Fourth International Conference on Materials Science and Manufacturing Technology (ICMSMT 2022)

IOP Conference Series: Materials Science and Engineering
Volume 1258

Online
8 – 9 April 2022

ISBN: 978-1-7138-6421-9
ISSN: 1757-8981

Printed from e-media with permission by:

Curran Associates, Inc.
57 Morehouse Lane
Red Hook, NY 12571

Some format issues inherent in the e-media version may also appear in this print version.

Institute of Physics
Dirac House, Temple Back
Bristol BS1 6BE UK

Phone: 44 1 17 929 7481
Fax: 44 1 17 920 0979

techtracking@iop.org

Additional copies of this publication are available from:

Curran Associates, Inc.
57 Morehouse Lane
Red Hook, NY 12571 USA
Phone: 845-758-0400
Fax: 845-758-2633
Email: curran@proceedings.com
Web: www.proceedings.com

TABLE OF CONTENTS

IOP Conf. Series: Materials Science and Engineering 1258 (2022) 011001 doi:10.1088/1757-899X/1258/1/011001

Preface

Materials science and manufacturing process is one of the significant symbols of modern day's high-tech revolution. The technological developments in the materials science paves the road for advancements in materials science and manufacturing technology across the world. Materials are the key factor for the emerging industries of all kind. In order for new material to get the actual application, exquisite preparation and manufacturing must be processed. No doubt manufacturing is critical in every walk of life. With rapid development computer technology, communications technology and network technology, traditional manufacturing process has evolved to intelligent manufacturing which is more intelligent, more precision and more efficiency. The Fourth International Conference on Materials Science and Manufacturing Technology 2022 (ICMSMT 2022) has provided a forum for researchers, scientists, academicians, scholars and practitioners around the world to present papers on recent developments in the broad fields of Materials Sciences and Manufacturing Technology. The ICMSMT series also provides a premier interdisciplinary platform for researchers, practitioners and educators to present and discuss the most recent innovations, trends, and concerns as well as practical challenges encountered and solutions adopted in the fields of Materials Science, Engineering and Manufacturing. The fourth edition of ICMSMT conference series (ICMSMT 2022) has been organized with partnered by Akshaya College of Engineering & Technology, Coimbatore and Diligentec Solutions, Coimbatore, India. The conference received potential number of submissions from across the world and the papers have been accepted by the editorial board of ICMSMT after the stringent screening and review process. ICMSMT is an annual technical event and has been aimed to be conducted in the second week of April every year. Due to the COVID-19, outbreak in all over the world followed by the national and international travel restrictions, the conference has been organized in online model. Each presenter were given ten minutes for presentation and three minutes for discussion on the queries. The entire meeting has been conducted using Google meet platform. The conference has been successfully completed with all the presentations and keynote sessions. Being the third edition, ICMSMT 2022 had received submissions from different regions of the world and scholars from 8 countries have registered and participated.

Dr. Ramya Muthusamy

Editor / Chair – TPC

ICMSMT 2022

IOP Conf. Series: Materials Science and Engineering 1258 (2022) 011001 doi:10.1088/1757-899X/1258/1/011001

ICMSMT 2022- Committeess

Executive Committee

Conference Chairs

Dr. Baharudin Bin Ismail, Universiti Malaysia Perlis, Malaysia

Dr. Kathiravan N, Dean(Academics), Akshaya College of Engineering and Technology, India

Organizing Chairs

Dr. Senthilkumar S, Diligentec Solutions, Coimbatore, India

Dr. Sivakumar R, Akshaya College of Engineering and Technology, India

Editorial Committee

Editor

Dr. Ramya Muthusamy, Diligentec Solutions, Coimbatore, India

(ramsthangam@gmail.com)

Members

Dr. Sivakumar R, Akshaya College of Engineering and Technology, India

Dr. Sendhil Kumar S, Akshaya College of Engineering and Technology, India

Dr. Anithasree A, Akshaya College of Engineering and Technology, India

Information Contact

Dr. Kapilan S, Akshaya College of Engineering and Technology, India

Publicity and Promotions

Mr. Sivakumar K, Diligentec Solutions, Coimbatore

IOP Conf. Series: Materials Science and Engineering 1258 (2022) 011001 doi:10.1088/1757-899X/1258/1/011001

Technical Program Committee Chairs

Dr. Joseph Xavier R, Advisor, Akshaya College of Engineering and Technology, India

Dr. Sabareesaan K J, Mechanical and Industrial Section, Nizwa College of Technology, Oman

Dr. Gopinth, S R Engineering College, Warangal, India

Dr. Rajamani R, PSG College of Technology, Coimbatore, India

Technical Program Committee

Dr. Jaroslaw Drelich, Michigan Technological University, USA
Dr. Chong Wen Tong, University of Malaya, Malaysia
Dr. Mohamed Arezki Mellal, M'Hamed Bougara University, Algeria
Dr. Khairel Rafezi Ahmad, University Malaysia Perlis, Malaysia
Dr. Mahanthi S D, Michigan State University, USA
Dr. Mohammed Henini, University of Nottingham, UK
Dr. Sharifah Shahnaz Syed Bakar, University Malaysia Perlis, Malaysia
Dr. Alexandru Bogdan Andreescu, Transilvania University of Brasov, Romania
Dr. Norinsan Kamil Othman, University Science Malaysia, Malaysia
Dr. Mitsuru Itoh, Tokiyo Institute of Technology, Japan
Dr. Sanjeev Kumar, Aalborg University, Denmark
Dr. Ramani Kannan, University Technology Petronas, Malaysia
Dr. Satyabrata Das, Indian Institute of Technology, Kanpur
Dr. Singh S P, Indian Institute of Technology, Delhi
Dr. Rajiv O. Dusane, Indian Institute of Technology, Bombay
Dr. Kamal K Kar, Indian Institute of Technology, Kanpur
Dr. Ashish Kumar Mishra, Indian Institute of Technology, Varanasi
Dr. Anand K, Indian Institute of Technology Madras, India
Dr. Sushil Mishra, Indian Institute of Technology, Bombay
Dr. Ajit Panigrahi, CSIR-Institute of Minerals and Materials Technology (IMMT), Bhubaneswar
Dr. Somnath Bhowmick, Indian Institute of Technology, Kanpur
Dr. Patil P, Indian Institute of Technology, Varanasi
Dr. Amrita Bhattacharya, Indian Institute of Technology, Bombay
Dr. Akhilesh Kumar Singh, Indian Institute of Technology, Varanasi
Dr. Krishnakant, Indian Institute of Technology, Delhi
Dr. Alankar Alankar, Indian Institute of Technology, Bombay
Dr. Vanitha C, National Institute of Technology, Warangal
Dr. Tanmoy Maiti, Indian Institute of Technology, Kanpur
Dr. Jitendra Kumar, Indian Institute of Technology, Varanasi
Dr. Rer. Nat. Ravikumar, Indian Institute of Technology, Madras
Dr. Bikash Kumar, CSIR-IMMT, Bhubaneswar
Dr. Ramkumar J, Indian Institute of Technology, Kanpur

IOP Conf. Series: Materials Science and Engineering 1258 (2022) 011001 doi:10.1088/1757-899X/1258/1/011001

Peer Review Statement

All papers published in this volume have been reviewed through processes administered by the Editors. Reviews were conducted by expert referees to the professional and scientific standards expected of a proceedings journal published by IOP Publishing Publishing.

- **Type of peer review:** Single Anonymous
- **Conference submission management system:** Morressier
- **Number of submissions received:** 102
- **Number of submissions sent for review:** 101
- **Number of submissions accepted:** 81
- **Acceptance Rate (Submissions Accepted / Submissions Received × 100):** 79.4
- **Average number of reviews per paper:** 2.7
- **Total number of reviewers involved:** 60
- **Contact person for queries:**
  Name: Dr. Ramya Muthusamy
  Email: ramsthangam@gmail.com
  Affiliation: Diligentec Solutions

Effect of annealing temperature on supercapacitive behaviour of manganese dioxide (MnO_2) thin-film prepared by electrodeposition technique and characterization

Priti N Jadhav[1], Dr. K V Sukhatankar*[1]

[1] Department of Physics, GogateJogalekar College, Ratnagiri, Maharashtra, India,

Email id:-pnjadhav609@gmail.com

Abstract: The purpose of this study is to demonstrate the construction of Manganese Dioxide (MnO_2) Thin-Film on a stainless-steel substrate (AISI304) using the electrodeposition process. A thin MnO_2 layer Potentiostatically formed by electrodeposition technique from $MnSO_4.H_2O$ (aqueous manganese sulphate monohydrated) solution was used to investigate the electrodes' super-capacitive capabilities. The impact of annealing temperature on the electrochemical characteristics of some materials at temperatures ranging from 100°C-500° C was investigated. The surface morphology of annealed material at 100° C, 200° C, 300° C, 400° C, and 500° C was investigated utilizing a SEM (scanning electron microscope). At 200° C, the cyclic Voltammogram is used to investigate the electrochemical characterisation. The highest value of specific capacitance was calculated from this data and was found to be 184.42F g^{-1} for 10 mV/s. Utilizing galvanostatic discharging and charging techniques, as well as electrochemical impedance spectroscopy, this research investigates the behaviour of super capacitive materials.

Keywords: Electrodeposition, Potentiostatic, Electrochemical, Supercapacitance.

1. Introduction:

As a result of its capacity to produce high power as well as resist high cycle counts, electrochemical capacitors are used as electrical energy storage devices. Mobile phones as well as other forms of personal devices, and also hybrid vehicles, are among the many possible practical uses. Effective charge storage in hybrid vehicles might have a positive influence on oil consumption patterns therefore reduce their effect on the climate if it is developed at a reasonable cost. Electrochemical capacitors can be used for a vast array of applications, including balancing electrical power loads and accelerating the charging and discharging processes of batteries [4]. Materials that are very dense and have a large surface area or materials that have many oxidation states within the solvent decomposition window, are used in the construction of supercapacitors. SCs store energy in either EDLC (electric double layer) or pseudo capacitive manner, depending on the charge storage systems. In the non-Faradaic EDLC technique, the electrolyte/electrode interface is separated via charge separation, that is largely based on carbon-based materials [2]. As electrode ions and materials in relevant potential window react with each other, pseudo capacitive (Faradaic) processes are generated [3].
Various kinds of conductive polymers, transition metal oxides, along with carbon are among the principal groups of materials used in supercapacitors [1]. As a result of their advantageous electrical, mechanical, morphological, and structural characteristics, transition metal oxides have acquired a lot of

interest in electrochemistry field. It is common for carbon counterparts to display pseudo capacitances; however, this is not the case here. Amongst numerous transition metal oxide electrode materials, for example IrOx, Co_3O_4, NiOx, and RuOx; MnO_2 has high energy density and specific capacitance with wide discharge/charge potential range, but has low abundance, toxicity, as well as ecologically benign nature; it has (~1370 F g^{-1}) high theoretical specific capacitance, which have drawn substantial interest for SCs as active electrode materials [2]. Since they are inexpensive, ecologically benign, and have better electrochemical performance than other transition metal oxides, MnO_2-based supercapacitors are now gaining a lot of attention. [3]. Electrochemical techniques were used to create a wide range of MnO_2 structures as well as morphologies, and their electrochemical characteristics were also studied.
Generally thin films deposition techniques are classified into two groups, Chemical methods and Physical method. Because thickness as well as particle size can be precisely controlled, chemical processes are far better suited for material production than physical ones. For the thin films' deposition, chemical procedures are easy, economical, and practical. Because of this, we know that with the electrodeposition method, we have three options for adjusting the electric field: varying the voltage and changing the current. [2,3]. The crystal structure, surface morphology, and therefore the supercapacitive characteristics of deposited material are all considerably influenced by the aforementioned 3 kinds of induced electric fields. Various MnO_2 morphologies and crystallographic forms have been shown to significantly influence performance [3].
Hydrous MnO_2 produced by Roger Amade[2] has a 104 F/g supercapacitance at 700^{0}C, Hydrous MnO_2 produced using a potentiostatic technique by D.P. Dubal [4] has 184 F g^{-1} supercapacitance, Hydrous MnO_2 generated using potentiodynamic electrodeposition approach has 392 F.g^{-1} supercapacitance, according to Pravin R. Jadhav [3]. MnO_2 generated using the galvanostatics approach has 240 F g^{-1} supercapacitance according to J.N. Broughton and M.J. Brett [5]. Chang and Tsai created hydrous MnO_2 supercapacitance via a potentiostatic technique [6].
The purpose of this study is to create thin films of MnO_2 by utilizing Potentiostatic mode at 1200 s. various annealing temperatures have been studied for the morphological, structural, along with supercapacitive characteristics of MnO_2 thin films. MnO_2 thin film charge–discharge in 1MNa_2SO_4 electrolyte has also been explored.

2. Experimental details:

AR grade chemicals were used in the preparation of a thin layer of MnO_2. The solution was adjusted to pH of 6.5 using potassium hydroxide (KOH) in an aqueous solution of 0.1 M $MnSO_4.H_2O$ (manganese sulphate) (molychem, A.R. Grade). Potentiostatic mode is used to deposit MnO_2 films for 1200 seconds. on stainless-steel foil After polishing using zero-grade polish paper as well as a detergent wash, the stainless steel foil was soaked in ethanol, acetone, along with HNO_3 before being ultrasonically cleaned and ready for depositing. MnO_2 was deposited in a three-electrode electrochemical cell with a standard three-electrode setup, including a stainless steel thin film as a working electrode, a graphite as a counter electrode, along with SCE (saturated calomel electrode) as reference electrode. At normal temperature, electrodeposition has been performed using 0.95 V dc voltages. This is followed by various annealing temperatures, such as 100° C, 200° C,300°C,400° C,500° C for all samples after deposition.
X-ray diffraction using a computer controlled Philips PW-3710 with CrK radiations was used to characterise the structure. SEM was used to examine the surface morphology. Metrohm autolab PGSTAT 204 was utilized to perform a super capacitor research, creating a 1M Na_2SO_4 electrolyte electrochemical cell with MnO_2 working electrode, graphite's counter electrode, a silicon carbide's reference electrode. Metrohm PGSTAT 204 has been used to conduct the charge–discharge research.

3. Results and discussion:

3.1 XRD(X ray diffraction method):-

Utilizing match 3 software, the XRD as well as JCPDS data were matched (reference No: 96-151-4235, No96-151-4236). MnO_2 XRD pattern, as illustrated in Figure1. This suggests that the film at 100^0 C and 200^0 C comprised entirely of MnO_2 colloidal particles, as there isn't any other diffraction peak other than the substrate in the XRD pattern (figs. 1a and 1b). The triangles serve as indexes to the steel-induced peak. Amorphous supercapacitor materials can be used because ions can easily penetrate the bulk of the active material. In "figure 1c and d" peak obtained at 37.38^0, 41.05^0, 64.58^0, 56.32^0, show crystalline phase of β MnO_2, the phase changes at 300 ^{0}C and 400 ^{0}C. Figure 1e shows crystalline structure and peak obtained at 18.70, 23.14^0, 32.95^0, 38.23^0, 55.18^0, 65.79^0 shows phase changes from MnO_2 to Mn_2O_3 at temperature of 500 ^{0}C [23-25]

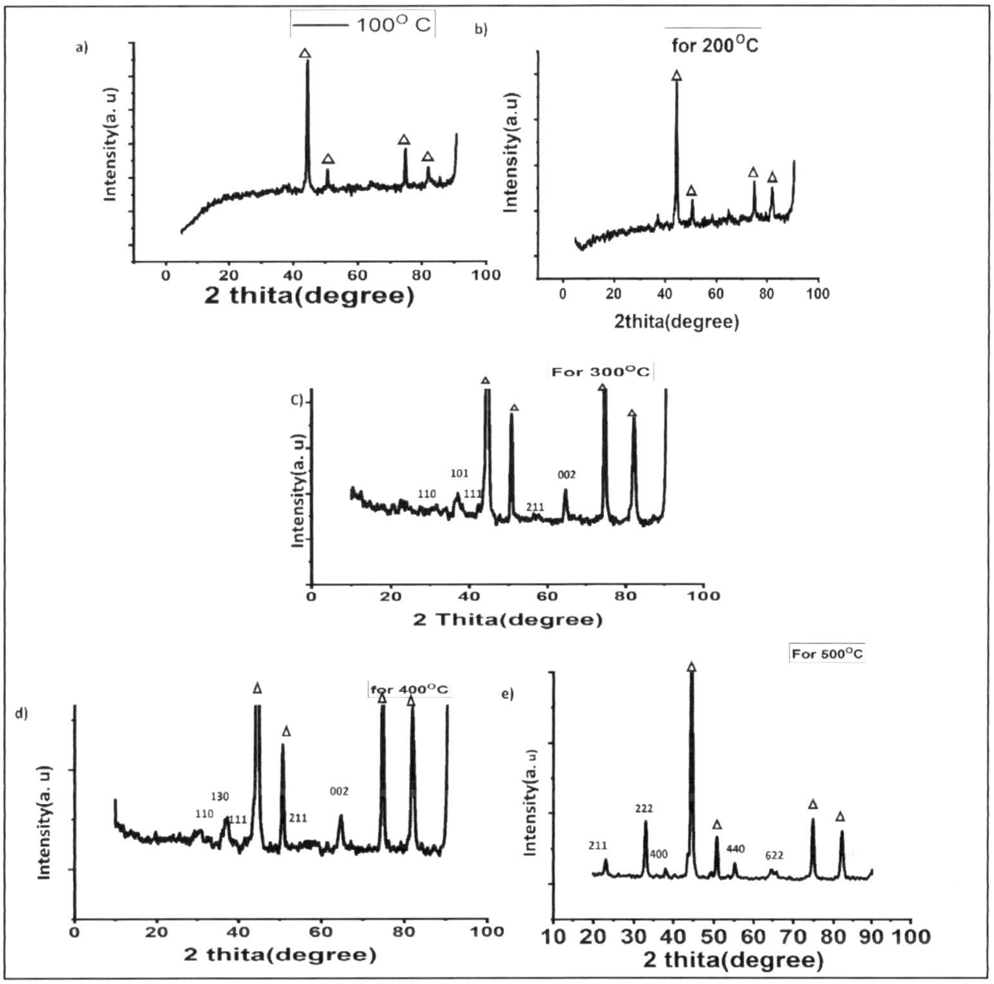

Figure 1. XRD of $MnO_2$for a) 100^0 C b) 200^0 C c) 300^0 C d) 400^0 C e) 500^0 C.

3.2 Raman Spectra:

Figure 2 demonstrates the MnO_2 Raman spectrum. Because of its high sensitivity to local material structure, Raman spectroscopy is an excellent tool for verifying the structure as well as phase of produced samples [16]. In the range of 500–700 cm^{-1}, various manganese oxide phases are identified [26]. Mn-O bond stretching mode in the octahedral units of [MnO_6] the sharp peaks at 625, 652, 648.77, 650.52 cm^{-1} are attributed in figure 2a, 2b, 2c, and 2d. Formation of MnO_2 and Mn_2O_3 is confirmed from Raman spectra.

IOP Conf. Series: Materials Science and Engineering 1258 (2022) 012001 doi:10.1088/1757-899X/1258/1/012001

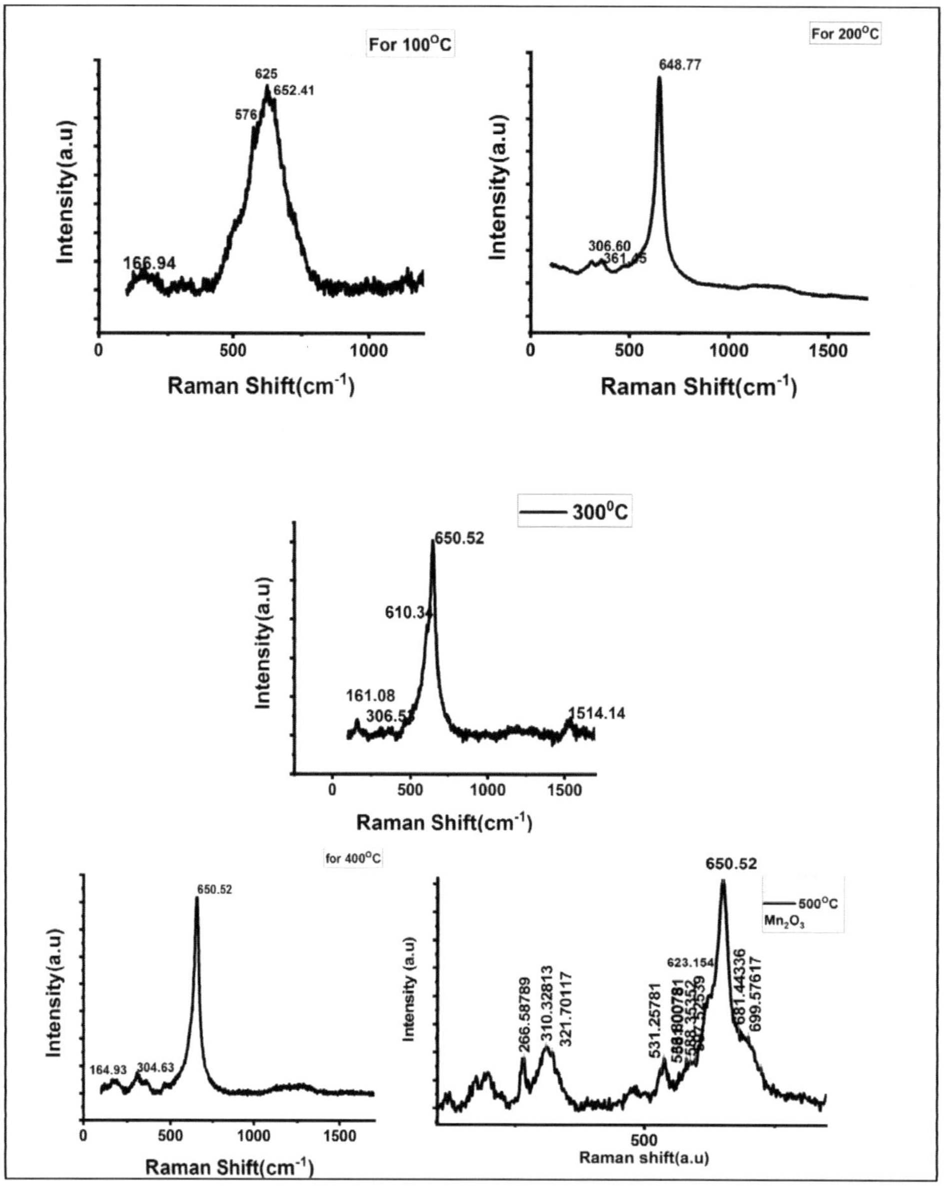

Figure 2. Raman spectra of MnO_2 for a) 100° C b) 200° C c) 300° C d) 400° C e) 500° C.

3.3 FTIR Spectra:

Figure 3 shows the FTIR transmittance spectrum of powdered form of MnO_2 thin film in the 350-4000 cm^{-1} region. Spectra contain 4 measure peak: at 3432 cm^{-1}, 3434 cm^{-1}, 3416 cm^{-1}, and 3440 cm^{-1}, that are achieved because of stretching of hydroxyl or water molecule group. Peak at 1632 cm^{-1}, 1621 cm^{-1}, 1608 cm^{-1}, 1627 cm^{-1} , 1631 cm^{-1}, can characterise bending of hydroxyl group or water molecule, whereas 392 cm^{-1}, 548 cm^{-1}, 611 cm^{-1}, 615 cm^{-1} , 724 cm^{-1} , 552 cm^{-1}, 585 cm^{-1}, peak may characterise Mn-O vibration in octahedral structure of MnO_6. From figure 3 it confirms that peak 585 cm^{-1}, 552 cm^{-1} shows γ MnO_2 Phase for 100° C and 200° C, Peaks 593 cm^{-1},615 cm^{-1}, 724 cm^{-1} shows β MnO_2 Phase for 300° C and 400° C, 548cm^{-1}, 392 cm^{-1} shows Mn_2O_3 Phase for 500° C.

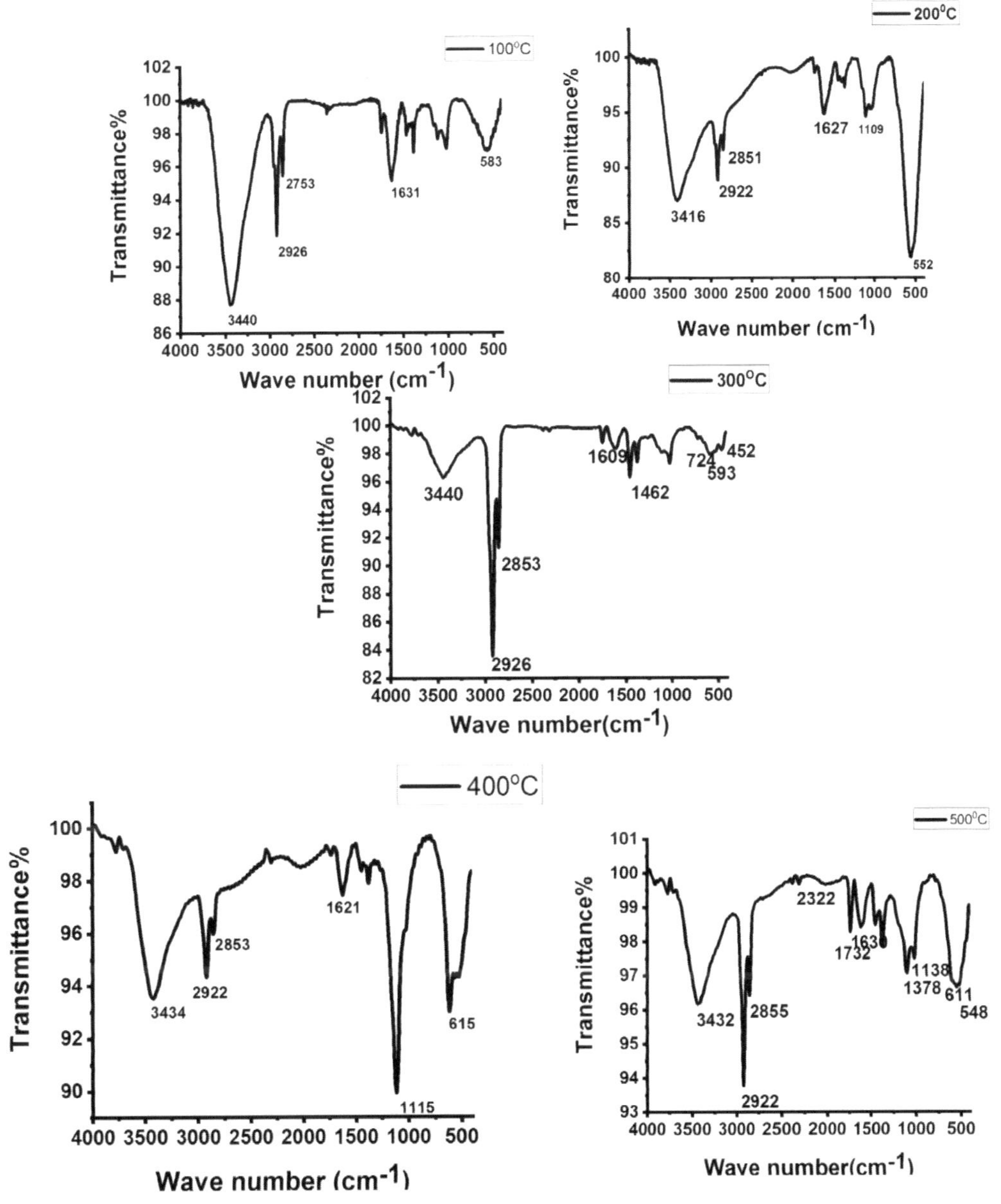

Figure 3. FTIR of MnO_2 for a) 100°C b) 200° C c) 300° C d) 400° C e)500° C.

3.4 Surface Morphology

Samples at temperatures of 100° C, 200° C, 300° C, 400° C, 500° C samples were examined under the SEM (Scanning electron microscope) as shown in figure 4 (a-e). MnO_2 nanoparticle nature can be seen in the SEM image, which reveals that the particles have been evenly distributed over the substrate. Also, altering the temperature of the annealing process might affect the manganese oxide base film's structure. Granular and Fibrous structures are generated when the annealing temperature is less than 200°C. When the temperature exceeds 300°C, the tangling of fibers leads morphology to evolve into a cluster-like structure; nevertheless, as the temperature rises, morphology may lead to the production of a flaky-like structure.

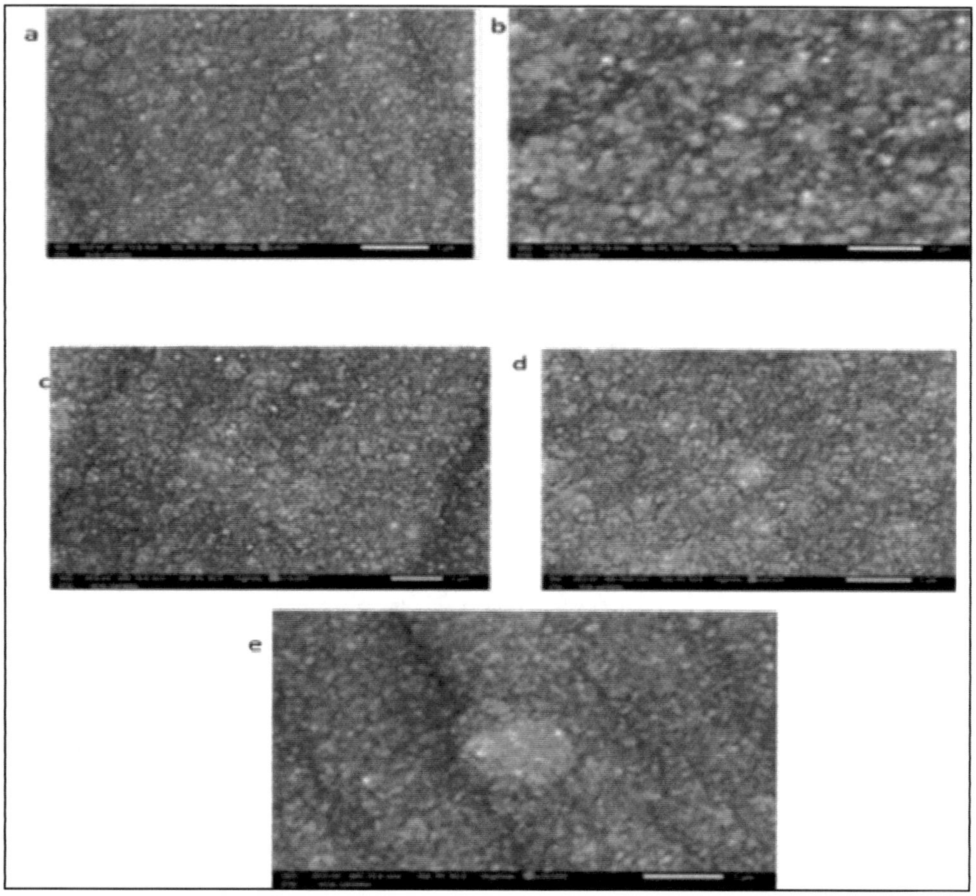

Figure 4. SEM Image of $MnO_2$for a)100° C b)200° C c)300° C d)400° C e) 500° C.

3.5 Electrochemical studies:

3.5.1. Cyclic Voltammetry: - CV Method has been used to investigate the electrochemical behavior of MnO_2. CV measurements were performed on all MnO_2 samples throughout the voltage range of 0 to 1 Volts. Figure 5a shows SCE in 1 M Na_2SO_4 electrolyte at 10 mV/s scan rate. The quasi-rectangular mirror images of all electrode curves are seen, and this is a hallmark of a perfect pseudo capacitive performance. The specific capacitance has been calculated by assessing area under the curve as well as calculating it using the formula

$$C_{sp} = \frac{\int I\, dv}{2mV}$$

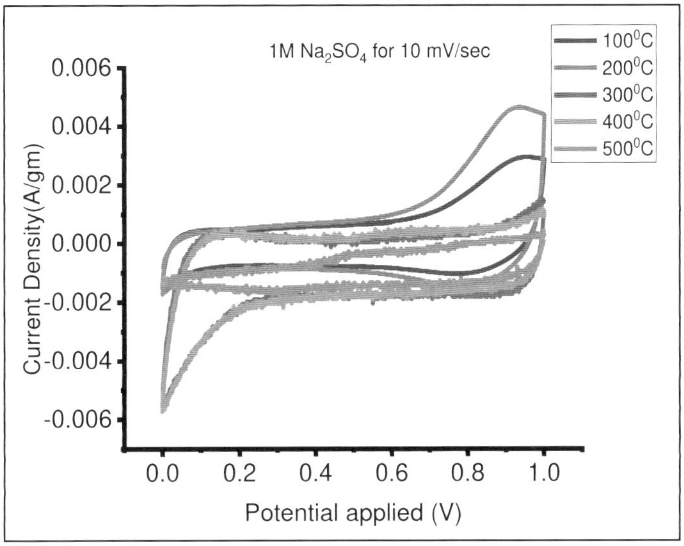

Figure 5(a). Cyclic Voltammetry of MnO_2 for 100° C, 200° C, 300° C, 400° C, 500° C in 1M Na_2SO_4

Where$\int$ I dv, is average charge deduced from CV, m is defined as the active material's mass that is deposited on working electrode along with scan rate that is defined by V. As indicated in Table 1, the CV technique's specific capacitance rises with temperature up to 200 °C because of the increased surface area, which is displayed. After then, specific capacitance reduces as the surface area drops. This is especially true for heat-treated materials, which are predicted to have reduced open porosity as well as smaller surface area, especially at temperatures above 200 °C.

Table 1.for measurement of specific capacitance, energy density, power density

Temperature (° C)	**Specific Capacitance by CV (F/g)**	**Specific Capacitance by charging discharging curve (F/g)**	**Energy density Wh/kg**	**Power density (W/kg)**
100	28.84	33.83	16.85	399.17
200	184.42	174.59	87.71	3007.08
300	175.00	172.81	85.64	2986.75
400	64.29	65.50	32.51	1281.06
500	29.69	93.62	46.90	563.07

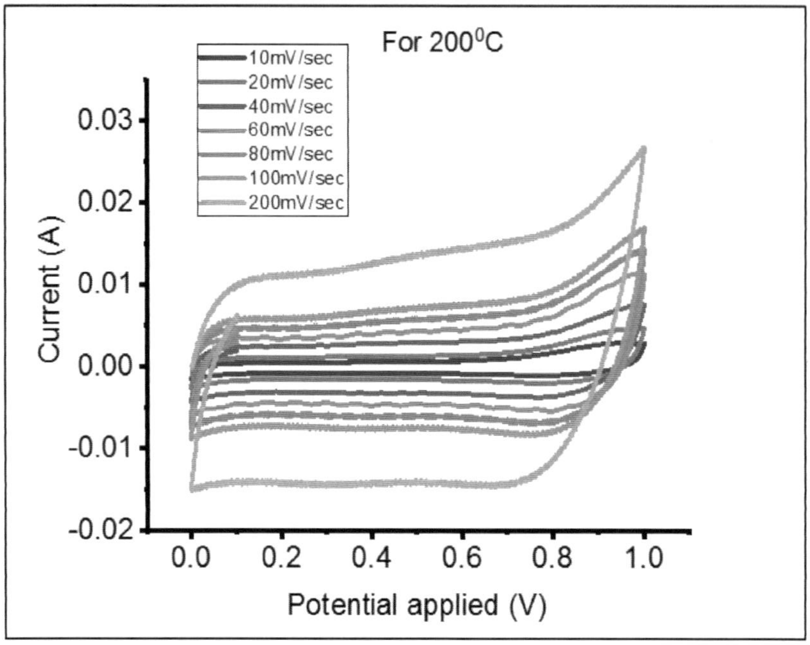

Figure 5(b) .Cyclic Voltammetry of MnO_2 at 200° C for scan rate 10 mV/s, 20 mV/s, 40 mV/s , 60 mV/s, 80mV/s, 100 mV/s, 200 mV/s in 1M Na_2SO_4

The scan rate has been varied from 10 to 100 mV/s to examine the influence on capacitance performance. The results are presented in figure 5b. It displays the CV curves for all of the samples that were captured at different scan rates. This deviation from ideal capacitive behaviour is due to a rise in area under the curve as well as an increase in scan speed [2]. Na^+ diffusion from the electrolyte can get access to practically all of the interior of the nanoparticles matrix at lower scan speeds, resulting in a full insertion reaction and thus a reduction procedure. During fast scan rates, the Na^+ ions only reached the electrode's outer surface layer whereas they were unable to exploit the internal pores of the nano particles matrix reducing the specific capacitance. The maximum capacitance for 200° C is 184.42 F/gm at 10 mV/s. Figure.5c shows stability curve of MnO_2 for 200° C at 100mV/s, it gives more specific capacitance stability with capacity retention 91.73% and average columbic efficiency is 99.99% up to 1000 cycle.

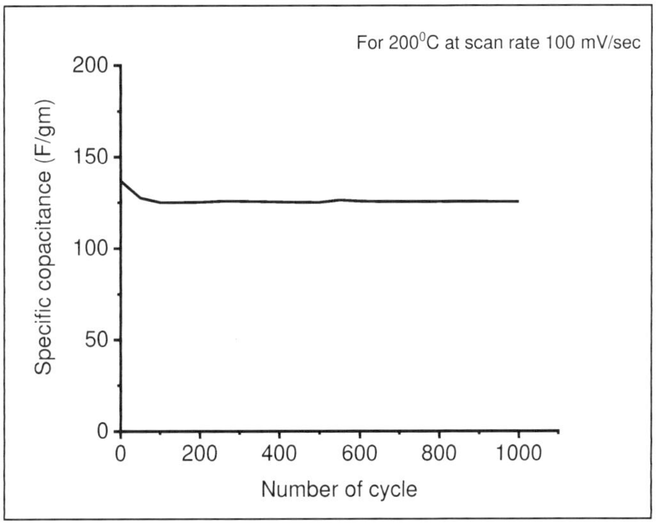

Figure 5(c). Cyclic Stability curve of MnO_2 at 200° C for 1000 cycles at 100 mV/s

3.5.2. Galvonostatic charging discharging studies:

In a galvanostatic charge-discharge experiment, MnO_2 supercapacitor devices shown to have excellent capacitive behaviour and a low internal resistance. In this experiment, the charge discharge curves were noted in potential ranges from 0 to 1 V while the current density was 1 mA. Figure 6 depicts a charge discharge curve with a slight voltage drop at the beginning of discharge curve. Small internal resistance is being shown among current collector as well as electrode.

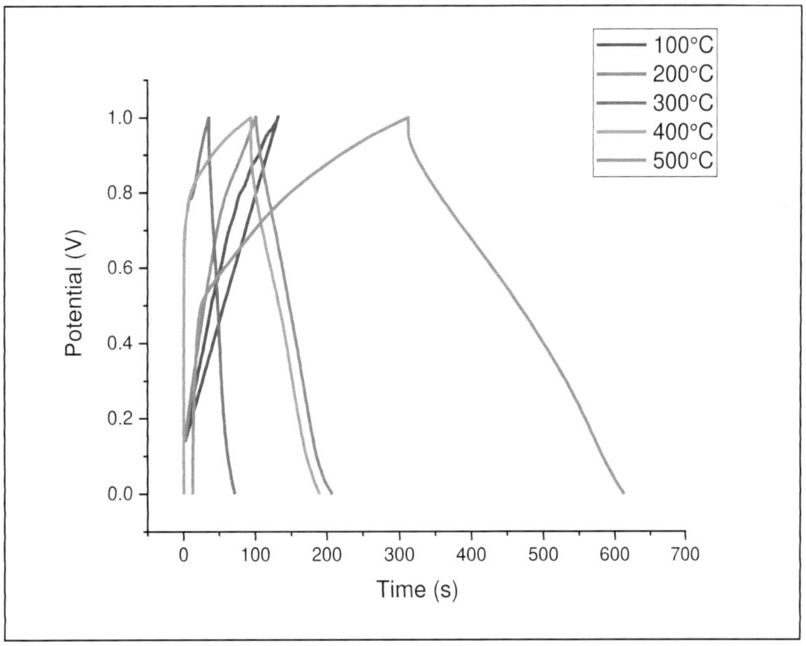

Figure 6. Charging discharging curve of MnO_2 for 100° C, 200° C ,300° C, 400° C, 500° C.

The following equation was used for calculating the specific capacitance based on the charging and discharging curve.

$$C_{sp} = \frac{I}{m \times dv/dt}$$

Here, m is deposited materials mass, I is defined as the applied current, dv/dt defined as discharge time from slope of galvonostatic charge discharge curve, 174.588F/g is maximum specific capacitance for 200^0 C at 1mA current density. The following equation may be used to determine the super capacitors energy density:

$$E = \frac{1}{2} C V^2$$

Pseudo capacitivity is demonstrated by the MnO_2 electrodes high energy density (87.7067 wh/kg) as well as high power density (3007.083 w/g).

3.5.3 Impedance Spectroscopy

Impedance Spectroscopy provides additional information on the electrochemical processes that take place at the interfaces between materials. And electrochemical impedance spectroscopy may be utilized for determining the bulk of electrode materials. In this study, the electrochemical impedance spectroscopy (EIS) of synthesised MnO_2 samples was investigated in the frequency range of 0.1 to 1 MHz at 50 mV as an alternating current (AC) amplitude using a 1M Na_2SO_4 electrolyte has been investigated. Figure.7 represents the electrode sample of Nyquist at 100° C, 200° C, 300° C, and 400° C and 500° C. A semicircle on the real axis has a high–frequency intercept on it, which represents the equivalent series or electrolyte resistance, as well as its diameter signifies the charge–transfer resistance (Rct) across the interface among electrode along with electrolyte [7]. A suitable pseudo-capacitive material for electrodes, MnO_2 has a Nyquist spectrum that shows a straight line, indicating a low charge transfer resistance [7]. The total ohmic resistance is included in the electrolyte resistance (Rs). Sample 200^0 C has a lower electrolyte resistance and charge transfer resistance than samples 100° C, 300° C, 400^0 C, and 500^0 C.

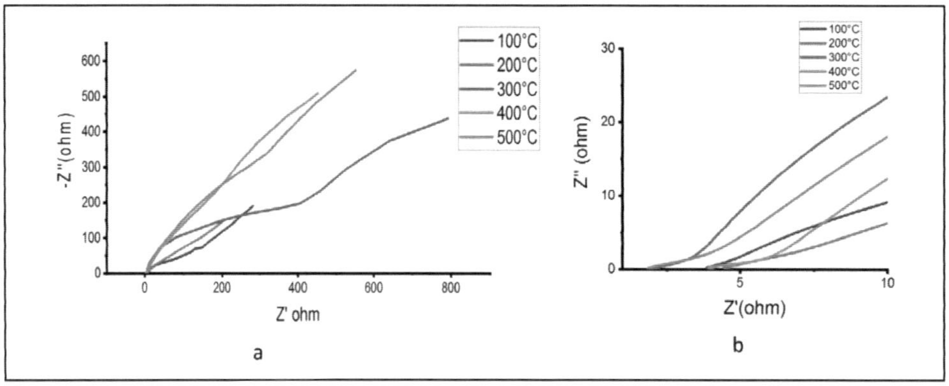

Figure 7: The Nyquist plot of sample 100° C, 200° C, 300° C, 400° C and 500° C electrode.

Additionally, the electrolyte resistance (Rs) for sample 500° C, 400° C, 300° C, 200° C, and 100° C, is discovered to be 1.79 Ω, 3.93 Ω, 2.35 Ω, 4.46 Ω, and 3.8Ω and charge transfer resistance (R_{ct})for 200° C electrode is zero. Nanostructured MnO_2 samples at 200° C, electrodes offer low resistance and easy access for intercalation and de-intercalation of charges between electrode and electrolyte, according to findings from ESI, GCD, and CV analysis. [7].

4. **Conclusion:-** In the potentiostatic mode, MnO_2 films were effectively manufactured. MnO_2 XRD analysis demonstrated the production of an amorphous structure (Figure 1). By using a suitable

annealing temperature, it is possible to convert the amorphousness into a crystalline phase. If the annealing temperature is increased to 300° C, the crystallinity begins to form. At 300° C crystal structure of MnO_2 was observed for the first time. Mn_2O_3 is formed when the temperature is raised to 500° C. Manganese oxide will operate better when a proton is included in the charge storage process. Low temperatures from 100° C to 200° C, improved the specific capacitance, at 200° C the maximum capacitance is 184.42 F/g and provides better cycling stability at 100 mV/s with high coulomb efficiency of 99.99%. At high temperatures above 200° C, the hydrous state is reduced and the specific capacitance is reduced with respect to high temperature.

5. Reference:

[1] Dose W M, Donne S W., Journal of Chemical Society,158(8)A905-A912 Dicipline of chemistry,University of Newcastle ,Callaghan,NSW 2308,Austalia.
[2] Amade R *, Muyshegyan-Avetisyan A , González J M ,2019,Materials,page no 483 volume 12.
[3] Jadhav P R, Suryawanshi M P, Dalavi D S, Patil D S , Jo E A, Kolekar S S , Wali, A A, Karanjkar M M., Kim J H, Patil P S,Development of nano fiber MnO_2 thin film electode and cyclic voltammetry behavior modeling using artificial neural network for super capacitor application Material Science in Semiconductor Processing 36(2015) 43-48.
[4] Dubal D P , Dhawale D S, Gujar T P , Lokhandea C D, Effect of different modes of electrodeposition on supercapacitive properties of MnO2 thin films Applied Surface Science Volume 257, Issue 8, 1 February 2011, Pages 3378-3382.
[5] Broughton * J N , Brett M J, Variations in MnO2 electrodeposition for electrochemical capacitors, ElectrochimicaActa 50 (2005) 4814–4819.
[6] Chang J K , Tsai W T , J.Powersource 158 (2006)1140.
[7] Kumar Y, Chopraa S, Gupta Y K, Ukec S J, Mardikard S P Materials Science for Energy Technologies Volume 3, 2020, Pages 566-574
[8] Pino A P, György E, Pascual E, Andújar J L , Serra E B , materials,MDPI.
[9] Chan Pei Yi and Majid S R, The Electrochemical Performance of Deposited Manganese Oxide-Based Film as Electrode Material for Electrochemical Capacitor Application, INTECH, http://dx.doi.org/10.5772/intechopen.71957.
[10] Raheem Z H, Abdul Kareem Ali M *,Department of chemistry, College of science, University of Baghdad., Baghdad, Iraq Corresponding Author: muraee@gmail.com[455,456], Sys Rev Pharm 2020;11(11):453-458A multifaceted review journal in the field of pharmacy.
[11] Yuqiu H, Hongcheng Z, Cathodic Potentiostatic Electrodeposition and Capacitance Characterization of Manganese Dioxide Film2010 International Conference on Nanotechnology and Biosensors,IPCBEE vol.2 (2011) © (2011) IACSIT Press, Singapore.
[12] Moulai F, Cherchour N , Messaoudi B & Zerroual, Electrosynthesis and characterization of nanostructured MnO2 deposited on stainless steel electrode: a comparative study with commercial EMD 2016 Springer-Verlag Berlin Heidelberg .
[13] Ying-Chu Chena, Yu-KueiHsub, Yan-Gu Lina, Yu-Kai Linc,d, Ying-Ying Horngc, Li-Chyong Chena, Kuei-HsienChena,c,*ElectrochimicaActa 56 (2011) 7124–7130.
[14] Rodrigues S *, Munichandraiahà N and Shukla A K *,A cyclic voltammetric study of the kinetics and mechanism of electrodeposition of manganese dioxide Journal Of Applied Electrochemistry 28 (1998) 1235±1241, Solid State and Structural Chemistry Unit* and Department of Inorganic and Physical Chemistry, Indian Institute of Science, Bangalore ± 560 012, India
[15] Conway B E, Electrochemical Supercapacitors: Scientific Fundamentals and Technological Applications, Kluwer-Plenum, New York, 1999.
[16] Wang C, Hsu H, Hu J, Power Sources, 249 (2014) 1.
[17] Choi D, Kumta P N, Electrochemist. Soc., 153 (2006) A22.
[18] Shinomiya T, Gupta V, Miura N, Electrochimica. Acta, 51 (2006) 44.
[19] Nayak P and .Munichandraiah N *,Simultaneous Electrodeposition of MnO2 and $Mn(OH)_2$ for Supercapacitor Studies Electrochemical and Solid-State Letters,12 _6_ A115-A119 _2009_1099-0062/2009/12_6_/A115/5/$25.00 © Electrochemical Society.

[20] Suhasini, Effect of deposition method and the surfactant on high capacitance of electrochemically deposited MnO2 on stainless steel substrate Journal of Electroanalytical Chemistry 690 (2013) 13–18.
[21] Das D, Sen P K, Das Kaushik *, Mechanism of potentiostatic deposition of MnO2 and electrochemical characteristics of the deposit in relation to carbohydrate oxidationElectrochimicaActa 54 (2008) 289–295.
[22] Fan JX, Wang X, Ge Li, Yu A, Chen Z, High-performance flexible electrode based on electrodeposition of polypyrrole/MnO2 on carbon cloth for supercapacitors Journal of Power Sources 326 (2016) 357e364.
[23] Yogesh Kumar, Seema Chopra, A. Gupta, Y. Kumar, S.J. Uke, S.P. Mardikar, Low temperature synthesis of MnO2 nanostructures for supercapacitor appli- cation, Materials Science for Energy Technologies (2020), doi: https://doi.org/10.1016/j.mset.2020.06.002
[24] Rahul Singhal1, Jonathan Lembeck1,2, Peter C. K. LeMaire III3, Daniel Pereira1, Peter LeMaire1*, Physical and Electrochemical Characterization of Nsutite, American Journal of Analytical Chemistry, 2021, 12, 440-445, DOI: 10.4236/ajac.2021.1211027.
[25] Yogesh Kumar , S J Uke , Ashwani Kumar , S P Merdikar , Meenal Gupta , A K Thakur , P Bocchetta , Anshu Gupta , Vinay Kumar and Yogesh Kumar, Triethanolamine–ethoxylate (TEA-EO) assisted hydrothermal synthesis of hierarchical β-MnO2 nanorods: effect of surface morphology on capacitive performance, Nano Express, 22 November 2021.
[26] Yanqing Xin, Zongcheng Ling*, Haijun Cao , Changqing Liu , Hongchun Bai, Morphology, Structural, And Spectral Characterizations Of Lab-Synthesized Manganese Oxides Relevant To Mars, 52nd Lunar and Planetary Science Conference 2021 (LPI Contrib. No. 2548)

Effect of ZnO on Structural, Physical and Optical Properties of Heavy Metal Oxides Doped Tellurite Glasses

R.P.Kumbhakar[1], Ghizal F. Ansari[1*], S. K. Dhiman[2], Afroj A. Khan[3]

[1]Physics Department, Madhyanchal Professional University, Bhopal (M.P.) 462044, India
[2] Department of Mechanical Engineering, Birla Institute of Technology, Mesra, Ranchi-835215, India
[3]Department of Physics, Integral University, Lucknow, 226026, India

**Corresponding author:ansarigf@rediffmail.com*

Abstract. A conventional technique has been used to prepare Heavy Metal Oxides (HMO) doped ternary tellurite glasses. The batch composition of samples were as (90-y)%TeO_2-10%WO_3-x%ZnO, (y=5%,10%,15%, and 20%). ZnO alter the optical and physical properties of the glasses are investigated. The thermal characteristics of the produced samples are studied using differential scanning calorimetry, and their structural conformation is determined using XRD. The existence of structural units is shown by Raman spectroscopy. Other parameters of glass samples have been calculated, including density, molar volume, polaron radius, and intermolecular distance. The absorption spectra of samples were measured in the spectrum range of 350nm to 1000nm, and the refractive index, direct and indirect optical energy band gaps, and Urbach's energy were determined.

Key Words: Heavy Metal Oxide; Ternary Tellurite Glass; Optical Properties

1. Introduction

In recent days glasses based on oxides of heavy metals, tellurite, and fluoride have been observed as generous concern of research that shows perspective to be used as nonlinear optical materials and applications in fiber optics as well as laser hosts [1]. The glasses composed of tellurium oxides attract the researchers, scientists towards the favorable outcomes in the applications of glass ceramics, nonlinear optical devices, and optical amplifiers [2,3]. They show low phonon energy [4], temperature of fusion [6],transition temperature and crystallization ability [7]; large 3rd order non-linear susceptibility [5], refractive indices [11], and dielectric constants [10]; as well as good chemical resistance [8], andinfrared rays transmission having comprehensive wavelength range [9]. A thorough analysis of the literature highlights the need for more study into the photophysical features of tellutite glasses that contain tungsten and Zinc Oxides. A TeO2–WO3–ZnO glass system was created in this work, and its properties were investigated using XRD, SEM-EDX, optical absorption, Raman, and DSC. The estimation of density, molar volume, polaron radius and intermolecular distance were conducted for glass samples. In addition the absorption spectra were recorded in the spectral range of 350-1000nm. Besides this the refractive index, direct and indirect optical energy band gaps and urbach's energy were also evaluated.Raman spectra were studied to the structural information.

2. Experiment

A classical melt and quench technique has been applied to synthesize Heavy Metal Oxides (HMO) doped tellurite glass samples.High purity chemical substances (Otto) TeO_2, WO_3 and ZnOwere taken in following composition:

I. TWZ1≡85%TeO_2-10%WO_3-5%ZnO
II. TWZ2≡80%TeO_2-10%WO_3-10%ZnO

III. $TWZ3 \equiv 75\%TeO_2\text{-}10\%WO_3\text{-}15\%ZnO$
IV. $TWZ4 \equiv 70\%TeO_2\text{-}10\%WO_3\text{-}20\%ZnO$

For appropriatemixing mortar and pestle were used. Thenthe blend was melted in alumina crucibles in a muffle furnace. Temperature wastaken as high as 800°Cto introduce molten vitreous property in it followed by physically stirring ensuring homogeneity in the mixture. It was further heated to 930°Cfor 30 minutes and then poured in stainless steel moulds to form pellets of diameter 10mmand thickness of 2 to 3mm.A hot oven maintained at 250 °C was used for annealing carried out for 60 minutes.X-RAY diffractogram of TWZ glasses were investigated for the endorsement of glassy nature. Differential Scanning Calorimetry for the analysis of thermal properties and the Raman Spectroscopy for the structural analysis was conducted. Absorption spectroscopy has been carried in the spectral range of 350 - 1000nm spectral range.

3. Results and Discussion

3.1 XRD

X-Ray diffractogram of TWZ1 and TWZ4 glasses shown in figure 1, carried by Rigaku, Japan, Smart Lab 9kW at normal temperature in 10° to 80°, twice of angle range. It is the signature diffractogram of tellurite base glass doped with HMO, It lacks a high peak and instead has a broad bump about 29°, which is owing to the lack of crystalline structure.

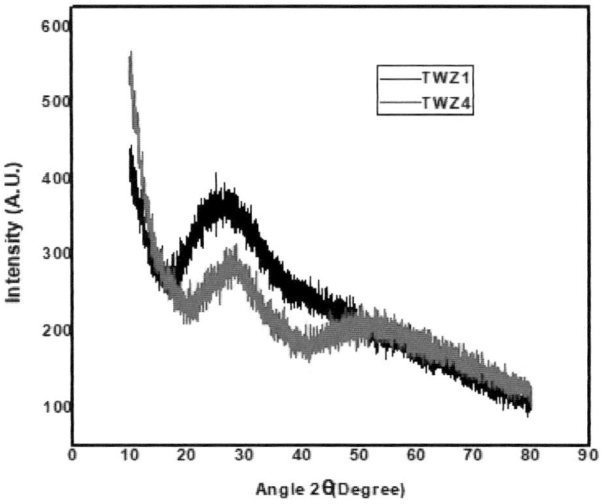

Figure 1.Diffractogram of TWZ1 and TWZ4 glass sample

3.2 DSC

The glass transition temperature of produced HMO doped tellurite glasses was determined using a differential scanning calorimetric graph utilising TA instruments' Q10 in the thermal range of 25°C to 460°C at a rate of 10°C/min. Figure 2 shows that the glass transition temperature of TWZ4 is 370°C.

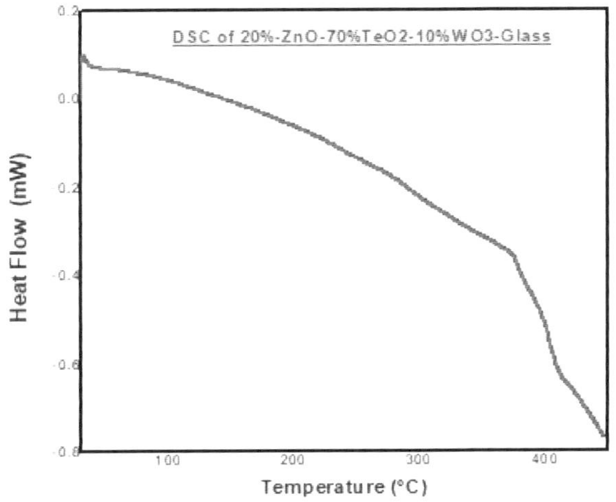

Figures 2. DSC plot of TWZ4 for glass transition temperature

3.3 SEM and EDS

The SEM allows for high-resolution inspection of tiny sample particles. This permits evaluation of a sample's surface, individual components, and focused study of the sample. SEM image of samples were carried by Joel, Japan; JSM-6390LV. Clear surface morphology of glass sample TWZ4 shown in Figure 3.

The chemical composition of TWZ4 glasses were analyzed by energy dispersive spectroscopy. EDS shown in figure 4 gives the information of elements of TWZ4.

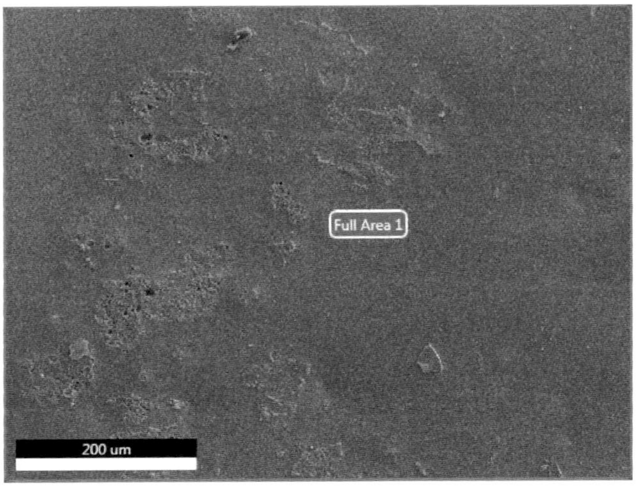

Figures 3. SEM of TWZ4

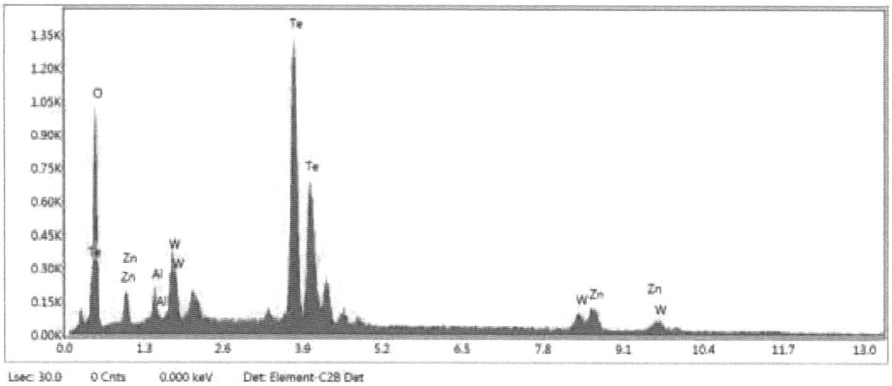

Figure 4.Energy dispersive spectrogram of TWZ4

3.4 Raman Spectra

Figure 5 shows the Raman Spectral characteristics of TeO2–WO3–ZnO glass with a mixture of 70% TeO2-10% WO3-20% ZnO (TWZ4) measured by Research India's RI-D2R-S Direct Coupled Raman Spectrometer in the range of 190–1000 cm1.

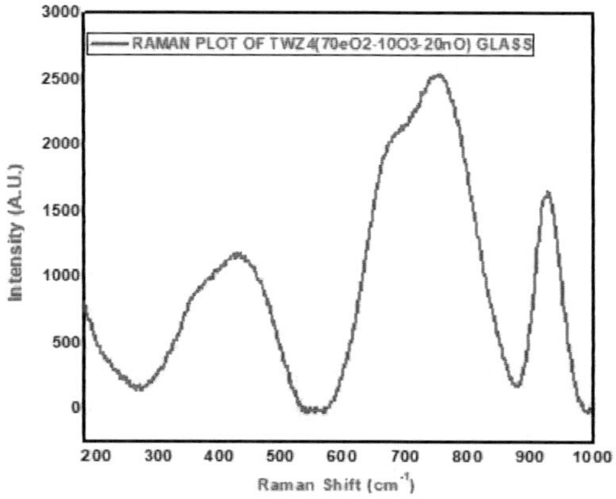

Figure 5. Raman plot of TWZ4 (70%TeO2-10%WO3-20%ZnO) glass

Raman spectra between 320 cm^{-1} to 480 cm^{-1} are credited to oxide group present in the glass as TeO_2, WO_3 and ZnO. Zinc – oxygen bond in zinc tetra oxide groups and tungsten – oxygen bond in tungsten hexa oxide units were in vibrating modes [12, 13]. Tellurium oxide gives strong line (near 660 cm^{-1}) and $[TeO3]^{2-}$ortho-anions gives strong vibration in interval 720–750 cm^{-1} [14-16]. Vibration bands at 850–900 cm^{-1} observed are is because of stretching vibrations of tungsten-oxygen bond in the tungsten tetra and hexa oxide units. The strong band centered at 960 cm^{-1} which is assigned to tungsten – oxygen – stretching vibrations in tungsten tetra and hexa oxides.

4. Physical Properties

Various parameters of HMO doped tellurite glass samples were determined, including density, Molar Mass(M), Molar Volume, Oxygen packing density, Zink ion concentration, polaron radius, and inter nuclear distance (Table 1)[17-19].

Table 1. The physical properties of TeO_2–WO_3–ZnO glass system

Name of the Sample	**TWZ1**	**TWZ2**	**TWZ3**	**TWZ4**
Batch composition in percent(TeO_2:ZnO: WO_3)	85:10:05	80:10:10	75:10:15	70:10:20
Molar Mass (M) gm/mol	163.37	159.43	155.5	151.56
ρ (g/cm^3)	5.122	5.184	5.238	5.285
V_m (cm^3/ mol)	31.895	30.756	29.686	28.677
OPD (mol/l)	62.704	65.032	67.370	69.741
N($\times 10^{23}$) (ions cm^{-3})	0.944	1.958	3.043	4.200
R_p (nm)	0.284	0.223	0.192	0.173
R_i (nm)	0.109	0.799	0.690	0.619
Rm(cm^3)	16.376	15.790	15.242	14.723

5. Optical properties

Ultraviolet-Visible-Infrared absorption spectra of synthesized tungsten tellurite glasses were performed in the wavelength range from 350nm to1000nm. If 'a' is the absorbance of a specimen of thickness 'd' the absorption coefficients α were obtained by using the equation [1]

$$\alpha = 2.303\, a/d \qquad (1)$$

Figure 6 shows a graph of the absorption coefficient as a function of wavelength. As the percentage of zinc oxide in the batch grows, the sample's absorption coefficient drops.
For amorphous materials, the relationship between and hν is stated as

$$\alpha = \frac{\{C(h\nu - E_{opt})^r\}}{h\nu} \qquad (2)$$

Where C is constant and E_{opt} is the optical energy gap [21]. Above equation can also be expressed as

$$(\alpha h\nu)^{\frac{1}{r}} = C(h\nu - E_{opt}) \qquad (3)$$

By extrapolating the straight region of the curve of $(\alpha h\nu)^{1/r}$ versus $h\nu$ to $(\alpha h\nu)^{1/r} = 0$, equations (3) were utilised to establish direct and indirect forbidden optical energy band gaps. In Tauc's figure, the value of index r is 2 for direct forbidden energy gap and 12 for indirect forbidden energy gap [22]. Figures 7 and 8 show Tauc's plots of TeO_2–WO_3–ZnO glass samples for direct and indirect optical band gaps.

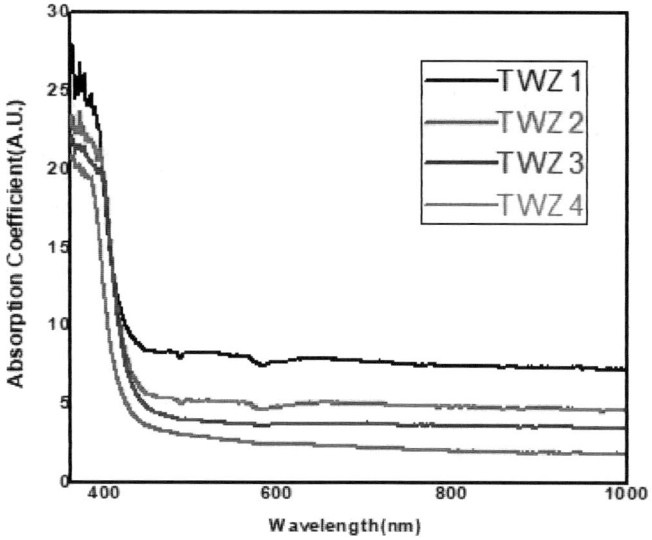

Figure 6. Absorption plot of TeO_2–WO_3–ZnO glass systems

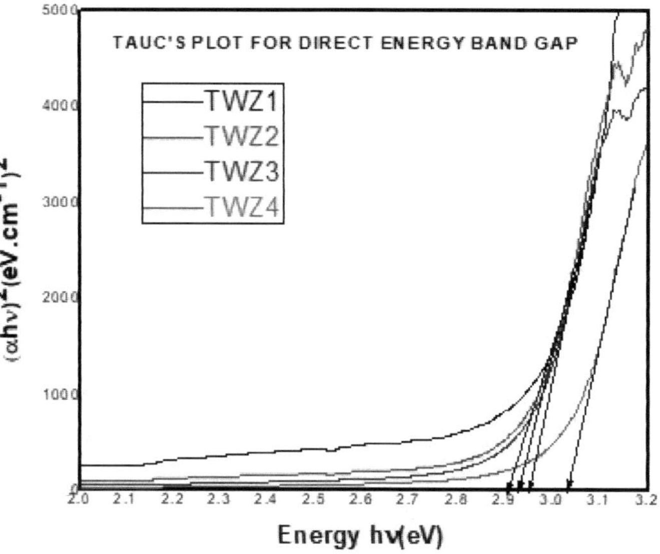

Figure 7. Absorption plot of TeO_2–WO_3–ZnO glass systems

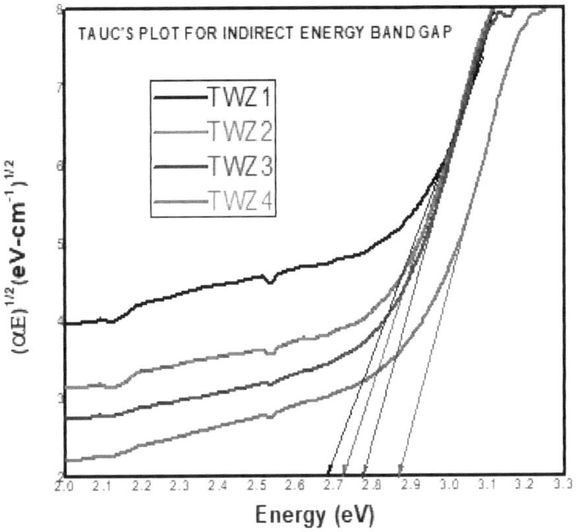

Figure 8. Absorption plot of TeO_2–WO_3–ZnO glass systems

Table 2 shows the values of prohibited energy gaps for direct and indirect transitions of TeO2–WO3–ZnO glass systems. With increasing ZnO contamination, Eopt values grow. This rise is due to a rise in zinc and tellurium non-bridging oxides [23].

Band tailing indicates disorder in amorphous substances such as glass, and it is known as Urbach's energy Eu and measured by the Urbach rule[24].

$$\alpha = \alpha_o e^{\frac{h\nu}{E_u}} \quad (4)$$

Where α_o is constant. Eq. (4) can be redraft as

$$ln\alpha = \frac{h\nu}{E_u} + constant \quad (5)$$

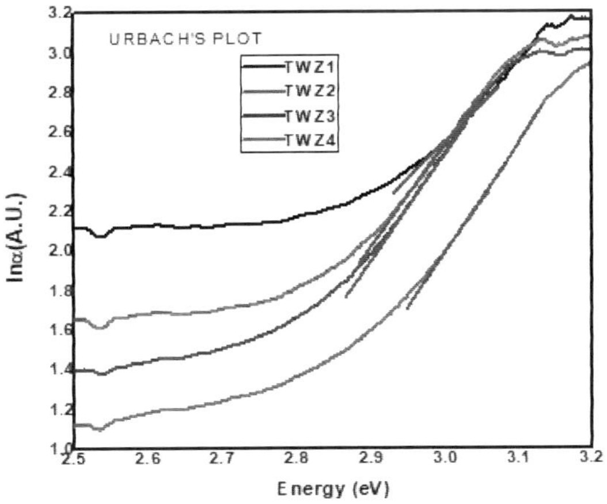

Figure 9.Urbach's plot of TeO_2–WO_3–ZnO glass systems

The Urbach plot of TeO2–WO3–ZnO samples is shown in Figure 9. Table 2 shows the calculated Eu values of TeO2–WO3–ZnO samples, which range from 0.288 eV to 0.205 eV.

Refractive indexes (n) of the synthesized TeO_2–WO_3–ZnO glasses were calculated by the equation (6).

$$\frac{(n^2-1)}{(n^2+2)} = 1 - \left(\frac{E_{opt}}{20}\right)^{\frac{1}{2}} \qquad (6)$$

Table 2. Optical properties of TeO_2–WO_3–ZnO glass systems

Name of the sample	TWZ1	TWZ2	TWZ3	TWZ4
Direct Eopt (eV)	2.90	2.92	2.95	3.02
Indirect Eopt(eV)	2.69	2.71	2.78	2.88
Refractive index (n)	2.424	2.418	2.411	2.391
Urbach energy (eV)	0.288	0.216	0.211	0.205

The values of E_{opt}increases with increase in mole percent of WO_3, in the E_{opt}increases whereas refractive index decreases.

6. Conclusion

The classic melt-quench process was used to make TeO2–WO3–ZnO glass systems using ZnO and WO3 as modifiers. XRD was used to validate the amorphous nature; nevertheless, no sharp peak could be seen in the diffractogram. Glass transition temperature was measured using differential scanning calorimetry when tellurite-based based glasses melted at low temperatures. Phonon energy of manufactured glass of batch composition 80% TeO2-10%WO3-20%ZnO investigate by Raman spectroscopy, which revealed that the tellurite glass sample is reasonably conducive to efficient visible and infrared emission.

Various physical features such as density, molar volume, OPD, polaron radius, inter ionic distance, and tungsten ion concentration were evaluated. The absorption coefficient, optical energy band gaps, and Urbatch's were examined using UV-VIS-IR absorption spectroscopy. The prepared samples were clear, with a direct energy band gap of 2.90 to 3.02 eV. The effects of zinc oxide mole percent on refractive index and energy band gap have been investigated. The findings of this study suggest that zinc oxide can be used as an HMO glass modifier with tellurite.

Acknowledgements
The authors are acknowledging the facility contribution of m/s Research India Bhopal, for the UV-VIS-IR and Raman characterization.

References

[1] Ansari G.F., Mahajan S.K., Journal of LuminescenceVolume 156,(2014), 97-101
[2] Sae-Hoon K., Toshinobu Y., J. Am. Ceram.Soc. 78 (1995) 106.
[3] Tatsumisago M., Minami T., Ceramics 28 (1993) 1227.
[4] Wang J.S., Vogel E.M., Snitzer E., Opt. Mater. 3 (1994) 187.
[5] Lin J., Huang W., Sun Z., Ray C.S., Day D.E., J. Non-Cryst. Solids 336 (2004) 189.
[6] Stanworth J.E., SocJ.. Glass Technol. 36 (1952) 217.
[7] Kosuge T., Benino Y., Dimitrov V., Sato R., T. Komatsu, J. Non-Cryst. Solids 242 (1998) 154.
[8] Shaltout I., Tang Y., Braunstein R., Shaisha E.E., J. Phys. Chem. Solids 57 (1996)1223.
[9] Waber M.J., Meyers J.D., Blackburn D.H., J. Appl. Phys. 52 (1981) 2944.
[10] Stanworth J.E., Nature 169 (1952) 581.

[11] El-Mallawany R.A., Tellurite Glasses Handbook – Physical Properties and Data,CRC Press, Boca Raton, FL, 2002.
[12] Edukondalu A., Hameed A., Kavitha B., Kumar R.V., Kumar K.S., Mater. Today. Process 2, 913–917 (2015)
[13] Taki Y., Shinozaki K., Honma T., Dimitrov V., Komatsu T., J. Solid State Chem. 220, 191–197 (2014)
[14] Cornette J., Merle-Mejean T., Mirgorodsky A.P., Colas M., Smirnov M., Masson O., Thomas P., J. Raman Spectrosc. 42 (2010) 758–764.
[15] Sekya T., Mochida M., Ohtsuka A., Tonokava M., J. Non-Cryst. Solids 144 (1992) 128–144.
[16] Champarnaud–Mesjard J.C., Blanchandin S., Thomas P., Mirgorodsky A.P., Merle-Mejean T., Frit `, J. Phys. Chem. Solids 61 (2000) 1499–1507.
[17] VeerannaGowda V.C., Effect of Bi3+ ions on physical, thermal, spectroscopic and optical properties of Nd3+ doped sodium diborate glasses, Physica B 426 (2013) 58–64.
[18] Alajerami Y.S.M., Hahim S., Muhamad W., MuhamadSaridan W., HassanW., Ramli A.T., Kasim A., Optical properties of lithium magnesium borate glasses doped with Dy3+ and Sm3+ ions, Physica B 407 (2012) 2398–2403.
[19] Yanlingwang S.D., Chen F., XuTiefeng, Nie Q., Physical properties and optical band gap of new tellurite glasses within the TeO2–Nb2O5–Bi2O3 system, Mater. Chem. Phys. 113 (2009) 407–411.
[20] Mott N.F., Davis E.A., Electronic processes in non-crystalline materials, Oxford University Press, Oxford, 1974.
[21] Mott N.F., Davis E.A., Electronic Processes in Non-Crystalline Materials, second ed., Clarendon Press, Oxford, 1979.
[22] Tauc J., Menth A., States in the gap, J. Non-Cryst. Solids 8e10 (1972) 569-585.
[23] RaoV.H., Prasad P.S., RaoP.V., Santos L.F., Veeraiah N., Journal of Alloys and Compounds 687 (2016) 898-905
[24] Urbach F., Phy. Rev. 92 (1953) 1324

IOP Conf. Series: Materials Science and Engineering 1258 (2022) 012003 doi:10.1088/1757-899X/1258/1/012003

Superconducting Properties of Ta Substituted ($Tl_{1-x}Ta_x$)(BaSr)$CaCu_2O_7$

A. B. P. Ilhamsyah[1], R. Abd-Shukor[1], Hafizal Yazid[2], Syahrul Humaidi[3] and A. N. Jannah[4]

[1]School of Applied Physics, Universiti Kebangsaan Malaysia 43600 Bangi, Selangor, Malaysia
[2]Malaysian Nuclear Agency, Bangi 43000 Kajang, Selangor, Malaysia
[3]FMIPA, Universitas Sumatera Utara, Jalan Biotechnology No 1, Medan 20155, Indonesia
[4]Faculty of Applied Sciences, Universiti Teknologi MARA, Cawangan Negeri Sembilan, Kampus Kuala Pilah, 72000, Kuala Pilah, Negeri Sembilan, Malaysia

Abstract. In this research, the effects of Ta substitution on the Tl-based superconducting materials have been investigated. The standard method of solid-state reaction was applied to prepare the ($Tl_{1-x}Ta_x$)(BaSr)$CaCu_2O_7$ superconductors of high temperature with $x = 0$, 0.4, 0.5 and 0.6. For all samples, temperature dependence measurements versus the electrical resistance demonstrated metallic normal state behavior. The X-ray diffraction measurement indicated coexistence of mixed phase with Tl-1212 as the major phase and Ta_2O_5 as an impurity for all substitution samples. A fraction with the highest volume of the Tl-1212 phase in Ta substitution samples is in $x = 0.4$ (67%), 30% comes from Ta_2O_5 and the other 3 % belongs to the Tl-1201. All samples with Ta substitution ($x \geq 0.4$) showed decreased in $T_{c\ onset}$, $T_{c\ zero}$ and T_p. The discussion on the effects of Ta_2O_5 in the Tl(BaSr)$CaCu_2O_7$ superconductors are also provided.

1. Introduction

Thallium based-cuprates $Tl_mSr_2Ca_{n-1}Cu_nO_{2n+m+2}$ with m = 1 and n = 2 or $TlSr_2CaCu_2O_7$ superconductor (Tl-1212) is one of a series which produces unique and interesting superconductivity properties. These superconductors are easily produced compared to other thallium systems and are easily stabilized with elements that are partially substituted at various sites in $TlSr_2CaCu_2O_7$ [1]. The elemental substitution or doping of different ions in the various site of cations in TlSr2CaCu2O7 has been known as one of the effective ways to optimize the hole concentration and hence stabilize the Tl-1212 phase and enhance the transition temperature. Among the factors which plays an important role in determiningthe phase formation of Tl-1212 and the critical temperature is the state of valence and the radius of ionof the substituted element [2, 3].

The effects of elemental substitutions on $TlSr_2CaCu_2O_7$ have been well studied [4-11]. F (CSCO) phase $Ca_{0.3}Sr_{0.7}CuO_2$ has been found in Bi-based superconductors as impurity phase [12]. This phase has also been discovered in other copper oxide superconductors where the main doping elements contain Sr and Ca [13]. The rare earth elements replaced at the Sr site are La, Ce, Pr, Nd, Sm, Eu, Gd, Tb, Dy, Ho, Er, Tm, Yb, and Lu. Lee et al. [14] have conducted a study of rare earth element substitution, R = La, Ce, Pr, Nd, Sm, Eu, Gd, Tb, Dy, Ho, Er, Tm, Yb and Lu in the $TlSr_{1.5}R_{0.5}CaCu_2O_7$ superconductor. The results showed that T_c between 80 to 90 K was successfully achieved for all substitution of earth elements which are rare except for Ce. The substitution of La yielded the highest onset transition temperature, T_c onset (90 K), while the substitution of Ce yielded the lowest T_c onset (5 K). The lattice parameter c was found to decrease inversely proportional to the atomic number except for Ce. However, Ce substitution in $TlSr_{1.65}Ce_{0.35}CaCu_2O_7$ have increased the T_c at 62 K [15].

Ta is a multivalent element with effective ionic radius (6 - fold coordination number) of 0.86 Å, 0.82 Å and 0.78 Å for Ta^{+3}, Ta^{+4} and Ta^{+5}, respectively. The ionic radius of Tl^{+1} (1.5 Å) and Tl^{+3} (0.885 Å) are smaller as compared to the ionic radius of Ta. It is essential to study how higher multivalent states with the substitution of element with smaller ionic radius affect the superconductivity of the Tl-1212 phase. Therefore, the aim of this research is to assess the effects of the Ta substitution at Tl site of $(Tl_{1-x}Ta_x)(BaSr)CaCu_2O_7$.

2. Experimental Details

A method which is known as solid-state reaction was applied to produce samples with the nominal starting composition $(Tl_{1-x}Ta_x)(BaSr)CaCu_2O_7$ with $x = 0.0$, 0.4, 0.5 and 0.6. An agate mortal was used to obtain an appropriate high purity of homogenous mixture (> 99.99%) SrO, CaO and CuO. Two intermittent grindings were applied while the precursor powders were heated at 900°C for 24 h in order to obtain a homogenous black powder. Then, Tl_2O_3 and Ta_2O_5 in the stoichiometric amounts were added into the precursor and were completely mixed. Next, the pellets of 1.3 cm in diameter and 0.2 cm thickness were produced by pressing the mixture. The loss of thallium during heating was compensated by adding an excess of 10% Tl_2O_3. The process was continued by heating the pellets for 4 min at 970°C in flowing oxygen. Finally, furnace cooling was applied to cool the pellets to room temperature.

The powder X-ray diffraction (XRD) pattern was obtained for identifying the resultant phase formation by using a diffractometer of Bruker model D8 Advance with CuKα radiation (λ =1.5418 Å).With the assumption that the amount of Tl-1201 and Tl-1212 phase is relative to the strongest line of each phase, the volume fraction of Tl-1201 and Tl-1212 phase was estimated from the highest peak intensity of each phase. The least square method was employed to estimate the lattice parameters of Tl-1212 phase.

The method of four-point with the contact of silver paste was applied to calculate dc electrical resistance from 40 to 300 K, combined with a closed cycle refrigerator from CTI Cryogenic Cycle Refrigerator Model 22 and a temperature controller from Lake Shore Temperature Controller Model 340. A current source between 1 and 100mA was constantly applied throughout the measurements. A Merlin Gemini scanning microscope was used to conduct field scanning electron microscopy (FESEM) for observing microstructure of the samples.

3. Findings and Discussion

Powder X-ray diffraction pattern of $(Tl_{1-x}Ta_x)(BaSr)CaCu_2O_7$ for $x = 0$, 0.4, 0.5 and 0.6 are shown in Fig. 1. All samples revealed the Tl-1212 phase with tetragonal unit cell (space group P4/mmm) as the main phase and Tl-1201 as the minor phase. Peaks (101), (102), (103), (110) and (200) demonstrated that the samples exhibited the main phase of Tl-1212. Peaks (105), (114) and (210) are normal in all substitute samples. The Tl-1201 peaks were denoted with *. Peaks due to the unreacted Ta_2O_5 were observed in all added samples indicated that Ta may not substituted in the crystal lattice.

According to the calculations, the fraction with highest volume of the Tl-1212 phase in Ta substituted samples is for $x = 0.4$ (67%), 30% comes from Ta_2O_5 and the other 3 % belongs to the Tl-1201. Moreover, the fraction of volume for Tl-1212 phase was 63 % and 44 % for $(Tl_{0.5}Ta_{0.5})(BaSr)CaCu_2O_7$ and $(Tl_{0.4}Ta_{0.6})(BaSr)CaCu_2O_7$, respectively. The fraction of volume for Tl-1212 phase dropped as Ta content was increased. Ta_2O_5 peak was also indicated in XRD pattern for all Ta-substituted $(Tl_{1-x}Ta_x)(BaSr)CaCu_2O_7$ samples [16]. According to to Table 1, no systematic change was observed in lattice parameters and the unit cell volume. The constrain of internal lattice was measured by using thelattice parameter *a*/*c* ratio. The *a*/*c* ratio ranged between 0.3047 and 0.3098.

Table 1. $T_{c\ onset}$, $T_{c\ zero}$, ΔT_c, T_p, lattice parameters, unit cell volume of $(Tl_{1-x}Ta_x)(BaSr)CaCu_2O_7$ for x = 0, 0.4, 0.5 and 0.6

x	$T_{c\ onset}$ [K]	$T_{c\ zero}$ [K]	ΔT_c [K]	T_p ($dR[dT$) [K]	a [Å]	c [Å]	V [Å^3]	a/c	V_{1212} [%]	V_{1201} [%]	V_{Ta2O3} [%]
0	98	88	10	92	3.825	12.554	183.7	0.3047	100	-	-
0.4	69	51	18	57	3.811	12.383	179.8	0.3078	67	3	30
0.5	76	51	25	55	3.811	12.303	178.7	0.3098	63	1	37
0.6	94/74	51	43	88/59	3.814	12.338	179.4	0.3091	44	8	48

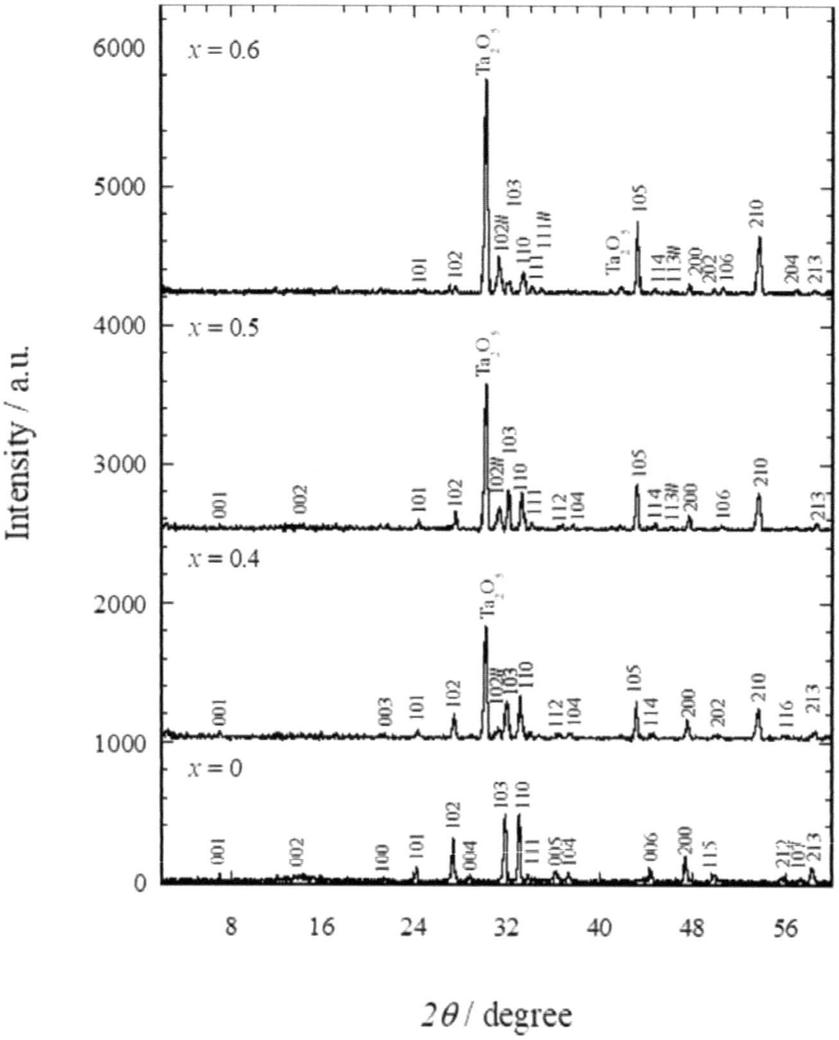

Figure 1 XRD pattern of $(Tl_{1-x}Ta_x)Sr_2CaCu_2O_7$ for x = 0, 0.4, 0.5 and 0.6. The Tl-1201 peaks are presented with *

Fig. 2 shows that the temperature dependence of electrical resistance for all samples was normal. A metallic behavior was indicated from the typical state of the samples until the transition temperature was achieved. In Fig. 2a, $T_{c\ onset}$ at 98 K and $T_{c\ zero}$= 88 K was observed for the non-substituted sample. The temperature curve versus resistance for samples at x = 0.4, 0.5 and 0.6 are displayed in Fig. 2b, 2c and 2d, respectively. The substitution of Ta significantly suppressed the temperature of zero-transition, $T_{c\ zero}$ which was 51 K for all substituted samples, as displayed in Fig. 2. $T_{c\ onset}$ declined from 98 K to 69 K for x = 0.4 sample and continuously decreased to 76 K for x = 0.5 sample. Sample with higher substitution sample (x = 0.6) demonstrated that the resistivity measurement had a double transition. This is the result obtained when the superconducting grains were decoupled [12].

Substitutions of Ta increased the width of transition, ΔT_c whereby ΔT_c of the non-substituted samples recorded the narrowest (10 K). For all substituted samples, the transition width rose from 18 K for x = 0.4 to 25 K for x = 0.5 and 43 K for x = 0.6 sample. This is due to inhomogeneities in transition temperatures of the individual superconducting grains increased. This results also showed that the presence of Ta as impurities at the grain boundaries contributed a crucial role to widen the transition temperatures. The peak temperature, T_p of dR/dT decreased with Ta content. Sample with x = 0.6 displayed two peaks in derivative curve which indicated the multiphasic behavior. $T_{c\ onset}$, $T_{c\ zero}$, ΔT_c, lattice parameters and unit cell volume for all samples are displayed in Table 1.

The linear fit in Fig. 2 represents the resistivity of background normal state. The superconducting fluctuations behavior (SFB) can significantly impact superconducting properties. At temperatures more and above the zero-resistance temperature, $T_{c\ zero}$, the fluctuation consists of the deviation of electrical resistivity from the linear metallic normal state behavior curve. It is proposed that SFB, resulting from the production of Cooper pairs at the early beginning, will dominate the excess conductivity zone. In general, Ta substitutions decreased the temperature at which the resistance moves away from linear resistive behavior as shown in Fig. 2. This indicates that superconducting fluctuations were reduced to a lower temperature as Ta was substituted.

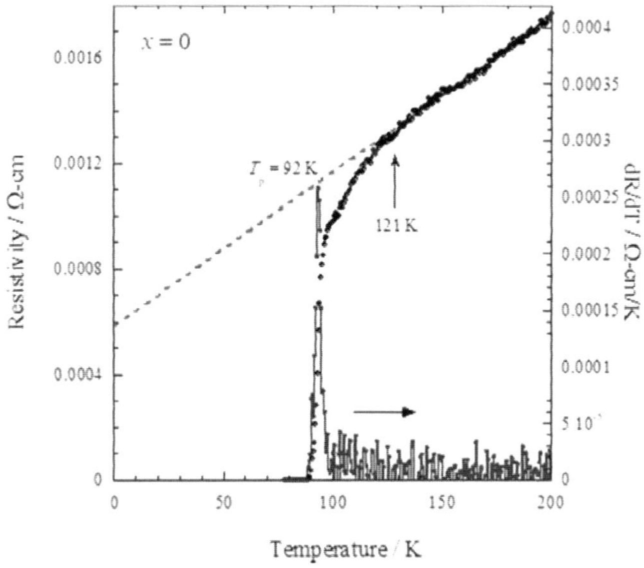

(a)

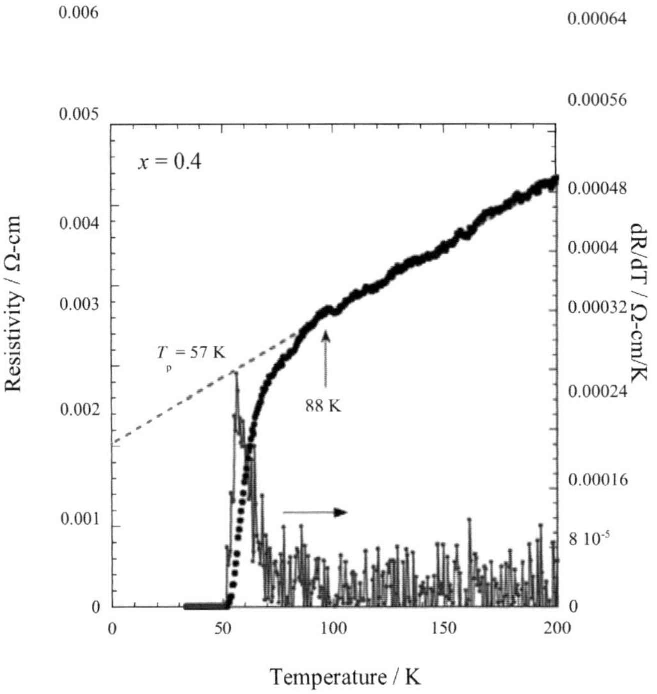
x = 0.4
T_p = 57 K
88 K
Resistivity / Ω-cm
dR/dT / Ω-cm/K
Temperature / K

(b)

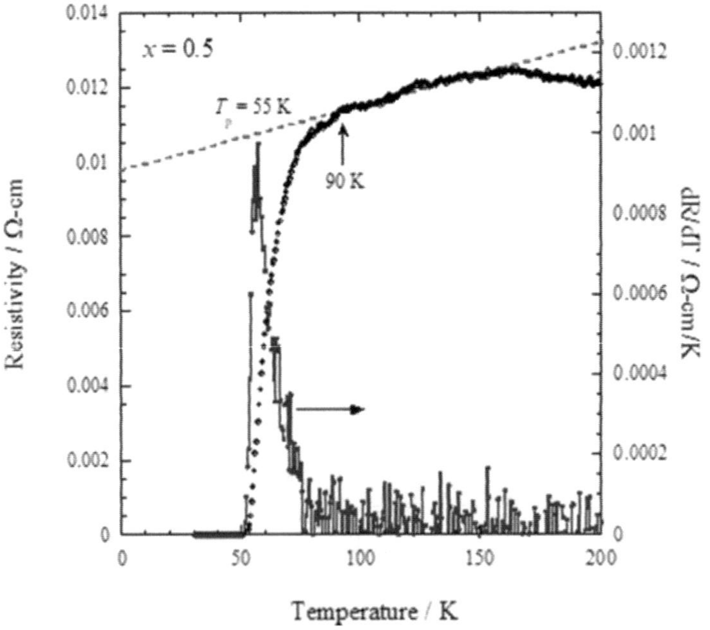
x = 0.5
T_p = 55 K
90 K
Resistivity / Ω-cm
dR/dT / Ω-cm/K
Temperature / K

(c)

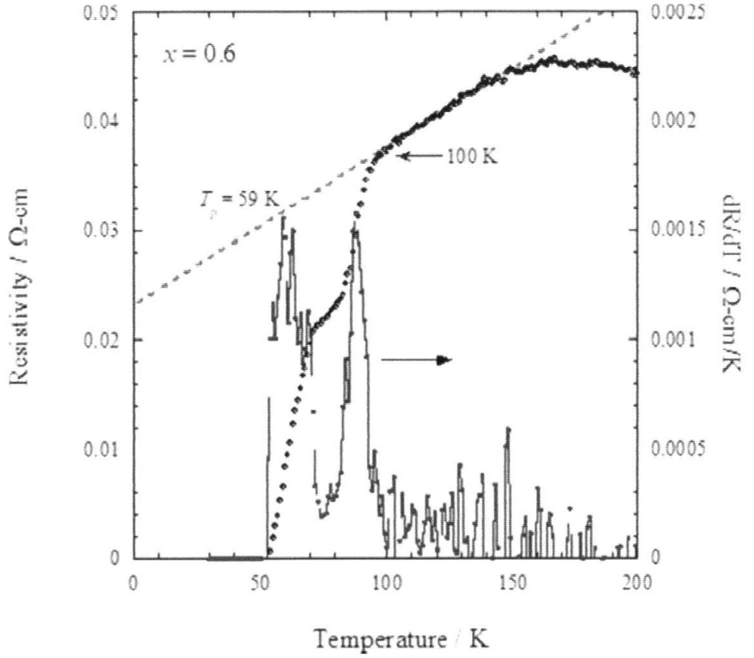

(d)

Figure 2 Normalized electrical resistance and the derivatives, dR/dT of $(Tl_{1-x}Ta_x)(BaSr)CaCu_2O_7$ for (a) $x = 0$, (b) $x = 0.4$ (c) $x = 0.5$ and (d) $x = 0.6$. Tp is the temperature at the peak of dR/dT.

The FESEM micrograph of $(Tl_{1-x}Ta_x)(BaSr)CaCu_2O_7$ for $x = 0$, 0.4, 0.5 and 0.6 are shown in Fig. 3. The non-substituted sample showed closed-packed plate and granular-liked grains structures with defined grained boundaries. The microstructure in the Ta substituted samples displayed grains without well-defined boundaries which were randomly oriented, indicating the evidence of partial melting.This has resulted in the lowering of T_p and $T_{c\ zero}$ which demonstrated that all Ta substituted samples had the weakening of grain coupling.

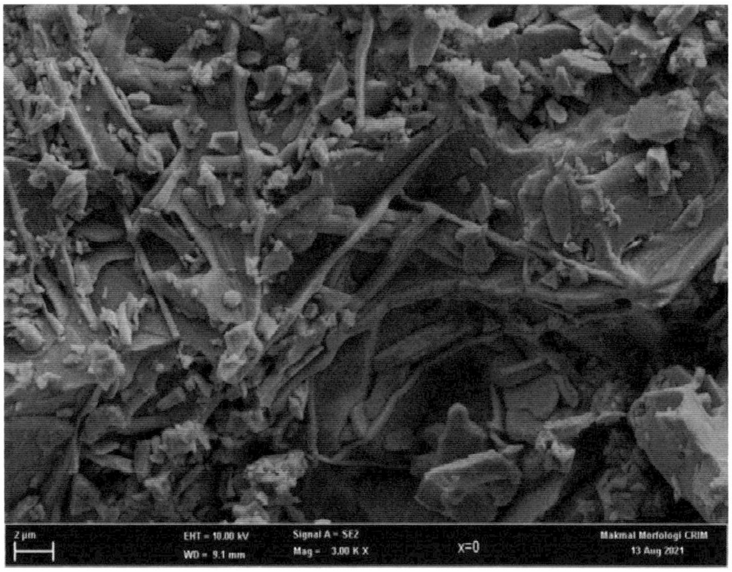
2 μm
EHT = 10.00 kV
WD = 9.1 mm
Signal A = SE2
Mag = 3.00 K X
X=0
Makmal Morfologi CRIM
13 Aug 2021

(a)

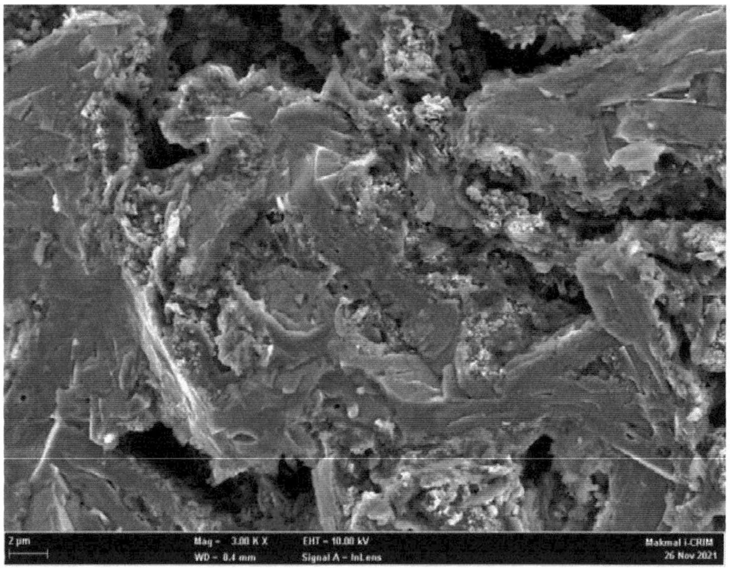
2 μm
Mag = 3.00 K X
WD = 8.4 mm
EHT = 10.00 kV
Signal A = InLens
Makmal i-CRIM
26 Nov 2021

(b)

IOP Conf. Series: Materials Science and Engineering 1258 (2022) 012003 doi:10.1088/1757-899X/1258/1/012003

(c)

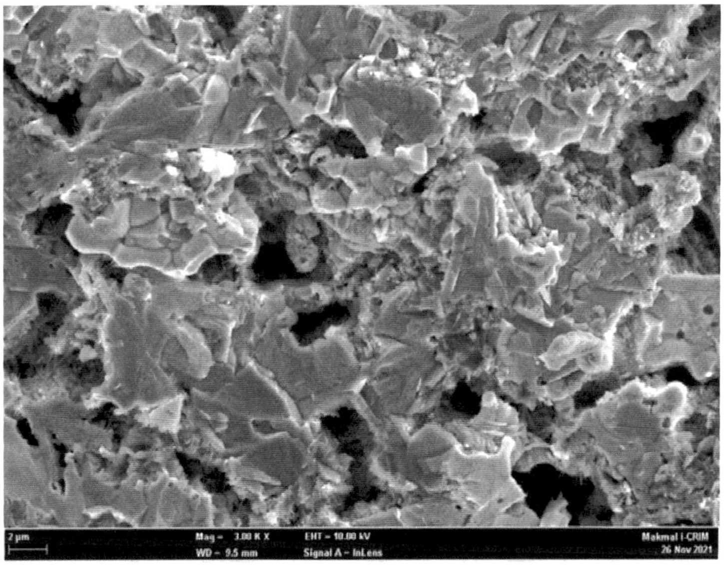

(d)

Figure 3. FESEM micrographs for $(Tl_{1-x}Ta_x)(BaSr)CaCu_2O_7$ for (a) $x = 0$, (b) $x = 0.4$, (c) $x = 0.5$ and (d) $x = 0.6$.

4. Conclusion

In conclusion, the investigation on how the Ta substitution affected $(Tl_{1-x}Ta_x)(BaSr)CaCu_2O_7$ for $x = 0$, 0.4, 0.5 and 0.6 was conducted. XRD pattern showed the existence of mixed phase with Tl-2212 as the major phase and Ta_2O_5 as an impurity for all substitution samples. Even though the Ta ionic radius is smaller than the Tl ionic radius, this research findings demonstrated that $T_{c\ onset}$, $T_{c\ zero}$ and T_p decreased in the sample with Ta substitution ($x \geq 0.4$). The sample with $x = 0$ with the highest T_c onset, $T_{c\ zero}$ and T_p was found to consist of the best superconducting properties. When Ta was substituted, the microstructure indicated partial melting. Further investigation on the effects of lower and higher Ta content should be conducted to assess the optimum stage of precursor preparation in order to enhance intergrain coupling and the final microstructure.

5. Acknowledgements

This study was funded by the Ministry of Higher Education of Malaysia under grant no. FRGS/1/2020/STG07/UKM/01/1. A. N. Jannah would like to thank UiTM (600-KNS (PJI(CRSF .5/3 (01/2022) and Ministry of Higher Education of Malaysia for post-doctoral grant.

References

[1] J. L. Jorda, Thallium-based superconducting cuprates, in: A. V. Narlikar (Eds.), *Frontiers in Superconducting Materials*, Springer, Berlin Heidelberg, 2005, pp. 833-868
[2] Abd-Shukor R and Quah C T 1994 *Mater. Sci. Eng.* B **23** 54
[3] Chen Z Y, Sheng Z Z, Tang Y Q, Li Y F and Pederson D O 1992 *Solid State Commun.* **83** 895
[4] Sheng Z. Z, Xin Y, Gu D X,. Meason J M, Bennett J and Ford D 1991, *Z. Phys.* B **84** 349
[5] Sheng Z Z, Gu D X, Xin Y, Pederson D O, Finger L W and Hadidiacos C G 1991 *Mod. Phys. Lett.* B **5** 635
[6] Akimov A I, Ksenofontov V, Lebedev S A and Tkachenka T M 2006 *Physica* C **443** 29
[7] Gerashchenko A P, Verkhovskii S V, Mikhalev K N, Piskunov Y V, Anan'ev A V, Okulova K A, Shustov L D, Yakubovskii A.and Yu A 1998 *Phys. Met. Metallogr.* **86** 43
[8] Abd-Shukor R 1993 *J. Mater. Sci. Lett.* **12** 1428
[9] Subramaniam M A, Calabrese J C, Torardi C C, Gopalakrishnan J, Askew T R, Flippen R B, Morrisey K, Chowdhry U and Sleight A W 1988 *Nature* **332** 420
[10] Salleh F M, Yahya A K, Hamadneh I and Abd-Shukor R 2002 *Solid State Sci. Technol.* **10** 30
[11] Jannah A N, Ilhamsyah A B P and Abd-Shukor R 2021 *Cryogenics* **113** 103222
[12] Misture S, Park C, Snyder R L, Jobst B and Seebacher B 1995 *Powder Diffr.* **10** 296
[13] Rao C N R, Nagarajan R and Vijayaraghaven R 1993 *Supercond. Sci. Technol.* **6** 1
[14] Lee W H, Yeh M R and Sung L R 1999 *Physica* C **311** 297
[15] Lee W H and Wang D C 1995 *Physica* C **253** 156

Equilibrium molecular dynamics simulation for investigation on the enhancement of thermal conductivity in ethylene glycol-based ZnO nanofluid

Shubhankar Sarkar * and Nanda Kumar Ghosh

Department of Physics, University of Kalyani, Kalyani-741235, WB, INDIA

*shubhankarphys21@klyuniv.ac.in

Abstract. The thermal conductivity of ethylene-glycol based ZnO nanofluid is computed using Equilibrium Molecular Dynamics Simulations in the Green-Kubo framework, taking into account the impact of nanoparticle volume fraction and nanofluid temperature. Mean square displacement (MSD) of the liquid phase for base fluid and nanofluid is also studied to investigate the thermal conductivity enhancement mechanism.

Keywords: Nanofluid, Base fluid, Thermal conductivity, Heat current autocorrelation function, Mean square displacement.

1. Introduction

Nanofluids are such a new type of fluid formed by mixing a small concentration of nanoparticles into a standard fluid, like water, ethylene glycol (EG), oils etc. Nanofluids have been widely utilised in a vast area of scientific and industrial domains, including heat transfer, cooling technologies, electrical systems, solar collectors, aircraft, and transportation. For heat transfer purposes nanofluids are used because their thermal conductivities are much greater than those of the base fluid [1]. Metal nanoparticles (Cu, Au, Ag...), metal oxide nanoparticles (Al_2O_3, ZnO, TiO_2, CuO), and carbon-based nanoparticles are the most commonly employed elements for heat transfer enhancement purposes. By using EMD simulation, the thermal conductivity increment effect on a number of graphene and hexagonal boron nitride nanosheets in water base fluid has been investigated [2]. M.J. Pastoriza-Gallego et al [3] experimentally shows the temperature and ZnO nanoparticle volume fraction effect on thermal conductivity for EG base fluid.

The goal of this study is to evaluate the thermal conductivity of an EG based ZnO nanofluid with variable nanoparticle concentration and temperature by using EMD simulation based on the Green-Kubo theorem.

2. Modelling and computational methods

In classical Equilibrium Molecular Dynamics (EMD) simulations atomic vibrations are computed and simulated by solving Newton's equation of motion. Here interactions between the atoms are modelled by various interatomic potentials and force fields. The OPLS-AA force field is used to describe the interaction between the EG molecules in the EG-ZnO nanofluid. In addition, pair-wise Lennard-Jones (LJ) potential is taken to describe the interatomic interaction.

$$U(r_{ij}) = 4\varepsilon_{ij}\left[\left(\frac{\sigma_{ij}}{r_{ij}}\right)^{12} - \left(\frac{\sigma_{ij}}{r_{ij}}\right)^{6}\right] \quad (1)$$

Where ε_{ij} indicates the depth of the potential well, σ_{ij} is the finite distance at which the pairwise potential zero (values are reported in Table 1), and the distance between atoms i and j is r_{ij}. The particle-particle-particle-mesh (PPPM) technique is used to calculate long-range Coulombic interactions. Lorentz-Berthelot combining rules have been used to determine Lennard-Jones parameters between two dissimilar interaction sites.

TABLE 1. Parameters for Lennard-Jones (LJ) interaction.

i-j	**C-C**	**C-H**	**O-H**	**Zn-O**	**Zn-C**	**Zn-H**
ε_{ij}(kcal/mol)	0.0665	0.0257	0.0258	0.0761	0.0759	0.0292
σ_{ij}(Å)	3.470	2.5025	3.07	2.745	2.965	2.65

In an EMD simulation, the Green-Kubo [4] theory is employed to relate the equilibrium heat current autocorrelation function to thermal conductivity.

$$\lambda = \frac{1}{3k_B V T^2} \int_0^{\infty} \langle J(0) \cdot J(t) \rangle \, dt \qquad (2)$$

Here λ, k_B, V, T, J denotes the thermal conductivity, Boltzmann constant, system volume, temperature, heat current vector respectively. The heat current autocorrelation function (HCACF) is $\langle J(0) \cdot J(t) \rangle$, where angular bracket represents the ensemble average.

EG molecules in the simulation box are filled in the shape of a simple cubic lattice to avoid overlapping in a random placement of molecules. To maintain the density of 1.101gm/cm^3, the simulation box size is taken 3.632 nm × 3.632 nm × 3.632 nm with 512 EG molecules. Since the simulations are run at a set temperature, the NVT ensemble is utilised. In which, dealing the whole period of the simulation, the system contains fixed number of total atoms, volume and obviously temperature. To implement NVT ensemble the Nose-Hoover thermostat is used. Velocity Verlet algorithm is used for the process of integration. The smallest integration time step is taken as 1 fs. Periodic boundary condition (PBC) is used in three dimensions of the simulation box with an interatomic interaction Cut-off radius of 9.25 Å. The Large-scale Atomic/Molecular Massively Parallel Simulator (LAMMPS) Package is used for all EMD simulations. To visualize the simulation, the Open Visualization Tool (OVITO) is used. In EMD simulation, mean square displacement (MSD) analysis is one of the most essential tools for understanding the dynamical characteristics of nanofluids. The average squared distance travelled by an atom is measured by the flowing equation

$$\text{MSD(t)} = \langle \{r_i(t) - r_i(0)\}^2 \rangle \qquad (3)$$

Where the vector distance travelled over a time period t by the ith particle is $\{r_i(t) - r_i(0)\}$. Angular bracket indicates average over all particles of this square vector distance travelled by the ith particle.

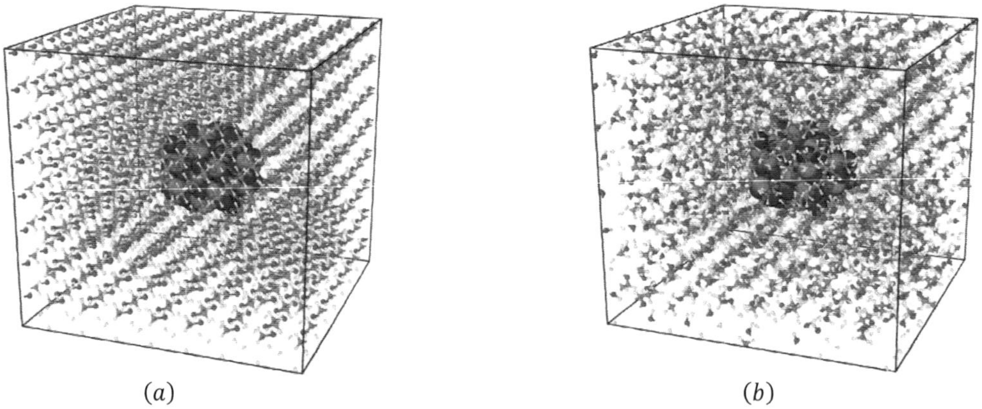

(a) (b)

Figure 1. (a) Initial simulation snapshot, and (b) Intermediate simulation snapshot of EG+ZnO nanofluid.

3. Results and Discussion

For various volume fractions (3.0%, 4.4%, 6.1%) of nanoparticle loading, thermal conductivity of base fluid and nanofluids is determined using Green-Kubo technique by EMD simulation, and these values are presented in Figure 2. From this figure it is observed that thermal conductivity of EG+ZnO nanofluid increases with nanoparticle volume fraction and temperature. These increments are similar to the previously published experimental data [3].

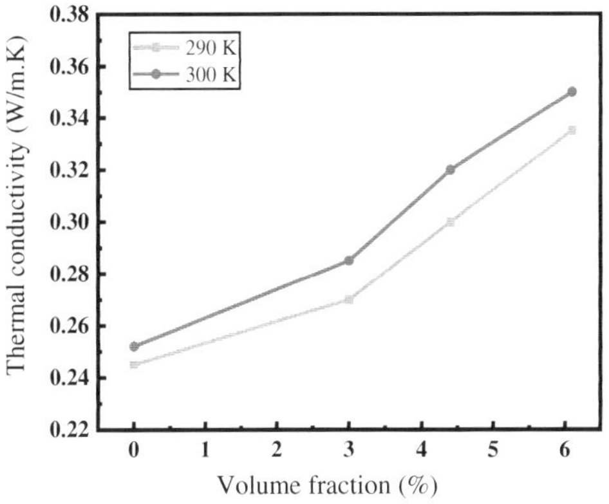

Figure 2. Change of thermal conductivity with nanoparticle volume fraction for the temperature of 290 K & 300 K.

Figure 3 showed HCACF for the nanofluid of volume fraction 3.0 %. In this case, HCACF declined to zero in 0.5 ps and the nature of this decay is oscillatory, which is due to back scattering of phonons carrying thermal energy with the solid nanoparticles. This type of behaviour also reported previously [5]. The integral of the HCACF is directly related to heat conductivity.

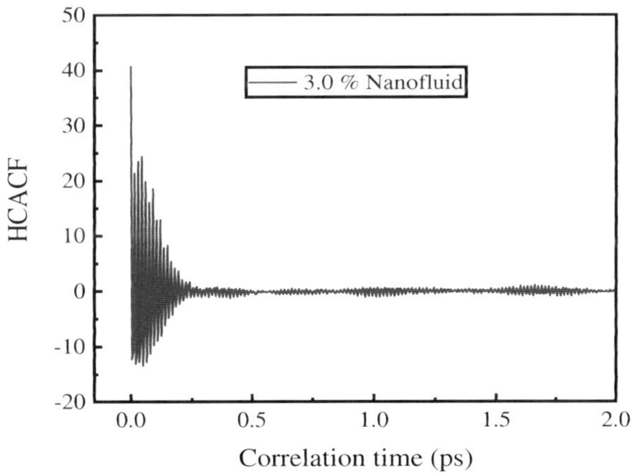

Figure 3. Change of heat current autocorrelation function with correlation time for the nanofluid of volume fraction 3.0%.

In the following paragraph of this section, MSD is investigated to explain the reasons of thermal conductivity increment in nanofluids by using EMD. MSD is the measurement of how far an atom travels on average. As MD can readily follow all atoms' displacements or motions in liquids and solids,

MSD of liquid atoms in nanofluid and base fluid is calculated and plotted. Figure 4, shows the variation of liquid phase atom's MSD in base fluid and nanofluid with simulation time for the particle volume fraction of 3.0%, 4.4% and 6.1%.

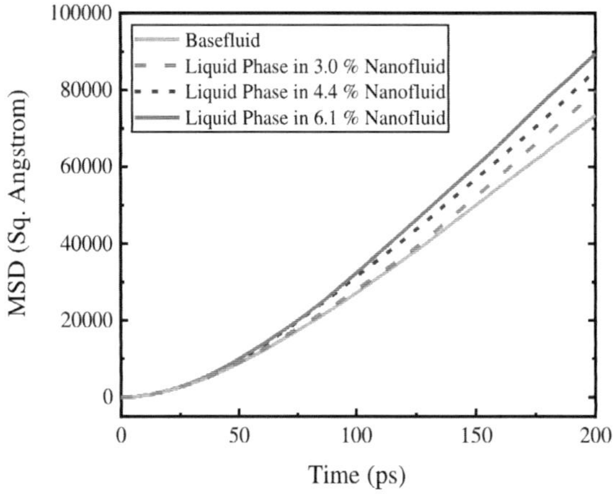

Figure 4. Variation of MSD of base fluid and liquid phase of nanofluid with time for different nanoparticle volume fractions at 300 K.

The MSD of the liquid phase of nanofluid is higher than the base fluid, and it increases as the particle volume percentage rises [6]. The enhanced mobility of liquid atoms is generated by interatomic interactions caused by the potential difference between metal oxide and liquid, as well as the heavier mass of nanoparticles. This enhancement of liquid phase MSD is significant because it increases heat transferability.

4. Conclusion

In this communication, EG based ZnO nanofluid systems is simulated to evaluate the thermal conductivity of the base fluid and nanofluid for various nanoparticle volume concentrations by using EMD simulation with the Green-Kubo formalism. Thermal conductivity of the nanofluids rises with particle volume fraction and temperature, but not in a linear way. From the mean square displacement (MSD) calculation, it can be concluded that the movement of liquid atoms in nanofluid increases dramatically compared to liquid atom movement in a base fluid, which is one of the main reasons for the increment of thermal conductivity.

Acknowledgement

S. Sarkar would like to acknowledge the support of Council of Scientific and Industrial Research (CSIR, Government of India) for a Junior Research Fellowship.

References

[1] Khamliche T, Khamlich S, Doyle T B, Makinde D and Maaza M 2018 *Materials Research Express* **5** 035020
[2] Shit S P, Pal S, Ghosh N K and Sau K 2021 *Journal of Molecular Structure* **1239** 130525
[3] Gallego M J P, Lugo L, Cabaleiro D, Legido J L and Piñeiro M M 2014 *The Journal of Chemical Thermodynamics* **73** 23-30
[4] Schelling P K, Phillpot S R and Keblinski P 2002 *Physical Review B* **65** 144306-12
[5] Khamliche T, Khamlich S, Moodley M K, Mothudi B M, Henini M and Maaza M 2021 *Journal of Molecular Liquids* **323** 114975
[6] Sarkar S and Selvam R P 2007 J. Appl. Phys. **102** 074302

Effect of next to next-nearest-neighbor (NNNN) hopping interaction in mixed valence systems and study of the thermodynamic properties

Piyali Ghosh*and Nanda Kumar Ghosh

Department of Physics, University of Kalyani, Kalyani-741235, WB, INDIA
*piyalighosh@klyuniv.ac.in

Abstract. The extended Falicov-Kimball model containing the next to next-nearest-neighbor hopping interaction (NNNN) is investigated using the exact diagonalization calculations in two dimensions. The f-electron density n_f with the energy of f-level E_f is investigated using the allowable exact solutions. The temperature dependency of thermodynamical parameters such as specific heat(C) and entropy (S) are also studied here. The significant observations are (i) a discontinuous valence transition (insulator-to-metal) at a certain energy of f-level (E_c) which varies with the value of NNNN interaction term (t_3), (ii) disorderedness of the system with the increasing value of t_3, (iii) the structures of single and double peaks in the specific heat versus temperature graphs varying with $t_3$and E respectively.

Keywords: Mixed valence systems, Falicov-Kimball model, hopping interaction, exact diagonalization calculations, valence transition.

1. Introduction

Several articles regarding the unusual physical phenomena of mixed valence compounds (i.e., SmB_6, *TmSe*, *SmS* etc.) [1-3] have already been published on the theoretical concept of Falicov-Kimball model (FKM) [4]. For example, non-local Coulomb interaction [5-7], correlated hopping [8], nearest neighbor hopping of d-electrons [9], next to nearest-neighbor (NNN) hopping interaction of *d*-electrons [10] etc. have been already studied. If an electron has sufficient kinetic energy to overcome the potential barrier of surrounding fields, it hops from one site to another site of the crystal. But the effect of next to next-nearest-neighbor (NNNN) hopping interaction on valence transition is not studied still now. If the hopping is strong enough, the electron can jump from one band (*d*-band) to another (*f*-band).

In this theoretical work, the extended FKM is further extended by addition of the third neighbor hopping interaction term (t_3). Actually, here the noticeable effect of this type of interaction on valence transition is investigated. The thermodynamic properties such as specific heat, entropy of mixed valence compounds are also examined through numerical simulations by exact diagonalization study.

2. Hamiltonian and formulation

Here, a two band, eight-site and the cluster of 2D square lattice is considered. The total Hamiltonian is taken as

$$\mathrm{H}=H_0+\mathrm{H}' \qquad (1)$$

Where, $H_0=\mathrm{E}\sum_{i\sigma}f_{i\sigma}^{+}f_{i\sigma}+\mathrm{U}\sum_{i}f_{i\uparrow}^{+}f_{i\uparrow}f_{i\downarrow}^{+}f_{i\downarrow}+\mathrm{G}\sum_{i\sigma\sigma'}f_{i\sigma}^{+}f_{i\sigma}d_{i\sigma'}^{+}d_{i\sigma'}+$
$$\mathrm{V}\sum_{<i,j>\sigma}(f_{i\sigma}^{+}d_{j\sigma}+d_{j\sigma}^{+}f_{i\sigma})-t_1\sum_{<i,j>\sigma}d_{i\sigma}^{+}d_{j\sigma}-t_2\sum_{<pq>\sigma}d_{p\sigma}^{+}d_{q\sigma} \qquad (2)$$

$$\mathrm{H}'=-t_3\sum_{<m,n>\sigma}d_{m\sigma}^{+}d_{n\sigma} \qquad (3)$$

Where, <*i*, j >, <p,q> and < m, n > are all pairs of NN, NNN and NNNN sites respectively on 2D square lattice; $f_{i\sigma}^{+}$ and $f_{i\sigma}$ are the creation and annihilation operators for an electron in the localized state at lattice site *i* with binding energy *E*, for a conduction band electron $d_{i\sigma}^{+}$ and $d_{i\sigma}$ are the creation and annihilation operators(σ,σ'=spin). The on-site Coulomb interaction is denoted by the letter *U*, the term *G* refers to the Coulomb type interaction between *f* and *d* electrons, between the *f* and *d* electrons the hybridization interaction term *V* is activated, t_1, t_2 and t_3 represent the itinerant *d*-electrons' quantum- mechanical hopping between NN, NNN and NNNN sites respectively.
The chosen eight-site basis state is in the form

$$|n_{1\uparrow}^{f}n_{1\downarrow}^{f}n_{1\uparrow}^{d}n_{1\downarrow}^{d}n_{2\uparrow}^{f}n_{2\downarrow}^{f}n_{2\uparrow}^{d}n_{2\downarrow}^{d}n_{3\uparrow}^{f}n_{3\downarrow}^{f}n_{3\uparrow}^{d}n_{3\downarrow}^{d} \dots\dots\dots n_{8\uparrow}^{f}n_{8\downarrow}^{f}n_{8\uparrow}^{d}n_{8\downarrow}^{d} > \quad (4)$$

The linear combination of basis vectors as in Eq. (4) produces the eigenvectors of H.

f-electron density $< n_i^f > = \frac{1}{N}\sum_{i\sigma} f_{i\sigma}^{+} f_{i\sigma}$

Where, the number of lattice sites is indicated by the letter N.

Per lattice site entropy is calculated using

$$S = \frac{1}{N}(k_B \ln Z + \frac{<H>}{T})$$

Specific heat at low temperature is defined by

$$C = k_B\beta^2 \frac{\partial^2}{\partial\beta^2}\ln Z$$

Where, $Z=\sum_a e^{-\beta E_\alpha}$,the total is calculated over all basis states, the related eigenvalues are E_a's and $\beta = \frac{1}{k_B T}$, k_B is thewell-known Boltzmann constant (the value of k_B is taken unity during calculations of my present articles).

3. Results and discussions

The outcome of various experimental observations supplies here the values of used parameters [11-13].

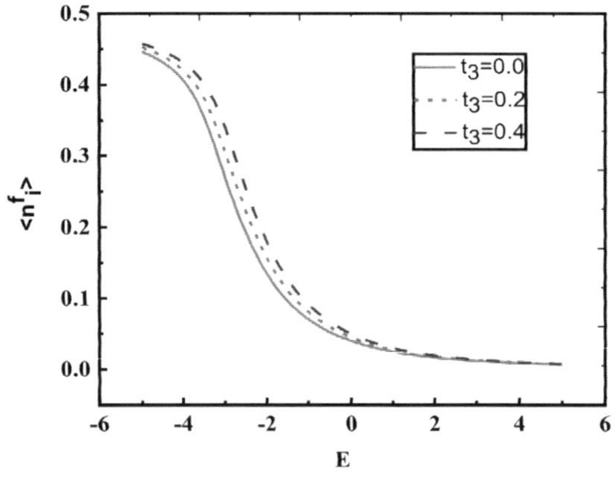

Figure 1. f-electron density $<n_i^f>$ vs. E for distinct values of NNNN hopping interaction term t_3. Here, V=0.5, G=1, U=5, t_1=1,t_2= -0.4.

Figure1 depicts the nature of *f*-electron density with *f*-level energy for distinct values of NNNN hopping interaction term (t_3). The valence transition of first order occurs here for lower values of t_3 at a certain value of E=E_c (it is a particular energy value of f-level at which valence transition happens). E_c is different for distinct values of t_3 and E_c shifts to higher E as t_3 is increased. So, a conclusion can be made that the energy of the *d*-level (the reference level energy) is shifted by the interaction t_3 caused by hopping of *d*-electrons. From a certain maximum value of $< n_i^f > \approx 0.45$ (approx.) it decreases with the energy of *f*-level *E* for distinct values of t_3.Then it becomes almost constant i.e., $<n_i^f> \approx 0$, as f-electrons jump to conduction band (*d*-band) due to increase in f-level energy (which parameterizes the pressure).

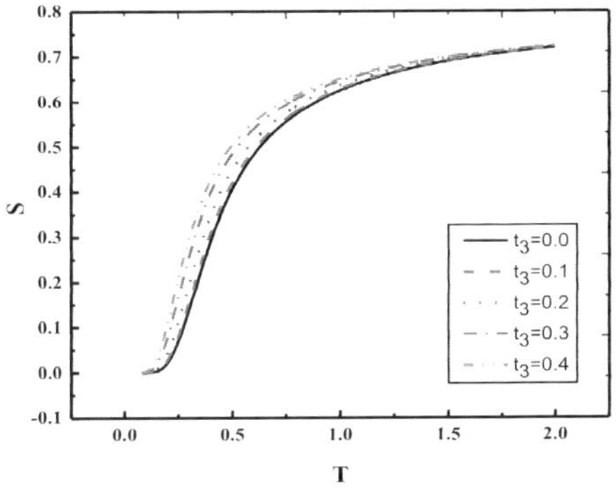

Figure 2. Entropy vs. Temperature graphs for distinct values of t_3. Here *E= -2, V=0.5, G= 1,* t_1*=1,*t_2*= -0.4, U=5.*

Figure 2 shows the nature of graph between entropy and temperature for different values of NNNN hopping interaction term t_3.As entropy measures the disorderedness of the system, from Figure 2 it can be concluded that the system goes to disordered state with the increasing value of t_3. It reveals the fact that the effect of next to next-nearest-neighbor of interaction of d-electrons exists there. The region 0<T<1.5 (approx.), where the splitting of graphs is observed, is identified as the transition temperature (T_c) region [14]. At

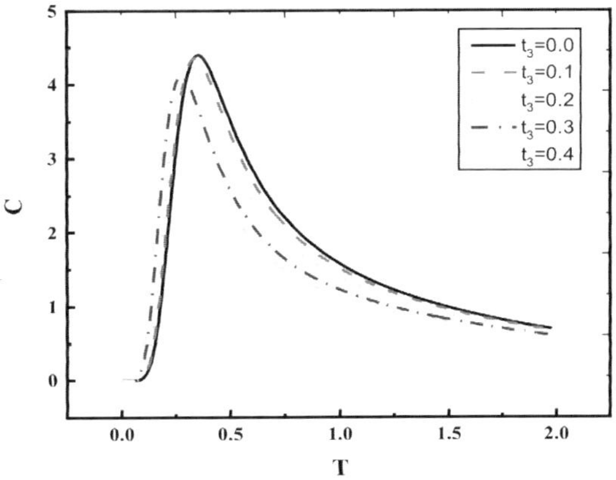

Figure 3. Graphs of specific heat *C* vs. Temperature for various t_3 values. Here, *E= -2, G= 1, V=0.5,*t_1*= 1,*t_2*= -0.4, U=5.*

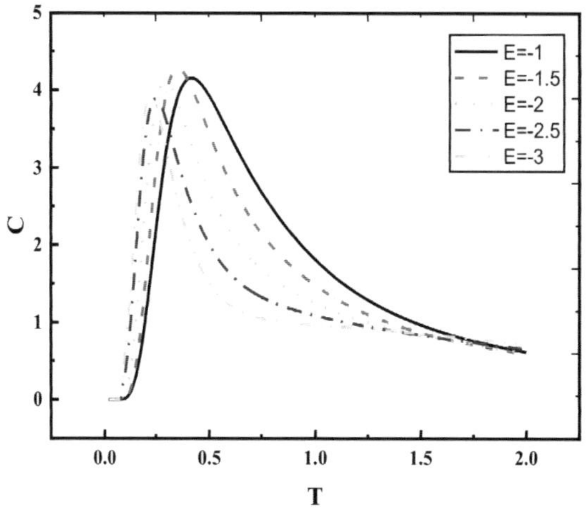

Figure 4. Temperature vs. Specific heat (*C*) graphs for distinct values of *E*. Here, *V=0.5, G=1, U=5,* t_1*=1,*t_2*=-0.4* t_3*=0.4.*

higher temperatures, i.e., T>1.5 (approximately), entropy tends to a limiting value (complete disorder in the system).

From Figure 3 and figure 4, an idea is emerged about the variation of low-temperature specific heat with temperature. It is noticed that the graphs of specific heat show either a single-peak (Figure 3) as well as double peak structure (Figure 4). The peak height (which is proportional to density of states) of sharp peaks (Figure3) decreases as t_3 increases and changes to a lower temperature region. The structure of double peak for E≤ E_C [Figure 4] is obtained. The first sharpened peak is caused by a significant number of degenerate many-body states with respect to the ground state. At higher temperatures, the wide second peak is usually Schottky-type.

4. Conclusion

In summary of the present work, there is no doubt that the effect of NNNN interaction is prominent in case of mixed valence system. An insulator to metal transition being discontinuity in nature is noticed at certain energy of f-level (E_c). The effect of inclusion of NNNN interaction is that E_c is different for different values of NNNN hopping interaction (t_3). The system becomes more disordered state with the increasing value of t_3. The specific heat vs temperatures curves exhibit single peak as well as double-peak structure. From our present study, we can conclude that the effect of NNNN hopping interaction should be taken into account to precisely design solar cells, Computer monitors etc. in the application side.

5. References

[1] Menth A, Buehler E and Geballe T H 1969 *Phys. Rev. Lett.* **22** 295.
[2] Nickerson J C, White R M, Lee K N, Bachmann R, Geballe T H and Hull G W 1971 *Phys Rev. B* **3** 2030.
[3] Allen J W, Batlogg B and Wachter P 1979 *Phys. Rev. B* **20** 4807.
[4] Falicov L M and Kimball J C 1969 *Phys. Rev. Lett.* **22** 997.
[5] Cencarikova H, Farkasovsky P and Zonda M 2008 *Acta Phys. Polon.* A **113** 287.
[6] Bhowmick S K and Ghosh N K 2012 *Pramana - J. Phys.* **78** 289.
[7] Farkasovsky P 2019 *Eur. Phys. J. B.* **92** 141.

[8] Ghosh N K, Bhowmick S K and Mondal N S 2011 *Pramana -J.Phys.* **76** 139.
[9] Ghosh N K and Mukherjee P 2016 *AIP Conference Proceedings* **1728** 020016.
[10] Bhowmick S K and Ghosh N K 2012 *Indian J. Phys.* **86(5)** 345.
[11] Varma C M 1976 *Rev. Mod. Phys.* **48** 219-238.
[12] Wohlfeld K L, Daghofer M, Oleś A M and Horsch P 2008 *Phys.Rev.* B **78** 214423(1-24).
[13] Farkasovsky P 1995 Phys. Rev. B **51** 1507.
[14] Mondal N S and Ghosh N K 2010 *Pramana – J. Phys.* **74** 1009.

Single Step Electrosynthesis of Co-Ferrite Thin Films from Alkaline Non-Aqueous Bath

M R Kale*
Department of Physics,
ASP College Devrukh, Dist: Ratnagiri, Maharashtra. India

* Corresponding author
Email: dr.mrkale@gmail.com

Abstract: Electrosynthesis in a single step was used to obtain high-quality Co-ferrite thin films from an alkaline non-aqueous bath at room temperature. Several preparative parameters were optimized to obtain smooth and adherent Co-ferrite thin films. From XRD studies, cubic crystal formation was confirmed. IR studies confirmed spinel structure formation in Co-ferrite thin films. From SEM and AFM studies, smooth, uniform, and compact film formation was observed. From magnetic measurements, the value of saturation magnetization was (304 emu/cm^3) obtained it is in good agreement with that of bulk value.

Keywords: Electrosynthesis, Non-aqueous bath, Co-ferrite thin films, structural and magnetic properties.

1. Introduction

For producing thin-film recording heads, magnetic thin films are significant to meet the future trends in high-density magnetic recording [1]. Cobalt ferrite (Co–Ferrite) which is a cubic ferromagnetic oxide, is a potential material for recording media with high-density. In general, inverse spinel structure has one-half of Iron $^{3+}$ ions on the 'A' site and the remaining, together with Cobalt^{2+} ions on the B site at room temperature. The thermal history of the sample is instrumental for the cation distribution in Co-Ferrite. [2-3]. Because of its physical properties such as greater recording density, reasonable saturation, more coercivity, hardness, and chemically stable behavior, thin films of Co-ferrite are prominent candidate for magneto-optical recording [4-6]. Bulk material as well as in thin films of, Co-ferrite has noteworthy applications in transformer cores, antenna rods, heads for recording, memory coils, magnetic refrigeration, microwave instruments, biomedical applications, and conversion of solar energy into H_2 [7-10].

Various chemical and physical techniques such as deposition with ion beam, Rf sputtering, sol-gel method, spray pyrolysis, ferrite plating, etc. have been used in the preparation of Co-ferrite thin films. [11-21]. The electrochemical reaction arises on the working electrodes, where either oxidation or reduction observed during the electrosynthesis.

Principally, electro-deposition can be done from aqueous and non-aqueous baths. The reduction of water is kinetically mostly observed at the electrode materials with the use of aqueous solutions. Hence aqueous solutions are problematic during electro-deposition. By using non-aqueous solvents, serious problems such as precipitation of ferric ($Fe(OH)_3$) can be avoided that were faced in case of aqueous solutions. [22]. Practically all-commercial electroplating baths contain one or more additive agents, which give smooth and bright deposits. During electro-deposition various electro-active species can be added to the electrolytes for improving coating characteristics. The addition of these electro-active species can also improve surface morphology and physical properties of deposits. Intercalation of oxygen can be possible in electrodeposited species and is a well-known phenomenon. The presence of oxygen in the electrodeposition bath gives oxidation of the deposited metal. An increase in pH near the cathode results in the hydrolysis phenomena and consequent precipitation of any iron group [E.g., Co (II), Ni (II)]. An alkaline bath generally depends on the presence of a complexing agent and electrodeposition from an alkaline bath helps incorporate the oxygen species [23-25].

Sartale et. al. [26] have deposited ferrite thin films by using an electrochemical route. The same is a promising substitute for other methods of deposition which also overcome the demerits occurred due to higher deposition temperature. The electrochemical route consisted of two steps during deposition. In the first step, electrodeposition of various alloy films such as MFe_2 (where M=divalent metal ion) and formation of ferrite thin films can be achieved by their subsequent electrochemical oxidation (anodization).

The present article is an outcome for the room temperature deposition of Co-ferrite thin films by understanding all the above factors during electrodeposition. In the electrochemical route, during the transfer of substrate, the possibility of contamination of alloy sample is possible after electrodeposition of alloy for anodization (oxidation). Thus, considering this drawback of the two-step electrochemical routes, there is a need to modify this two-step method into single step for the ferrite thin films formation. Using electrosynthesis from an alkaline non-aqueous bath, this two-step electrochemical route is modified into a single step. The Co-ferrite thin film was deposited from a mixture of cobalt sulfate, iron sulfate, citric acid (complexing agent), and sodium hydroxide using ethylene glycol as an organic solvent. Various preparative parameters were optimized during the deposition of Co-ferrite thin films to deposit good quality, well adherent, and smooth film formation. Post-heat treatment was introduced for enhancement of the properties. Structural characterizations of thin films of Co-ferrite were done with the help of X-ray diffraction (XRD), infrared spectroscopy (IR), scanning electron microscopy (SEM), and atomic force microscopy (AFM) techniques. Magnetic parameters for thin films of Co-ferrite were conducted with the use of Vibrating Sample Magnetometer (VSM) technique.

2. Experimental

2.1 Co-Ferrite thin films preparation

2.1.1 Electrolytes used:

All the chemicals used for the deposition of Co-ferrite thin films were of Analytical Grade (AR) and used as such supplied by S. D. Fine Chemicals Ltd.; India. All the solutions, except the solution of sodium hydroxide, were prepared in ethylene glycol to make the bath non-aqueous. The required amount of chemicals was weighed and dissolved in a definite volume.

2.1.2 Electrodeposition of the Co-ferrite thin films:

The deposition of the thin film of Co-ferrite was done from an electrolyte-containing mixture of 0.1 M $CoSO_4$, 0.1 M $FeSO_4$, 0.1 M citric acid, and 1 M NaOH solutions in various proportions. Bath composition was confirmed with the use of atomic absorption spectroscopy (AAS) technique. Three-electrode system was used for the electrosynthesis of thin films of Co-ferrite, graphite was used as a counter electrode, the substrate acts as a working electrode and saturated calomel electrode (SCE) acts a reference electrode. The 20 mL of an unstirred solution of this mixture was used for the deposition of Co-ferrite thin films. The electro analytical measurements and deposition of thin films of Co-ferrite were carried out by using potentiostat/galvanostat model 263A connected to personnel computer viva GPIB electrochemical software M 270, in potentiostatic mode. Measurements of deposition potentials were carried out using polarization curves obtained with the help of Linear Sweep Voltammetry (LSV). Preparative parameters for depositing smooth, uniform co-ferrite thin films are tabulated in

Table 1. After deposition, thin films of Co-ferrite were washed with double distilled water and used for their further characterization.

2.1.3 Post-heat treatment

The as deposited films may contain defects, the air annealing helps to remove such types of defects and improvement in crystallinity. Annealing of electrodeposited Co-ferrite thin films was carried out in air at 773 K for 4 hours. Above this temperature, Co-ferrite thin films got peeled off from the surface of a substrate. It may be happened due to a thermal incompatibility between substrate and film. The heat treated Co-ferrite thin films were slowly cooled in the furnace.

2.2 Characterization of Co ferrite film

2.2.1 Composition of Co ferrite film

The composition of films was carried out using the atomic absorption spectroscopy (AAS) spectroscopic technique with a Perkin Elmer model 3030B spectrophotometer. At room temperature, Co-ferrite thin films were deposited onto a stainless-steel substrate. For detaching film from the surface, pre-electroforming treatment [25] was given before the deposition of Co-ferrite thin films. Normally all solutions were examined with the concentration range of 1-5 ppm using four standards. After the deposition films were dissolved into concentrated nitric acid and further that solution was diluted into known volume for measurement of the weight percentage of metal deposited in the film.

2.2.2 Structural characterization

Philips X-ray diffractometer PW 3710 was used to obtain X-ray diffraction patterns of as deposited and heat treated Co-ferrite thin films deposited onto stainless steel substrates. X-ray diffractometer with a computer controlling system and having CrKα radiations was used for analyzing X-ray diffraction patterns of co- ferrite thin films.

Perkin Elmer model 783 was used for recording IR spectra of cobalt ferrite thin films with the spectral range of 200-800 cm^{-1}. Powder of Co-ferrite thin film was scratched from the surface of the film and mixed with KBr for making pellets and IR spectra of these pellets were recorded.

2.2.3 Surface Morphological Characterization

The SEM images of Co-ferrite thin films were obtained using the Scanning electron microscope model JSM 5600. The AFM images of thin films of Co-ferrite were carried out using Digital Nanoscope (II) operating in contact mode.

2.2.4 Magnetic measurements

Magnetic measurements were carried out with a fully automated [27] vibrating sample magnetometer (VSM) by applying a magnetic field in the plane of the film. The curves of magnetization (M) versus magnetic field (H) for Co-ferrite thin films deposited over copper substrates were obtained at room temperature.

3. Results and discussion

Hydrogen evolution is an important factor in electro-deposition, that affects the kinetics, composition and iron group metals properties. For acidic solution, it can be described as, [28]

$$2H^{+} + 2e \Leftrightarrow H_2 \qquad (1)$$

Non-aqueous baths can overcome this problem of hydrogen evolution. According to Dahms and Croll [24], an increase in the pH at the cathode causes the hydrolysis and consequent precipitation of the metal ions such as Co(II), Ni(II), and Fe(II), etc., are observed.

By considering this, we have carried electrodeposition of Co ferrite thin films from the alkaline non-aqueous bath. Citric acid was added to avoid precipitate formation during the alkaline bath. As oxide formation in electrodeposition is possible in a strong alkaline bath by applying higher negative potential than reduction potential. For the deposition of co-ferrite thin films from the non-aqueous bath, sodium hydroxide solution was prepared in water. From that solution, water acts as a complexing agent which means it does not take part in the chemical reaction but helps in reducing the rate of reaction.

3.1 Bath composition

Co-ferrite thin films were deposited from the bath made up of mixture of 0.1 M $CoSO_4$, 0.1 M $FeSO_4$, 0.1 M citric acid and 1 M NaOH solutions. In this solution total quantity is constant of 20 ml, variation in quantities of 0.1 M $CoSO_4$ and 0.1 M $FeSO_4$ solutions was made in different ratios. Graph 1 shows plot of percentage of iron in deposited film with volume of 0.1 M $FeSO_4$ solution in the fixed quantity of electrolyte. It is depicted that for (10cc) 0.1 M $CoSO_4$ + 0.1 M $FeSO_4$ (10cc) bath solution, the deposited film contained atomic percentage of Co ~34% and Fe ~66% composition.

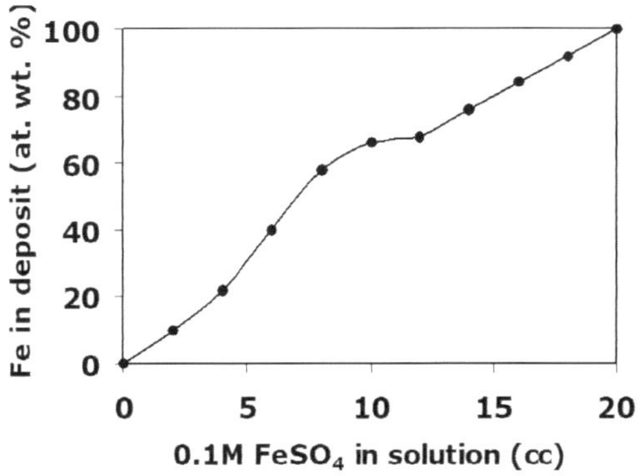

Fig.1 A Plot of Fe in deposit against quantity of 0.1 M $FeSO_4$ solution in bath.

3.2 Potential for deposition

To know the potentials for the deposition of Co-ferrite thin films linear sweep voltametry (LSV) technique was used for obtaining polarization curves for co-ferrite thin films at scan rate 50 mV/SCE and in between voltage range 0 to –1.5 V/SCE. Fig. 2 shows plot of current against voltage for (a) 0.1 M $CoSO_4$+ 0.1M $FeSO_4$ (b) 0.1M $CoSO_4$ + 0.1 M $FeSO_4$+ 0.1M citric acid and (c) 0.1M $CoSO_4$ + 0.1 M $FeSO_4$+ 0.1M citric acid+ 1 M NaOH solutions. From fig 2, it is clear that after addition of citric acid solution the deposition potential was shifted toward negative side and further by making bath alkaline, it was slightly shifted towards negative side. Thus due to the shifting of deposition potential towards more negative side, possibility of oxide formation of both these metals is considered. Various characterizations techniques confirmed the formation of co-ferrite thin films and various preparative parameters are tabulated in Table 1.

Table 1 Parameters for thin films of Co-ferrite deposited from non-aqueous bath

Optimized preparative parameters	For Co-ferrite thin films deposited from non-aqueous bath at pH 12
Bath composition	0.1 M $CoSO_4$ (10 cm^3) + 0.1 M $FeSO_4$ (10 cm^3) + 0.1 M Citric acid (in ethylene glycol) + 1 M NaOH (in distilled water)
Deposition potential, V/SCE	-0.65
Deposition time (min)	20

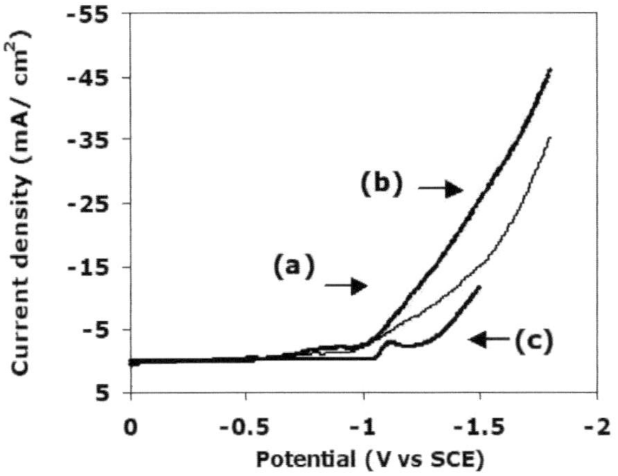

Fig.2 Polarization curves for cobalt ferrite thin films from non-aqueous acidic and alkaline baths

3.3 X-ray diffraction results

Fig. 3 shows X-ray diffraction patterns for as deposited and heat treated co-ferrite thin films. Well-defined peaks of $CoFe_2O_4$ films for as deposited and heat treated films were observed. After heat treatment preferred orientation along (311) plane was observed, as like bulk $CoFe_2O_4$ [29], and one more peak (422) was observed, which confirms formation of spinel cubic $CoFe_2O_4$ phase. Lattice parameters were calculated and are found to be matches with standard value [8] for $CoFe_2O_4$. After heat treatment, it was observed that the value of lattice parameter was greater for as deposited film than heat treated film. This can be be owing to the increase of strain in the film after undergoing the heat treatment. Observed and standard'd' values [30] matched well with each other and are tabulated in Table 2. The value of lattice parameter for cubic crystal structure was calculated using the formula,

$$d(hkl) = \frac{a}{\sqrt{(h^2 + k^2 + l^2)}}$$

Table 2. X-ray diffraction studies for thin films of Co-ferrite

Films	Standard 'd' values	Observed 'd' values	Plane	Lattice parameters
As deposited Co-ferrite	2.531	2.531	(311)	8.3943
	2.08	2.0866	(400)	8.3044
Heat treated Co-ferrite thin films	2.531	2.531	(311)	8.3995
	2.08	2.076	(400)	8.3464
	1.713	1.793	(422)	8.3256

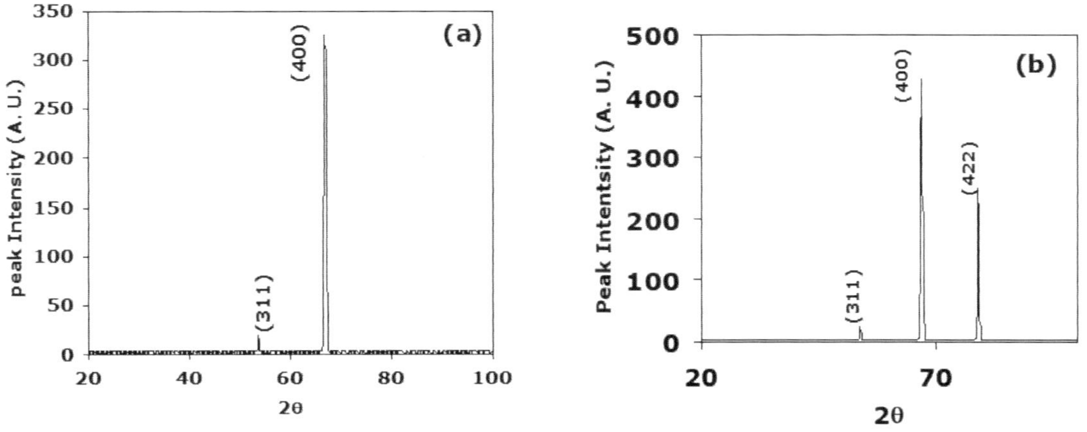

Fig. 3 XRD patterns for (a) as deposited and (b) heat treated cobalt ferrite thin films obtained from non-aqueous bath.

3.4 Infrared spectroscopy

Fig. 4 (a) and (b) shows IR spectra for deposited an annealed Co-ferrite thin films. From fig., it has been detected that for low frequency band observed at ν_1 = 407and 462 cm^{-1} and for high frequency at ν_2= 590 and 618 cm^{-1} for as deposited and heat treated co-ferrite thin films respectively. Also from fig a broad spectrum is observed for both the co-ferrite thin films and such broadening is commonly observed in inverse spinel ferrite (MFe_2O_4). Thus from IR spectra two band observed, confirmed the intrinsic vibration at tetrahedral and octahedral sites [31] and spinel $CoFe_2O_4$ formation. Shoulder and the band splitting also seen in IR spectra for thin films of co-ferrite, can be due to the occurrence of Jahn-Teller Iron^{2+} ions in the ferrite film.

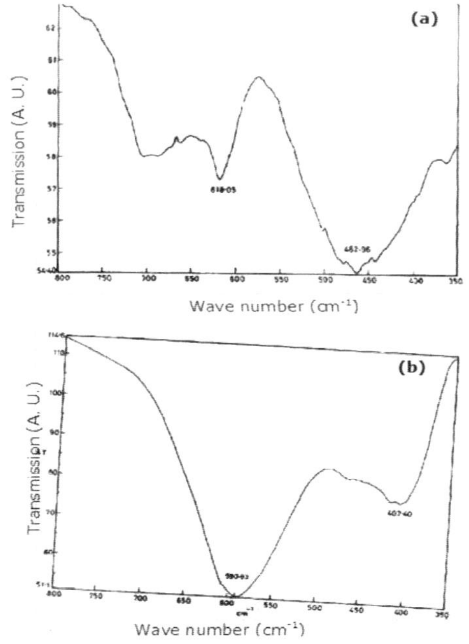

Fig. 4 IR spectra for (a) as deposited and (b) annealed cobalt ferrite thin films.

IOP Conf. Series: Materials Science and Engineering 1258 (2022) 012006 doi:10.1088/1757-899X/1258/1/012006

3.5 Surface morphology

Figures 5 (a) and (b) show SEM pictures of as deposited and heat treated Co-ferrite thin film. From images it has been observed that films are crack free, uniform and smooth. Heat treated co-ferrite thin films showed an increase in grain size and compactness of film. Also films are grown in the orientation of surface of substrate. Thus from SEM studies, necessity of post annealing is confirmed and well covered, completely grown co-ferrite thin film formation was observed.

Fig. 5 SEM images for (a) as deposited and (b) heat treated cobalt ferrite thin films.

3.6 Atomic force microscopy studies

As X-ray diffraction and SEM studies, confirmed that complete growth of $CoFe_2O_4$ takes place after heat treatment so that AFM pictures of heat treated Co-ferrite thin films were carried out. Fig. 6 (a) and (b) shows topographical 3-dimensional AFM images of deposited and heat treated thin films of co-ferrite. RMS value for surface roughness of Co-ferrite thin film was calculated from three-dimensional view of AFM image [32]. From fig. 6 (b) it is found to be around about 20 nm which confirmed co ferrite thin films are smooth and uniform.

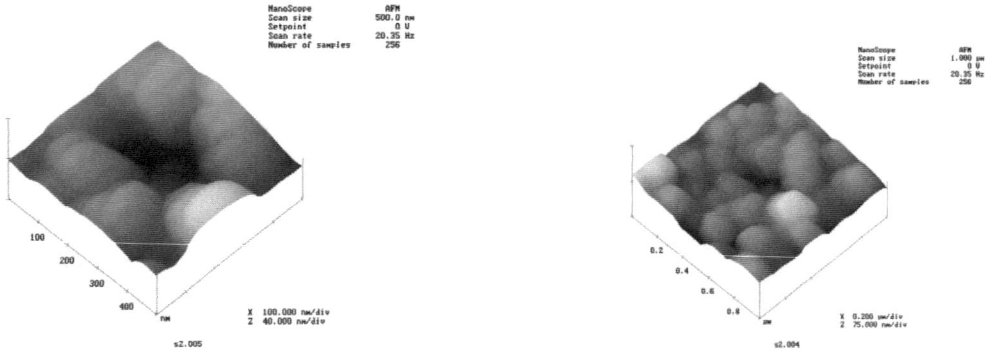

Fig. 6 three dimensional AFM images of (a) as deposited and (b) annealed cobalt ferrite thin films

3.7 Magnetic studies

Fig. 7 shows hysteresis curve for annealed Co-ferrite thin films, which shows normal behaviour observed for inverse spinel ferrite. This curve shows that the, saturation magnetization is 295 emu/cc, which is in consonance with the bulk value of saturation magnetization [33] for Co-ferrite. Value of residual magnetization found to be 100 emu/cc and coercive force was 0.05 Koe.

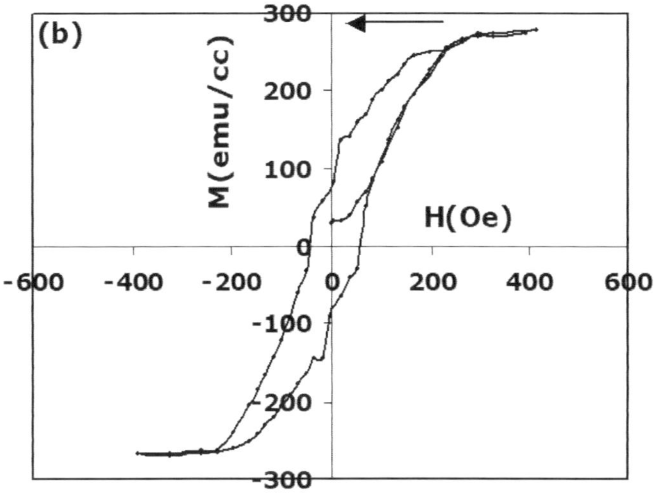

Fig. 7 plot of magnetization (M) against magnetic field (H) for annealed cobalt ferrite thin film.

4. Conclusions

Thin films of Co-ferrite were deposited at room temperature in a single step electro-deposition. Non-aqueous bath escaped the hydrogen evolution problem and gives an enhanced quality of thin films of Co-ferrite. With the use of alkaline bath, introduction of oxygen in solid film was achieved. It results into thin films of Co-ferrite formation at room temperature. Small-scale post heat treatment enabled to achieve good quality of Co-ferrite thin films, which showed comparable results to that of the bulk. In structural investigations, XRD studies showed formation of cubic spinel Co-ferrite thin films with preferred orientation along (311) plane. From IR studies, spinel structure formation is confirmed. With the help of SEM and AFM images surface morphology was examined, which showed smooth, well covered growth to surface of substrate, uniform and compact Co-ferrite thin film formation. From magnetic measurements, value of saturation magnetization found to be comparable to the bulk cobalt ferrite. To conclude, it can be summed up that better thin films of Co-ferrite are electrosynthesised at room temperature from alkaline non-aqueous bath. Thus electrodeposition is an eco-friendly method and does not require sophisticated instrumentation and high temperature, which avoids many environmental hazardous events, and it give ferrite film formation at room temperature too.

References

[1] E. S. Murdoch, 1992 *IEEE Trans. Magn.* **28** 3078
[2] G. D. Rieck, J. J. M. Thijssen, 1968 Acta *Crystallogr.* **B**. 982
[3] G. A. Sawatzky, F. Vander Woude, A. H. Morish,1968 *J. Appl. Phys.* **39** 1204.
[4] K. C. Patil, S. Sundar Moharan, D. Gajapathy, in N. P. Cheremisinoff 1990 (Ed). *Handbook of Ceramics and composites,* **Vol. 1** Synthesis and properties, Marcel Dekker, Inc New York.
[5] V. S. Darshane, S. S. Lokegaonkar, S. G. Oak, 1997 *J. Phys. IV France* **7** C1-683.
[6] Y. Tamara, M. Kojima, N. Hasegawa. M. Tsuji, K. Eherensberger, A. Steinfeld, 1997 *J. Phys; IV France* **7** C1- 673
[7] F. Bodker, S. Morup, S. Linderoth, 1994 *Phys. Rev. Lett.* **72** 282.

[8] M. Abe, T. Itoh, y. Tamaura, 1992 *Thin Solid Films* **216** 155.
[9] C. V. Gopal Reddy, S. V. Manorama, V. J. Rao, 2000 *J. Matter Sci. Lett.* **19** 715.
[10] R. K. Ahrenkiel, t. J. Couburn, 1975 *IEEE Trans. Magn.* **11** 1103.
[11] J. W. D. martens, W. L. Peeters, H. M. Van Noort, M. Errnan, 1985 *J. Phys. Chem. Solids.* **46** 411.
[12] S. N. Okuna, S. Hashimoto, K. Inomata, 1992 *J. Appl. Phys.* **71** 5926.
[13] R. Valenzuela, 1994 *Magnetic Ceramic,* Cambridge University Press, Cambridge.
[14] N. Matsushita, S. Nakagawa 1992 *IEEE Trans. Magn.* **2850** 3108.
[15] M. Naoe, n. Matsushita, 1996 *J. Magn. Magn. Mater.* **155** 216
[16] P. C. Dorsey, P. Lubitz, D. B. Chirsey, J. S. Horwiz, 1996 J. *Appl. Phys.* **79** 6338.
[17] H. Y. Zhang, B. X. Gu, H. R. Zhai, M. Lu, H. B. Haung, 1995 *J. Magn. Magn. Mater.* **140** 699.
[18] D. M. Schleich, y. Zhang, 1995 *Mater. Res. Bull.* **30** 447.
[19] J. A. Na, 1996 *J. Appl. Phys.* **79** 4893.
[20] J. Lee, J. Y. Park, Y. Oh, C. S. Kim, 1998 *J. Appl. Phys.* **84** 2801.
[21] E. S. Murdock, R. F. Simmons, R. Davidson, 1992 *IEEE Trans. Magn.* **28** 3078.
[22] M. Abe, Y. Tamaura, Jpn. 1983 J. *Appl. Phys. Lett.* **22** 511.
[23] M. Abe, Y. Tamaura,1984 J. *Appl. Phys.* **55** 2614.
[24] V. Surve, V. Puri, 1998 *Bull. Electrochem.* **14** 151.
[25] S. S. Bellad, C. D. Lokhande and C. H. Bhosale 1997 *Ind. J. Pure Appl. Phys.* **35** 565.
[26] S. D. Sartale, C. D. Lokhande, 2002 *Ceramic International* **28** 467.
[27] R. K. Pandey, S. N. Sahu, S. Chandra, 1996 *Handbook of Semiconductor Electrodeposition,* Marcel Dekker, Inc. New York.
[28] H. Dahms and I. M. Croll, 1965 *J. Electrochem. Soc.* **112** 771.
[29] A. Brenner, 1963 *Electrodeposition of Alloys,* ***Vols. 1 and 2,*** Academic press, New York.
[30] JCPDS data file card no. 22-1086.
[31] R. V. Krishnan and A. Banerjee, 1999 *Rev. Sci. Inst.* **70** 85.
[32] S. G. Eills, 1967 J. Appl. Phys. **38** 2906.
[33] S. S. Djokie, 1994 in *Electrochemistry and Material Science of Cathodic absorption and adsorption,* B. E. Conway and G. Jerkirwiez (Ed), the electrochemical Proceeding Series New York.

Structural and electrical properties of bismuth sodium titanate ceramic

B L Kuanar[1]*, H S Mohanty[1], B Dalai[1] and D Behera[2]

[1]Department of Physics, School of Science, GIET University, Gunupur, Rayagada, Odisha-765022
[2]Department of Physics and Astronomy, National Institute of Technology, Rourkela, Odisha-769008
*Corresponding author e-mail address: bkkuanar@giet.edu

Abstract. Single-phase lead-free $Bi_{0.5}Na_{0.5}TiO_3$ (BNT) perovskite ferroelectric ceramic was produced using a solid-state reaction method. A detailed investigation of the structural and electrical properties of BNT ceramic is conferred. According to phase analysis employing Rietveld refined X-Ray diffraction, the crystal structure is single phase with rhombohedral (R3C) symmetry. Raman spectra measurement also confirms the rhombohedral structure of BNT by originating numerous peaks from the TiO_6 octahedron. The ferroelectric character of the BNT sample was established by a hysteresis loop measurement of polarisation vs electric field (P-E). The remnant polarisation (P_r) and coercive field (E_c) have characteristic values of 1.63 μC/cm^2 and 29.91 kV/cm, respectively. The ferroelectric phase change is seen in the temperature-dependent dielectric research, with a transition temperature of 323 ^oC. The compound possessed a low value of tanδ even at a high temperature (500 ^oC) at 1 MHz. The prepared sample exhibited excellent dielectric characteristics from room temperature to high temperatures, making it ideal for various applications.

Key Words: BNT Ceramic, X-Ray diffraction, Raman Scattering, FTIR study, Dielectric properties.

1. Introduction

The finding of new efficacious piezoelectric ceramic materials that enable mechanical and electrical energy interconversion has resulted in a significant increase in the number of research investigations for a wide variety of applications. In recent years, the development of high energy storage materials has gotten a lot of attention due to the ever-increasing demand for vital passive components. Furthermore, the stored energy in the ceramics can be released in a relatively short time to form an intense power pulse, broadening their applications in modern commodities, such as tuneable microwave devices, computer memory (NVRAMs), optical waveguides, pyroelectric detectors, micropumps, ultrasonic transducers, printing devices, micro-positioning devices, micro-motors, and so on. The fact that lead-based has good electromechanical properties, as well as a high piezoelectric coefficient and coupling factors, has prompted us to construct devices using it. Moreover, due to the implementation of rigorous environmental legislation, concerns about lead concentration have limited the wide range of applications. [1-3]. As a result, the development of new lead-free ceramics is critical. Ferroelectric researchers search for alternative potential lead-free systems, which should show inferior functional properties as lead-based ceramic. Various lead-free ceramics are $BaTiO_3$, $KBiTiO_3$, $NaTiO_3$, $KNbO_3$, $BiFeO_3$ $BiScO_3$, $BaZrTiO_3$, and $NaNbO_3$ etc [4]. Non-linear dielectrics, with piezoelectric, pyroelectric, and ferroelectric properties are known as ferroelectrics. Because of the strong coupling of their polarization with the electric and stress fields, ferroelectric perovskites, such as those of the ABO_3 type, are not only of basic interest but also of technological importance. Their sensitivity to stress and strain fields makes them valuable in micro-electromechanical systems, and their switchability of polarization direction with applied electric field makes them useful in producing non-volatile memories. Due to its excellent piezoelectric and ferroelectric characteristics at ambient temperature, high residual polarisation (Pr =38 $\mu C/cm^2$) and high Curie temperature, Bismuth Sodium Titanate (BNT) is recognised as one of the most prime ferroelectric materials. [5-7]. BNT shows improved properties near its morphotropic phase boundary (MPB). Smolenskii *et al.* investigated bismuth sodium titanate (BNT) for the first time in 1961 [8]. BNT is a relaxor ferroelectrics with the common formula $(A'A'')BO_3$ with rhombohedral symmetry at ambient temperature that transitions to tetragonal symmetry at almost 320 °C with bismuth (Bi^{3+}) and sodium (Na^{1+}) cations at the A-sites and titanium cation (Ti^{4+}) at the B-site. In an ABO_3 perovskite, sodium and bismuth cations occupy the corners, oxygen anions occupy the face centres, and a titanium cation sits in the centre of the oxygen octahedra produced. On the other hand, it is a three-dimensional cubic network of 8 corner-sharing TiO_6 octahedral with bismuth and sodium cations at the centre of the cube formed by the octahedral. Several methods of preparation, which include mechanical, ball milling, the chemical method, sol-gel, wet-dry, solid-state reaction, and auto combustion, etc are there. However, each approach has its own set of pros and disadvantages when it comes to achieving a high-quality product in terms of purity, homogeneity, reactivity, particle size, and so on. By virtue of performance, dependability, repeatability, and economy, the solid-state reaction was found to be the most approachable and least expensive method among the other accessible ways. This research describes the synthesis of BNT ceramic via a solid-state reaction method, as well as its structural and electrical analysis. Temperature-dependent dielectric studies were carried out to investigate the different phase transitions involved in the system.

2. Experimental Procedure

A Solid-state reaction was used to make the $Bi_{0.5}Na_{0.5}TiO_3$ ceramic. Bi_2O_3 (99%), was chosen as the beginning material, followed by Na_2CO_3 (99.5%), and TiO_2 (99.5%). To maintain the stoichiometric ratio, the raw components were weighed in proportion and then mixed in acetone medium for 6 hours with an agate motor. The dry powder was calcined in an alumina crucible at 950 °C for 3 hours using a typically programmed furnace. To compress the lumps created during calcination, the calcined powder was crushed again for 1 h. After that, using a hydraulic press and a tiny amount of 3.5wt% polyvinyl alcohol (PVA) as a binder, the fine powder of the foregoing ingredients was condensed into pellets of 1 mm in thickness and 10 mm in diameter. The pellets were sintered in a closed alumina crucible for 3 hours at 1150 °C, then cooled to room temperature. Silver paint was used as electrodes for measuring electrical properties, and the pellets were dried at 100 °C for 10 minutes. The sample's X-ray diffraction (XRD) pattern was produced using an X-ray powder diffractometer (Model Rigaku Japan ultimav IV) with Cu-Kα radiation (λ=1.54178Å) over a 2θ range of 20° to 80° with a step size of 0.05 degrees per minute. The Raman scattering measurements were performed in PL Micro Raman Spectrometer with 532 nm frequency stabilized, narrow slit of a laser diode. The spectra were collected in the range of 100-1000 cm^{-1}. A PerkinElmer FTIR spectrometer was used to conduct spectroscopy studies on ceramic particles in a KBr medium. A computer-controlled impedance analyser (Model HIOKI IM 3570) was used to measure impedance, capacitance, dielectric loss and other relevant characteristics in the frequency range of 10^2-10^6 Hz from room temperature to 500 °C. The measurement of ferroelectric properties was performed at a fixed frequency of 10 Hz by using Radiant Ferroelectric Tester.

3. Results and Discussions

3.1. X-Ray Diffraction

Figure.1 elucidates the room temperature XRD pattern of BNT ceramic. The diffraction pattern reveals that at ambient temperature, the considered ceramic has only a single-phase layer with a pure perovskite structure having no impurity phase. The optical diffraction pattern of BNT ceramic as displayed in Figure 1, shows successive reflections at different Bragg angles of 22.77°, 32.43°, 40.06° and 46.51° which are the characteristic peaks of the BNT powder. The diffraction peaks are assigned to the rhombohedral phase. It is exactly matched with the reported literature by H.S. Mohanty et. al., [9]. A contradictory report on the crystal structure of BNT reported by several authors that it is crystallized into reduced symmetry with the monoclinic phase [10, 11]. Rietveld refinement of XRD data was carried out, to confirm the crystal structure, which shows the rhombohedral structure with R3C space group symmetry for the considered sample.

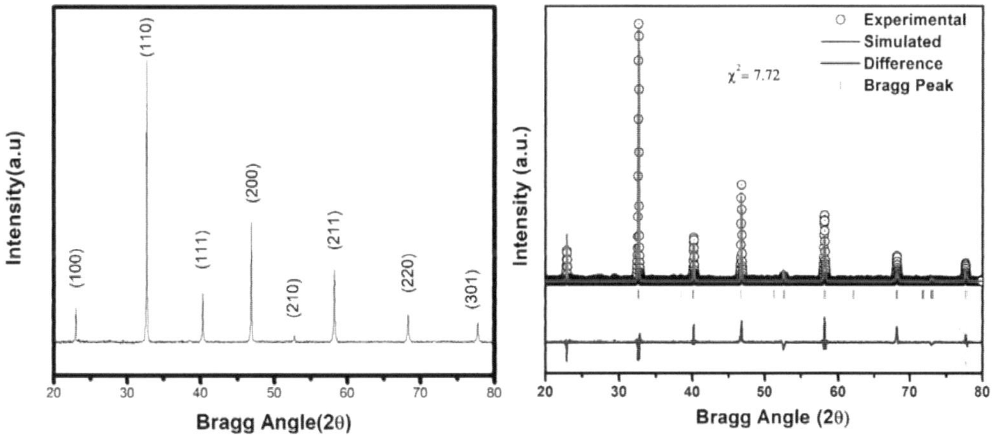

Figure 1. (a) X-ray Diffraction pattern and (b) Rietveld refined XRD pattern of BNT ceramic.

In the diffraction pattern of the pure BNT system, a small, weak super-lattice reflection corresponding to Miller indices ½ (311) approximately 2 θ = 38.5° was discovered, as seen in Figure 1. (b). This weak reflection is the signature of rhombohedral crystal structure [12]. The lattice parameters are determined to be a =b =5.4921 (7) Å and c =13. 4808 (5) Å, respectively, and the unit cell volume is determined to be 352.14 $Å^3$.

3.2. Raman Study

The Raman scattering spectra of BNT ceramic were taken at room temperature shown in Figure 2. The sample has a little bit broad Raman peak, which could be attributable to disorder in the A-site of the rhombohedral structure or deformed octahedral BiO_6 and NaO_6 clusters. The detected phonon modes at 140, 274, 546, and 753 cm^{-1} are quite similar to those previously described [13]. The internal modes of TiO_6 can be categorised as BNT vibrational modes. The deformed octahedral BiO_6 and NaO_6 clusters and vibration correspond to the mode at 140cm^{-1}. The mid-frequency regions 274 cm^{-1} arise from the external vibration of TiO_6 octahedra, high-frequency regions 546 cm^{-1}, 753 cm^{-1} and 840 cm^{-1} correspond to internal Ti-O bond vibration and TiO_6 octahedral stretching vibration, respectively [14,15]. These modes are well known in perovskites to involve oxygen atoms vibrations, while cations are almost at rest. The appearance of all the above-mentioned modes signifies the presence of a perovskite structure.

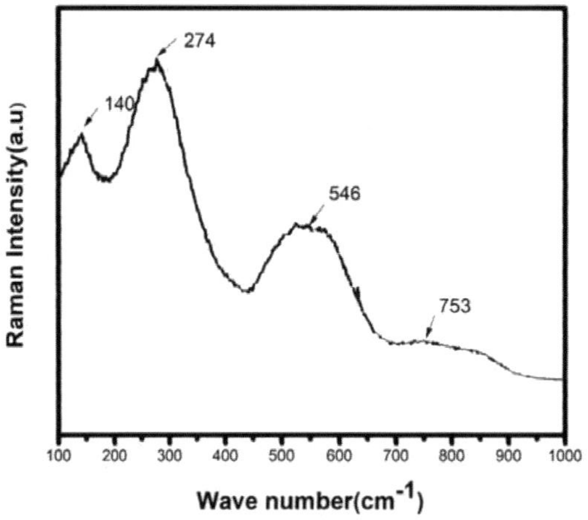

Figure 2. Raman spectrum of BNT ceramic

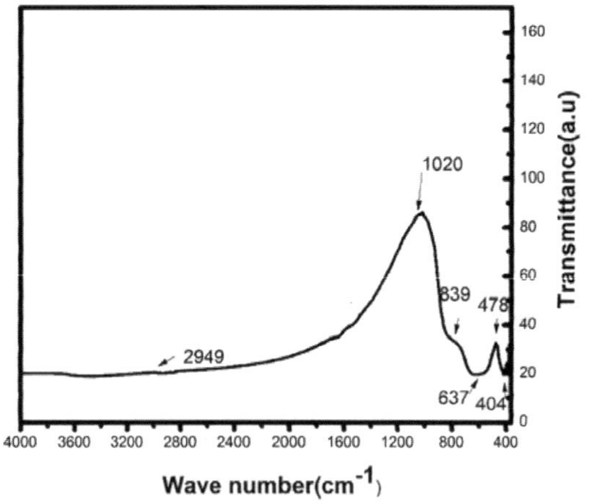

Figure 3. FTIR spectra of studied sample

3.3. FTIR Spectroscopy

FTIR spectroscopy is another tool for studying the structural and vibrational properties of ceramic materials, in addition to XRD and RAMAN. The FTIR spectra of BNT samples were investigated at room temperature over a wave number range of 400-4000 cm^{-1}, as shown in Figure 3. In the investigated sample, several forms of vibrations occur at 404, 478, 637, 839, 1020, and 2949 cm^{-1}.

The Ti-O octahedral group in the perovskite structure is assigned to the asymmetric band of approximately 400-1020 cm^{-1} [16]. The intrinsic vibration bands of BNT are located near 637 and 839 cm^{-1} [17]. The band corresponding to 2949 cm^{-1} was assigned as the stretching mode of O-H [18]. CO vibration may be responsible for the peak near 1020 cm^{-1} [19]. This pattern resembles the Raman spectra, which confirm the structure of the ceramic.

3.4. Electrical Properties

The temperature dependence of the relative dielectric permittivity and dielectric loss of BNT ceramics are illustrated in Figures 4(a) and 4(b), for varied frequencies of 1, 10, 100, and 1000 kHz. At temperatures ~190 °C (T_d) and ~ 320 °C (T_m), two dielectric abnormalities were discovered in the composition. The phase transition between ferroelectric (FE) and antiferroelectric (AFE) phases is responsible for T_d. Tm is the temperature at which the transition from the antiferroelectric (AFE) to the paraelectric (PE) phase occurs. The thermal dependency of the dielectric constant of BNT ceramic agrees well with what has previously been reported in the literature [20].

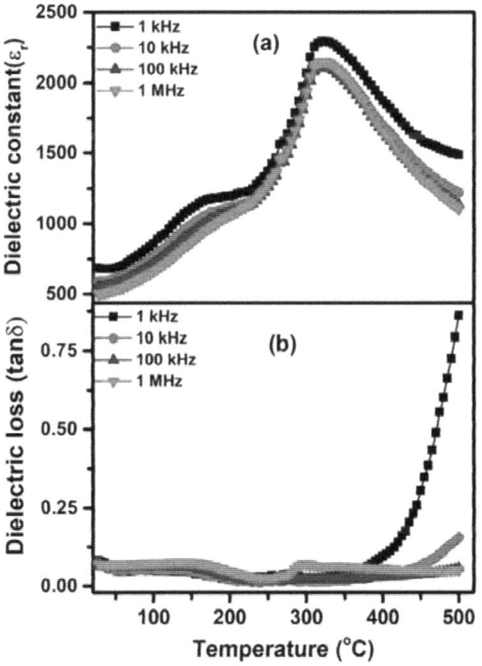

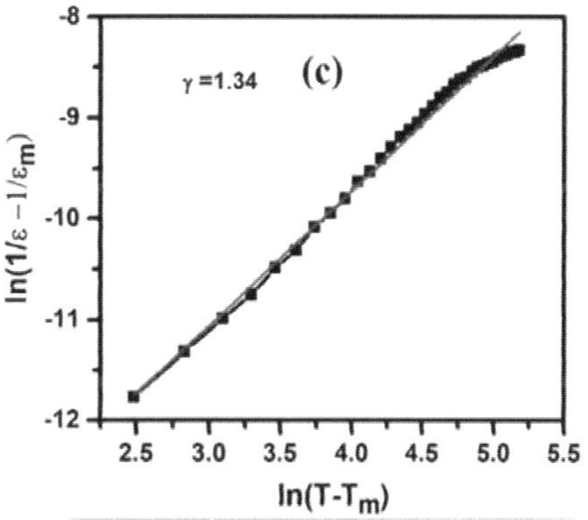

Frequency (kHz)	Maximum Dielectric constant (ε_r)	tanδ
1	2315	0.087
10	2167	0.163
100	2148	0.062
1000	2167	0.055

Figure 4. (a) Temperature Dependent Dielectric constant, (b) Dielectric loss and (c) Order of diffusivity of BNT ceramic.

Again, the transition temperature is nearly constant across all frequencies, indicating that BNT belongs to the typical ferroelectric group. From room temperature to 400 °C, the loss tangent (tanδ) is very low and nearly constant, then increases slightly with increasing temperature for low frequencies. The tan values at room temperature imply minimal loss materials. Defects such as oxygen vacancies, which act as space charges in contributing to the electrical polarisation of perovskite materials, can produce dielectric loss [21]. Materials with a high dielectric constant and low dielectric loss are ideal for capacitor applications. The dielectric constant and tangent loss for different frequencies are displayed in tabular form in figure 4.

The thermally dependent dielectric research reveals a large peak at phase transition temperature, confirming a diffuse type phase transition. The order of diffusivity or disorder ness that correlates to the ferroelectric to paraelectric phase transition (Tm) temperature in the sample may be determined using the modified Curie-Weiss law:

$$\frac{1}{\varepsilon} - \frac{1}{\varepsilon_m} = \frac{(T - T_m)^{\gamma}}{C'} \quad (1)$$

where γ is the diffusivity, ε is dielectric constant , c' is the Curie– Weiss constant at a given temperature T and ε_m is its maximum value of permittivity at Tm. In general, the diffusion factor is between 1 and 2, reflecting the standard ferroelectric phase transition and diffuse phase transition. The plot of $\ln(1/\varepsilon - 1/\varepsilon_m)$ as a function of $\ln(T-T_m)$ is shown in the figure 4.c. The value of diffusivity (γ) after linear fitting is found to be 1.34 for the frequency 1 kHz.

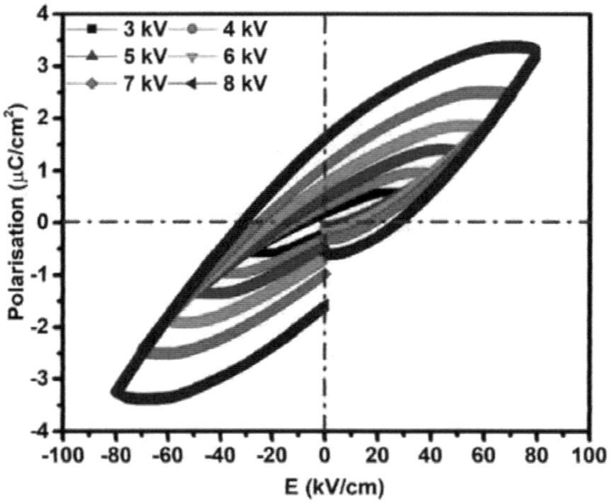

Figure 5. Polarization Vs Electric field loop of BNT ceramic

As illustrated in figure 5. the ferroelectric nature of BNT ceramic is investigated in a room temperature P-E loop under varied electric fields ranging from 30 kV/cm to 80 kV/cm. Because of the experimental setup's constraints, the saturation polarisation is much lower than planned. It is observed that the remnant polarization and coercive field increases with increase in applied electric field. Under various electric fields, the remnant polarisation (P_r), the saturation polarisation (P_s), and coercive field (E_c) are observed below.

Table 1. Remnant polarization (P_r), saturation polarization (P_s) and coercive field (E_c)

E (kV/cm)	*P_r (μC/cm^2)*	*P_s (μC/cm^2)*	*E_c (kV/cm)*
30	0.254	0.67	9.045
40	0.322	1.08	9.775
50	0.509	1.449	12.26
60	0.720	1.950	16.57
70	1.045	2.56	21.36
80	1.63	3.385	29.91

4. Conclusions

In conclusion, the structural, electrical, and temperature-dependent dielectric properties of BNT ceramic have been elucidate in detail. Rietveld refinement, Raman spectroscopy, and FTIR spectroscopy were used to confirm a single phase of perovskite structure. The phase transition with a curie temperature of 323 °C is found from dielectric investigation. The modified Curie-Weiss law was used to get the diffusivity parameter. The diffusivity parameter confirmed the relaxor ferroelectric behaviour of BNT ceramic. The ferroelectric feature is supported by a ferroelectric hysteresis loop in addition to dielectric studies. All of these findings suggest that BNT ceramics could outperform lead-based ceramics in high-temperature industrial applications such as capacitors and storage devices.

5. References

[1] Parida R, Parida B, Bhuyan R K, and Parida S K 2021 *ferroelectrics* 571 162–174.

[2] Nilkhao S, Sumang R, Charoonsuk T, and Prasertpalichat S 2021 *Integrated ferroelectrics* 214 79–89.

[3] Panda P K 2009 *J Mater Sci.* 44 5049–5062.

[4] Kuanar B, Mohanty H S, Behera D, Nayak P and Dalai B 2021 *Engineering and Applied Science Research* 49 288-299.

[5] Muhammed K R, Scrimshire A, Sterianou I, Bell A M T and Bingham P A 2020 *Journal of the Australian Ceramic Society* 56 1441–1449.

[6] Piewpan P, Eittsayeam S and Pengpat K 2021 *Integrated ferroelectrics* 214 37–45.

[7] Hajra S, Sahoo S, Das R and Choudhary R N P 2018 *Journal of Alloys and Compounds* 750 507-514.

[8] Sahoo S, Hajra S, De M and Choudhary R N P 2017 *ceramics International* 44 4719-4726.

[9] Mohanty H S, Kumar A, Sahoo B, Kulriya P K and Pradhan D K 2018 *J. Mater. Sci. Mater. Electron* 29 6966.

[10] Aksel E, Forrester J S, Jones J L, Thomas P A, Page K and Suchomel M R 2011 *Phys. Lett.* Appl. 98 152901.

[11] Rao B N and Ranjan R 2012 *Phys. Rev. B* 86 134103.

[12] Mohanty H S, Dam T, Borkar H, Pradhan D K, Mishra K K, Kumar A, Sahoo B, Kulriya P K, Cazorla C, Scott J F and Pradhan D K 2019 *J. Phys.: Condens. Matter* 31 075401.

[13] Behara S, Ikeda K and Thomas T 2021 *ceramic international* 47 12870-12878.

[14] Badapanda T, Sahoo S and Nayak P 2017 *IOP Conf. Series: Materials Science and Engineering* 178 012032.

[15] Barick B K, Mishra K K, Arora A K, Choudhary R N P and Pradhan D K 2011 *J. Phys. D: Appl. Phys*. 44 355402.

[16] Jr A F, Banerjee P and Romanholo P L 2018 *Journal of Alloys and Compound* 764 122-127.

[17] Selvadurai A P B, Pazhnivelu V, Vasanth B K, Jagadeeshwaran C and Murugaraj R 2015 *J of Materials Science: Mat. in Elect.* 26 7655-7665.

[18] De Manojit, Mukherjee A and Tewari H S 2015 *Processing and Application of Ceramics* 9 193–197.

[19] Urlaub R, Posset U and Thull R 2000 *J. Non-Cryst. Solids* 265 276–284.

[20] Pal V, Dwivedi R K and Thakur O P 2013 *Materials Research Bulletin* 54 189-196.

[21] Anem S, Rao K S, Rao K H 2016 *ceramic International* 42 15319-15326.

Preparation and Study the Structural Properties for Compound $Tl_{0.9}Hg_{0.1}Sr_2Ca_2Cu_3O_{8+\delta}$ Using the Modified Scherer Equation

Firas Kadhim Nsaif[1] , Jaafar Sadeq Mohammed[2] , Intehaa Ahmed Mohammed[1] and Kareem Ali Jasim[1]

[1]Department of Physics , College of Education for Pure Science / Ibn Al-Haitham , University of Baghdad , Baghdad , Iraq
[2]Department of Physics , College of Science , University of Diyala , Diyala , Iraq

Email : firas.k.n@ihcoedu.uobaghdad.edu.iq

Abstract : In this paper , the compound $Tl_{0.9}Hg_{0.1}Sr_2Ca_2Cu_3O_{8+\delta}$ was produced using solid status reaction procedure . The crystal construction of specimen was studied using x-ray deviation device . X-ray diffraction (XRD) study of the crystalline nature confirmed that the sample of compound has a lattice structure from a mixture for different phases , it is has a tetragonal structure . The crystal size was calculated and investigated from the traditional Debye-Scherer method and the modified method for it . It is note the approximate values of crystal size and crystallinity in percentage between the modified Scherer method and the traditional Debye-Scherer method change depending on the peaks intensity of XRD diagram . Furthermore , the sample was taken into account to determine the physical and microstructure factors such as crystal strain .

Keywords : x-ray diffraction , structural properties , Debye-Scherer equation , modified Scherer equation , crystal strain .

1. Introduction

By the comparative study of the similar twosome of Hg-1223 and Tl-1223, confirmed that T detours of the two identical when schemed at the lowered temperature [$1/(T_c/T)$] , the stages of Tl-superconductors are commonly specified by $Tl_1Ba_2Ca_{n-1}Cu_nO_{2n+4-\delta}$ where n is count of the serial planes for Cu-O [1,2] . The elevated T_c was 113 K at $Tl_{0.6}Pb_{0.4}Ba_{1.5}Sr_{0.5}Ca_2Cu_3O_{8.62}$. All specimens were produced using flux of O_2 . Analysis of x-ray deviation have indicated a tetragonal construction with an increment the lattice constant (c-direction) for specimens which were containing Hg as compared to the ones don't have Hg [3,4] .

Chemically , replacment at locations of Tl lead to extra metallicity in the bulk planes between conducting planes of Cu-O [5] . Hg-base group stills gets the maximum attention in HTS groups and investigation of unresolved mechanism of high T_c superconductivity . Zero resistance with elevated value of T_c at 135 K was noted in Hg-1223 , while Hg-1223 doped with Tl still holds the record until now (138 K) [6,7] . There is a study that studied the partial replacement of mercury with silver and chromium , also studied Scherer method , other methods for calculating the crystal size and crystallinity , the full width half maximum (FWHM) and lattice constants [8] . A survey was offered effect of the fractional replacement of Cu with Ni in the superconducting compound $Tl_1Ba_2Ca_2Cu_{3-x}Ni_xO_{8-\delta}$ (x = 0 to 1.0). By solid status reaction procedure, the specimens were produced with 860 °C temperature of sintering for 24 hrs. . The constitution of specimens were examined using x-ray deviation , the preferable temperature of transition was got at the specimen $Tl_1Ba_2Ca_2Cu_{2.4}Ni_{0.6}O_{8-\delta}$ and equivalent to 136.4 K with a majority phase is 1223 [9] . By solid status reaction procedure , supreme temperature superconductors called ($Tl_{0.9}Pb_{0.1}$-1223) , ($Tl_{0.9}Sb_{0.1}$-1223) and ($Tl_{0.9}Cr_{0.1}$-1223) have been successfully produced and temperature of sintering was (855-860) °C [10] . Effect of the copper substitution with nickel in unperspicuous stages for superconducting compound $Tl_{0.5}Pb_{0.5}Ba_2Ca_2Cu_{3-x}Ni_xO_{2n+3-\delta}$ (x = 0 to 1.0) was studied [11] .

The specimens were produced using solid status reaction procedure at temperature of sintering 860 °C for 24 hours and the four-probe manner was used to calculate temperature of transition . The x-ray deviation analysis displayed that the composition for perspicuous and unperspicuous stages was tetragonal with weakly-T_c stage 1202 and elevated-T_c stage 1223,1212 and addition to existence of several contaminated stages . In previous search solid status reaction procedure was used to make three different high temperature superconductors (HTS) , namely ($Tl_{0.85}Cd_{0.15}$-1201) , ($Tl_{0.85}Cd_{0.15}$-1212) and ($Tl_{0.85}Cd_{0.15}$-1223) . The optimum calcinations were performed at 800 °C and the sintering temperature was within (855-860) °C [12] .

In this study , a modern way was explored for produced (Tl,Hg) base by investing a quite perfect dominance of reaction for the compound $Tl_{0.9}Hg_{0.1}Sr_2Ca_2Cu_3O_{8+\delta}$. The purpose of this partial substitution between thallium and mercury was to reduce the interatomic distances between the mercury atoms from TlO to HgO , and then study the crystal size and crystal strain by using the traditional Debye-Scherer method and the modified Debye-Scherer method .

2. Theoretical part

X-rays were used to find out the distance between the details in the matters relating to the interior model [13] . The sample was examined using x-ray diffraction and the spacing between the atomic planes (d_{hkl}) was calculated from the diffraction angle using Bragg's law [14] :

$$d = \frac{n\lambda}{2\sin\theta} \quad (1)$$

Where d : is the spacing between atomic planes ,θ : Bragg's angle ,n : is an integral numbers order , λ : the wavelongitude of x-ray ≈ 1.541 Å .

The total domain half maximum to symbol (β) was calculated by using the origin program (2019b-64bit) which this led to determine the crystal size from Scherer method (D_{Sh}) and modified Scherer method (D_{modSh}) . The crystallinity and crystal strain calculated by using some x-ray diffraction analysis methods . Scherer method is one of the methods used to analysis the results of XRD , this method was used to detect about the crystal size with XRD , the equation for calculating the crystal size by equation of Scherer is [15] :

$$D_{Sh} = \frac{K\lambda}{\beta\cos\theta} \quad (2)$$

Where D_{Sh} : is the crystal size by Scherer equation , θ : is Bragg's angle in Radian unit , λ : wavelongitude of x-ray ≈ 1.541 Å , K : is the factor of Scherer = 0.89 , β : is line broadening at total domain half maximum peaks with Radian unit .

The purpose of the modified Scherer equation is to present a new type of method for using Scherer's equation, so that someone can apply the least squares technique in order to reduce the sources of errors in pervious method [16,17] . The errors is reduced from calculating size of the crystal by formula of the modified Scherer equation [18] :

$$\ln\beta = \frac{\ln K\lambda}{D_{modSh}} + \ln\frac{1}{\cos\theta} \quad (3)$$

Where D_{modSh} : is the crystal size by the modified Scherer m method .

The advantage of modified Scherer method is that it is possible to calculate the crystal strain by it , and this cannot be done in the traditional Scherer method . Also , when the peaks in pattern of x-ray deviation have large values of the full width half maximum (FWHM) , calculation of the crystal size using the modified Scherer method gives a more accurate result than is in the case using the traditional Scherer method .

The crystallinity in percentage was calculated using equilibrium (4) [17] :

$$\text{The crystallinity} = \frac{\textbf{Area of the crystalline peaks}}{\textbf{Area of all the peaks}} \times 100 \qquad (4)$$

3. Experimental part

Using the standard solid status reaction procedure , the $Tl_{0.9}Hg_{0.1}Sr_2Ca_2Cu_3O_{8+\delta}$ component was produced from element oxides (TlO , HgO , SrO , CaO , and CuO) . The oxides of this compound were weighted and their proportions enumerated conformity to the next equation :

$$\frac{9}{10}\,TlO + \frac{1}{10}\,HgO + 2SrO + 2\,CaO + 3CuO \rightarrow Tl_{0.9}Hg_{0.1}Sr_2Ca_2Cu_3O_{8+\delta} \qquad (5)$$

In a furnace with a programmable sintering controller , the powder was placed for the calcination process which is the heating treating of degassing the powder . In this procedure , the mix was heated up to 800 °C for three hrs. . The powder was re-milled again and pressed into a hydraulic press under 7 $(\mathrm{ton/cm^2})$ pressure to make pellet with (3 to 3.5) mm highness and 20 mm radius . The pellet was put in capsules then placed in the furnace with a programmable sintering controller and was sintered with 860 °C for 24 hours at average of 100 °C per an hour , and then as the heating procedure it was cooled down to temperature of room . By employ x-ray deviation measurements the construction of the produced sample was got , in range $\theta = 20^{\circ}$ to 60°. A computer programme depended on Cohen minimal quadratic procedure [14,19] was used to calculate lattice parameters a , b and c .

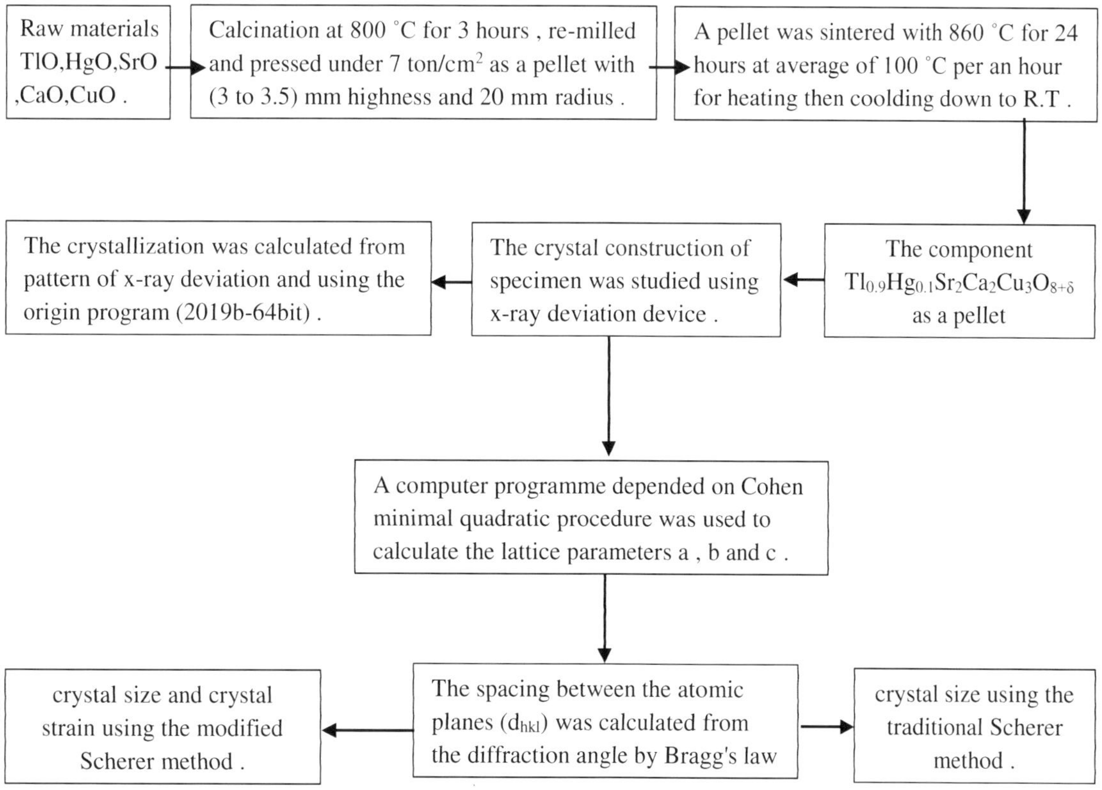

Fig. 1 : Summary of the practical part .

IOP Conf. Series: Materials Science and Engineering 1258 (2022) 012008 doi:10.1088/1757-899X/1258/1/012008

4. Consequences and argumenation

4.1 The x-ray deviation results

Pattern of x-ray deviation was taken for sample of the compound $Tl_{0.9}Hg_{0.1}Sr_2Ca_2Cu_3O_{8+\delta}$. Figure (2) shows the sharp peaks , It is noted from this diagram that the crystal structure of the sample is a tetragonal , in addition to that , there is more than one phase in the sample , but the dominant phase is 1223 (High phase) and a small percentage of the phases 1201 and 1212 (Low phase) and some impurities (SrCuO and CaCuO) phases are also noted at end of the diagram.

The crystal lattice constants a , b and c were calculated for the sample using Bragg's law (equation1) and the crystal lattice constants were (3.6518 , 3.652 and 15.763) Å respectively . The partial substitution results in a clear change in the crystal lattice. The percentage of the high phase in the compound was 75.64%, while total of the rest of the phases did not exceed 25% .

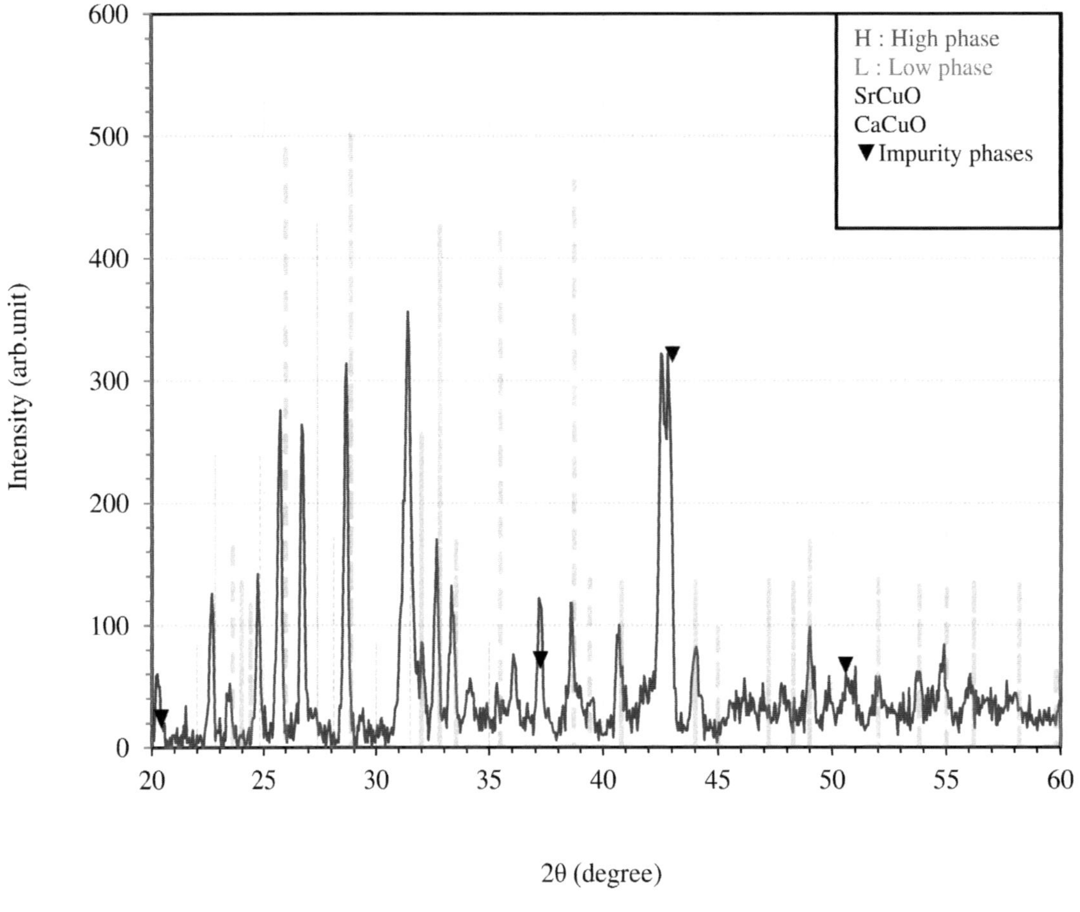

Fig. 2 : Represented XRD for the compound $Tl_{0.9}Hg_{0.1}Sr_2Ca_2Cu_3O_{8+\delta}$.

4.2 Crystal size by using Scherer equation

Values of 1/β in the y-axis and values of cos θ in the x-axis were graphed from Scherer method (equation 2) and the best straight line was drawn using the origin program (2019b-64bit) . Slope of the line in this method was used to calculate the crystal size as shown in the figure (3) , it was 403.56 nm and the result was recorded in the table (1) .

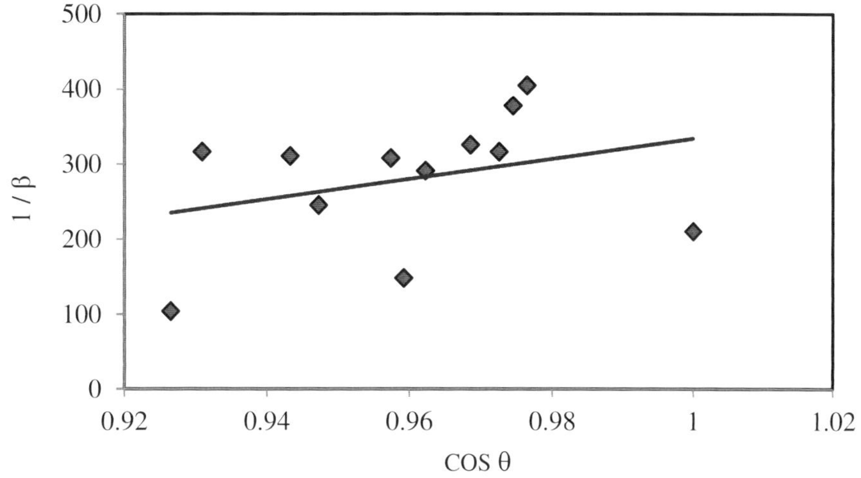

Fig. 3 : The relation between values of 1/β in the y-axis and values of cos θ in the x-axis for the compound $Tl_{0.9}Hg_{0.1}Sr_2Ca_2Cu_3O_{8+\delta}$ by using Scherer equation .

4.3 Crystal size by using the modified Scherer equation

As for calculating the crystal size and the crystal strain using the modified Scherer method (from equation 3) , the relation between values of ln (1/cos θ) in y-direction and values of ln (β) in x-direction were drawn using origin program (2019b-64bit) [20] . The best straight line was drawn and slope of the line represents the crystal strain as shown in the figure (4) , it was found that it is equal to 6.51466 and the result is laid out in the table (1) .

As for calculating the crystal size and the crystal strain , someone can be obtained them from the modified Scherer method . By taking value of the cut for the straight line which drawn in the figure (4) with y-axis , the crystal size was calculated which is equal to 403.56 nm and the result was recorded in the table (1) .

Through the two crystal size values which were calculated by the two methods and recorded in the table (1) , it was observed that the crystal size values were identical between the traditional and modified Scherer method . This indicates that this compound has a narrow top , so this match occurred [17,20] .

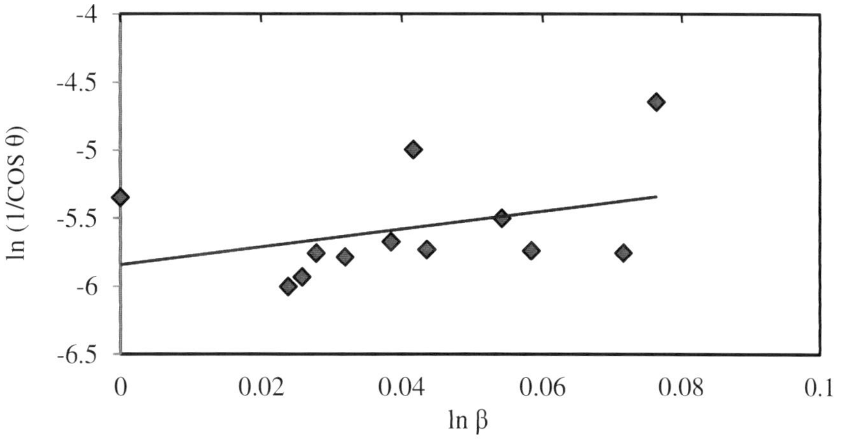

Fig. 4 : The relation between values of ln (1/cos **θ**) in y-direction and values of ln (**β**) in x-direction for compound $Tl_{0.9}Hg_{0.1}Sr_2Ca_2Cu_3O_{8+\delta}$ using the modified Scherer equation .

4.4 Crystallinity

Degree of the crystallinity can be assessed by using the x-ray crystallography and many materials can be prepared as mixtures of crystalline and amorphous regions . In such this status , crystallinity is generally observed as a proportion between the phase size for crystalline substance to the phase size for crystalline and amorphous for the prepared substance .

However , stage of the completeness in construction can alter even within fully crystalline materials . The crystallization was measured by using the equation (4) and the result was recorded in the table (1) . While the full width half maximum (FWHM) of the symbol (β) was calculated using the original program (2019b-64bit) .

Table 1 : Shows values of the crystal size using Scherer method , the crystal size and crystal strain using the modified Scherer method and the crystallinity .

Scherer method	modified Scherer method		crystallinity (%)
crystal size [nm]	crystal size [nm]	crystal strain	
403.56	403.56	6.51466	53.86

5. Conclusions

By using x-ray diffraction , the crystal size , crystal strain , crystallinity , area of all peaks , FWHM and Miller's indices for the compound $Tl_{0.9}Hg_{0.1}Sr_2Ca_2Cu_3O_{8+\delta}$ were calculated . Value of the crystallinity was equal to 53.86 % and the crystal strain 6.51466 . It was observed that the crystal size values (403.56 nm) were identical between the two traditional and modified Scherer methods . This indicates that this compound has a narrow top , so this match occurred .

References

[1] S. Parkin et al. , Bulk superconductivity at 125 k in $Tl_2Ca_2Ba_2Cu_3O_y$ compound , in ten years of superconductivity : 1980-1990 , Springer , 5 (1988) 309-312 .

[2] T. Nabatame , Y. Saito , K. Aihara , T. Kamo and S. Matsuda , Comparison of transport properties between Tl-(1223) and Tl-(2223) phases of Tl-Ba-Ca-Cu-O systems , Jpn. J. Appl. Phys. , 32 (1993) 484 .

[3] A. A. Gapud , J. Z. Wu , B. W. Kang , S. L. Yan , Y. Y. Xie and M. P. Siegal , Superfluidity and superconductivity giant T_c shift in $HgBa_2CaCu_2O_{6-\delta}$ and $TlBa_2CaCu_2O_{7-\delta}$ superconductors due to Hg-Tl exchange , Phys. Rev. B-Condensed Matter , 59 (1999) 203-206 .

[4] K. A. Jasim and T. J. Alwan , The effect of simultaneous substitution of strontium at the barium site of $Tl_{0.6}Pb_{0.4}Ba_{2-x}Sr_xCa_2Cu_3O_{9-\delta}$ superconductors , Journal Supercond. Nov. Magn. , 22 (2009) 861-865 .

[5] H. Maeda , Y. Tanaka , M. Fukutomi and T. Asano , A new high T_c oxide superconductor without a rare earth element , Jpn. J. Appl. Phys. , 27 (1988) 209 .

[6] A. Iyo et al. ,Variation of T_c in multilayered cuprates of $HgBa_2Ca_{n-1}Cu_nO_y$, Phys. C, Supercond. and it is Appl. , 460 (2007) 436-437 .

[7] A. Iyo et al. , T_c vs Relationship for Multilayered High T_c Superconductors , J. Phys. Soc. Japan , 76 (2007) 9711 .

[8] R. A. Hamood , Structural and Electrical Properties of Substitution Elements on the Hg-Base Compound in High Temperature Superconductor , Thesis in M.Sc. of phys. , College of education for pure science / Ibn Al-Haitham , University of Baghdad , (2021) .
[9] L. A. Mohammed and K. A. Jasim , Improvement the Superconducting Properties of $TlBa_2Ca_2Cu_{3-x}Ni_xO_{9-\delta}$ Superconducting Compound by Partial Substitution of Copper with Nickel Oxide , Energy Procedia , 157 (2019) 135-142 .
[10] K. A. Jasim , S. A. Makki and A. abud Almohsin , Comparison Study of Transition Temperature between the Superconducting Compounds $Tl_{0.9}Pb_{0.1}Ba_2Ca_2Cu_3O_{9-\delta}$, $Tl_{0.9}Sb_{0.1}Ba_2Ca_2Cu_3O_{9-\delta}$ and $Tl_{0.9}Cr_{0.1}Ba_2Ca_2Cu_3O_{9-\delta}$, Phys. Procedia , 55 (2014) 336-341 .
[11] L. A. Mohammed and K. A. Jasim ,Synthesis and study the structural,electrical and mechanical properties of high temperature superconductor $Tl_{0.5}Pb_{0.5}Ba_2Ca_{n-1}Cu_{n-x}Ni_xO_{2n+3-\delta}$ substituted with nickel oxide for n = 3 , Ibn AL-Haitham J. Pure Appl. Sci. , 31 (2018) 26-32 .
[12] K. A. Jasim and T. J. Alwan , Effect of Oxygen Treatment on the Structural and Electrical Properties of $Tl_{0.85}Cd_{0.15}Sr_2CuO_{5-\delta}$, $Tl_{0.85}Cd_{0.15}Sr_2Ca_2Cu_2O_{7-\delta}$ and $Tl_{0.85}Cd_{0.15}Sr_3Ca_2Cu_3O_{9-\delta}$ Super conductors , J. Supercond. Nov. Magn. , 30 (2017) 3451-3457 .
[13] R. Habla and S. Khawla , Preparation and determination of the crystalline structure of the compound $SrFe_{0.7}Mo_{0.3}O_{3-\delta}$ " , (2018) .
[14] T. Nguyen , E. Sierra , H. Eguiraun and E. Lizundia , Iridescent cellulose nanocrystal films:the link between structural colour and Bragg's law , Eur. J. Phys. , 39 (2018) 45803 .
[15] M. K. Kamil and K. A. Jasim , Investigation the Crystalline Size and Strain of Perovskite $YBa_2Cu_3O_6$ by variant method , Test Engineering and Management , 7 (2020) 8719-8723 .
[16] M. K. Kamil and K. A. Jasim , Calculating of crystalline size , strain and degree of crystallinity of the compound $HgBa_2Ca_2Cu_3O_{8+\delta}$ by different method , IOP Conference Series:Materials Science and Engineering , 928 (2020) 1-7 , doi : 10.1088/1757-899X/928 /7/072109 .
[17] K. M. Wadi , K. A. Jasim , A. H. Shaban , M. K. Kamil and F. K. Nsaif , The effects of sustainable manufacturing pressure on the structural properties of the $Pb_2Ba_2Ca_2Cu_3O_{9+\delta}$ compound , Journal of Green Engineering , 10 (2020) 6052-6062 .
[18] A. K. Saadon , K. A. Jasim and A. H. Shaban , Superconducting Compound $Hg_{0.8}Sb_{0.2}Ba_2Ca_2Cu_3O_{8+\delta}$ Compared with $Hg_{0.8}Sb_{0.2}Ba_2CaCu_2O_{6+\delta}$ to Evaluate Transition Temperature , Neuro Quantology , 18 (2020) 14-18 .
[19] A. Monshi , M. R. Foroughi and M. R. Monshi , Modified Scherrer equation to estimate more accurately nanocrystallite size using XRD , World J. Nano Sci. Eng. , 2 (2012) 154-160 .
[20] N. Rani , S. Chahal , A. S. Chauhan , P. Kumar , R. Shukla and S. K. Singh , X-ray analysis of MgO nanoparticles by modified Scherer's Williamson-Hall and size-strain method , Mater. Today Proc. , 12 (2019) 543-548 .

Synthesis and characterization of thermally treated $Co_{1-x}Fe_{2+x}O_4$ (x = 0.0 –0.8) spinel nano ferrite

C Parmar[1], R Verma[1], S S Modak[2], F Mazaleyrat[3] and S N Kane[1,*]
[1]Magnetic Materials Laboratory, School of Physics, D. A. University, Khandwa road, Indore-452001, India
[2]Physics Department, Jaypee University of Engineering and Technology, Raghogarh, Guna 473226, India
[3]SATIE, ENS Paris-Saclay, 4 Avenue des Sciences, 91190 Gif-sur-Yvette, France

[*]E-mail: kane_sn@yahoo.com(S. N. Kane)

Abstract. We report the effect of surplus-Fe-content, and thermal annealing on the properties of sol gel auto-combustion synthesized $Co_{1-x}Fe_{2+x}O_4$ (x = 0.0–0.8) spinel ferrites. X-ray diffraction (XRD), magnetic measurements, Fourier transform infrared (FTIR) spectroscopy, Ultraviolet–visible (UV–Vis) are used to monitor the effect of excess Fe-content, and thermal annealing dependence of properties, and correlation among them. XRD validates presence of spinel nanoferrite (31.54 – 44.89 nm) phase, minor presence of α-Fe_2O_3 phase was also detected. Results reveal that variation of Fe-content leads to: i) variation of lattice-parameter, re-distribution of Fe^{3+}. Co^{2+} ions on A, B-site, ii) alteration of inversion degree, disorder, iii) modification of saturation magnetization due to spin disorder at surface, iv) variation of coercivity, and related grain size imply that studied ferrites are in single-multi domain overlap regions, while squareness-ratio indicates variation of inter-grain interaction, isotropic behaviour of multi-domain particles; iv) fine-tuning of bandgap (1.35 – 1.5 eV) via changes in cationic-distribution and lattice-parameter, v) variation of Debye temperature, suggests modification of lattice vibrations. Current work implies strong correlation between bandgap magnetic, structural properties.

1. Introduction

Ferrimagnetic spinel-ferrites are represented by formula $Me^{2+}O.Fe^{3+}{}_2O_3$, display face centred cubic 'fcc' structure, with Fd_3m space group, and have two inter-penetrating sub-lattices tetrahedral (A), octahedral (B) [1, 2, 3]. Ferrites are vital materials for their use in microwave devices, high frequency devices, etc. For their usage in various devices, magnetic properties should to be optimized [4, 5]. By choosing: i) a particular method: e.g. – sol gel [6, 7], hydrothermal [8], solid state reaction [9], co-precipitation [10] etc., ii) compositional variation [6, 7, 11], and iii) thermal annealing /or no thermal annealing [12], ion irradiation [13] etc. Literature also reports tuning of bandgap, structural, magnetic properties [14, 15, 16]. $CoFe_2O_4$ has inverse spinel structure, and Fe^{3+} ions are dispersed nearly equally on A (tetrahedral), B (octahedral) site, while Co^{2+}ions on B site. Although, $CoFe_2O_4$ is a well studied ferrite, exhibits interesting properties including excellent chemical stability, high Curie temperature, high saturation magnetization, coercivity [17], and widely used in electronic devices, magnetic-resonance-imaging (MRI) [18, 19], targeted drug delivery, for cancer treatment (hyperthermia) [20], transformer cores, high density storage devices, magnetic fluids, , humidity, and gas sensors, etc. [4, 21]. However, studies on influence of excess amount of Fe (*where there is excess Fe i.e. more than 2*) in Co-based $Co_{1-x}Fe_{2+x}O_4$-type spinel ferrites are less reported [22]. $Co_{0.8}Fe_{2.2}O_4$ system *(here excess Fe is 0.2)*, reported in [22] shows magnetic field induced phase transition [23]. Kane et el [14] report the study of bandgap, magnetic, structural, properties, in dry gel $Co_{1-x}Fe_{2+x}O_4$ (x = 0.0–0.8) spinel nano ferrites, where excess Fe-content has considerable effect on aforementioned properties. In spinel ferrites, compositional variation, thermal annealing etc. is known to affect structural, magnetic properties, bandgap e. g. – [6, 7, 11, 14, 24] etc. So study of effect of thermal annealing in $Co_{1-x}Fe_{2+x}O_4$(x = 0.0–0.8) spinel ferrite, would be interesting to obtain the information on thermal annealing induced changes in structural, magnetic properties, bandgap and correlation between them.

Thus, objective of the present work is to study the influence of surplus-Fe-content, and thermal annealing on the properties of sol gel autocombustion synthesized $Co_{1-x}Fe_{2+x}O_4$(x = 0.0–0.8) spinel ferrites. Effect of excess Fe-content, and thermal annealing dependence of properties was monitored by XRD, magnetic measurements, UV-Vis, FTIR spectroscopy.

2. Material synthesis, material characterization, and data analysis

Preparation of $Co_{1-x}Fe_{2+x}O_4$ (x = 0.0 – 0.8) spinel nano ferrites was done by sol-gel autocombustion method, utilizing AR grade nitrate-citrate precursors, During synthesis solution pH was maintained at 7, by adding Ammonia solution [NH_4OH], and metal salt to citric acid ratio was kept as 1:1. More details on synthesis can be seen in [14]. Prepared dry gel samples were annealed in muffle furnace (in air) at 600 °C for 3 hours, and were used for characterization.

Structural characterization of the samples was done at room temperature by X-ray diffraction (XRD) measurements using Bruker D8 diffractometer utilizing Cu-K_α radiation (wavelength 'λ' = 0.1541 nm) in $\theta - 2\theta$ configuration (step size: 0.02°). Vibrating sample magnetometer (Lake Shore 7310 model), at room temperature was used to get magnetization measurements. Room temperature Ultraviolet-Visible (UV-Vis) measurements (in reflection mode) were done by Perkin-Elmer spectrometer (Model: Lamda 950) to get the bandgap. Absorption mode Fourier transmission infrared (FTIR) measurements (in KBr matrix) were done by Perkin Elmer spectrometer.

MAUD (Material Analysis Using Diffraction), Reitveld refinement software [25] is employed to get full profile analysis of the XRD pattern, which also verified the formation of fcc spinel phase. Analysis of XRD data as described in [6, 26, 27], yields structural parameters: lattice parameter (a_{exp}), Scherrer's grain diameter (D_S), X-ray density (ρ_{XRD}), specific surface area (S), dislocation density (ρ_D). Using X-ray diffraction line intensities, cationic distribution of each composition was estimated via employing Bertaut method [27, 28], was used to compute theoretical lattice parameter (a_{th}), oxygen parameter (u), inversion parameter (δ), Néel magnetic moment 'n_N' or '$M_{S(th)}$' (theoretical magnetization at 0 K) and, canting angle ($\alpha_{Y\text{-}K}$) [27]. Hysteresis loops were used to obtain coercivity (H_c), saturation magnetization (M_S), reduced remanence (M_r/M_S) anisotropy constant (K_1). Energy bandgap was computed by using Kubelka-Munk (K-M) function via following equations: F(R) = $[1-R^2] / 2R$, and $\alpha h\nu = A\ (h\nu - E_g)\ n$, where R – reflectance and F(R) is Kubelka-Munk function, is proportional to incorporation of the adsorbate molecules, while hν, α, A, and n represent respectively the energy of the photon, the absorption coefficient, the material parameter, and E_g– the bandgap [14, 27, 29]. FTIR data was analyzed to obtain absorption band positions, force constants for tetrahedral, octahedral sites (K_T, K_O), and Debye temperature (θ_D). Error shown in table(s), figure(s) is '*Standard Deviation*' obtained from the data.

3. Results and discussions

X-ray diffractograms of the studied samples are shown as figure 1 (a), confirm the formation of nano-crystalline (grain diameter: 31.54 nm to 44.89 nm) cubic spinel structure. Small fraction of Fe_2O_3 was also detected, ascribed to moderate thermal annealing temperature: 600 °C, as was also observed earlier [27, 30]. Figure 1 (b) illustrates the representative Rietveld refined plot for x = 0.2 sample. Figure 2 (a) shows compositional dependence of Scherrer's grain diameter 'D_S', while inset depicts

variation of $a_{exp.}$. Based on compositional dependence of D_S, $a_{exp.}$, following experimental relations can be written: i) D_S = 44.90268 -16.68858 [Fe content], ii) $a_{exp.}$= 0.83852 – 0.00313 [Fe content]. Aforementioned experimental relations show association among composition, and structural parameters: D_S, $a_{exp.}$. Observed variation of $a_{exp.}$is ascribable to slight variation of ionic radii of Co^{2+} (0.065 nm), and Fe^{3+} (0.067 nm), and allocation of Fe, Co ions on A, B site (see table 1), as was also reported earlier [14] where minor changes in $a_{exp.}$were observed, and no thermal treatment was used. Agreement between theoretical lattice parameter $a_{th.}$ (range between: 0.8306 – 0.8414 nm), and $a_{exp.}$ (0.8354 – 0.8390 nm), shows that obtained cationic distribution (see table 1) is close to actual one. Observed grain diameter (range between 31.54 – 44.89 nm) values are higher as compared to earlier report (range between 28.26 – 36.67 nm) [14], are ascribable to annealing assisted growth of grains [31]. Figure 2 (b) depicts the compositional dependence of dislocation density, ascribable to compositional dependent variation of Fe, Co ions allocation on site A, B (shown in table 1). Figure 2(b) inset depicts grain diameter dependence of ρ_D. Based on figure 2, and its inset, the following experimental relations are written, showing strong correlation between composition, and structural parameters: i) ρ_D = 0.47215 + 0.623 [x], and
ii) ρ_D = 24.85 exp.($-D_S$/6.17) + 3.99 exp.($-D_S$/18.66) + 0.12.

Table 1 depicts the cation distribution, oxygen parameter (u), inversion parameter (δ), theoretical lattice parameter (a_{th}), specific surface area (S), and x-ray density (ρ_{xrd}). Table 1 shows that ρ_{XRD} range between 5272.17 – 5297.48 kg/m^3, shows that with reducing Co-content, there is a shrinkage of unit cell (evidenced by reduction of $a_{exp.}$) leading to increase of ρ_{XRD} , which is slightly higher than ρ_{XRD} values reported in earlier work [14], where ρ_{XRD} varies between 5268.99 – 5292.15 kg/m^3. Observed higher ρ_{XRD} values are ascribable to thermal annealing. Specific surface area (S) varies between 25.35 – 36.08 m^2/g, which is higher than as reported in [14]: 16.23 30.03 m^2/g. Higher S value is useful for heterogeneous catalysis, owing to the higher surface area. Table 1 shows that thermal annealing leads to re-distribution of cations (Co^{2+}, Fe^{3+}) on site A, B, affecting degree of inversion (δ) revealing mixed

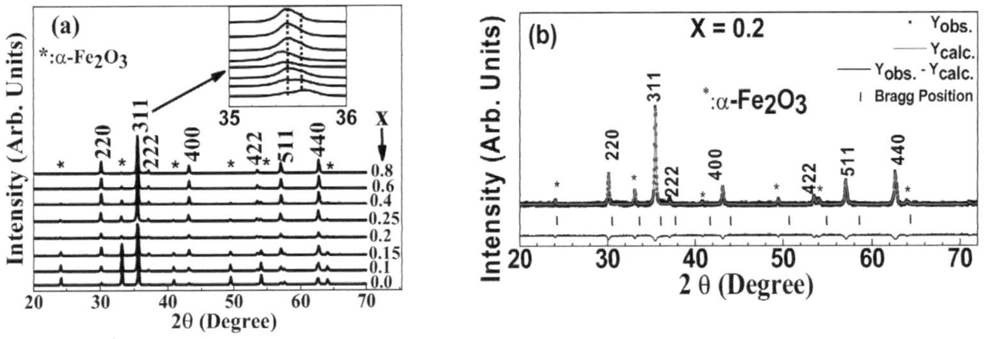

Figure 1. (a) XRD patterns of $Co_{1-x}Fe_{2+x}O_4$ samples, Inset: zoom part of (311) peak, where peak shift is shown by dotted lines, (b) Representative Rietveld refined XRD pattern for x = 0.2 (R_p = 10.47 %, R_{wp} = 14.18 %, R_{exp} = 10.47 %, G.O.F. = 1.74).

Table 1. Iron content (x) dependence of the cation distribution, oxygen parameter (u), inversion parameter (δ), theoretical lattice parameter (a_{th}), specific surface area (S) and, x-ray density (ρ_{xrd}) of annealed samples of $Co_{1-x}Fe_{2+x}O_4$.

x	Cation Distribution *A site*	*B site*	u (±0.0017)	δ (±0.27)	a_{th}(nm) (±0.0036)	S (m^2/g) (±04.10)	ρ_{xrd}(kg/m^3) (±12.15)
0.0	$(Co^{2+}_{0.23}Fe^{3+}_{0.77})$	$[Co^{2+}_{0.77}Fe^{3+}_{1.23}]$	0.3803	0.77	0.8414	25.59	5297.48
0.1	$(Co^{2+}_{0.00}Fe^{3+}_{1.00})$	$[Co^{2+}_{0.09}Fe^{3+}_{1.10}]$	0.3789	1.00	0.8399	25.81	5292.62
0.15	$(Co^{2+}_{0.17}Fe^{3+}_{0.83})$	$[Co^{2+}_{0.68}Fe^{3+}_{1.32}]$	0.3799	0.83	0.8393	26.32	5292.86
0.2	$(Co^{2+}_{0.04}Fe^{3+}_{0.96})$	$[Co^{2+}_{0.76}Fe^{3+}_{1.24}]$	0.3789	0.96	0.8386	29.74	5263.56
0.25	$(Co^{2+}_{0.75}Fe^{3+}_{0.25})$	$[Co^{2+}_{0.00}Fe^{32}_{2.00}]$	0.3834	0.25	0.8383	25.35	5272.17
0.4	$(Co^{2+}_{0.60}Fe^{3+}_{0.40})$	$[Co^{2+}_{0.00}Fe^{3+}_{2.00}]$	0.3827	0.40	0.8362	29.87	5282.03
0.6	$(Co^{2+}_{0.40}Fe^{3+}_{0.60})$	$[Co^{2+}_{0.00}Fe^{3+}_{2.00}]$	0.3815	0.60	0.8335	36.08	5273.28
0.8	$(Co^{2+}_{0.08}Fe^{3+}_{0.92})$	$[Co^{2+}_{0.12}Fe^{3+}_{1.88}]$	0.3797	0.92	0.8306	33.90	5289.52

ferrite as well as oxygen parameter (u), and shows cationic distribution alteration as compared to earlier results [14]. Obtained oxygen parameter (u) range between 0.3789 - 0.3834, which is greater than its ideal value u_{ideal} - 0.375 [1], is used to quantitative measurement of oxygen distortion in the structure and, its enhancement suggests higher structural disorder.

Figure 3 (a) illustrates Fe^{3+} content dependence of bond angles (θ_1, θ_2, θ_3, θ_4, θ_5), which are closely related to the strength of A-O-B, B-O-B and, A-O-A super-exchange interaction [1]. Perusal of figure 3 (a) reveal that increase of Fe-content of B–B interaction is strengthened, while weakening of A–B, A–A interaction weakens, which will have effect on magnetic properties, as was also observed earlier [14]. Linear variation of oxygen parameter (u) with inversion parameter (δ) (see fig. 3 (b)) reveal variation of disorder, as observed in [14, 32]. Using figure 3 (b), following experimental relation can be written: $u = 0.38507 - 0.00615\ [\delta]$, showing strong connection between u and δ.

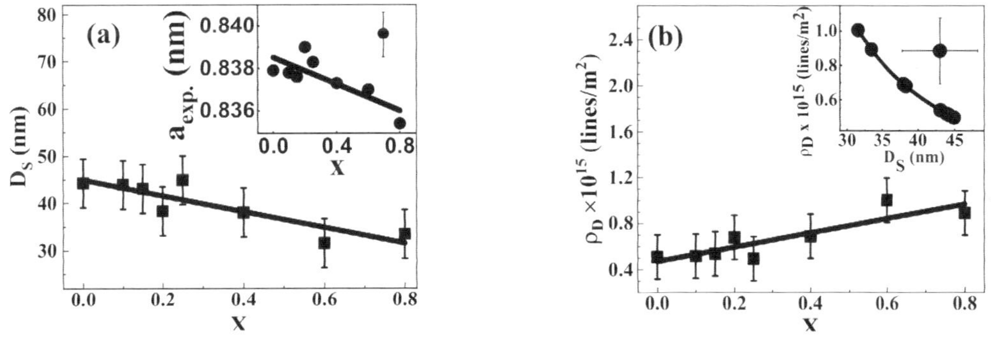

Figure 2. (a) Dependence of D_S with Fe content, Inset: Dependence of $a_{exp.}$ with Fe content, error bar (s) are also shown. **(b)** Variation of ρ_D with Fe content, Inset: D_S dependence of ρ_D, error bar is also shown. Continuous lines: fit to data.

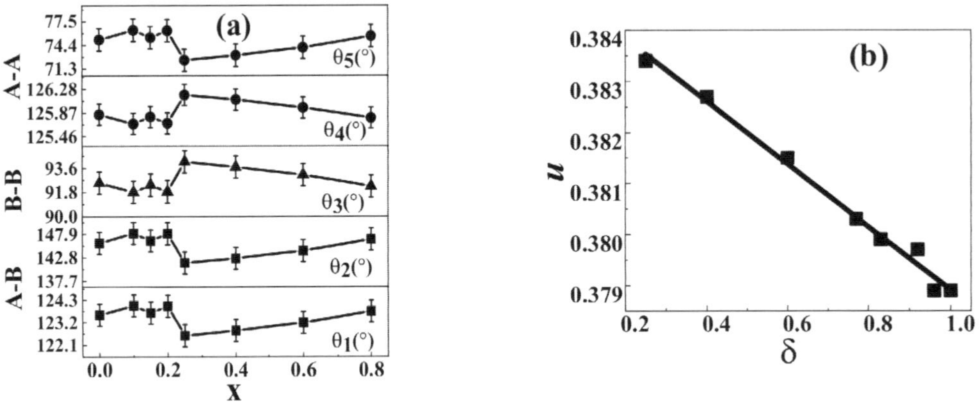

Figure 3. (a) Fe content dependence of bond angles ($\theta_1^{A\text{-}O\text{-}B}$, $\theta_2^{A\text{-}O\text{-}B}$, $\theta_3^{B\text{-}O\text{-}B}$, $\theta_4^{B\text{-}O\text{-}B}$, $\theta_5^{A\text{-}O\text{-}A}$). Line connecting points are guide to eye. Error in the estimation of bond angles range between ± 0.18 – ±2.40 degree, **(b)** Variation of u with δ. Continuous line is linear fit.

Figure 4 (a) depicts the hysteresis loops, and its perusal of reveal noticeable considerable changes in hysteresis loops with varying Fe-content, and is reflected in magnetic parameters, given in table 2. Inset of figure 4(a) reveal dependence of $M_{S(exp.)}$ on $\alpha_{Y\text{-}K}$. Perusal of figure 4(a) inset shows linear decrease of saturation magnetization '$M_{S(exp.)}$' with increasing canting angle '$\alpha_{Y\text{-}K}$', as was also observed in earlier work on samples without ant post-preparation thermal treatment [14]. Based on figure 4 (a) inset, following experimental relation can be written:

$M_{S(exp.)}$= 93.22– 1.06 [$\alpha_{Y\text{-}K}$], reveal correlation between structural ($\alpha_{Y\text{-}K}$), and magnetic properties. Observed behavior in figure 4(a) inset shows that magnetization is described by three sub-lattice model given by Yafet-Kittel [14, 33,], and the saturation magnetization reduction is because of surface spin disorder in spinel nano particles.

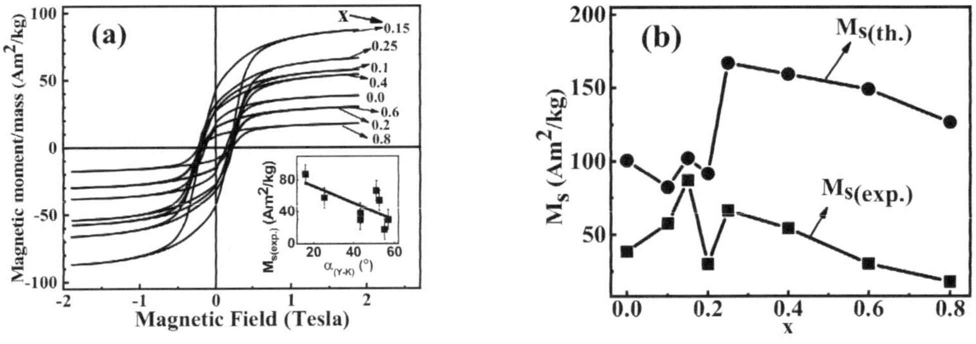

Figure 4. (a) Hysteresis loops of $Co_{1-x}Fe_{2+x}O_4$ (x=0.0-0.8) samples, Inset: dependence of $M_{S(exp.)}$ on $\alpha_{Y\text{-}K}$, linear fit is shown by continuous line, **(b)** Dependence of $M_{S(exp.)}$ and $M_{S(th.)}$ on x, line connecting points are guide to the eye.

Table 2. Iron content (x) dependence of coercivity (H_c), remanence (M_r), experimental saturation magnetization ($M_{s(exp.)}$), theoretical magnetization at 0K ($M_{S(th.)}$), squareness ratio (M_r/M_s), canting angle (α_{Y-K}), Anisotropy constant (K_1) of the studies of $Co_{1-x}Fe_{2+x}O_4$ samples.

x	Sample composition	H_c (Oe) (±0.0325)	M_r (Am^2/Kg) (±11.14)	$M_{s\,(exp.)}$ (Am^2/Kg) (±22.79)	$M_{s\,(th)}$ (Am^2/Kg) (±32.74)	$M_r/M_{s(exp.)}$ (±0.009)	(α_{Y-K}) (Degree) (±14.84)	$K_1 \times 10^4$ (erg/cc) (±02.71)
0.0	$Co_1Fe_2O_4$	1394.7	18.26	38.30	100.61	0.476	43.37	2.89
0.1	$Co_{0.9}Fe_{2.1}O_4$	1893.3	28.13	57.48	082.13	0.489	25.32	5.88
0.15	$Co_{0.85}Fe_{2.15}O_4$	1895.8	42.68	86.88	102.15	0.492	16.06	8.90
0.2	$Co_{0.8}Fe_{2.2}O_4$	1628.4	15.11	29.88	091.49	0.506	43.29	2.61
0.25	$Co_{0.75}Fe_{2.25}O_4$	2325.0	32.05	66.26	166.83	0.484	51.14	8.29
0.4	$Co_{0.6}Fe_{2.4}O_4$	2284.4	27.13	54.11	159.13	0.501	52.69	6.66
0.6	$Co_{0.4}Fe_{2.6}O_4$	1984.3	14.80	29.97	148.87	0.494	56.98	3.20
0.8	$Co_{0.2}Fe_{2.8}O_4$	2210.7	8.66	17.72	126.45	0.489	55.155	2.11

Figure 4(b) elucidates variation of $M_{S(exp)}$, $M_{s(th)}$ with varying Fe-content. Non - similar trend of $M_{S(exp)}$, $M_{S(th)}$ observed, shows that magnetization behavior is directed by three sub lattice model [14, 33], as explained earlier. Table 2 shows that obtained H_c values are consistent with computed K_1 values. Obtained H_c values, equivalent grain size (range between: 31.54 – 44.89 nm) suggest that studied ferrites are in single-multi domain overlap regions [34]. Obtained M_r/M_s between 0.27 – 0.47, indicates variation of inter-grain interaction, and isotropic behavior of multi domain grains, and does not have favored magnetization direction is consistent with literature [24, 27, 35].

Figure 5 (a) depicts the representative TAUC plot for x= 0.4 sample, while figure 5 (a) inset shows variation of bandgap with a_{exp}. Based on figure 5 (a) inset, the following experimental relation is written: Bandgap = 75.10 - 87.93 [$a_{exp.}$], show relationship among bandgap, structural changes i. e. $a_{exp.}$ alteration due to variation of Fe^{3+} content, and bandgap, as was also reported in [14, 15]. Figure 5 (b) depicts the Fe-content dependence of bandgap. Perusal of figure 5 (b), shows exponential increase of bandgap with increase in Fe-content, attributable to variation of cationic distribution (*changes in Fe^{3+}, Co^{2+} ions on A, B site*) [15, 16, 36, 37], and using figure 5 (b) following experimental relation can be

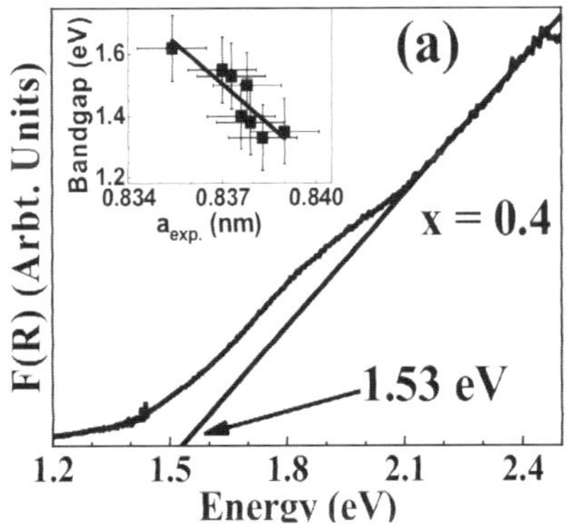

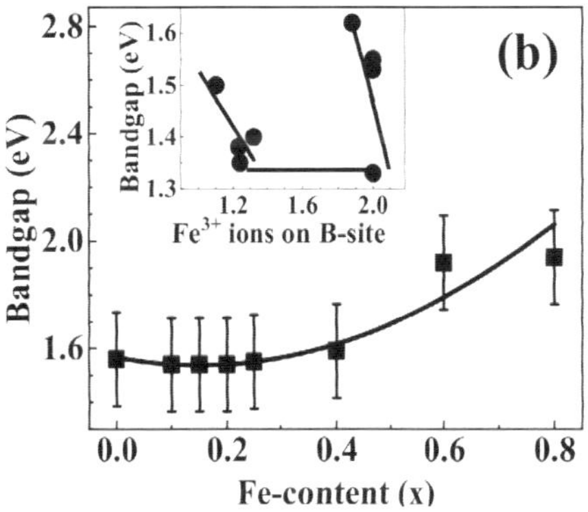

Figure 5. (a) Representative TAUC plot for $Co_{0.9}Fe_{2.1}O_4$ sample (x=0.4), Inset: bandgap variation on $a_{exp.}$, continuous line is linear fit),**(b)** Fe content (x) dependence of bandgap for the studied $Co_{1-x}Fe_{2+x}O_4$ samples, continuous line is exponential fit to the data; Inset: Variation of bandgap with Fe^{3+} on B-site, continuous lines are guide to eye.

IOP Conf. Series: Materials Science and Engineering 1258 (2022) 012009 doi:10.1088/1757-899X/1258/1/012009

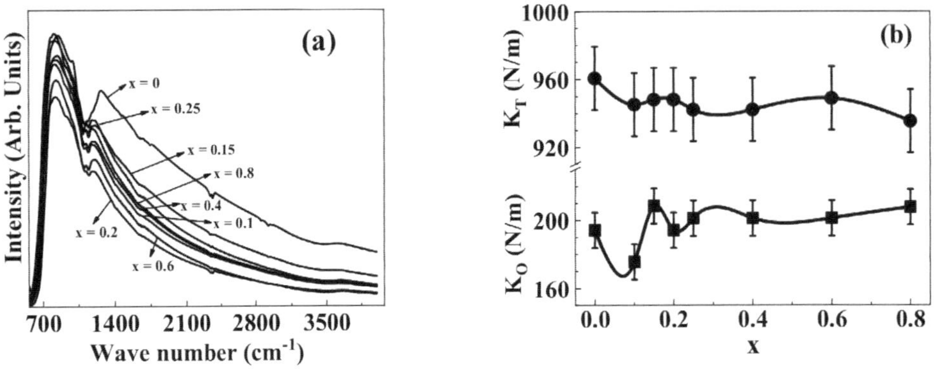

Figure 6. (a)FTIR data of studied $Co_{1-x}Fe_{2+x}O_4$ (x= 0.0-0.8) samples, **(b)** Variation of tetrahedral force constant K_T and octahedral force constant K_O with Fe content for the studied sample.

written: Bandgap = 65.24 exp.(– [Fe-content] / 4.92) – 63.67 + 12.86 [Fe-content]. Figure 5 (b) clearly shows that the variation of Fe-content in the studied ferrites, can be used for fine-tuning of the bandgap (range between 1.35 eV to 1.5 eV) useful semiconductor device fabrication. Aforementioned explanation of figure 5 (a) inset, is understood in terms of variation of cationic distribution, and figure 5 (b) inset, depicts the bandgap dependence of B-site Fe^{3+} ions, corroborate with the cationic distribution dependence of bandgap, and explained as follows: perusal of figure 5 (b) inset shows that Fe^{3+} ions occupancy on B-site up to 1.33 the bandgap and shows sharp linear decrease, beyond 1.33, and till 1.98 it is almost constant; and beyond 1.98 again a sharp linear increase of bandgap is observed, showing strong correlation between structural properties, and bandgap.

FTIR spectra of the studied $Co_{1-x}Fe_{2+x}O_4$ samples are depicted in Figure 6 (a), while absorption bands *(v_1, v_2)*, tetrahedral (K_T), octahedral (K_O) force constants, and Debye temperature (θ_D) are depicted in table 3. FTIR spectra reveal two main characteristic absorption bands of spinel crystal lattice in the frequency range 500- 4000 cm^{-1}: the first one is high frequency 'v_1' is in the range 800–853 cm^{-1}, while the second one is low frequency 'v_2' is in the range 482–504 cm^{-1} as was also reported in [14, 38]. These bands v_1, v_2 are ascribed respectively to vibrations of MeA-O and MeB-O vibrations. The difference in the band positions of v_1 and $v_2$as a function of Fe-content is ascribable to the variation in the distances of the MeA-O, and MeB-O complexes respectively [14, 39]. It is found that minor shift of band positions with the function of Fe–content (see table 3) are attributable to the

Table 3. Iron content (x) dependence of IR absorption bands, (K_T and K_O) force constants and Debye temperature (θ_D) of annealed samples of $Co_{1-x}Fe_{2+x}O_4$.

x	Sample composition	Tetrahedral complex v_1 (cm^{-1})	Octahedral complex v_2(cm^{-1})	K_T (N/m) (±18.6)	K_O (N/m) (±10.4)	θ_D (K) (±07.3)
0.0	$Co_1Fe_2O_4$	852.6	486.3	530.7	194.2	960.6
0.1	$Co_{0.9}Fe_{2.1}O_4$	835.2	482.2	507.8	175.6	945.2
0.15	$Co_{0.85}Fe_{2.15}O_4$	817.7	503.8	487.8	208.4	948.1
0.2	$Co_{0.8}Fe_{2.2}O_4$	835.6	486.3	508.0	194.2	948.1
0.25	$Co_{0.75}Fe_{2.25}O_4$	817.7	495.6	491.1	201.2	942.2
0.4	$Co_{0.6}Fe_{2.4}O_4$	817.7	495.6	490.3	201.2	942.2
0.6	$Co_{0.4}Fe_{2.6}O_4$	826.9	495.6	500.3	201.2	948.8
0.8	$Co_{0.2}Fe_{2.8}O_4$	800.3	503.4	466.7	207.7	935.4

variation of ionic distances FeB-O, and FeA-O [14, 40], explains the stronger covalent bonding of Fe^{3+}ions at the tetrahedral than at octahedral sites. Perusal of table 3, figure 5 (b) depicts that the K_T decreases (*for the studied thermally annealed samples, with increasing Fe content*) is found to increase as a_{exp} decreases, suggests the strengthening of inter-atomic bonding, while the force constants K_O is found to increase as $a_{exp.}$ decreases, shows inter-atomic bond weakening [14, 38]. Perusal of table 3 shows that Debye temperature (θ_D) decreases with increasing Fe-content, suggests variation of lattice vibrations, and agrees well with the values reported in literature [14, 40].

4. Summary
Nanocrystalline spinel $Co_{1-x}Fe_{2+x}O_4$ ferrites were prepared via sol gel autocombustion technique. Formation of spinel phase is validated by XRD. Results reveal modification of – lattice parameter, dislocation density, cationic distribution, degree of inversion, A–A, B–B, A–B, super-exchange interactions, saturation magnetization, coercivity, anisotropy, bandgap, and lattice vibrations. Current work noticeably reveal strong connection between structural properties, bandgap, and magnetic properties.

Acknowledgments
Authors thank Dr. M Gupta, and Dr. U. P. Deshpande, UGC-DAE CSR, Indore for respectively for XRD, and UV-Vis, FTIR measurements. One month hospitality to S. N. Kane, during June 2018 as invited professor at Ecole Normale Superieure de Cachan, University Paris-Saclay, Cachan (France), gratefully acknowledged.

References
[1] Smit J and Wijn H P J 1959 *Ferites* (Eindhoven, The Netherlands: Philips Technical Library)
[2] Sertkol M, Köseoglu Y, Baykal A, Kavas H and Toprak M S 2010 *J. Magn. Magn. Mater.***322** 866-71
[3] Costa A C F M, Silva V J, Cornejo D R, Morelli M R, Kiminami R H G A and L. Gama 2008 *J. Magn. Magn. Mater.***320** e370-72
[4] Willard M A, Kurihara L K, Carpenter E E, Calvin S and Harris V G 2004 *Int. Mater. Rev.* **49** 125-70
[5] Shanmugavel T, Gokul Raj S, Rajarajan G and Ramesh Kumar G 2014 *Procedia Mater. Sci.* **6** 1725-30
[6] Satalkar M, Kane S N, Ghosh A, Ghodke N, Barrera G, Calegato F, Coisson M, Tiberto P and Vinai F 2014 *J. Alloys Comp.* **615** S313-16
[7] Panchal S,Raghuvanshi S, Gehlot K, Mazaleyrat F and Kane S N 2016*AIP Adv.* **6** 055930-1-6
[8] Feng X, Huang Y, Chen X, Wei C, Zhang X and Chen M 2018 *J. Mater. Sci.* **53** 2648–57
[9] Chen D H and He X R 2001 *Mater. Res. Bull.* **36** 1369–77
[10] Wang X, Hou X, Mao J, Gao Y, Ru Q, Hu S and LamKh 2016 *J. Mater. Sci.* **51** 5843–56
[11] Barrera G, Coisson M, Celegato F, Raghuvanshi S, Mazaleyrat F, Kane S N and Tiberto P 2018 *J. Magn. Magn. Mater.* **456** 372–80
[12] Raghuvanshi S, Mazaleyrat F and Kane S N 2018 *AIP Adv.* **8** 047804-1-11
[13] Raghuvanshi S, Tiwari P, Kane S N, Avasthi D K, Mazaleyrat F, Tatarchuk T and Mironyuk I2019 *J. Magn. Magn. Mater.* **47** 521-28
[14] Kane S N, Tiwari P, Deepti, Verma R, Deshpande U P and Mazaleyrat F 2020 *Mater. Today: Proc.* **32** 358-64
[15] Verma R, Mazaleyrat F, Deshpande U P and Kane S N 2020 *Mater Today Proc.* **32** 329–33
[16] Devi E C and Soibam I 2018 *J. Supercond. Nov Magn.* **31** 1183–88
[17] Cedeno-Mattei Y, Perales-Perez O, Uwakweh O N C and Xin Y 2010 *J. Appl. Phys.***107** 09A741-1-3
[18] Na H B, Song I C and Hyeon T 2009 *Adv. Mater.***21** 2133-48
[19] Ahmad T, Rhee I, Hong S, Chang Y and Lee J 2011 *J. Nanosci. Nanotech.* **11** 5645-50

[20] Zhang W, Zuo X, Niu Y, Wu C, Wang S, Gun S and Silva S R P 2017 *Nanoscale* **91** 3929-37
[21] Kharabe R G, Devan R S, Kanamadi C M and Chougule B K 2006 *Smart Mater. Struct.* **15** N36-39
[22] Raghuwanshi S, Kane S N, Lalla N P and Reddy V R 2016 *J. Phys.: Conf. Ser.* **755** 012049-1-5
[23] Kriegisch M, Ren W, Sato-Turtelli R, Muller H, Grossinger R and Zhang Z 2012 *J. Appl. Phys.* **111** 07E308-1-3
[24] Tiwari P, Verma R, Modak S S, Reddy V R, Mazaleyrat F and Kane S N 2021 *Hyperfine Inter.* **242** 51-1-15
[25] Lutterotti L and Scardi P 1990 *J. Appl. Cryst.* **23** 246-52
[26] Verma R, Kane S N, Tiwari P, Modak S S, Tatarchuk T and Mazaleyrat F 2018 *Mol. Cryst. Liq. Cryst.* **674** 130-41
[27] Verma R, Kane S N, Deshpande U P and Mazaleyrat F 2021 *Mater. Today: Proc.* **46** 2205-11
[28] Bertaut E F 1950 *Comptes Rendus Hebdomadaires des Seances de l'Academie des Sciences* **230** 213-15
[29] Kubelka P and Munk F Z 1931 *Tech. Phys.* **12** 593-601
[30] Mallesh S, Prabu D and Srinivas V 2017 *AIP Adv.* **7** 056103-1-8
[31] Yadav R S, Kuřitka I, Vilcakova J, Havlica J, Masilko J, Kalina L, Tkacz J, Švec J, Enev V and Hajdúchová M 2017 *Adv. Nat. Sci.: Nanosci. Nanotechnol.* **8** 045002-1-14
[32] Barrera G, Coisson M, Celegato F, Raghuvanshi S, Mazaleyrat F, Kane S N and Tiberto P 2018 *J. Magn. Mgn. Mater.* **456** 372-80
[33] Murthy N S S, Natera M G, Youssef S I, Begum R J and Srivastava C M 1969 *Phys. Rev.* **181** 969-77
[34] Kolhatkar A G, Jamison A C, Litvinov D, Willson R C and Lee T R 2013 *Int. J. Mol. Sci.* **14** 15977-16009
[35] Shirsath S E, Toksha B G and Jadhav K M 2009 *Mater. Chem. Phys.* **117** 163-168
[36] Chakrabarty S, Pal M and Dutta A 2015 *Mater. Chem. Phys.* **153** 221–28
[37] Taufiq A, Hidayat A, Sunaryono, Hidayat N and Soontaranon S 2019 *IOP Conf. Series:Mater. Sci. Eng.* **515** 012047-1-11
[38] Tiwari P, Verma R, Kane S N, Tatarchuk T and Mazaleyrat F 2019 *Mater. Chem. Phys.* **229** 78–86
[39] Zipare K V, Bandgar S S and Shahane G S 2018 *J. Rare Earths* **36** 86-94
[40] Modi K B, Raval P Y, Shah S J, Kathad C R, Dulera S V, Popat M V, Zankat K B, Saija K G, Pathak T K, Vasoya N H, Lakhani V K, Chandra U and Jha P K 2015 *Inorg. Chem.* **54** 1543–55

Template assisted green synthesis of nickel oxide nanoparticles for photocatalytic applications

OK Nimisha, A P Reena Mary, Rajita Ramanarayanan*

Department of Physics, Government Victoria College,

Palakkad, Kerala, India-678001

rajithar@gvc.ac.in

Abstract:

Nickel oxide nanomaterials are a highly researched class of magnetic nanomaterials showing tremendous applications in catalysis, sensing, batteries, solar cells and magnetic devices. This work shows the green synthesis of nickel oxide nanoparticles using vermicelli as template and black tea as reducing agent. The obtained product was analysed by XRD, UV and SEM measurements. The sample showed good photocatalytic behaviour when tested under solar cell illumination in the degradation of Methylene Blue dye. Further the nanosheet morphology of the material has potential to be used in energy generation and storage applications.

Keywords: green synthesis, template, nanosheet

1. INTRODUCTION

Metal oxide nanoparticles have been a great research interest to scientists owing to their outstanding physical, chemical and optical properties. Among metal oxides ,Nickel oxide nanoparticles (NPs) is one of the prominent transition metal oxide due to its varied applications in catalysis, sensing, photovoltaic and antibacterial properties [1]. NiO NPs are of great interest to researchers worldwide as vital semiconductors with a wide band gap and magnetic properties. Among the various transition-metal oxides, nickel oxide has remarkable electrochemical performance to be used in supercapacitors[2].
Inspite of great advances by physical and chemical methods, the green synthesis of nanomaterials has assumed great significance due to its environment friendly approach along with its simple approach in the recent years. Nickel oxide nanoparticles have been synthesized through plant mediated methods as reported by previous literature with remarkable results [3, 4, 5].

In this investigation, we have used black tea extract as reducing and capping agent and vermicelli as template to synthesize nickel oxide nanoparticles. The major components of tea extract as per previous reports are tannins/gallic acid and flavonoids which have hydroxyl /carbonyl groups for reduction and stabilisation of nanoparticles[6,7]. The aim to use vermicelli template is to attempt to synthesize nanoparticles of different shapes with a simple affordable lab method. The obtained nanoparticles have sheet/shell like morphology which is promising due to its increase in surface area making it suitable for a wide variety of applications. Further the synthesized sample was tested for its Photocatalytic behaviour using Methylene Blue (MB) dye.

2. MATERIALS AND METHODS

2.1Materials

Nickel nitrate [Ni $(NO_3)_2.6$ H_2O] was purchased from Merck Chemicals. The black tea and vermicelli were collected from local market. The extract of the tea was prepared using the method employed in our previous report [7] by boiling a specified amount of tea in distilled water. The obtained filtered black tea extract was used as the reducing agent in this work.

2.2 Synthesis of nickel oxide

90ml of Nickel nitrate solution (precursor) was taken in a beaker and 10 ml of black tea extract was added to the stock solution as reducing agent .To this solution 100gms of vermicelli (template) was added and stirred for 30 minutes using magnetic stirrer at 60°C (Figure 1(a) and (b)). The liquid part was removed and vermicelli transferred to a Petri dish (Figure 1(c)). Petri dish was heated at 60°C in an oven for drying. The dried template along with nanoparticles deposited on it was heated in a muffle furnace for 450 °C for two hours. After cooling, the synthesized sample was found to be a greyish black powder as shown in Figure 1(d). The vermicelli which is primarily composed of starch is used as template in this work and gets removed from sample when heated above 400 °C[8].

Figure 1 (a) Precursor Solution (b) Precursor –template-extract mixture (c) Vermicelli after drying (d) powder obtained after calcinations

The proposed scheme of the reaction mechanism indicating the formation of NiO nanoparticles is illustrated in Figure 2. In this mechanism, nickel ions (Ni^{2+}), in the presence of plant reducing agent gallic acid, undergo reaction to form stabilised Ni nanoparticles which further on calcination become NiO nanoparticles.

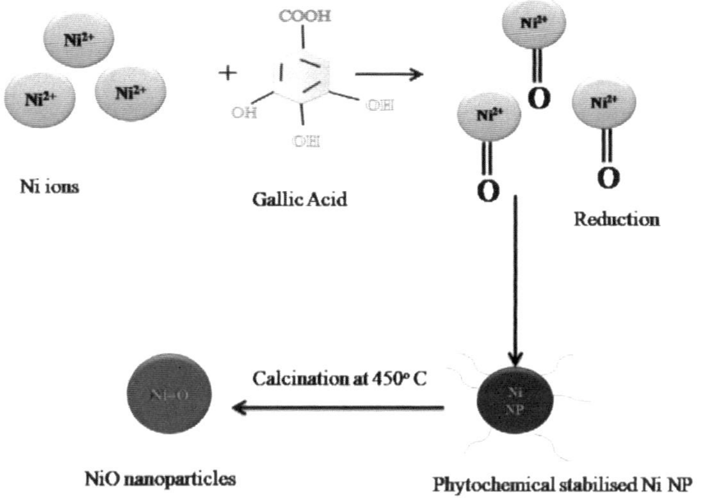

Figure 2 Proposed scheme of reaction mechanism of NiO nanoparticles

2.3 Photocatalysis

5 mg of the nickel oxide nanopowder was mixed with 25 ml of 20 ppm Methylene Blue (MB) dye stirred for two minutes and placed in sunlight (870-880 Watt/m^2) for 180 minutes during which at intervals of 60 minutes, the supernatant solution was taken out and UV absorbance spectrum was taken and compared with the original dye solution.

2.4 Measurements

The UV-vis absorption measurements were recorded using spectrophotometer (Shimadzu) in the range of 200-900 nm. X-ray diffraction (XRD) measurements were performed using Rigaku Miniplex–X Ray diffractometer with monochromatic Cu-Kα radiation (λ=1.54 Å). The sample morphology was observed through Scanning Electron Microscope (Jeol).

3. RESULTS AND DISCUSSION

The phase and crystal nature of the synthesized samples were investigated by the XRD pattern as shown in Figure 3.

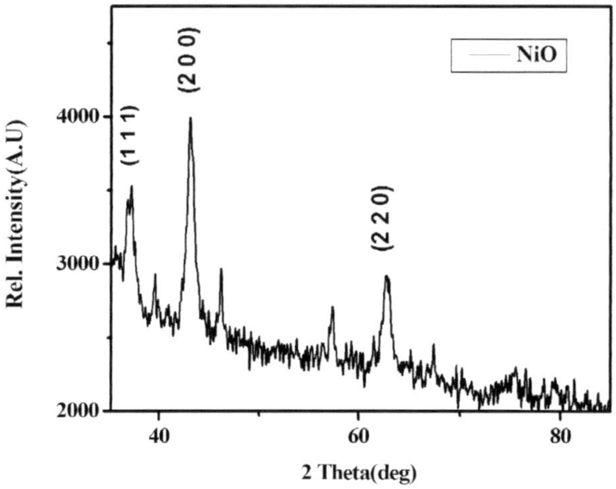

Figure 3 XRD of NiO nanoparticles

The peaks in the pattern for NiO nanoparticles are consistent with that of the JCPDS reference file for face-centered cubic (FCC)Nickel Oxide (JCPDS card number 47-1049).The dominant diffraction peaks at approximately 2θ= 37.01°, 43.06°, 62.61° can be related to (111), (200) and (220) crystal planes respectively. The crystallite size of NiO nanoparticles was obtained from Debye-Scherrer formula (D=0.9λ/β Cosθ) where λ= 1.54056A^O is the wavelength of the X-ray, β= FWHM (full width at half maximum) and is angle obtained from 2θ values corresponding to maximum intensity peak in XRD pattern (200).The crystallite size was calculated to be 18 nm which agrees with the reported literature [9, 10].

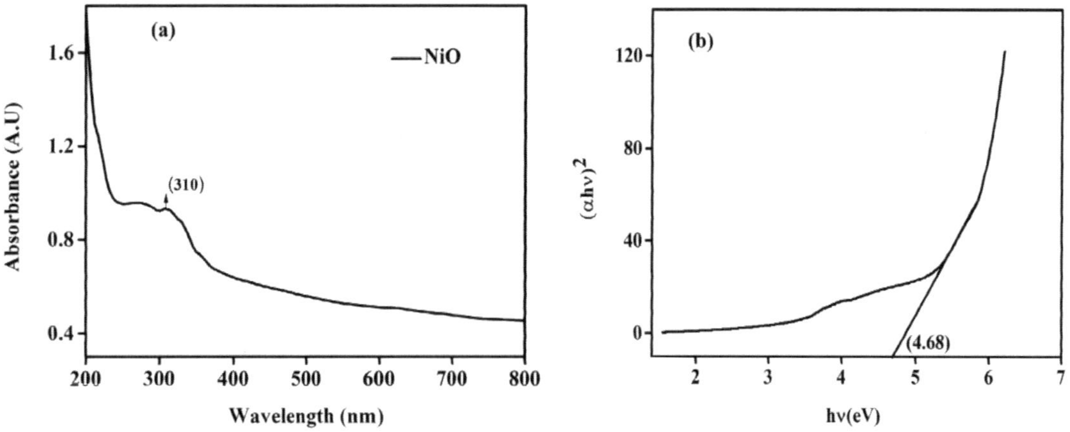

Figure 4 (a) Absorbance (b) Optical band gap of NiO nanoparticles

The absorbance graph of NiO nanoparticles shows absorbance in UV region as shown in Figure 4(a) and a peak centered at 310 nm is observed in agreement with the previous reports [10]. The presence of absorption at UV regions shows the potential for NiO nanoparticles to be used as a photocatalyst under solar light illumination. Considering the direct band transition in NiO NPs, a plot between $(\alpha h\upsilon)^2$ versus photon energy (hυ) is drawn in Figure 4(b). It is observed from that the optical band value 4.68 eV is higher than bulk value of 3.65 eV indicating quantum confinement [11].

Morphology of the synthesized Nickel oxide nanoparticles is studied using scanning electron microscope and shown in Figure 5. It shows that the synthesized nickel oxide nanoparticles have several shapes such as Nano sheets, Nano shells and they appear as Nano clusters.

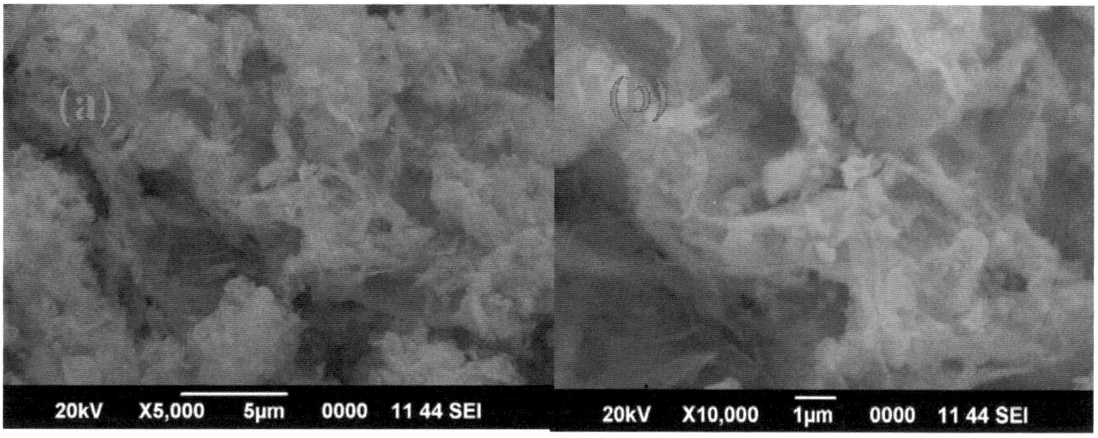

Figure 5(a) and (b) SEM images of NiO with different magnification

To understand the photocatalytic activity of the NiO nanoparticles in the degradation of Methylene Blue dye, the plot of C/C_0 versus time was drawn as shown in Figure 6 (a), where C_0 is the initial concentration and C is the concentration of MB dye at regular intervals.

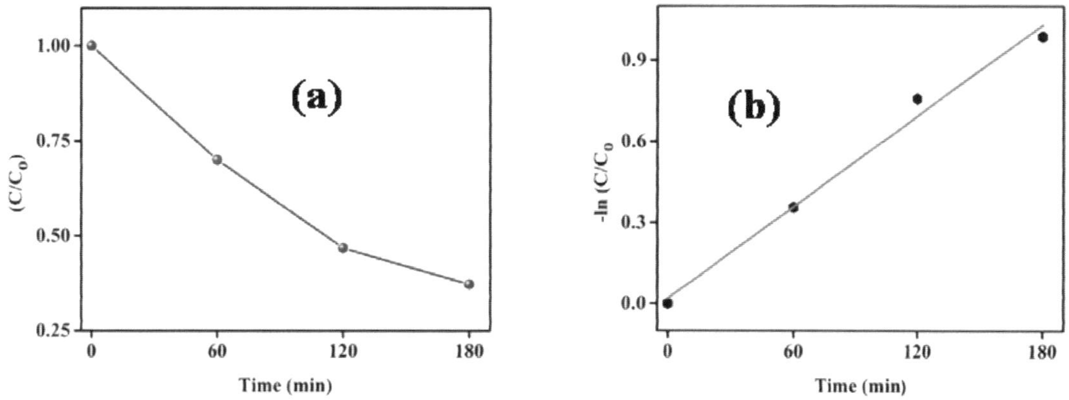

Figure 6 (a) C/C_0 versus time (b) $-\ln(C/C_0)$ versus time

The degradation efficiency (%) is calculated using the formula $(1-C/C_o)$ and it was observed that the sample showed 63.3% degradation efficiency for a period of 180 minutes which is promising and comparable to the previously published reports as indicated in Table 1. The Figure 6(b) shows the plot of $\ln(C/C_o)$ versus irradiation time for Methylene Blue dye. It is observed that the graph shows a linear fit (linear correlation coefficient R^2 =0.983) indicating a good fit. The Figure 6(b) shows that the reaction is a pseudo first order degradation with a rate constant k showing a value of 0.00561 min^{-1} in agreement with previous reports [10].

Catalyst	Dye	Degradation %	Reported Work

NiO	MB dye	80 (270 min)	10
NiO	MB dye	80 (300 min)	12
NiO	MB dye	90.43(360 min)	13
PANi-NiO	MB dye	76(300 min)	14
NiO	MB dye	62 (360 min)	15
NiO	MB dye	63.3(180 min)	Present Work

Table 1 Comparative study of Photocatalytic behaviour of NiO nanoparticles

The synthesized NiO nanoparticles in this work have shown interesting morphology and good Photocatalytic behaviour. However to effectively use them in various applications their absorbance can be extended to visble region using metal nanoparticles /doping etc. The role of template in assisting photocatalysis may be studied bẏ reducing calcinations temperatures below 400 degrees and testing their respective Photocatalytic behaviour. The morphology/shape studies with various solvents and temperatures are likely to give controlled two dimensional nanostructures with profound applications in various fields of technology and medicine. Overall this work has been successful in the simple and cost effective synthesis of NiO nanostructures with a great scope for further research and applications.

4. CONCLUSION

A template assisted synthesis to obtain NiO nanoparticles from black tea extracted was successfully conducted in this investigation. The template used was the low cost food material vermicelli which is easily available in market. The XRD studies showed a crystallite size of 18 nm indicating their nano regime. This data is further supported by absorbance around 310 nm which is unique for NiO nanoparticles. The morphology studied using SEM showed nanocluster with nanosheets, nanoshells and other irregular structures. The sample showed good Photocatalytic behaviour with 63.3% degradation for 180 minutes with Methylene Blue dye suggesting its potential for waste water treatment studies.

ACKNOWLEDGEMENTS

The authors acknowledge Sophisticated Test and Instrumentation Centre (STIC) Cochin University of Science and Technology for SEM analysis. The authors also acknowledge Department of Physics, University of Calicut for XRD analysis.

References:

[1] Jaji N-D, Lee H L, Hussin M H, Akil H M, Zakaria M R and Othman M B H 2020 *Nanotechnology reviews* **9** 1456-80.

[2] Purushothaman K K, Babu I M, Sethuraman B and Muralidharan G 2013 *ACS Appl. Mater. Interfaces* **5** 10767−773.

[3] Gebretinsae HG, Tsegay MG and Nuru ZY 2021 *Materials today proceedings* **36** 566-70.

[4] Ezhilarasi A A, Vijaya J J, Kaviyarasu K, Kennedy L J, Jothiramalingam R and Al-Lohedan H A 2018 *Journal of Photochemistry & Photobiology, B: Biology* **180** 39-50.

[5] Prabhu S , Thangadurai T D , Vijai Bharathy P and Kalugasalam P 2022 *Results in Chemistry* **4** 100285.
[6] Yadav K C, Ashok P, Bishnu B K and Shiwakoti L D **2020** *Hindawi Journal of Food quality* 1-13.
[7] Ramanarayanan R and Swaminathan S 2021 *IOP Conf. Series: Materials Science and Engineering* **1166** 012041.
[8] Lu H, Zheng M, Chen J, Li N, Xue L and Cao J 2011 *Integrated Ferroelectrics* **127** 128-133.
[9] Al-Sehemi A G, Al-Shihri A S, Kalam A , Du G and Ahmad T 2014 *Journal of Molecular Structure* **1058** 56-61.
[10] Sabouria Z , Akbaria A , Hosseini H A and Darroudi M 2018 *Journal of Molecular Structure* **1173** 931-36.
[11] Vijaykumar P, Jafar Ahamed A and Karthikeyan M 2019 *SN Applied Sciences* **1** 1-15.
[12] Sabouria Z, Fereydouni N, Akbari A, Hosseini H A , Hashemzadeh A , Amiri M S, Oskuee R K, Darroudi M 2020 *Rare Metals* **39** 1134-44.
[13] Nasser M Y, Aly H M, Abdelrahman E A and Mousthafa M E 2017 *Journal of Molecular Structure* **1143** 462 – 471.
[14] Jayaramu V and Balamurugan P 2019 *Mater. Res. Express* **6** 1-22.
[15] Ramesh M 2021 *SN Applied Sciences* **3** 1-13

Synthesis, structural and optical properties of colloidal 2D organometal halide $(C_{12}H_{25}\text{-}NH_3)_2PbI_4$ nanoparticles

Kshetra Mohan Dehury, G. Vijaya Prakash*

Nanophotonics Lab, Department of Physics, Indian institute of Technology Delhi, Hauz Khas, New Delhi 110016

prakash@physics.iitd.ac.in

Abstract. The self-assembled IO hybrid semiconductors have been widely explored due to their unique characteristics of the formation of different dimensionalities crystal structural packing. The room temperature Mott-type excitons are formed as a result of quantum confinement and dielectric contrast effects. The excited free electron and hole bound together through strong coulomb interaction which is known as excitons and the excitons energy levels are present slightly below the conduction band. The large dielectric constant of extended inorganic network structure reduces the strong coulomb interaction between the excited free electron and holes known as Mott-type excitons. The Mott-type excitons has larger Bohr radius (30-100 A°) as compared to lattice spacing and having binding energies~10-30 meV. The different crystal structural packing are formed due to the structural reorganization of organic moiety (based on shape, size and position of amine group) within the extended inorganic layered structure. A systematic study of synthesis, structural and optical properties of 2D $[(C_{12}H_{25}\text{-}NH_3)_2PbI_4]$ nanoparticles are discussed and presented.

1. Introduction

The self-assembled 2D IO hybrid semiconductors are represented by general formula $(R\text{-}NH_3)_2\text{-}MX_4$ (R is the organic moiety, 'M' is the divalent metal (Pb^{2+}, Sn^{2+}, Cu^{2+} etc.) and 'X' is the halide-ion (Cl^-, Br^-, I^-) [1-3]. The multiple quantum well (MQW) structure are formed due to the quantum width limit of both quantum well and quantum barrier. The charge carriers are confined in the lowest band gap of extended inorganic network structure which is known as quantum confinement effects. The large difference in the dielectric constant mismatch ($\varepsilon_{org} < \varepsilon_{inorg}$) between the organic moiety and extended inorganic network structure induces strong coulomb intercalation between the excited free electron and hole (dielectric screening/contrast effect). The dielectric screening effects depend upon the dielectric constant value of organic and inorganic network structure and thickness of barrier. The room temperature excitons are formed as a result of quantum confinement and dielectric contrast effects [4, 5]. The extensive studied of optical exciton emission features of 2D $[(R\text{-}NH_3)_2PbI_4]$ and 3D $(R\text{-}NH_3PbI_3)$ IO hybrid semiconductors are already reported [6-12]. In this study, we have successfully synthesized the 2D nanoparticles and explore the structural and optical exciton emission features.

2. Experimental

2.1 Characterization technique

The colloidal 2D $(C_{12}H_{25}NH_3)_2PbI_4$ (hereafter, C12PI) nanoparticles (NPs) are synthesized by using chemical synthesis method [13]. The XRD patterns of 2D nanoparticles are performed by using Cu-K_α radiation (1.5418 Å) with diffraction angle (2θ) from 3° to 60°. The TEM micrographs images of colloidal nanoparticles are recorded by using FEITecnai TF20 microscope. The HRTEM micrograph images are recorded by using Phillip CM 12 high resolution electron microscope. The optical absorption spectra of 2D nanoparticles solutions are carried out by using UV-VIS-NIR spectrophotometer. The photoluminescence spectra and real time PL monitoring are recorded in a reflection mode using advanced modified optical microscopes.

3. Results and Discussion

3.1 Structural and optical properties of C12PI colloidal 2D nanoparticles

The crystal structure packings and multiple quantum well (MQW) structure of 2D IO hybrid semiconductors are shown in Figure 1(a). In 2D crystal structure packing, the organic ammonium cation is interdigitized between the extended inorganic layered structures with corner sharing MX_6 octahedra

[1-3]. The schematic of exciton formation in IO hybrid semiconductors are shown in Figure 1(b). The conventional absorption and emission spectra of 2D nanoparticles are shown in Figure 1(c). The 2D nanoparticles are showing strong exciton absorption peak at 494 nm and exciton emission peak at 497 nm. These colloidal 2D nanoparticles are showing strong green colour emission which is the characteristics of their emitting materials. The XRD pattern of 2D nanoparticles shows strong diffraction peak along *004l* (*l*=1, 2, 3) demonstrate the crystalline nature (Figure 1(d). The TEM micrograph images of 2D nanoparticles are shown in Figure1 (e). The d spacing (~0.667 nm) are obtained from HRTEM micrographs images.

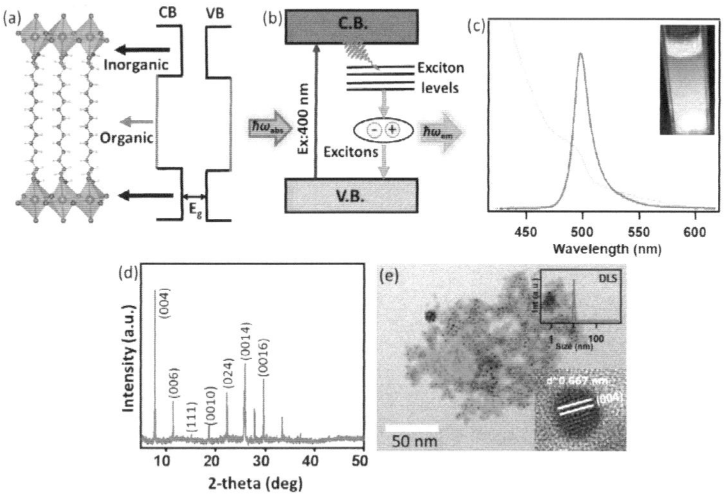

Figure 1. (a) 2D crystal structure packing in which an organic moiety is interdigitized between two extended inorganic layered structure along with MQW structure (b) schematic of exciton formation (c) Exciton absorption and photoluminescence spectra (Ex:400 nm CW laser). Inset images are 2D nanoparticles under laser induced strong green color emission (d) typical XRD pattern and (e) is the TEM micrograph images of colloidal 2D nanoparticles. The insert images are DLS particles size distribution and HRTEM micrograph images (d~0.667 nm).

3.2 Real time PL monitoring of 2D nanoparticles

The laser light interaction with these 2D Vander wall IO hybrid semiconductors demonstrated linear and nonlinear optical process depends upon the laser excitation wavelength and intensities. The real time PL monitoring of C12PI colloidal 2D nanoparticles are performed in a reflection mode to confirm/check the PL peak stability and laser threshold (Figure 2). The real time PL monitoring of IO hybrid semiconductors are influenced by cumulative local heating effect, material degradation and laser ablation [11, 12]. The real time PL monitoring of 2D nanoparticles are shown in Figure 2(a). The collected PL (PL_{max}~497 nm) spectra are shown in Figure 2(b). The PL peak intensity plotted with laser exposure time are shown in Figure 2(c).

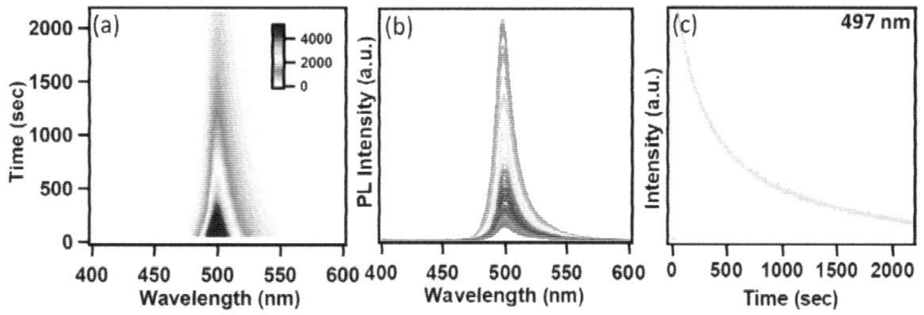

Figure 2. (a) Real time vs. PL spectral mapping along with scale bar which is just representation for relative intensity (b) collected PL spectra and (c) is the PL peak intensity plotted with laser exposure time. Ex: 400 nm CW laser.

4. Conclusions

IO hybrid semiconductors are widely explored due to the formation of different dimensional crystal structure packing along with their large exciton binding energy. The XRD pattern demonstrate the crystalline nature of C12PI 2D nanoparticles along *004l* (*l*=1, 2, 3) diffraction modes. The laser induced strong green colour emission of C12PI 2D nanoparticles demonstrate the highly emitting nature of the characteristics materials. The real time PL monitoring demonstrate the PL peak stability and laser threshold. The strong optical exciton emission features of colloidal 2D nanoparticles solutions are favourable for various optoelectronic device applications such as such as Light emitting diodes (LED), Thin film transistors (TFT) and nonlinear optical device application (multiphoton imaging, optical limiting etc).

Acknowledgements
The authors acknowledge DST-INSPIRE research fellowship for financial support and central research facility (CRF) at IIT Delhi.

References

[1] Saparov B and Mitzi D B 2016 *Chem. Rev.* **116** 4558-96.

[2] Billing D G and Lemmerer A 2007 *Acta. Crystallogr. Section B. Struct. Sci.* **63** 735-47

[3] Mitzi D B 1999 *J. Solid State Chem.* **145** 694-704.

[4] Shimizu M, Fujisawa J I and Ishi-Hayase J 2005 *Phys. Rev. B* **71** 205306.

[5] Ishihara T, Hong X, Ding J and Nurmikko A V 1992 *Surf. Sci.* **267** 323-6.

[6] Billing D G and Lemmerer A 2007 *CrystEngComm.* **9** 236-44.

[7] Vijaya Prakash G, Pradeesh K, Ratnani R, Saraswat K, Light M.E and Baumberg J J 2009 *J. Phys. D: Appl. Phys.* **42** 185405.

[8] Pradeesh K, Rao Kotla N, Ahmad S, Dwivedi V. K and Vijaya Prakash G 2013 *J. Nanoparticles* **2013**, 1-3.

[9] Adnan M, Rao K N, Acharyya J N, Kumar D, Dehury K M and Vijaya Prakash G 2019 *ACS omega* **4** 19565-72.

[10] Pradeesh K, Baumberg J J and Vijaya Prakash G 2009 *Appl. Phys. Lett.* **95** 283.

[11] Adnan M, Baumberg J J and Vijaya Prakash G 2020 *Sci. Rep.* **2615** 1-0.

[12] Adnan M, Dehury K M, Kanaujia P K and Vijaya Prakash G 2020 *J. Appl. Phys.* **128** 023104.

[13] Dehury K M, Acharyya J N, Adnan M and Vijaya Prakash G 2022 *Opt. Mater.* **124** 111969.

Structural, morphological and optical attributes of ZnO thin films deposited via spray pyrolysis process: Impact of molarity variation

S V Nikam[1], B T Jadhav[2], S M Chivate[3], S M Nikam[4], S H Mujawar[5] and A P Torane[6]

[1,2]Electronics, Rayat Institute of Research and Development
[3]Dr. Vishvanath Karad MIT World Peace University, Pune
[4]Nanoscience and Technology, Y. C. Institute of Science Satara
[5]Physics, Y. C. Institute of Science Satara
Email: sushamanikam12@gmail.com

Abstract. Undoped thin films of ZnO were prepared over glass substrates using a simple chemical spray pyrolysis process. The solution of aqueous 'Zinc acetate dehydrate' was used as precursor and sprayed over preheated glass substrate at 400 °C. The concentration of solution changed from 0.2M to 1 M with the step of 0.2 and its impact on structural, morphological and optical properties has been studied with the help of XRD, SEM, UV-Vis Spectroscopy and four-probe technique. The films are polycrystallines, have a hexagonal wurtzite structures, and have a preferred orientation in the (002) plane. SEM analysis indicates the dependency of film morphology on precursor concentration. Optical Transmittance decreases with increasing precursor concentration. Maximum optical transmittance reaches up to 51% with film obtained from 0.2M concentration. Decreasing band gap values from 3.18 eV to 3.04 eV are noted as carrier concentration increases with precursor concentration.

1. Introduction

Metal oxide semiconductors are extensively studied by many researchers for the last few decades due to their interesting optoelectronic properties. As compared to other metals ZnO has wide bandgap energy 3.37eV, resistivity, large 60meV excitation energy, thermal stability, high transmittance, non-toxicity and inexpensive make it auspicious or reliable material for various applications [1], [2]. It can be used in various applications such as gas sensing, transparent conducting electrodes, LEDs, LCDs [3], optoelectronic applications, solar cells [4], and photocatalysts [5]. ZnO has number of applications in microelectronic devices, energy storing devices [6], magnetic devices, spintronics devices [7], electrochemistry, antibacterial, and food packaging,[8], etc. ZnO forms in different structures according to application needs such as nanorods used in solar cell and biomedical applications [9], [10], nanoplatelets for sensors [11], nanowires in optoelectronics, photonics, and actuators [12], nanotubes, thin films, etc [13]. There are various methods that can be used to synthesized ZnO some of them are PLD, hydrothermal, solvothermal, electrospinning, CBD, CVD, sol-gel process, spray pyrolysis technique, etc.[14], [15].

The spray pyrolysis process is widely used, easy and inexpensive technique for large area as well as thin film deposition [16]. In this process, the precursor solution is sprayed over the heated substrate under the control of various preparation parameters such as spraying rate, substrate temperature, precursor concentration, substrate to nozzle distance, and gas pressure of the carrier. Structural properties, morphological and optoelectrical properties of films depend upon process parameters [17]–[20]. In this presented work, thin film of ZnO is prepared onto the glass surface with varying molar precursor concentration using a spray pyrolysis method, as well as the impact of varying precursor concentration on structural, morphological and optical properties of ZnO thin films was studied.

2. Experimental

The first step involve in thin film deposition was substrate cleaning. It is the most important part before thin film deposition to get the uniform and adhesive film all-over the surface area. In this study, simple glass micro slides were used as a substrate. It is further kept in chromic acid for 15 min at 100°C and then washed with laboline and acetone. These slides were ultrasonically cleaned by using double distilled water for 30 min.

The chemical spray pyrolysis method was implemented to form ZnO thin films to a preheated substrate of glass. The powder of zinc acetate dehydrate was dissolved in 50 ml of double distilled water to form aqueous solution of precursor. The films are deposited at 0.2, 0.4, 0.6, 0.8 and 1 molar concentration. All the films are deposited by keeping other preparative parameters constant. The process parameters has 400°C substrate temperature, nozzle–substrate distance at 34cm, and solution spray rate was 3 ml/min and compressed air as carrier gas. All films are kept at 200°C for annealing period of 1 hr. These sample films were further investigated with the help of XRD technique for structural properties, SEM for morphological studies, and UV-Vis spectroscopy for optical properties.

3. Results and Discussion

3.1 Structural properties

The XRD technique used to analyze the impact of molar concentration variation on structural properties of zinc oxide films. Following Figure. 1 Demonstrates XRD data plots in 2θ range from 20°-80° for the ZnO films deposited at various molar concentrations. XRD patterns indicates that, all the films are polycrystallines, hexagonal wurtzite structure, preferred diffraction peaks at (100), (002), (101), (110) and it is verified with standard JCPDS card N0-36-1451. It is also observed that the more intense (002) peaks for the films deposited at 0.6, 0.8 and 1 molar concentrations were as at 0.2 and 0.4 molars (100) peaks are more intense [21]. XRD plot also indicates that with increasing molar concentration of precursor solution peak intensity at (002) plan also goes on increasing because of decreasing grain size.

The interplanar distance calculated according to 'Bragg's law'

$$2dsin\theta = n\lambda \quad (1)$$

Where, interplanar distance 'd', Bragg's angle of diffraction 'θ', diffraction order 'n' and X-ray wavelength 'λ'.

Crystallite size was calculated with the help of more intense peaks of diffraction patterns using Scherer formula

$$0.9\lambda/2\beta Cos\theta \quad (2)$$

Where, crystallite size 'D', X-ray wavelength 'λ' (1.5406 A°), angle diffraction 'θ' and Full-width half-maxima intensity (FWHM) of peak 'β'. Table 1 given bellow demonstrate XRD analysis data.

IOP Conf. Series: Materials Science and Engineering 1258 (2022) 012012 doi:10.1088/1757-899X/1258/1/012012

Table 1. XRD analysis

Molar Concentration (M)	2θ Degree	h k l	FWHM (degree)	Crystallite Size D (nm)	d spacing (nm)
0.2	31.53	100	0.5199	16.58334	2.835191
0.4	31.93	100	0.6938	12.43909	2.800581
0.6	34.565	002	0.5131	16.93567	2.592876
0.8	34.565	002	0.5134	16.92578	2.592876
1	34.703	002	0.693	12.54395	2.58288

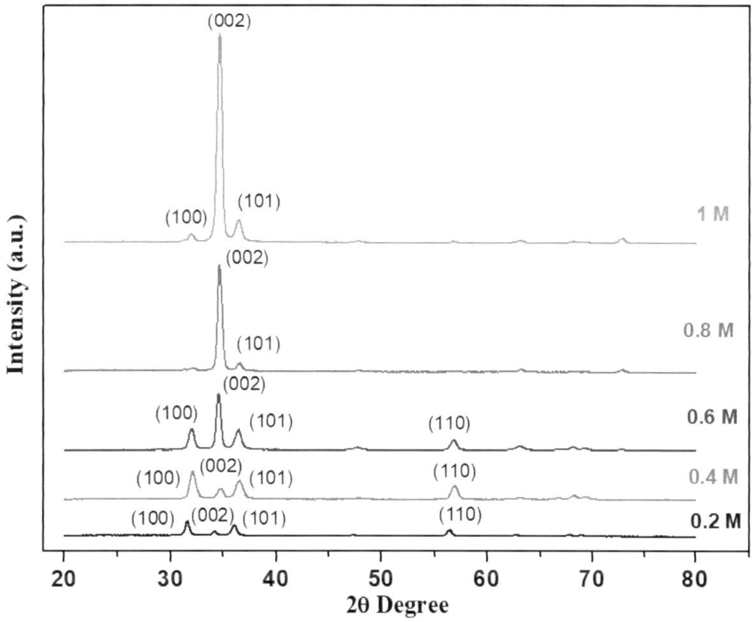

Figure1. X-ray diffraction patterns of Zinc Oxide films deposited at various molar concentrations of the precursor solution.

3.2 Morphological properties:

The prepared ZnO films have been investigated using SEM technique to reveal morphological properties. SEM morphology of ZnO films synthesized at various molar concentrations is depicted in Figure 2 given below. In figure 2 SEM morphology indicates that film morphology depended on precursor molar concentration [22]. ZnO films deposited at 0.2 molar concentration shows Nanopatelets like morphology in fig 2 (a), due to their large surface area it has potential applications in sensor technology [23]. The nanorods like dense structure was seen for the film synthesized using precursor solution of 0.4 M. For the 0.6 molar concentration of precursor solution ZnO film shows an elongated root-like structure [24]. The film deposited at 0.8 M concentration nano flex like morphology reveals, whereas with 1M concentration agglomerated Nanosheets like morphology [25]. This change in morphology occurs due to increasing grain size with increasing molar concentration.

IOP Conf. Series: Materials Science and Engineering 1258 (2022) 012012 doi:10.1088/1757-899X/1258/1/012012

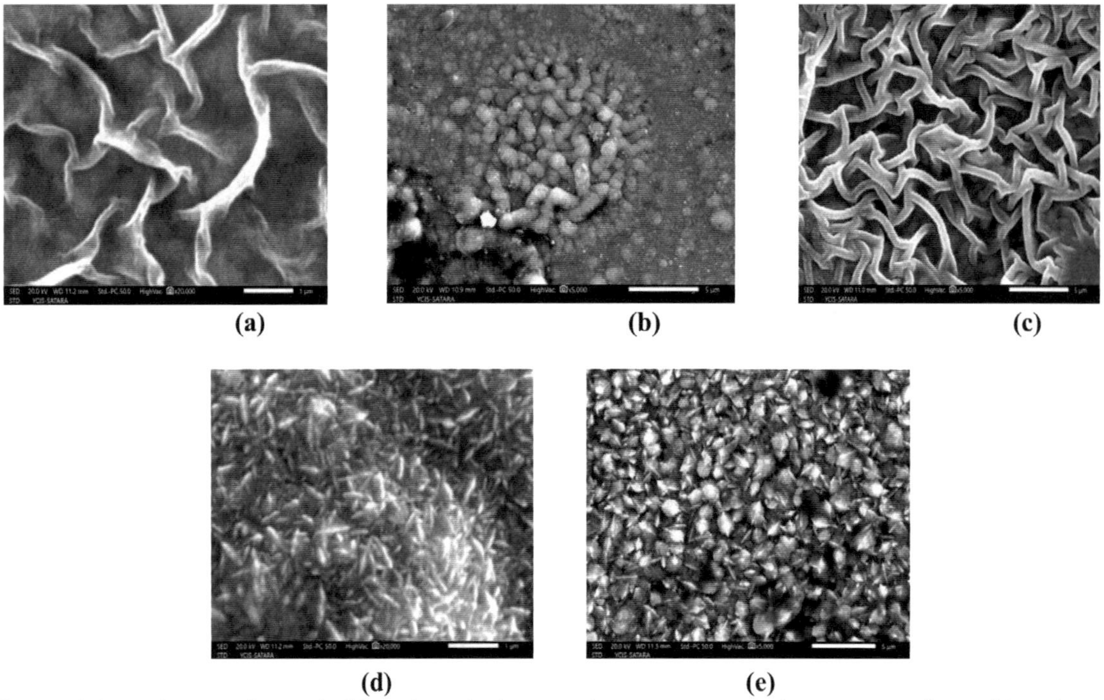

Figure 2 SEM images for ZnO films deposited at varying precursor molar concentrations. (a) at 0.2 M nanopatelets, (b) 0.4 M dense structure of nanorods, (c) 0.6 M elongated root-like structure, (d) 0.8 M nanoflex like structure, (e) 1 M agglomerated nanosheets

3.3 Optical Properties:

The UV-Vis spectrophotometer was utilized to explore optical properties. The transmittance vs wavelength characteristics with respect to different molarities are shown in Figure 3. In figure 4 absorption plot was shown.

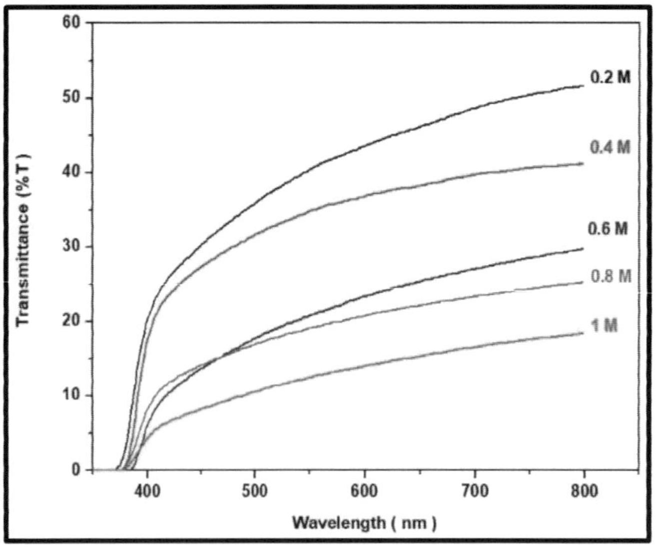

Figure 3. transmittance vs wavelength characteristics with respect to different molarities

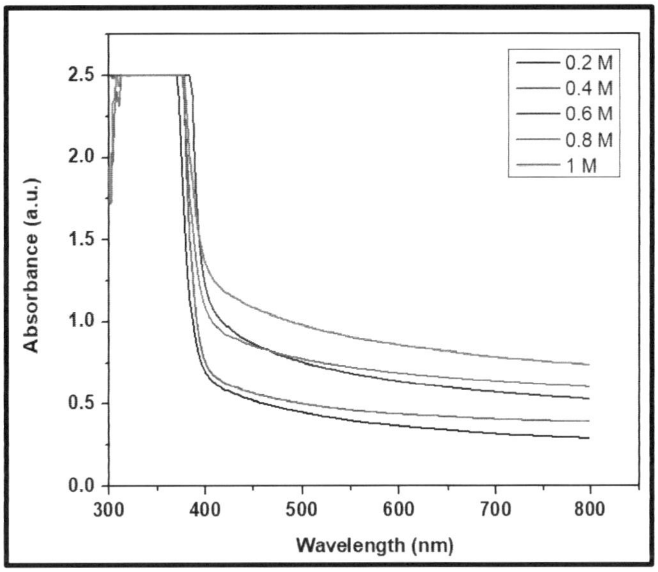

Figure 4. Absorbance vs wavelength characteristics with respect to different molarities

It is clearly indicated that, transmittance goes on decreasing as precursor molar concentration increases. The maximum transmittance was reported at 0.2M concentration and it is about 51 %. Further it decreases upto 18% for the film deposited with 1M concentration. The optical bandgap is calculated by using absorbance spectra. From absorption plot, sharp absorption adge near 375 nm was observed as well as red shift appears with increaseing molar concentration. In figure 5 tauc plots of corresponding absorption spectra are reported and summarized in table 2 given below. It shows that the bandgap value decreases with increasing precursor molar concentration. It is because of increasing carrier concentration.

Table 2: Calculated bandgap values with respective molar concentration

Molar Concentration(M)	0.2	0.4	0.6	0.8	1
Bandgap (eV)	3.18	3.17	3.10	3.08	3.04

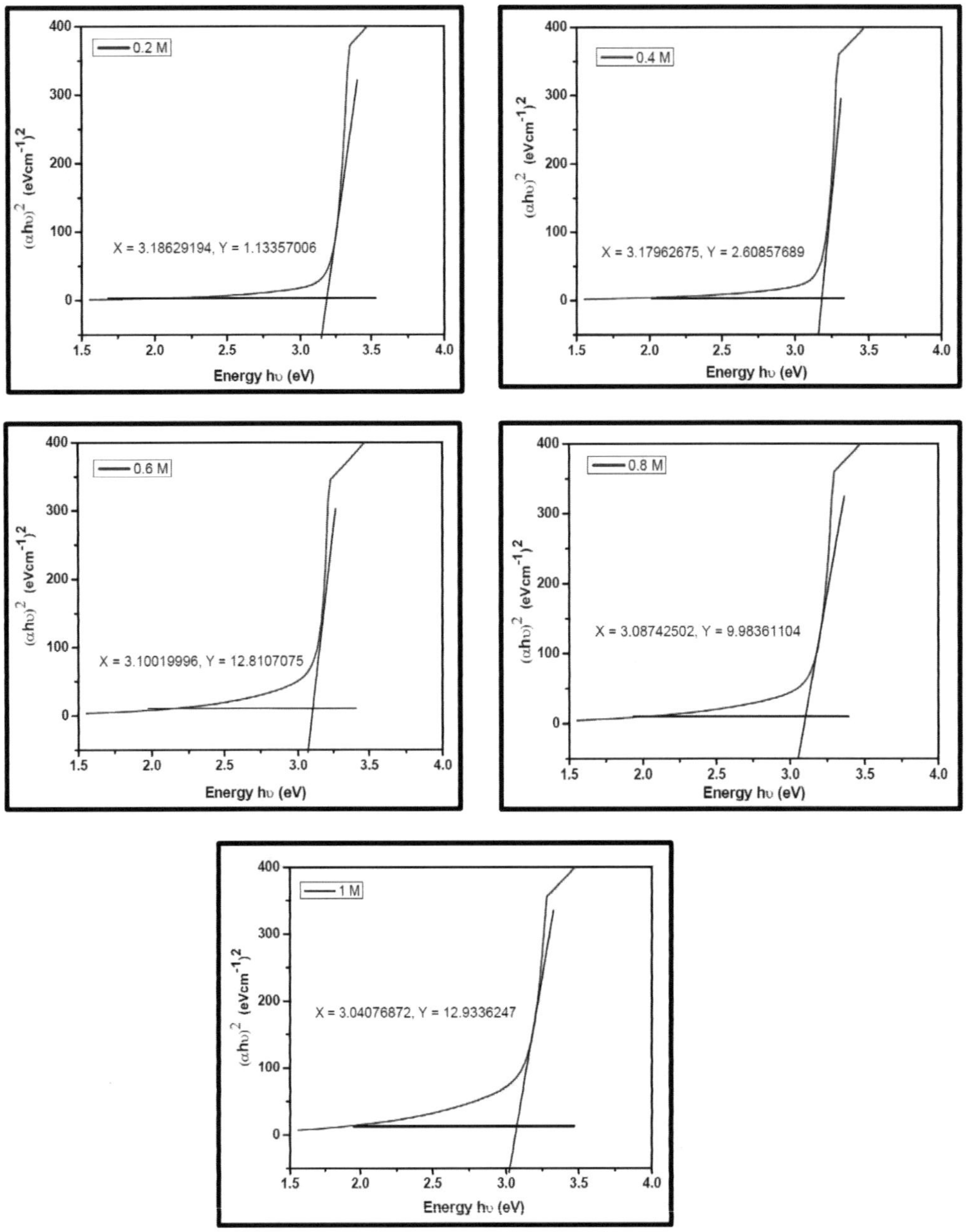

Figure 5. Tauc plots from absorbance characteristics of ZnO films with respect to varying precursor molar concentrations.

4. Conclusion:

The thin films of aqueous zinc acetate dehydrate of varying molarities has been successfully synthesized over glass substrate. X-ray diffraction studied reveals that films are polycrystallines, 'hexagonal wurtzite structure' and (002) plane oriented in preferred direction. The increasing molar concentration of precursor leads to decreasing crystallite size. The optical transmittance decreases from 51% to 18% with increasing molar concentration from 0.2M to 1M. Band-gap energy decreases from 3.18 to 3.04 eV with increasing concentration of precursor. SEM morphology reveals that it completely depends on precursor concentration. For varying molar concentrations nanoplatelets, nanorods, Elongated roots, and agglomerated nanosheets like morphologies are observed.

References

[1] Alami Z Y, Salem M and Gaidi M Effect of Zn concentration on structural and optical proprieties of ZnO thin films deposited by spray pyrolysis Oct. 2015 *Adv. Energy: An Int.J.* **2(4)** pp. 11–24.

[2] Gaikwad, R.S., Patil, G.R., Shelar, M.B. et al. Nanocrystalline ZnO films deposited by spray pyrolysis: Effect of gas flow rate Jul. 2012 Int. J Self-Propag. High-Temp. Synth. **21(3)** pp. 178– 182.

[3] Lee S D, Nam S H, Kim M H, and Boo J H, Synthesis and photocatalytic property of ZnO nanoparticles prepared by spray-pyrolysis method 2012 Phy. Procedia **(32)** pp. 320–326.

[4] Keis K, Magnusson E, Om H L, Lindquist S E, Hagfeldt and Anders A 5% efficient photoelectrochemical solar cell based on nanostructured ZnO electrodes 2002 *Solar Energy Materials and Solar Cells* **(73)** pp 51-58.

[5] Ong C B, Ng L Y and Mohammad A W A review of ZnO nanoparticles as solar photocatalysts: Synthesis, mechanisms, and applications 2018 *Renewable and Sustainable Energy Reviews* **(81)** pp. 536–551.

[6] Fahimi Z and Moradlou O Fabrication of ZnO@C foam: A flexible free-standing electrode for energy storage devices Apr. 2020 *Materials and Design* **(189)** 108525.

[7] Pearton S J, Norton D P, Heo Y W, et al. ZnO spintronics and nanowire devices 2006 *J. Electron. Mater*. **(35)** 862–868.

[8] Espitia P J P, Soares N de F F, Coimbra J S dos R, Andrade N J de, Cruz R S and Medeiros E A A Zinc oxiden anoparticles: synthesis, antimicrobial activity and food packaging applications Jul. 2012 *Food and Bioprocess Technol* **(5)** pp. 1447–64.

[9] Hames Y, Alpaslan Z, Kosemen A, San S E and Yerli Y Electrochemically grown ZnO nanorods for hybrid solar cell applications Mar. 2010 *Solar Energy* **84 (3)** pp. 426–431.

[10] Gopikrishnan R, et al.,Synthesis, characterization and biocompatibility studies of zinc oxide (ZnO) nanorods for biomedical application Mar. 2010 *Nano-Micro Letters* **2 (1)** pp. 31–36.

[11] Mani G K and Rayappan J B B, Novel and facile synthesis of randomly interconnected ZnO nanoplatelets using spray pyrolysis and their room temperature sensing characteristics Jul. 2014 *Sensors and Actuators, B: Chemical* **(198)** pp. 125–133.

[12] Cui J Zinc oxide nanowires Feb. 2012 *Materials Characterization* **(64)** pp. 43–52.

[13] Theerthagiri J, et al. A Review on ZnO nanostructured materials: energy, environmental and biological applications Jul. 2019 *IOP publication* Nanotechnology **30 (39)** pp 392001.

[14] Gili M B, Chu R and Balela M Effects of molarity variation on the optical property and topography of ZnO thin films deposited via spray pyrolysis Apr. 2019 *J. Phys.: Conf. Ser.* **1191** 012050.

[15] Sivalingam K, Shankar P, Mani G K and Rayappan J B B Solvent volume driven ZnO nanopetals thin films: spray pyrolysis Nov. 2014 *Materials Letters* **(134)** pp. 47–50

[16] Nikam S V and Jadhav B T Jadhav Review on the effect of preparative parameters of spray pyrolysis technique on optical and electrical properties of Zinc Oxide thin film 2021 *IJARIIT* **7(4)** pp 1238- 44.

[17] Islam M N, Hakim M O and Rahman H The effects of deposition variables on spray-deposited ZnO thin film prepared from Zn (C2 H3O2) 2 Apr. 1987 J. Mat. Sci. **(22)** pp 1379-84.

[18] Reguig B A, Regragui M, Morsli M, Khelil A, Addou M and Bernede J C Effect of the precursor solution concentration on the NiO thin film properties deposited by spray pyrolysis Jun. 2006 *Solar Energy Materials and Solar Cells* **90 (10)** pp 1381 - 92.

[19] Perednis D and Gauckler L J Thin Film Deposition Using Spray Pyrolysis Mar. 2005 *J Electroceramics* **(14)** pp 103–111.

[20] Patil P S Versatility of chemical spray pyrolysis technique 1999 *Mater. Chem. Phys* **59 (3)** pp

185–198.
[21] Hacini N, Ghamnia M, Dahamni M A, Boukhachem A, Pireaux J J and Houssiau L Compositional, structural, morphological, and optical properties of ZnO thin films prepared by PECVD technique Feb. 2021 *Coatings* **11(2)** pp. 1–11.
[22] Baneto M, Enesca A, Lare Y, Jondo K, Napo K and Duta A Effect of precursor concentration on structural, morphological and opto-electric properties of ZnO thin films prepared by spray pyrolysis 2014 *Ceramics International* **40 (6)** pp 8397–8404.
[23] Zeming Li et al. Preparation and optical properties of ZnO nanostructures 2010 *3rd Int. Nanoelectronics Conf (INEC)* (2010): pp 953-954.
[24] Sutanto H, Durri S, Wibowo S, Hadiyanto H and Hidayanto E Rootlike morphology of ZnO: Al Thin Film Deposited on Amorphous Glass Substrate by Sol-Gel Method 2016 *Phy. Research Int.*, vol. 2016, 4749587.
[25] Kumar N, Mittal H, Reddy L, Nair P, Ngila J C and Parashar V Morphogenesis of ZnO nanostructures: Role of acetate (COOH-) and nitrate (NO3-) ligand donors from zinc salt precursors in synthesis and morphology dependent photocatalytic properties 2015 *RSC Advances* **5(48)** pp. 38801–809.

Acknowledgments: The authors are thankful to RUSA and Y. C. Institute of Science (Autonomous) Satara, Maharashtra, India for financial assistance.

Effect of Diethylenetriamine on Mechanical and Electrical Properties of Epoxidized Natural Rubber Thin Films

K Phatcharasit[1] and T Sittiruk[2]

[1] Department of Rubber and Polymer Engineering, Faculty of Engineering, Thaksin University, Phatthalung Campus, Thailand.
[2] Department of Mechatronics Engineering, Faculty of Engineering, Thaksin University, Phatthalung Campus, Thailand.

Email: kritsada.p@tsu.ac.th

Abstract. A novel thin film material was fabricated from epoxidized natural rubber latex (ENRL) and chemicals for rubber. The effects of different diethylenetriamine (DETA) contents filled composition on ENRL compounds were investigated on mechanical properties, such tensile strength, water absorption dielectric properties and morphology analysis. The method ENRL thin films were prepared by using solution casting technique and diethylenetriamine as a hardener. The optimal diethylenetriamine content was found to be 6 phr which demonstrated the highest tensile strength due to crosslinking between the epoxy molecules and the water absorption of thin film of increased with the increase of diethylenetriamine content due to the highly porous and hydroxyl groups in the structure. Furthermore, dielectric properties increased with the diethylenetriamine content, it is shown that the conductivity of DETA without supporting electrolyte was very low. The phase morphology of homogeneous compatibility in ENRL phase can be regulated by the crosslinking degree of ENRL and DETA.

Keywords: Diethylenetriamine, Surface coating, Epoxidized natural rubber, Thin film

1. Introduction

Surface coating layer of modern technology has gained a great interest and is on the increase as coatings become ever more diverse and demands on them rise, both in range of applications and quality of performance [1, 2]. Coatings are used both to remedy deficiencies of substrates and to confer additional properties such as biologically, chemically, electrically, etc [2]. Surface coatings are needed to understand of the wetting, adsorption and adhesion is required to begin to understand the adhesion of coating to substrates. Surface coatings are natural polymer or macromolecular materials. Most of them are also composite materials, although is just two dimensional ones. There are many other compound ingredients used amino groups such as triethylenediamine (TEDA), diethylenetriamine (DETA) and ethylenediamine (EDA) it is well known that strong interactions with the surface atoms of many materials. Diethylenetriamine (DETA) has been used for this study due to use in the hardener and coordinated chemistry.

Epoxidized natural rubber (ENR) has been commercially prepared from natural rubber (NR) by reacting with natural rubber with peroxide [3, 4]. Epoxy groups of ENR improves the high polarity of rubber [3, 5]. ENR help to increasing hydrophilicity and chemical resistance. Polyvinyl alcohol (PVA) is high water solubility with large hydroxyl group [6]. PVA polymer has many advantages, mechanical strength, chemical stability and good thin film-forming ability [7]. PVA can be mixed with optical fibers, electronic sensors. A previous study, PVA based soft electronics possess it is found that to increase the mechanical strength and form a conductive network that enables PVA composite film [8].

This study targets to elucidate the influence of these novel diethylenetriamine (DETA) contents in epoxidized natural rubber based on the mechanical properties, water absorption, dielectric properties and morphology analysis by fabricated using solution casting technique.

2. Material and Methods

2.1 Material

Epoxidized natural rubber latex (ENRL) containing 50 mol% of epoxy groups was supplied by Muangmai Guthrie Public Co.,Ltd. Thailand. Diethylenetriamine (DETA) and Polyvinyl alcohol (PVC) was supplied Sigma Aldrich. All chemicals for vulcanization in including activators (zinc oxide, ZnO), accelerator (Zinc 2-mercaptobenzothiazole, ZMBT), antioxidant N-(1,3-dimethylbutyl)-N'-phenyl-p-phenylenediamine, 6PPD) and sulfur is a vulcanizing agent purchased from Kij Paiboon Chemical Ltd., Thailand.

2.2 Methods

2.2.1 Formulation and film fabrication

Epoxidized natural rubber latex formulation for the rubber films preparation consisted of activating agent, sulfur curing agent, antioxidant agent, accelerator agent, PVA and DETA, as shown in Table 1. the preparation process was initiated by pre-dispersing the curing agent, activating agent, antioxidant agent, and accelerator agent in water through homogenization. After adding PVC solution at 5%w/w and DETA mixture to formulation.

Table 1. Formulation compositions for rubber latex

Components	Parts per hundred (phr)				
	1	2	3	4	5
ENRL50	100	100	100	100	100
DETA	0	2	4	6	8
5% PVA	5	5	5	5	5
50% ZnO	3	3	3	3	3
50% ZMBT	2	2	2	2	2
50% 6PPD	1	1	1	1	1
50% Sulfur	2.5	2.5	2.5	2.5	2.5

The latex formulation was casted on the glass molds to produce a uniform thickness film. After drying at room temperature, the films were cured at 100 °C for 2 h, as shown in Fig. 1.

Figure 1. Preparation of thin films by solution casting technique.

2.2.2 Thin film characterization
Tensile strength. The rubber films were cut into strips of specific dimensions (70 mm × 10 mm) in accordance with ASTM D882-18. Tensile testing was performed on five specimens with a strain rate 500 mm/min using a LLOYD instrument tensile tester using 100 N load cell.

Water absorption testing: This experiment was repeated 3 times to obtain an average value. Samples measuring 25 mm × 12 mm were dried in an oven at 80 °C prior to testing. After weighing the dry samples, the sample is immersed in a beaker filled with 100 mL of water for 5 days. The percentage of water absorption was calculated as following Eq. (1)

$$\text{Water absorption (\%)} = [(W_t - W_0)/W_0] \times 100 \qquad (1)$$

Where W_t represents the weight (g) of the sample before immersion in water. W_0 is the weight (g) after immersion.

Dielectric properties: To obtain the dielectric properties of the analysed polymer composites, the LCR dielectric meter (Agilent 4285A) was used. The operating conditions were the frequency range of 75 kHz to 30 MHz, the capacitance C and the loss tangent (tan δ) were measured at room temperature. The three samples used to cut into round discs. The dielectric constant ε′ and dielectric loss ε'' of the tested materials obtain information about the behaviour of the dielectric and electrical components for the thin films, calculated as following Eq. (2) and (3) [6].

$$\varepsilon' = \frac{C \times I}{\varepsilon_0 \times A} \qquad (2)$$

$$\varepsilon'' = \tan\delta \times \varepsilon_1 \qquad (3)$$

Where A and I are the impedance, C and ε_0 is the measured capacitance of the material and the permitting of the vacuum respectively. The frequency dependence for the dielectric constant ε′

Morphology analysis: The ENRL/DETA thin films morphology study was preformed using field emission scanning electron microscope (FE-SEM, Quanta 450 FEG) operating at an acceleration voltage of 5 kV. All samples were coated with a thin layer of gold.

3. Results and discussions

3.1 Tensile strength
The tensile strength of the thin films were evaluated to study the effect of the DETA as a hardener filled with ENRL films. The tensile strength curves of ENRL/DETA mixture with different DETA contents and times to crosslink from casted film formulations are shown in Fig. 8. The maximum of tensile strength improvement was noted at 8 phr of DETA loading, increased about 40% in tensile strength from 12.52 MPa to 18.02 MPa. It was evident that there was increased with the DETA addition and long drying. In addition, DETA has an aliphatic ethylene amine with two primary and one secondary amine groups. It is used as a catalyst where the reaction takes place in the epoxide group, this results in crosslinking between molecules [9]. The long drying employed in the casted films provides the particles ample time to interact with each other of solution led to the observed enhancement in the tensile strength [10].

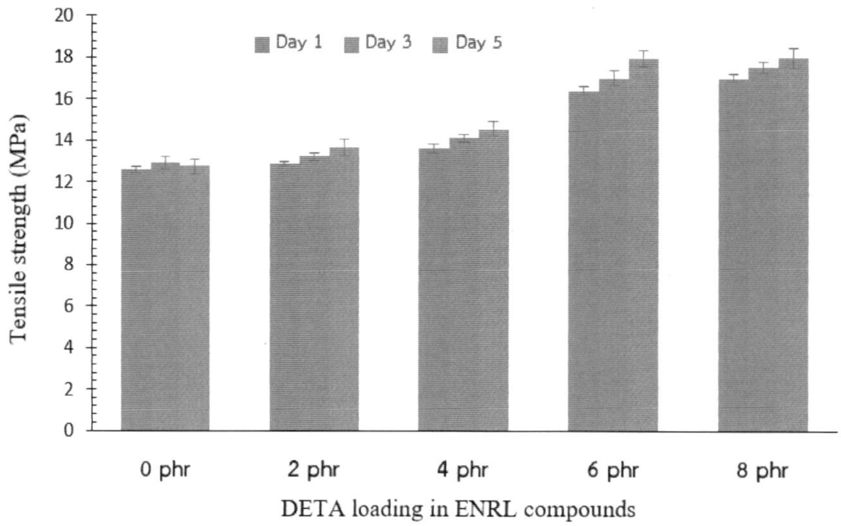

Figure 2. Tensile strength of ENRL/DETA mixture with different DETA contents.

3.2 Water absorption testing

Water absorption studies show that equilibrium absorption of ENRL/DETA was reached after 5 days of water immersion, and the data are recorded in Fig. 3. The water absorption of ENRL without DETA was 25%. Due to PVA was added into the ENRL because of the increased hydrophilicity, the hydrophilic interactions between water molecules and hydrophilic chain segments [11]. Incorporating DETA into the ENRL shifted the higher values compared to samples without DETA. At the amount of DETA increased form 6 phr to 8 phr, the results showed that the water absorption increase as the amount of DETA increased because of more chemical interactions between the DETA and ENRL reduced the water absorption of the sample [11]. In a previous study, aqueous mixtures of diethylenetriamine (DETA) have three amine groups, two terminal primary amino groups and one secondary central amino group. Due to that, they are expected to exhibit high absorption capacity [9].

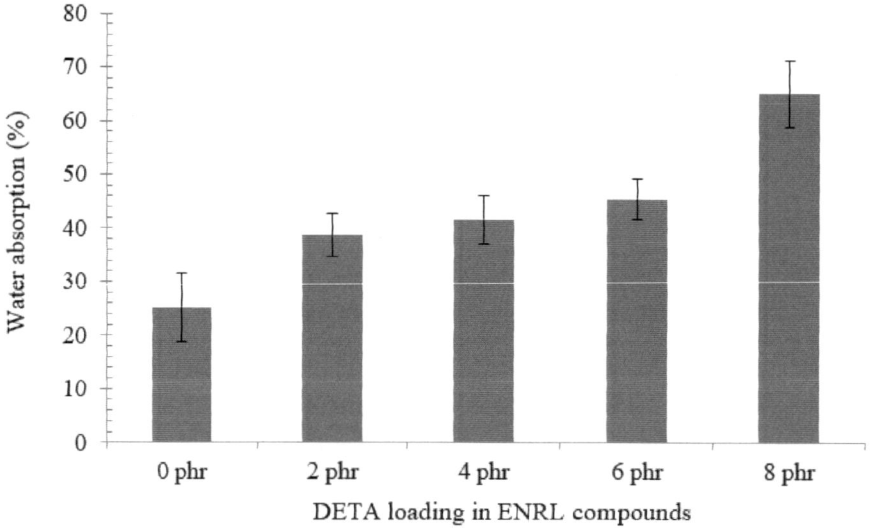

Figure 3. Water absorption of ENRL/DETA mixture with different DETA contents.

3.3 Dielectric properties

The dielectric constant is inverse proportion to the incident electric field, it means a higher values of the incident electric decreases dielectric constant. The ENRL/DETA thin films at room temperature are shown in Fig.4. It can be seen that dielectric constant takes higher values were carried out in all cases, which gradually decreases with increasing frequency due to interfacial polarization. In addition, DETA is medium of solvent oxidation at the electrode due to primary polyamine electron rich nitrogen, which is importance for a fundamental understanding of electrochemical reactions.

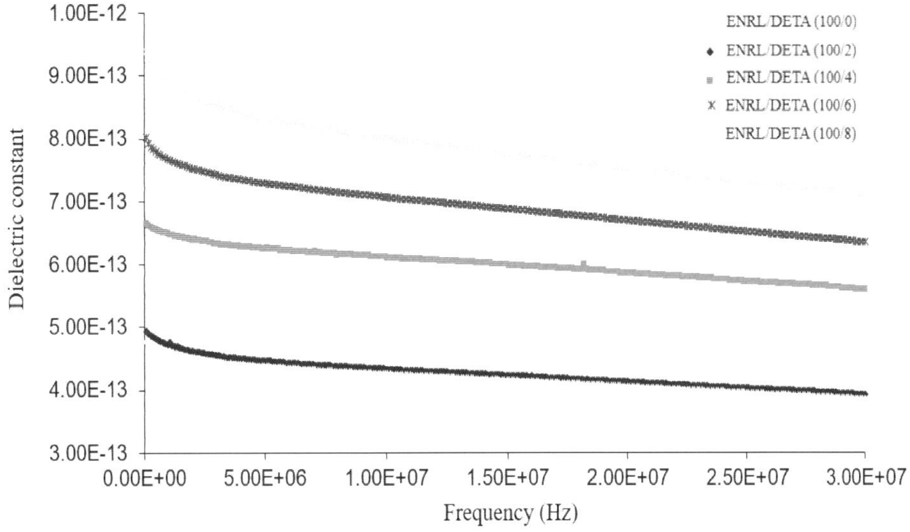

Figure 4. Dielectric constant of ENRL/DETA mixture with different DETA contents.

3.4 Morphology analysis

The polymeric solution of ENRL/DETA thin films were observed by SEM and the results are recorded in Fig. 5 The original ENRL/DETA film (a-c) has dense thin film. which could be explained by mechanical properties. Especially, among in all samples, the ENRL/DETA (100/6) (b) had the highest tensile strength, which was related to the higher crosslinking density of the thin film. Simultaneously, pore small cavities (c) were observed on the surface morphology, that not phase separated occurred in thin film. This result could be attributed to the increase of free motion in the thin film and solution compatibility. However, the more pore developed can deduced the weak interaction between ENRL with DETA [12]. These results were agreement with the results obtained from water absorption properties.

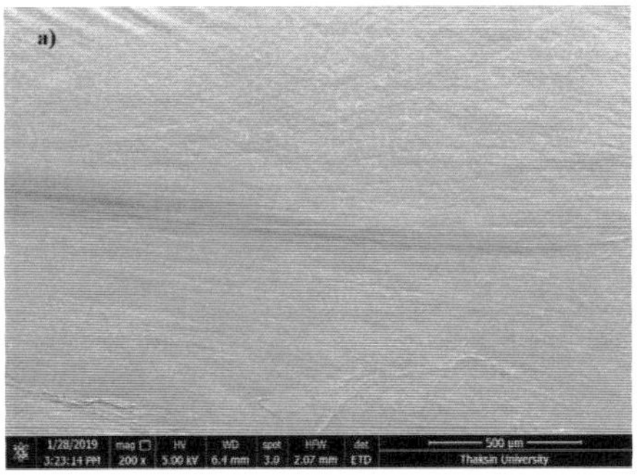

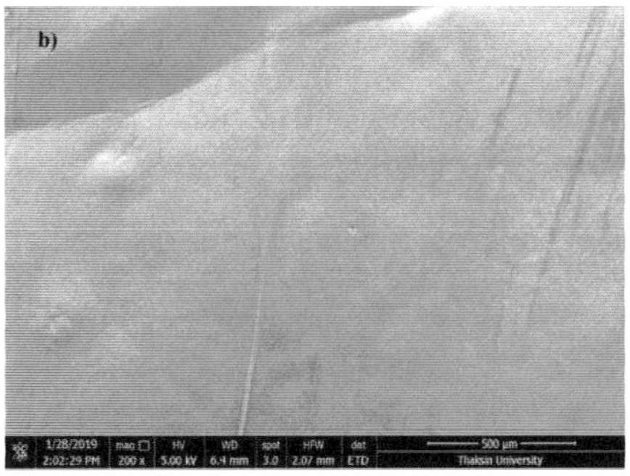

Figure 5. SEM images of ENRL/DETA blends with different DETA contents. (a) DETA 0 phr, (b) DETA 6 phr and (c) DETA 8 phr

4. Conclusion

In this study, the ENRL thin films were successfully prepared by incorporating with different DETA content using solution casting technique. The tensile strength of ENRL was enhanced with DETA content whereas resulted in increased tensile strength, results in crosslinking between molecules and homogeneous compatibility. The water absorption of ENRL/DETA thin films increased with increasing DETA because of chemical interactions. The electrical properties of ENRL/DETA thin films as verified demonstrated improvement in electrical conductivity and dielectric properties due to interface polarization thin films. Furthermore, the morphology was found that good polymer compatibility among ENRL and DETA.

Acknowledgements

This work was supported by Nation Higher Education, Science, Research and Innovation Policy Council, Thaksin university (Research project grant) Fiscal Year 2022. Authors would like to thank Department of Rubber and Polymer Engineering, Faculty of Engineering, Thaksin University.

References

[1] Wen C 2015 *Surface Coating and Modification of Metallic Biomaterials* (New York-Elsevier)
[2] Ouattara L, Diaco T and Bokra Y 2006 *Bull. Chem. Soc. Ethiop.* **20** 269-277
[3] Ruksakulpiwat C, Nuasaen S, Poonsawat C and Khansawai P 2008 *Adv. Mater. Res.* **47-50** 734-737
[4] Zhang C, Wang W, Huang Y, Pan Y, Jiang L, Dan Y, Luo Y and Peng Z 2013 *Mater. Des.* **45** 198-205
[5] Zhang H, Yuan W Q, Weng Y, Zeng J B and Li Y D 2022 *Indus. Crops Prod.* **180** 114707
[6] Mohammed M I, Mahmoud S and Yahia S 2022 *Optik* **253** 168582
[7] Rodica D and Silvia P 2013 *Appl. Surf. Sci.* **285** 56-64
[8] Cai G, Wang J, Qian K, Chen J, Li S and Lee P S 2017 *Adv. Sci.* **4** 1600190
[9] Fragkiskos T, Athanasios I P, Panos S and Ioannis T 2021 *Chem. Thermo. Therm. Anal.* **3-4** 100017
[10] Adibi A, Kim J, Mok J, Lenges C, Simon L and Mekonnen T H 2021 *Carbohydr. Polym.* **267** 118234
[11] Sa-Ad R 2022 *Indus. Crops Prod.* **177** 114415
[12] Samad N A, Othaman R and Abdullah I 2014 *Int. J. Mater. Eng. Inno.* **5** 61-69

IOP Conf. Series: Materials Science and Engineering 1258 (2022) 012014 doi:10.1088/1757-899X/1258/1/012014

Noble metal nanoparticles dispersed on nanocellulose: a green platform for catalytic organic transformations

Roya Thach-Nguyen[1,2], Trung Dang-Bao[1,2,*]
[1]Faculty of Chemical Engineering, Ho Chi Minh City University of Technology (HCMUT), 268 Ly Thuong Kiet Street, District 10, Ho Chi Minh City, Vietnam
[2]Vietnam National University Ho Chi Minh City, Linh Trung Ward, Thu Duc District, Ho Chi Minh City, Vietnam

Corresponding author's e-mail: dbtrung@hcmut.edu.vn

Abstract. In recent years, plant-derived biomaterials, typically cellulose, acting as catalytic supports have a great impact on heterogeneous catalysis thanks to their biodegradability, non-toxicity, low-cost, availability and easy-implementation. As the most abundant biopolymer found in nature, cellulose consists of repeating cellobiose units which are built up from two anhydroglucose rings and linked by a β-1,4 glycosidic bond. The term of "nanocellulose" has been widely used to describe cellulose nano-objects, involving cellulose nanocrystals (CNCs), cellulose nanofibrils (CNFs) and bacterial cellulose (BC). Nanocellulose features high specific surface area and controllable surface chemistry, high crystalline structure, superior mechanical strength and thermal stability, resulting in its applications in food, cosmetics, pharmaceutical, biomedical and paper industries. Concerning to catalytic support application, the nanocellulose surface possesses the hydroxyl (in nature) or the sulfate ester groups (modified via acid hydrolysis), facilitating metal ions reduction towards the corresponding metal nanoparticles. In addition, the supramolecular structure of cellulose permits to disperse metal nanoparticles and prevent their agglomerations. In this context, nanocellulose is introduced as matrices for immobilizing noble metal nanoparticles and then applied to catalytic organic transformations.

1. Introduction

Metal-based catalysts play an important role in many chemical areas, covering organic synthesis, petrochemicals and environmental remediation; in particular, homogeneous catalysis and heterogeneous catalysis are considered to be the two most widely studied fields. Homogeneous catalysts are designed based on metal salts or complexes dissolved in reaction media, which often show high activity and selectivity; however, the catalytic stability is relatively low and difficult to recover and recycle. In contrast, heterogeneous catalysts are often designed based on solid catalysts/carriers with good durability, easy recovery and recycling; however, such reactions often require relatively harsh conditions, in addition to the disadvantages caused by mass transfer problems [1]. Since the 90s of the last century, along with the development of nanoscience and nanotechnology, nanocatalysis (based on nanometer-sized structured materials, in the nanoscale range of 1–100 nm) has brought a landmark revolution, combining the advantages of both traditional catalytic systems, increasing their catalytic efficiency. More importantly, catalytic reactivity of nano-objects can be tuned by controlling their size, morphology and chemical compositions, via synthetic conditions, solvents or supports [2]. In general, in order to avoid the agglomeration of nanoparticles and to facilitate the recovery and reuse of nanocatalysts, metal nanoparticles are often dispersed and immobilized on solid carriers, typically carbon-derived materials [3–5], silica [5, 6], metal oxides [4, 5] and polymers [4, 5]. The achievements in the catalysis field show that the dispersion of metal nanoparticles on high surface area carriers is a major direction for the catalysis industry and organic synthesis applications.

In recent years, natural-origin materials, typically biopolymers, have been highly appreciated as catalyst carriers, with outstanding factors such as biodegradability, low-cost, ubiquity and ease of implementation. As the most abundant biopolymer on earth, cellulose-based materials have been paid great attention to a variety of industrial applications. Cellulose consists of repeating cellobiose units (a dimer of glucose) and each glucose is linked by a β-1,4 glycosidic bond. In nature, microfibrils are built up from cellulose chains, and then bundled together into microfibers/fibers in a hierarchical structure oriented by hydrogen bonding [7]. The natural cellulose has high crystallinity; however, the studies on its properties have been challenging, especially in determining the typical size of single crystals of cellulose. Intriguingly, hydrolysis of cellulose structure (using strong acids) is only capable of breaking the most accessible glycosidic bonds in the circuit and releasing nanometer-sized cellulose crystals, namely nanocellulose. The most common hydrolysis methods mainly use sulfuric acid because of its ability to modify some sulfate ester groups on the obtained cellulose, making them readily dispersed in water [8]. Besides, mechanical purification and hydrolysis using HCl or enzymes have been also reported; in general, such techniques do not affect the chemical surface of nanocellulose [9].

Nanocellulose is usually rod-shaped or ribbon-shaped, 50–1000 nm in length and 3–50 nm in width, so the length/width ratio of 10–100 [10, 11]. The term of "nanocellulose" refers cellulose nanocrystals (CNCs, extracted from plants by acid hydrolysis), cellulose nanofibrils (CNFs, extracted from plants by mechanical treatment) or bacterial cellulose (BC, derived from bacteria) [10]. The size and the morphology of nanocellulose mainly depend on the cellulose source, the fabrication conditions (acid nature and concentration, acid/cellulose ratio, treatment temperature and time). Nanocellulose has high specific surface area, high crystallinity, outstanding mechanical strength and controllable chemical surface. In addition, cellulose-based materials are renewable, biodegradable, non-toxic, thermally stable, and applicable in industrial scale [10, 11]. Various applications of nanocellulose have been widely reported in paper, cosmetics, food, pharmaceutical and biomedical industries [11–16]. Modestly, the feasibility of nanocellulose as a catalyst carrier has been recently concerned with the following highlights: (i) high specific surface area, good thermal stability and ability in surface functionalizing; (ii) natural surface activity induced by hydroxyls and sulfate esters, which play a role in the reduction of metal ions; (iii) good dispersion of nanocellulose in water, favoring the design of supported catalysts in two-phase systems; (iv) biodegradability, non-toxicity and large-scale applicability [16, 17]. In comparison with other carbon-related materials (typically carbon nitride and carbon nanotube) mainly obtained from chemical resources (carbon-sources) for a wide range of industrial purposes [18–20], nanocellulose has been mainly extracted from plant resources but less exploited in catalysis fields.

In this context, the strategies of the synthesis of metal nanoparticles supported on nanocellulose will be briefly described. Since the comprehensive contribution of Moores *et al.* previously reported in 2016 [17], the present work will update some catalytic behaviors of noble metal nanoparticles (gold, silver, palladium) contributed since 2017, with the aim of establishing the scope of such nanocatalysts in organic transformations.

2. Synthetic strategies for nanocellulose-supported metal nanoparticles

In principle, metal nanoparticles can be prepared by the in-situ reduction of metal ions in the presence of nanocellulose in the same reaction medium [17]. It should be emphasized that the reduction of metal ions towards zero-valent metal nanoparticles must be carried out in the assistance of both reductant and stabilizer. The roles of both agents can be separately functionalized by using different agents or jointly functionalized on a same entity. In all cases, nanocellulose acts as a carrier for dispersing and immobilizing metal nanoparticles thanks to its supramolecular structure [10], preventing the particle agglomeration. Furthermore, the highly chemical reactivity on the surface of nanocellulose promises an efficient and flexible reducing agent, ruling out the use any external co-reductants. Briefly, the synthesis of metal nanoparticles supported on nanocellulose can be categorized into 2 approaches: (i) nanocellulose acting as a reducing agent thanks to functionalized groups, and (ii) using additional reducing agents, as illustrated in Figure 1.

(i) Nanocellulose acting as a reducing agent: the reduction of metal ions towards the corresponding metal nanoparticles could be occurred thanks to surface functionalized groups on nanocellulose, likely the hydroxyls (in nature) or the sulfate esters (modified during acid hydrolysis) [17]. No doubt, the adsorption of metal ions onto the surface oxygen atoms of the hydroxyls can be rationally foreseen via the electrostatic attraction, followed by the corresponding chemical reductions. Consequently, the functionalized groups onto nanocellulose take part in the both roles: stabilizing and reducing. More interestingly, in some cases, the introduction of functionalized groups with negative surface charges (like carboxylate [21]) facilitates the deposition of metal species onto the surface of nanocellulose via electrostatic interactions.

(ii) Using additional reducing agents: in order to ensure the complete reduction of metal ions, the external co-reductants have been introduced, typically hydrides, hydrazine, hydrogen gas or ascorbic acid, *etc.* [17]. In general, the higher the reduction rate is, the smaller the nanoparticles are; therefore, the use of a strong reducing agent facilitates the nucleation during the formation of metal nanoparticles. By this way, in spite of using an external hazard chemical for the reduction of metal ions, the well-understanding insights on the preparation of metal nanoparticles dispersed on diverse supports [3–6] have promoted the catalytic applications in a variety of organic transformations, in comparison with the former synthetic strategy.

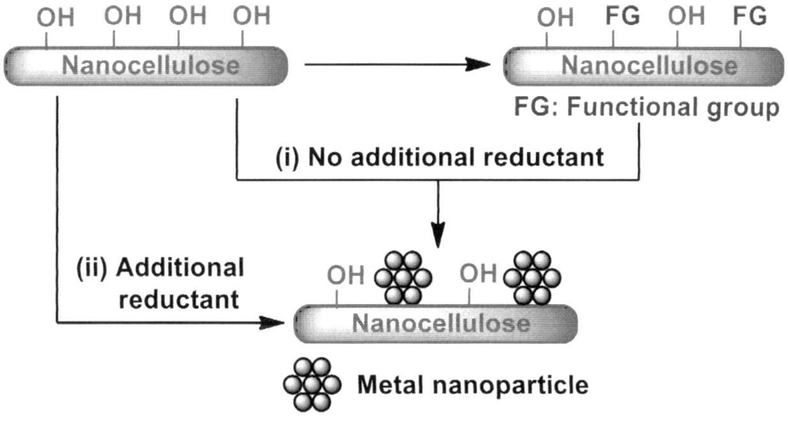

Figure 1. Schematic illustration for synthesis of metal nanoparticles supported on nanocellulose [17].

3. Noble metal nanoparticles supported on nanocellulose for catalytic organic transformations

3.1. AuNPs and AgNPs supported on nanocellulose for reduction of 4-nitrophenol

According to the first synthetic strategy, Fu *et al.* reported the preparation of cellulose nanocrystals from cotton linters by H_2SO_4 hydrolysis and then ultra-fine gold nanoparticles supported on as-prepared cellulose nanocrystals (AuNPs/CNCs) coming from $HAuCl_4$ [22]. The formation of spherical AuNPs (face-centered cubic crystals, < 8 nm in diameter) on CNCs (rod-like shape with 100–300 nm in length and 10–30 nm in width) was observed, accompanying with the oxidation to form the carboxyl groups on the surface of nanocellulose by Au(III) ions. On the other hand, the surface modification of nanocellulose can be carried out prior to accelerating the chemical reduction of gold ions as reported by Zhang *et al.* [23]. In this work, cellulose nanofibrils were surface modified via periodate oxidation to yield dialdehyde nanocellulose, which can act as a reducing and stabilizing agent for the further preparation of AuNPs from $HAuCl_4$. The presence of aldehyde groups onto nanocellulose was confirmed by FT-IR analysis (with the C=O stretching at 1732 cm^{-1}) and titrimetric method (its content of 3.8 mmol/g). The spherical AuNPs (d_{mean} = 29.61 nm) dispersed on cellulose nanofibrils (5–20 nm in diameter) were observed thanks to the reduction of metal ions by aldehyde groups (forming zero-valent metal species and carboxylate groups). Furthermore, the carboxylate groups on cellulose nanofibrils (from 2,2,6,6-tetramethylpiperidinyl-1-oxy (TEMPO) oxidation in the extraction of nanocellulose) possessed the negative charges, leading to well-dispersion of AuNPs due to the electrostatic interaction.

By the second approach to gain more contributions, the chemical reduction of metal ions to form metal nanoparticles was performed in the assistance of external reducing agents. Eisa *et al.* used ascorbic acid to reduce Au(III) ions towards AuNPs (18–25 nm in diameter, [Au] = 5%) onto cellulose nanocrystals (needle shape, 100–170 nm in length and 15–20 nm in width) at the solid state [24]. As the most frequently used reducing agent, $NaBH_4$ was widely introduced for the preparation of AuNPs (d_{mean} = 10.5 nm, [Au] = 9.7 wt% [25] and d_{mean} = 4.3 nm [26]) from Au(III) ions onto anionic cellulose nanofibers (surface modified by carboxyl) [25, 26]. In the same frameworks reported in the same groups, AgNPs/CNCs (6–35 nm in diameter, [Ag] = 10%) using ascorbic acid [24] and AgNPs/CNFs (d_{mean} = 17.3 nm, [Ag] = 22.6 wt%) using $NaBH_4$ [25] were also obtained by the reduction of $AgNO_3$. As above-mentioned, metal ions were firstly adsorbed on the hydroxyl groups, followed by their reduction to form Au(0) or Ag(0) by adding an external reducing agent. In particular, the anionic groups induced the electrostatic interaction with positively-charged metal ions, resulting the better dispersion of formed nanoparticles on the surface of nanocellulose.

In some cases, instead of anchoring on the hydroxyls of nanocellulose, metal nanoparticles could be positioned and stabilized by using a cationic surfactant, like hexadecyltrimethylammonium bromide (CTAB), thanks to non-covalent interactions [27]. In the assistance of CTAB, AgNPs ($d < 10$ nm) were well-dispersed on cellulose nanocrystals (200–250 nm in length, 15–20 nm in width) by the reduction of $AgNO_3$ using $NaBH_4$, in contrast to the bigger aggregates of AgNPs ($d > 15$ nm) observed in the absence of CTAB, resulting in their different catalytic behaviors. In fact, CTAB constructed a bilayer consisting of two molecules with hydrophobic interactions (the first hydrophilic head interacts with hydroxyl of nanocellulose and the second hydrophilic head interacts with metal nanoparticles); therefore, both steric effect and electrostatic interaction could be concerned with the better dispersion of AgNPs in the system.

In general, the selective catalytic reduction of 4-nitrophenol towards 4-aminophenol using $NaBH_4$ as a reducing agent has been frequently investigated in the framework of nanocellulose-supported AuNPs or AgNPs (Scheme 1) and some corresponding achievements [24–27] are summarized in Table 1. According to the Langmuir–Hinshelwood mechanism, 4-nitrophenolate ions firstly adsorb on the metal sites and the electron transfer then occurs from metal hydrides (based on reducing agents, like $NaBH_4$) to nitro groups, thus generating 4-aminophenol as the sole product of the reduction [17]. Practically, the reaction rate obeys the pseudo-first-order kinetics, depending on the concentration of 4-nitrophenol. The current challenges on such a reaction are to minimize the amount of reducing agents, to increase the TOF (turnover frequency), to be able to separate and recover the nanocatalysts for further cycles. The use of nanocellulose as a catalytic support permits to well-disperse the metal nanoparticles thanks to its anchor sites, resulting in their smaller sizes; therefore, the catalytic reactivity of nanocellulose-supported metal nanoparticles is expected to be higher, examined via the reaction rate and the TOF.

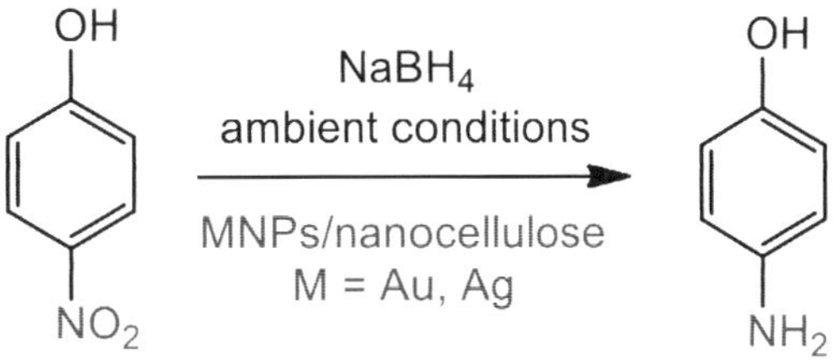

Scheme 1. Reduction of 4-nitrophenol towards 4-aminophenol catalyzed by nanocellulose-supported AuNPs or AgNPs [24–27].

Table 1. Reports on the reduction of 4-nitrophenol towards 4-aminophenol catalyzed by nanocellulose-supported AuNPs or AgNPs [24–27].

Metal nanoparticles (size)	Nanocellulose	$NaBH_4$/4-NP/MNPs (mole ratio)	Reaction rate, k (s^{-1}) TOF (h^{-1})	Ref.
(a) Gold nanoparticles				
AuNPs (18–25 nm)	CNCs	12/0.3/1	$k = 3.31\times10^{-3}$ TOF = 531	[24]
AuNPs (10.5 nm)	CNFs	100/1/1	$k = 9.78\times10^{-2}$	[25]
AuNPs (4.3 nm)	CNFs	125000/125/1	$k = 6.75\times10^{-3}$ TOF = 750	[26]
(b) Silver nanoparticles				
AgNPs (6–35 nm)	CNCs	9/0.23/1	$k = 4\times10^{-3}$ TOF = 420	[24]
AgNPs (17.3 nm)	CNFs	21/0.21/1	$k = 9.05\times10^{-2}$	[25]
AgNPs (< 10 nm)	CNCs	15000/150/1	$k = 1.6\times10^{-3}$ TOF = 545	[27]

3.2. PdNPs supported on nanocellulose for C-C cross-coupling reactions

During the last decade, palladium nanoparticles (PdNPs) have been extensively explored in a variety of catalytic organic transformations, typically C-C cross-coupling reactions, thanks to their high reactivity and versatility [28]. As awarded Nobel Prize in 2010, Pd-catalyzed C-C cross-coupling reactions involving Suzuki, Heck and Sonogashira have been paid great attention thanks to a large application in the synthesis of carbon-based molecules [29]. In such efforts of using heterogeneous Pd-based catalysts instead of classical homogeneous catalysts, functionalized nanocellulose-supported Pd-based catalysts have been currently reported on such C-C cross-coupling reactions, though still modest compared to others. Inherited from the previous studies on traditional supports [3–6], Pd(II) complexes loaded onto modified nanocellulose were mainly developed, such as Pd(II) catalyst supported on cellulose Schiff-base for Suzuki reaction under microwave assistance with 8 cycles [30], Pd(0) and Pd(II) hybrid supported on N-Methylimidazole functionalized carboxymethylcellulose for Suzuki reaction with 4 cycles [31]. Practically, Pd(0) nanoparticles supported on nanocellulose have been less reported due to more challenges in tuning their reactivity [28]; in the scope, the straightforward synthesis of PdNPs/nanocellulose for such C-C cross-coupling reactions will be briefly addressed, excluding the contributions on nanocellulose-supported Pd(II) hybrid catalysts.

IOP Conf. Series: Materials Science and Engineering 1258 (2022) 012014 doi:10.1088/1757-899X/1258/1/012014

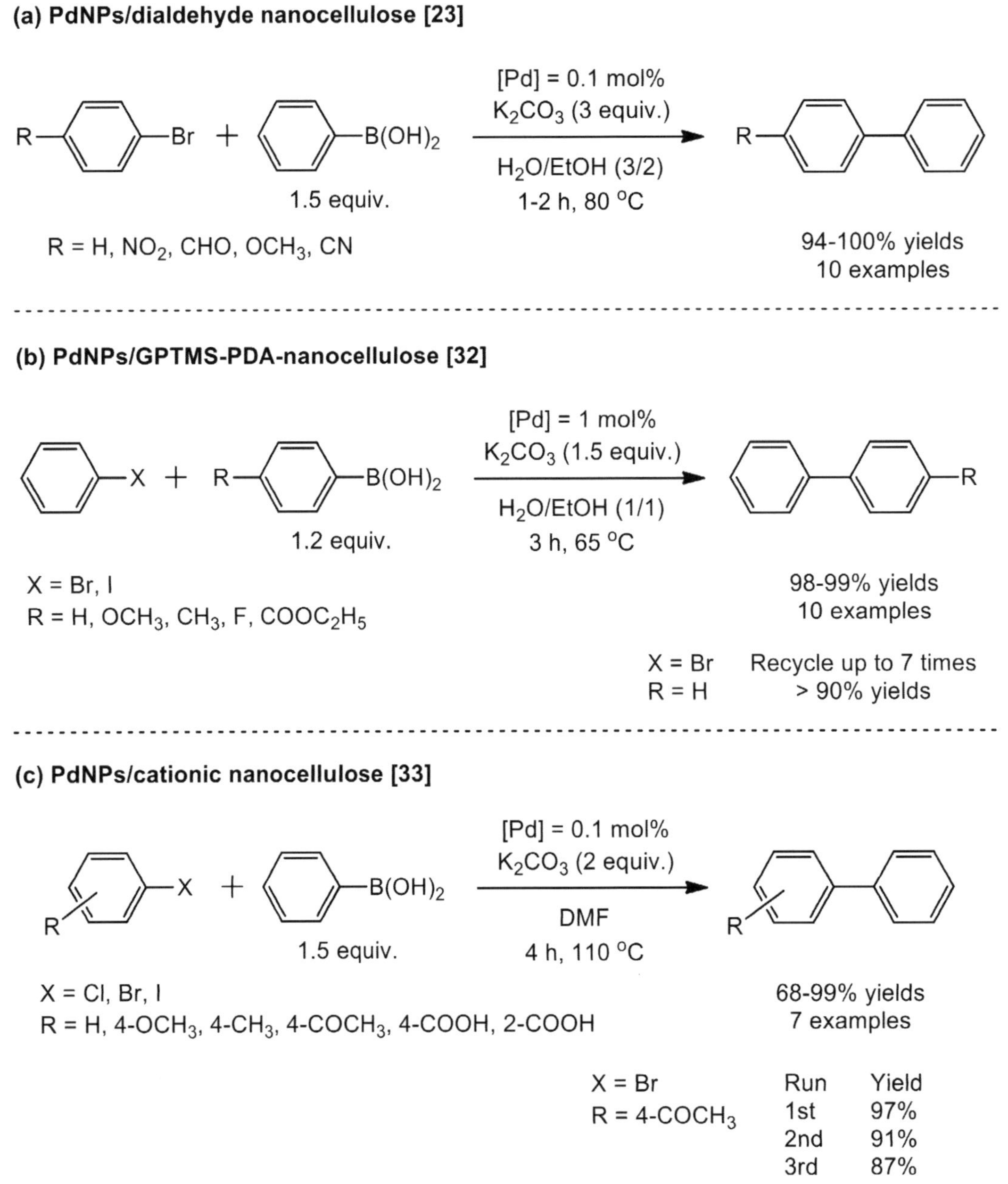

Scheme 2. Suzuki cross-coupling reactions catalyzed by PdNPs/nanocellulose [23, 32, 33].

On the same strategy to prepare AuNPs or AgNPs supported on nanocellulose as above described, the hydroxyl groups have been flexibly functionalized prior to reducing Pd(II) ions to corresponding Pd(0) species. For instance, 2,3-dialdehyde nanocellulose was obtained by the periodate oxidation of cellulose nanofibrils, and then applied as a reducing and stabilizing agent for the chemical reduction of $[Pd(NH_3)_4]^{2+}$ towards PdNPs, without any external reductants [23]. The aldehyde groups onto cellulose nanofibrils (5–20 nm in diameter) contributed to the reduction of Pd(II) ions, forming the corresponding zero-valent metal nanoparticles and carboxylate groups. Interestingly, a higher PdNPs concentration with a smaller size could be obtained at a higher concentration of dialdehyde nanocellulose (the mean sizes of spherical nanoparticles of 19.33, 18.28 and 15.86 nm using 0.1, 0.2

and 0.4 wt.% dialdehyde nanocellulose, respectively). Besides, the carboxylate groups on cellulose nanofibrils (from 2,2,6,6-tetramethylpiperidinyl-1-oxy (TEMPO) oxidation in the extraction of nanocellulose) possessed the negative charges, resulting in well-dispersion of PdNPs due to the electrostatic interaction. The as-prepared PdNPs/nanocellulose (PdNPs with mean size of 18.28 nm and [Pd] = 1.7 wt.%) was applied to the Suzuki reactions between aryl bromides and phenylboronic acid (Scheme 2a), affording the excellent yields of the corresponding biaryls (> 94% within 1 hour and > 98% within 2 hours for all the products). Mao and Sui *et al.* reported the modification of cellulose nanofibers via dual-crosslinking with γ-glycidoxypropyltrimethoxysilane (GPTMS) and polydopamine (PDA), and the chemical reduction of $PdCl_2$ to form spherical PdNPs (d_{mean} = 2.5 nm) thanks to catechol moiety of polydopamine without any external reducing agents [32]. Its catalytic reactivity was performed on the Suzuki reactions between aryl halides and arylboronic acids giving the corresponding biaryls with excellent yields (Scheme 2b) and the Heck reactions between aryl halides and aryl ethylene (Scheme 3). The catalyst could be reused for 7 times remaining both activity and selectivity, and without Pd leaching during the Suzuki cross-coupling reaction.

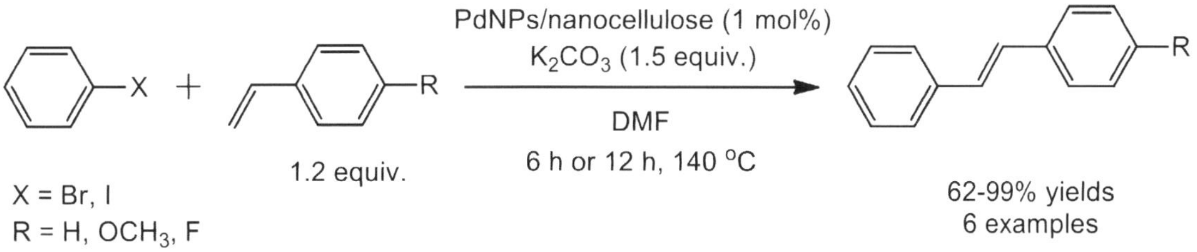

Scheme 3. Heck cross-coupling reactions catalyzed by PdNPs/nanocellulose [32].

In the case of using an external reducing agent, $NaBH_4$ was typically introduced to reduce $PdCl_2$ towards PdNPs in the presence of cationic cellulose nanofibrils [33]. In this work, nanocellulose was firstly extracted from Marram grass stems, and then modified with trimethylammonium chloride groups, anchoring Pd(II) species and resulting in the better dispersion of PdNPs (from 3 to 9 nm in sizes) on the surface of nanocellulose. The as-prepared catalyst was applied to the Suzuki reactions between aryl halides and phenylboronic acid in DMF (Scheme 2c), affording biaryls in moderate to excellent yields. Importantly, the nanocatalyst could be used for three cycles, preserving the catalytic reactivity. In most cases, nanocellulose-supported PdNPs were evidenced to be highly efficient in the Suzuki and Heck cross-coupling reactions, even using a low Pd loading for a short time, covering on aryl iodides, bromides and chlorides.

Besides the C-C cross-coupling reactions, the catalytic systems based on PdNPs/nanocellulose have been also applied to hydrogenation processes, such as the reduction of 4-nitrophenol towards 4-aminophenol [34, 35] and the selective hydrogenation of 3-nitrostyrene towards 3-aminostyrene or 3-nitroethylbenzene [35]. The efficient catalytic reactivity of such PdNPs/nanocellulose has been practically evidenced, promising to widen its scope in catalytic organic transformations.

In general, functionalization of nanocellulose (such as dialdehyde groups [23], dual-crosslinking with γ-glycidoxypropyltrimethoxysilane and polydopamine [32], trimethylammonium chloride [33]) is beneficial to the reduction of metal ions, even better dispersing metal nanoparticles on the surface nanocellulose thanks to the electrostatic interaction, resulting in smaller sizes. More challenging but full of opportunities, the catalytic activity and selectivity of such nanocellulose-supported metal nanoparticles can be rationally tuned via the accessible functionalized groups or stabilizers/ligands. As a key point towards a sustainable catalyst, the reuse of such catalysts should be further improved due to the possible metal leaching and agglomeration, leading to an decrease in their catalytic reactivity during recycling. The idea for this issue is to fabricate nanocellulose containing gel-trapped metal nanoparticles.

4. Conclusions and Outlooks

In the past 5 years, noble metal nanoparticles have been increasingly developed based on diverse solid supports for a variety of catalytic organic transformations. In comparison with other carbon-derived materials, cellulose possessing hydroxyl groups in nature is capable of functionalizing its chemical surface to anchor metal ions prior to reducing them; therefore, metal nanoparticles are well-dispersed on the surface. Mainly focused on cellulose nanofibrils and cellulose nanocrystals, the chemical reduction of metal ions towards metal nanoparticles (gold, silver, palladium) could be occurred with or without an external reducing agent. In most cases, nanocellulose can act as a reducing and stabilizing agent, permitting achieving the smaller nanoparticles and preventing their agglomeration. Their high efficiency has been practically evidenced in the reduction of 4-nitrophenol towards 4-aminophenol catalyzed by AuNPs, AgNPs or PdNPs immobilized on nanocellulose, the C-C cross-coupling reactions catalyzed by PdNPs immobilized on nanocellulose.

The promising results offer the alternative strategy for the preparation of metal nanoparticles by exploring diverse functionalized groups on nanocellulose, resulting in a variation in size and shape of metal nanoparticles. As a consequence, the catalytic reactivity can be tuned and the scope of application can be widened.

Acknowledgments
We would like to thank Ho Chi Minh City University of Technology (HCMUT), VNU-HCM for the support of time and facilities for this study.

References

[1] Thomas S J M 2010 *ChemCatChem* **2** 127–132.
[2] Philippot K and Serp P 2012 Concepts in Nanocatalysis *Nanomaterials in Catalysis* ed P Serp and K Philippot (Wiley-VCH) pp 1–54.
[3] Serp P and Figueiredo J L 2008 *Carbon Materials for Catalysis* (Wiley-VCH).
[4] White R J, Luque R, Budarin V L, Clark J H and Macquarrie D J 2009 *Chem. Soc. Rev.* **38** 481–494.
[5] Gawande M B, Goswami A, Felpin F X, Asefa T, Huang X, Silva R, Zou X, Zboril R and Varma R S 2016 *Chem. Rev.* **116** 3722–3811.
[6] Dhiman M, Singh B and Polshettiwar V 2021 Silica-Supported Nanoparticles as Heterogeneous Catalysts *Nanoparticles in Catalysis: Advances in Synthesis and Applications* ed K Philippot and A Roucoux (Wiley-VCH) pp 215–238.
[7] Klemm D, Heublein B, Fink H P and Bohn A 2005 *Angew. Chem. Int. Ed.* **44** 3358–3393.
[8] Beck S, Bouchard J and Berry R 2012 *Biomacromolecules* **13** 1486–1494.
[9] Sacui I A, Nieuwendaal R C, Burnett D J, Stranick S J, Jorfi M, Weder C, Foster E J, Olsson R T and Gilman J W 2014 *ACS Appl. Mater. Interfaces* **6** 6127–6138.
[10] Moon R J, Martini A, Nairn J, Simonsen J and Youngblood J 2011 *Chem. Soc. Rev.* **40** 3941–3994.
[11] Habibi Y, Lucia L A and Rojas O J 2010 *Chem. Rev.* **110** 3479–3500.
[12] Noremylia M B, Hassan M Z and Ismail Z 2022 *Int. J. Biol. Macromol.* **206** 954–976.
[13] Yu S, Sun J, Shi Y, Wang Q, Wu J and Liu J 2021 *Environmental Science and Ecotechnology* **5** 100077.
[14] Trache D, Tarchoun A F, Derradji M, Hamidon T S, Masruchin N, Brosse N and Hussin M H 2020 *Front. Chem.* **8** 392.
[15] Reshmy R, Philip E, Paul S A, Madhavan A, Sindhu R, Binod P, Pandey A and Sirohi R 2020 *Rev. Environ. Sci. BioTechnol.* **19** 779–806.
[16] Thomas B, Raj M C, Athira K B, Rubiyah M H, Joy J, Moores A, Drisko G L and Sanchez C 2018 *Chem. Rev.* **118** 11575–11625.
[17] Kaushik M and Moores A 2016 *Green Chem.* **18** 622–637.
[18] Ramasamy V, Murahari P, Banapurmath N R and Ramesh K 2021 *Carbon Trends* **5** 100079.

[19] Ismael M 2020 *J. Alloys Compd.* **846** 156446.
[20] Rathinavel S, Priyadharshini K and Panda D 2021 *Mater. Sci. Eng. B* **268** 115095.
[21] Lam E, Hrapovic S, Majid E, Chong J H and Luong J H T 2012 *Nanoscale* **4** 997–1002.
[22] Guo Z and Fu S 2019 *BioResources* **14** 3057–3068.
[23] Zhang K, Shen M, Liu H, Shang S, Wang D and Liimatainen H 2018 *Carbohydr. Polym.* **186** 132–139.
[24] Eisa W H, Abdelgawad A M and Rojas O J 2018 *ACS Sustain. Chem. Eng.* **6** 3974–3983.
[25] Gopiraman M, Deng D, Saravanamoorthy S, Chung I M and Kim I S 2018 *RSC Adv.* **8** 3014–3023.
[26] Chen Y, Chen S, Wang B, Yao J and Wang H 2017 *Carbohydr. Polym.* **160** 34–42.
[27] An X, Long Y and Ni Y 2017 *Carbohydr. Polym.* **156** 253–258.
[28] Reina A, Dang-Bao T, Guerrero-Ríos I and Gómez M 2021 *Nanomaterials* **11** 1891.
[29] Yin L and Liebscher J 2007 *Chem. Rev.* **107** 133–173.
[30] Baran N Y, Baran T and Menteş A 2017 *Appl. Catal. A: Gen.* **531** 36–44.
[31] Dong Y, Wu X, Chen X and Wei Y 2017 *Carbohydr. Polym.* **160** 106–114.
[32] Li Y, Xu L, Xu B, Mao Z, Xu H, Zhong Y, Zhang L, Wang B and Sui X 2017 *ACS Appl. Mater. Interfaces* **9** 17155–17162.
[33] Jebali Z, Granados A, Nabili A, Boufi S, Rego A M B, Majdoub H and Vallribera A 2018 *Cellulose* **25** 6963–6975.
[34] Chan H, Shi C, Wu Z, Sun S, Zhang S, Yu Z, He M, Chen G, Wan X and Tian J 2022 *J. Colloid Interface Sci.* **608** 1414–1421.
[35] Meng J, Liu Y, Shi X, Chen W, Zhang X and Yu H 2021 *Sci. China Mater.* **64** 621–630.

Recent development in electron transport layers for efficient tin-based perovskite solar cells

R Rani, K Monga, S Chaudhary*

Department of Applied Sciences, Punjab Engineering College (Deemed to be University) Chandigarh-160012, India

rekharani@pec.edu.in, shilpichaudhary@pec.edu.in

Abstract. Hybrid organic-inorganic tin (Sn)-based perovskite materials became a promising choice as an alternative to lead-free perovskite solar cells (PSCs) due to their outstanding optical and electrical properties. But, so far, a power conversion efficiency (PCE) of only 13% has been achieved for Sn-based PSCs. To achieve highly efficient and stable PSCs, not only the properties of the active layer but the charge selective contacts (electron and hole transport layers) should be selected wisely. The interfaces between the perovskite active layer and charge transport layers play an important role in achieving the better performance of PSCs. In the present review, the spotlight is on the recent developments made on the optimization of electron transport layers (ETLs) for the efficient Sn-based hybrid organic-inorganic PSCs. Further, we comprehensively discuss the significance and the impact of the lowest unoccupied molecular orbital level of electron transport material on the charge transport, which additionally affects the photovoltaic performance of the device. In summary, with continuous research on the Sn-based hybrid organic-inorganic perovskite materials as an absorbing layer, conventional ETLs (metal oxides) cannot be used. Thus, the optimum candidate for befitted ETLs must be explored and investigated in detail for efficient PSCs.

1. Introduction

Halide perovskite material has emerged as a star material since the breakthrough in 2012 [1-3], because of its exceptional optoelectronic properties such as tunable band gap, high absorption coefficient (10^4-10^5) [4], low exciton binding energy (37-50 eV) [5], high mobility of charge carriers (for electrons 164 ± 25 $cm^2V^{-1}S^{-1}$ and for holes 24.8 ± 4.1 $cm^2V^{-1}S^{-1}$) [6], the long transmission distance of charge carriers (approximately 100 nm or greater than this) [7]. Thus, perovskite materials have the potential to become an alternative material for active layers in photovoltaic devices. A general formula for perovskite material is ABX_3, where A is monovalent cation, B is divalent cation and X is the anion. The crystal structure of perovskite is shown in figure 1 [8,9].

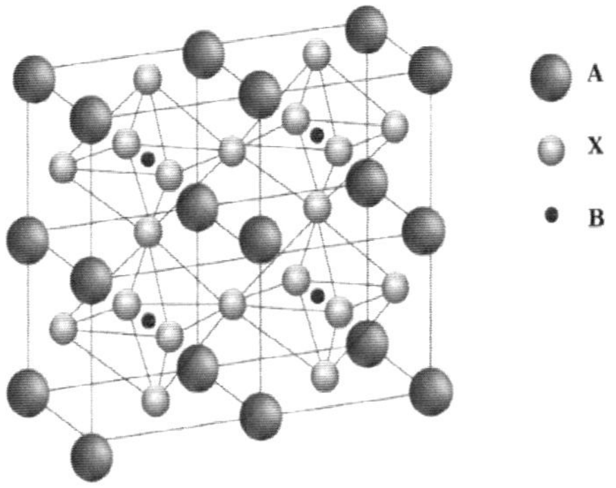

Figure 1. Basic structure of halide perovskite

Miyasaka et al. reported the photovoltaic outcomes for the perovskite materials for the first time [10]. They have demonstrated a power conversion efficiency (PCE) of 2.2% by using $CH_3NH_3PbBr_3$, as a light-absorbing material [11]. Further, a PCE of 3.8% was achieved by replacing bromine with iodine [12]. Recently, Ulsan National Institute of Science and Technology and the Swiss Federal Institute of Technology Lausanne claim 25.6 % efficiency in a single junction perovskite solar cell. Despite these marvellous outcomes, the toxicity of lead (Pb) is a cause of concern for lead-based perovskite materials, which is limiting the commercialization of perovskite solar cells (PSCs). Furthermore, according to Shockley- Quissor limit theoretical PCE of lead-based PSCs is limited to 32% due to its band gap which is greater than 1.5 eV [13]. Therefore, researchers are currently working toward lead free hybrid perovskite solar cells with the intention of replacing lead with the less toxic suitable candidate. Out of the various substitutes, tin (Sn) is regarded as a more suitable substitute, as Sn and Pb belong to the same group and have a similar ionic radius (1.19 Å for Pb^{2+} and 1.18 Å for Sn^{2+}) [14]. The PCE of tin halide perovskite solar cells can be greater than 33% due to the smaller band gap of tin (1.2-1.4 eV) [15], which enables it to extend its absorbance in the infrared region leading to higher theoretical efficiency. Moreover, Sn-based perovskite materials also exhibit excellent optoelectronic properties for e.g., the small value of exciton binding energy, higher value of absorption coefficient, higher values of charge carrier mobility [16]. However, the experimental efficiency of Sn-based PSCs is affected by the instability of Sn^{2+}. The electrons in 5s orbital of tin are more active as compare to electrons in the 6s orbital of lead, due to this Sn^{2+} tends to oxidize in Sn^{4+} [17, 18]. Due to the presence of Sn^{4+}, a process called self doping take place in Sn-based perovskite materials. It causes the enhancement in the hole concentration which may lead to severe recombination of charge carriers in the solar cell [19]. Based on the thermogravimetric analysis and in situ X-ray diffraction measurements Leijten et al. provide the oxidation and degradation path of tin-based perovskites. The degradation equation of tin perovskite can be given as [20].

$$2ASnI_3 + O_2 \qquad\qquad 2AI + SnO_2 + Sn$$

In 2014, Hao et al. reported a PCE of 5.73% from a solution-processed solid-state photovoltaic device based on methylammonium (MA) tin perovskite [21]. In the same year, a PCE of 6.4% was achieved by using $MASnI_3$ perovskite independently [19]. Since then, a lot of efforts were executed to enhance the stability and efficiency of Sn-based PSCs [22-28]. So far, a maximum PCE of only nearby 13 % has been achieved for Sn-based PSCs [29-30]. One of the main causes of poor performance of Sn-based PSCs is the presence of energy barriers and mismatch band alignment at the interface between electron transport layer (ETL)/ perovskite and perovskite/ hole transport layer (HTL). Since tin perovskite exhibit shallower conduction band (CB) and valence band (VB) as compared to lead perovskite [31], the mismatch between CB of Sn perovskite/ETL and VB of Sn perovskite/HTL take place, which leads to poor charge transport and hence less efficiency of the device. To achieve the highly efficient and stable PSCs, not only the properties of the active layer (high absorption coefficient of the order 10^4-10^5, high mobility, and long charge carrier diffusion length) but the efficient charge transport is also an essential condition. Thus, the charge transport structure engineering to obtain optimized band alignment between ETL/perovskite/HTL plays an important role in obtaining better charge transport, for efficient solar cells. In this review, we emphasize the recent developments made on the optimization of electron transport layers (ETLs) for the efficient Sn-based hybrid organic-inorganic perovskite solar cells. Here, we comprehensively discuss the significance and the impact of the lowest unoccupied molecular orbital (LUMO) level of electron transport material on the charge transport, which additionally affects the forward-reverse scan current density-voltage (J&V) characteristics, open-circuit voltage (V_{oc}), PCE. Figure 2 shows the various ETLs used in Sn-based PSCs and their corresponding PCE and V_{oc}. However, to the best of our knowledge, no comprehensive review on various ETLs used in Sn-based PSCs presenting the whole picture has been reported so far. The other possibilities for achieving efficient transport layers include increasing mobility and varying charge transport layer thickness. Further, charge transport layers can be doped with other materials to tune the LUMO/HOMO level of charge transport layers to match with the CB and VB of the active layer, which can lead to better photovoltaic performance.

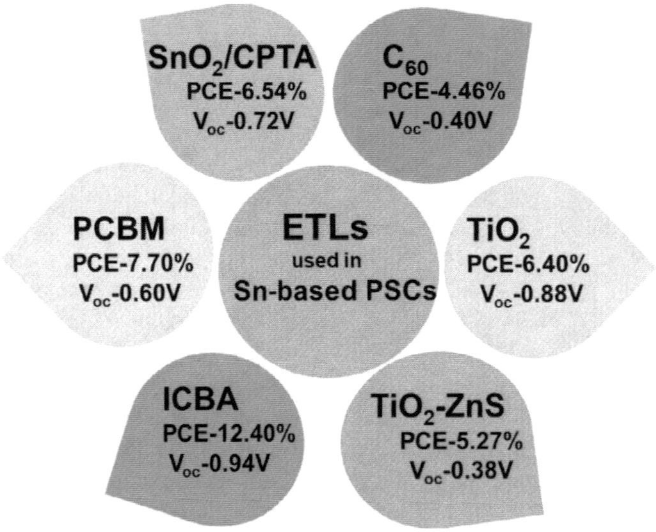

Figure 2. Electron transport layers contributing important role in the enhancement of performance of tin perovskite solar cells

2. Working principle of PSCs

PSCs are typically made in two different structures: planar n-i-p, and planar inverted p-i-n structure, which consists of transparent conductive oxide (TCO) (fluorine doped tin oxide (FTO) and Indium doped tin oxide (ITO)), ETL, active layer, HTL, and a metallic electrode. When light falls on the perovskite layer, electron and hole pairs are generated. These generated electrons and holes are transported to their respective ETL and HTL. Both charge transport layers efficiently transfer charge carriers through them. In the end, both charge carriers are transported to the electrode and generate current [32]. ETL and HTL in the inverted p-i-n structure are flipped compared to the planar n-i-p structure. Standard orientation in the inverted structure is glass/TCO/HTL/perovskite active layer/ ETL/Metal electrode as compared to planar structure of configuration glass/ETL/perovskite active layer/ HTL/Metal electrode.

For the efficient transfer of electrons and holes, LUMO and the highest occupied molecular orbital (HOMO) of ETL and HTL should match with the CB and VB, respectively. The performance of PSCs can be defined by the interface characteristics between the perovskite layer and adjacent charge transfer layers. PSCs have additional features of tunable bandgap through the compositional engineering process. Compositional engineering can be done by partially substituting organic and inorganic elements of similar properties as A, B, and X ion. According to Marchiore et al. electron-hole pair separation take place at two heterojunction interface 1) ETL / perovskite 2) HTL / perovskite and it is followed by the injection of electrons into ETL and holes in HTL as shown Figure 3, process (1) and (2) respectively [33]. At the same time, there is also a probability of reverse transmission of electrons and holes (shown in process (3) and (4)), exciton annihilation also takes place, as shown in process (5) which leads to a decrease in the efficiency of a solar cell. There is also a probability of recombination of electrons and holes at the interface of ETL and HTL as shown in process (6). For the efficient transfer of electrons and holes, LUMO of ETL and HOMO of HTL should match with the CB and VB of the active layer, respectively.

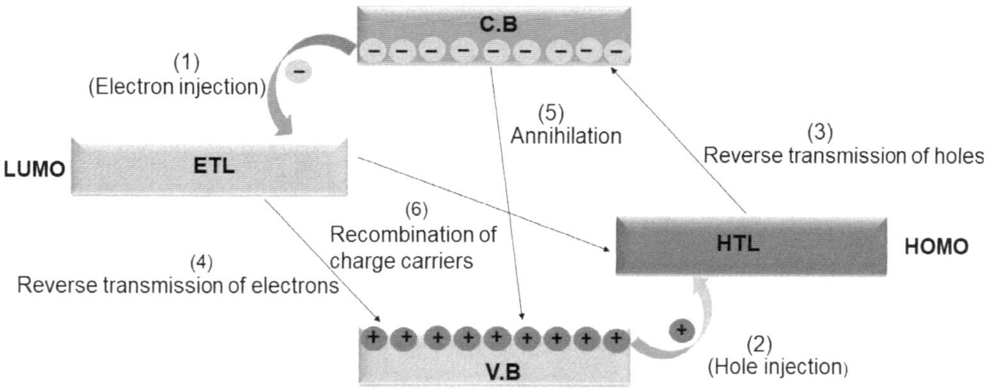

Figure 3. Pictorial depiction on the working of perovskite solar cell.

3. Progress in ETLs for Sn-based hybrid organic-inorganic perovskite (Sn-HOIP) solar cells

The basic function of ETL is to efficiently transfer photogenerated electrons from the perovskite layer to the respective electrode. It also prevents hole migration in order to reduce the recombination of charge carriers. ETL fabricated with different compositions and techniques can be employed in PSCs to achieve high PCE. In general, electron transport material should have high carrier mobility and band structure should match with the active layer (perovskite material). Here, two broad categories of ETL materials have been discussed based on recent developments targeting efficient Sn-based PSCs.

3.1 Metal oxides and their composite with inorganic/organic compound

Transparent TiO_2 has been widely studied ETL due to the large band gap [34-36]. In the beginning, PSCs with n-i-p device structure used TiO_2 as an electron selective contact, however, the high-temperature treatment of TiO_2 leads to J-V hysteresis [37]. The other choice of material as ETLs can be fullerene (C_{60}), which does not require high-temperature treatment [38]. When C_{60} or its derivative is employed as ETL in tin-based PSCs a considerable efficiency has been achieved [39]. In 2014 Noel et al. [19] fabricated the first lead-free perovskite solar cell using Sn-based hybrid organic-inorganic perovskite material (methylammonium tin iodide, $MASnI_3$) with device structure FTO/compact-TiO_2/mesoporous(meso) TiO_2/$MASnI_3$/Spiro-OMETAD/Au. It was concluded that the device constructed on meso TiO_2 scaffold provides better PCE and other photovoltaic parameters than the device constructed on meso Al_2O_3. They achieved PCE of 6.4 %, open-circuit voltage (Voc) 0.88 V, short circuit current density (J_{sc}) 16.8 mA cm^{-2}, fill factor (FF) 42% by using meso TiO_2 of thickness 400 nm. In 2015 Koh et al. [40], found that variation of thickness of meso TiO_2 as ETL has a significant impact on the photovoltaic parameters of Sn-based PSCs. As shown in figure 4(a) highest current density has been achieved by using 500 nm thick meso TiO_2 as ETL with device structure FTO/TiO_2 blocking layer/meso TiO_2/$FASnI_3$/spiro-OMeTAD/Au. An increment in the thickness of meso TiO_2 leads to the decrease in the capping layer of formamidinium tin iodide ($FASnI_3$) on the meso TiO_2 which leads to the excellent pore filling and reduces the charge recombination in bulk capping layer. This further improves the charge injection and collection. Miyamoto et.al.[41] investigated the effect of structure of ETL on the electron extraction process using time resolved microwave conductivity (TRMC). They employed nano-porous niobium oxide (np-Nb_2O_5) ETL with different pore size to obtain efficient charge extraction from perovskite material by shortening electron extraction length. TRMC showed that the small pore size of

ETL improves the device performance due to a decrease in the distance between photogenerated electron and ETL. Smaller pore size ETL extract electrons effectively as compare to larger pore size ETL. Charge carrier recombination also gets suppressed with the use of nano-porous ETL. Solar cell device based on np-Nb_2O_5 as ETL pore size 5 nm with device structure ITO-ATO/c-Nb_2O_5/np-Nb_2O_5/$FA_{0.83}PEA_{0.17}SnI_3$/PTAA/Au exhibit a PCE of 7.0%, J_{sc} 24 mAcm^{-2} and FF 63%, here ATO and PTAA represents antimony doped tin oxide, poly(bis(4-phenyl) (2,4,6- trimethylphenyl) amine, respectively. These results show that np-Nb_2O_5 can be used as efficient ETL in Sn-based PSCs.

In 2016 Ke et al. [42], investigated the effect of the ZnS interfacial layer on the performance of the $FASnI_3$ (with bandgap ~1.4eV) perovskite solar cell. $FASnI_3$ was used as light absorbing material, the structure is shown in figure 4(b). In their study, TiO_2-ZnS cascade was employed as ETL in Sn-based PSCs with device structure FTO/c-TiO_2/ZnS/$FASnI_3$/ PTAA/Au. As shown in figure 4(b), the conduction band minima (CBM) of $FASnI_3$, ZnS and TiO_2 lies at ~ 3.3eV, ~ 3.5 eV and ~ 4.2eV, respectively. Clearly, the CBM or LUMO of TiO_2 is much lower than that of $FASnI_3$, which restricts the efficient injection of the electron in TiO_2. Since the LUMO of ZnS ~ 3.5 eV lies in between the $FASnI_3$ and TiO_2, the electron can efficiently transfer from $FASnI_3$ to TiO_2, resulting in the high value of V_{oc} and other photovoltaic parameters. Charge recombination would effectively be blocked as a consequence of energy barrier at TiO_2/ZnS interface charge, which increases the quasi-fermi level for electrons and results in the high value of V_{oc}. The J-V characteristics of PSC with and without ZnS are shown in figure 4(c). By using TiO_2/ZnS as an ETL, PCE 5.27%, V_{oc} 0.38 V, J_{sc} 23.09 mA cm^{-2}, FF 60% and for TiO_2 as an ETL PCE of 4.90%, V_{oc} 0.31 V, J_{sc} 23.28 mA cm^{-2}, FF 54% were reported. Further, the use of TiO_2/ZnS as an ETL displays more significant quenching than TiO_2 alone (shown in figure 4(d)), which indicates the efficient transfer of electrons between perovskite and ETL. Thus, the study demonstrated that ZnS modified meso-TiO_2 can serve as an efficient ETL for $FASnI_3$ perovskite solar cells.

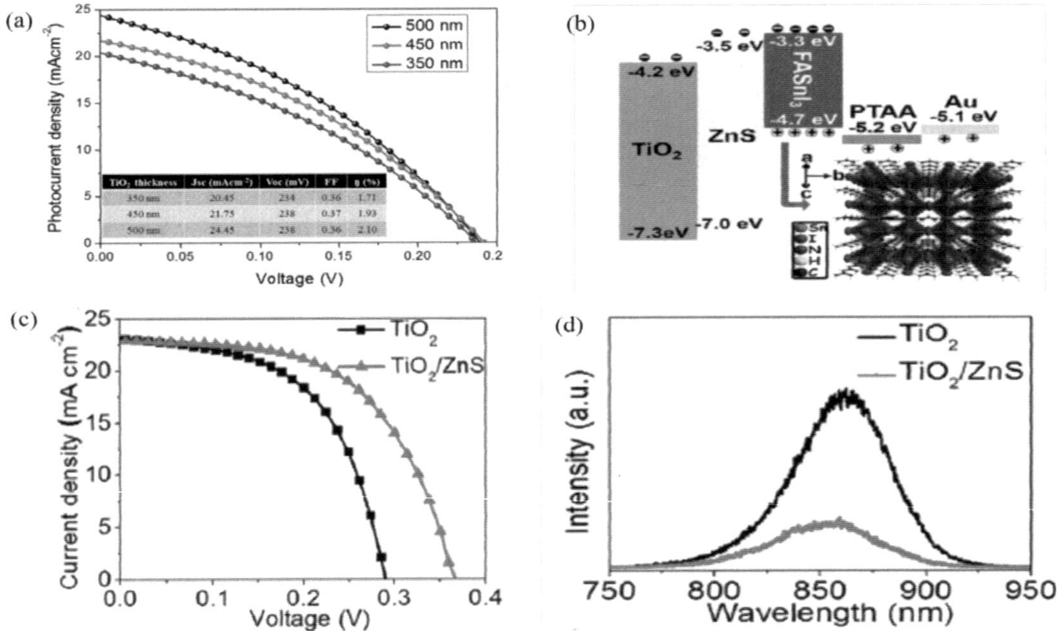

Figure 4. (a) Variation of current density with voltage at the different thickness of TiO_2 [32]. Copyright, Journal of Material Chemistry A. (b) energy band diagram of $FASnI_3$ based perovskite solar cell and crystal structure of perovskite, (c) variation of current density with voltage with neat and ZnS modified mesoporous electron transport layer, (d) PL spectra. Copyright, Journal of the American chemical society [42]

SnO_2-C_{60} pyrrolidine tris-acid (CPTA) as ETL in Sn-based PSCs has been investigated by Yang et al. [43]. Sn-based PSC with device structure FTO/SnO_2/CPTA/ethylenediamine(en)$FASnI_3$/PTAA/Au has been fabricated based on (en)$FASnI_3$ perovskite and SnO_2-CPTA as an ETL. As it is clear from the energy band diagram (shown in figure 5) LUMO of CPTA (-3.9 eV) is very much shallower than LUMO of SnO_2 (-4.5 eV), which results in less energy loss during electron transfer across SnO_2/(en)$FASnI_3$ interface. Moreover, electron transfer to SnO_2 might be prone to recombination with photogenerated holes inside the active layer as VBM of (en)$FASnI_3$ (-4.8 eV) is close to LUMO of SnO_2. The introduction of CPTA layer leads to a decrease in a decay lifetime from 1.5 ns (with SnO_2) to 1.2 ns (with SnO_2 -CPTA) which indicate faster electron transfer from the active layer to charge selective contacts and it also enhances the degree of quenching (quenching is a measure of charge separation). The energy barrier at SnO_2/CPTA interface suppresses the charge reunite which leads to a higher PCE of 6.54% and also device acquires high stability in air. J-V hysteresis also gets significantly reduced by employing SnO_2-CPTA as ETL.

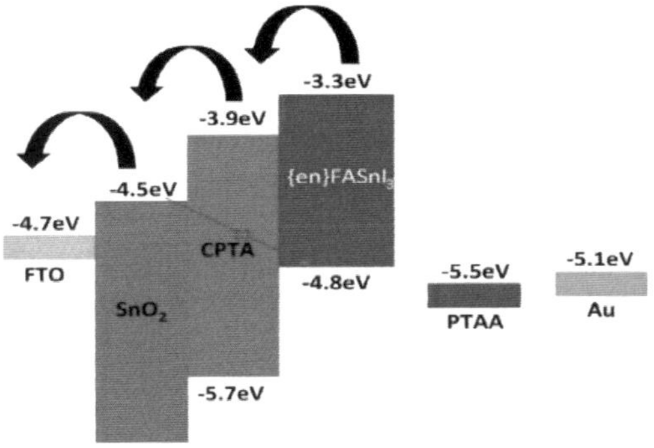

Figure 5. Energy band diagram of {en}$FASnI_3$ based perovskite solar cell, the red arrow shows the recombination at SnO_2/(en)$FASnI_3$ interface (redrawn) Copyright, Advanced Functional Materials [43]

3.2 Fullerene and it's derivative

Organic ETLs such as C_{60}, phenyl-C_{61}-butyric acid methyl ester (PCBM), CPTA, and Indene – C_{60} -bis adduct (ICBA) are also employed for highly efficient Sn-based PSCs. Even a record high PCE of 12.4% has been obtained by using organic material ICBA, as electron selective contact [44]. The chemical structures of efficient ETLs used in Sn-based PSCs are shown in figure 6. Here we will discuss the impacts of these ETLs on the PCE and other photovoltaic parameters of Sn-based PSCs.

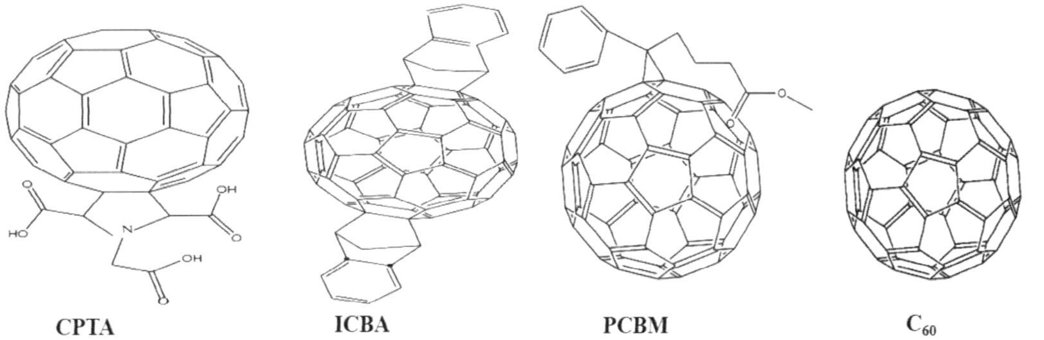

Figure 6. Chemical structure of organic electron transport material used in Sn-based PSCs

Zhang et al. [39] explored C_{60} and PCBM for efficient ETLs in Sn-based PSCs with $FASnI_2Br$ (bandgap ~1.68eV) as an absorbing layer. In case of PCBM as an ETL, with device architecture ITO/PEDOT-PSS/$FASnI_2Br$/PCBM/Ca/Al exhibits PCE = 1.43 %, J_{sc} = 7.27 mA cm^{-2}, V_{oc} = 0.36 V and FF = 53 %.

Whereas in the case of C_{60} as an ETL with device structure ITO/PEDOT-PSS/$FASnI_2Br$/C60/Ca/Al resulted in the hysteresis-less PCE = 1.72%, J_{sc} = 6.82 mA cm^{-2}, V_{oc} = 0.46 V, and FF = 54%. Although the efficiency of PSC with PCBM based device retains 60% of its initial efficiency after 30h and efficiency of C_{60} quickly drops within a few hours. The stability of PCBM based devices in nitrogen atmosphere has been reported to be much higher as compared to C_{60}.

A recent study by Jiang et al. investigated the potential of indene – C_{60} -bis adduct (ICBA) as ETL material instead of conventionally used PCBM [44]. The perovskite film was fabricated by tuning the concentration of phenethylammonium ($C_6H_5CH_2CH_2NH_3^+$=PEA) and by adding ammonium thiocyanate (NH_4SCN) as an additive to enhance the film growth. The PSC with ICBA as an electron selective contact leads to a PCE of 12.4%, V_{OC} of 0.94 V, J_{SC} 17.4 mA cm^{-2}, and FF 75 %. However, the PSC with PCBM as ETL resulted in lower PCE (7.7%) and V_{oc} (0.60 V). ICBA not only resulted in the ultrahigh open-circuit voltage (0.94 V) owing to its shallower energy level but also decrease recombination of interface carriers. It may be pointed out that the CBM of PEA15-SCN perovskite lies at -3.69 eV. As shown in figure 7(a) LUMO level of ICBA lies at -3.74eV shallower than LUMO of PCBM (lies at -3.91 eV) offers an ultra-high value of V_{oc} and PCE. Further, it was concluded that the ICBA based device shows less hysteresis as compared to PCBM, shown in figure 7(b). Figure 7(c) shows less interface recombination in ICBA based device as compared to PCBM as injection of electrons from iodide to PCBM take place due to deep-lying LUMO level of ETL and results in high electron density, whereas the shallow level of ICBA prohibits electron injection from iodide. Interface carrier recombination has been studied by performing TRPL measurement, PCBM containing perovskite film shows faster recombination at the interface. The high electron density of PCBM could inflame interface carrier recombination.

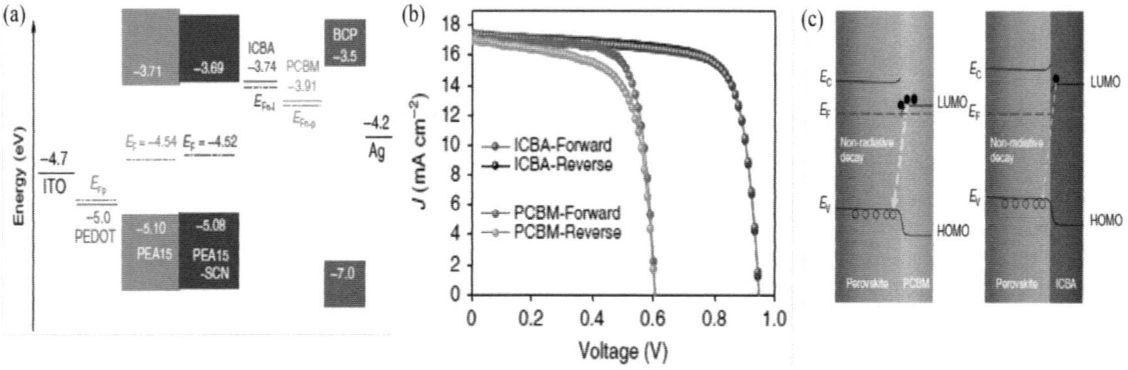

Figure 7. (a) Schematic illustration of energy levels. Dashed lines represent the quasi-Fermi level of ICBA (EFn–I), PCBM (EFn–P), and PEDOT (EFP), (b) J-V curves for the PEA15-SCN device with ICBA and PCBM, (c) Schematic diagram for the interfacial recombination of the device with ICBA and PCBM.

Later on, highly dependency of V_{oc} and PCE of Sn-based PSCs on the LUMO level of ETL was demonstrated by Lee et al. [45]. Here ICBA is compared with generally used PCBM and C_{60}. They used $FA_{0.9}PEA_{0.1}SnI_3$ as a light-absorbing material. The addition of PEA can improve the morphology of $FASnI_3$ film. ICBA suitably matches the energy level of $FA_{0.9}PEA_{0.1}SnI_3$ as compare to PCBM and C_{60} because of their deep-lying LUMO level. The energy level alignment between $FA_{0.9}PEA_{0.1}SnI_3$ and ETLs was investigated by using ultraviolet photoelectron spectroscopy (UPS) as shown in figure 8(a). The CBM of $FA_{0.9}PEA_{0.1}SnI_3$ lies at -3.63 eV and LUMO level of ICBA, PCBM, C_{60} lies at -3.73eV,-3.93eV,-4.56eV respectively. As the LUMO level of ICBA lies upper than PCBM and C_{60}, it gives, energy offset ΔE 0.10 eV less than PCBM (0.30 eV) and C_{60} (0.93eV). ΔE act as a driving force for the extraction of the electron. ICBA gives the smallest value of energy offset, it gives the longest PL lifetime also in the light-absorbing material, so this clearly demonstrates that ICBA can extract electrons efficiently from light-absorbing material. By using ICBA as an ETL in Sn-based PSC with device structure ITO/PEDOT-PSS/$FA_{0.9}PEA_{0.1}SnI_3$/ICBA/Al significant-high PCE (7.05%), V_{oc} (0.65) were obtained as compared to PCBM and C_{60}. Figure 8(b) shows the Voc value for the device by using the above-mentioned ETL Figure 8(c) shows the J-V characteristics for Sn-based PSC by using ICBA, PCBM, C60 as ETLs.

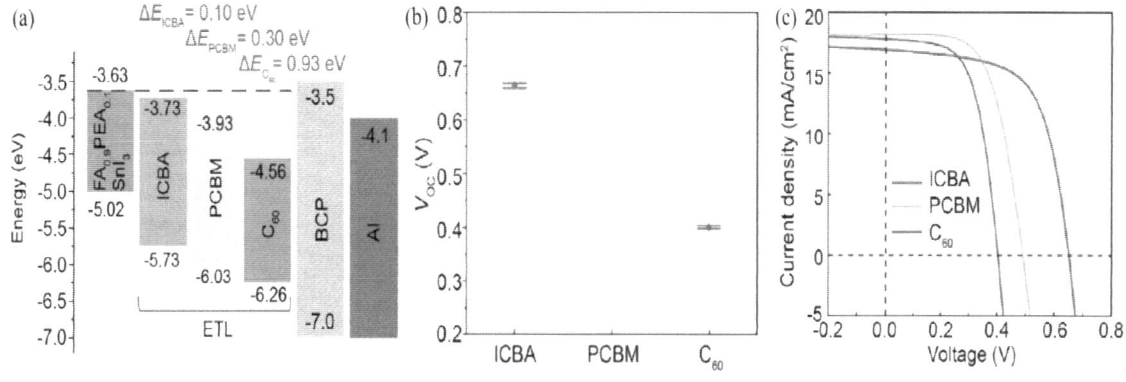

Figure 8. (a) Schematic representation of energy levels of $FA_{0.9}PEA_{0.1}SnI_3$, ETLs and of other constituents of the device (b) V_{oc} values for the PSC by using ICBA, PCBM, C_{60} (c) J-V curves for the PSC by using ICBA, PCBM, C_{60}. Copyright, ACS Applied Energy Materials [45]

Table 1 provides a sum upon the above discussed two categories of ETL materials used in various Sn-based PSCs along with their photovoltaic parameters for the past few years. Additionally, the LUMO levels of above discussed ETL materials based on the metal oxides, metal-oxide-composites, fullerene, and its derivatives used in Sn-based PSCs are shown in figure 9. Nowadays, device modelling also emerging as a convenient method for the optimization of solar cell devices. Although there are some discrepancies in the experimental and theoretical estimated results due to factors like fabrication technique, process paraments, and solvents used. Roy et.al. [46] investigated the effect of TiO_2, ZnO, and PCBM as ETL by using device modelling method SCAPS-1D software. They found that TiO_2 and ZnO based PSCs provides almost identical performance, whereas PCBM based device deliver low PCE but provide stability on temperature variation. Devices based on TiO_2 ETL resulted in PCE of 23.24%. Lastly, it is noteworthy to discuss a theoretical study by Gagandeep et al. [47], where doped graphene was employed as ETL and HTL in Sn-based PSC with device structure n-graphene (doped with nitrogen atom)/$CH_3NH_3SnI_3$/p-graphene (doped with boron atom). In their study, the structure of Sn-based PSCs with n-graphene as ETL and p-graphene as HTL was simulated. Boron doping of graphene results in shifting of Fermi level below the Dirac point and nitrogen doping shifts the Fermi level above the Dirac point. Change of Fermi level accounts for the change in the work function of graphene, which further affects the position of HOMO and LUMO level of graphene. The LUMO of n-graphene and HOMO of p-graphene lies at -4.25 eV and -5.47 eV, respectively. After the optimization of layer parameters best possible PCE of 10.67 %, V_{oc} 0.68 V, J_{sc} 19.93 $mAcm^{-2}$, and FF of 77.92 have been achieved.

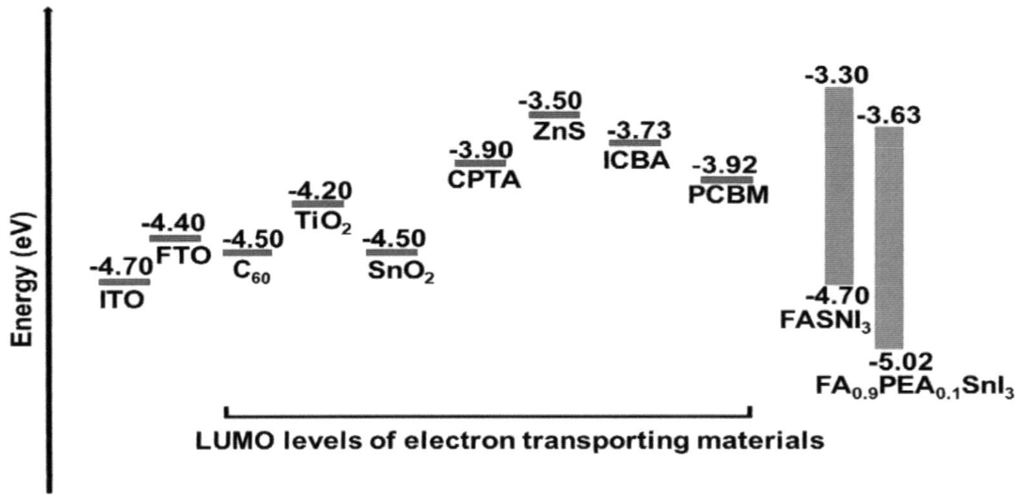

Figure 9. Summary of LUMO levels of various ETLs, conduction band and valence band of perovskite materials

Table 1. Performance summary of Sn-based PSCs utilizing different electron transport layers.

S.No	Device Structure	ETL	V_{oc} (V)	J_{sc} ($mAcm^{-2}$)	PCE (%)	FF (%)	Year	Ref
1	FTO/c-TiO_2/MesoTiO_2/$CH_3NH_3SnI_3$/Spiro-OMeTAAD	TiO_2	0.88	16.8	6.4	42	2014	[19]
2 (i)	ITO/PEDOT-PSS/$FASnI_2Br$/C_{60}/Ca/Al	C_{60}	0.46	6.82	1.72	54	2016	[39]
(ii)	ITO/PEDOT-PSS/$FASnI_2Br$/PCBM/Ca/Al	PCBM	0.36	7.27	1.43	53		
3 (i)	ITO-ATO/c-Nb_2O_5/ $FA_{0.83}PEA_{0.17}SnI_3$/PTAA/Au	c-Nb_2O_5	0.40	10.4	1.7	41	2020	[41]
(ii)	ITO-ATO/c-Nb_2O_5/np-Nb_2O_5/$FA_{0.83}PEA_{0.17}SnI_3$/PTAA /Au	c-Nb_2O_5/np-Nb_2O_5	0.46	24	7	63		
4 (i)	FTO/c-TiO_2/ZnS/$FASnI_3$/PTAA/Au	TiO_2-ZnS	0.38	23.09	5.27	60	2016	[42]
(ii)	FTO/c-TiO_2/ $FASnI_3$/PTAA/Au	TiO_2	0.72	13.42	4.90	54		
5 (i)	FTO/SnO_2/CPTA/en($FASnI_3$)/PTAA/Au	SnO_2-CPTA	0.72	13.42	6.54	67	2019	[43]
(ii)	FTO/SnO_2/en($FASnI_3$)/PTAA/ Au	SnO_2	N/A	N/A	N/A	N/A		
6 (i)	ITO/PEDOT-PSS/PEA15-SCN/ICBA/Ag	ICBA	0.94	17.40	12.40	75	2020	[44]
(ii)	ITO/PEDOT-PSS/PEA15-SCN/PCBM/Ag	PCBM	0.60	17.10	7.70	74		
7 (i)	ITO/PEDOT-PSS/$FA_{0.9}PEA_{0.1}SnI_3$/ICBA/Al	ICBA	0.65	16.88	7.05	64	2020	[45]
(ii)	ITO/PEDOT-PSS/$FA_{0.9}PEA_{0.1}SnI_3$/PCBM/Al	PCBM	0.49	17.37	5.54	62		
(iii)	ITO/PEDOT-PSS/$FA_{0.9}PEA_{0.1}SnI_3$/C_{60}/Al	C_{60}	0.40	17.16	4.46	62		

4.Conclusion

In this review, the fundamental aspects of ETL engineering for the optimization of efficient ETLs in Sn-based PSCs have been discussed. Further, this review summarizes the recent significant developments on different ETL materials based on the metal oxides, metal-oxide-composites, fullerene, and its derivatives used in Sn-HOIP solar cells. It is pointed out unanimously that efficient ETLs can be selected by adjusting LUMO levels of ETLs to minimize the V_{oc} loss at the interface and by achieving optimal driving force to extract electrons from the perovskite layer. This can help to meet the requirements for the theoretical limit of PCE in Sn-based PSCs. The efficient charge extraction/ transfer and low recombination rate at the interface are the consequences of improved band alignment at the interface of ETL/perovskite materials. Significant improvements in the PCE of Sn-based HOIP solar cells have been achieved by implementing of the new approach of cascading ETLs such as TiO_2-ZnS, SnO_2-CPTA. Charge transport properties of TiO_2 can be significantly improved by varying its thickness and interfacial modification with the materials having shallower LUMO level leads to the high performance of the device. Further, organic ETLs, such as C_{60}, PCBM, and ICBA, are proven to be promising candidates as ETLs in Sn-based PSCs. Recently, an organic electron selective contact ICBA reported a record high efficiency of 12.40% with ultra-high V_{oc} of 0.94V in Sn-based PSCs.

It can be summarized that optimization of ETLs is an efficient method to enhance the photovoltaic parameters such as PCE, V_{oc}, J_{sc}, FF and also to reduce hysteresis in Sn-based PSCs. It can be pointed out that continuous research efforts are being made to improvise the existing and exploration of new perovskite materials for highly stable and efficient PSCs. Thus, the conventional ETL materials, in particular, metal oxides cannot serve as efficient ETLs with desirable properties with respect to the active layers. This demands the careful selection of ETL materials with an in-depth understanding of the electron transfer mechanism between ETL and the active layer.

Acknowledgements

We gratefully acknowledge the grant from the Science & Engineering Research Board (SERB), Department of Science and Technology (DST) under Start up Research Grant, Government of India (SRG/2020/000777). Authors also acknowledge the Punjab Engineering College (Deemed to be University) for research initiation grant No. PEC/DSR&IC/4636 for the financial support.

References

[1] Roy P, Sinha N.K, Tiwari S and Khare A 2020 *Sol. Energy* **198** 665-8
[2] Shamsi J, S. Urban A, Imran M, De Trizio L and Manna L 2019 *Chem. Rev.* **119** 3296-3348
[3] Kim H-S, Lee C-R, Im J-K, Lee K-B, Moehl T, Marchioro A, Moon S-J, Humphry-Baker R, Yum J-H, Moser J, Grätzel M and Park N-G *Sci. Rep.* 2012 **2** 591
[4] Qiu X, Cao B, Yuan S, Chen X, Qiu Z, Jiang Y, Ye Q, Wang H, Zeng H, Liu J andG.Kanatzidis M 2017 *Sol. Energy Mater Sol. Cells* **159** 227-34
[5] Miyata A, Mitioglu A, Plochocka P, Portugall O, Tse-Wei Wang J, D. Stranks S, J. Snaith H and J. Nicholas R 2015 *Nat. Phys.* **11** 582-87
[6] Dong Q, Fang Y, Shao Y, Mulligan P, Qiu J, Cao Land Huang J 2015 *Science* **347** 967-70
[7] Ibn-Mohammed T, C.L.Koh S, M.Reaney I, Acquaye A, Schileo G, B.Mustapha K, Greenough R 2017 *Renew. Sustain. Energy Rev.* **80** 1321-44
[8] Yi Z, Ladi N.H, Shai X, Li H, Shen Y and Wang M 2019 *Nanoscale Adv* **1** 1276-89
[9] Lu H, Krishna A, M.Zakeeruddin S, Grätzel M and Hagfeldt A 2020 *iScience* **23** 101359
[10] A. Green M, Ho-Baillie A and J. Snaith H 2014 *Nat. photonics* **8** 506-14
[11] Kojima A, Teshima K, Miyasaka T, and Shirai Y 2006 *J. Electrochem. Soc.* 397
[12] Kojima A, Teshima K, Shirai Y and Miyasaka T 2009 *J. Am. Chem. Soc.* **131** 6050-51

[13] Anttu N 2015 *ACS Photonics* **2** 446-53
[14] Zhou Y and Zhao Y 2019 *Energy Environ. Sci.* **12** 1495-1511
[15] Cao J and Yan F 2021 *Energy Environ. Sci.* **14** 1286-1325
[16] Wu T, Liu X, He X, Wang Y, Meng X, Noda T, Yang X and Han L 2020 *Sci. China Chem.* **63** 107–15
[17] Li H, Wei Q and Ning Z 2020 *Appl. Phys. Lett.* **117** 060502
[18] Gao F, Li C, Qin L, Zhu L, Huang X, Liu H, Liang. L, Hou Y, Lou Z, Hu Y and Teng F 2018 *RSC Adv* **8** 14025-30
[19] K. Noel N, D. Stranks S, Abate A, Wehrenfennig C, Guarnera S, Haghighirad A, Sadhanala A, E. Eperon G, K. Pathak S, B. Johnston M, Petrozza A, M. Herz L and J. Snaith H 2014 *Energy Environ. Sci.* **7** 3061-68
[20] Leijtens T, Prasanna R, Gold-Parker A, F. Toney M and D. McGehee M 2017 *ACS Energy Lett.* **2** 2159-65
[21] Hao F, C. Stoumpos C, Hanh Cao D, P. H. Chang R and G. Kanatzidis M 2014 *Nat. Photonics* **8** 489-94
[22] Konstantakou M and Stergiopoulos T 2017 *J. Mater. Chem. A* **5** 11518-49
[23] Liang L, Gao P 2017 *Adv. Sci.* **5** 1700331
[24] Wang F, Jiang X, Chen H, Shang Y, Liu H, Wei J, Zhou W, He H,Liu W and Ning Zhigun 2018 **2** 2732-43
[25] Jokar E, Chien C-H, Fathi A, Rameez M, Chang Y-H and and Wei-Guang Diau E 2018 *Energy Environ. Sci.* **11** 2353-62
[26] Wang C, Gu F, Zhao Z, Rao H, Qiu Y, Cai Z, Zhan G, Li X, Sun B, Yu X, Zhao B, Liu Z, Bian Z, Huang C 2020 *Adv. Mater.* **32** 1907623
[27] Nakamura T, Yakumaru S, Anh Truong M, Kim K, Liu J, Hu S, Otsuka K, Hashimoto R, Murdey R, Sasamori T, Do Kim H, Ohkita H, Handa T, Kanemitsu Y and Wakamiya A 2020 *Nat. Commun.* **11** 3008
[28] Nishimura K, Akmal Kamarudin M, Hirotani D, Hamada K, Shen Q, Likubo S, Minemoto T, Yoshino K and Hayase S 2020 *Nano Energy* **74** 104858
[29] Afaq Ali Shah S, Hassan Sayyad M, Khan K, Guo K, Shen F, Khan Tareen A, Gong Y and Guo Z 2020 *Energies* **13** 5092
[30] Jiang X, Li H, Zhou Q, Wei Q, Wei M, Jiang L, Wang Z, Peng Z, Wang F, Zihao, Xu K, Hou Teale S, Zhou W, Si R, Gao X, H. Sargent E and Ning Z 2021 *J. Am. Chem. Soc.* **143** 10970-76
[31] Shi Z, Guo J, Chen Y, Li Q, Pan Y, Zhang H, Xia Y and Huang W 2017 *Adv. Mater.* **29** 160505
[32] Baikie T, Fang Y, M. Kadro J, Schreyer M, Wei F, G.Mhaisalkar S, Graetzel M and J. White T 2013 *J. Mater. Chem. A* **1** 5628-41
[33] Marchioro A, Teuscher J, Friedrich D, Kunst M, Van de Krol R, Moehl T, Grätzel M and Moser J-E 2014 *Nat. Photonics* **8** 250-55
[34] Zhou Di, Zhou T, Tian Y, Zhu X and Tu Y 2018 *J. Nanometer* **2018**
[35] Shahiduzzaman M, Sakuma T, Kaneko T, Tomita K, Isomura M, Taima T, Umezu S and Iwamori S, 2019 *Sci. Rep.* **9** 19494
[36] Murugan Arivunithi V, Kim S, Choi J, Hun Sung J, Deog Yoo H, Shin E, Noh Y, Gal Y, Lee H and Jin S, 2020 *Org. Electron.* **86** 105922
[37] Suk Jung H, Park N-G 2015 *Small* **11** 10-25
[38] Roldán-Carmona C, Malinkiewicz O, Soriano A, Espallargas Mínguez G, Garcia A, Reinecke P, Kroyer T, Dar M, Khaja Nazeeruddin M and Bolink H. J 2014 *Energy Environ. Sci.* **7** 994-97
[39] Zhang M, Lyu M, Yun J-H, Noori M, Zhou X, A. Cooling N, Wang Q, Yu H, C. Dastoor P and Wang L 2016 *Nano Res.* **9** 1570-77
[40] Ming Koh T, Krishnamoorthy T, Yantara N, Shi C, Lin Leong W, P. Boix P, C. Grimsdale A, G. Mhaisalkar S and Mathews N *J. Mater. Chem. A* **3** 14996-15000
[41] M Yumi, K Shohei, Y Tomoyasu, N Yu, M Taisuke, K Takashi, N Ryosuke, S Akinori and K Yukihiro, 2020, *ACS Appl. Nano Mater.* **3** 11650–57
[42] Ke W, C. Stoumpos C, Leigh Logsdon J, Leigh Logsdon M, Yan Y, Fang G and G. Kanatzidis M, 2016 *J. Am. Chem. Soc.* **138** 14998-15003

[43] Yang Z, Zhong M, Liang Y, Yang L, Liu X, Li Q, Zhang J and Xu D 2019 *Adv. Funct. Mater.* **29** 1903621
[44] Jiang X, Wang F, Wei Q, Li H, Shang Y, Zhou W, Wang C, Cheng P, Chen P, Chen L, Ning Z 2020 *Nat. Commun.* **11** 1245
[45] Lee M, Kim D, Kyu Lee Y, Koo H, Tae Lee K and Chung I 2020 *ACS Appl. Energy Mater.* **3** 5581-88
[46] Roy P, Tiwari S and Khare A, 2021 *Results Opt.* **4** 100083
[47] Gagandeep, Singh M, Kumar R and Singh V 2019 *Mater. Res. Express* **6** 115611

Synthesis characterization and Luminescence Properties of $B_2BiMg_2V_3O_{12}$ based phosphors with rare earth activated Dy^{3+} phosphor for solid state lighting

S P Tatte[1], N S Dhoble[2], G C Mishra[3] and S J Dhoble[4]

[1]Department of Chemistry, O. P. Jindal University, Raigarh, 496109, India
[2]Department of Chemistry, Sevadal Mahila Mahavidhyalaya, Nagpur,440009,India
[3]Department of Physics, O. P. Jindal University, Raigarh, 496109, India
[4]Department of Physics, R.T.M. Nagpur University, Nagpur-440033 (India)

Corresponding author: sonaltatte89@gmail.com

Abstract
Highly new efficient vanadate based phosphor $B_2BiMg_2V_3O_{12}$ (B= K & Na) material incapacitated through rare- earth Dysprosium (Dy) and it was characterized by high temperature solid-state synthesis. Equipped phosphor was established by X-ray diffraction method. In Photo-luminescence measurements shows that the prepared phosphors doped by Dy is excited by near UV ultraviolet light reaching after 300 nm to 380 nm efficiently acceptable to recognize the emission in visible spectrum (in the range 400 nm –570 nm. The highly efficient prepared phosphor doped with lanthanide doped with Dy phosphor likewise showed the exact emission point at 487 nm and 571 nm at the excitation point at 325 nm. Hence, these prepared phosphors can find numerous applications as green emitting phosphor in the field of solid-state lighting area.

Keywords: Solid state synthesis method; X-Ray diffraction (XRD); Photoluminescence (PL); Concentration quenching.

1. Introduction

A vanadate is a purely complex compound. It consists of oxyanion of vanadium. Vanadium compound is good corrosion inhibitor & it is widely used in different types of industry like paint & pigment system. The compound vanadium specially monitor in different types of forms like metavanadate, vanadate, pyrovanadate & polyvanadate. Out of these vanadate groups, mostly metavanadate is a stable compound. In generally way, the vanadate based phosphor is not steady and hydrolysis responses quickly happen hence development of pyrovanadate result out.

Phosphors are a type of luminescence and it transforms the dissimilar kinds of excitation energy alike light, heat, acoustics, electricity, X-ray radiation, be contingent on their configuration nature. Phosphor mostly self-possessed of host crystal as well as small quantity like activators. [1-5]. Mostly in oxide materials attracted great interested by way of higher corrosion resistance and low toxicity associated to the sulfides compound which reduce them eco-naturally materials.[3-6]. At this time many oxide host constituents like Y_2O_3, Lu_2O_3, Sc_2O_3, La_2O_3, In_2O_3, Gd_2O_3, & yttria-stabilized zirconia (YSZ) [7-14].out of these, Y_2O_3 is an very important host materials (Er) doped component subsequently of resemblances between Y^{3+} & Er^{3+} in ionic radii in addition to the similar crystal structures of Er_2O_3& Y_2O_3 [7-15]. Moreover it consumes expansive transmittance (230 to 8000 nm), higher melting point at 2400°C and the larger index of refraction (1.91), small value of phonon energy in 380 cm^{-1}, larger the band gap i.e.5.8 eV & higher the chemical stability [16-17]. The resultant powders formed by a huge variation of different methods like sol-gel synthesis, solvo-thermal [16] synthesis of the combustion [17] Hydrothermal synthesis [18] Liquid membrane-Emulsion system [19] Laser ablation Synthesis [20] spray pyrolysis method [21]. There are large numbers of application like in optical sensing devices [17] infrared to discernable light converters [19], thermometry synthesis [22] bio labeling [23] solar-cells system [24] in optical-heaters process [25] in lasers [26] & in the scintillators [27]. In term of Scintillators, it can be well-defined as phosphorus constituents that permit the adaptation of higher energy-ionizing radioactivity in to the ultraviolet light or visible light.

In rare-earth activated doped based vanadate phosphors it can radiate numerous colors light due to higher luminescence substantial yields originating from the f–f level transition have been extensively studied [28-35]. In the field of nanomaterial various investigators have equipped the nanoparticles of vanadate materials and it's used to convert ultra-violet pumping light of the III-N type semiconductor LED hooked on the band of wavelength in visible mode. [36]. in complex-ion VO_4^{3-} group, the central V-metal ion corresponding with IV O^{2-} ions in tetrahedral proportion and it has been recognized as competent luminescence centre [37]. Commercially available soli state lighting sources usage commercially available blue LED light- emitting diodes i.e. max 450–480 nm based on GaN semiconductor technology.

We have described the synthesis-characterization and photoluminescence characteristics of prepared phosphor by solid-state synthesis. Equipped phosphors were characterized by the process of X-ray diffraction process & also investigated their luminescent properties.

2. Experimental

$B_2BiMg_2V_3O_{12}$: Dy (B=K & Na) phosphor have been synthesized by modified solids state synthesis. There are two different phosphor like $K_2BiMg_2V_3O_{12}$: Dy and $Na_2BiMg_2V_3O_{12}$: Dy phosphor. Synthesis procedures are same for prepared phosphors. High- purity i.e. 99.99% preliminary ingredients were $BaCO_3$, Na_2CO_3, BiO_2, MgO_2, K_2CO_3 and NH_4VO_3. Stoichiometric quantities of substances were systematically mixed and crushed by using agate mortar paster. The reaction and combination was occupied in purity the crucible like silica crucible then it gradually excited in oven (Furnace) at 600^{0}C aimed at 6 hr & 800^{0}C for 24 hr. Then sample were cooled despondent to the room temperature for obtaining white crystal powder.

Phase structure & composition remained categorized by X-ray diffraction arrangement by by Analytical diffractrometer. Excitation and emission spectra were characterized by exhausting Shimadzu RF -5301 PC-spectrofluoro-photometer at room temperature. PL emission & excitation peak spectra were verified by exhausting spectral slit width of 1.5 nm.

3. Result

3.1. XRD Pattern of B2BiMg2V3O12 (B=K & Na)

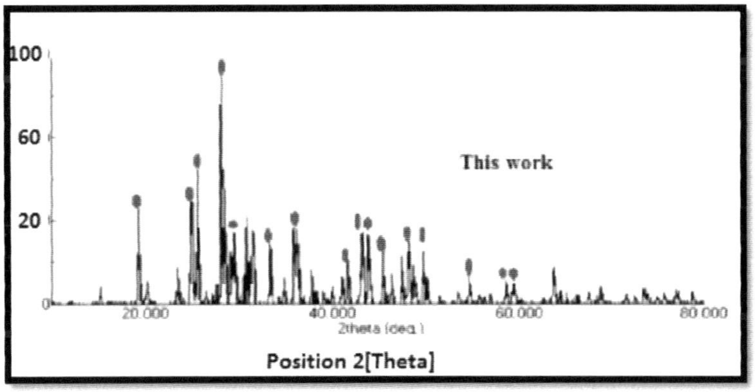

Figure 1. (a): X-ray diffraction arrangement of $K_2BiMg_2V_3O_{12}$ Phosphor

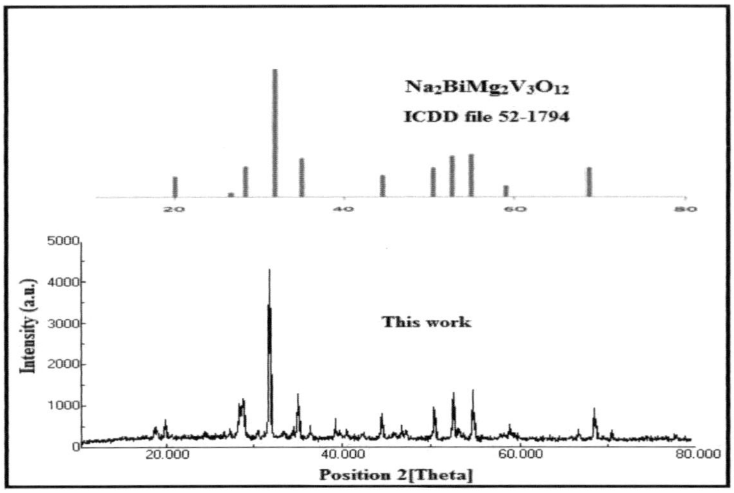

Figure 1 (b): X-ray diffraction arrangement of $Na_2BiMg_2V_3O_{12}$ Phosphor

With the help of Fig. 1 (a) it can be shown that, X-ray diffraction spectrum of $K_2BiMg_2V_3O_{12}$ phosphor powder, the sample displayed single phase, entirely the peaks were originate to characterize complete formation of the prepared compound. Standard file is not at all obtainable for the comparison so the XRD pattern was not associated by any statistics. From these figure it observed that the high intense sharp lines of our prepared material confirm the formation of compound excepting some lower intense lines these lines are aroused may be because of the impurity present in preliminary materials used.

X-ray diffraction pattern of as-prepared $Na_2BiMg_2V_3O_{12}$ is shown in fig. 1(b). An excellent match was observed with ICDD file 52-1794. The X-ray diffraction pattern did not indicate presence of constituent carbonates, which is indirect indication for development of the desired compound. Lattice constants of $Na_2BiMg_2V_3O_{12}$ were determined through Millar Indices of (2 2 0), (4 0 0), (4 2 0), (4 2 2), (6 1 1), (4 4 4), (6 4 0), (6 4 2), (8 4 2). Prepared phosphor gives the cubic structure with space group Ia3d with unit cell a=b=c= 12.52. Resultant result indication that, the final manufactured product was designed in uniform-form with unit cell by body center.

3.2. PL- Photoluminescence properties of $B_2NaMg_2V_3O_{12}$: Dy (B= Na & K)

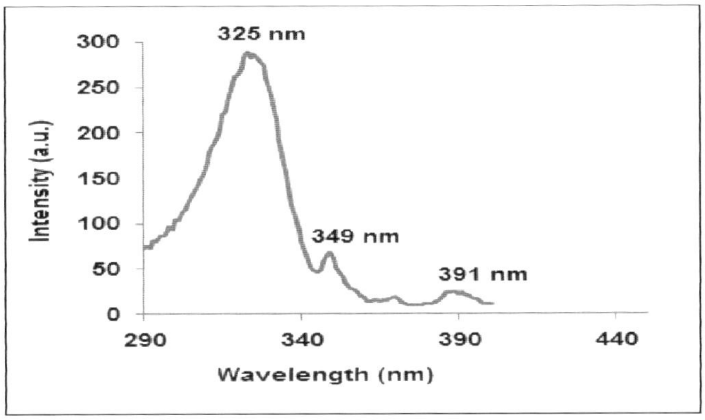

Figure 2 (a) Excitation Spectra of $Na_2BiMg_2V_3O_{12}$: Dy

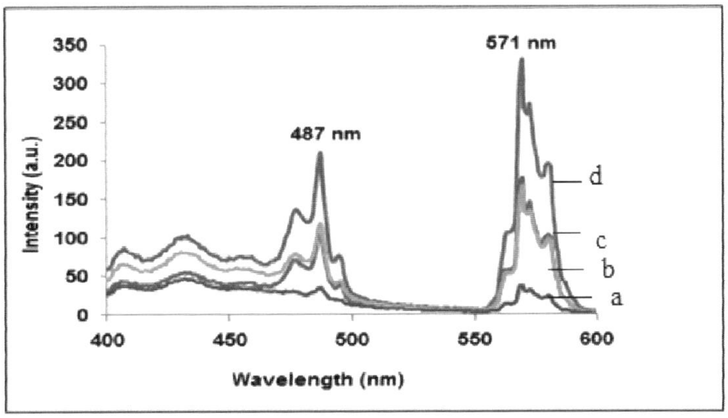

Figure 2 (b) Emission Spectrum of $Na_2BiMg_2V_3O_{12}$: Dy

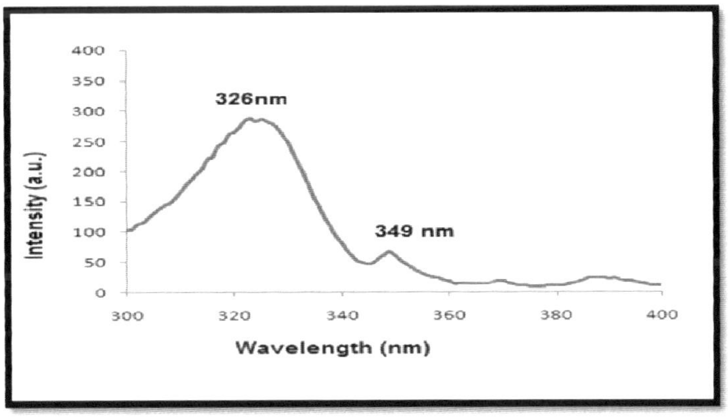

Figure 3 (a) Excitation Spectra of $K_2BiMg_2V_3O_{12}$: Dy

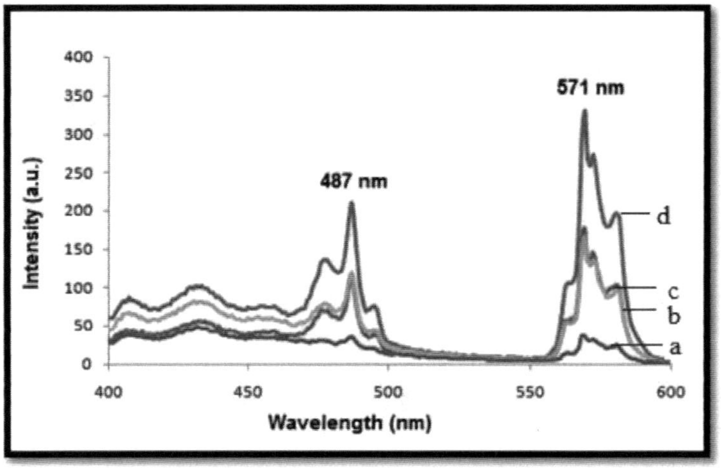

Figure 3 (b) Emission Spectrum of $K_2BiMg_2V_3O_{12}$: Dy

Fig.2 (a) shows broad-band between 290 & 340 nm is recognized to band structure by host absorption (HAB) and state of charge (CTS) owing to Dy^{3+} – O^{2-} collaborations. Longer wavelength state area (350–370 nm) weaker band of excitation likewise appears originated after transition of f–f. Fig.2 (b) & Fig.3 (b) indicate the corresponding emission spectrum of $Na_2BiMg_2V_3O_{12}$:Dy^{3+} like (a=0.1, b=0.3, c=0.5 & d=1mol %) & $K_2BiMg_2V_3O_{12}$:Dy^{3+} (a=0.1, b=0.3, c=0.5 & d=1mol %) with altered concentration of Dy^{3+} ions and it can be measured in series of 400–700 nm. It is comprehended that they have analogous features. In term of upon excitation process in the 325 nm band and in term the emission band have two broad-strong bands by 487 nm & 571 nm, respectively.

3.3. The emission transition and change in energy levels of Dy^{3+} ion:

In prepared synthesized emissions band relate to characteristic emission of Dy^{3+} outcome of transitions of ${}^4F_{9/2} \rightarrow {}^6H_{15/2}$ & ${}^4F_{9/2} \rightarrow {}^6H_{13/2}$ respectively. I found that, in luminescent properties are strongly be contingent proceeding the doping concentration in host-lattices. If concentration of dopant like Dy^{3+} ion increases up to scale of 0.5 mol% then the peak intensity are strongest & slowly peak decreases at the higher Dy dysprosium concentrations called concentration-Quenching of lanthanide-doped structure due to process of communal Dy^{3+} to Dy^{3+} interactions.

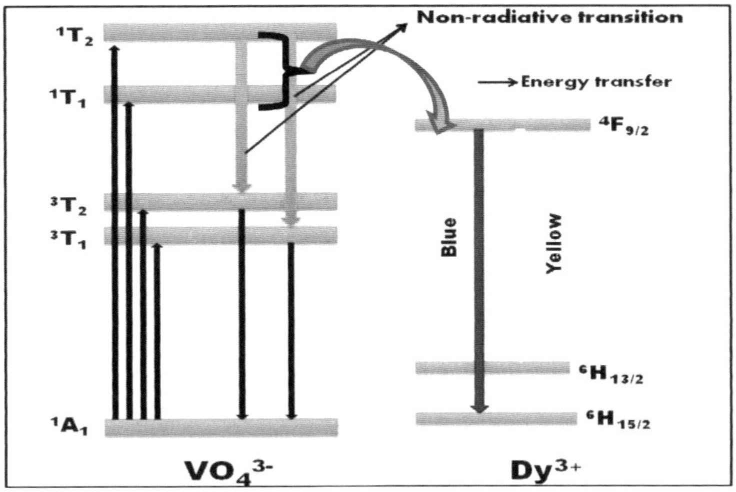

Figure 4. Energy levels diagram of Dy^{3+} ion concentration & emission-transitions.

Figure 4 represent energy levels diagram of Dy^{3+} ion and transition of emission. In this figure, it indicates the states elaborate in luminescence procedure and prospects of transition for dopant Dy^{3+} ions in $Na_2BiMg_2V_3O_{12}$ host. When Eu^{3+} ion comes in to the vicinity of VO_4^{3-} group the energy levels of both were at the same distance. According toward this model structure first excited the system since $^1A_1$ level of ground- state of VO_4^{3-} group to $^1T_2$, $^1T_1$, $^3T_2$ and $^3T_1$ excited states. At these excited levels VO_4^{3-} group transfers its energy to $^5D_0$ excited level of Dy^{3+} ion and then it comes to ground state non-radioactively. Eu^{3+} ion uses this energy which was transferred by VO_4^{3-} group for its transition after $^4F_{9/2}\rightarrow{}^6H_{15/2}$ & $^4F_{9/2}\rightarrow{}^6H_{15/2}$ levels giving blue and yellow emission lines. It indicates that Dy^{3+} emission nearby 482 nm ($^4F_{9/2}\rightarrow{}^6H_{15/2}$) is of magnetic-dipole origin & 576 nm ($^4F_{9/2}\rightarrow{}^6H_{13/2}$) is electric-dipole moment. In transition $^4F_{9/2}\rightarrow{}^6H_{15/2}$ is predominant but in only after Dy^{3+} ions are placed at very lower symmetrical sites with no transposal centers [38].

In lower symmetrical center position of doped Dy^{3+} outcomes in predominate emission state of $^4F_{9/2}\rightarrow{}^6H_{15/2}$ transition band. Meanwhile emission level at 482 nm is prime hence it can recommend that, here is very slight deviance from inversion symmetry in this matrix. In optical characteristics property of resources are frequently influenced by structure of matrix & synthesis technique [39]. Therefore, in ration of yellow and blue ratios are called as the asymmetry- ratio of Dy^{3+} and it fluctuates while discovering in dissimilar host-lattices. According to Lopez [40] et al. indicated that, for energy transfer process, oxygen vacancies doing as sensitizers to ion of rare-earth due to mixing of charge transfer states. On the other hand, excess oxygen vacancies in host would unavoidably destroy the crystalanity, which leads to quenching of the luminescence. Hence in fig.4 shows, the additional important reason is that, more Dy^{3+} entered the lattice, which can enhance the $^4F_{9/2}\rightarrow{}^6H_{15/2,\ 13/2}$ transitions of the samples greatly.

4. Conclusion

In the present study, a series of $B_2BiMg_2V_3O_{12}$:xDy^{3+} (x = 0, 0.1, 0.3, 0.5, 0.7, 1.0 mol %) (B= K & Na) phosphors were synthesized by solid state synthesis. The high purity (99.99%) starting materials were $BaCO_3$, K_2CO_3, Na_2CO_3, BiO_2, MgO_2, and NH_4VO_3. Stoichiometric amounts of reagents were thoroughly mixed and ground using an agate mortar. In Phase purity of the prepared phosphor was analysed through XRD analysis. Further, luminescence properties analysed using PL technique. The Photoluminescence excitation & the photoluminescence emission spectra of without rare earth ions doped host material shows self-luminescence properties. With the doping of Dy^{3+} ions emission intensity varied however nature of emission band is similar. In the emission spectra of photo-luminescence, it has two characteristics peaks they are situated at different region. One peak found in blue region & another peak in in yellow region of wavelength. The excitation peak is a mercury (Hg) free wavelength. Hence in this excitation peak is relevant for the development of ecofriendly lamp phosphors for purpose of solid state lighting. Hence entire investigation suggested, Synthesized Dy^{3+} doped $B_2BiMg_2V_3O_{12}$ phosphor have huge potential for future solid state lighting application.

References

[1] Unal F, Kaya C and Kazmanli K 2019 *A statistical approach Ceram. Int.* **45**, 14-19.

[2] Minami T, Kobayashi Y, Miyata T.and Yamazaki M. 2003 *Thin Solid Films*. 443, 1-2.

[3] Krishna R.H..2013 *J. Phys. Chem. C* **117**, 4-9.

[4] Jacobsohn L.G., Bennett B.L. Muenchausen R.E. Smith J.F. and Cooke D.W. 2006 *Proc. SPIE 6321,Nanophotonic Materials III, San Diego.*

[5] Pflieger R., Gravier L. Guillot G. Ashokkumar and M. Nikitenko S.I.; 2018 *Ultrason. Sonochem.* **46**, 10.

[6] Pitale S.S.,Kumar, V. Nagpure, I.M. Ntwaeaborwa, O.M. Coetsee, E.and Swart H.C. 2011 *J. Appl. Phys.* **109**,1.

[7] Unal F. and Kazmanli K. 2019 *Powder Metall. Met. Ceram.* **58**, 3-4.

[8] Zhang X. Zhao, H.L. Gao, S.and Zeng, Q.F. 2021, *J. Rare Earths*, **39**, 4.

[9] Permin D.A, Balabanov S.S. Snetkov I.L. , Palashov O.V. ,Novikova, A.V. ,Klyusik, O.N. and Ladenkov, 2020 *I.V. Opt. Mater.* **100,** 60-65.

[10] Gangwar B.P., Irusta S. and Sharma, S. 2020 *J. Lumin. 219*, 14.

[11] Wang Y.Z. ,Wen Z.C., Ye W.G., Feng Z., Zhao C. Zuo, C.D. Li Y.B. Cao Z.Q., Cao Z.J. and Cao Y.G. 2020 *J. Lumin. 221.*

[12] Ćirić, A. and Stojadinović S. 2020, *Opt. Mater* **99.**

[13] Li, Y.Z. , Cai, L.X., Yang, S.T., Guo, D. ,Peng, X.F., Zhao and Liu, Y.Z. 2020 *Sens. Actuators B* **308**,6.

[14] Yang, L.X. , Peng, D. , Shan, X. , Guo, F.W. , Liu, Y.Z. , Zhao, X.F. and Xi ao, P. 2018 *Sens. Actuators B* **254**, 578

[15] Van T.T. Bargar J.R and Chang, J.P. 2006 *J. Appl. Phys.* **100**, 2.

[16] Guo, H. and Qiao, Y.M.2009 *Opt. Mater.* **31**, 4.

[17] Rai, V.K. Pandey, A. and Dey, R. 2013 *J. Appl. Phys.* **113**, 8.

[18] Lu, Q.P., Hou, Y.B.,Tang, A.W. Liu, X.J.and Teng, F. 2013 *Mater. Lett.* **99**, 115.

[19] Hirai, T., Orikoshi, T and Komasawa, I. 2002 *Chem. Mater.* **14**, 8.

[20] Nunokawa, T. Odawara, O.and Wada, H. 2014 *Mater. Res. Express.* **1**, 3.

[21] Lojpur, V., Mancic, L. ,Vulic, P. , Dramicanin, M.D. ,Rabanal, M.E. and Milosevic, O. 2014 *Ceram. Int.* **40**, 2.

[22] Eldridge J.I. 2019 *J. Lumin.* **214**, Art. No. 116535.

[23] Mi, C.C. ,Zhang, J.P. ,Gao, H.Y.,- Wu, X.L. ,Wang, M, Wu, Y.F. ,Di, Y.Q.- Xu, Z.R. , Mao, C.B.,and Xu, S.K.

2010 *Nanoscale*.**2**, 7.

[24] Meng, F.L. Luo, Y. Zhou, Y.L. Zhang, J.W. Zheng, Y.Z. Cao, G.Z. and Tao, X. J. 2016 *Power Sources*. **316**, 207.

[25] Dey, R. Pandey, A.and Rai, V.K. 2014 *Sens. Actuators B*. **190**, 512.

[26] Yin, D.L. Wang, J Wang, Y. Liu, P. Ma, J. Xu, X.D. Shen, D.Y. Dong, Z.L. and Kong, L.B. 2020 *J. Eur. Ceram. Soc.* **40**, 2.

[27] Yanagida, T.2013 *Opt. Mater.* **35**, 11.

[28] Natarajan, V, Doable, A.R and Lu C.H. 2009 *J. Lumin.* **129**, 290.

[29] Chung, J.W.,Yang, H.K., Moon, B.K., Choi, B.C.,Jeong, J.H.and Bae Kim, J.S.2009 *Curr.Appl. Phys.* **9**,222.

[30] Singh, N.S, Ningthoujam, R.S., Singh, S.D., Viswanadh, B, Manoj, N.and Vasta, R.K. 2010 J.Lumin. **130**, 2452.

[31] Wang, N., Chen, W., Zhang, Q.F.and Dai, Y.2008 *Mater. Lett.* **62**,109.

[32] Li, X., Yu, M., Hou, Z.Y,Li, G.G., Ma, P.A., Wang, W.X.,and Cheng, Z.Y. 2011 *J. Solid State Chem.* **184**,141.

[33] Xu, Z.H., Kang, X.J., Li, C.X.,Hou, Z.Y, Zhang, C.M., Yang, D.M.and Li, 2010 *J. Inorg.Chem.* **49**,6706.

[34] Jia, G., Song, Y.H., Yang, M, Huang, Y.J.; Zhang L.H.and You, H.P. 2009 *Opt. Mater.* **31**, 1032.

[35] He ,F., Yang, P.P., Niu, N.,Wang, W.X. Gai, S.L.and Wang, D. 2010 *Colloids Interface Sci.* **343**,71.

[36] Qiao, Y.B. Liu, X.F. Zhang, Q. Chen, D.P. and Wang, Y.W. 2010 *Mater Letter.* **64**, 1306.

[37] Tomaszewicz, E., Kaczmarek, S.M.,nd Fuks, H. 2010 *Mater Chem Phys*. **122**, 595.

[38] Mohameda, R.M., Harrazb, F.A.;and Mkhalida, I.A. 2012 *J. Alloys and Compounds*. **532**, 55.

[39] Wang, P.L.and Chen, L.M. 2012 *Mater.Charecterisation.* **69**, 108.

[40] Yu, D.C., Huang, X.Y., Ye, S., Peng, M.Y.and Zhang, Q.Y. 2012 *Solar Energy Materials & Solar Cells.* **101**,303.

A Review on Barrier Properties of Nanocellulose and Polylactic acid Composites

Parvathy S U[1], Hema S[1], Malavika Sajith[1], Rashid Sulthan[1], Sreelekshmi C, Sreedha Sambhudevan[a*] and Balakrishnan Shankar[b]

[a] *Department of Chemistry, Amrita Vishwa Vidyapeetham, Amritapuri, Kerala, 690525, India*
[b] *Department of Mechanical Engineering, Amrita Vishwa Vidyapeetham, Amritapuri, Kerala, 690525, India*

E-mail: sreedha@am.amrita.edu

Abstract. Bio-based polymers, which are environmentally benign and contribute to a range of applications, are in great demand right now. The relevance of nanocellulose polylactic acid polymers, their benefits, production processes, and the wide variety of applications given by their barrier property are highlighted in this paper. Cellulose can be defined as the final product obtained as a result of biosynthesis of plants, animals, or bacteria, and nanocellulose refers to cellulosic take-outs or materials with definite nanoscale structural dimensions. Poly (lactic acid) or polylactide (PLA) is a thermoplastic polyester, which has the ability to replace petrochemical based polymers. Poly (lactic acid), often known as polylactide (PLA), is a thermoplastic polyester that can replace petrochemical-based polymers

1. Introduction

For the last two decades of twentieth century there has been a fundamental change from biostable materials to biodegradable materials for a broad range of applications including packaging, barrier applications, medical applications etc. Green composites are biomaterials made up of natural reinforcing and matrix components acquired from natural sources. [1]. Green composites such as nanocellulose-polylactic acid composites is an excellent example. Nanocellulose is a nanostructured cellulose made out of natural fibre and polylactic acid, the most widely used biopolymer [2]. The most common natural polymer is cellulose, which is a chain of glucose residues that may be easily acquired from nature. It has a lot of distinct characteristics. It has many unique properties [3]. Poly lactic acid is a biodegradable polymer that is widely utilised in a variety of applications. It is a member of the aliphatic polyester family, which is made up mostly of alpha-hydroxy acids such as polyglycolic acids [4]. PLA may be made by either condensing polylactic acid directly or polymerizing cyclic lactide polymer directly [4].

PLA and nanocellulose are referred to as "green composites," which have a significant environmental impact since they are both renewable and biodegradable. The addition of nanocellulose to a polylactic acid substrate can induce notable changes in mechanical, thermal, and thermomechanical characteristics. [5][6]. Processing a layered system of nanocellulose and PLA might balance out each other's drawbacks, resulting in packaging with better barrier qualities than the individual components. The nanosized cellulosic fillers were found to be an excellent component for barrier applications due to their strong hydrogen bonding, which forms an impenetrable network inside the composites, making it difficult for molecules to permeate through them. [7] [8]

2. Classification of Nanocellulose

Cellulose is one of the abundant natural polymers and has the strength to overcome the challenges faced by it because of its biodegradability, renewability, cost and energy. It is also known as a nontoxic material and mainly found in trees, plants, fruits, barks and leaves which constitutes the structural component of cell walls in plant tissues [9]. It is reported that the Braconnot and Payen done the first investigations on cellulose, but the clear structure was demonstrated by Staudinger in 1962 [10].

2.1 Cellulose Nanofibrils (CNFs)

It can be referred as a natural cellulosic material and comprises of high volume of cellulose fibers with large surface area having 20-60 nm diameter and micrometer lengths. They can be well distinguished through their structure due to the elongated masses of fibrils containing crystalline and amorphous domains [9]. Without regard to other cellulosic sources, they can be manufactured from fiber pulp by mechanical methods, which overcomes the inter fibrillar hydrogen bonds and produces CNFs [11]. The properties shown by the product is mainly based on the source of cellulosic material.

2.2 Cellulose Nanocrystals.

CNCs were isolated from wood pulp and cotton in the 1940s and given the names crystallites and cellulose micelles. Acid hydrolysis was used to isolate the samples. Due to features such as stiffness, optical qualities, nature, chemical purity, and most significantly, their ability to totally re-disperse from dry powder, CNCs occupy a separate area than the other two forms of nanocellulose [3]. Hongxiang Xie and colleagues investigated the use of mineral acid hydrolysis to prepare CNCs from raw cellulosic materials, with the end product demonstrating good quality and inherent properties.

2.3 Bacterial Nanocellulose.

Bacterial nanocellulose apart from cellulose nanofibrils and cellulose nano crystals is extracted from several bacterial species like Glucano acetobacter (Acetobacter xylinium) [12]. The specialties of BNCs include high purity, a nanofiber network structure and 99% water content in the form of mechanically and thermally stable hydrogel bodies [3]. BNCs are extracted by keeping these bacterial species in an aqueous cultural media for several days which contain carbon and nitrogen sources [12] [13].

3. Isolation of Nanocelluloses

Chemical, mechanical, and enzymatic approaches are the most common ways for isolating nanocellulose. Vern Yi Tan and Pui May Chou et al. evaluated the effects of chemical and mechanical approaches on nanocellulose extraction from Kapok Husk in a research. Mechanical methods were used to create fibrils with a diameter of 50-100 nm and a length of a few micrometers, while chemical and mechanical methods were used to create fibrils with a diameter of 5-20 nm. [14]. Milena Martelli et al. used enzymatic hydrolysis to separate nanocellulose from soybean straw [15] Cellulose nanofibers with a thickness of 10 nm and a length of 300 nm were created. Several research [16] [17] have employed high pressure homogenization to isolate nanocelluloses. This approach was utilized by Antonia Norio Nakagaito et al. to investigate the morphological changes in plant fiber-based composites after Nano size fibrillated cellulose was mixed on them. [18]

3.1 Alkali Hydrolysis/Bleaching

This is the most favored pre-treatment applied for hemi cellulose, lignin and pectin solubilization before the synthesis cellulose nano fibrils occur. This happens by three steps- Fibers are soaked in 12-17.5 w t% solution of sodium hydroxide for about 2 hours, then the fibers are soaked in 1 M solution of HCL at 60-80 ^{0}C. After solubilization it is again treated with 2 wt % of sodium hydroxides about 2 hours so that the lignin structure and the bonds between carbohydrate and lignin can break. This treatment overall increases the cellulose yield by 43% to 84% [19].

3.2 Mechanical Process

There are many mechanical methods for converting cellulose fibers into nanocellulose which includes high pressure homogenization, micro fluidization, grinding, high intensity ultrasonication, etc. In high pressure homogenization the cellulosic pulp is made to passthrough a small nozzle at high pressure which then decrease the size of cellulose fibers to a nanoscale one. Microfluidizers work similar to HPH, they use an intensifier pump and the interaction chamber shears and impact forces to defibrillate the fibers [5].

3.3 Enzyme Hydrolysis

Since the cellulosic fibers contain many organic compounds, a simple enzyme cannot be used for the extraction. Cello-bio-hydrolases are used for crystalline cellulose and Endoglucanases are used for disordered structure of cellulose [19].

3.4 Chemical Hydrolysis

Mainly sulfuric acid is used in chemical hydrolysis for extraction of nanocellulose because this commences the esterification on the facet of cellulose and encourages the growth of anionic sulphate ester groups. Recently TEMPO Oxidation method is dragging interaction in this field [20]. TEMPO oxidized nanocellulose are synthesized from 2,2,6,6tetramethylpiperidine-1-oxyl mediated oxidation [21] [14].

4. Properties of Nanocellulose

4.1 Thermal, Physical and Chemical Properties

Comparing with other traditional cellulose fibre, nanocellulose shows high surface area, less density and immense physical strength. The young's modulus of nanocellulose is attractive due to crystal nature and interconnection of amorphous and crystalline regions [22]. Properties of electro spun nanofibers can be improved by the pre-treatment of them called as electrospinning which is a simple technique and thus can be used in many applications [10]

4.2 Biological Properties

Nanocellulose has attained an improved desire not only because of its mechanical and thermal properties but also due to biological properties. Biocompatibility means the lack of inflammation, cyto toxicity or a negative reaction to body cells, tissues, or organs in in vivo. Another property includes cyto toxicity which is also satisfied by nanocellulose. Biodegradability is also another important property which is also 99% satisfied by the nano cellulosic materials [14].

4.3 Barrier Properties

Nanocellulose materials have ability to form stable self-standing films or coatings which has great barrier applications. It shows resistance to gas transport when it is converted to a film or when combined with plastics or paper or other materials which results in a dense multi-layer structure. Gas barrier properties of nanocellulose has great importance when it comes to packaging applications. The resistance of oxygen transport shown by these films were nearly equal to those shown by other synthetic materials. Nanocellulose shows barrier properties because of its characteristic structure which comprises of extensive inter and intra molecular hydrogen bonding. Apart from paper and plastic, when nanocellulose combines with other materials such as PLA, it gives rise to an impermeable surface for oxygen, carbon dioxide and nitrogen [23]. Barrier properties for oxygen are very high for nanocellulose when compared to those for water vapour properties since they tend to be too slow for unmodified of cellulose materials [24].

5. Polylactic Acid (PLA)

PLA is thermoplastic, biologically degraded which contributes them as a good biological material in many fields. The company, which is popular in the field of development of PLA was Cargill Incorporated. PLA is one of the developed biopolymers which has a promising characteristic for many applications [25]. It is an aliphatic polyester which is prepared solely from renewable resources [8]. It comprises of linear polymeric chains [26].

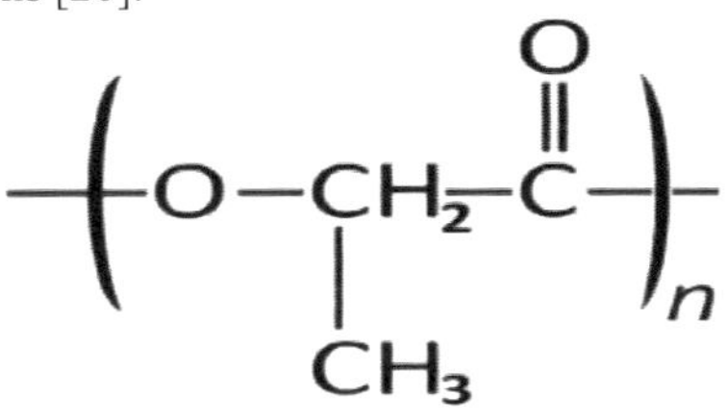

Figure 1. Structural Formula of PLA [25].

Lactic acid being a chiral molecule and has two stereoisomeric forms (L and D) [25]. Lactic acid obtained through fermentation has 90% of L isomer and by the formation of cyclic lactide dimer intermediates occur in 3 forms. The D and L forms are optically active and the D, L or meso form is inactive [9].

5.1 Preparation Methods.

Polylactic acid can be synthesized by two methods such as Direct condensation of lactic acid and Ring opening of lactide dimer.

5.1.1 Direct Condensation
This is a simple method where lactic acid monomers undergo esterification in the presence of hydroxyl and carboxyl groups and leads to polymerisation which produces water as by product. Even though water can be removed by the help of high temperature and vacuum, it again produces low molecular intermediate because of the presence of other impurities. Therefore, it was believed that this method is not possible for producing high molecular weight polymers [27].

5.1.2 Ring Opening Polymerization
This method is used to produce high molecular weight polylactic acid. The resulting product is formed from the residual mixture (by removing water) due to oligo-polymerisation. The lactic acid formed is then undergone ring opening polymerisation which uses metal alkoxides as catalysts and thus produces efficient PLA [27].

5.2 Properties of Polylactic Acid

PLA obtained and the properties exhibited by them depend upon the mixture of enantiomers of lactic acid from which the final form of PLA is synthesized. When racemic mixture is used for the formation Of PLA the meso lactide is produced first and then converted to D, L polymerlactide and then at last forms to an amorphous polymer. Most of the properties of PLA is influenced by the ratio Of D and L polymers used [27].

5.2.1 Physical Properties
The physical properties of polylactic acid is illustrated in the below table 1.

Table 1, Physical properties of PLA [27]

PROPERTY	L-PLA	DL- PLA
Glass transition temperature (Tg)	60-65	50-60
Melting Point	184^0 C	Amorphous
Specific Gradient	1.24	1.25
Tensile Strength (MPa)	55.2-82.7	27.6-41.4
Elongation (%)	5-10	3-10
Modulus (MPa)	2758-4137	1379-2758
Inherent Viscosity(dl/g)	0.90-1.2	0.55-0.75

5.2.2 Barrier Properties

According to various analysis the barrier properties shown by polylactic acid is appreciable than any other common polymers and thus can be used for many applications. Barrier properties include the permeability against gases, water vapour and atoms. The comparison below is from the study conducted by Lehermeier and group which was mentioned in Thermal and rheological Properties of commercial grade Polylactic Acids [28].

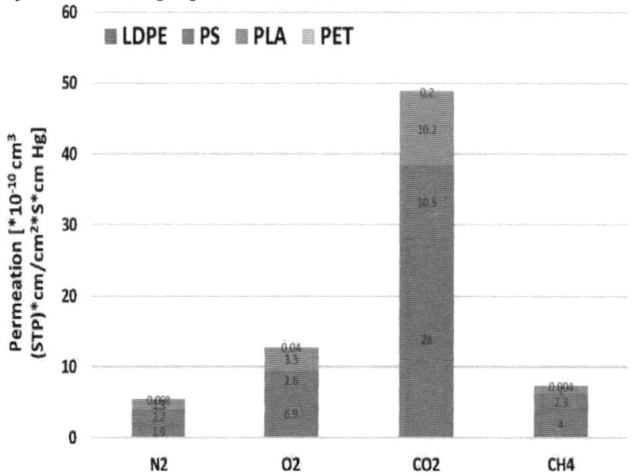

Figure 2. Comparison of barrier properties of common polymers at 30 Degree Celsius [28].

6. Barrier Properties of Nanocellulose- Polylactic Acid Composites.

Nowadays there is a great demand for the usage of biomaterials for barrier applications. But the discouraging factor behind it is most of the materials lack several characteristics needed for their adoption as one. Nanocellulose is such a material which falls under this category. Even though it has ambulant barrier for certain oils, grease and oxygen, but have a poor resistance for water vapour which makes it impossible to cope up with humidity. PLA is another such material which has barrier against water vapour but poor resistance to wards oxygen. So, by combining these two materials into coatings can supress both the shortcomings and can be raised to a product which shows fabulous barrier properties [7].

A study conducted by Rajesh Koppulu et al; [7] is focussed on combining nanocellulose and polylactic acid into multilayer coatings on a paper board. Here micro fibrillated cellulose or cellulose nanocrystals was coated to a pigment coated paper board continuously by the help of a slot die which was customised. Again, the material was coated with PLA by a plot -scale extrusion coater thus forming a multilayer coating on the paper board. LDPE coated on nanocellulose by extrusion method was taken as the reference to compare the barrier properties with those with polylactic acid -based coatings. The water vapour transmission rate was observed to be lower than those with PLA and LDPE coatings even at 90 % RH. When compared with Just PLA, nanocellulose/PLA multilayer coatings were found to reduce OTR by 98%. Similarly, a reduction occurred with HVTR also when compared with baseboard. In addition, the grease barrier of these coatings was increased by 5-fold than those with nanocellulose only and 2-fold with Polylactic acid only. Hence the perspective of processing of multilayer coatings of nanocellulose and polylactic produced a paperboard with fabulous barrier properties [7]

Another study involved the combination of cellulose nanofibrils or nanocrystals with bio based polylactic acid. Three phase multi-layered materials were made by a heat pressing process in which a dry CNF OR CNC based films was complexed with two polylactic acid sheets. CNFs and CNCs were modified with the rosin based nano emulsion even before it was processed into films. These three-phase materials showed great barrier properties against both oxygen and water vapour.
By placing nanocellulose as an inner layer between the two polylactic acid layers, the oxygen barrier property was increased. This decreased the oxygen permeability and compressed in between 84% to 96% for nanocelluloses and about 44% to 50 % for nanocelluloses with rosin emulsion which is used as the inner layer. The final product which include rosin combined nanocellulose showed high anti oxidative property and scavenging rate up to 20% [29]

Bich Nam Jung on another study focused to study the optimal properties obtained when triethyl citrate was combined with nanocellulose and polylactic acid to produce films which is also bio based. In order to find the optimum plasticizer content an analysis was done based on mechanical properties shown by the triethyl citrate when combined with PLA. Then a solvent mixture of triethyl citrate and ethanol was prepared and CNF was combined with varying values (0.25,0.5,1 w t %). Melt mixing process for about 7 minutes was used to prepare the CNF/PLA nano composites. By the results produced after conducting a Zeta potential analysis it was proved that the presence of tri ethyl citrate, CNF was dispersed more efficiently and the rate of dispersion of CNF in polylactic acid while melt mixing process shown increase also.On the other hand, the oxygen barrier property of t-CNF/PLA of 1 w t% of CNF lessen to 47.3 % when compared with neat PLA and thus proved that apart from coating method, melt processing method can also be used to produce oxygen barrier films from CNF. The tensile strength of nanocomposites was increased by 11-20% and the young's modulus by 13-68% when compared with t – PLA and hence the material produced by the study was bio based and has a fabulous barrier property also [30].

Studies by D Rigotti et al; was the explanation of barrier properties shown by polylactic acid nanocomposites by dispersing lauryl functionalised cellulose nanocrystals. Here the films were prepared by solvent casting method. Gas permeation technique was used to study the gas transport properties of composites with different filler content. Thermal desorption spectroscopy was used to analyse the released solvent materials (CHCL3). In this study they correlated the two processes and thus found out that the both barrier properties of nanocomposites and kinetics of residual solvent dispersion get controlled by LNC-PLA layer interface. The gas barrier properties were then again improved by the critical filler content of 6.5 w t% which helps to form gas permeable PLA. The results from TDS spectra analysed that the residual CHCL3 molecules get trapped in interface regions and only at high temperature they got released apart from those which are dispersed in PLA matrix. Hence the final showed great potential in barrier applications [31] Pickering emulsion was a next approach to nanocellulose polylactic acid composites. Here CNC were thoroughly dispersed into PLA by emulsion method. CNCs were prepared by HCL vaporisation method and the concentration was analysed gravimetrically which was found to be 1.01 w t % for final dispersion [32].

The composites were prepared then by Pickering emulsion in which the CNC dispersion was diluted by deionised water with concentrations 0.25,0.580,7.5. Then CHCL2 solution of PLA was added. The precipitate was vacuum filtered and composites were prepared and the solvent casting was done. The TG measurements revealed that CNC slightly increased the thermal decomposition of PLA. The resultant product was found to be thermally, mechanically, rheologically fabulous and thus, have various applications. The study carried out by E. Fortunate, and team was based on barrier and migration property of Nanocellulose Polylactic acid composites. The composites were made up by solvent casting method. CNCs and surface modifies CNC s was used and the effect caused by the modification and amount were also studied. TEM analysis was conducted, and it showed a s-CNC has a good dispersion in nanoscale level a d this indicates that the reason behind dispersion was the addition of this surfactant.

Even though there was a 34% reduction in 1 w t %of CNC s, better oxygen barrier properties were shown by all the composites formed and the nanocomposites with 2 and 3 w t % of CNCs exhibited high oxygen and water barrier properties [33]. Ali Abdul Khani et al; conducted a study on properties of Nanocellulose/ PLA Composites where the team prepared composites one with acetylated cellulose nanofiber and one with microcrystalline cellulose by solvent casting method and was compared. When taking account of these material in applying packaging Water vapour permeability and water vapour transmission rate was measured. Here PLA was dissolved in chloroform and was agitated for 8 hours at 25 degree. Then the solution is taken into grease coated mould and dried. Then the PLA nanocomposites films are prepared. The micro structure was analysed By SEM analysis of CNF- AC and MCC films [34].

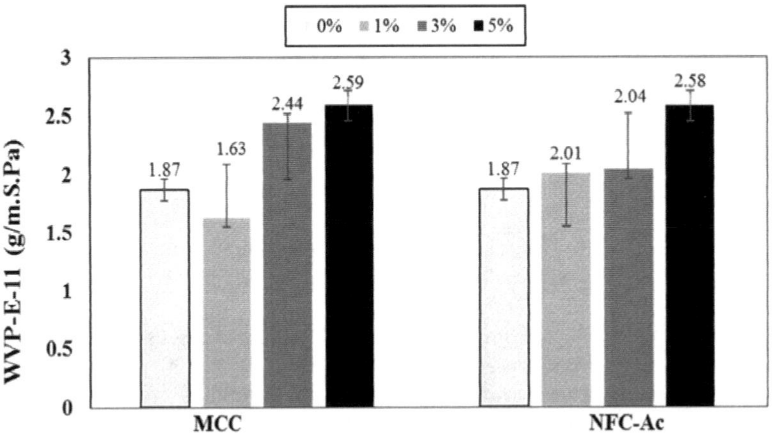

Figure 3: Impact of CNF – AC (w t%) on WVP of PLA composites [34]

From the figure 3 and 3% CNF-Ac have even distribution over PLA matrix and can thoroughly fill up the pores. Also, a noticeable increase in WVP of about 38.5 % for 5% of CNF-Ac films was observed but when compared with this WVP decreased when 1% of MCC was added to the matrix. Therefore, the reinforcement of CNF-Ac to PLA thus produced a material with appreciable barrier properties [34]. Christian Aulin and his team conducted another study on the gas barrier properties shown by the transparent films of nanocellulose combined with PLA. But in this study for the preparation of NFC, layer by layer (Lbl)method was used apart from usual procedures such as solvent casting method, filtering, or other coating techniques to produce films with excellent oxygen barrier properties. Here NFC (nano fibrillated cellulose) or CMC (carboxy methyl cellulose) was layered as multi layered thin films using PEI (polyethyleneimine) as a cationic material which is branched on PLA [35].

The rate of adsorption of components to the PLA substrate was analysed through QCM- D technique. When focussing on the barrier property shown that is both gas and water, the oxygen permeance and water vapour permanence of the films (PEI/CMC & PEI / NFC) on PLA was studied at 23 degree Celsius and 50% RH. It was observed that there is a decrease in oxygen permanence with the increase of number in layers deposited. The oxygen permanence shown by PLA coated with 20 layers of PEI/NFC was found to be 49.3cm^3 /m^2. day. atm whereas for uncoated PLA it was 151.5 cm^3 /m^2. day.atm. The permanence was again decreased when the deposition increased to 50 layers and was recorded as 9.0cm^3 /m^2. day. atm. PEI/CMC layers also exhibited similar results as that of PEI/NFC. When water permanence was analysed PEI/NFC with 20 bilayers showed 4.8 g /m^2. day.

atm. but further decreased for a 50 bilayer up to 50%. These trends were shown by PEI/CMC [33]. Thus, the resistant product that is the Lbl coated PLA / Nanocellulose film showed an increased barrier property both oxygen and water vapour properties. From the experiment value it is proved that PEI/CMC and PEI/NFC showed an order of oxygen permanence low than those with uncoated PLA and also improved the water permanence also [35].

Sandeep S Nair and his group conducted where polylactic acid composites was reinforced by nanocellulose fibrils with high lignin content in order to study the properties shown by the resultant bio composites. In this experiment Lignin was used in order to create a strong compatibility between Nanocellulose fibrils with high lignin content (NCFHL) and PLA polymer. The NCFHL/PLA composites was prepared by adding or mixing different content of NCFHL suspensions (around 5-20 t %) to PLA matrix when it is wet itself. The film was further prepared by solvent casting and hot pressing. The bio composites showed a noticeable increase in barrier, mechanical and thermal property when 5-10R% NCFHL additions was done. The addition caused decrease of 52% of WVPTR when compared to PLA alone [36]. The WVT was analysed by ASTM E-96-95 water method. The water vapour barrier efficiency showed a noticeable increased up to the addition of 10% of NCFHL in neat PLA. It is then reduced to half for the further additions. This is due to network caused by the fibrils over the films. Due to the addition it showed the presence of cellulose crystalline peak of around 22 degree within the peak of neat PLA. and this study proved that the resultant films can have a noticeable need in packaging field due to its amazing barrier, thermal and mechanical property [34].

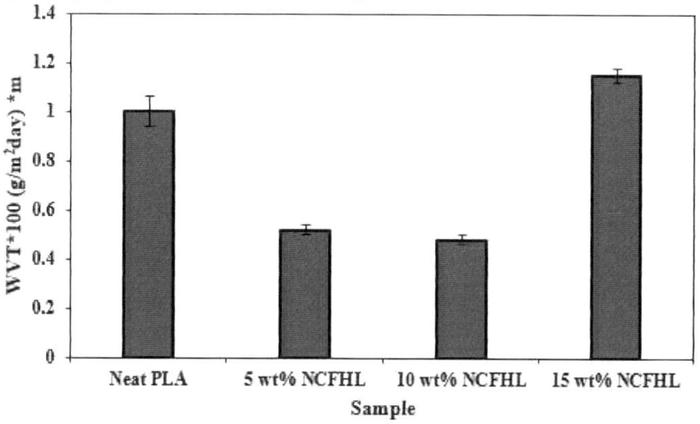

Figure 4: WVPTR of neat PLA and PLA/NCFHL composites [36].

Ritesh Kumar and his team conducted a study on barrier and mechanical properties exhibited by PLA when nano cellulosic fibrils get combined to it. In this work the films were prepared by adding specific amounts of CNF to PLA such as 1, 3, 5 w t% by solvent casting method. The green composites prepared were then analysed by methods such as SEM, X ray method and thermogravimetric method etc. It was found that embedment of CNF in PLA improved the barrier and mechanical properties of the composites. Especially the 5-w t % of CNF addition to PLA contributed the main improvements [37] From the analysis it was proved that when CNF is loaded poorly that barrier property exhibited is also very low and this can lead to low crystallinity. An improvement in barrier property was shown after the addition of 5 w t% which also improved crystallinity. Oxygen permeability also decreased with the addition of CNF and thus lead to a composite with high barrier property [37]

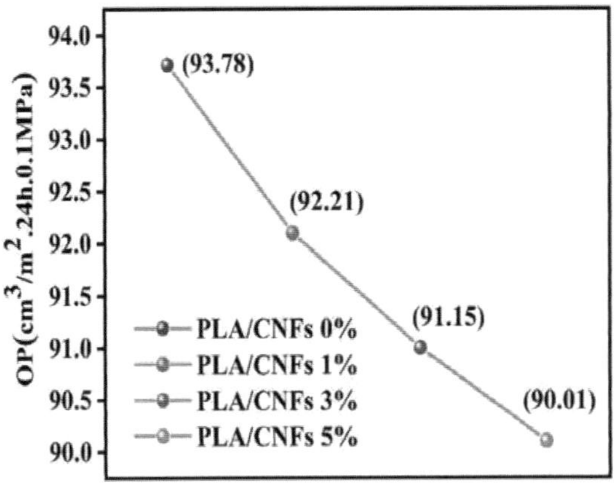

Figure 5: Oxygen permeability shown by composites at various addition of CNF [37].

Jon Trifol and team conducted a study on the barrier properties shown by hybrid polylactic acid combined with nanocellulose and nano clay to form composites. Here nanocellulose (partially acetylated CNFs or NCC) was combined with nano clay in A proportion such as 1 or 5 w t % of nanocellulose with 1,3 and 5 w t% of nano clay. And the results showed that the combination of nano cellulose and nano clay reduced the OTR up to 90% and WVPTR to 76%. The composites also showed a high film transparency due to further additions [38]. Nanocomposites with 5 w t% of C30B and 5 w t% of CNF exhibited a 90% reduction in the OTR and 76% of WVTR than the neat PLA. But when observed even at lower content filling that is at 1 w t% of C30B and CNF there was a reduction in OTR of about 74% and WVTR about 57%. And thus, makes this composite with high barrier property and makes it more advantageous in packaging field. D Rigotti, R Checchetto and team members conducted a study on properties shown by Polylactic acid lauryl functionalised nanocellulose composites. The composites were made by combining (1 to 20 w t%) of nanocellulose esterified with lauryl chains with PLA matrix. And then the gas transport and thermomechanical properties were studied. It was observed that addition of LNC particles, the gas barrier properties of PLA vs Deuterium, nitrogen and carbon dioxide increases up to 6.5 w t % of concentration of LNC and gas permeability of composites were lower than those of neat PLA alone [39].

The improvement of gas barrier properties was established by the formation of robust regions with c size around 8-10 nm surrounding the LNC particles. All the composites thus prepared exhibited good transparency and high oxygen barrier properties and was suitable for packaging applications [39]. Leire Urbina and group conducted a study by impregnating PLA to Bacterial Cellulose Membranes and analysed the barrier properties and transparency caused. Here the PLA was infused to BC. After the preparation of Bacterial Cellulose from Glucan acetobacter medellinensis, the BC/PLA-PEG films were prepared by solvent casting. PEG was used it improve the bonding between PLA and BC [40]. The water permeability of the composites was analysed by Sartorius BP210D gravimetric cell. mechanical properties were also tested by A 200 N load cell.

In order to study the barrier property of films, the effect of addition of PLA to BC substrate was analysed and WVTR was taken. Contact angle measurements was also carried out in additional to measure the barrier property shown by the films. It was found that the addition increased the hydrophobicity of the surface of the films also. Thus, impregnation of PLA-PEG to bacterial cellulose, films with noticeable barrier and mechanical properties was produced [40].

Chunhui Xiang and Nuria C. Acevedo in a work, studied about the properties shown by composites produced by combining BC with PLA/ Lipid Nanofibers (electro spun). Their study was the first to report the production of nanocomposites with in situ self -assembled BC and electro spun PLA/Lipid Fibers. In the study it was founded that the toughness of the composites increased by 52% due to the addition of Electro spun PLA to BC. It lowered the moisture gain and thus reduced water vapour transmission also. The composites were undergone for certain analysis such as SEM, XRD, WVTR etc in order to study all the properties [41]. WVTR was analysed by ASTM E 96-95 (1995b) gravimetrically [41]. From the results it was proved that materials coated with BC and electro spun PLA/ Lipid fibers will protect it from moisture and will make the nanocomposites brittle and fragile [41]

7. Conclusion

Green Composites which are nowadays a central attraction of research are the combination of natural resins with plant fibers. Green composites are preferred due to their specialities or abilities such as recyclability, renewability, flexibility, degradability, cost friendly etc.

Nanocellulose is one of the most abundant bio materials found in nature and have a wide variety of applications. When it is combined with polymers like Polylactic Acid the nature of the whole material can be changed and can be converted to films, composites or materials that has the property of both PLA and Nanocellulose and can be used for various applications. This study mainly focusses on Barrier Properties provided by these composites.

Barrier property as a whole can be defined as the barrier against oxygen, water, atoms, thermal energy or any other foreign bodies. This review contains a compilation some of the works carried out in this area by many researches and institutions which have been proved to be very successful in the aims. Due to this barrier property established by Nanocellulose Polylactic acid composites, a variety of applications are there in such as packaging field, industrial field, medical field etc. Studies are continuing in these areas and thus in upcoming future there will be a huge rise in the field of green composites and its implementations in all fields.

REFERENCES

[1] Abhemanyu, P. C., Prassanth, E., Kumar, T. N., Vidhyasagar, R., Marimuthu, K. P., & Pramod, R. (2019, March). Characterization of natural fiber reinforced polymer composites. In *AIP Conference Proceedings* (Vol. 2080, No. 1, p. 020005). AIP Publishing LLC.

[2] Murali, M., Finosh, G. T., & Jayabalan, M. (2013). Studies on biodegradable polymeric nanocomposites based on sheet molding compound for orthopedic applications. *Advances in Materials Science and Applications*, *2*, 73-87.

[3] Hema, S., Krishnan, A., Akther, A., Suresh, A., Sambhudevan, S., & Shankar, B. (2021). Green nanocomposites based on natural rubber latex containing xylan from sugarcane bagasse–Synthesis, characterization and dye absorption studies. *Materials Today: Proceedings*, *46*, 2950-2954.

[4] R. Z. Khoo, H. Ismail, W.S. Chow, 2016, Thermal and Morphological Properties of Poly (lactic acid), Procedia Chemistry 19, 788-794.

[5] Mokhena, T.C.; Sefadi, J.S.; Sadiku, E.R.; John, M.J.; Mochane, M.J.; Mtibe,2018, A. Thermoplastic processing of PLA/Cellulose nanomaterials composites. Polymers10, 1363

[6] Hossain, K.M.Z.; Hasan, M.S.; Boyd, D.; Rudd, C.D.; Ahmed, I.; Thielemans, W., 2014, Effect of cellulose nano whiskers on surface morphology, mechanical properties, and cell adhesion of melt-drawn polylactic acid fibers. Biomacromolecules 15, 1498–1506.

[7] Rajesh Koppolu, Johanna Lahti, Tiffany Abitbol, Agne Swerin, Jurkka Kuusipalo, Martti Toivakka,2019, Continuous Processing of Nanocellulose and Polylactic Acid into multilayer Barrier coatings, ACS Applied materials and interfaces,11, 11920-11927

[8] Jianxiao Lu, Chuanyue Sun, Kexin Yang, Kaili Wang, Yingyi Jiang, Rogers Tusiime, Yun Yang, Fan Fan, Zeyu Sun, Yong Liu, Hui Zhang, Keqing Han, Muhuo Yu, 2019, Properties of Polylactic Acid Reinforced by Hydroxyapatite Modified Nanocellulose, Polymers,11, 1009.

[9] Katrina Pul Yee Shak, Yean Ling Pang, Shee Keat Mah, 2018, Nanocellulose: Recent advances

and its prospects in environmental remediation, BEILSTEIN Journal of Nanotechnology,9, 2479-2498

[10] Trache D, Tarchoun AF, Derradji M, Hamidon TS, Masruchin N, Brosse N, Hussin MH, 2020, Nanocellulose: From Fundamentals to Advanced Applications, Journal of Frontiers in Chemistry,8:392.

[11] Yan Xue, Zihao Mou, Huining Xiao,2017, Nanocellulose as sustainable biomass material: Structure, Properties, Present Status and Future Prospects in Biomedical applications, Royal society of Chemistry.

[12] Amjad Farooq, Mohammed Kayes Patoary, Meiling Zhang, Hassan Mussana, Mengmeng Li, Muhammad Awais Naeem, Muhammad Mushtaq, Aamir Farooq, Lifang Liu, 2020, Cellulose from sources to nanocellulose and an overview of synthesis and properties of nanocellulose/zinc oxide nanocomposite materials, International Journal of Biological Macromolecules,154,1050-1073.

[13] D. Klemm, F. Kramer, S. Moritz, T. Lindström, M. Anker forks, D. Gray, et al., 2011, Nanocelluloses: a new family of nature-based materials, Angew. Chem. Int. Ed. 50 , 5438–5466.

[14] Vern Yi Tan, Pui May Chou, 2020, Effectiveness of Using Different Mechanical Methods for Extraction of Nanocellulose from Kapok Husk, AIP publishing, 978-0-7354-1992-6.

[15] Milena Martelli-Tosia, Maraiane M. Massonb, Natália C. Silvab, Bruno S. Espostob, Taís T. Barrosc, Odílio B.G. Assisc, Delia R. Tapia-Blácidob,2018, Soybean straw nanocellulose produced by enzymatic or acid treatment as a reinforcing filler in soy protein isolate films, Carbohydrate Polymers 198 ,61–68

[16] Iwamoto S, Nakagaito AN, Yano H, Nogi M, 2005, optically transparent composites reinforced with plant fiber-based nanofibers. Appl Phys A Mater Sci Process, 81:1109–1112.

[17] Eldho Abraham, B. Deepa, L.A. Pothen, J. Cintil, S. Thomas, M.J. John, R. Anandjiwala, S.S. Narine, 2013, Environmentally friendly method for the extraction of coir fiber and isolation of nanofiber. Carbohydrate polymers 92, 1477-1483

[18] Nakagaito AN, Yano H, 2004, The effect of morphological changes from pulp fiber towards nano-scale fibrillated cellulose on the mechanical properties of high-strength plant fiber based composites. Appl Phys A Mater Sci Process, 78:547–552

[19] Mohammed Nasir, Rokiah Hashim, Othman Sulaiman, Mohd Asim, 2017, Nanocellulose: Preparation methods and applications, Cellulose – reinforced Nanofiber Composites, 261-276

[20] Mohammed Nasir, Rokiah Hashim, Othman Sulaiman and Mohd Asim, Nanocellulose; Preparation methods and applications, Cellulose-Reinforced Nano fibre Composites.

[21] Shen, X.J, Huang, P.L, Chen, J.H, Wu, Y.Y, Liu, Q.Y and Sun, R.C, 2017, Comparison of acid-hydrolysed and TEMPO – oxidised nanocellulose for reinforcing alginate fibers, BioResources 12(4),8181-8198.

[22] Neftali L.V. Careeno, Ananda M. Barbosa, Bruno S. Noremberg, Mabel M.S. Salas, Susana C.M. Fernandes, Jalel Labidi,2017, Advances in Nanostructured Cellulose-based Biomaterials, Springer Briefs in Applied Sciences and Technology.

[23] Ali H. Tayeb, Mehdi Tajvidi, Douglas Bousfield ,2020, Sustain. Chem, 1, 198–208 [29] Ana Ferrer, Lokendra Pal, Martin Hubbe, 2016, Nanocellulose in packaging: Advances in barrier layer technologies, Industrial Crops and Products.

[24] Ana Ferrer, Lokendra Pal, Martin Hubbe, 2016, Nanocellulose in packaging: Advances in barrier layer technologies, Industrial Crops and Products.

[25] Zeeshan Sheikh, Shariq Najeeb, Zohaib Khurshid, Vivek Verma, Haroon Rashid, Michael Glogauer, 2015, Biodegradable Materials for Bone repair and Tissue Engineering Applications, Materials ,8,5744-5794

[26] Ray E, Drumright, Patrick R, Gruber, David E. Henton, 2000, Polylactic Acid Technology, Advanced Materials, 12, No. 23.

[27] Pranas Vitkevicius, 2017, Synthesis of polylactic acid, Master's Thesis in Chemical Engineering.

[28] Shady Farah, Daniel G. Anderson, Robert Langer, 2016, Physical and mechanical properties of

PLA, and their functions in widespread applications – A comprehensive review, Advanced drug delivery reviews 107,367-392.
[29] Manon Le Gars, Benjamin Dhuiege, Aurore Delvart , Mohamed N. Belgacem, Karim Missoum, and Julien Bras, 2020,High barrier and antioxidant poly(lactic) acid/ Nanocellulose multi layered materials for packaging , ACS omega ,2020,22816-22826
[30] Bich Nam Jung, Hyun Wook Jung, Dong Ho Kang, Gi Hong Kim, Miji Lee, Jin Kie Shim, Sung Wook Hwang, 2020The fabrication of flexible and oxygen barrier cellulose nanofiber/polylactic acid nanocomposites using cosolvent system, Journal of applied polymer chemistry.
[31] D. Rigottia, A. Pegorettia, A. Miotellob, R. Checchettob, 2020, Interfaces in biopolymer nanocomposites: Their role in the gas barrier properties and kinetics of residual solvent desorption, Applied Surface Science 507,145066. Yunchong Zhang, Lu Cui, Hong Xu Xueling Feng, Bijia Wang, Béla
[32] Pukánszky,Zhiping Mao, Xiaofeng Sui, 2019, Poly (lactic acid)/nanocellulose nanocrystal composites via the Pickering emulsion approach: Rheological, Thermal and mechanical properties, International Journal of Biological Macromolecules,137,197-204.
[33] E. Fortunati, M. Peltzer, I. Armentano, L. Torre, A. Jiménez, J.M. Kenny, 2012, Effects of modified cellulose nanocrystals on the barrier and migration properties of PLA nano-bio composites, Carbohydrate Polymers,90,948-956.
[34] Ali Abdulkhani, Jaber Hosseinzadeh, Saeed Dadashi and Mohammad Mousavi, A Study Of Morphological, Thermal, Mechanical and Barrier properties of PLA based bio composites prepared with micro and nanosized cellulosic fibers.
[35] Christian Aulin, Erdem Karabulut, Amy Tran, Lars Wagberg and Tom Lindström, 2013, Transparent Nano cellulosic Multilayer Thin Films on Polylactic Acid with Tunable Gas Barrier Properties, ACS Applied Materials, Interface
[36] Sandeep Sudhakaran Nair, Heyu Chen, Yao Peng, Yanhui Huang, and Ning Yan, 2018, Polylactic acid bio composites reinforced with nanocellulose fibrils with high lignin content for improved mechanical, thermal and barrier properties, ACS Sustainable Chemical Engineering
[37] Ritesh Kumar, 2019, Effect of nano-cellulosic fiber on mechanical and barrier properties of Polylactic acid (PLA) green nanocomposite, Mater. Res. Express
[38] Jon Trifol, a David Plackett, Cecile Sillard, Peter Szabo, Julien Brasc and Anders E Daugaarda, 2016, Hybrid poly (lactic acid)/nanocellulose/ nanoclay composites with synergistically enhanced barrier properties and improved thermomechanical resistance Polymer Interfaces ,65: 988–995
[39] D. Rigotti, R. Checchetto, S. Tarter, D. Caretti, M. Rizzuto, L. Fambri, A. Pegoretti, 2019, Polylactic acid-lauryl functionalized nanocellulose nanocomposites: Microstructural, thermomechanical and gas transport properties, eXPRESS Polymer Letters, Vol.13, No.10,858-876.
[40] Leire Urbina, Itxaso Algar, Clara Garcıa-Astrain, Nagore Gabilondo, Alba Gonzalez, Ma Angeles Corcuera, Arantxa Eceiza, Alona Retegi, 2016, Biodegradable composites with improved barrier properties and transparency from the impregnation of PLA to bacterial cellulose membranes , Journal of applied Polymer science
[41] Chunhui Xiang and Nuria C. Acevedo, 2017, In Situ Self-Assembled Nanocomposites from Bacterial Cellulose Reinforced with Eletrospun Poly (lactic acid)/Lipids Nanofibers, Polymers 9, 179.

A Framework Towards Realization of Smart Manufacturing Systems

Sachin Karadgi
KLE Technological University, BVB Campus, Hubballi – 580031, India
E-mail: sachin.karadgi@kletech.ac.in

Abstract. Germany's National Academy of Science and Engineering (acatech) published its proposals for implementing the key initiative Industry 4.0 in 2013, requiring horizontal integration and vertical integration within and across multiple enterprises and end-to-end digital integration across the product lifecycle. Likewise, a smart manufacturing system emphasizes enhancing the capabilities of manufacturing enterprises considering multiple objectives, like resource utilization and productivity, necessitating the realization of the business cycle for supply chain management, product development lifecycle, and production system lifecycle. However, realizing these individual lifecycles and integrating them as part of a smart manufacturing system is not straightforward due to manifold reasons (e.g., difficult to define the interface points necessary to interact with the various systems associated with these lifecycles). The current article elaborates a systematic framework considering these lifecycles to realize a smart manufacturing system. The framework is divided into different layers starting from the process layer at the bottom all the way up to the smart layer at the top. Finally, a use case from the end-to-end additive manufacturing process has been discussed that employs the previously elaborated framework.

1. Introduction

Germany's National Academy of Science and Engineering (acatech) published its proposals for implementing the key initiative Industry 4.0 in April 2013, which laid the path for the adoption of Industry 4.0 by numerous industries [1]. The technological evolution (e.g., embedded systems, cloud computing) has a decisive influence on manufacturing, designated as Industry 4.0 [1]. Industry 4.0 was introduced to focus on data to efficiently manage the business and manufacturing processes. These require bringing CPS to production and logistics, i.e., Cyber-Physical Production Systems (CPPS) [1, 2]. Similarly, the Internet of Things (IoT) was introduced as the Industrial Internet of Things (IIoT) to enhance the industrial processes, especially located on the shop floor.

A manufacturing enterprise can be viewed as consisting of hierarchical levels. These levels can be categorized as strategic, tactical, operational, and resource levels [3], as illustrated in Figure 1. Each of these levels has pre-defined functionalities. Successful implementation of Industry 4.0 requires the following features considering the hierarchical levels [1]:

- Horizontal integration through value networks: A value network consists of multiple enterprises to fulfill customers' requirements. It is crucial to integrate various IT systems, especially located at a strategic level. These systems manage different stages of the business and manufacturing processes.

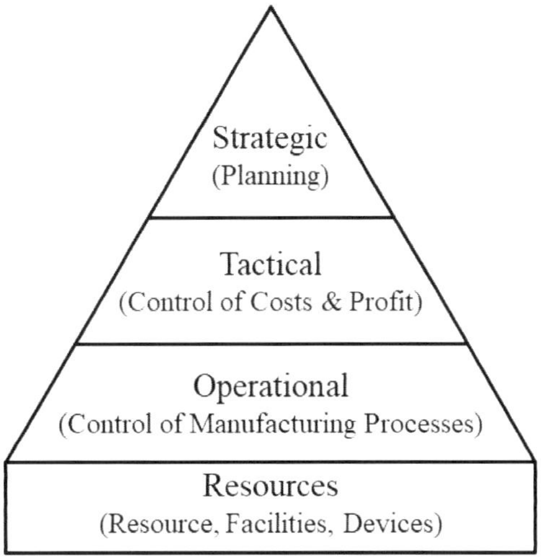

Figure 1. Hierarchical layers of an enterprise, adopted from [3].

- End-to-end digital integration of engineering across the entire value chain: It is indispensable to realize end-to-end digital integration of the digital world and real-world encompassing the value chain. Here, the integration starts with the engineering process and ends with services.
- Vertical integration and networked manufacturing system: Industry 4.0 requires integrating various IT and automation systems at different hierarchical levels of an enterprise or across several enterprises to realize end-to-end services with the focus to fulfill customer needs.

A cloud-based IoT platform was suggested [4], as illustrated in Figure 2. Likewise, Reference Architecture Model for Industrie 4.0 (RAMI 4.0) has been proposed that addresses the features mentioned above [5]. Finally, there is a discussion about fully developing the digital economy encompassing new business models and ecosystems and fully autonomous systems [4]. A smart manufacturing system, similar to Industry 4.0, focuses on maximizing the capabilities of manufacturing enterprises considering multiple objectives [6, 7]. These require implementing the business cycle for supply chain management, product development lifecycle, and production system lifecycle [7].

Industry 4.0 has been widely researched and adopted since its introduction under different synonyms. However, companies fail to appreciate the benefits of Industry 4.0, which is mainly due to the lack of transparency in identifying the benefits. This lack of transparency is also true in the case of Small and Medium-Sized Enterprises (SMEs) with the additional issue of lack of access to investments [8, 9]. Likewise, realizing the individual lifecycles of a smart manufacturing system and further integrating them as part of a smart manufacturing system is not straightforward due to manifold reasons [6, 7]. Subsequently, the Industry 4.0 maturity index has been introduced to assist companies in determining the maturity of their transformation journey [8]. Likewise, a readiness report for the future of production assessment has been carried out for different countries[10, 11].

The current article aims to present underlying concepts of smart manufacturing systems and subsequently present a framework to combine these concepts to realize smart manufacturing systems. The article is structured as follows. Section 2 introduces various concepts of smart manufacturing systems, more specifically from the perspective of smart manufacturing systems lifecycles. Next, Section 3 elaborates a framework to realize smart manufacturing systems. Next, a use case that employs the previously elaborated framework from the end-to-end additive

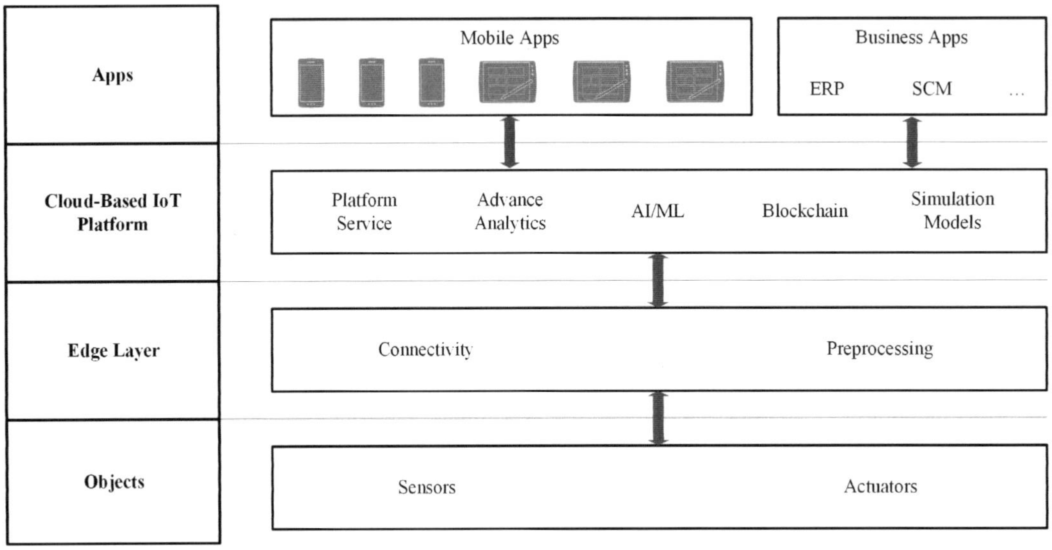

Figure 2. Cloud-based IoT platform, adopted from [4].

manufacturing process is discussed in Section 4. Finally, conclusions and possible next steps in smart manufacturing systems are highlighted in Section 5.

2. Smart Manufacturing Systems - Lifecycles and Dimensions

Smart manufacturing systems focus on using ICT and intelligent systems to enhance the use of resources, materials, labor, and so forth, and at the same time, respond to fluctuation in the market to fulfill customers' needs [7]. A smart manufacturing system is an "attempt to maximize those capabilities by using advanced technologies that promote rapid flow and widespread use of digital information within and different manufacturing systems" [7].

A smart manufacturing ecosystem is identified based on productivity, agility, quality, and sustainability capabilities, with an ecosystem encompassing business, product development, and production system lifecycles [7]. Similarly, the functional dimensions coverage on a smart manufacturing system consists of the digital thread, smart factory, and value chain management dimensions [6]. A smart manufacturing system's lifecycles and dimensions overlap, as depicted in Figure 3. Furthermore, these lifecycles converge at operations, more specifically when the value gets added [7, 6]. The end-to-end digital integration of the real-world and digital world encompassing the value chain is crucial. This requires integrating Enterprise Resource Planning (ERP) systems along the business cycle, Product Life Cycle (PLM) system along the digital chain, and Manufacturing Execution Systems (MES) along the production lifecycle to provide the necessary context for manufacturing enterprises [6].

2.1. Business Lifecycle and Value Chain Management

The focus of the business lifecycle is on control of profits and costs. This lifecycle interacts with different stakeholders (e.g., suppliers, customers) to support various management processes (e.g., supplier management) [6, 7]. Furthermore, standards play a critical role during interactions with various stakeholders (e.g., APICS SCOR [12], IEC 62264-3 [13]). ERP systems supported by Supply Chain Management (SCM) and Customer Relationship Management (CRM) systems are employed to fulfill the processes and functionalities of the business lifecycle. Integrating these systems and their interaction will assist in realizing horizontal integration at the strategic level.

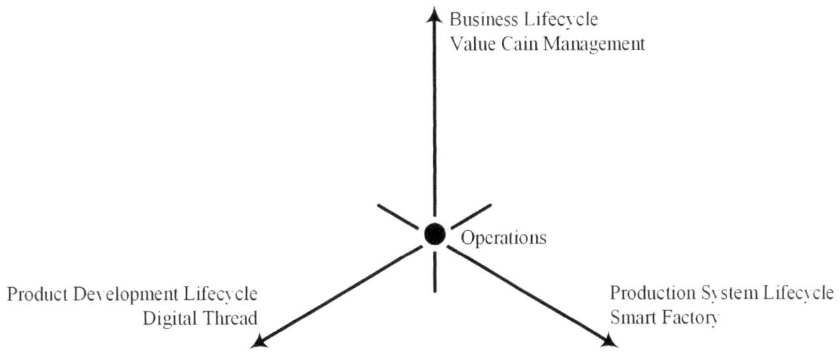

Figure 3. Three dimensions of smart manufacturing systems.

The business lifecycle provides necessary context, which can be related to the product development and production system lifecycle. This context can be employed to realize traceability, and computing performance metrics, among others. However, this realization is difficult, mainly because interface points for information exchange between the ERP system and other systems need to be exclusively defined. These points, in turn, depend on various factors (e.g., processes employed, manufacturing strategies). Nevertheless, standards support exchanging information between various systems but are rarely implemented. For example, IEC 62264-2 lists the possible information exchange between the ERP system and MES [14].

2.2. Product Development Lifecycle and Digital Thread

The product development lifecycle focuses mainly on design planning, engineering, use, service, end-of-lifecycle, and recycling [7, 15]. These activities are supported by various systems, like Computer-Aided (CAx) Technologies and Product Lifecycle Management (PLM). These activities can be assigned either to specifications management or product service management [6]. Industry 4.0 highlights the significance of end-to-end digital integration of engineering across the entire value chain, represented as digital thread, and will provide a single source of truth about products [6].

The digital definition of each product is made available and updated through feedback from different activities of specification and product service management [6]. These activities are performed by heterogeneous systems, stressing the need for product data exchange standards among various CAx and manufacturing systems. Here, multi-part ISO 10303 or STandard for the Exchange of Product model data (STEP) standards is available to address these issues [16]. Nonetheless, proprietary context information is not supported during the exchange.

2.3. Production System Lifecycle and Smart Factory

The production system lifecycle focuses on designing machine tools, building, commission, operations, maintenance, decommission, and recycling [7]. Similarly, the smart factory dimension focuses on operations and related resources available on the shop floor [6]. The main functionalities include data acquisition, traceability, performance measurement, monitoring and control of manufacturing processes, quality, inventory, and maintenance [6, 13]. Enterprise integration plays a crucial role in fulfilling of delivering the right information to the right persons at the right place and right time [17]. For example, the digital thread along with the real-time acquired data from physical assembly line assist in realizing the digital twin, which in turn support semi-physical commissioning of assembly line [18].

Research has been carried out to define frameworks to realize enterprise integration [19, 20]. The integration needs to be realized horizontally and vertically in a given enterprise, and MES

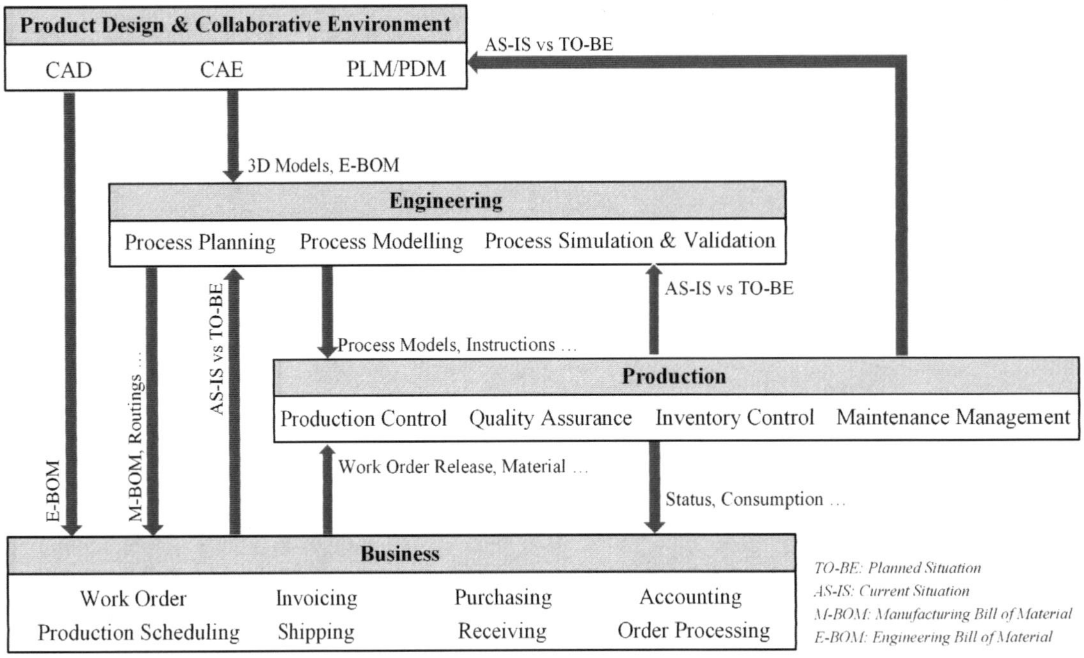

Figure 4. Flow of information among different enterprise applications, adopted from [24].

supports this integration. In contrast, Industry 4.0 discusses horizontal integration and vertical integration across the value chain and value networks [1]. The research on enterprise integration has cumulated in defining ISO 15704 [21]. ISO 15704 lists the requirement for enterprise reference architecture and methodologies with necessary guidelines for realizing enterprise integration. This standard defines particular and partial reference models to realize an integrated enterprise [14]. IEC 62264 (or ISA-95) defines reference architecture for discrete industry [13, 14, 22] and IEC 61512 (or ISA-88) defines reference architecture for batch/process industry [23].

3. Methodology

A smart Manufacturing ecosystem encompasses the business, product development, and production system lifecycles [7]. The information flow between various systems associated with these lifecycles is complicated, as illustrated in Figure 4 (adopted from [24]). The main reason would be that it is difficult to define the interface points necessary to interact with the systems located on different dimensions compounded by temporal and semantic challenges. The definition of interface points gets complicated when different production strategies are involved (e.g., make-to-order, make-to-stock). Similarly, it is cumbersome to manage these points as the business and manufacturing processes are modified to reflect the internal and external situations.

A layered framework has been envisioned to realize a smart manufacturing system, as depicted in Figure 5. For simplicity, arrows showing flow information and control data are not depicted. The following paragraphs elaborate on the individual layers.

- Process Layer: Process can denote the process itself, operations, activities, tasks, and so forth. The external environment is dynamic, which requires their internal processes to be flexible. However, enterprises address this situation by haphazardly modifying their internal processes. These processes will be untenable after some time. Nonetheless, managers take these processes along the transformation journey. The only way to proceed is to simplify and standardize the processes before embarking on a smart manufacturing system journey. It is

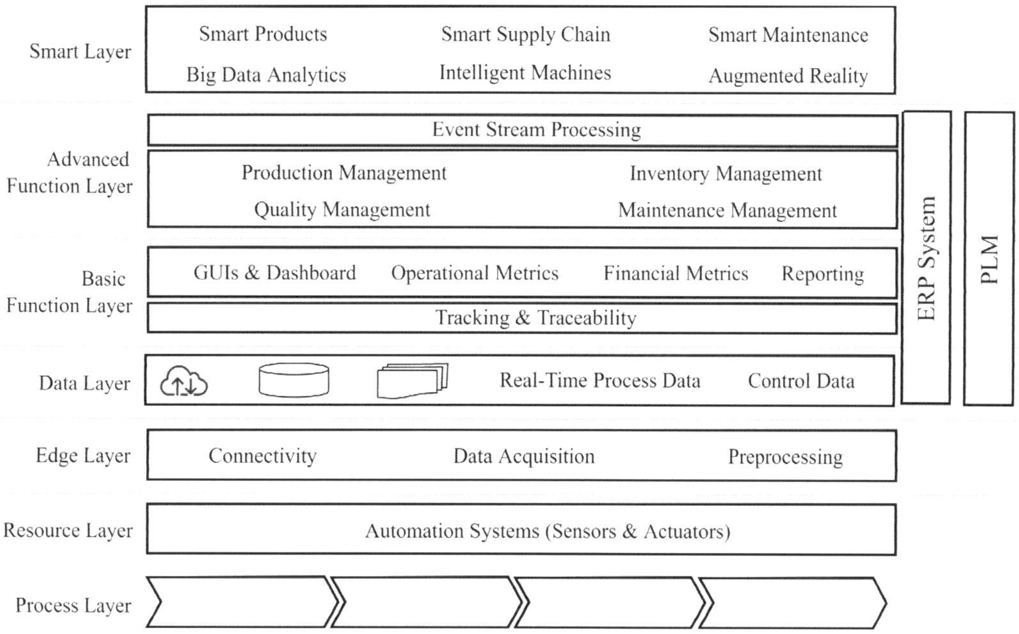

Figure 5. Framework for the realization of smart manufacturing systems.

crucial to model the processes using suitable techniques (e.g., ARIS, Event-Driven Process Chain (EPC)) [25]. If required, process analysis and re-design need to be carried out, including Business Process Reengineering (BPR) [26]. Likewise, data models using Unified Modeling Language (UML), Data Flow Diagrams (DFDs), Entity Relationship Diagrams (ERDs), and so on need to be developed to understand the business and manufacturing processes [27]. The data and process models are necessary to define the interface points for integrating smart manufacturing system lifecycles.

- Resource Layer: Resources indicate automation systems encompassing sensors and actuators. These systems can also contain industrial PCs, Programmable Logic Controllers (PLCs), and so forth. Additionally, these systems implement communication protocols for bi-directional communication. Automation systems implement protocols (e.g., OPC (Open Platform Communication) [28], TCP/IP, MTConnect [29], MQTT [30]), which might be standard or proprietary. It is crucial to rely extensively on standards to overcome the interoperability issues toward realizing integrated enterprise.
- Edge Layer: Resources need to be connected to other business and manufacturing applications. In many cases, the connection is established through edge nodes. The cost to realize the connectivity through edge nodes is directly dependent on automation systems and underlying controllers, which depend on the communication protocols employed. Edge nodes provide three-fold functionalities – pushing sensor data to higher-level systems, transferring control data from higher-level systems to controllers, and establishing a short-closed feedback loop.
- Data Layer: Data collected from the edge layer need to be stored to realize various downstream functionalities (e.g., product tracking and genealogy, computation of performance metrics). Additionally, there can be a possibility of collecting data from an ERP system, PLM, and other business and manufacturing applications that depend on the previously discussed interface points between applications. These applications can be located in different enterprises. Overall, the collected data from various sources need to be

stored either in a local database or in the cloud to provide a good context and avoid data islands. Also, control data must be stored before data is transmitted to controllers. It is crucial to design a data model relying extensively on available standards. The design of the data model has an impact on numerous functionalities of previously mentioned layers and upcoming layers and the performance of the overall system.

- Basic Function Layer: This layer provides a holistic overview of an enterprise or the value network through various functionalities. Tracking and traceability are important as they establish numerous relationships between enterprise entities (e.g., machines, quality) [31, 32]. This relationship will assist in performing performance analysis, recall, and so forth. Enterprises should have a strong analytic culture, which plays a vital role in defining financial and operations metrics as in ISO 22400 [33, 34], VDMA 66412 [35, 36], or any other standards. The data can be utilized to present the situation using graphical components. Similarly, detailed offline reports can be generated considering roles and responsibilities. Overall, the basic function layer provides a holistic overview of an enterprise or the value network.
- Advanced Function Layer: Actions need to be initiated, mostly by operators and supervisors, depending upon the situation. As mentioned earlier, IEC 62264-3 has identified production, maintenance, quality, and inventory operations management [13]. Hence, it is necessary to provide tools (or user interfaces) to manage production, quality, inventory, and maintenance. Additionally, it is necessary to automate the previously mentioned actions, which can be taken near real-time. Here, Complex Event Processing (CEP) [37, 38], instead of Rule-Based System (RBS) [39], can play a major role. One of the major issues in realizing vertical integration is the temporal gap, which CEP attempts to address.
- Smart Layer: The layers above provided opportunities to fine-tune the processes and activities, but with the involvement of operators, supervisors, and so on. However, there is a limit to which operators, supervisors, and so on can influence the processes and activities. This is where Industry 4.0 and smart manufacturing systems come into the picture to optimize internal processes, focusing on customers and adapting to external market conditions [7]. Here, huge strides in algorithms and applications have been made to analyze structured, semi-structured, and unstructured data. Furthermore, the situation in the internal and external environments keeps on changing. Thus, enterprises must have access to different smart services.

4. Use Case - End-to-End Additive Manufacturing Process

The different processes of Additive Manufacturing (AM) are depicted in Figure 6. Part design deals with fulfilling the functional requirements and is supported with different CAx (e.g., CAD – Computer-Aided Design) and PLM [40]. The outputs of this process are design parameters (e.g., wall thickness, orientation) and act as inputs for the downstream processes. In parallel, material needs to be designed to complement the part design and its functional requirements. The outputs of this process are material specifications (e.g., chemical composition, grain size).

The design parameters, material specifications, and selected 3D printer are the inputs for the process plan. The outputs of the process plan are process parameters (e.g., layer deposition details, nozzle temperature). Here, software tools provided by the 3D printer vendors are employed. Also, commercial simulation tools can be used to simulate the build process. The build process involves setting up the 3D printer as per the process parameters, printing, and post-processing. The outputs of the build process are real-time process data. This data can be classified as 1D (e.g., scanning speed), 2D (e.g., optical images), and 3D (e.g., tomography) data depending upon the sensors available with the printer [41]. Furthermore, the data can be categorized as structured, quasi-structured, and unstructured [42]. The printed part is subjected

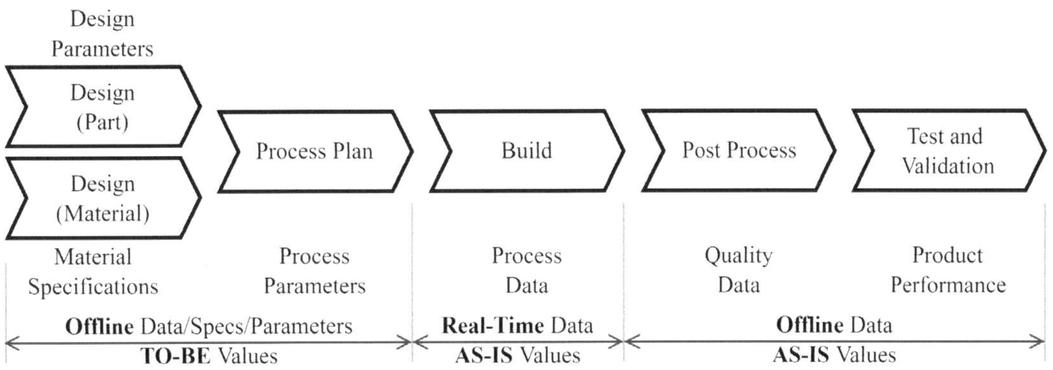

Figure 6. End-to-End additive manufacturing process.

to various quality checks in the post-process that generate quality data. Finally, the product will be tested and validated to check the part performance.

The above-elaborated reference framework for realizing a smart manufacturing system can be adapted to realize in-situ monitoring and control of the end-to-end AM process. The process layer indicates the end-to-end additive manufacturing process. The resource layer encompasses 3D printers and other metrological/laboratory devices, specifically fulfilling the build, post-process, and test and validation processes. The printers and devices must be equipped with suitable sensors and drivers. The corresponding data can be read and written in real-time using available communication protocols. The edge layer is concerned with establishing bi-direction communication with the printers and devices in the resource and data layers.

The end-to-end AM process generates a huge amount of data denoting TO-BE values (e.g., parameters, specifications, process data) and AS-IS values (e.g., temperature, bed temperature). These data can be obtained offline (e.g., specifications), and the remaining data (e.g., process data) can be acquired in real-time/online. It is crucial to define relationships among different data types, as illustrated in Figure 7. This will support traceability, compute performance metrics, and generate reports for root cause analysis. Data can also be acquired from manufacturing support systems, especially part designs and material designs.

The basic function layer will display the operational metrics and actual process data (i.e., AS-IS values) from printers in real-time and the planned parameters and specifications (i.e., TO-BE values). This will provide operators with a comprehensive view of the printing processes. The rules-of-thumb and pre-defined events can be encoded as patterns in CEP to realize in-situ monitoring and control of the end-to-end AM process. For example, the incoming real-time process data can be analyzed with the help of patterns to modify the nozzle temperature and bed temperature proactively. Likewise, the incoming real-time process data can be analyzed to predict the quality of the build process and proactively imitate the corrective actions to improve the build quality.

The end-to-end AM process generates enormous data, encompassing parameters, specifications, process data, and quality data. AI/ML algorithms can generate valuable insights when employed on these data [40, 43, 44]. As part of the smart layer, numerous services can be provided to address the different issues related to the end-to-end AM process. For example, topology optimization can be made available that employs a neural network. Similarly, the data and suitable AI/ML algorithms can predict a new material for given requirements. Likewise, it should be possible to predict the quality of the printed part [41] or predict the degradation of material using Deep Learning (DL) models [43]. Overall, the data can be employed along with a suitable combination of AI/ML to assist the various stages of the end-to-end AM process.

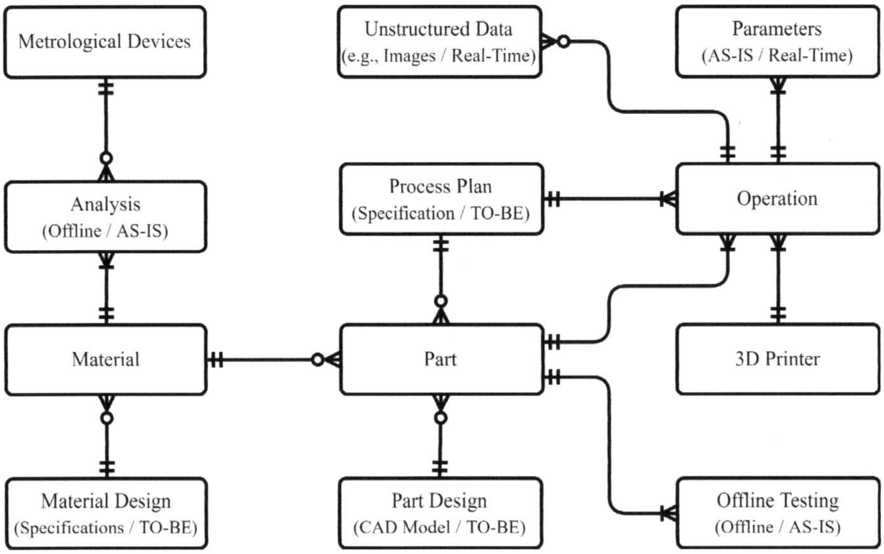

Figure 7. Relationship among various entities of end-to-end AM process.

5. Conclusion and Future Work

Industry 4.0 has been widely researched and adopted since its introduction in 2013 in developed countries, which requires implementing horizontal and vertical integration within and across multiple enterprises and end-to-end digital integration of products. A smart manufacturing system emphasizes the realization of the business cycle for supply chain management, product development lifecycle, and production system lifecycle, similar to the implementation requirements of Industry 4.0. The implementation of these three lifecycles and integrating them is not straightforward. Subsequently, a reference framework has been elaborated to help realize the three lifecycles of a smart manufacturing system.

The elaborated reference framework provides the necessary guidelines for realizing smart manufacturing systems. It is critical to start with the process layer and move to the smart layer. It is crucial to simplify the existing processes and nurture analytic culture among the employees to realize smart manufacturing systems. Likewise, feedback control loops are indispensable within and across multiple layers. The implementation of reference layers assists in acquiring data and the necessary context from multiple sources. This data is an essential input for the realization of the smart layer. A use case was elaborated to realize the reference framework for in-situ monitoring and controlling the end-to-end AM process.

The framework provides a layered approach to realizing smart manufacturing systems by elaborating various concepts, technologies, and so forth. However, getting all the pieces together (i.e., integration) is really challenging. Researchers, in most cases, work in isolation focusing on addressing a particular theme of smart manufacturing systems but missing the bigger picture. The lack of (data) context will have a negative impact on the insights derived from the application of AI/ML techniques in the smart layer of the framework. Thus, effort needs to be put into realizing the individual lifecycles of smart manufacturing systems as per the elaborated framework.

References

[1] Kagermann H, Wahlster W and Helbig J 2013 Recommendations for implementing the strategic initiative industrie 4.0 Tech. Rep. April acatech

[2] Fraser J 2016 Manufacturing software for Industry 4.0 - embracing change and decentralization for success Tech. rep. Iyno Advisors Inc.
[3] Grauer M 1997 *Concise international encyclopedia of business and management* ed Warner M (London: International Thomson Business Press) pp 296–305
[4] Kagermann H 2018 Industrie 4.0 - perspektiven für den standort deutschland die zweite welle der digitalisierung deutschlands zukunftsprojekte Tech. rep. acatech
[5] Adolphs P, Bedenbender H and Dirzus D 2015 Status report: Reference achtitecture model industrie 4.0 (RAMI4.0) Tech. Rep. July VDI
[6] Leiva C 2018 Three functional dimensions converge on smart manufacturing Tech. Rep. Whitepaper 59 MESA International
[7] Lu Y, Morris K and Frechette S 2016 *Current Standards Landscape for Smart Manufacturing Systems* vol 8107 (National Institute of Standards and Technology) ISBN 1069600690287
[8] Schuh G, Anderl R, Gausemeier J, ten Hompel M and Wahlster W 2018 Industry 4.0 maturity index Tech. rep. acatech STUDY
[9] Vijaya Kumar N, Karadgi S and Kotturshettar B B 2020 *IOP Conference Series: Materials Science and Engineering* **872** 012006 ISSN 1757-899X
[10] Kearney A 2018 Readiness for the future of production report 2018 Tech. rep. World Economic Forum
[11] Martin C, Samans R, Leurent H, Betti F, Drzeniek-Hanouz M, Geiger T and Schulz O 2018 *World Economic Forum's System Initiative on Shaping the Future of Production* pp 89–91
[12] Association for Supply Chain Management 2021 SCOR - supply chain operations reference model https://scor.ascm.org/ accessed on 2022-03-09
[13] IEC 62264–3 2016 Enterprise-control system integration – part 3: activity models of manufacturing operations management Standard IEC
[14] IEC 62264–2 2015 Enterprise-control system integration – part 2: model object attributes Standard IEC
[15] Heynitz H, Bremicker M, Amadori D and Reschke K 2016 The factory of the future Tech. rep. KPMG
[16] ISO/TC 184/SC 4 Secretary 1994 Overview and fundamental principles Standard ISO
[17] Kosanke K, Vernadat F and Zelm M 1999 *Computers in Industry* **40** 83–97 ISSN 01663615
[18] Leng J, Zhou M, Xiao Y, Zhang H, Liu Q, Shen W, Su Q and Li L 2021 *Journal of Cleaner Production* **306** 127278 ISSN 0959-6526
[19] Ho L T and Lin G C I 2004 *International Journal of Production Research* **42** 3731–3742 ISSN 0020-7543
[20] Karadgi S 2014 *A Reference Architecture for Real-Time Performance Measurement: An Approach to Monitor and Control Manufacturing Processes* (Cham: Springer International Publishing) ISBN 978-3-319-07007-0
[21] ISO 15704 2000 Industrial automation systems - requirements for enterprise-reference architectures and methodologies Standard ISO
[22] IEC 62264–1 2013 Enterprise-control system integration – part 1: models and terminology Standard IEC
[23] IEC 61512–1 1997 Batch control-part 1: models and terminology Standard IEC
[24] Slansky D 2008 Integrating PLM with the shop floor validating as-built to today's business requirements Tech. rep. ARC Advisory Group
[25] Aguilar-Savén R S 2004 *International Journal of Production Economics* **90** 129–149 ISSN 0925-5273 production Planning and Control
[26] Chaffey D and Wood S 2005 *Business Information Management: Improving Performance using Information Systems* (FT Prentice Hall, Harlow) ISBN 9780273686552
[27] Chaffey D, Hickie S, Greasley A and Bocij P 2006 *Business Information Systems: Technology, Development and Management for the E- Business*
[28] OPC 2021 OPC Foundation https://opcfoundation.org/ accessed on 2022-03-09
[29] MTConnect 2021 MTConnect https://www.mtconnect.org/ accessed on 2022-03-09
[30] MQTT 2021 MQTT: The standard for IoT messaging https://mqtt.org/ accessed on 2022-03-09
[31] Grauer M, Karadgi S S and Metz D 2011 *10th International Conference on Wirtschaftsinformatik* pp 282–291
[32] Karadgi S, Grauer M and Metz D 2014 *Tagungsband Multikonferenz Wirtschaftsinformatik 2014, MKWI 2014* pp 351–362 ISBN 9783000453113
[33] ISO 22400–1 2014 Automation systems and integration – key performance indicators for manufacturing operations management – part 1: overview, concepts and terminology Standard ISO
[34] ISO 22400-2:2014 2014 Automation systems and integration – key performance indicators (kpis) for manufacturing operations management – part 2: Definitions and descriptions Standard International Organization for Standardization (ISO)
[35] VDMA 66412-1 2009 Manufacturing execution systems (MES) – kennzahlen Standard VDMA
[36] VDMA 66412-2 2010 Manufacturing execution systems (MES) – kennzahlen-wirkmodell Standard VDMA
[37] Etzion O and Niblett P 2010 *Event Processing in Action* (Manning Publications) ISBN 9781935182214
[38] Grauer M, Seeger B, Metz D, Karadgi S and Schneider M 2011 *Towards a Service-Based Internet.*

ServiceWave 2010 Workshops ed Cezon Micheland Wolfsthal Y (Berlin, Heidelberg: Springer Berlin Heidelberg) pp 180–187 ISBN 978-3-642-22760-8
[39] Drools JBoss Community 2021 Drools JBoss https://www.drools.org/ accessed on 2022-03-09
[40] Razvi S S, Feng S, Narayanan A, Lee Y T T and Witherell P 2019 *Proceedings of the ASME Design Engineering Technical Conference* **1**
[41] Qi X, Chen G, Li Y, Cheng X and Li C 2019 *Engineering* **5** 721–729 ISSN 20958099
[42] Butt J 2020 *Designs* **4** ISSN 2411-9660
[43] Bhuvaneswari V, Priyadharshini M, Deepa C, Balaji D, Rajeshkumar L and Ramesh M 2021 *Materials Today: Proceedings* **46** 3263–3269 ISSN 2214-7853 international Conference on Materials, Manufacturing and Mechanical Engineering for Sustainable Developments-2020 (ICMSD 2020)
[44] Anbuchezhian N, Priyadharshini M, Devarajan B, Priya A and Rajeshkumar L 2020 *Solid State Technology* **63** 12310–12319

Telescopic handlers: applications, challenges and scope

Ayush Verma[1]

Student , Department of Mechanical Engineering, NSUT, New Delhi-110078, India

Abstract. Telescopic Handlers, also known as Telehandlers or teleporters or boom lifts are small human-operated machines extensively used by the agricultural, construction and landscaping industries. Different types of attachments are being used with these telehandlers which make them extremely versatile, safe, swift and costumer friendly. They prove out to be very efficient when used with various attachments designed as per the market requirements. Common attachments include forks, lifting jibs, man-basket or work platforms, buckets etc which are attached to the telescopic boom of the telehandlers as per use. However, these attachments should be chosen with utmost care keeping in mind that it does not pose any threat on the operation safety. While, these telehandlers have numerous advantages and applications, they also have certain limitations which slightly restricts their use in some cases. The present work is a review on telescopic handlers, their applications, limitations and future scope of research on their improvement to expand the domain of the usage.

Keywords: boom lifts; attachments; forks; buckets; operation; domain.

1. Introduction

Material Handling, lifitng loads, workspace cleanings have been some of the areas where humans have invested a considerable amount of their time and capital since the time when industrialisation started. Until, in the 19th century, when the Clark company which used to manufacture axles, started using 5 ton (5T) truclifts shown in Fig. 1 below for material handling in their plants. In the early 1990s, various industrial companies started using these truclifts. This was the time, when these small machines started to be used by various industries for loading, unloading materials, also for lifting objects for transporting them from one place to another [3]. Telehandlers were first used in the US to replace rough terrain forklifts to allow loads to be lifted upto four-five storey buildings with Pettibone and Koehring as the pioneers. In 1977, JCB launched the first ever loadall offering various interchangeable attachments like jibs, forks etc. 36 telehandler manufacturers grew by 1990s with over 25,000 units manufactured per year [4]. Since, then a number of features, and technologies are being added, and now a number of lifting machies like telehandlers, forklifts, rough-terrain loaders, skid steers and pick and carry cranes have been developed and put into use. Telehandlers are widely used by businesses following their extreme versatile nature. Examples of telehandlers in the market are JCB 530-70 Telehandler (shown

IOP Conf. Series: Materials Science and Engineering 1258 (2022) 012019 doi:10.1088/1757-899X/1258/1/012019

in Fig. 2 below) ,MANITOU MRT-X 1645 Telehandler, Haulotte HTL5210, SkyTrak 3013 Telehandler etc.

Figure 1. 5T Clark truclifts [1]

Figure 2. JCB 530-70 Telehandler [2]

Nowadays, compact telehandlers have also come into existence, which are easy to transport and can serve variety of purposes in compact and inaccessible regions. The telehandlers are heavy duty and normally have diesel run engines. These can be used in a number of terrains, to serve purposes such as load handling, snow ploughing, farming, waste handling etc. Telehandlers prove to be extremely cost efficient and operationally versatile when used which attachments like pallet forks, buckets, mulchers, graples etc. However in the current markets, despite of having so many features, that enable high yield productivity and efficiency, telehandlers are unable to create a wider market footprint, which shows a need for marketing strategies that should be adopted by manufacturers worldwide [8]. According to industry estimates, the current market size of telehandlers in India is around 350 to 400, which is very low when compared with the world's advanced markets. While rental opportunities for telehandlers in a mature market are immense, but India's rental market is small, and the rates quite high as there is a limited number of telehandlers available.Going forward, the telehandler segment in India is set to gather momentum. Large corporates engaged in construction and infrastructure projects have started using them, though their percentage is still very small [20].

2. Types of Telehandlers

Telescopic handlers are divided into two categories:

2.1 Non-rotating

A telescopic boom is attached and pivots on a motorised wheeled chassis on these machines. The boom can be raised to nearly vertical height from below the horizontal. A fork carriage is attached to the boom's outer end for handling unit loads. When raising or lowering a load, the forks automatically level as the boom's elevation changes, ensuring perfect balance. Non-rotating handlers can transfer items to various parts of a project and position them at the proper height [19].

2.2 Rotating

These telehandlers have all of the characteristics of non-rotating telehandlers, plus a slewing or rotating superstructure on which the boom and operator's cab are placed. Outriggers at either end of the chassis on rotating telehandlers allow the entire chassis to be lifted entirely off the ground for maximum stability.

In comparison to non-rotating telehandlers, rotating telehandlers have the following advantages:

- The chassis is small and compact.
- Stability has improved
- Lifting height has been increased
- Placement of loads without shifting the chassis is simple [19].

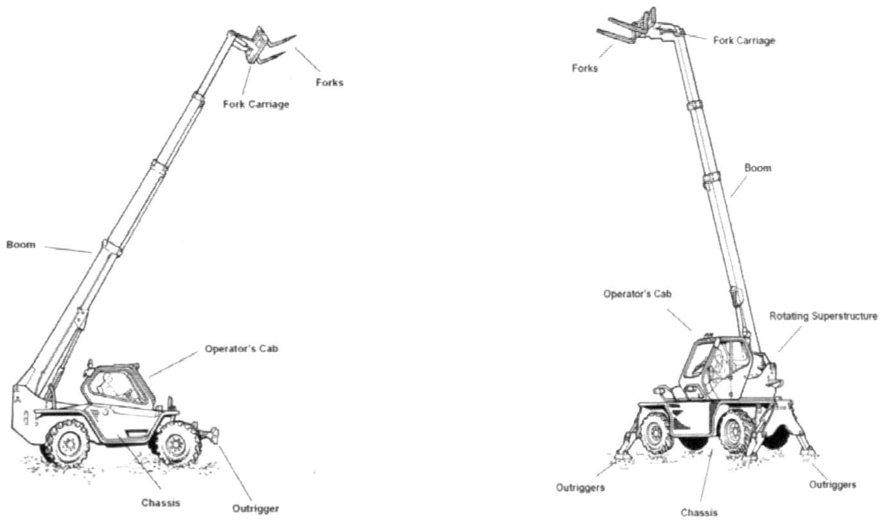

Figure 3. Parts of a non-rotating and rotating telehandler [23].

3. Applications of Telehandlers

In the construction industries, the workmen often require heavy loads to be lifted to and from places. Sometimes, they can also need to remove massive loads from the back of the delivery trucks. This is where telehandlers with a pallet fork attachment find their major usage. In the winters, businesses often use telehandlers with a snow plough attachment to remove snow from their workspaces, thus preventing it to slow them down. Telescopic Handlers are also widely being used for clearing sites using buckets as the attachment, reparing tall buildings or statues using a man-basket attachment as shown in Fig. 3 below.

Figure 4. Telehandler with a man-basket attachment

Tree surgeons often use telescopic handlers so as to access the parts of the trees that are inaccessible. By doing this, they can easily check the health, safety of the tree, prune branches and also can fill cavities that have appeared in a tree trunk. Some telehandler models are also fitted with breakers which can be used to down-size the materials before they are put to crushers. In the construction industry, telehandlers are majorly used for the movement of palletised or loose materials across the job site or to altitudes, roofing, renovation, cladding etc [8,9,10].
Other applications can be seen in Airports, Oil & Gas, Tyre, Tunneling, Container Stuffing and De-Stuffing, waste managements etc. At the mining sites, telehandlers are used for lifting tyres of large dump trucks, large cylinders and cable reels. Farmers also make use of telescopic handlers to tranport bricks, blocks, heavy boxes and bales of hay [8].

4. Compact Telehandlers

Compact Telehandlers can be thought of as a combination of a skid steer and a fully-sized telehandler as shown in Fig.4 below. It can efficiently fulfill a variety of tasks in tight and not-easy-to-operate-on workspaces. It can also be used as a loader and can be called as a replacement for a skid steer [18].

Figure 5. A compact telehandler with bale-handler attachment

Compact telehandlers are popular because of their size, low weight, mobility and attachment versatility which makes these an attractive and lucrative alternative to rough-terrain forklifts. Compared to other rough terrain vehicles, a compact telehandler:

- Weighs less.
- They are easier to move.
- They can operate in areas inaccessible to rough-terrain forklifts.
- They are easier to transport and most of them can be hauled behind a one-ton truck.

5. Advantages of telehandlers over conventional mobile cranes and loaders

Telescopic handlers are available with a wide range of attachments, including pallet forks, winches, jibs, and cranes, as well as re-handling multi-purpose buckets, tyre handlers, glass handlers, slewing access platforms, concrete skips, and mixers. The list is dynamic and ever-changing as new application areas are discovered over time. Manufacturing companies believe that telehandlers outperform pick-and-carry cranes and have a wide range of applications in Indian markets [9]. Telehandlers have a significantly superior load centre, and when combined with all of the safety features, they show to be highly stable, capable of loading heavier weights at greater heights. Normal pick-and-carry cranes, due to their design, tend to tip over in specific locations, but telescopic handlers are quite stable. Telescopic handlers and their many attachments are used in a variety of sectors, including manufacturing, mineral processing, mining, construction, oil and gas, and agriculture. Mobile cranes are only capable of lifting items using the crane hook, but telehandlers may be equipped with a variety of attachments to handle both solid and loose loads. Telehandler is an all-rounder that combines the functions of both cranes and forklifts while being significantly less expensive than cranes and forklifts [8].

The key technological aspects that differentiate the various models are the lifting capacity and the maximum height attained by the telescopic boom. There are several options available on the market. Compact or "mini" lifts are ideal for small producers such as greenhouses, nurseries, and horticultural businesses where adaptability and manoeuvrability in compact places are critical. A footprint (less than 2 metres broad and about 4 metres long) and mass of no more than 5,000 kg become critical. The highest lifting capability is around 2,600 kg, and the maximum boom height is 5-6 m. Engine power is restricted (50-60 kW) [21].

These versions adapt well to the most diverse operating challenges: total size grows (2 m in width and 4.5-5 m in length), maximum capacity ranges between 3,000 and 4,000 kg. The engine has a variable power range of 55 to 80 kW. The maximum capacity reaches and occasionally exceeds 6,000 kg, while the maximum boom height exceeds 10 m, with a few machines reaching as high as 15 m. The maximum engine power ranges from 100 to 130 kW [21].

Costumers mostly look out for standard telehandlers containing forks used for moving or shifting palletised loads, However, depending upon the type of weight to be lifted, the user can use additional and optional attachments like buckets (used for loose load, sand, gravel, stone chips, loose metals etc.), tyre handlers (for handling tyres), hooks jibs and winch and man-basket for people lifting (shown in Fig. 3).

Figure 6. A compact telehandler with different attachment options

Telehandlers have a safety aspect which is better than most of the present machines. Manufacturers claim that telehandlers are very safe to operate with their load gauges, hose burst protections, FOPS and ROPS etc, and the 11m telehandler also have stabilizers for making it work more safely on site [11,12]. Maintou claims that it's telehandlers are reliable and highly productive with a major focus on operator's and machine's safety with overall visibility of the site leading to less fatigue, high ground clearance. Handling of loads within restricted and narrow areas have been made very simple with different steering modes such as the Crab mode, 4 Wheel Drive (WD), and 2WD, braking on all 4 brakes, inbuilt safety systems and allowances, and hydraulic cut offs for safe management of loads [8,10].

A single telehandler may do many load handling operations, improving productivity significantly. According to Manitou, a Load Management System (LMS) is required for any telehandler like the Load Stability Indicator provided by JLG Industries. Additionally, many Telehandlers have features such as pothole guards with a cut-off system preventing major dangerous movements and a limited slip on the front axle and the cabin, as well as an inching pedal for smooth and precise forward movements, even when the engine is at full throttle [8].

Figure 7. Load Indication Systems for Telehandlers [22].

6. Limitations of Telehandlers

Telehandlers offer a number of uses and advantages, but they also have certain limits. While carrying a load, the boom serves as a lever, causing the trucks to become progressively unstable, despite the counterweights installed at the back. This means that as the working radius rises, the lifting capacity diminishes rapidly. The distance between the front of the wheels and the centre of the load is measured as the working radius. However, the operator is furnished with a load chart that allows him to decide if a certain duty is feasible or not, taking into account weight, boom angle, and height [9].

Moreover, the telehandlers can't be used for handling senstive materials like flammable or inflammable liquids, glass objects and other medical or surgical equipments due to its high raising speed. They can't be used for handling finished products in warehouses or manufacturing plants. They also cannot be used for handling nuclear waste as it cannot be remotely or distance operated.

7. Challenges and Scope

The introduction of telehandlers into the material handling business, with its safe technology, adaptability, and requirement for fewer people, has sparked fast growth in the Indian market. Because of their wide range of interchangeable attachments, these machines are finding use in a number of industries, and they should be marketed as multi-utility machines capable of doing a variety of jobs. With new legislative mandates requiring safer work environments on construction sites, the industry will increasingly adopt telehandlers as a more dependable and efficient investment than other conventional devices [8]. Telehandlers are very popular equipment worldwide and are extensively used globally. In 2018, 66,408 new units of telehandlers were sold worldwide and for 2019 the numbers are expected to be around 73,000. Although the demands vary from market to market based on maturity, the opportunities are quite encouraging even in developing markets where awareness is slowly increasing.

According to Manitou's Hafeez Khan, performance that is technologically driven and reliable, as well as total cost of ownership, will have a far greater value of money in the years to come. Users seeking solutions at a somewhat higher cost will find telescopic handlers to be a significant value addition, as the entire cost and utilisation in projects will justify the customer's original investment. According to Jasmeet Singh of JCB, telehandlers are adaptable equipment with immaculate performance, and they may overlap the performance capacities of other machines. However, because they are one-of-a-kind and application-specific, new application areas should be developed so that they can function and, as a result, create a sizable demand potential for themselves [8].
Telescopic Handlers are widely utilised for material handling and lifting in industrialised countries. As a result, this should occur in Indian markets as well. Despite increased awareness, there is still more work to be done in marketing telehandlers' applications, advertising them, and encouraging users not to be put off by the expense. All lifting tasks must be performed by cranes; this thinking must be corrected. In India, telehandlers are currently in the idea stage. Telehandlers' sluggish popularity in Indian markets might be attributed to their expensive cost in comparison to other mobile cranes. As a result, the Indian market for telehandlers is still in its early stages, with little awareness in various industries. As a result, it's past time for manufacturers to start pushing telehandlers by offering various new attachments [8]. As the telehandlers are man-operated, the possible reach and modification can be possible by making them unmanned or remotely operated from a certain distance wired if the material is non-hazardous and unwired when material to be handled is hazardous like nuclear waste or bio-hazardous material in decomposition plants [3].

8. Conclusions

Telehandlers are ideally suited for material handling and load lifting since they are cost effective and have the option of replaceable attachments. Telehandlers, on the other hand, cannot be utilised for delicate material handling such as medical, surgical, flammable/inflammable materials, liquid materials, or glass products for a variety of reasons. As a result, unique telehandlers must be built for this purpose, which will increase the present market need for telescopic handlers. Furthermore, the use of compact telehandlers with the same capabilities as conventional telehandlers should be pushed since certain clients require them to be used in tight, compact task sites. Although the benefits of telescopic technology are still confined to a narrow customer base, demand is growing, particularly in hazardous environments where safety is an issue. As a result, producers such as JCB, Manitou, Escorts, and others must advertise and sell telehandlers while also raising awareness of the machine's safety features. Companies must continue to investigate new application areas for telehandlers in order to expand their market share. Their aim should be to identify new industries by tapping into market need for safe, cost-effective, adaptable, and time-saving technologies such as telehandlers. Manufacturing companies like JCB, Manitou should also lay emphasis on operator training to ensure optimum utilization of their telescopic machines by forming dedicated teams. After-sale services should be given uttermost importance. The telehandler market in India is steadily expanding. As awareness increases and the industry and market mature further, the demand for telehandlers of various capacities is set to rise up in the near future.

9. References

[1] https://www.trucknews.com/transportation/8-technologies-that-transformed-trucking/1003093667/attachment/clark-tructractor/ (accessed on 01/02/2022 at 0800 hours).

[2] https://www.indiamart.com/proddetail/jcb-loadall-530-70-telescopic-handlers-21846332355.html (accessed on 01/02/2022 at 0800 hours).

[3] Dr. V.R. Gandhewar1 , Rahul R. Ulmale2 , Gaurav P. Ghodeswar3 , Pinki S. Turankar4. Application of Forklift for Sensitive Material Handling, *International Journal of Engineering Science and Computing, April 2018,* **Volume 8 Issue No.4.**

[4] https://www.cranestodaymagazine.com/features/telehandlers-rising-6940129/ (accessed on 01/02/2022 at 0800 hours).

[5] Richard L. Neitzel , Noah S. Seixas & Kyle K. Ren (2001) A Review of Crane Safety in the Construction Industry, *Applied Occupational and Environmental Hygiene*, **16:12**, 1106-1117

[6] A systematic review of scholarly works addressing crane safety requirements Sanaz Sadeghi a Nazi Soltan mohammad loua,PayamRahnamayiezekavatb.

[7] https://www.nbmcw.com/interviews/manitou-south-asia.html (accessed on 05/02/2022 at 0830 hours).

[8] https://www.nbmcw.com/equipment-machinery/material-handling-cranes-awp/telehandler-the-all-rounder.html (accessed on 08/02/2022 at 0830 hours).

[9] https://en.wikipedia.org/wiki/Telescopic_handler (accessed on 08/02/2022 at 0930 hours).

[10] https://dozr.com/blog/telehandler (accessed on 08/02/2022 at 0930 hours).

[11] https://www.adaptalift.com.au/blog/what-is-a-telehandler (accessed on 08/02/2022 at 0930 hours).

[12] https://sbskills.co.uk/what-is-a-telehandler-and-what-are-they-used-for/ (accessed on 08/02/2022 at 1230 hours).

[13] Hyun, H.; Park, M.; Lee, D.; Lee, J. Tower Crane Location Optimization for Heavy Unit Lifting inHigh-RiseModularConstruction. *Buildings* **2021**, *11*,121.

[14] Mr. Sagar Gajanan Kashid, Prof. Mr. Pandit Rangrao Sawant, 2014, Design and Development of Material Handling Crane, *international journal of engineering research & technology (ijert)* **Volume03, Issue 10 (October 2014),**

[15] Beavers, J. E., Moore, J. R., Rinehart, R., & Schriver, W. R. (2006). Crane-related fatalities in the construction industry. *Journal of Construction Engineering and Management*, **132(9)**, 901-910.

[16] Neitzel, R. L., Seixas, N. S., & Ren, K. K. (2001). A review of crane safety in the construction industry. *Applied Occupational and Environmental Hygiene*, **16(12),** 1106-1117.

[17] Shapira, A., & Lyachin, B. (2009). Identification and analysis of factors affecting safety on construction sites with tower cranes. *Journal of Construction Engineering and Management*, **135(1),** 24-33.

[18] What is a Telehandler and What are they Used For? - MTeevan (mteevanhire.co.uk) (accessed on 14/02/2022 at 0930 hours).

[19] What Are Telehandlers Used For? – Dwight Crane Ltd. (accessed on 14/02/2022 at 0930 hours).

[20] India's Telehandler Market: Slow & Steady Rise (nbmcw.com) (accessed on 14/02/2022 at 0930 hours).

[21] https://www.mondomacchina.it/en/the-telehandler-top-level-multi-functionality-c2575 (accessed on 25/06/2022 at 1337 hours).

[22] https://www.jlg.com/en (accessed on 25/06/2022 at 1430 hours).

[23] https://www.highwayssafetyhub.com/uploads/5/1/2/9/51294565/2d_guidance_on_safe_use_of_telehandlers.pdf (accessed on 25/06/2022 at 1530 hours).

Modelling of Surface Roughness effect on the Dry Turning of Duplex Stainless Steels

Naresh Deshpande[1] and Hari Vasudevan[2]

[1]Research Scholar, Mechanical Engineering, Dwarkadas J. Sanghvi College of Engineering, Vile Parle (West), Mumbai - 400056, India.
[2]Principal, Dwarkadas J. Sanghvi College of Engineering, Vile Parle (West), Mumbai - 400056, India.

[1]ncdeshpande72@yahoo.co.in ; [2]principaldjs@gmail.com

Abstract. Surface integrity features such as roughness, induced by machining Duplex Stainless Steel (DSS) components, is an important factor affecting their functional performance. This work used a multi-regression modelling method to state the relationship between turning parameters & roughness average for standard DSS components. Four process parameters, viz. cutting speed, rate of feed, cutting depth & radius of tool nose curvature were varied between three levels each and their effect on roughness average was noted. Experiments were performed using Definitive Screening Design (DSD). The surface roughness parameter was measured using Taylor Hobson branded Talysurf-5 profilometer. A Step Wise regression approach with forward selection of terms and alpha value of 0.25 resulted in a quadratic model with R square value of 99.61%. Validation experiment with new specimens yielded prediction accuracy of 88%. For minimization of roughness, cutting speed should be set at 100 m/min, feed rate should be set at 0.1 mm/rev, cutting depth should be set at 1.5 mm and nose radius should be set at 0.8 mm. These settings yielded minimum value of roughness average as 0.0845 microns.

1. Introduction

Use of Duplex Stainless steel (DSS) is found in several applications across different industrial sectors such as, mining & hydrometallurgy, chemical processing plants, oil & gas, bulk materials handling equipment, desalination, shipbuilding & marine and pollution control equipment, to name just a few. Many products in these fields of application are processed using conventional machining. DSS is categorised as difficult to cut material because it has low thermal conductivity, higher tendency for work hardening & built-up edge formation. It also displays higher tensile strength & toughness. DSS components produced by machining, should have strong corrosion resistance & higher fatigue life [1]. It is established from extant literature that surface integrity features such as roughness average are strongly related with these properties. However, the influence of turning parameters like cutting speed, rate of feed, cutting depth, type of tool & cutting environment on surface integrity of DSS is not sufficiently explored.

Other types of stainless steels have been extensively studied for surface integrity features similar to roughness average. Kaladhar M. et al. [2] investigated the effect of carbide inserts coated by application of physical vapour (PVD) & chemical vapour deposition (CVD) methods for cutting AISI 202 austenitic stainless steels. They observed improvement in roughness average Ra, when machined with CVD coated inserts. The rate of feed & radius of tool nose curvature were the most influential parameters. Korkut I. et al. [3] found out optimum cutting speed, during turning of AISI 304 austenitic type stainless steels. They observed that roughness decreases with increase in cutting speed. Anthony Xavior M. and Adithan M. [4] compared the influence of various cutting fluids for reduction of tool wear & roughness while turning of AISI 304 austenitic type stainless steels. Fernandez Abia A. I. et al. [5] studied the influence of very high speed on roughness while turning of austenitic stainless steels without using cutting fluids. Mahavinejad R. and Saeedy S. [6] inferred that the rate of feed is most influential in causing variation in roughness. The roughness decreases with decrease in feed. Rajguru R. and Vasudevan H. [7] also found rate of feed to be the most influential for the variation of roughness.

Compared to traditional response surface experimental designs, like standard central composite designs, Definitive Screening Designs (DSD) with three levels have superior properties [8]. DSD provide more reliable, cost effective and efficient one step approach for problems involving twelve factors. These designs are efficient in estimating the main effects, interactions involving two factors & unbiased quadratic effects with minimum number of experiments [9]. Omar Ahmed et al.[10] used DSD to investigate the short term creep deformation of PC-ABS under different fabrication conditions. Yunpeng Yan et al. [11] used combination of DSD and discrete elemental methodology for studying the effect of the five parameters & two operating conditions on the sliding wear of a transfer chute. Omar Ahmed Mohamed et al. [12] used integrated approach of DSD and ANN (Artificial Neural Networks) for modelling and optimization of fused deposition modelling fabricated parts. R. Mahendran et al. [13] studied the effect of cryogenic treatment on tool life of multilayer coated carbide inserts.

The novelty of the present research work lies in the use of Definitive Screening Design for investigation of surface roughness effects in dry turning of Duplex Stainless Steels with multilayer coated carbide inserts.

2. Materials and Method

The material selected for this study is Standard Duplex Stainless Steel S32205 (UNS). Duplex stainless steel's structure consists of approximately fifty percent austenite and fifty percent ferrite. The physical properties of these group of steels combine the properties of ferritic and the austenitic grades. However, due to their high yield & tensile strength, higher ability to work harden, lower values of thermal conductivity, stronger tendency for the formation of built-up edge. They also possess higher values of fracture toughness and hence are classified as difficult to machine. DSS composition is as indicated in table 1.

Table 1. DSS composition.

C % max.	Cr %	Ni %	Mo %	N %	Mn %
0.03	22.0	4.5 - 6.5	3.0 - 3.5	0.14 - 0.20	2.0

Hot rolled circular rods with 34 mm diameter and 150 mm length were turned after annealing on MTAB Max Turn Plus + CNC turning centre. This machine with FANUC controller with 6000 rpm maximum spindle speed & 7.5 KW power rating. Taylor Hobson branded Talysurf-5 roughness profilometer was applied for roughness parameter measurement. Gaussian filter & cut-off length of 0.8mm with five cut-offs, and total traverse length 4 mm was used for data acquisition. SESURF software was used for processing and data analysis purpose. The turning parameter settings are as per table 2. These values are chosen based on recommendations by cutting tool & machine tool manufactures and International Molybdenum association. Current marine components industry practise where DSS machining is carried out is also considered for selection of these parameters.

Table 2. Turning variables and their set values.

Code	Parameter	Levels		
		-1	0	1
A	Speed of cutting (m/min)	80	100	120
B	Rate of feed (mm/rev)	0.1	0.2	0.3
C	Cutting depth (mm)	0.5	1	1.5
D	Radius of nose curvature (mm)	0.8	1.2	1.6

Chemical vapor deposited triple layer coated, i.e., Titanium carbonitride, aluminium oxide and titanium nitride, carbide cutting tool inserts of Sandvik make with designation SNMG 120408 MR 2220, SNMG 120412 MR 2220 and SNMG 120416 MR 2220 were used for the turning operation. The tool holder was of Echain brand with designation ESDNN 2525 M12. The Definitive Screening Design used for this study and the observed values of surface roughness are given in table 3.

Table 3. Experimental plan and observations.

Std Run	Run order	A (m/min)	B (mm/rev)	C (mm)	D (mm)	Ra (microns)
1	J1	0	1	1	1	1.45
4	J2	0	-1	-1	-1	0.9
5	J3	1	0	1	-1	1.675
9	J4	-1	0	-1	1	0.81
12	J5	1	1	0	1	2.115
2	J6	-1	-1	0	-1	0.625
3	J7	1	-1	1	0	2.035
6	J8	-1	1	-1	0	1.035
10	J9	1	-1	-1	1	3.02
11	J10	-1	1	1	-1	3.89
7	J11	1	1	-1	-1	1.155
8	J12	-1	-1	1	1	2.47
13	J13	0	0	0	0	1.435

3. Results and Discussion

A multi-regression model using stepwise procedure and forward selection of terms with alpha to enter as 0.25 was generated using the popular statistical software, Minitab. The analysis of variance for the model is as given in table 4.

Table 4. ANOVA table for regression model for surface roughness.

Source	DF	Adj SS	Adj MS	F-Value	P-Value
Model	9	10.8131	1.20146	84.55	0.002
Linear	4	6.0291	1.50728	106.07	0.001
A Speed	1	0.0001	0.00006	0.00	0.951
B Feed	1	4.5765	4.57652	322.07	0.000
C Depth of cut	1	0.1166	0.11664	8.21	0.064
D Nose radius	1	1.3359	1.33590	94.01	0.002
Square	2	0.4136	0.20678	14.55	0.029
A Speed*A Speed	1	0.3615	0.36146	25.44	0.015
D Nose radius*D Nose radius	1	0.0343	0.03427	2.41	0.218
2-Way Interaction	3	3.5165	1.17217	82.49	0.002
A Speed*B Feed	1	0.0962	0.09616	6.77	0.080
B Feed*D Nose radius	1	2.2205	2.22050	156.27	0.001
C Depth of cut*D Nose radius	1	0.5375	0.53754	37.83	0.009
Error	3	0.0426	0.01421		
Total	12	10.8558			

This analysis of variance resulted in R^2 value of 99.61 %. Table 5. below represents the values of various coded coefficients of the model terms.

Table 5. Coded Coefficients

Term	Coef	SE Coef	T-Value	P-Value	VIF
Constant	1.439	0.118	12.24	0.001	
A Speed	0.0025	0.0377	0.07	0.951	1.00
B Feed	0.6765	0.0377	17.95	0.000	1.00
C Depth of cut	0.1080	0.0377	2.87	0.064	1.00
D Nose radius	-0.3655	0.0377	-9.70	0.002	1.00
A Speed*A Speed	0.536	0.106	5.04	0.015	1.84
D Nose radius*D Nose radius	-0.1455	0.0937	-1.55	0.218	1.43
A Speed*B Feed	0.1648	0.0634	2.60	0.080	2.26
B Feed*D Nose radius	-0.6318	0.0505	-12.50	0.001	1.44
C Depth of cut*D Nose radius	0.3740	0.0608	6.15	0.009	2.08

Following is the resultant regression model for roughness average Ra in coded terms.

$$Ra = 1.439 + 0.0025\,A + 0.6765\,B + 0.1080\,C - 0.3655\,D + 0.536\,A{*}A - 0.1455\,D{*}D + 0.1648\,A{*}B - 0.6318\,B{*}D + 0.3740\,C{*}D$$

Figure 1 shows that the residuals are distributed normally, which validates one of the important assumptions of the regression model. The significance of various terms of the model is quite clear from Figure 2. It was found out that the rate of feed has the highest effect on roughness. subsequently nose radius and cutting depth also affected the response variable.

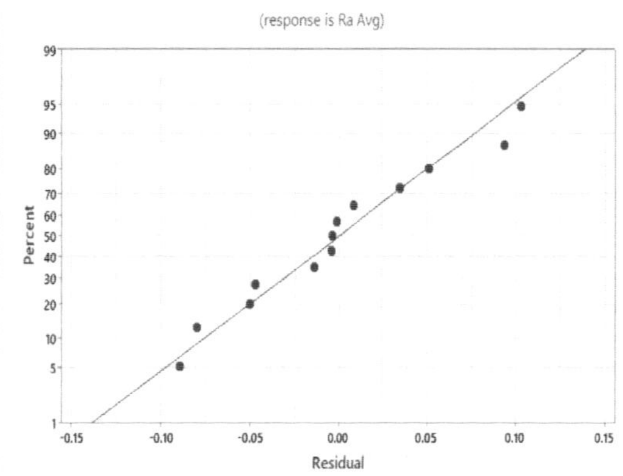

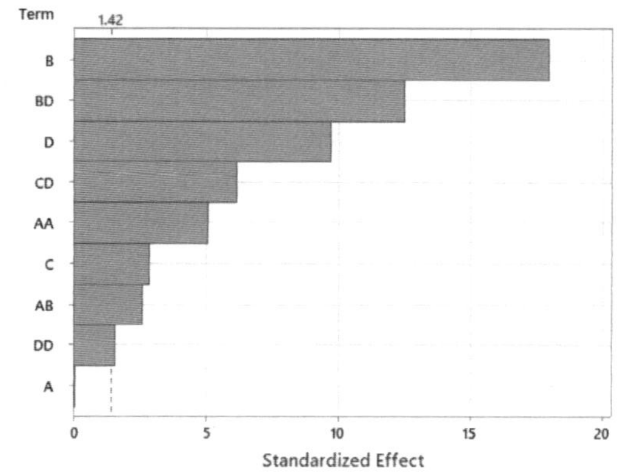

Figure 1. Normal Probablity Plot of Residuals.

Figure 2. Pareto chart of standardised effects

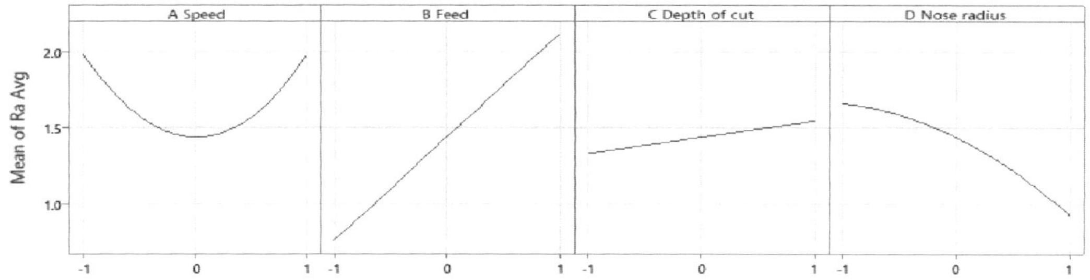

Figure 3. Main Effects Plot

The effects of nose radius and speed are quadratic as observed from Figure 3, by the curvature in the main effects plot. Figure 4, shows the response surface for roughness Ra, when feed and nose radius are varied, while keeping speed & cuting depth at level zero. The interaction effect of rate of feed & radius of nose curvature is plotted in Figure 5. A significant interaction between these two process parameters is quite evident, since the plots for various levels of nose radius are crossing each other.

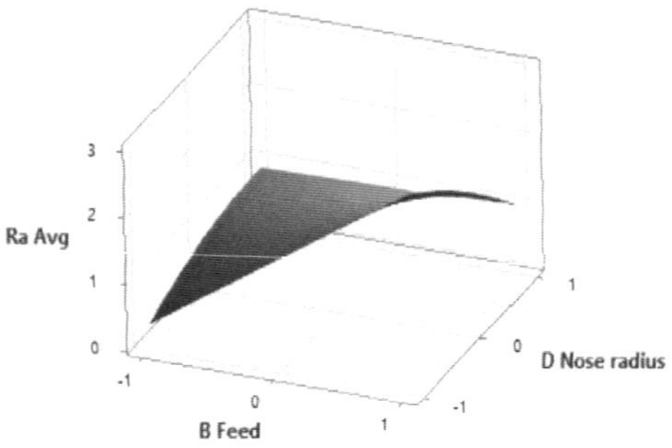

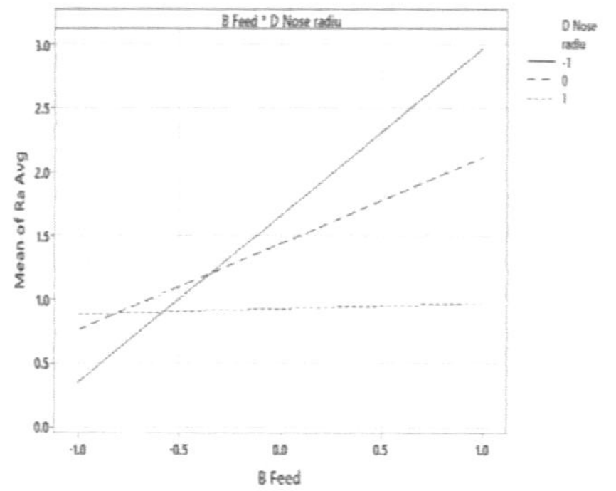

Figure 4. Response surface for Ra Vs B x D

Figure 5. Interaction effect for Ra Vs B x D

Figure 6, shows the response surface for roughness Ra, when cutting depth & radius of tool nose are varied, while keeping speed and feed at level zero. The interaction effect of cutting depth & nose radius is plotted as shown in Figure 7. A significant interaction between these two process parameters is quite evident, since the plots for various levels of nose radius are crossing each other.

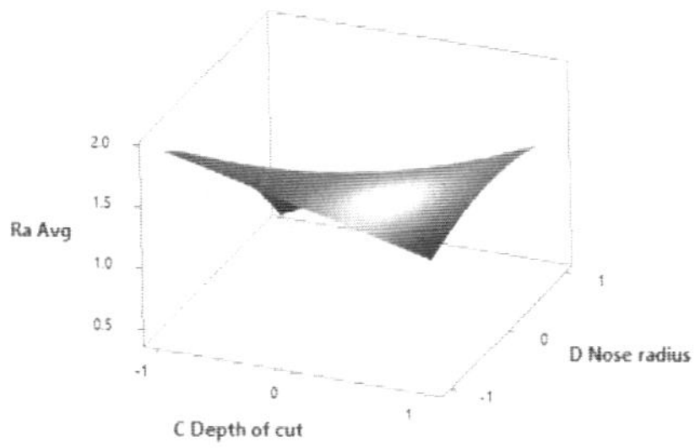

Figure 6. Response surface for Ra Vs C x D

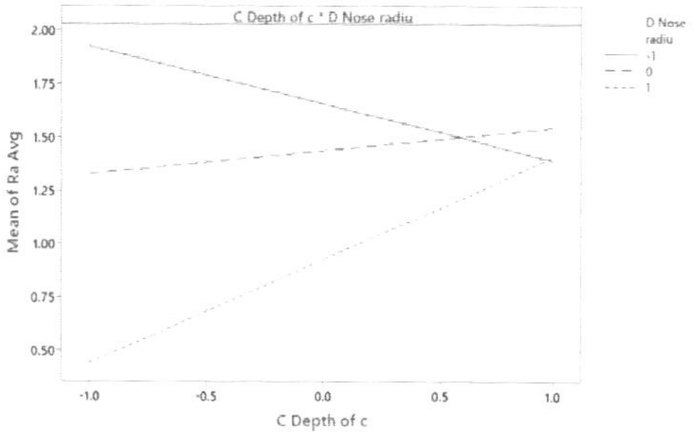

Figure 7. Interaction effect for Ra Vs C x D

Figure 8 below, shows the response surface for roughness Ra, when speed and feed are varied, while keeping depth of cut and nose radius at level zero. The interaction effect of cutting speed and rate of feed is plotted in Figure 9. This interaction is not very strong, but statistically significant as the graphs for different levels of feed will eventually converge and intersect towards lower values.

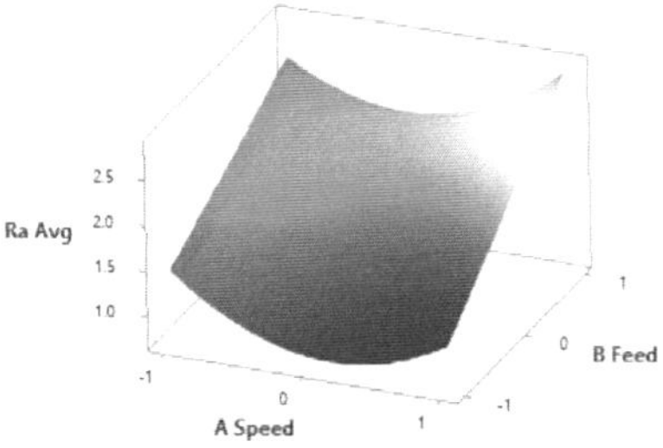

Figure 8. Response surface for Ra Vs A x B

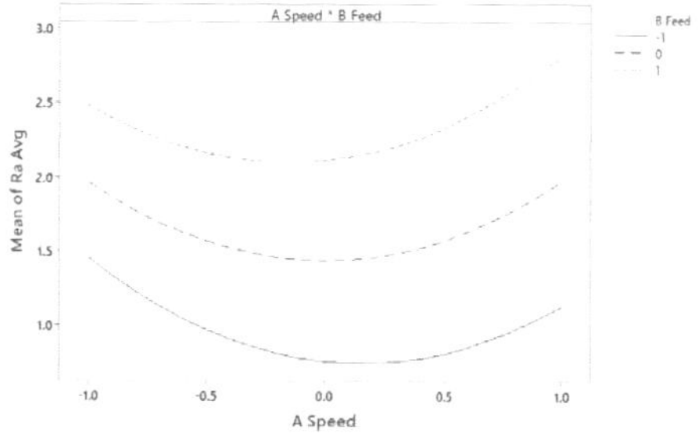

Figure 9. Interaction effect for Ra Vs A x B

Optimization for minimum value of roughness average was carried out using desirability. Figure 10, shows the optimization plot, and it indicates that for minimization of roughness, cutting speed should be set at level 0, i.e., 100 m/min, feed rate should be set at -1 level i.e., at 0.1 mm/rev, cutting depth should be set at 1 level i.e., 1.5 mm and nose radius should be set at level -1 i.e., 0.8 mm. These settings yielded minimum value of roughness average as 0.0845 microns. The composite desirability score for these optimum settings was 100 %.

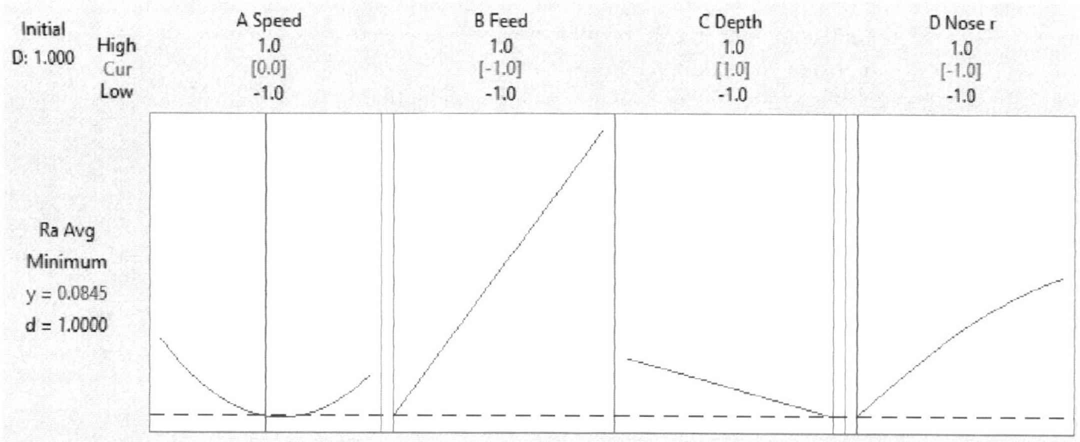

Figure 10. Optimization Plot for minimization of Ra.

Cross validation of the regression model was done by testing four new samples as given in table 6.

Table 6. Cross Validation Process Parameters and calculations

Sample	A (m/min)	B (mm/rev)	C (mm)	D (mm)	Predicted Ra	Measured Ra	% Delta	% Accuracy
E1	120	0.3	1.5	1.6	2.158	1.866	15.64	84.35
E2	120	0.1	1.5	1.6	1.739	0.593	193.25	93.25
E3	100	0.1	0.5	1.6	0.4013	0.3725	7.73	92.26
E4	100	0.1	1.5	0.8	0.7625	0.6555	16.32	83.67
			Average accuracy					88.38

4. Conclusion

A multi-regression model using stepwise procedure and forward selection of terms with alpha to enter as 0.25 was generated for surface roughness Ra during the dry turning of standard duplex stainless steels. ANOVA for the generated model resulted in R^2 value of 99.61 %. Rate of fed has the highest influence on roughness. It is followed by cutting depth & lastly radius of nose curvature. Cross validation with four new samples resulted in prediction accuracy of around 88%. For minimization of Ra, speed should be set at level 0, i.e., 100 m/min, feed rate should be set at -1 level i.e., at 0.1 mm/rev, cutting depth –should be set at 1 level i.e., 1.5 mm and nose radius should be set at level -1 i.e., 0.8 mm. These settings yielded minimum value of roughness average as 0.0845 microns. The composite desirability score for these optimum settings was 100 %.

Acknowledgments

The authors express their gratitude to Prof. U. A. Dabade for permitting to use the machining facilities at Walchand College of Engineering Sangli, Maharashtra, India. Authors are also thankful to Salbro Engineers, Mumbai, Maharashtra, India, for providing access to roughness measurement facilities for the measurement of roughness parameters, that too at a concessional rate.

References

[1] Deshpande N, Vasudevan H. Machining Induced Surface Integrity Aspects of Stainless Steels: A Review. Mater Today Proc 2020;22:1499–506.

[2] Kaladhar M, Venkata Subbaiah K, Srinivasa Rao C, Narayana Rao K. Process parameters optimisation and coatings influence on the surface quality during turning of AISI 202 austenitic stainless steel. Int J Mach Mach Mater 2012;11:371–84.

[3] Korkut I, Kasap M, Ciftci I, Seker U. Determination of optimum cutting parameters during machining of AISI 304 austenitic stainless steel. Mater Des 2004;25:303–5.

[4] Xavior MA, Adithan M. Determining the influence of cutting fluids on tool wear and surface roughness during turning of AISI 304 austenitic stainless steel. J Mater Process Technol 2009;209:900–9.

[5] Fernández-Abia a. I, Barreiro J, Lacalle LNL De, Martínez S. Effect of very high cutting speeds on shearing, cutting forces and roughness in dry turning of austenitic stainless steels. Int J Adv Manuf Technol 2011;57:61–71.

[6] Mahdavinejad RA, Saeedy S. Investigation of the influential parameters of machining of AISI 304 stainless steel , Sadhana 2011;36:963–70.

[7] Rajguru R, Vasudevan H. Investigating the effect of cutting conditions and tool geometry on surface roughness in dry end milling of Inconel 625 using TiAlSiN ultra hard coated solid

carbide tool. Adv Mater Process Technol 2020;001–10.
[8] Jones B, Nachtsheim CJ. A class of three-level designs for definitive screening in the presence of second-order effects. J Qual Technol 2011;43:1–15.
[9] Jones B, Nachtsheim CJ. Definitive screening designs with added two-level categorical factors. J Qual Technol 2013;45:121–9.
[10] Mohamed OA, Masood SH, Bhowmik JL. Experimental investigation of creep deformation of part processed by fused deposition modeling using definitive screening design. Addit Manuf 2017;18:164–70.
[11] Yan Y, Helmons R, Wheeler C, Schott D. Optimization of a convex pattern surface for sliding wear reduction based on a definitive screening design and discrete element method. Powder Technol 2021;394:1094–110.
[12] Mohamed, O.A., Masood, S.H. & Bhowmik, J.L. Modeling, analysis, and optimization of dimensional accuracy of FDM-fabricated parts using definitive screening design and deep learning feedforward artificial neural network. Adv. Manuf. 2021; 9:115–129.
[13] Mahendran, R., Rajkumar, P., Nirmal Raj, L. et al. Effect of deep cryogenic treatment on tool life of multilayer coated carbide inserts by shoulder milling of EN8 steel. J Braz. Soc. Mech. Sci. Eng.2021; 43:378

Numerical Investigation on Vortex-Induced Vibration over Circular Cylinder to be used for Energy Harvesting Application

Sonawane Chandrakant[1], Naina Chowdhury[1], Mehul Rajhans[1], Movva Sai Chaitanya[1], Eshan Sarkar[1], Atul Kulkarni[2], Anand Pandey[1]

[1] Symbiosis Institute of Technology, Symbiosis International Deemed University, Pune, India.
[2] Symbiosis Center for Nano-science and Nano-technology, Symbiosis International Deemed University, Pune, India.

Abstract. Renewable energy is energy generated through inexhaustible processing resources, like sunlight, wind, etc. Existing and popular wind energy harvesting technology is wind turbine farms that have very high installation costs and demand strong wind flow to deliver sufficient power. Thus, wind energy generation based on the vortex-induced vibration-based energy harvester has become a more popular field of interest for many researchers. The vortex shedding produced by the unsteady incompressible flow over a bluff body has been used to excite the piezoelectric plate, generating electricity by its vibration. Hence generating an optimised vortex shedding and obtaining the optimised oscillation frequency. This paper investigates the computational fluid dynamics-based flow analysis over a smooth cylinder carrying a plate. The Reynolds number, Re, is 100 used for unsteady incompressible flow and the plate density ratio is varied from 5 – 40. The variation in lift and drag forces and frequency of vibration, and maximum vibration amplitude achieved by the piezoelectric plate is observed and determined. The results presented here help design a better and more efficient vortex-induced energy harvester.

1. Introduction

Our knowledge of Renewable energy sources makes it apparent that they are becoming vital due to the cumulative rise in energy demands coupled with the exhaustion of natural resources. Tremendous research has been done on renewable energy sources, which includes the generation of energy through solar energy, wind energy, hydro energy, biomass, and other viable renewable energy sources. As per the report, wind power is a major factor in determining the future economic stability of India is India's self-reliance in the power generation sector. India currently has the 4th most wind capacity installed, at 39.25 GW (as of March 2021) [1]. The Indian Government is rapidly promoting development in this sector by giving fiscal and financial incentives to wind power projects. Hence, the focus on renewable energy, typically on wind energy, is increased. However, the current technology of wind turbines with big blades requires high capital investments and large land spaces to produce sufficient power. Vortex-induced vibration (VIV) based energy generation is one of the approaches explored by many authors due to its simplicity, low maintenance cost and ability to work with low wind speed. Here a detailed review of recent work on wind energy by vortex-induced vibration-based energy harvesters is presented.

One of the early works by Barrero-Gil et al. [2] analysed the vortex-induced vibration over a smooth circular cylinder and found that this can be a potential source for harvesting wind energy. The most common parameters that have been studied were the mass ratio, the mechanical damping coefficient, and therefore the Reynolds number. Soti et al. [3] have numerically studied the vortex-induced vibrations over a cylindrical surface, keeping $Re < 200$ where the flow is two-dimensional. To generate electricity, they attached a magnet to the cylinder and made it free to oscillate in the direction transverse to the airflow. When VIV causes the cylinder to oscillate, electrical energy gets induced inside the coil. On further rectifying these pulses, a load was connected, and the total power produced was measured. They generated electrical power without using the concept of piezoelectricity. Hence the effect of Reynolds on the overall power output was studied and observed that the output power increased by 30% with an increase in Re from 100 to 150 and by 45% with an increase in Re from 150 to 200. Zhang et al. [4] numerically investigated vortex-induced vibrations of two cylinders stationed in an adjacent arrangement. They observed stronger vortex shedding patterns between 0^0 to 30^0 attack angles. They achieved the max possible energy conservation with 21.7% efficiency. Zunlong et al. [5] try to improve the problem of low bandwidth and low harvester efficiency of the vortex-induced vibrations based piezoelectric energy harvester. They modified the cylindrical surface by adding pits on the surface and found that the locking area lies between 2 to 3.5 m/s of wind velocity. Their experiments also recommended the smooth surface of a cylinder to obtain the most efficient vortex for varying wind velocities.

Chin et al. [6] developed PVEH (piezoelectric vibration-based energy harvester) plates made of brass that is compressed between two Pb-zirconate-titanate (PZT) plates. They analysed different wind speeds and recommended that much research be carried out to increase efficiency. Ippilli et al. [7] have developed an eco-friendly Pb-free methylammonium tin-iodide (MASnI3) compound, which has a high piezoelectric coefficient and generates a current density of 0.35 μAcm^2. Mohammad et al. [8] prepared Poly Vinylidene fluoride (PVDF) composite nanofibers and investigated their electrical outputs. They observed that adding a very less amount of Graphene (0.1%wt) increased the electrical output of the composite. Olivier et al. [9] focused on the linear stability of the plates used in piezoelectric coupling. They observed that the waves or other modes of disturbance maximise energy efficiency conversion. The whole effect of piezoelectric flags was achieved by using the piezoelectric patches and attaching them to the face of the flags. Lee et al. [10] used additive manufacturing with a polymeric poling process to manufacture piezoelectric PVDF film. The advantage of this process is that it keeps the piezoelectric polymer dipoles symmetric and unvarying with respect to a large area. They also introduced plasma-assisted polling to increase piezoelectricity, which helps fabricate new multi-functional polymers in a convenient, fast, and precise manner. Sukumaran et al. [11] used PVDF piezoelectric film to convert mechanical energy into electrical energy. The piezoelectric coefficient (d33) is used to measure the current generated when subjected to mechanical stress. The electronic circuit design is presented to improve the overall efficiency of the harvester.

Lee et al. [12] have developed a silicon chip integrated VIV based micro-electrochemical energy harvester. The maximum power output of MEMS was observed to be 1.6 nW. It was observed that the VIV concept goes well with MEMS compared to other alternators that work on electromagnetic induction. Narendran et al. [13] have discussed a vortex-induced vibration-based linear generator subjected to Reynolds number above 10^5. Their setup had a peak efficiency of 90%. They have given the output in the form of power density ratio, which is the amount of power generated per unit volume. The power generation density of their setup was 3,29,950 mW/Kgm^3, whereas, for a diesel generator, the power density is said to be 2,50,00,000 mW/Kgm^3. Liu et al. [14] developed a piezoelectric-based wind energy harvester that uses microfiber composites (MFC) as the piezoelectric material. Using a copper hinge, they demonstrated the experiment in a wind tunnel by fixing the MFC on a substrate with a triangular leaf linked to the same substrate. They observed that the leaf flutters at certain wind speeds. The flutter is responsible for generating electrical power. However, the substrate's thickness and stiffness affect the harvester's performance by altering the fluttering of the substrate. Muhterem et al. [15] discussed using the electro-spinning process with a rotating drum collector to prepare pure PVDF and doped it with 10%, 20%, and 30% lead zirconate titanate (PZT). The piezoelectric nanogenerator (PEN) is made using PVDF or PZT nanotechnology fibres as a non-conducting material between the two conductive plates. Such PEN converts vibration energy into electrical energy. These nanogenerators need not be externally polarised to work.

Jung et al. [16] developed a piezoelectric PVDF-based energy harvester for roadway applications. They fabricated a flexible energy harvester using PVDF and found it to perform well for many repeated bending cycles. The harvester allows up to 60 parallel connections that make large amounts of power extraction possible. The instantaneous power output was 200 mW across a 40 k ohms resistor at speed - 8km/h and applied weight of 250 kgf. Michelin et al. [17] created a flag-like material, added piezoelectric patches, and tested its efficiency at different speeds. The wind flows in an axial direction, and the formed deformation is converted to electricity. They studied the effect of damping; it increases the dissipating energy capacity but reduces the robustness of the structure with velocity flow velocity fluctuations. Kundu et al. [18] analysed the vibrations induced by flow (FIV) on a splitter plate by coupling two solvers. They used Immersed boundary-flow solver and FEM based solver, which is structural coupled by implicit partitioning. The coupling is implemented with the dynamic under-relaxation parameter for faster computation and numerical stability compared to constant under-relaxation.

From the above literature, it can be seen that the vortex-induced vibration-based energy generation and related developments like flow physics, various materials involved, structural requirements, etc., have been gaining attention across the global research community. Hence, in our present investigation, the structural and flow dynamics characteristics of vortex-induced Vibrations flow over a circular cylinder having a splitter plate is explored. This Splitter plate is analogous to the piezoelectric structural element specifically kept under the influence of vortex-induced Vibrations to harness wind energy while optimising on spatial and installation concerns. The multi-physics numerical simulation is performed using COMSOL software [19 – 22]. Here a splitter plate is allowed to move along a transverse direction to the flow hence the fluctuations in lift force and/or thrust causes the oscillation in the body, i.e. vortex-induced vibration [23 - 26]. These mechanical vibrations are converted to alternating current using piezoelectricity, which means electricity generated from pressure and latent heat. Here, present paper work simulates and studied the flow at Reynolds number of 100. The density ratio is varied from 5 – 40, and its effect is observed on the amplitude and lift and drag forces produced.

2. Mathematical formulation and Theory of Vortex generation

This section presents the basics of vortex-induced vibration generated behind a cylindrical object to identify the ideal location to set up piezoelectric plates.

2.1. Mathematic equation: Navier-Strokes Equation

The two dimensional, unsteady viscous incompressible flow Navier–Stokes equation for a Newtonian fluid [19 – 22] flowing over a circular cylinder can be given as
can be written as

$$\frac{\partial U}{\partial X}+\frac{\partial V}{\partial Y}=0 \tag{1}$$

$$\frac{\partial U}{\partial t}+U\frac{\partial U}{\partial X}+V\frac{\partial U}{\partial Y}=-\frac{\partial P}{\partial X}+\frac{1}{Re}\nabla^2(U) \tag{2}$$

$$\frac{\partial V}{\partial t}+U\frac{\partial V}{\partial X}+V\frac{\partial V}{\partial Y}=-\frac{\partial P}{\partial Y}+\frac{1}{Re}\nabla^2(V) \tag{3}$$

where the dimensionless velocity (U and V), dimensionless time (t), dimensionless temperature difference (θ), and dimensionless pressure (P) are,

$$U=\frac{u}{u_\infty};V=\frac{v}{u_\infty};\ t=\frac{t_r u_\infty}{D};\qquad P=\frac{p-p_\infty}{\rho u_\infty};\qquad X=\frac{x}{D};\ Y=\frac{y}{D}; \tag{4}$$

where v represents flow velocity in the respective x, y directions, p denotes pressure, T_w is cylinder wall temperature, T_∞ represents far-field fluid's temperature. u_∞ is the reference far-field fluid velocity, D is the cylinder's diameter (characteristic dimension), ρ, μ represents the fluid density and dynamic viscosity respectively, ν, β_t and α is the kinematic viscosity, thermal coefficient of expansion and thermal diffusivity of the fluid respectively.

2.2. Vortex Generation concept

Vortex is a whirlpool that one can witness in our daily life. The basic theory of vortex is when the air hits a bluff body (cylinder and or plate), it will start shedding vortex on the downstream side of the cylinder. Here, the aim is to convert these vortices generated into usable electricity. The concept of vortex generation is related to two major parameters: Reynold's number and Strouhal's number [26 - 28]. Vortex can be generated from a particular unsteady incompressible flow, mostly for Reynold number higher than 40 till to a large value of 3.5 x10^6. The Reynolds number (*Re*) can be given as

$$Re = \frac{\rho\, V\, D}{\mu} \tag{5}$$

where, ρ is fluid density, V is fluid flow velocity, D is characteristic dimension and μ is the fluid viscosity.

The parameter Strouhal number (*St*) can be given as

$$St = \frac{fD}{V} \tag{6}$$

where f represents frequency for vortex shedding.

2.3. Piezoelectric effect

The piezoelectric effect refers to the formation of electric charges on the outer surface of insulating materials in proportion to the influence of mechanical stress. The converse also stands true, known as the converse piezoelectric effect. The mechanical stress is proportional linearly to the amount of electric charge generated, and the constant of this proportionality is the piezoelectric constant (d) and given as

$$d = k\sqrt{\varepsilon_0 k^T S^E} \tag{7}$$

where d = density of charge in open circuit/stress applied, k is electro-mechanical coupling coefficient, k^T represents relative dielectric constant, and S^E represents elastic compliance (10 m/N) at a constant electrical field.

The PVDF (polyvinylidene fluoride plate) is made using polyvinylidene fluoride, a type of non-reactive thermoplastic and is available in various forms like tubes, sheets, etc. It is synthesised by controlled radical polymerisation using vinylidene fluoride (VDF) monomer. PVDF is usually available in the non-piezoelectric phase and must be annealed to get it in the piezoelectric phase. However, PVDF thin films have stresses between thin films, sufficient to get into the piezoelectric phase.

2.4. Structural modelling of flow-induced vibration of a circular cylinder Problem

For the structural modelling of the vibrating splitter plate, the dynamic equation of the structural system can be written [29 - 30] as

$$\rho_S \frac{\partial^2 X}{\partial t^2} = \frac{\partial \sigma_{ij}}{\partial x_j} + \rho_S f_i \tag{8}$$

Where ρ_S structural density, X is displacement, σ_{ij} Tensor (Cauchy Stress) and f_i is the force component. Since for elastic material, the structure has geometric nonlinearity, Saint Venant–Kirchhoff law is employed, and thus the stress and strain relationship is based on Green Lagrangian tensor for strain (E) and 2nd Piola–Kirchhof tensor for stress, *S(E)* as

$$S(E) = \int F^{-1}\sigma\, F^{-T} \tag{9}$$

IOP Conf. Series: Materials Science and Engineering 1258 (2022) 012021 doi:10.1088/1757-899X/1258/1/012021

3. Simulation Results and Discussion

Figure 1 shows the schematic of the cylinder-elastic plate geometry and an outer domain considered for the current study. The domain has a size of 3 x 1.2 and is divided into three parts. The first one with dimensions 0.6 x 1.2 starts from the inlet to the point of contact or half of the cylinder. The second one with dimension 0.55 x 1.2 includes the plate and the other half of the cylinder; this domain is very important. The fluid-solid interactions for our simulation happen here, and meshing in this domain is finer than in other domains. The third one with dimensions 1.85 x 1.2. A cylinder has a diameter of 0.1 units, and a plate has a length of 0.35 units with 0.02 unit width and is fixed with the cylinder.

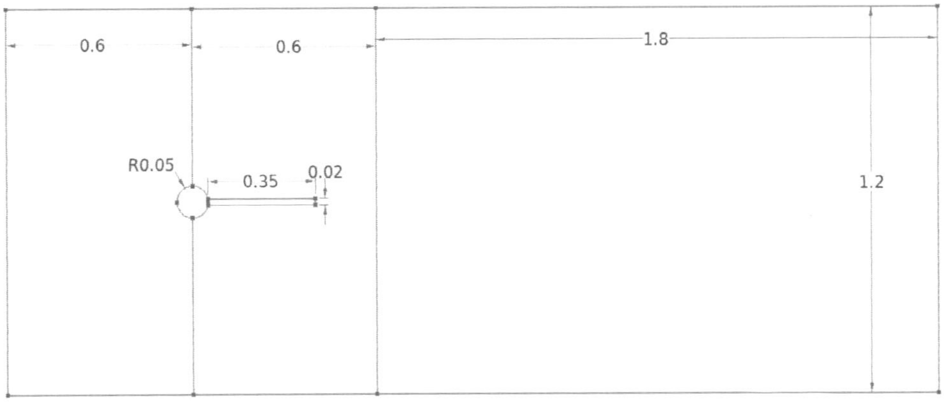

Figure 1. Schematic Diagram of the domain.

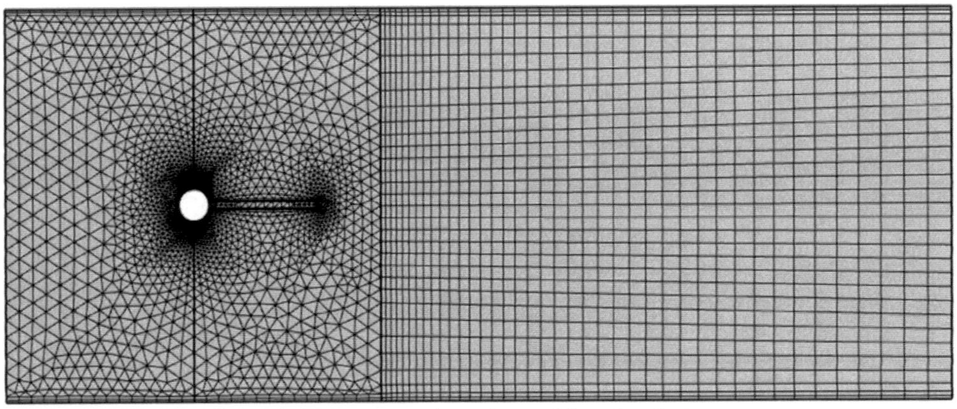

Figure 2. Unstructured mesh used for simulation.

Table 1: Mesh details

Description	Value
Minimum element quality	0.1516
Average element quality	0.9137
Triangle	9691
Quad	4518

Table 2: Flexible Plate: Material properties

Properties	**Variable**	**Expression**	**Unit**
Young's Modulus	E	2.27	GPa
Poisson's Ratio	Y	0.225	1
Density	ρ_s	1780	kg/m^3

Table 3: Fluid properties

Properties	Variable	Expression	Unit
Density	ρ_f	2.27	GPa
Dynamic Viscosity	μ_f	0.225	Pa.s

Figure 2 shows the unstructured mesh used for the simulation. The unstructured mesh contains the 14209 mixed; quad and triangular elements with 12876 total nodes. Table 1 shows the details of mesh quality. This mesh is selected based on the mesh convergence study conducted. The mesh convergence is not presented here for the sake of brevity. As per as boundary conditions are concerned; The fluid enters from the left side with fluid flow velocity corresponding to the Reynolds numbers of 100. The right, top and bottom sides are given as pressure outlet conditions. The cylinder is fixed, and no-slip boundary conditions are provided. The moving plate's tip vibration/oscillation movement is evaluated using equation (9) and allowed to be displaced as per fluid forces.

Table 2 shows the flexible plate material properties, whereas table 3 shows the fluid–air properties. Here, during simulation, the plate material properties such as Young's Modulus and Poisson's Ratio taken same as of [18]. The density of the plate is varied from 5 to 40 kg/m^3 in 9 different cases, and due to this, the mass ratio changed in every case. Here the effect of mass ratio on vortex shedding capacity is studied. The bending stiffness is kept the same during this process, and the Reynolds number is maintained as 100. Also, note that the Kb has been constant due to constant/same values for young's modulus and dimensions. During simulation the density is increased for each case, which led to increases in the mass ratio (M). The value of Ur changes with changes in M and Kb. AYtip is the oscillation amplitude in the Y direction.

Now, the simulated model demonstrates the working of proposed wind energy harvester in real outdoor conditions. Figure 3 shows the velocity contours obtained after 200 s for the case 1 scenario. It can be seen that the vortex shedding is generated downstream of the cylinder creating the unsteady flow. The unsteady flow causes the time-varying lift and drag forces, as shown in figure 4. The displacement attained by the PVDF flexible plate is shown in figure 5. It is observed the PVDF film will vibrate with an amplitude of 138.23mm on each side. Hence, an alternating current could be generated. Depending upon the application, an electronic circuit needs to be connected to the output terminals of PVDF.

Table 4 shows the simulation parameters and the results obtained, especially for the oscillation amplitude of the flexible PVDF plate. It is observed the flexible plate displacement increases with increasing density ratio. However, after a certain density ratio (larger than 60), A_{YTip} starts to decrease as density increases, this is because the wind speed is not strong enough to oscillate the plate. The natural frequency of oscillation decreases with an increasing density ratio. When the ratio F_p/F_{n2} gets closer to 1, the lock-in condition is achieved, which means that resonance occurs between vortex shedding and the plate. This lock-in condition makes the oscillations continuous with a large and constant amplitude.

IOP Conf. Series: Materials Science and Engineering 1258 (2022) 012021 doi:10.1088/1757-899X/1258/1/012021

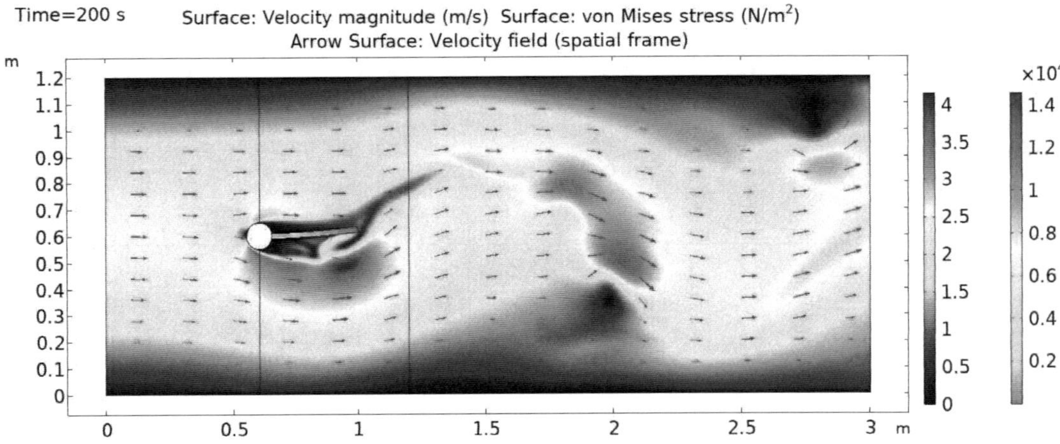

Figure 3. Velocity contours for case-1 obtained after time 200 s.

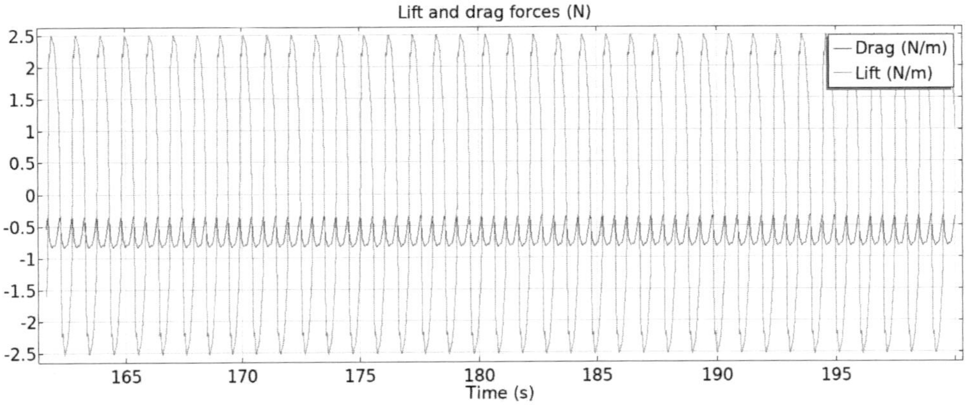

Figure 4. Variation of lift and drag forces over a flexible vibrating plate for case-1.

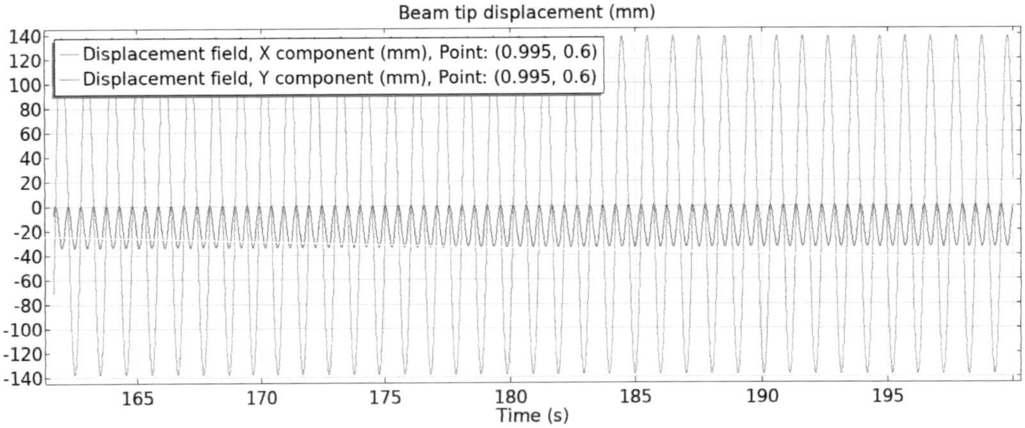

Figure 5. Displacement of a flexible vibrating plate for case-1.

Table 4: Simulation results for various density ratio

Case	Density Ratio	K_b	Fn_2	M	U_r	Simulated results		
						A_{YTip}	F_p	F_p/F_{n2}
1	5	0.0218	0.276	0.286	3.623	0.74	0.213	0.769
2	7.5	0.0218	0.225	0.429	4.437	0.885	0.196	0.867
3	10	0.0218	0.195	0.572	5.123	1.018	0.18	0.919
4	15	0.0218	0.159	0.859	6.275	1.25	0.158	0.989
5	20	0.0218	0.138	1.145	7.246	1.474	0.14	1.012
6	25	0.0218	0.123	1.431	8.101	1.67	0.127	1.026
7	30	0.0218	0.112	1.717	8.874	1.775	0.116	1.02
8	35	0.0218	0.104	2.003	9.585	1.818	0.106	1.013
9	40	0.0218	0.097	2.289	10.24	1.975	0.098	1.00

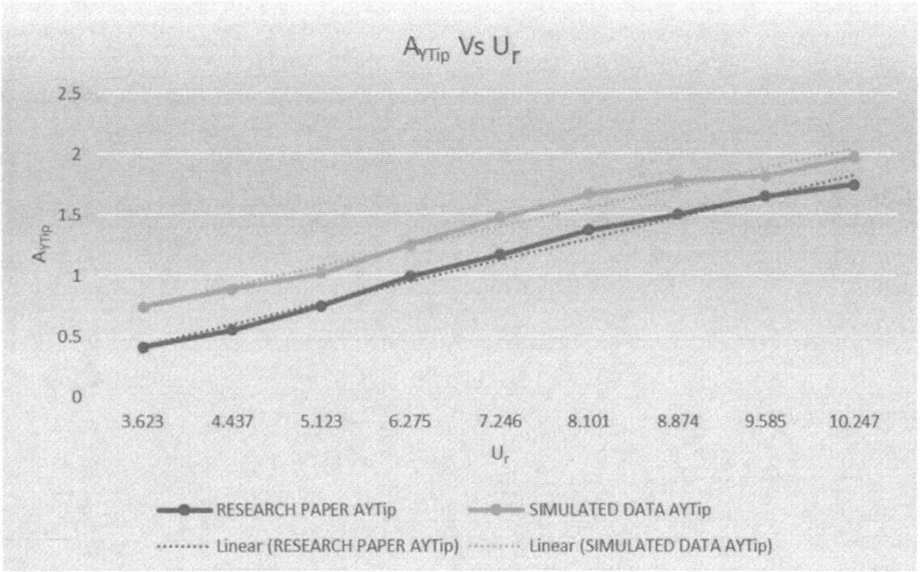

Figure 6. Comparison of displacement of a flexible vibrating plate with literature [17] results

4. Conclusion

This paper successfully simulates the model presented by literature [18] using COMSOL multi-physics software. The density ratio varies from 5 – 40 values, and the fluid flow characteristic is observed. The plate displacements are the major parameter for electricity generations using the piezoelectric concept. The average difference of 0.27 m in tip displacement value, Ay_{tip} is seen, refer to figure 6, compared to literature results. The K_b and mass ratio, M, are the major factors that affect the vortex shedding capacity of a plate or splitter. The Polyvinylidene fluoride (PVDF) film of length 35 cm and thickness 2 cm, when used as the piezoelectric material for our energy harvester, shows a displacement of 13.823 cm when subjected to the wind velocity of 1.67 m/s, which is the average wind velocity of Pune. This occurs when it is mounted on a cylinder of 10 cm diameter. The robustness of the PVDF plate is such that it can withstand extreme outer conditions due to its material property. It is suitable for all climatic conditions. As per the results, the oscillating frequency is 0.98 Hz with an amplitude of 138mm will increase Ay_{tip} and hence the power output. The current model of wind energy harvester is easy to manufacture, inexpensive, compact, eco-friendly, and lightweight compared to the conventional turbine-based harvesters. It can be placed in rural areas where electricity is scarce. This model can be used where there is a constant scarcity of electricity.

5. References

[1] Website of Ministry of New and Renewable energy : https://mnre.gov.in/wind/current-status/

[2] Antonio Barrero-Gil, Santiago Pindado, Sergio Avila, Extracting energy from Vortex-Induced Vibrations: A parametric study, Applied Mathematical Modelling, Volume 36, Issue 7, 2012, Pages 3153-3160, https://doi.org/10.1016/j.apm.2011.09.085.

[3] Atul Kumar Soti, Mark C. Thompson, John Sheridan, Rajneesh Bhardwaj, Harnessing electrical power from vortex-induced vibration of a circular cylinder, Journal of Fluids and Structures, Volume 70, 2017, Pages 360-373, https://doi.org/10.1016/j.jfluidstructs.2017.02.009.

[4] Zhang, L., Mao, X., & Ding, L. (2019). Influence of attack angle on vortex-induced vibration and energy harvesting of two cylinders in side-by-side arrangement. Advances in Mechanical Engineering. https://doi.org/10.1177/1687814018822598

[5] Zunlong Jin, Guoping Li, Junlei Wang, Zhien Zhang, "Design, Modeling, and Experiments of the Vortex-Induced Vibration Piezoelectric Energy Harvester with Bionic Attachments", Complexity, vol. 2019, Article ID 1670284, 13 pages, 2019. https://doi.org/10.1155/2019/1670284

[6] Chin, W.K.; Ong, Z.C.; Kong, K.K.; Khoo, S.Y.; Huang, Y.-H.; Chong, W.T. Enhancement of Energy Harvesting Performance by a Coupled Bluff Splitter Body and PVEH Plate through Vortex Induced Vibration near Resonance. *Appl. Sci.* 2017, *7*, 921.

[7] Swathi Ippili, Venkatraju Jella, Ji-Ho Eom, Jaegyu Kim, Seungbum Hong, Jin-Seok Choi, Van-Dang Tran, Nguyen Van Hieu, Yun-Jeong Kim, Hye-Jin Kim, Soon-Gil Yoon, An eco-friendly flexible piezoelectric energy harvester that delivers high output performance is based on lead-free MASnI3 films and MASnI3-PVDF composite films, Nano Energy, Volume 57, 2019, Pages 911-923, https://doi.org/10.1016/j.nanoen.2019.01.005.

[8] Mohammad Mahdi Abolhasani, Kamyar Shirvanimoghaddam, Minoo Naebe, PVDF/graphene composite nanofibers with enhanced piezoelectric performance for development of robust nanogenerators, Composites Science and Technology, Volume 138, 2017, Pages 49-56, https://doi.org/10.1016/j.compscitech.2016.11.017.

[9] Olivier Doaré, Sébastien Michelin, Piezoelectric coupling in energy-harvesting fluttering flexible plates: linear stability analysis and conversion efficiency, Journal of Fluids and Structures, Volume 27, Issue 8, 2011, Pages 1357-1375, https://doi.org/10.1016/j.jfluidstructs.2011.04.008.

[10] Lee, C., Tarbutton, J.A. Polyvinylidene fluoride (PVDF) direct printing for sensors and actuators. Int J Adv Manuf Technol 104, 3155–3162 (2019). https://doi.org/10.1007/s00170-019-04275-z

[11] Sunija Sukumaran, Samir Chatbouri, Didier Rouxel, Etienne Tisserand, Frédéric Thiebaud, et al.. Recent advances in flexible PVDF based piezoelectric polymer devices for energy harvesting applications. Journal of Intelligent Material Systems and Structures, SAGE Publications, 2020, 32 (7), pp.746-780. ⟨10.1177/1045389X20966058⟩. ⟨hal-02994589⟩

[12] Lee, Y.J., Qi, Y., Zhou, G. et al. Vortex-induced vibration wind energy harvesting by piezoelectric MEMS device in formation. Sci Rep 9, 20404 (2019).

[13] K. Narendran, K. Murali, V. Sundar, Investigations into efficiency of vortex induced vibration hydro-kinetic energy device, Energy, Volume 109, 2016, Pages 224-235, https://doi.org/10.1016/j.energy.2016.04.110.

[14] *Liu, Jianjun et al. "Wind energy harvesting using piezoelectric macro fiber composites based on flutter mode." Microelectronic Engineering 231 (2020): 111333.* https://doi.org/10.1016/j.mee.2020.111333

[15] Muhterem Koç, Levent Paralı, Osman Şan, Fabrication and vibrational energy harvesting characterisation of flexible piezoelectric nanogenerator (PEN) based on PVDF/PZT, Polymer Testing, Volume 90, 2020, 106695, https://doi.org/10.1016/j.polymertesting.2020.106695.

[16] Inki Jung, Youn-Hwan Shin, Sangtae Kim, Ji-young Choi, Chong-Yun Kang, Flexible piezoelectric polymer-based energy harvesting system for roadway applications, Applied Energy, Volume 197, 2017, Pages 222-229, https://doi.org/10.1016/j.apenergy.2017.04.020.

[17] Michelin, S., & Doaré, O. (2013). Energy harvesting efficiency of piezoelectric flags in axial flows. Journal of Fluid Mechanics, 714, 489-504. doi:10.1017/jfm.2012.494

[18] Kundu, A., Soti, A.K., Garg, H. et al. Computational modeling and analysis of flow-induced vibration of an elastic splitter plate using a sharp-interface immersed boundary method. SN Appl. Sci. 2, 1110 (2020). https://doi.org/10.1007/s42452-020-2876-z

[19] C. R. Sonawane, Y. B. More, Anand Kumar Pandey, "Numerical simulation of vortex-induced vibrations of a circular cylinder: Isothermal and Heat transfer cases", E3S Web of Conferences, 128, 10010, 2019. https://doi.org/10.1051/e3sconf/201912809001

[20] C. R. Sonawane, Y. B. More, Anand Kumar Pandey, "Numerical Simulation of unsteady channel flow with moving indentation", E3S Web of Conferences, 128, 09001, 2019. https://doi.org/10.1051/e3sconf/201912810010

[21] Chandrakant R. Sonawane, Yogesh B. More & Anand Kumar Pandey (2022) Numerical Simulation of Unsteady Channel Flow with a Moving Indentation Using Solution Dependent Weighted Least Squares Based Gradients Calculations over Unstructured Mesh, Heat Transfer Engineering, 43:3-5, 300-313, DOI: 10.1080/01457632.2021.1874661

[22] Sonawane Chandrakant, Hitesh Panchal & Kishor Kumar Sadasivuni (2021) Numerical simulation of flow-through heat exchanger having helical flow passage using high order accurate solution dependent weighted least square based gradient calculations, Energy Sources, Part A: Recovery, Utilization, and Environmental Effects, DOI: 10.1080/15567036.2021.1900457

[23] Jadav C. Mandal & Chandrakant R. Sonawane (2014) Simulation of Moderator Flow and Temperature Inside Calandria of CANDU Reactor Using Artificial Compressibility Method, Heat Transfer Engineering, 35:14-15, 1254-1266, DOI: 10.1080/01457632.2013.876802

[24] Mandal, J.C. and Sonawane, C.R. (2013), "Simulation of flow inside differentially heated rotating cavity", International Journal of Numerical Methods for Heat & Fluid Flow, Vol. 23 No. 1, pp. 23-54. https://doi.org/10.1108/09615531311289097

[25] J.C. Mandal, C.R. Sonawane, A.S. Iyer, S.J. GosaviInamdar, Incompressible flow computations over moving boundary using a novel upwind method, Computers & Fluids, Volume 46, Issue 1, 2011, Pages 348-352, https://doi.org/10.1016/j.compfluid.2010.08.011.

[26] Sonawane Chandrakant R., Priyambada Praharaj, Anand Pandey, Atul Kulkarni, High order accurate numerical simulation of vortex induced vibration of a cooled circular cylinder case using solution dependent weigted least square gradients calculations, E3S Web of Conferences, Vol 321 04005, 2021. https://doi.org/10.1051/e3sconf/202132104005

[27] C R Sonawane, J C Mandal and S. P. Roa, "High-Resolution Incompressible Flow Computations over Unstructured Mesh using SDWLS Gradients" Journal of The Institution of Engineers (India): Series C, Feb 2019, Vol 100, issue 1, pp 83-96 (DOI: 10.1007/s40032-017-0390-x).

[28] Sonawane Chandrakant R., Atul Kulkarni, Rohan V. Sawant, Anand Kumar Pandey, Hitesh Panchal, Kishor Kumar Sadasivuni & Anil Kumar (2021) Numerical simulation of cascaded cyclone separator for nanosize aerosol, Energy Sources, Part A: Recovery, Utilization, and Environmental Effects, DOI: 10.1080/15567036.2021.1974127

[29] Chandrakant R. Sonawane, Kuldeep Tolia, Anand Pandey, Atul Kulkarni, Hitesh Punchal, Kishor Kumar Sadasivuni, Anil Kumar & Mohammad Khalid (2021) Experimental and numerical analysis of heat transfer and fluid flow characteristics inside pulsating heat pipe, Chemical Engineering Communications, DOI: 10.1080/00986445.2021.1974413

[30] Chandrakant Sonawane, Priyambada Praharaj, Anand Pandey, Atul Kulkarni, Ketan Kotecha, Hitesh Panchal, Case studies on simulations of flow-induced vibrations of a cooled circular cylinder: Incompressible flow solver for moving mesh problem, Case Studies in Thermal Engineering, Volume 34, 2022, 102030, https://doi.org/10.1016/j.csite.2022.102030.

Optimization and S-N life cycle analysis of helical gear with nickel alloy MP35N

Amit D Modi[1,2,4], **K R Gawande**[3]

[1]Mechanical Engineering Department, C U Shah University, Wadhwan -363030, Gujarat, India
[2]Mechanical Engineering Department, Government Polytechnic, Amreli-365601, Gujarat, India
[3]Mechanical Engg. Deptartment, C U Shah University, Wadhwan City-363030, Gujarat, India

a_d_modi@yahoo.co.in (Amit D Modi)

Abstract : Gears are most critical components in mechanical power transmission systems. Helical gear is used to transmit power or torque at high speed, heavy loads with smooth and silent operation. It involves many variables including specific materials with consideration of force/stress analysis and its mechanical properties. The optimal design and its lead curve is the primary objective of the present work. The bending and surface strength of the gear tooth is considered to be the main contributors to gear failure. Thus, stress analysis on gear is vital to reduce the failures for optimal design. Analytical design for helical gear has been done with five different sets of parameters and then 3D modelling and FEM carried out with 30 cases and optimization was carried out to derive the best selection so that the fillet stress can be reduced and the life can be improved. The optimum lead curve is determined by the design of the experiment, transmission error calculation and response surface method. The results obtained in analytical design and FEM have been validated and it has been observed that by means of life it's clear that Nickel alloy MP35N is more valuable and better material for manufacturing helical gear.

Keywords: Life cycle analysis, Safety factor, FE Analysis, Equivalent stress, Moment-load analysis

1. Introduction

Helical gear transmission is largely used in mechanical transmission systems due to smooth operation, long life, low noise, economic performance and drive reliability[1]. Gear optimization and modification is an emerging technology which is more widely used in recent years, includes the modification along the tooth profile, axial and three-dimensional modification to reduce the vibration of gear transmission device[2].The optimal design of the gear and its lead curve is the primary objective of the present work. The optimum lead curve is determined by the design of experiment, transmission error calculation, and response surface method[3]. Further, the working life span of helical gear was found, which shows promising results. It was understood that the transmission error or stress fatigue results in wear and thus reduces the gear lifetime[4,5].The material selection in the present work is the main factor while the other critical parameters are constant as of the other materials selected. It is observed that the optimum input values of Nickel alloy MP35N result in a significant life factor of helical gear, designed by standard physical parameters based on AGMA standards. All the elements in MP35N have relatively high modulus [6].

Fatigue, life stress, and moment are the crucial parameters affecting the gear design, which need to be optimized[7]. Finite Element Analysis (FEA) using ANSYS is compared with analytical values, which shows satisfactory results in terms of the moment of the gear. Gear specification used in this analysis is shown in Table 1..

Table 1. Parametric chart for analytical design (Set-5)

Nomenclature	**Notation**	**Units**	**Helical Gear** No of Teeth Z_1=28 and Z_2=31	**Output Value**
Normal Module	m_n	mm	$m_n = \frac{2a\cos\alpha}{Z_1+Z_2}$	4
Transverse Module	m_t	mm	$m_t = \frac{m_n}{\cos\alpha} = \frac{2a}{Z_1+Z_2}$	4.14
Center Distance	a	mm	$a = \frac{m_n}{\cos\alpha}\frac{Z_1+Z_2}{2}$	122.13
Height Factor	f_o	mm	$f_o = 1$ (for Special Cases $f_o = 0.8$)	0.8
Bottom Clearance	c	mm	$c = 0.25m_n$	1
Tooth Depth	h	mm	$h = 2.25m_n$	9
Pitch Diameter or Reference Diameter	d	mm	$d_1 = \frac{m_n}{\cos\beta}Z_1$ $d_2 = \frac{m_n}{\cos\beta}Z_2$	115.92 128.34
Tip Diameter	d_c	mm	$d_{c1} = \left(\frac{Z_1}{\cos\alpha}+2f_o\right)m_n$ $d_{c2} = \left(\frac{Z_2}{\cos\alpha}+2f_o\right)m_n$	122.35 134.77
Root Diameter	d_f	mm	$d_{f1} = d_{c1} - 2c$ $d_{f2} = d_{c2} - 2c$	121.35 132.77
Helix Angle	α	mm	For Helical $\alpha = 8^{\circ} to 25^{\circ}$ (Std) $\cos\alpha = \frac{m_n(Z_1+Z_2)}{2a}$	0.966 $\alpha = 15^{\circ}$

Lead error, lead crowning, and input torque are selected as the candidate design variables. To eliminate unimportant variables, the effect of transmission error analysis on the candidate ones for a screening experiment has been investigated using Analysis of Variance[3].

2. Data Sheet of Analysis

The optimization datasheet in Table 2shows the various materials used for the analysis and experiments. The analysis was carried out over equivalent stress and factor of safety to get the optimized results. The stress analysis and safety factor are important for analyzing the performance and life of any gear [8]. It is obvious that there are various metals that are not being used to manufacture the gear, because of their cost and machining difficulties. But still, there are sufficient materials being available that is left out due to uncertainness and lack of experimentation. Those types of materials are listed in Table 2.

Table 2. Optimization data sheet

Case No.	Material	Torque (N-mm)	Equivalent Stress (Mpa)			Safety Factor		
			Min	Max	Avg	Min	Max	Avg
Case 1	AISI 4140 Alloy Steel	95000	0.000264	194.83	3.5603	3.3619	15	14.895
Case 2	AISI 4140 Alloy Steel	159000	0.000443	326.08	5.9588	2.0087	15	14.746
Case 3	AISI 4140 Alloy Steel	191000	0.000532	391.71	7.1581	1.6722	15	14.681
Case 4	AISI 4140 Alloy Steel	127000	0.000353	260.45	4.7596	2.5148	15	14.818
Case 5	AISI 4140 Alloy Steel	209000	0.000582	428.62	7.8327	1.5282	15	14.648
Case 6	304 Alloy Steel	95000	0.000221	197.55	3.5733	1.2301	15	14.568
Case 7	304 Alloy Steel	159000	0.000369	330.63	5.9806	0.73495	15	14.362
Case 8	304 Alloy Steel	191000	0.000444	397.18	7.1843	0.61182	15	14.286
Case 9	304 Alloy Steel	127000	0.000295	264.09	4.777	0.92014	15	14.451
Case 10	304 Alloy Steel	209000	0.000485	434.61	7.8613	0.55913	15	14.248
Case 11	AISI 6061 Aluminum Alloy	95000	0.000234	189.56	3.5387	1.3663	15	14.596
Case 12	AISI 6061 Aluminum Alloy	159000	0.000391	317.26	5.9227	0.81635	15	14.39
Case 13	AISI 6061 Aluminum Alloy	191000	0.00047	381.12	7.1147	0.67958	15	14.315
Case 14	AISI 6061 Aluminum Alloy	127000	0.000313	253.41	4.7307	1.022	15	14.48
Case 15	AISI 6061 Aluminum Alloy	209000	0.000514	417.03	7.7852	0.62105	15	14.277
Case 16	Nickel Cobalt Chromium Alloy	95000	0.000264	191.51	3.5461	4.287	15	14.941
Case 17	Nickel Cobalt Chromium Alloy	159000	0.000441	320.52	5.9351	2.5614	15	14.82
Case 18	Nickel Cobalt Chromium Alloy	191000	0.00053	385.03	7.1296	2.1323	15	14.761
Case 19	Nickel Cobalt Chromium Alloy	127000	0.000352	256.01	4.7406	3.2068	15	14.882
Case 20	Nickel Cobalt Chromium Alloy	209000	0.00058	421.32	7.8015	1.9487	15	14.731
Case 21	High Carbon Steel	95000	0.000264	194.83	3.5603	3.906	15	14.926
Case 22	High Carbon Steel	159000	0.000443	326.08	5.9588	2.3338	15	14.795
Case 23	High Carbon Steel	191000	0.000532	391.71	7.1581	1.9428	15	14.734
Case 24	High Carbon Steel	127000	0.000354	260.45	4.7596	2.9218	15	14.86
Case 25	High Carbon Steel	209000	0.000582	428.62	7.8327	1.7755	15	14.703
Case 26	MP35N Alloy (AMS 5844)	95000	0.000263	174.46	3.4913	9.0854	15	14.998
Case 27	MP35N Alloy (AMS 5844)	159000	0.00044	291.98	5.8433	5.4284	15	14.968
Case 28	MP35N Alloy (AMS 5844)	191000	0.000529	350.75	7.0193	4.5189	15	14.942
Case 29	MP35N Alloy (AMS 5844)	127000	0.000352	233.22	4.6673	6.7962	15	14.988
Case 30	MP35N Alloy (AMS 5844)	209000	0.000579	383.8	7.6808	4.1297	15	14.925

The various optimization results of these materials show that some of them are useful for the manufacturing of helical gears. The best result is obtained when the Nickel alloy MP35N is analyzed. Based on performance and life span with a minimum stress factor (3.4913 MPa) and maximum factor of safety(9.0854), it is indeed that the material is suitable. Further, the results of the ANSYS FEA simulation are found satisfactory compared with analytical values of the gear design and mechanical properties. Figure 1 shows the variation of safety factor minimum with respect to life minimum.

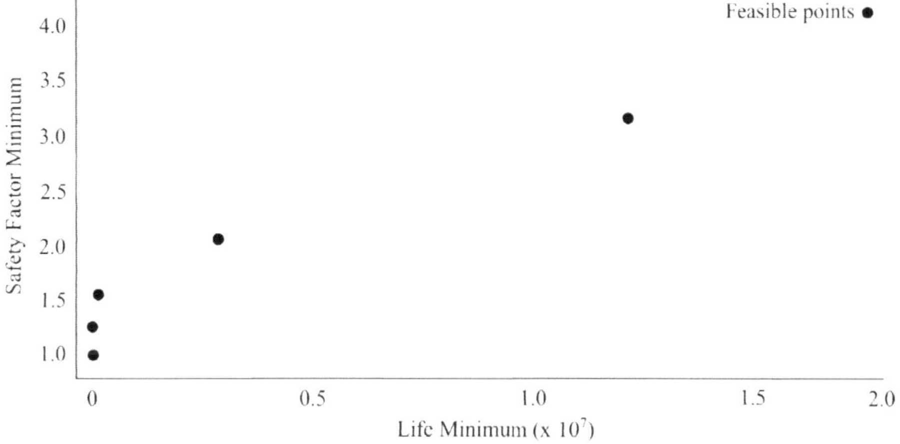

Figure 1. Result of optimization

The final optimization is performed separately over Nickel alloy MP35N, which was identified as the most suitable material. The analysis has been carried out with the consideration of stress, the factor of safety, moment, and working life. To ensure the best output, additional moments and stress values are also evaluated and put into the analysis. The resulting force and stress lead to wear and cause decreased life span. The optimization shows us the overall life, stress, and moment curves from the Figure 2.

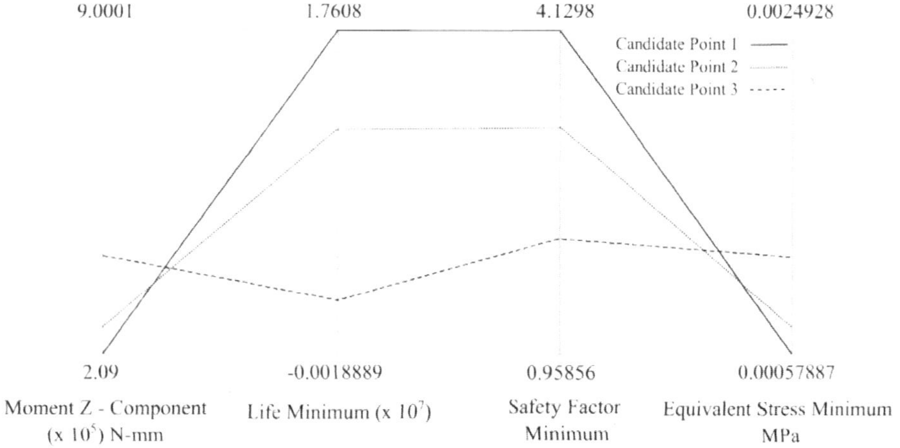

Figure 2. Outline of optimization result

The analytical value of the moment at which the gear made of Nickel alloy MP35N is matching with the ANSYS result. This also proves that the gear in general working situations will outperform the other standard materials and alloys used. Considering the stress generated upon the gear, the moment(M) is calculated as,

$$M = \sqrt{(M_1)^2 + (M_2)^2}$$

$$= \sqrt{(559800)^2 + (56000)^2}$$
$$= 562594 \text{ N} - \text{mm}$$

The bending moment and bending stress on the gear are also optimized so that chances of failure are avoided[9]. The Turning Moment of the gear is also satisfactory($600161 \text{ N} - \text{mm}$) with the analysis result. The alloy MP35N itself has better ductility and corrosion, heat, and wear-resisting nature[10]. Further, the material becomes even more suitable for almost any manufacturing operation, if age hardened[11].Till now it is used for fasteners, springs, and high ductile operations and not used for gear manufacturing because of its composite structure and the extra process of hardening or cold drawing. Indeed, it is much more reliable and better for the purpose when compared to many standards and non-standard materials available.

3. S-N Life Cycle Analysis

Constant amplitude load is made to be reversed, and reverse loading analysis is performed to understand the rotational movement and deformation of Nickel alloy MP35N that can occur during same as gears face both parallel load and reversible load in their operational life. Figure 3shows the load Applied to calculate the life cycle of the gear.

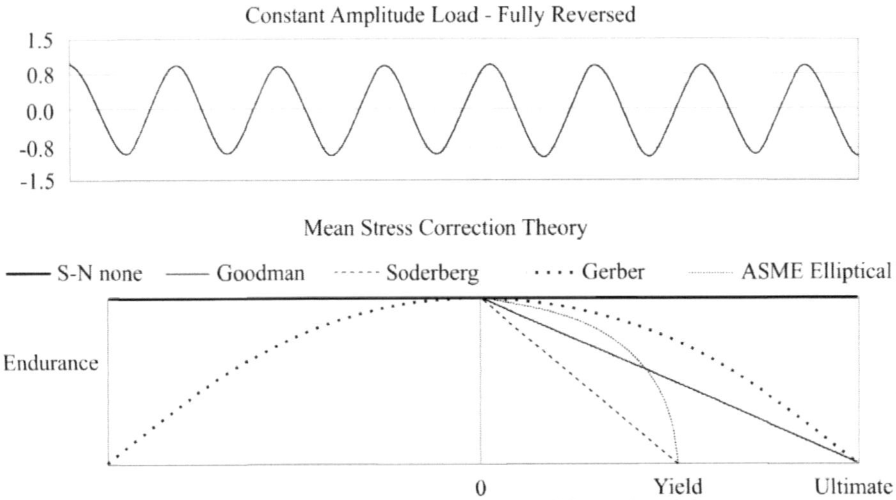

Figure 3. Load applied to calculate the life cycle of the gear

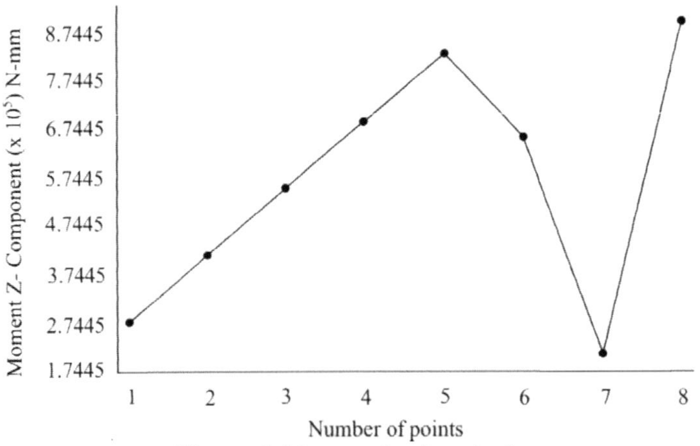

Figure 4. Moment-load analysis

Figure 4 shows the Moment to load analysis. Moments that are calculated analytically as the form of torque are applied here to understand the load variation. The load variation shows the deformation tendency of the gear which includes wear, shear, bending, misalignment, and tip error. Torque (T) is calculated as,

$$T = \frac{P \times 60}{2\pi N}$$

$$= \frac{35 \times 10^3 \times 60}{2 \times \pi \times 1600}$$

$$= 209 \times 10^3 \text{ N} - \text{mm}$$

The best value can be achieved with results of analytical calculation and valuation Set-5 where torque is maximum ($209000 \text{ N} - \text{mm}$).The stress acting upon the helical gear due to transmitted power, moment, and load. There is Normal shear stress, Principal shear stress, Bending stress, and Sustainable stress (yield stress). The overall principal stress is acting along with the turning moment of the axis of the helical gear are the types of stresses that can act on the gear when load and moment are applied.

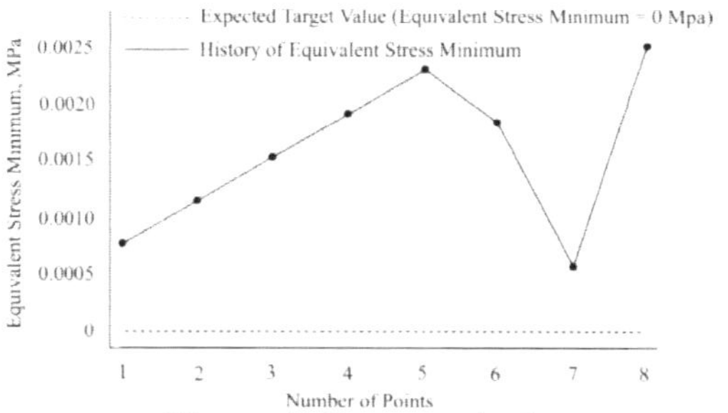

Figure 5. Stress analysis

The stress life of gear directly affect the wear of gear and shear limits and so the torque and power input need to be regulated to have a longer working life of the gear. Generally, the life of helical gear is greater than of the spur gear for its uniform load distribution due to the curve along with the teeth and thus having a larger axis contact ratio. But here, the trial has been performed using different materials and different torque referring to the standard possible dimension and parameters as per AGMA. Nickel alloy MP35N is giving the best results at a torque up to $209000 \text{ N} - \text{mm}$ and the stress life is maintained at a higher level. So, when stress is low, and the torque is high, the working life of the helical gear increases significantly[12].Thus, the S-N curve from the optimization analysis performed in ANSYS shows the superiority of the material over other standard materials. Stress analysis and life cycle analysis are shown in Figure 5 and Figure 6 respectively.

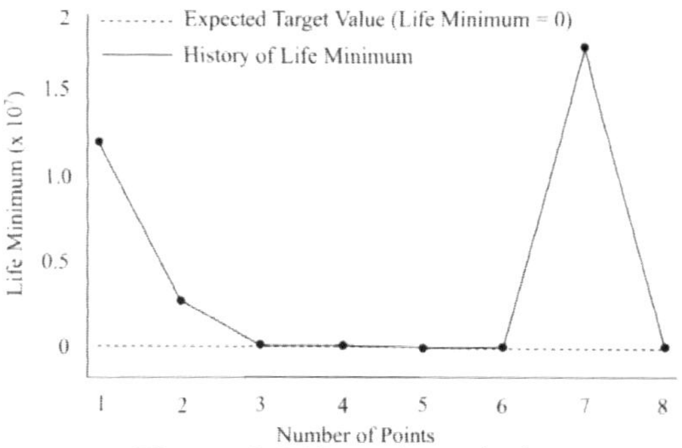

Figure 6. Life cycle analysis

While performing the analysis of the working life of the gear, all materials showed some variations in their effective work life. Among the variations, again Nickel alloy MP35N showed the most improvement of the gear life due to low wear, low fatigue, and fewer corrosion effects[10].

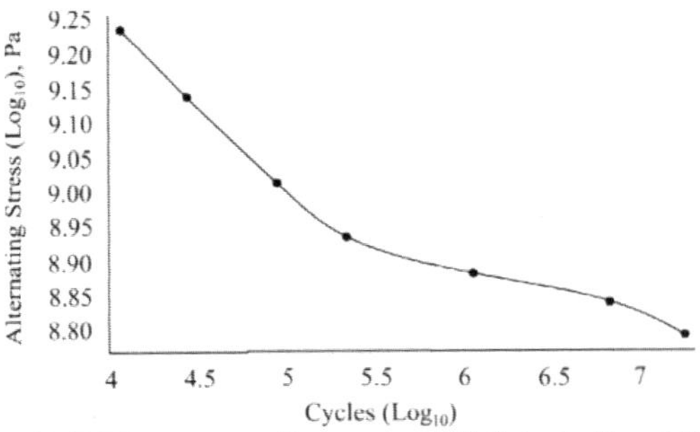

Figure 7. S-N curve performance of Nickel alloy MP35N

The age-hardened MP35N is even better when performing the analysis. The alternating stress, and cycle time analysis is performed in the last optimizing process to understand the highest reachable sturdy nature of the material, and result is shown in Figure 7.

4. Results and Discussions

By comparing the values of equivalent stress and safety factor of different materials from it is clear that Nickel alloy MP35N is the highest performing material under different loading conditions, and is much better material for manufacturing helical gear.

Table 3.Data of case studies

Case No.	Material	Torque (N-mm)	Equivalent Stress (Mpa)			Safety Factor		
			Min	Max	Avg	Min	Max	Avg
Case 1	AISI 4140 Alloy Steel	95000	0.000264	194.83	3.5603	3.3619	15	14.895
Case 2	AISI 4140 Alloy Steel	159000	0.000443	326.08	5.9588	2.0087	15	14.746
Case 3	AISI 4140 Alloy Steel	191000	0.000532	391.71	7.1581	1.6722	15	14.681
Case 4	AISI 4140 Alloy Steel	127000	0.000353	260.45	4.7596	2.5148	15	14.818
Case 5	AISI 4140 Alloy Steel	209000	0.000582	428.62	7.8327	1.5282	15	14.648
Case 6	304 Alloy Steel	95000	0.000221	197.55	3.5733	1.2301	15	14.568
Case 7	304 Alloy Steel	159000	0.000369	330.63	5.9806	0.73495	15	14.362
Case 8	304 Alloy Steel	191000	0.000444	397.18	7.1843	0.61182	15	14.286
Case 9	304 Alloy Steel	127000	0.000295	264.09	4.777	0.92014	15	14.451
Case 10	304 Alloy Steel	209000	0.000485	434.61	7.8613	0.55913	15	14.248
Case 11	AISI 6061 Aluminum Alloy	95000	0.000234	189.56	3.5387	1.3663	15	14.596
Case 12	AISI 6061 Aluminum Alloy	159000	0.000391	317.26	5.9227	0.81635	15	14.39
Case 13	AISI 6061 Aluminum Alloy	191000	0.00047	381.12	7.1147	0.67958	15	14.315
Case 14	AISI 6061 Aluminum Alloy	127000	0.000313	253.41	4.7307	1.022	15	14.48
Case 15	AISI 6061 Aluminum Alloy	209000	0.000514	417.03	7.7852	0.62105	15	14.277
Case 16	Nickel Cobalt Chromium Alloy	95000	0.000264	191.51	3.5461	4.287	15	14.941
Case 17	Nickel Cobalt Chromium Alloy	159000	0.000441	320.52	5.9351	2.5614	15	14.82
Case 18	Nickel Cobalt Chromium Alloy	191000	0.00053	385.03	7.1296	2.1323	15	14.761
Case 19	Nickel Cobalt Chromium Alloy	127000	0.000352	256.01	4.7406	3.2068	15	14.882
Case 20	Nickel Cobalt Chromium Alloy	209000	0.00058	421.32	7.8015	1.9487	15	14.731
Case 21	High Carbon Steel	95000	0.000264	194.83	3.5603	3.906	15	14.926

Case 22	High Carbon Steel	159000	0.000443	326.08	5.9588	2.3338	15	14.795
Case 23	High Carbon Steel	191000	0.000532	391.71	7.1581	1.9428	15	14.734
Case 24	High Carbon Steel	127000	0.000354	260.45	4.7596	2.9218	15	14.86
Case 25	High Carbon Steel	209000	0.000582	428.62	7.8327	1.7755	15	14.703
Case 26	MP35N Alloy (AMS 5844)	95000	0.000263	174.46	3.4913	9.0854	15	14.998
Case 27	MP35N Alloy (AMS 5844)	159000	0.00044	291.98	5.8433	5.4284	15	14.968
Case 28	MP35N Alloy (AMS 5844)	191000	0.000529	350.75	7.0193	4.5189	15	14.942
Case 29	MP35N Alloy (AMS 5844)	127000	0.000352	233.22	4.6673	6.7962	15	14.988
Case 30	MP35N Alloy (AMS 5844)	209000	0.000579	383.8	7.6808	4.1297	15	14.925

The higher corrosion resistance, ductility and thermal effect gives an overhand for the material than compared to other materials. It has been already proved that the Nickel alloy MP35N offers the best results for different combinations and set of parameters and now from the Table 4, it is clear and obvious that Set-5 has the minimum percentage error between analytical design and ANSYS FEA values, and other values are even obtained within the permissible range of percentage error.

Table 4. Comparison of analytical and ANSYS design values of MP35N

Sr. No.	Valuation Set	Equivalent Stress (MPa)		Percentage Error
		Design Value	ANSYS Value	
1	Set - 1	3.3000	3.4913	5.80 %
2	Set - 2	5.0000	5.8433	16.87 %
3	Set - 3	6.6400	7.0193	5.71 %
4	Set - 4	3.8000	4.6670	22.82 %
5	Set - 5	7.3460	7.6808	4.56 %

The contact stress analysis of the helical gear modelled using the finite element[5] is to show accurate results. The element is broadly grouped under the tetrahedral, hexahedral family of the element. In this analysis, the 3D eight nodded brick element is chosen from the hexahedral family for the element representation. This three-dimensional iso-parametric formulation element is chosen because it takes care of a shift in cross-section along the axis of the gear. To get better and accurate results, mapped-meshing is employed instead of free-meshing. The CAD model of helical gear is designed in Solid Works and is analyzed in ANSYS. The load sharing ratio of the helical gear pair teeth in one mesh cycle is normally calculated by the average load shared by teeth. The load shared among the contact element of the paired teeth is often obtained by applying the force of 10 MPa on the nodes of the inner radius of the rim in the tangential direction and constrained within the remaining direction of the pinion.

Figure 8. FEM Analysis of assembly for equivalent stress

Therefore, the gear is additionally constrained in each direction due to which the contact load is generated in each period of the contact element. The result for Von-Mises stress calculated with ANSYS for Set-4 is shown in Figure 8. The total stress analysis is done by using overall principal stresses and loads acting upon the gears (Axial load, Radial load, and Tangential load.). This helped to develop the life log of the gear in both static and dynamic conditions and allowable fatigue stress is also increased by the design of the gear. Deformation that occurred on the gear surface due to fatigue load and stress is further analyzed in the ANSYS, shown in Figure 9and the result is found satisfactory when compared to other materials selected.

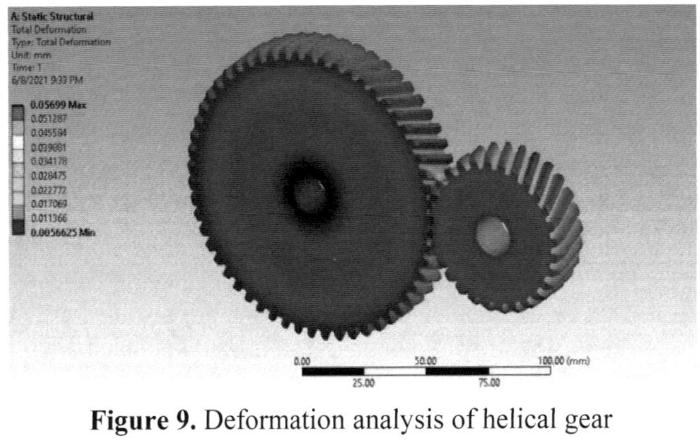

Figure 9. Deformation analysis of helical gear

The deformation noticed on the surface bending process while applying ultimate stress and load ranged from 0.005 to 0.06 mm(5 – 60 microns).Still, the deformation value goes above 50 microns when the stress is way more excessive and load is uneven, which means point loads along with uniform load are applied to check circumstantial effects on the gear.

Figure 10. Fatigue safety factor

Overall bending moment is significantly under the level of operation and fatigue is less than the value of yielding. Strain overload is achieved under the analysis of stress and strain, shown in Figure 11. The configuration of stress is $50\,\text{MPa}$ or, $50\,\text{N/mm}^2$ acting on the gear and pinion while in static condition.The results show higher bending moment and stress, due to the satisfactory strain level. The overall stress is lesser than that of the strain, thus the theoretical design satisfies the technical design calculated by means of gear life cycle under both ISO and AGMA standards.

Figure 11. Strain factor analysis

The moment reaction of the designed model of gear is analyzed from different tangential axes and in both clockwise and anti-clockwise rotation of the gear where the standard acceleration due to gravity, $g = 9.81$ is applied.The moment reaction values are then compared and configured by tangential load optimization over different time-lapses, shown in Figure 12.

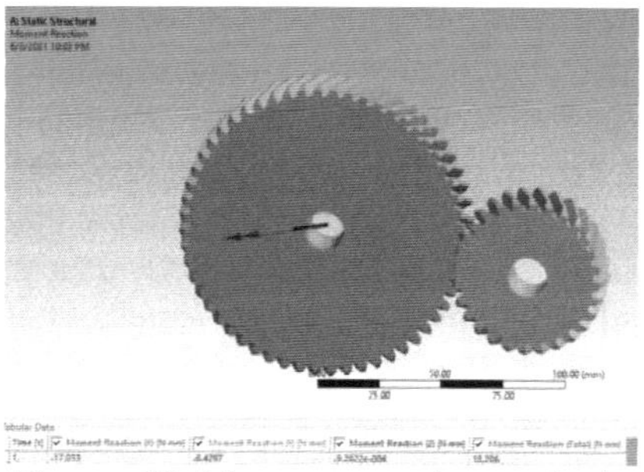

Figure 12. Moment reaction analysis

Moment Reaction at X-axis $= -17.033\ \text{N} - \text{mm}$

Moment Reaction at Y-axis $= -6.4297\ \text{N} - \text{mm}$

Moment Reaction at Z-axis $= -9.2622\ \text{e} - 004\ \text{N} - \text{mm}$

Total Moment Reaction $= -18.206\ \text{N} - \text{mm}$

Therefore, the moment reaction is modified by the twisting and turning method of value modification which includes static structural analysis. This moment reaction is satisfactory with the gear design methodology and further can be experimented upon by the same twist and turn method for future optimization with non-standard parameters of gear. The total moment of the gear is under satisfactory and safe numbers, where the chances of deformative structural failure are null. Even with the twisting moment, $M_T = 600161\ N - mm$, the structure of gear in static and dynamic conditions is safe and more efficient by means of working life cycle laps and span. However, the overall efficiency of gear is not changed by a large amount, but a very small figure of improvement is observed as in static operational condition while the load is applied. The structural analysis of the gear with the twist and turn values with twisting moment acting in static analysis is optimized and results are evaluated to match the necessary coordination, which is shown in Figure 13.

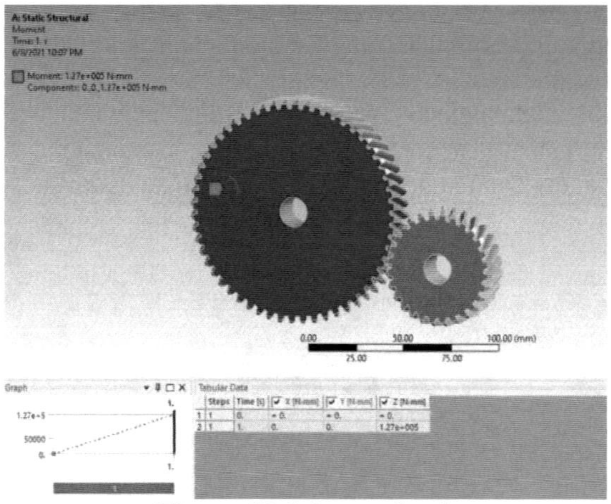

Figure 13. Moment load

The surface sustainability values are higher than that of the other materials and yield, and stress factors are safe to operate.

5. Conclusion

Analytical method utilizes a variety of basic assumptions to calculate the peak stress concentrations. This research work describes the use of computational approach to the helical equipment's surface and shear stress. A parametric study is additionally taken under consideration with a change in facial size and helix angle to justify their effect on the performance of the helical gear. Analytical design for helical gear has been done with five different sets of parameters being selected and then 3D Modelling and FEA Analysis have been carried out with 30 different cases. Further, optimization was carried out to derive the best selection of parameters so that the fillet stress can be reduced, and the life of the helical gear can be improved. The results obtained in analytical design and FEA Analysis have been validated, and it has been observed that by means of life it is clear that Nickel alloy MP35N is more of a valuable and much better material for manufacturing helical gear. Different materials have been investigated to derive the best and optimum material to increase the life of helical gear which can be used in various domains. It is very distinct that Nickel alloy MP35N is a much more valuable and efficient material for increased life. The material properties like higher corrosion resistance, ductility, and thermal effect on a higher side give an overhand for the material compared to the other materials being investigated. The field of operation has proven that the time and principal amount invested to manufacture the gear will prove to be more effective in the future this will obviously be a better option. There are certainly some scopes available for further experiments and research that can open many more doors for higher efficiency by using this material under the design parameters suggested or simply under all standard design parameters and properties.

6. Reference

[1] Sabarinath, P., Thansekhar, M. R., Jeganathan, R., and Saravanan, R. "Multi Objective Design Optimization of Helical Gear Pair Using Adaptive Parameter Harmony Search Algorithm." *Applied Mechanics and Materials*, Vols. **813–814, 2015**, pp. 1032–1036. https://doi.org/10.4028/www.scientific.net/AMM.813-814.1032.

[2] Wang, C. "Multi-Objective Optimal Design of Modification for Helical Gear." *Mechanical Systems and Signal Processing*, Vol. **157, 2021**, p. 107762. https://doi.org/10.1016/j.ymssp.2021.107762.

[3] Park, C. IL. "Multi-Objective Optimization of the Tooth Surface in Helical Gears Using Design of Experiment and the Response Surface Method." *Journal of Mechanical Science and Technology*, Vol. **24, No. 3, 2010**, pp. 823–829. https://doi.org/10.1007/s12206-010-0117-x.

[4] Rai, P., and Barman, A. G. "An Approach for Design Optimization of Helical Gear Pair with Balanced Specific Sliding and Modified Tooth Profile." *Structural and Multidisciplinary Optimization*, Vol. **60, No. 1, 2019**, pp. 331–341. https://doi.org/10.1007/s00158-019-02198-7.

[5] Kang, J. S., and Choi, Y.-S. "Optimization of Helix Angle for Helical Gear System." *Journal of Mechanical Science and Technology*, Vol. **22, No. 12, 2008**, pp. 2393–2402. https://doi.org/10.1007/s12206-008-0804-z.

[6] Ke Han, Ishmaku, A., Yan Xin, Garmestani, H., Toplosky, V. J., Walsh, R., Swenson, C., Lesch, B., Ledbetter, H., Kim, S., Hundley, M., and Sims, J. R. "Mechanical Properties of MP35N as a Reinforcement Material for Pulsed Magnets." *IEEE Transactions on Appiled Superconductivity*, Vol. **12, No. 1, 2002**, pp. 1244–1247. https://doi.org/10.1109/TASC.2002.1018627.

[7] Yavuz, S. D., Saribay, Z. B., and Cigeroglu, E. "Nonlinear Dynamic Analysis of a Drivetrain Composed of Spur, Helical and Spiral Bevel Gears." *Nonlinear Dynamics*, Vol. **100, No. 4**, 2020, pp. 3145–3170. https://doi.org/10.1007/s11071-020-05666-8.

[8] Godwin Raja Ebenezer, N., Ramabalan, S., and Navaneethasanthakumar, S. "Design Optimisation of Mating Helical Gears with Profile Shift Using Nature Inspired Algorithms." *Australian Journal of Mechanical Engineering*, **2020**, pp. 1–8. https://doi.org/10.1080/14484846.2020.1761007.

[9] Chen, Z., Zeng, M., and Fuentes-Aznar, A. "Computerized Design, Simulation of Meshing and Stress Analysis of Pure Rolling Cylindrical Helical Gear Drives with Variable Helix Angle." *Mechanism and Machine Theory*, Vol. **153, 2020**, p. 103962. https://doi.org/10.1016/j.mechmachtheory.2020.103962.

[10] Yang, Y., Radhakrishna, U., Hunter, J. F., Eagar, T. W., and Lang, J. H. "An Electromagnetic Translational Vibration Energy Harvester Fabricated in MP35N Alloy." *Journal of Microelectromechanical Systems*, Vol. **29, No. 6, 2020**, pp. 1518–1522. https://doi.org/10.1109/JMEMS.2020.3026057.

[11] Ishmaku, A., and Han, K. "Characterization of Cold-Rolled and Aged MP35N Alloys." *Materials Characterization*, Vol. **47, No. 2, 2001**, pp. 139–148. https://doi.org/10.1016/S1044-5803(01)00162-0.

[12] Liang, D., Luo, T., and Tan, R. "Generation and Meshing Analysis of a New Type of Double Helical Gear Transmission." *Mathematical Problems in Engineering*, Vol. **2018, 2018**, pp. 1–10. https://doi.org/10.1155/2018/1539849.

Machinability Studies on Hybrid Aluminum Metal Matrix Composites

Shrinivasa D[1] and G V Naveen Prakash[2]

[1]Faculty of Mechanical Engineering, Vidyavardhaka College of Engineering, Mysuru. Karnataka, India
Email: shrinivasashetty@vvce.ac.in

[2]Faculty of Mechanical Engineering, Vidyavardhaka College of Engineering, Mysuru. Karnataka, India
Email: gvnp@vvce.ac.in

Abstract: Hybrid metal matrix composites are new invention in material science and Engineering. The hybrid metal matrix composites (Hybrid MMC's) generally consist of two or more reinforcements in it. These hybrid MMC's have several benefits over the single reinforcement. The presence of reinforcements in it exhibits improved mechanical, tribological and machinability properties. Aluminium-based MMC's have made attention in recent years as engineeringmaterials. But in some Aluminium-based MMC's the use of single reinforcement may sometimes lead to weakening in its physical properties. Conversely, to overcome the problem of single reinforced composites, the idea of the use of two different types of reinforcements is being explored in the Aluminium matrix. In the hybrid MMC's, generally one of the reinforcements will be a hard phase and the other reinforcement being a soft lubricating phase. Hard reinforcements such as TiO_2, Al_2O_3, SiC, etc. will improvethe hardness and abrasive wear resistance of Aluminium but it has a negative effect on the machinability of Aluminium. To balance these effects, reinforcements like graphite which is a solid lubricant can be dispersed in Aluminium along with hard reinforcements. In the present study, the paper reviews on the effect of processing techniques, effect of reinforcement and effect of machining parameters on the machinability of hybrid aluminium metal matrix composites.

Keyword: MMC, tribological, machinability, hybrid, graphite

1. Introduction

Machinability is the most important manufacturing process to convert the casted part into the desired size and shape. Machinability refers to the easiness of removing metal in transforming raw material into a finished product. Such a comprehensive measure requires information from cutting tool or tool wear tests, cutting force tests, surface finish tests, etc.

From the industrial or manufacturing point of view, the machining operations have become challenging during the manufacturing process. Production engineers always put considerable importance for the machinability of materials. Since MMC'S are relatively new materials, comprehensive tool wear analysis data are to be established and thus research on tool wear monitoring during turning of MMCs, is gaining popularity in recent years. The influence of reinforcing particles and also machining parameters have significant effect on the machinability.

2. Machinability of Al alloys

Machinability studies on hybrid metal matrix composites (AMMC) is being researched by number of good researchers employing various machining processes. But there was a lack of clarity how the reinforcement particles will affect the machining of aluminium MMC and the effect of machining parameters during the machinability process Following are some of the research studies that have been assessed and reported categorically.

2.1 Effect of processing techniques

The most common methods for fabricating AMMC are liquid casting and powder metallurgy [1-2] and these methods are typically cost effective [3]. Before casting and solidification, the reinforcements particles are thoroughly spread over the liquid metal in the liquid casting procedure [4]. Stir casting is the simplest and cheapest method for fabricating composite materials yet stir casting composites have poor reinforcing particle integration and dispersion. This is mainly due to the reinforcement particle size, which results in showing agglomeration tendency and reduced wettability of particles within the melt. In contrast to composites made by stir casting, composites made by compo casting have a more uniform dispersion of reinforcing particles in the solidified matrix, as well as higher hardness and lower porosity [5].

2.2 Effect of reinforcements

Al 2014 aluminum alloy often used in the aerospace industry and machinability of Al 2014 alloys are determined by surface roughness and the volume % of reinforcement. Graphite and SiC are most commonly used reinforcing materials with Al 2014 due to its enhanced properties. Volume % of Graphite has major influence on the machinability of Al based MMC. Increase in graphite decreases cutting forces and surface roughness [6]. When graphite reinforced with aluminium 2219 and particularly focused on tool wear in turning operation, it is found that presence of graphite in aluminium MMC reduces tool wear in all the cutting conditions and reduces the coefficient of friction between cutting tool & graphite reinforced with aluminium alloy and also between tool & chip interface [7]. Silicon and Graphite reinforced with Aluminium LM25 hybrid MMC, it was found that presence of graphite particles in the hybrid MMC provide better machinability properties [8]. In Silicon carbide and TiB_2 reinforced with aluminium 6061 MMC, the increase in weight percentage of reinforcement (Silicon carbide 10 % & TiB_2 5%) leads to formation of cluster and from wear analysis it was observed that TiB_2 particles increase wear resistance property of hybrid MMC. It was also noticed that the weight % of TiB_2 reinforcement is the most important factor on surface roughness [9]. In Al 6061 material, the presence of TiC reinforcements increased the hardness of the material and also observed that increasing TiC reinforcement increases surface roughness of Al 6061 MMC and flank wear of the cutting tool [10]. Presence of graphite in the aluminum reinforced with silicon and zinc hybrid MMC improves surface finish due to reduction of friction between tool and surface of the composites which leads to improvements in the machinability [11].

2.3 Effect of Machining Parameters

The current study focuses on optimizing machining process parameters for aluminum alloys and their impact on surface roughness and material removal rate. Because aluminum alloy is a ductile material, we set cutting speed, feed rate and depth of cut for better results when machining it. During the machining of silicon carbide reinforced with aluminum alloy it was noticed that flank wear rate during the turning operation is high at low cutting speed and with the experimentation results it was found to have high cutting speed, low feed and low depth of cut to achieve good surface finish [12]. When comes to surface quality of these hybrid MMC in the machining operation, it was observed that cutting speed has major effect on the tool wear among all machining parameters on further exploration second influential machining parameter was feed rate and higher feed rates increases tool wear [13]. In the machining of Aluminium 6061 reinforced with silicon carbide and aluminium oxide hybrid MMC, there was an increase in surface roughness with increase in feed rate and surface roughness reduce with the increase in cutting speed [14]. In the drilling of Al 6061 reinforced with silicon carbide and graphite hybrid MMC using coated carbide drill as cutting tool, it is observed that feed rate has important effect on thrust force and surface finish of the hybrid MMC and also found that surface roughness increases with the increase in feed rate and decreases with increase in spindle speed [15]. During turning process, the feed rate had a greater impact on surface roughness, roundness and MRR than cutting speed or depth of cut while machining the Al 6061 [16]. During the machining of aluminium reinforced with silicon and zinc hybrid MMC, surface roughness depends on feed rate and then by cutting speed was observed [11]. Machining of aluminium MMC reinforced with silicon carbide particles using tungsten carbide inserts revealed that surface roughness decreases when the increase in cutting speed, and surface roughness increases as the depth of cut is increased at a constant feed rate. [17-18]. Furthermore, machining Silicon Carbide and Boron Carbide reinforced aluminium hybrid MMC, surface roughness is found to be higher at higher feed rates and lower at higher cutting speeds. Cutting force and surface roughness increased as the weight percentages of Silicon Carbide particle increases. [19].

3. Conclusions

From the above research work, the researchers observed that there is an essential need to select an appropriate percentage of reinforcements and machining process parameters for effective machining of hybrid aluminium metal matrix composites. And also in the research works, it is studied that machining of hybrid aluminium MMC is one of the major challenges. The weight percentage of reinforcements is the most significant factor which have most influence on surface roughness of hybrid MMC followed by speed, feed, depth of cut, with respect to cutting force applied. Hybrid composites are better substitutes to reinforced metal matrix composites with single reinforcement material as these shown improved mechanical, tribological and machinability properties.

References

[1] William C Harrigan 1998 *Materials Science and Engineering* A 244 pp 75–79

[2] Prakash K S, Moorthy R S, Gopal, P. M and Kavimani V 2016 *Int. Journal of Refractory Metals and Hard Materials* pp 223–229

[3] Kerti I and Toptan F 2008 *ELSEVIER Materials Letters 62* pp 1215–18

[4] Moses J J, Dinaharan I and Sekhar S J 2016 *Transactions of Nonferrous Metals Society of China* **26(6)** pp 1498–1511

[5] Glasdon J A K, Sheriff N M, Dinaharan I and Selvam J D R 2015 *Transactions of Nonferrous Metals Society of China* **25(3)** pp 683–691

[6] G Haribabu, K Satyanarayana and M Ratna Deepika 2018 *Journal of Innovation in Mechanical Engineering* Vol. 1(2) pp 2581-89

[7] S.Basavarajappa 2009 *Taylor and Francis, Materials and Manufacturing Processes* pp 484–487

[8] P Suresh, K Marimuthu, S. Ranganathan and T Rajmohan 2014 *Elsevier, Trans. Nonferrous Met. Soc.* China pp 2805–14

[9] Johny James 2014 *Procedia Engineering* pp 1018 – 26

[10] D. Sai Chaitanya Kishore K. Prahlada Rao and A Mahamani 2014 *3rd Int. Conf. on Materials Processing and Characterisation (ICMPC 2014)*

[11] Anthony Xavior Ma and Ajith Kumar J P 2017 *Elsevier, Procedia Engineering* pp 1110 – 18

[12] Manna A and Bhattacharyya B 2003 *Journal of material processing technology* Vol. 140 pp 711-716

[13] Ferial Hakami 2016 *SAGE J Engineering Manufacture IMechE*

[14] T.Sasimurugan and K.Palanikumar 2011 *Journal of Minerals & Materials Characterization & Engineering* Vol. 10, No.13 pp 1213-24

[15] Dr. A Muniaraj 2017 *IJMER* Vol. 7,Issue 3, ISSN: 2249–6645 pp 34-40

[16] P Jayaraman and L Mahesh Kumar 2014 *Proc. of 2nd Int. Conference on Advances in Industrial Engineering Applications (ICAIEA 2014)* Anna University Chennai p 309

[17] Chintada Shoba, Nallu Ramanaiah and Damera Nageswara Rao 2015 *Proc. Materials Science 10* pp 220 – 229

[18] Manna and B. Bhattacharayya 2003 *Journal of Materials Processing Technology* pp 711– 716

[19] K Venkatesan, R Ramanujam, J Joel, P Jeyapandiarajan, M Vignesh,Darsh, Jiten Tolia and R.Venkata Krishna 2014 *Procedia Engineering 97* pp 677–686

A Review on Biodiesel Production from Various Feedstocks by Transesterification

Sridevi Veluru[1], Husam Talib Hamzah[*2], Tukaram Bai M[3], Venkata Rao Poiba[4] and Husam Salah Mahdi[5]

[1,2,3,4]Department of Chemical Engineering, College of Engineering, Andhra University, Visakhapatnam, India.
[5]Department of Computer Science and System Engineering, College of Engineering, Andhra University, Visakhapatnam, India.

Graphical abstract

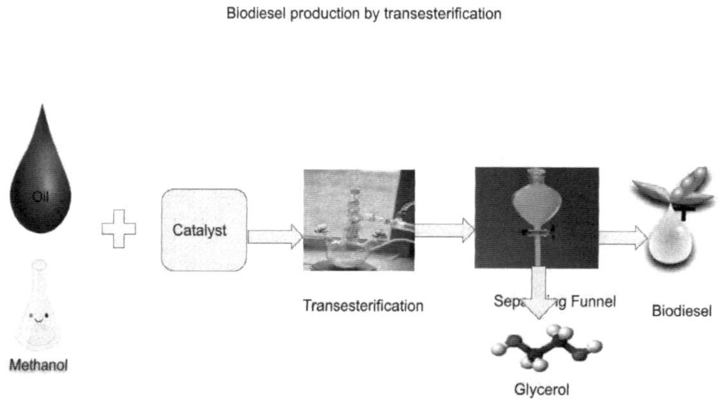

Abstract. Since fossil fuel emissions will continue indefinitely, we must find a suitable and long-term alternative, owing to the fact that it is biodegradable, non-toxic, and eco-friendly, biodiesel an excellent substitute for diesel engines. EASAC classifies the evolution of biodiesel into four generations. Biodiesel feedstocks and their advantages and disadvantages for different generations of the fuel are thoroughly analysed in this article. An in-depth investigation is provided in this article, of the benefits and drawbacks of various feedstocks used in the manufacturing process of different generations of biodiesel.
In terms of the production of biodiesel, transesterification is the best method because it produces high-yield biodiesel with comparable properties to diesel, making it an ideal choice. As far as economics are concerned, this process is also viable. It is possible to meet the energy requirements of the future by blending different oil feedstocks. The system used and the cost of feedstock have the most significant impact on the cost of biodiesel production. Characteristics of biodiesel such as the oxidation stability, cold flow and cetane number, viscosity, and density, are some of the most

important characteristics of biodiesel. Biodiesel's performance in diesel engines was also discussed in this paper, and it was suggested that biodiesel is safer for the environment than Petro-diesel. Unlike Petro-diesel, it degrades four times faster and has with a higher flash point, making storage and handling easier. It's also nontoxic and causes less irritation to the skin than soap and water. The paper also looked at the production of biodiesel using feedstocks from the first through the fourth generation.

Key Words: Biodiesel, Feed stock, Transesterification, Properties, Catalysts, Performance.

1. Introduction

Biomass was the primary energy source prior to the nineteenth century, but as technology advances, the search for new energy-efficient sources of energy continues. For cooking and heating, dried wood was the most common fuel, ethanol and vegetable oil on the other hand were the most commonly used fuels for lighting. Early humans also used wood to heat their homes and to heat-treat clay. [1]Industrialization's need for energy led to a shift over many years, the transition from wood to coal, and then from coal to fossil fuels, has occurred. [2] As a result, fossil fuels have come to dominate the world's energy supply, accounting for about 48% of total demand. [3]

As a result of growing concerns about environmental safety, as well as an ever-increasing global demand of fuel for usage in transportation and the industrial sector have led to an ever-demand for energy is on the rise. Because in the event of the depletion of fossil fuel reserves, has sparked a search for new renewable energy sources and Liquid fuel sources that are friendly to the environment. [4] All forms of economic and social growth are dependent on the availability of energy, in the field of a variety sources of alternative energy, the natural phenomena process of converting feedstock into useful forms of energy takes place. [5] The socioeconomic benefits of using renewable energy sources instead of conventional power generation methods are significant. It is imperative that biofuels be developed for four primary reasons: (1) combating climate change (2) responding to rising energy consumption (3) ensuring a reliable supply of energy (4) and maximizing the use of limited resources. Combating global warming and exploring low-carbon fuel alternatives are two of the most important tasks we face today. The world's population is expected to grow to 8 billion or even 10 billion people by the year 2050, according to estimates. With the rise of emerging economies, the demand for energy is expected to rise significantly. As fuel resources are dispersed across the globe, meeting future energy demand will be a major challenge. Another critical aspect of our long-term survival is cutting down on the amount of energy we waste and making the most of our precious natural resources, all of these issues can be solved by biofuels. [6]

When compared to commercial diesel fuel, the Biodiesel is considered to be one of the preferable possible sources of biofuel because it has no effect on engine performance parameters. Another benefit of using biodiesel is that it is non-toxic and oxygenated to reduce engine emissions, as well as being biodegradable and free of Sulphur and minerals. [7]

The biochemical process by which biodiesel is produced is known as transesterification by using biological components such as animal or vegetable fats, for example. However, a century ago Rudolf Diesel experimented with the use of engine fuel derived from vegetable oil, several problems arise in the case of vegetable oils directly used in diesel engines, many issues, such as a result of incomplete combustion and excessive smoke emission as well as deposition of carbon on the engine's components as a result of its high viscosity; of SVO's, straight vegetable oils. In addition, vegetable oil has a high flash point, which reduces the oil's volatility. Using transesterification, biodiesel has been formulated to overcome these issues. Biodiesel esters and glycerol are formed when Alcohols react with fats or oils to produce biodiesel. In order to accelerate the reaction time and yield increase, an alkaline or acidic catalyst is used. [8]Acid alkyl esters

IOP Conf. Series: Materials Science and Engineering 1258 (2022) 012024 doi:10.1088/1757-899X/1258/1/012024

from fatty acids are formed during the transesterification of triglycerides is the goal of the process of transesterification Transformation of triglycerides into fatty acid alkyl esters is the goal of the transesterification process, by reacting an alkali or an acid with an alcohol, such as methanol, and resulting in glycerol as a waste product. [9]

Renewable fuel research has focused on biodiesel because of its ability to be produced from biomass, Non-toxic and non-explosive properties, non-flammable, higher flash point, Application in diesel engines without the need for large changes is possible and Reduced reliance on fossil fuels. [10] Biodiesel can be used as a "clean energy source" in place of fossil fuels. It can reduce CO2, SO2, CO, and HC emissions. In contrast to the biodiesel combustion process, plants absorb more CO2 than are released. Thus, compared to the use of fossil fuel, using biodiesel can more effectively reduce CO2 emissions, protect the environment, and maintain ecological balance. [11]

Despite the fact that using edible oils to make biodiesel helps reduce our dependence on fossil fuels, there has been widespread criticism of this strategy due to the conflict between food and fuel. Despite the fact that the world's population continues to expand, there will be more people who will need to be fed. Using food crops to make biodiesel could lead to food shortages and price spikes if food demand outpaces supply. [12]

2. Feedstock

Vegetable oil, for example, can be utilised as a feedstock in the production of biodiesel, algae, microbial oil and animal fats. There are significant differences in the purity and compositions of biodiesel obtained from different feedstocks, [13] A biodiesel plant's feedstock is selected as the first step in the manufacturing process, which has a significant impact on a number of variables, Such as biodiesel purity, price, composition, and production. Classifying edible and non-edible as well as waste-based biodiesel sources is dependent on the feedstocks' availability and type. [14] Some additional considerations must also be made when selecting sources of biodiesel feedstock that are locally accessible and competitively priced and easily accessible as well as technically feasible. [15]

3. Types of feedstocks

3.1. First generation biodiesel

Biodiesel from the first generation is made from food-based feedstocks, edible feedstocks include things like rapeseed and soybean oils as well as coconut and palm oils as well as mustard and olive oils, etc. [12]

At the dawn of the biodiesel era, the production of biodiesel from edible feedstocks was widespread. Plants are readily available, and the conversion process is relatively simple, making first-generation feedstocks an attractive option. Because these feedstocks raise the cost of food products, there is a risk of food shortages. [16] Adaptability to environmental conditions, for biodiesel production from edible feedstock, high costs and having a small area to cultivate is also an issue. These drawbacks compelled biodiesel producers to look for additional sources of fuel. [17]

3.2. Second generation biodiesel

biodiesel of Second generation is made from (nonedible)feedstocks, Nagchampa oil, Jatropha oil, Neem oil, Calophylluminophyllum oil, Karanja oil, Rubber seed oil, and Mahua Indica oil examples of non-edible feedstocks. [18,19]Non-edible feedstocks are being researched because of first generation feedstocks drawbacks of. In addition to being environmentally friendly and cost-effective, second generation biodiesel also eliminates food inequity and reduces the need for farmland. [12-19]

There is no need to replant food crops and no need for agricultural land when using second generation biodiesel oils. Second-generation fuels have some drawbacks, such as lower yields from (nonedible) plants like Jojoba oil, Jatropha oil, and Karanja oil, which are the main non-edible sources of these fuels.

3.3. Third generation biodiesel

It is referred to as "third generation biodiesel" when it is produced using microalgae and discarded cooking oils. [19] decreased emissions of greenhouse gases are one of the most significant advantages of third-generation biodiesel, increased productivity and growth, less competition for farming land, more oil, and less impact on the food supply. The main drawbacks are the high costs, the need for sunlight, the difficulty of mass production, and the difficulty of extracting oil. [20-23] Currently, research is being carried out in order to increase the production rate of biodiesel from algal biomass as well as the extraction process. Fish oil and animal fat are the primary sources of third-generation biodiesel, micro algae, waste cooking oil etc.[24]This third generation of biodiesel's feasible sources overcome the problems of earlier generations' feedstocks, which affect the food chain availability and the environment. conditions of the environment, economic viability, As needed, it can be adapted to suit the requirements.

3.4. Fourth generation biodiesel

solar-powered fuels derived from photobiological processes are being considered for the biodiesel's fourth generation. By converting solar energy into biodiesel from raw materials, solar biofuels can be created, an entirely new area of study is being conducted on this conversion procedure. A wide variety of raw materials can be found. unlimited supply at a low cost. Such a transformation is made possible by the use of synthetic biology. [25] For long-term growth and development, need to look for solutions that are brand new to nature that make synthetic living firms and modern microorganisms to directly convert solar energy and effectively converted into a source of power. Similarly, metabolically engineered microbial fuel development with inorganic water splitting catalysts is an expanding method for liquid fuel production and storage.[26]

4. Parameters to be checked for feed stock

4.1. Moisture

Soaps and emulsions can form during the process of transesterification when moisture interacts with the catalyst. [67] Less than 0.050 wt. % should be expected.

4.2. Free Fatty Acid (FFA)

The interaction of FFA in Sodium methoxide catalyst and feedstock emulsions may be formed making the process of biodiesel separation more difficult; resulting, perhaps, in a reduction in yield. Additionally, emulsions can increase costs by requiring additional cleaning and filter replacement. To decrease the amount of soap which is generated as a result of a chemical reaction, FFA in the feedstock should be reduced to 0.5 weight percent or less as a goal. [67]

4.3. Kinematic Viscosity

Velocity measures a substance's resistance shear or allow to flow; the Movement of liquid near a solid border, like the piping's interior walls, which described by the formula, which, to begin with, is strongly influenced by temperature.

4.4. FAC Color

Members of (FAC) a Fat Analysis Committee, the color technique compares the oils and fats to color standards to determine their color.

4.5. Value of Saponification

a measure of the amount of saponification, is defined as (KOH) potassium hydroxide, in milligrams, it is necessary in order to saponify a grame of fat or oil in the specified conditions. [70] In terms of molecular weight, and length of the fatty acid chain in a given fatty acid that existing is used to calculate the value of saponification. The average chain length of fatty acids can be compared.

4.6. Volatile Matter and Moisture

Volatile matter in a feedstock that found, may lead to loss in fatty acid methyl ester yield either by diluting the feedstock, or by reacting with the catalyst.

4.7. Insoluble Impurities

Measurement of solids is done using an insoluble impurities tests, that are kerosene and petroleum, are insoluble in it. Sand and dirt are two examples of these solids, and in the situation of vegetable oil seed fragments and Animal fats or used cooking oil can contain bone and gum fragments. [68]

4.8. Unsaponifiable Matter

Organics are the primary constituents of unsaponifiable matter, in which the base does not react in order to produce soap. A Higher molecular weight of alcohols, waxes, sterols, pigments, and hydrocarbons are all included. [69] After the transesterification reaction, these non-polar components are still present in the biodiesel.

4.9. Moisture, Insoluble, and Unsaponifiable (MIU)

It's the fat or oil that MIU represents, those that it is not possible to change to mono alkyl fatty esters by transesterification or esterification.

4.10. Oxidation Stability

Based on oxidation stability, age or the conditions in which it was stored of the oil or fat that can be determined and is able to foretell if the feedstock is accomplishing a goal of the lowest necessities for the stability of biodiesel oxidation, as stated by (ASTM D6751). Two factors influence the stability of oxidation, first the fact that hydrogen atoms exist, next to double bonds of carbon-carbon, in which oxidation can take place. [69] Second is the presence of naturally occurring antioxidants in the raw material, can avoid of the triglyceride moleculesoxidation.[70]

4.11. Sulfur

The maximum Sulphur contented in biodiesel is (15 ppm, by ASTM D6751).

4.12. Magnesium and Phosphorous, Calcium

The needs of phosphorous in biodiesel be partial to (10 ppm) (0.001 % mass maximum), and It is expected that the total amount of magnesium and calcium will be less than (5 ppm), minor components such as calcium, phosphorus, and magnesium are typically associated with gums and phospholipids that have the potential to function as emulsifiers or produce sediment, transesterification yields can be reduced. [67]

5. Transesterification

Vegetable oils can be transformed into diesel-like fuels using a variety of methods. Transesterification, blending, cracking, micro emulsification, and pyrolysis are all examples of these processes. [58-62]Biodiesel is most commonly produced through the transesterification process. [59,62,63] When compared to other techniques, this method is more efficient. it has a wide range of benefits. e.g., Biodiesel of high quality is produced with a high yield in normal conditions. [38,39] Over the last few years, transesterification has gained widespread acceptance for the converting vegetable oils into fuel-like products that are more compatible from a technical standpoint. To make biodiesel, transesterification is a crucial step in the production process so that it can bring the feedstock-vegetable oils closer to the viscosity of Diesel fuel that is derived from fossil fuels. [27] Among organic reactions, transesterification is an important one This occurs as a result of the exchange of the alkoxy moiety as a result of the conversion of one ester into another as per Fig.1. Transesterification is a steady-state process, describing the

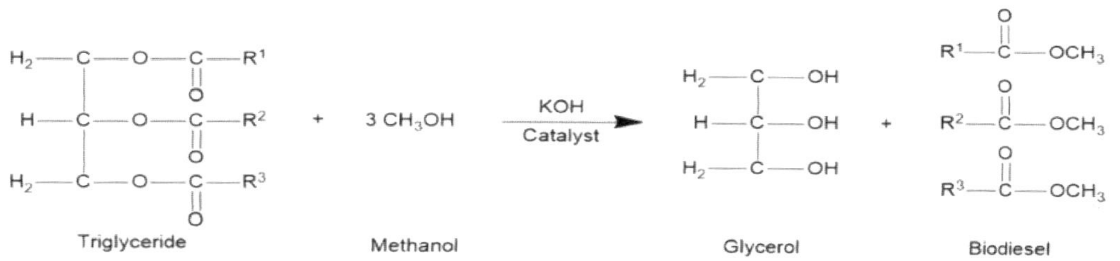

Fig.1. Transesterification process.[117]

alcoholics of carboxylic esters usually performed in the in the presence of a standard catalyst, e.g., NaOH and KOH, for valued accelerate the rate at which the equilibrium adjustment is made to increase the amount of esters produced. [28-30] An appropriate catalyst is needed for transesterification, in which alcohol is mixed with vegetable oil and reacted. Because ethyl and methyl alcohols are so widely used, ethyl-methyl esters are created. As a result of the response, liquids in two layers of distinct, i.e., ethyl-methyl ester and glycerin are appeared, which are separated. There are two types of transesterification: catalytic and noncatalytic processes. [31]Acid and base catalyzed transesterification processes can be subdivided. Depending on the catalyst phase, Homogeneous, heterogenous, and enzymatic, the acid or base catalysts can be subclassified further.[32,33] Homogeneous catalyzed transesterification conversion efficiency is improved when the (FFA) is fewer than (1 mg of KOH/g) of oil. (FFA)A free fatty acid values when larger than (1 mg of KOH/g) of oil require a two-step transesterification process. [34] Fig.2. shown the basic transesterification technology.

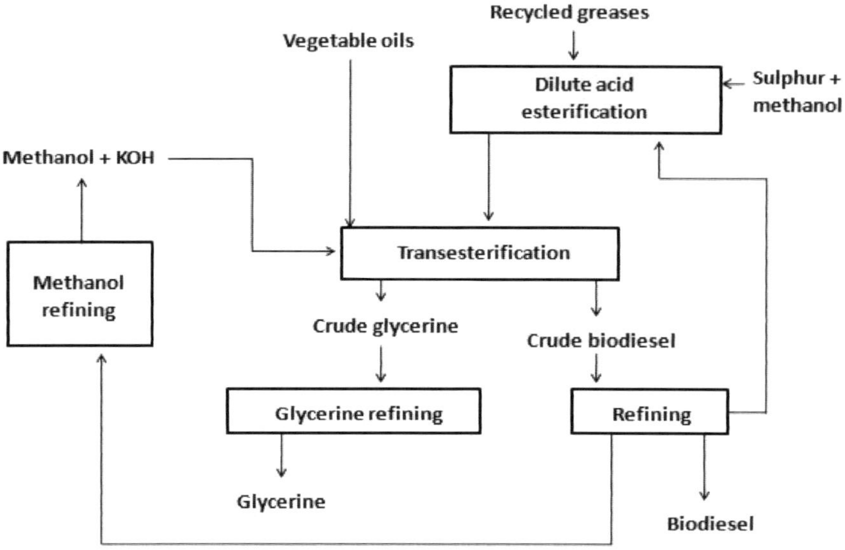

Fig.2. Basic transesterification technology.[118]

There are many variables that affect the rate of biodiesel production via transesterification, of these, the most critical are type and amount of alcohol, type and amount of catalyst, reaction time and temperature. The yield is also affected by the feedstock's content of moisture and the free fatty acids amount. Generally, Reaction time is 2 hours at a temperature between 50 and 75 degrees Celsius. Gravity or centrifugal force is used to separate biodiesel from glycerol. [63,64]

While transesterification has many benefits, it also has a few drawbacks to consider. Vegetable oil transesterification takes a long time to complete. and Only a few post-treatment steps, such as separation, are required, the use of water for washing and heating purposes. To purify biodiesel, more water is needed. [36,35]

6. Catalyst

Catalysts for biodiesel production generally fall into one of three categories: alkalis, acids, or enzymes.[37-39] the used in biodiesel production Alkali and acid catalysts are more frequently than enzyme catalysts. [40] Homogeneous and heterogeneous catalysts were the next classifications made. Enzyme catalysts, on the other hand, have recently gained popularity due to their ability to reduce soap formation and their ease of purification. Despite this, they aren't widely used because of their slower response times and higher costs. Despite this, they aren't widely used because of their slower response times (longer reaction times), and costs higher. Some researchers recently, new biocatalysts have been created, effort to reduce costs. These biocatalysts don't require purification, which is a benefit. [41,42]

7. Homogeneous base-catalyst

Using homogeneous alkaline catalysts for transesterification is more common due to the slower rate of the acid counterpart. [43]Potassium hydroxide is the most commonly used basic catalyst (KOH), ($KOCH_3$) potassium methoxide, (NaOH) sodium hydroxide, ($NaOCH_3$) sodium methoxide, and sodium ethoxide ($NaOCH_2CH_3$).[44-46] As a result of their abilities, to catalyze reactions at atmospheric pressure and low temperature, these catalysts are widely used, fast and efficient conversion rates, as well as a low cost. [44]

($KOCH_3$) potassium methoxide and ($NaOCH_3$) Sodium methoxide, are improved catalyst than (NaOH) sodium hydroxide and (KOH)potassium hydroxide, as a result of the dissociation into CH_3O^- and Na^+ and $CH_3O$and K^+ respectively when comparing on the yield of biodiesel.[46]Commercial biodiesel production typically employs an alkaline catalyst The reason for this is that it does not generate water during the transesterification reaction. The one that is most frequently used homogeneous base catalysts in the production of biodiesel are NaOH and KOH. [47]

8. Homogeneous acid-catalyst

in waste oils the Free fatty acids can't be converted to biodiesel using an alkaline catalyst. Ester separation will be prevented by soap derived from these FFAs, washing water, and glycerin, Once, with the help of an alkaline catalyst,[48] the (liquid acid-catalyzed transesterification), is offered as a means of overcoming a lot of the conundrums related with liquid base catalysts. Sulfonic acid, hydrochloric acid, Sulfuric acid, organic sulfonic acid, for transesterification, acids containing ferric sulphate are the most widely used catalysts in chemical reactions. [46]

9. Heterogeneous base-catalyst

Heterogeneous catalysts have been the subject of numerous investigations in production of biodiesel in order to get around the drawbacks of homogeneous catalysis. the vast majority of biodiesel production heterogeneous catalysts are whichever (alkaline oxide) or (alkaline earth metal oxide), that large area was successfully supported. The activity of heterogeneous basic catalysts is greater than that of heterogeneous acid catalysts. Their homogenous counterparts are quite similar.[49] solid alkaline catalysts, for instance, (CaO) calcium oxide provide numerous benefits such as long catalyst life times, higher activity, and Running in a moderate reaction condition is possible. [50] Nonetheless, CaO as catalyst can also be used to Reduce the rate of reaction in the biodiesel synthesis process.[51] Fig.3. shown the Diagrammatic representation of the biodiesel manufacturing process using ostrich bones derived heterogeneous catalyst.

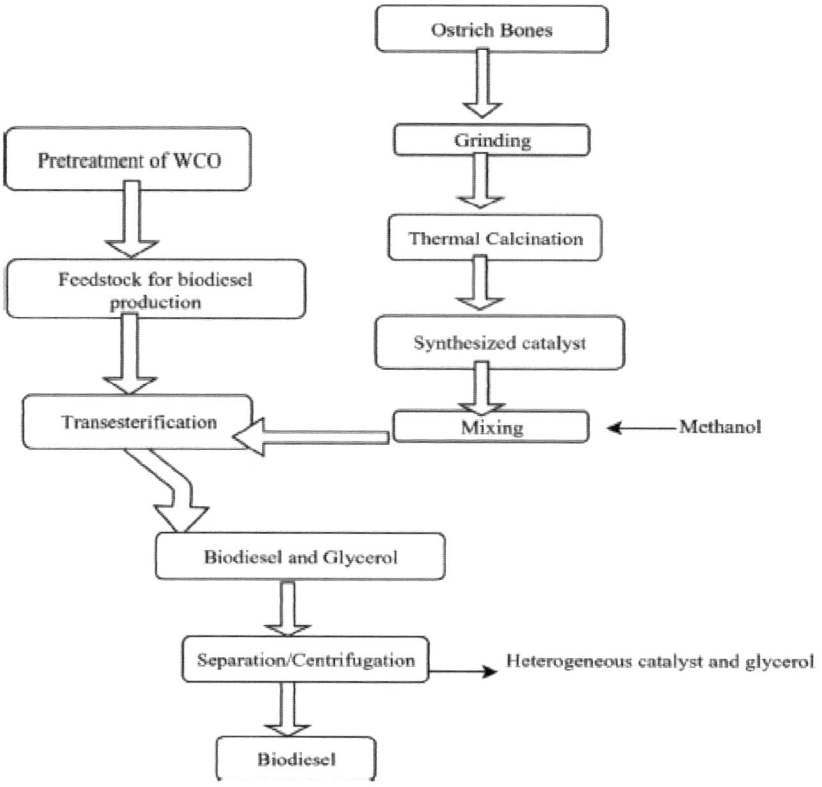

Fig. 3. Diagrammatic representation of the biodiesel manufacturing process using ostrich bones derived heterogeneous catalyst.[119]

10. Heterogeneous acid-catalyst

However, despite the homogeneous acid catalyst's efficiency, can lead to problems contamination that necessitate effective separation and procedures for removing impurities from a product. [52] As a result, production costs will rise. Heterogeneous acid catalysts are a viable alternative to homogeneous acid catalysts and have the potential to be more efficient. the Heterogeneous acid catalysts have some advantages, insensitive to content of (FFA), can simultaneously conduct the esterification and transesterification, remove the (washing step) from the biodiesel production process , a more efficient way to separate of catalyst from product, reusing and regenerating the catalyst, is possible and also reduce the issues with corrosion [53-55] Solid acid catalysts are preferable over liquid acid catalysts in many applications, They contain a number of different sites as a result of this, with different strength of Bronsted, or Lewis acidity.[46,56] A significant advantage of bronsted acid catalysts based is that it promotes the simultaneous esterification and transesterification of fatty acids with the utilization of fewer expensive feedstock with an increased concentration of free fatty acids.

11. Enzyme (Biocatalyst) catalyst

Researchers have focused on enzymatic transesterification, as a result of an issue with downstream processing posed through the process of transesterification in the lab. Massive amounts of wastewater are being produced, and glycerol recovery is problematic for a variety of reasons, that it will eventually rise the total cost of producing biodiesel and not being ecofriendly, In contrast, enzyme catalysis happens without the generation of (by-products), insensitive to high (FFA oil),easy recovery product and mild condition of reaction, and It is possible to re-use the catalyst.[56] enzyme catalyzed biodiesel production has a process with a high probability of being environmentally friendly and a talented alternative to the chemical process. Biodiesel production was aided by the enzyme has some limits particularly when implemented at a large-scale manufacturing facility as a result of the high cost of enzyme, and enzyme deactivation and slow reaction rate.[57]

12. Biodiesel Properties

Cetane number and cold flow characteristics are define biodiesel's properties, viscosity, oxidation stability, and fuel density.[71]A fuel's ability to be pumped and filtered is its cloud point has influenced it and cold filter plugging point. [72] regardless of Biodiesel sources, has a cloud point that is generally higher in comparison with diesel.[73] Saturated fatty acid content of biodiesel mayrise the cloud point.[74] saturated fatty acids with long chains lead to biodiesel with higher cloud points.[75]the biodiesel derived from less saturated fatty acid-containing sources has lower cloud points. Additives, on the other hand, the cold flow enhancers or transesterification procedures with branches of alcohol, can be put to use to biodiesel's cold flow properties should be improved .[76]

is a measure of the fuel's cetane number is an important factor in determining its cold flow properties. The High cetane number fuels have ignition and combustion are separated by a short period of time.[77] In spite of the fact that a low cetane number would not be ideal, this may also cause fuel to burn near the injection line if it has a high cetane number, which results in a clogged nozzle. [78] Generally, the cetane number of conventional diesels is lesser than that of biodiesel, Cetane number can rise in the existence of Fatty acids with a long chain and a saturated fatty acids high content. [79] For example, biodiesel from (UCO), has a higher cetane rating comparative to (SO) biodiesel.[80,81] a link exists between biodiesel's cetane number and its viscosity. [82] Therefore, the use of the (grease and AF) with high viscosity can result in biodiesel with a high cetane number [83,78] Combustion of an engine with a high cetane number may result in an increase in gaseous pollution. [83] PM and CO emissions are still a concern, may decrease if preferable to use biodiesel with a high cetane number is used in the combustion process. NOx emissions are rising, because of Biodiesel with higher oxygen content. [84]The viscosity of biodiesel is another important factor in determining its quality. The injection and atomization processes will be impacted by high viscosity. [85]Because of the large droplets that form, and poor mixing with air during the injection process, it may result in decreased combustion efficiency. [85,86]The high viscosity of biodiesel is attributed to its saturated fatty acids at high content and long chains of carbon. [87]Fuel injection and combustion in an engine are both impacted by density, and viscosity. Slow injection can be caused by a high density. Slow injection can be caused by a high density, as a result, the atomization and combustion process are less efficient. [88] Further, Pollutant emissions could be accelerated as a result, (i.e., NOx, PM, CO) from the engine.[89]Generally speaking, the density of biodiesel is higher than that of conventional diesel fuel. [90]Unsaturation at a high level, the number of double bonds, in carbon chains, and the density of biodiesel is amplified by the attendance of hydroxyl esters in the fatty acids. [91,92]

Biodiesel's oxidation stability is yet another essential quality. The storage of biodiesel for an extended period of time can result in changes in the biodiesel quality due to the fatty acid's oxidation. [93] Oxidation results in the formation of gums. which has a significant impact on the fuel's physiochemical properties (e.g., viscosity, cetane number, etc.) and This will lead to a less efficient combustion. [94,95] During the process of burning oxidized fuel, carbon deposits form in the combustion chamber. [96]Nonoxidized biodiesel produces less NOx than oxidized biodiesel. [97]A biodiesel's oxidation tendency rises as the unsaturation level of fatty acids rises. [98] Consequently, biodiesel derived with linoleic and linolenic acid concentrations in vegetable oils above normal, such as SO biodiesel, has a low oxidation stability. " [99] The temperature point at which the fuel supply begins to ignite is known as the "flash point," and it is closely linked to the volatility of the fuel. the fuel with a high flash point indicates delay combustion as a result of in which form does carbon deposition take place, and large-scale emissions of polluting gases (e.g., NO_x), Long carbon chain fatty acids are present in biodiesel, make it less volatile and a higher flash point, compared to diesel fuel in its traditional form. [100]

13. Performance of biodiesel in diesel engine

Biodiesel it has shown promise as an alternative diesel engine fuel in numerous studies. However, For the most part, vegetable oils or oil mixes when utilised directly are considered undesirable and unworkable, for both direct and indirect injection diesel engines. Viscosity, acidity, and the presence of free fatty acids of such oils, along with the formation of gums, because of the oxidation and polymerization that occurs during storage and A few of the more obvious issues include combustion, oil thickening, carbon deposits.

Combustion of Biodiesel in a Diesel Engine, several promising characteristics have been demonstrated, reduction of exhaust emissions is also included. Because of their similar properties to diesel, trans esterified oils are a promising alternative to the fossil fuel. Petroleum-based diesel fuel has the same chemical and physical properties as biodiesel, Because of this, it is suitable for use in diesel engines. Existing diesel engines can use biodiesel as a diesel fuel replacement with no modifications needed. In accordance with the results, Compression ignition engines can use biodiesel without making any adjustments, in place of diesel. Fueled by the same principles as petroleum diesel, Compression-ignition engines are required for biodiesel use. Blends of up to 20 biodiesels mixed with petroleum diesel fuels, can used in a wide range of diesel engines, and compatible with a wide range of storage systems and distribution systems [65].

14. Reviews from Previous studies:

Md Mofijur Rahman et al., (2017)[101] investigated the wear and frictional properties of native Australian species, biodiesels of the first and second generation using a four-ball tribotester. It was decided to conduct the tribological test at 1800 rpm with various temperatures and loads. The shape of the ball's exterior and pattern on the surface was tested by (SEM)scanning electron microscope and (EDX) energy dispersive X-ray spectroscopy analysis. The results of the experiments demonstrated that the usage of biodiesel has (COF) a lower frictions coefficient, and lower (WSD) wear scar diameter up to (83.50% and41.28%), respectively when compared to regular diesel. The results of the worn-surface area measurements revealed that A biodiesel fuel must contain a minimum amount of carbon and oxygen, except Fe, compared to diesel. Also discovered was a worn surface area for diesel that (2.20%–27.92%) to be higher than biodiesel. The results of this investigation showed that the tribological performance of examines the total cost of production of the biofuel blends; was better more so than diesel, and among the various types of biodiesel fuel, macadamia biodiesel displayed enhanced lubrication properties.

Azad et al., (2013)[102] investigated experimentally the use of a diesel direct-injection engine, with1st generation biofuel blends with fossil fuel, like kerosene. Trans-esterification reactions have not been used to produce the first generation biofuel (mustard oil). M20, M30, M40, and M50 are the approximate volume percentages used in the blend with kerosene oil. The biofuel's characteristics blends were tested in accordance with a variety of ASTM standards. It was determined and compared how temperature changes the properties of various biofuels. After that, a four-stroke test engine, single cylinder, in place of the bed of the dynamometer used to measure a variety of parameters of the engine, like break house power, and break specific fuel consumption, break mean effective pressure, and lube oil temperature, temperature and volume of exhaust gas etc. A thorough investigation into the performance of various biodiesel blends was conducted with kerosene and performance of diesel fuel in engines. This paper also discusses the biofuel blends' total cost of production per unit and their economic impact.

Emilio A. Viornery Portillo et al., (2020)[103] carried out a comparative assessment of Environmental impacts of (B25) biodiesel blend production and use, and Using ultra-low Sulphur fuel in a power generator of (33 kW) at 100 percent of the rated impact and load analysis based on the lifecycle methodology, Based on primary data for each stage of processing and utilization. The model's final output, that (B25) overtakes low Sulphur diesel in every class. Abiotic depletion was found to have the greatest impact reductions, potentially harmful to human health, acidification potential, Classifications of eutrophication and global warming with (39.48%), (39.44%), (39.24%), (38.73%), and (35.77%) respectively, Exhaust emission measurements carried out in the laboratory a rise was observed in (CO) (52%) and a decrease in (NOx) (41.54%) for (B25), As compared to diesel with ultralow Sulphur content. The findings demonstrate that biodiesel is beneficial and a viable alternative fuel.

Hirani et al., (2018)[104] focused on the most recent developments in scientific study of the second-generation biofuel production from diverse feedstocks. reducing emissions of greenhouse gases and the dependence on fossil fuels are two of the primary goals of renewable sources of energy. Biofuel industries, particularly ethanol and biodiesel, have been rapidly expanding alongside agricultural crop production for more than a decade. Agriculture crops are heavily relied upon in the production of the first-generation biofuels such as corn, sugarcane, sugar beets, soybean, and canola. As a result, for the past few years, the inherent competition between food and fuel has been up for debate in our society. Next-generation biofuels, many different feedstocks can be used to produce, including agricultural waste products, recent technological advances in research and development, high-yielding grass species' cellulosic biomass and crop residues.

Cristiano Varrone et al., (2017)[105] investigation of various mixed microbial cultures' performance (MMC), capable of fermenting glycerol crude produced using biodiesel produced from the fat of animals, to produce (1,3) propanediol (1,3PDO), and During non-sterile conditions (butyric acid). There were eight different CSTRs, each with a different inoculum and growth medium. Each CSTR was set up in a different way. The metabolic product distribution was studied under varying operating conditions. The kinetic properties of each MMC were analyzed, and stoichiometric reactions were designed for each of the reactions in the dataset. The use of Next Generation Sequencing (NGS) allowed researchers to examine changes in microbial populations. The maximum rate of degradation of the substrate was about (101 g/L/d) of glycerol, with a output of (38 g/L/d and 11 g/L/d) for 1,3 PDO and butyric acid respectively, the results were obtained using a hydraulic retention duration of (12 h) and 60 g/L input. A concentration that is very close to the maximum of feedstock was achieved, (90 g/L), resulting in an incomplete degradation of the substrate.

Dariusz Kurczyński et al., (2021)[106] studied the exhaust gas composition as a result of second-generation biofuels, from (1104D-44TA) a Perkins diesel engine. Diesel fuel and first-generation biofuel were used as a standard for comparison. Biofuel of the second generation, the babassu butyl esters (BBuE) are created through transesterification with the use of a sulphuric acid catalyst, and butyl alcohol. Rapeseed oil methyl esters were used as the biofuel of first generation in the tests (RME). A component of the engine's combustion process as a concentrations of carbon monoxide, carbon dioxide, hydrocarbons and amount of particulate matter in the exhaust was lowered in the engine, as a comparison that powered by either (RME or DF). By using BBuE as the engine's power source, lower nitrogen oxide concentrations were found in comparison to its powering with the (RME). Since (BBuE) contains several short chain esters that are not present in (RME). This has led to the development of a brand-new type of biofuel powered engine has fewer potentially harmful components emitted into the atmosphere as exhaust gas than do engines powered by first generation biofuel or diesel oil.

Wenjun Zhong et al., (2016)[107] compared second-generation biodiesel (SGB) with the (FGB)the first generation of biodiesel, there has been very little published research about (SGB). optical methods can be used to examine their combustion processes. and the emissions-related features. The (PLIF)planar laser induced fluorescence technique was practical to Try experimenting with various blend ratios, to see how the flame's OH distribution changes, diesel-SGB in a chamber with a constant volume. A variety of operating conditions can cause ignition delays. by using high-speed imaging, it was discovered. After that, the various fuel blending ratios and their impact on emissions were examined in engine with direct injection, which is a four-cylinder diesel engine through the ESC testing cycle. Biodiesel blends with higher percentages resulted in lesser ignition delays, while blending ratios that are higher resulted in greater LIF signal areas. Additionally, Fuel for IV B20's blending process has improved both in terms of emissions and economic performance as compared to other fuels, and the sensitivity of (HC) emissions to the speed of the load is relatively low, in contrast to (NOx) emissions at the same load. The speed of the engine was a major factor.

Janske van Eijck et al., (2014)[108] reported that the Net Present Value (NPV), and cost as a whole of production for Biofuels of the first and second generations, calculations are made in (74 settings), covering8 feedstock types, and 5 fuel output types, There are 12 countries and 8 different combinations of agricultural management systems in this study, between 2010 and 2030. a wider selection and improved methods of crop management, the yields are expected to rise. Quite a high level (NPVs) (This refers to profitable production), are calculated in relation to cassava, up to (16,000 $/ha, and palm production, up to nearly 7000 $/ha), However, cassava can have a detrimental impact (NPV), in which it is stated that there is some risk associated with a project's investment, The estimated of the net present value (NPV) of jatropha ranges from (900 to 2000 $/ha), The estimated NPVs forsoy and sugarcane. The net present value (NPV) is always positive (2500–5000 $/ha and 200–3000 $/ha respectively), and gainful as well.

Enrico Mattarelli et al., (2015)[109] conducted studies in order to study the combustion mechanism when used in a small industrial engine powered by Algae biodiesel is blended with conventional diesel fuel. For the most part, most applications make use of the blend composition, i.e., 20% of the fuel is biodiesel, while the remaining 80% is diesel (B20). As a means of providing a precise reference point, all of the tests were re-performed using only pure Diesel fuel, adding to the mix 20 percent commercial rapeseed biodiesel, one of the best-known examples of a first-generation biofuel. Using a 1.4-liter, IDI 4-cylinder, naturally aspirated engine, the researchers ran their tests. The algal B20 fuel conversion efficiency was found to be slightly improved, as compared to the normal Diesel version when compared to the standard Diesel version. Finally, the lower heating value of algal B20 necessitates in order to compensate, a larger mass flow rate fuel must be used.

Anoop Singh et al., (2011)[110] studied that the First- and second-generation biofuels have a number of drawbacks. Third generation biofuels do not have these drawbacks, competition between food and fuel, as well as Change in land use etc., and is therefore regarded as a viable source of alternate sources of energy. With regards to sustainability in all its forms, environmental, social and economic, assessment of the entire life cycle is most important for avoiding problems in regard to shifting problems. The operation of carbon dioxide, organic waste, and other gases in flue gases for the production of biomass additional rises the third-generation biofuels sustainability, because it reduces greenhouse gas emissions and waste problems.

Oliveira, L.E et al., (2019)[111] reported that It's possible to produce a fourth-generation biofuel. As a result of their strong economic development and productivity growth, the brown and red strains were chosen and subjected to a 6-percent KOH treatment for extraction of carrageenan. The results of this residue's chemical analysis display that there is significant amount of fermentable sugar that is present in the residue of carrageenan extraction Both strains were able to convert glucan to glucose at a 100% rate, and the glucose concentrations from thered and brown strains, were (13.7 and 11.5 g L^{-1}) respectively, results show that generating data is a highly efficient process An important bioproduct of carrageenan residue, When used as a food thickening, Carrageenan has been utilised for hundreds of years, and this would save it.

Hoofar Shokravi et al., (2021)[112] reviewed that the fourth-generation biofuel (FGB), where the algae biomass used to produce biofuel is genetically modified, Academic and industrial circles have given it a lot of attention. Many obstacles remain in the way of replacing fossil fuels with FGB. However, GM algal biomass has received hardly a sliver of attention. Only a few studies have examined the A look at the current state and future directions of FGB's algae genetics, the reason for this investigation's existence is in order to close this gap in the literature by reiterating the findings of recent research and achievements in a safe, and (FGB) production is facilitated by the efficient genetic manipulation, the most critical issues to consider and the aspects related to Highlights are given to the genome editing of algal strains, It was determined that the problems with the technology had been remedied, and FGB's biosecurity considerations, furthermore, there is the issue and variety of corresponding rules, legitimacy concerns, environmental risks, and health, are some of the most popular significant difficulties, that need a strong commitment on both the national and international levels to reach a worldwide understanding.

M. Mofijur et al., (2021)[113] this study, it can be summarised that Nanocatalysts can be utilised in processes that need lower temperatures, and the utilisation of these catalysts in transesterification reactions can speed up the reaction process. In addition, the presence of free fatty acids and the amount of water in feedstocks had no impact on the performance of nanocatalysts. Because of their ability to be reused, nanocatalysts offer significant cost savings when compared to other types of catalysts. However, a greater quantity of alcohol is required for a satisfactory yield, and the preparation of sufficient catalysts might be an expensive endeavour. The development of catalysts that are effective, economical, and environmentally friendly is essential to overcoming the issues that are now being faced. Therefore, the successful development of a catalyst with these qualities that allows for successful transesterification would be a landmark in the fuel sector. In addition, in order to solve the ongoing difficulties associated with the manufacture of biodiesel in a manner that is energy-efficient , A comprehensive understanding of the mechanisms behind surface catalytic reactions should be the primary focus of research efforts, This is essential for the development of logical ideas for new catalysts with predetermined improvements in catalytic efficiency for target processes.

Abd El-Fatah Abomohra et al., (2021)[114] This study evaluated *in-situ* co-transesterification of wet spent coffee ground (SCG)/microalgae mixture for enhanced biodiesel production. The SCG had a lipid level of 16.0 wt percent, while the microalgae had a lipid value of 23.6 wt percent. At various temperatures, durations, and solvent ratios, a total of 27 transesterification runs were carried out using wet SCG: algae (1:1, w/w) as the reactant mixture. At a variety of SCG/microalgae blend ratios, the biodiesel yield from pure microalgae was found to be greatest, coming in at 20.15 weight percent. An increase in the SCG ratio caused a considerable decrease in the amount of biodiesel produced, achieving the lowest possible amount of 11.2 weight percent by utilising just pure SCG. However, on the other hand, SCG demonstrated superior biodiesel qualities compared to microalgae in regard of iodine value, cetane number, and oxidation stability. The findings presented here provide further evidence that the addition of algae to SCG has the dual effect of improving both the quantity and quality of the biodiesel produced, in contrast to the process of individual transesterification.

Shashi Kant Bhatia et al., (2021)[115] reported that a technique that makes use of non-edible oils and waste oils (such as animal fat and WCO) for the creation of biodiesel appears to be a potential method for the production of biodiesel. It does not appear that the efficient recovery of oils using single methods will be sufficient, and the employment of several combined approaches demonstrated a high recovery of oil. In order to accomplish transesterification, the application of heterogeneous catalysts with biomass appears to be an eco-friendly and economic viewpoint, as the majority of the procedures detailed and discussed in this article demonstrated a biodiesel output and conversion of greater than 95 percent. A wide variety of PI technologies have the potential to provide a high biodiesel yield in a short reaction time at a lower alcohol: oil ratio. The economics of the biodiesel manufacturing process can be further improved by integrating the production process with the recovery process and converting the by-products into valuable products. This will allow for a more efficient use of resources.

Adeyinka Sikiru Yusuff et al., (2021)[116] This research showed that the zinc-modified anthill catalyst could be employed effectively in the methanolysis of low-grade feedstock such as WCO. The biodiesel evaluation based on the results of the gas chromatography, exhibited a high amount of FAME content. The findings of the optimization showed that all of the analysed variables had a substantial impact on the biodiesel yield. These variables included temperature, catalyst loading, and the methanol/oil molar ratio. In addition, the findings demonstrated that a biodiesel output of 83.16 percent, which was the greatest possible, was accomplished under the optimal conditions for the reaction, which are as follows: (catalyst loading = 0.5 wt%, methanol/ WCO molar ratio = 17.99:1 and reaction temperature = 66.54 °C). By combining WCO-biodiesel with diesel fuel derived from petroleum, extraordinary qualities can be conferred upon diesel fuel. According to the findings of reusability and stability tests conducted on the catalyst, the ZnMA composite is relatively stable after multiple cycles of reuse. On the basis of the FAME yield that was obtained, zinc oxide supported anthill is thought to be acceptable as a heterogeneous catalyst for the generation of renewable fuel, and as such, it ought to be tested with various kinds of oil.

15. Conclusion

However, the high cost of these oils and fats adds up to a higher total cost of production. The improvement of biodiesel quality with less harm to the environment has been the primary focus of successive generations' research. Biodiesel feedstocks for the fourth generation are still in the early stages of development. Uses synthetic biology technology to produce biodiesel in the fourth generation, which will be the next generation of biodiesel. This generation uses metabolic engineering, biological tools should be added to the enhancement of biodiesel's quality and quantity from various feedstocks. The content energy of biodiesel, from the fourth generation is especially high, rapid of feedstock growth, and amplified capacity for CO_2 absorption during the process of creating feedstock, as an alternative, several research projects have been carried out, in order to advance catalysts at a reasonable price, that reduce the cost of production. a variety of catalysts, such as homogeneous or heterogeneous acid catalysts, the Researchers have explored, and used biocatalysts(enzymes), and heterogeneous-homogeneous base catalysts for biodiesel synthesis. In commercial production, base-catalyzed transesterification is frequently employed, because of the high yield of FAME, in low activation energy requirements with a short reaction time as compared to acid-catalyzed transesterification. a method of transesterification that uses an acid catalyst has been reported requirements higher molar ratios of alcohol to oil, it is the most effective way when a poor quality product of oil is used From the investigation, This means that It is possible to utilise biodiesel as a substitute for diesel fuel in a compression ignition engine, without modifying the engine in any way. In the same way that petroleum diesel works, Compression-ignition engines can run on biodiesel.

16. Future prospects for biodiesel

Rapid growth can be seen in the production of biodiesel, prompted by concerns about energy security as well as other environmental issues. Given geographic disparities between demand and supply potential and supply cost, expanding commerce in biodiesel makes sense. The production of biodiesel's potential on a worldwide scale is not entirely understood, On the other hand, it has the potential to fulfil a sizeable portion of the need for fuel to power transportation. As the number of people living on Earth continues to rise, there will be a higher demand for agricultural land to cultivate in order to meet the rising demand for food. Because there are not enough lands available, this could result in an increase in the cost of producing biodiesel feedstock. These issues are already present in Asia, a region where the cost of vegetable oil is relatively high and continues to rise. The identical pattern will eventually take place in the other parts of the world. Therefore, the use of non-edible oil, genetically altered plants, and microalgae feedstock are all potential answers to this issue, and they have the potential to assure the long-term sustainability of the production of biodiesel in the future. It is essential to have rules in place to facilitate research on biodiesel and bring its costs down to a level where they can compete with those of other traditional energy sources. The economic viability of the biodiesel manufacturing process as a whole will be improved by further improvements in the usage of by-products.

References

[1] Guo, Mingxin, Song, Weiping, Buhain and Jeremy 2015Bioenergy and biofuels: history status and perspective *Renew Sustain Energy Rev J* **42(C)** 712–725.

[2] Mosarof M H, Kalam M A, Masjuki H H, Ashraful A M, Rashed M M, Imdadul H K and Monirul IM 2015 Implementation of palm biodiesel based on economic aspects,performance, emission, and wear characteristics *Energy Convers Manag* **105** 617–629.

[3] EmodiNnaemeka and Vincent 2016 Energy Policies for Sustainable Development Strategies *(Springer)*1-8.

[4] Diniz A P M M, Sargeant R and Millar G J 2018 Stochastic techno-economic analysis of the production of aviation biofuel from oil seeds *Biofuels* **11** 161.

[5] KalogirouS A 2004 Solar thermal collectors and applications *Prog Energy Combust Sci* **30** 231–295.

[6] Lin L, CunshanZ, Vittayapadung S, Xiangqian S and Mingdong D 2011 Opportunities and challenges for biodiesel fuel *Appl Energy* **88** 1020–1031.

[7] Issariyakul T and Dalai AKJR 2014 Biodiesel from vegetable oils *Renew Sustain Energy Rev* **31** 446–471.

[8] Fukuda H, Kondo A and Noda H 2001 Review: biodiesel fuel production by transesterification of oils *Journal of Bioresource and Bioengineering* **92** 405–416.

[9] Palligarnai T Vasudevan and Michael Briggs 2008Biodiesel production—current state of the art and challenges *J Ind Microbiol Biotechnology* **35**421–430.

[10] P Qiang Tan, Z Yuan Hu, D Ming Lou and Z Jun Li 2012 Exhaust emissions from a light duty diesel engine with Jatropha biodiesel fuel *Energy* **39** 356-362.

[11] Lapuerta M, Armas O amd Fernandez J R 2008 Effect of Biodiesel Fuels on diesel engine emissions *Progress in Energy and Combustion Science* **34** 198-223.

[12] HwaiChyuan Ong, Jassinnee Milano, Arridina Susan Silitonga and Masjuki Haji 2019 Biodiesel production from Calophylluminophyllum-Ceiba pentandra oil mixture: Optimization and characterization *Journal of Cleaner Production* **219** 183-198.

[13] Mahdavi M, Abedini E and Darabi AH 2015 Biodiesel synthesis from oleic acid by nano catalyst (ZrO_2/Al_2O_3) under high voltage conditions *RSC Adv* **5** 55027–55032.

[14] Demirbas A 2009 Production of biodiesel fuels from linseed oil using methanol and ethanol in noncatalytic SCF conditions Biomass *Bioenergy* **33** 113–118.

[15] Anwar M, Rasul M G and Ashwath N 2018 Production optimization and quality assessment of papaya (Carica papaya) biodiesel with response surface methodology*Energy Convers Manage* **156** 103–112.
[16] Aransiola EF, Ojumu TV, Oyekola OO, Madzimbamuto T F and IkhuOmoregbe DIO 2014 A review of current technology for biodiesel production: state of the art *Biomass Bioenergy* **61** 276–297.
[17] Tariq M, Ali S and Khalid N 2012 Activity of homogeneous and heterogeneous catalysts, spectroscopic and chromatographic characterization of biodiesel: A Review *Renew Sustain Energy Rev* **16** 6303–6316.
[18] Shameer PM, Ramesh K, Sakthivel R and Purnachandran R 2017 Experimental evaluation on oxidation stability of biodiesel/diesel blends with alcohol addition by rancimat instrument and FTIR spectroscopy *J.Mech Sci Technol* **31** 455–463.
[19] Mohamed Shameer P and Ramesh K 2017 Green technology and performance consequences of an ecofriendly substance on a 4-stroke diesel engine at standard injection timing and compression ratio *J Mech Sci Technol* **31** 1497–1507.
[20] Hannon M, Gimpel J, Tran M, Rasala B &Mayfield S 2010 Biofuels from algae: challenges and potential *Biofuels* **1**7 63–84.
[21] Paul Abishek M, Patel J and Prem Rajan A 2014 Algae Oil: a sustainable renewable fuel of future *Biotechnology Res Int* **27** 2814.
[22] Puneet Verma, Sharma M P &Dwivedi Gaurav 2016 Impact of alcohol on biodiesel production and properties *Renew Sustain Energy Rev* **56** 319–33.
[23] Cameron DE, Bashor C J and Collins JJ 2014 A brief history of synthetic biology *Nature Rev/Microbiol* **12** 381–90.
[24] Digambar Singh, Dilip Sharma, S LSoni, Sumit Sharma, Pushpendra Kumar Sharma and Amit Jhalani 2020 A review on feedstocks, production processes, and yield for different generations of biodiesel *Fuel* **262** 116553.
[25] Math M C and Chandrashekhara KN 2016 Optimization of alkali catalyzed transesterification of safflower oil for production of biodiesel *Journal of Engineering* **6** 1-7.
[26] Rashid U and Anwar F 2008 Production of biodiesel through base-catalyzed transesterification of safflower oil using an optimized protocol *Energy and Fuels* **22** 1306-1312.
[27] Mumtaz MW, Adnan A, Anwar F, Mukhtar H, Raza M A and Ahmad F 2012 Response surface methodology: An emphatic tool for optimized biodiesel production using rice bran and sunflower oils *Energies* **5** 3307-3328.
[28] Ulf S, Ricardo S and Roge´rio MV 1998 Transesterification of vegetable oils: A review *Journal of Brazilian Chemical Society* **9** 199-210.
[29] Khan TMY, Atabani AE, Badruddin IA, Badarudin A, Khayoon M S and Triwahyono S 2014 Recent scenario and technologies to utilize non-edible oils for biodiesel production *Renew Sustain Energy Rev* **37** 840-851.
[30] Shin HY, Lee SH, Ryu JH and Bae SY 2012 Biodiesel production from waste lard using supercritical methanol *J Supercrit Fluids* **61** 134-138.
[31] Schorken U and Kempers P 2009 Lipid biotechnology: Industrially relevant production processes *Eur J Lipid Sci Technol* **111** 627-645.
[32] Karmakar A, Karmakar S and Mukherjee S 2010 Properties of various plants and animals' feedstocks for biodiesel production *BioresourTechnol* **101** 7201-7210.
[33] Xie W and Li H 2006 Alumina-supported potassium iodide as a heterogeneous catalyst for biodiesel production from soybean oil *J Mol Catal A Chem* **255** 1-9.
[34] Yin J Z, Xiao M and Song J B 2008 Biodiesel from soybean oil in supercritical methanol with co-solvent *Energy Convers Manag* **49** 908-912.
[35] M Canakci and J Van Gerpen 1999 Biodiesel Production via Acid Catalysis *Trans Am Soc Agric Eng* **42** 1203–1210.

[36] L Nelson, T Foglia and W Marmer 1996 Lipasecatalyzed production of biodiesel *J Am Oil Chem Soc***73** 1191–1195.
[37] Y Shimada, Y Watanabe, T Samukawa, A Sugihara, H Noda, H Fukuda and Y Tominaga 1999 Conversion of vegetable oil to biodiesel using immobilized Candida antarctica lipase *J Am Oil Chem Soc* **76** 789–793.
[38] F Ma and M A Hanna 1999 Biodiesel production: A review *bioresource technology volume* **70** 1-15.
[39] H Fukuda, A Kondo and H Noda 2001Biodiesel fuel production by transesterification of oils *J Biosci Bioeng* **92** 405–16.
[40] M Kaieda, T Samukawa, T Matsumoto, K Ban, A Kondo and Y Shimada 1999 Biodiesel fuel production from plant oil catalyzed by Rhizopus oryzae lipase in a water-containing system without an organic solvent *J Biosci Bioeng* **88** 627–631.
[41] E Lotero, Y Liu, D E Lopez, KSuwannakarn, D A Bruce and J G Goodwin 2005Synthesis of Biodiesel via Acid Catalysis *Ind Eng Chem Res* **44** 5353–5363.
[42] M K Lam, K T Lee and A R Mohamed 2010 Homogeneous, heterogeneous and enzymatic catalysis for transesterification of high free fatty acid oil (waste cooking oil) to biodiesel: A review *Biotechnology Adv* **28** 500–518.
[43] K Narasimharao, A Lee and K Wilson 2007 Catalysts in Production of Biodiesel: A Review *J Biobased Mater Bioenergy***1** 1-12.
[44] I M Atadashi, M K Aroua, A R Abdul Aziz and N M N Sulaiman 2013 The effects of catalysts in biodiesel production: A review *J Ind Eng Chem* **19** 14-26.
[45] Y C Sharma and B Singh 2008 Development of biodiesel from karanja, a tree found in rural India *Fuel* **67**1740–1742
[46] M Canakciand J Van Gerpen 1999Biodiesel Production via Acid Catalysis *Trans Am Soc Agric Eng* **42** 1203–1210.
[47] G Arzamendi, I Campoa, E Arguinarena, M Sanchez, M Montes and L M Gandia 2007 Synthesis of biodiesel with heterogeneous NaOH/alumina catalysts: Comparison with homogeneous NaOH *Chem Eng J* **134** 123-130.
[48] M Kouzu, T Kasuno, M Tajika, Y Sugimoto, S Yamanaka and J Hidaka 2008 Calcium oxide as a solid base catalyst for transesterification of soybean oil and its application to biodiesel production *Fuel* **87** 2798–2806.
[49] T L Chew and S Bhatia2008Catalytic processes towards the production of biofuels in a palm oil and oil palm biomass-based biorefinery *Bioresour Technol* **99** 7911–7922.
[50] E Lotero, Y Liu, D E Lopez, K Suwannakarn, D A Bruce and JG Goodwin 2005 Synthesis of Biodiesel via Acid Catalysis *Ind Eng Chem Res* **44** 5353–5363.
[51] M G Kulkarni and A K Dalai 2006 Waste Cooking Oils an Economical Source for Biodiesel: A Review *Ind Eng Chem Res* **45** 2901–2913.
[52] D E López, J G Goodwin, D A Bruce and E Lotero 2005 Transesterification of triacetin with methanol on solid acid and base catalysts *Appl Catal A Gen* **295** 97-105.
[53] M Di Serio, R Tesser, LPengmei and E Santacesaria 2008 Heterogeneous Catalysts for Biodiesel Production *Energy and Fuels* **22** 207-217.
[54] Z Helwani, M R Othman, N Aziz, J Kim and W J N Fernando 2009 Solid heterogeneous catalysts for transesterification of triglycerides with methanol: A review *Appl Catal A Gen* **363** 1-10.
[55] A Bajaj, P Lohan, P N Jha and R Mehrotra 2010 Biodiesel production through lipase catalyzed transesterification: An overview *J Mol Catal B Enzym* **62** 9-14.
[56] Sinha S, Agarwal AK and Garg S2008 Biodiesel development from rice bran oil: Transesterification process optimization and fuel characterization *Energy Conver Manage* **49** 1248–1257.
[57] Ramadhas S, Jayaraj Sand Muraleedharan C 2004 Use of vegetable oils as I C engine fuels – A review *Renew Energy* **29** 727–742.

[58] Demirba A 2002 Biodiesel from vegetable oils via transesterification in supercritical methanol *Energy Convers Manage* **43** 2349–2356.
[59] Schwab AW, Bagby MO and Freedman B1987 Preparation and properties of diesel fuels from vegetable oils *Fuel* **66** 1372–1378.
[60] Sharma YC and Singh B 2008 Development of biodiesel from karanja, a tree found in rural India *Fuel* ***87*** *1740–1742.*
[61] Meher LC, Sagar DVand Naik SN 2006 Technical aspects of biodiesel production by transesterification – A review *Renew Sustain Energy Rev* **10** 248–68.
[62] Shahid EM and Jamal YSept 2009 Performance of direct injection compression ignition engine with cotton seed oil biodieselin *3rd Int Conf. on sustainable and energy protection.*
[63] Hamed M, Mashad E, Zhang R, Roberto J and Bustillos A 2008 A two-step process for biodiesel production from salmon oil Biosyst *Eng* **99** 220–227.
[64] Nagarhalli M V, Nandedkar V M and Mohite K C 2010 Emission Performance Characteristics of Karanja biodiesel and its blends in a C I Engine it's Economics *ARPN Journal of Engineering and Applied Sciences* **5** 52-56.
[65] Knothe G, Van Gerpen JandKrahl J 2005 The Biodiesel Handbook *Champaign, IL: AOCS Press.*
[66] Keith F W, Blachly F E and Sadler F S J1954*Am Oil Chem Soc* **31** 298-302.
[67] McCormick R L, Ratcliff M, Moens L and Lawrence R 2007 several factors affecting the stability of biodiesel in standard accelerated tests *Fuel Processing Technology* **88** 651–657.
[68] AOCS Ca 6a-40 1998 Unsaponifiable Matter Official Methods and Recommended Practices of the AOCS *5th Edition AOCS Press, Champaign, IL.*
[69] Benjumea P, Agudelo J and Agudelo A 2008 Basic properties of palm oil biodiesel–diesel blends *Fuel* **87** 2069–2075.
[70] Srikanth H V, Venkatesh J, Godiganur S, Venkateswaran S and Manne B2017Bio-based diluents improve cold flow properties of dairy washed milk-scum biodiesel *Renewable Energy* **111** 168–174.
[71] Joshi R M and Pegg M J2007Flow properties of biodiesel fuel blends at low temperatures *Fuel* **86** 143–151.
[72] Abe M, Hirata S, Komatsu H, Yamagiwa K and Tajima H 2016 Thermodynamic selection of effective additives to improve the cloud point of biodiesel fuels *Fuel* **171** 94–100.
[73] Dunn R O, Ngo H L and Haas M J 2015 Branchedchain fatty acid methyl esters as cold flow improvers for biodiesel *Journal of the American Oil Chemists Society* **92** 853–869.
[74] GiraldoS Y, Rios L A and Suarez N2013Comparison of glycerol ketals, glycerol acetates and branched alcohol-derived fatty esters as cold-flow improvers for palm biodiesel *Fuel* **108** 709–714.
[75] Knothe G, Matheaus A C and Ryan T W2003 Cetane numbers of branched and straight-chain fatty esters determined in an ignition quality tester *Fuel* **82** 971–975.
[76] Canakci M and Sanli H2008Biodiesel production from various feedstocks and their effects on the fuel properties *Journal of Industrial Microbiology & Biotechnology* **35** 431–441.
[77] Pala EnN 2012 On-road measurement of NOx and CO2 emissions from biodiesel produced from different feedstocks. Doctoral Thesis *the University of Texas at Arlington.*
[78] SaydutA,Duz M Z, Kaya C, Kafadar A B and Hamamci C 2008 Transesterified sesame (Sesamum indicum L) seed oil as a biodiesel fuel *Bioresource Technology* **99** 6656–6660.
[79] Zaharin M S M, Abdullah N R, Najafi G, SharudinH, and Yusaf T 2017 Effects of physicochemical properties of biodiesel fuel blends with alcohol on diesel engine performance and exhaust emissions: A review *Renewable and Sustainable Energy Reviews* **79** 475–493.
[80] Knothe G2005Dependence of biodiesel fuel properties on the structure of fatty acid alkyl esters *Fuel Processing Technology* **86** 1059–1070.
[81] IssariyakulT , Kulkarni M G, Dalai A K and Bakhshi N N 2007 Production of biodiesel from waste fryer grease using mixed methanol/ethanol system *Fuel Processing Technology* **88** 429–436.

[82] Mofijur M, Atabani A, Masjuki H, Kalam M and Masum B2013A study on the effects of promising edible and non-edible biodiesel feedstocks on engine performance and emissions production: A comparative evaluation *Renewable and Sustainable Energy Reviews* **23** 391–404.
[83] Park S H, Kim, H J Suh, H K and Lee C S 2009 A study on the fuel injection and atomization characteristics of soybean oil methyl ester (SME) *International Journal of Heat and Fluid Flow* **30** 108-116.
[84] Sallevelt J, Gudde J, Pozarlik A and Brem G 2014 The impact of spray quality on the combustion of a viscous biofuel in a micro gas turbine *Applied Energy* **132** 575–585.
[85] Ong HC, Masjuki H, Mahlia T, Silitonga A, Chong W and Leong K 2014 Optimization of biodiesel production and engine performance from high free fatty acid Calophylluminophyllum oil in CI diesel engine *Energy Conversion and Management* **81** 30–40.
[86] Ghazali W N M W, Mamat R, Masjuki H H and Najafi G 2015 Effects of biodiesel from different feedstocks on engine performance and emissions: A review*Renewable and Sustainable Energy Reviews* **51** 585–602.
[87] Ejim C, Fleck B and Amirfazli A 2007 Analytical study for atomization of biodiesels and their blends in a typical injector: surface tension and viscosity effects *Fuel* **86** 1534–1544.
[88] Lapuerta M, Armas O and Rodrıguez Fernandez J2008Effect of biodiesel fuels on diesel engine emissions *Progress in Energy and Combustion Science* **34** 198–223.
[89] Ramırez Verduzco L F, Rodrıguez Rodrıguez J E D and Rayo Jaramillo Jacob A 2012 Predicting cetane number, kinematic viscosity, density and higher heating value of biodiesel from its fatty acid methyl ester composition *Fuel* **91** 102–111.
[90] Oner C and Altun SS 2009 Biodiesel production from inedible animal tallow and an experimental investigation of its use as alternative fuel in a direct injection diesel engine *Applied Energy* **86** 2114–2120.
[91] Sanford S D, White J M, Shah P S, Wee C, Valverde M A and Meier G R 2009 Feedstock and biodiesel characteristics report *Renewable Energy Group November 17th, 2009Ames, IA* **5** 0010-0888
[92] Xin J,Imahara H and Saka S2009 Kinetics on the oxidation of biodiesel stabilized with antioxidant *Fuel* **88** 282–286.
[93] Monyem A and Van Gerpen J H 2001 The effect of biodiesel oxidation on engine performance and emissions *Biomass and Bioenergy* **20** 317–325.
[94] Jakeria M, Fazal M and Haseeb A 2014 Influence of different factors on the stability of biodiesel: A review *Renewable and Sustainable Energy Reviews* **30** 154–163.
[95] Ranganathan S V, Narasimhan, S L and Muthukumar K 2008 An overview of enzymatic production of biodiesel *Bioresource Technology* **99** 3975–3981.
[96] Kalam M A and Masjuki, H H 2008 Testing palm biodiesel and NPAA additives to control NOx and CO while improving efficiency in diesel engines *Biomass and Bioenergy* **32** 1116–1122.
[97] Atabani A E, Silitonga A S Badruddin I A, Mahlia T M I, Masjuki H H, and Mekhilef S 2012 A comprehensive review on biodiesel as an alternative energy resource and its characteristics *Renewable and Sustainable Energy Reviews* **16** 2070–2093.
[98] Karmakar A, Karmakar S and Mukherjee S 2010 Properties of various plants and animals' feedstocks for biodiesel production *Bioresource Technology* **101** 7201–7210.
[99] Xue J, Grift T E and Hansen AC 2011 Effect of biodiesel on engine performances and emissions *Renewable and Sustainable Energy Reviews* **15** 1098–1116.
[100] Bozbas K 2008 Biodiesel as an alternative motor fuel: production and policies in the European Union *Renewable and Sustainable Energy Reviews* **12** 542–552.
[101] Md Mofijur Rahman, Mohammad Rasul and Nur Md Sayeed Hassan 2017 Study on the Tribological Characteristics of Australian Native First Generation and Second-Generation Biodiesel Fuel *Energies* **10** 55.

[102] AzadaA K and Ameer UddinS M 2013 Performance study of a diesel engine by first generation bio-fuel blends with fossil fuel *an experimental study, Journal of Renewable and Sustainable Energy* **50** 13118.
[103] Emilio A. Viornery-Portilloa, Brenda Bravo-Diaza and Violeta Y 2020 Mena-Cervantes, Life cycle assessment and emission analysis of waste cooking oil biodiesel blend and fossil diesel used in a power generator *Fuel* **281** 118739.
[104] Arvind H, Hirani, Nasir Javed, Muhammad Asif,Saikat K. Basu and Ashwani Kumar 2018 A Review on First- and Second Generation Biofuel Productions *Biofuels: Greenhouse Gas Mitigation and Global Warming* 141–154*(springer)*.
[105] Cristiano Varrone, GeorgisFloriotis, TonjeM. B. Heggeset, Simone B. Le, Sidsel Markussen, IoannisV.Skiadas and Hariklia N. Gavala 2017 Continuous Fermentation and Kinetic Experiments for the Conversion of Crude glycerol derived from Second-Generation biodiesel into 1,3 Propanediol and Butyric Acid, *Biochemical Engineering Journal* **128** 149-161.
[106] Dariusz Kurczyńskia, Piotr Łagowskiaand Grzegorz Wcisło2021 Experimental study into the effect of the second-generation BBuE biofuel useon the diesel engine parameters and exhaust composition *Fuel* **284** 118982.
[107] Wenjun Zhong, Tiemin Xuan, Zhixia He, Qian Wang, Da Li, Xin Zhang and Huang Yue Yin 2016 Experimental study of combustion and emission characteristics of diesel engine with diesel/second generation biodiesel blending fuels *Energy Conversion and Management* **121** 241–250.
[108] Janske van Eijck, Bothwell Batidziraiand André Faaij 2014 Current and future economic performance of
first and second-generation biofuels in developing countries *Applied Energy* **135** 115–141.
[109] Enrico Mattarelli and Carlo Alberto Rinaldini2015 Combustion Analysis on an IDI CI Engine Fueled by
Microalgae *SAE Technical Paper* **24** 2484.
[110] Anoop Singh, Stig Irving Olsena and Poonam Singh Nigam 2011 A viable technology to generate third generation biofuel *J Chem Technol Biotechnol* **86** 1349–1353.
[111] Oliveira LE1, Cedeno R F, Chavez EG, Gelli VC and Masarin 2019 Red Macroalgae Kappaphycusalvarezii as feedstock for nutraceuticals, pharmaceuticals and fourth generation biofuel production *International Conference on Renewable Energies and Power Quality (ICREPQ'19) Tenerife (Spain)*.
[112] HoofarShokravi, Zahra Shokravi, Mahshid Heidarrezaei, Hwai Chyuan Ong, Seyed Saeid Rahimian Koloor,
Michal Petrů,Woei Jye Lau and Ahmad Fauzi Ismail 2021 Fourth generation biofuel from genetically, modified algal biomass: Challenges and future directions *Chemosphere* **285** 131535.
[113] M. Mofijur, Sk. Yasir Arafat Siddiki, Md. Bengir Ahmed Shuvho, F. Djavanroodi, I.M. Rizwanul Fattah, Hwai Chyuan Ong, M.A. Chowdhury, T.M.I. Mahlia 2021 Effect of nanocatalysts on the transesterification reaction of first, second and third generation biodiesel sources- A mini-review *Chemosphere* **270** 128642.
[114] Abd El-Fatah Abomohra, Xuying Zheng, Qingyuan Wang, Jin Huang, Reham Ebaid 2021 Enhancement of biodiesel yield and characteristics through *in-situ* solvo-thermal co-transesterification of wet microalgae with spent coffee grounds *Bioresource Technology* **323** 124640.
[115] Shashi Kant Bhatia, Ravi Kant Bhatia, Jong-Min Jeon, Arivalagan Pugazhendhi, Mukesh Kumar Awasthi, Dinesh Kumar, Gopalakrishnan Kumar, Jeong-Jun Yoon, Yung-Hun Yang 2021 An overview on advancements in biobased transesterification methods for biodiesel production: Oil resources, extraction, biocatalysts, and process intensification technologies *Fuel* **285** 119117.

[116] Adeyinka Sikiru Yusuff, Afeez Olayinka Gbadamosi , Lekan Taofeek Popoola 2021 Biodiesel production from transesterified waste cooking oil by zinc-modified anthill catalyst: Parametric optimization and biodiesel properties improvement *Journal of Environmental Chemical Engineering* **9** 104955.
[117] Inam Ullah Khan, Zhenhua Yan and Jun Chen 2020 Production and Characterization of Biodiesel Derived from a Novel Source Koelreuteria paniculata Seed Oil *Energies* **13** 791.
[118] Ejaz M. Shahid*, Younis Jamal 2011 Production of biodiesel: A technical review *Renewable and Sustainable Energy Reviews* **15** 4732– 4745.
[119] Haris Mahmood Khan, Tanveer Iqbal, Chaudhry Haider Ali, Ansar Javaid, Izzat Iqbal Cheema 2020 Sustainable biodiesel production from waste cooking oil utilizing waste ostrich (Struthio camelus) bones derived heterogeneous catalyst *Fuel* **277** 118091.

A State of Art Review on Grass Fibre Reinforced Polymer Composite

Hrithik Kumar[1], Debangshu Jaiswal[1], Hemalata Jena*[1]
[1]School of Mechanical Engineering, KIIT Deemed to be University, Bhubaneswar, India
*hemalata.jenafme@kiit.ac.in

Abstract. The environment awareness and cost cutting are the main beginning to think about eco-friendly materials. Of course, natural fibre as reinforcement in polymer matrix is one of the best option due to its biodegradability, recyclability, low cost and high specific strength/stiffness. It motivates the researchers to find new and easily available bio-fibre which may replace with non-biodegradable material and maintain the proper balance for the ecosystem. Now a day's, researchers search on new plant fibre that facilitate easy and viable extraction that does not mess up the properties of the fibres. There has been an incredible development in polymer matrix composites of various plant fibre. Grass fibre is one of the important plant fibres which occurs in different parts of plants, mainly in the stems and leaves. Present paper studied on different grass fibres and its extraction process. Different surface treatment of grass fibre has been reviewed to extract the knowledge of the adhesion property of the fibre with polymer and its mechanical property. The present paper has also reviewed the mechanical property of the grass fibre composite. The scope of the grass fibre-reinforced composite material in various structural and engineering application are also addressed here.

1. Introduction

Polymer composite materials are widely used in numerous mechanical regions to oblige the low density and high strength property prerequisites for specific applications. Wide popularity of glass fibre reinforced composite in various application has resulted into its disposability and recyclability issue. Hence alternate finding of glass fibre motivates the researchers to try biodegradable fibre. Agricultural by-products, for example, sugar cane, rice husk, coir and grass are arising as novel and appealing materials due its inexpensive, eco-friendly, carbon dioxide sequestration, economical, recyclable, low density, and biodegradable in nature. Despite of various advantage of natural fibre, its application in polymer matrix is limited due to its poor interfacial bonding with matrix, chemical incompatibility of highly polar nature of fibre with nonpolar polymer [1]. This results into lower the mechanical strength of composite material. Dimensional instability is another problem associated with natural fibre composite. To overcome the problem related to natural fibre, different techniques have been adopted like surface treatment of fibre, hybridisation of natural fibre with synthetic fibre/ other natural fibre [2-3] and matrix added with filler etc [4-6]. Mohanty et al [7] shows the different surface treatment method of natural fibre [7]. The treatment is a typical strategy to remove cellulose, lignin, wax etc. from the fibre and change the fibre surface to bring down surface strain and upgrade interfacial bond between hydrophilic fibre and polymer matrix. Hybridization of natural fibre composites have been studied by Karthi et. al [8]. They have prepared the hybrid composite made of coir and napier grass fibre in unsaturated polyester and found that hybridization improves the tensile, flexural and impact properties of the composite.

In present study, grass fibre has been taken into consideration as fibre reinforcement in polymer matrix composite. It is a strengthening tissue in a plant elongated cells which take place in different parts of plants i.e., stems and leaves. Grass fibres are lignocellulosic in nature which consist of layers of lignin (outermost layer), hemicellulose (inner layer), cellulose (innermost layer). Different grass fibres have different percentage of lignin, hemicellulose, cellulose which is shown in Table 1 along with its density and diameter of the fibre. Density of the fibre is an important physical property which is required to calculate the fibre volume fraction and weight fraction of the composite. Fibre density can be calculated by using various method like Truong method or pycnometer. Similarly, fibre diameter is another important property which decides the mechanical property of the fibre. Table 1 shows all these physical and chemical property of different types of grass fibres. Different types of grass fibres are there which are discussed below and shown in the Figure 1.

1.1. Wild cane grass:
Wild cane grass belongs to the Poaceae family (Saccharum spontaneum). It a is a tall perennial cane which is found in fresh or moderately saline damp soils. It is mostly found in eastern and southern Asia. It is also known as Spanish cane, Giant reed, Arundo, Carrizo, Arundo donax, and Colorado River Reed [9].

1.2. Napier grass:
Napier grass (Pennisetum purpureum) belongs to the Poaceae family. It is also known as elephant grass. It is a fast-growing perennial grass [10].

1.3. Broom grass:
Broom grass belongs to Poaceae family (Thysanolaena maxima). Its flowering culms are basically used for making broom sticks. The suitable place for growing broom grass is at humid ravine banks, river banks and
slopes of hill. It is basically grown from 4.2 ft. to 5.1 ft height mostly found in north-east of India [11].

1.4. Snake Grass:
Snake grass fibre is vernacular name of Sansevieria ehrenbergii which belongs Acanthaceae family. This is a medicinal plant used for skin, snake bites etc.

1.5. Kusha Grass:
Kusha fibre (Desmostachya bipinnata) belongs to Poaceae family. It is abundantly available in south asian countries. Traditional use of kusha fibre in yoga and religions institutions which now used in material science for its wide application.

(a)Wild cane grass[9]

(b)Napier grass[12]

(c)Snake grass [13]
Figure 1. Different types of grass fibres

Table 1. Physical and chemical property of different grass fibre

Grass Fibre	Physical Properties			Chemical composition			
	Density(gr/cm3)	Diameter (mm)	Length (mm)	Cellulose (%)	Hemi cellulose (%)	Lignin	Ref
Kusha Grass	1.102	0.07-0.1	-	70.58	-	14.35	14
Napier grass	0.358	0.15-0.55	142	47.12	31.27	21.63	15-16
Corn husk	0.34	1.86*10-4	-	46.15	33.7	3.92	17
Sisal	0.76	0.122-0.135	-	78	10	8	18
Belulang Grass	1.213	0.081-0.178	-	45.8	32.29	7	19
Pennisetum orientale grass	1.045	0.213		60.3	16	12.45	20

Kush grass	1.1	0.320	37	70.6		14.4	21
Snake grass	0.88	0.045-0.250					13
Sansevieria cylindrica				79.7		3.8	22

2. Extraction process of grass fibres

Grass fibre must be extracted from its raw from and then it can be act as reinforcement polymer matrix. Haameem et. al. [16] shows the steps of making napier grass fibre composite from its the extraction to its composite fabrication as shown in Figure 2. Figure 2.a shows the extraction of fibre by using a very common water retting process. In water retting process, stem of grass fibre is dipped in water and this process helps in producing high-quality fibres. Before doing retting process, grass must be taking away from the internodes of stem and then stems are crushed with a mallet. After this, crushed stems immersed in a water tank for about three to four weeks. Figure 2.b shows the extraction of each fibre from immersed napier grass fibre stem using a roll-out machine. Subsequently, the extracted raw fibres are dried under sun to remove excess water. Figure 1.c. shows the alkali treatment of fibre in different alkali concentrations of 5, 10, 15, and 20%. Figure 1.d shows the fibres are cut into pieces of having length 1 to 3 cm. Figure 1.e. shows the napier grass fibre with polyester matrix of having 10, 15, 20, 25, and 30 vol% of fibre. Table 2 shows the extraction process of grass fibre which can be classified into three i.e. (i) water retting process (ii) mechanical process and (iii) the combination of mechanical processes with water retting. The common extraction of grass fibre in water retting and mechanical process. Kusha grass fibre, elephant grass fibre, broom grass fibre etc. can be splitted during the water rating process. Some of the fibre like belung grass, broom grass etc can be extracted after water retting process.

Table 2. Extraction of different grass fibre

Different grass fibre	Extraction process	Ref
Sansevieria cylindrica	1.Mechanical Process- By using decorticator	22
Kusha Grass	1. Water rettin-28 days 2. Drying- At ambient temperature for 7 days	14
Mendong grass	1. Water retting-7 days 2. Drying- Wind dry	23
Elephant grass	1. Water retting-4 days 2. Drying- Air dry for 8 hours and oven dry at 160^{0}C for 1 hr	24
Snake grass	1. Water retting-10 days 2. Drying- Under the Sun for one week	25
Arindo donax L	1.Mechanical Process- By using decorticator	26 83.
Belung grass	1. Water retting-8 days 2. Drying- At ambient temperature for 24 hr 3. Separation of fibre- By using Plastic comb	19
Sansevieria ehrenbergii	1.Mechanical Process- Decorticator	27
Corn husk	1. Water retting-8 days 2. Separation of fibre- By using Plastic comb 3.Drying- Air dry	17

Broom grass	1. Water retting-16 days 2. Drying- Sun dry for 7 days and oven dry for 2 hr at 70^0C 3. Separation of fibre- By manually	28
Sisal	1. Water retting 2.Scraping 3.Mechanical Process- Decorticator	29

3. Mechanical property of grass fibre composite

Grass fibre as reinforcements in polymer matrix composites have been researched for finding its potential to replace traditional fibres. Grass fibre is the alternate solution for the high-cost composites. Various attempts have been made on treatment of grass fibre to improve the interaction between the fibres and the matrix and hence increase the mechanical property [30]. Liu et al [31] have tried to improve the mechanical property of Indian grass fibre by alkali solution and found the improvement in tensile, flexural properties and impact strength. Figure 3 (a), (b), (c), (d) shows the micrograph of grass fibre alkali treated with 5% for 2 hr, 10% for 2 hr and 10% for 4 hr respectively. Lignocellulosic fibres have cellulose, hemicellulose, lignin and wax are present. This is the reason of hydrophilic in nature of fibre and improper adhesion of matrix with fibre. From the SEM image, it is observed that smaller individual single fibres are combined together by the inter-fibrillar material and form the grass fibres. After alkali treatment of 5% for 2 hr, 10% for 2 hr and 4 hr, inter-fibrillar region is decreases. At 5 or 10% alkali solution for 2 hr treatment carbon atomic ratio in fibre surface improved which in turn changes the surface structure of the fibre. At 10% alkali solution for 4 hr treatment carbon atomic ratio in fibre surface largely increases that decreases the content of lignin and hemicelluloses as compared to 5 or 10% alkali solution for 2 hr. Maximum decrement of inter-fibrillar region is at 10% concentration of alkali for 4hr due to more removal of lignin, wax and others. This is the cause of improve the surface roughness and improve the mechanical locking between fibre and matrix.

Length of the fibre play important role in fibre reinforced polymer composite. It is also observed that long fibre composites have superior mechanical properties than short fibre composites [10,16]. Sathishkumar et. al. [13]. have studied the different fibre length of snake grass and different fibre volume fraction and concluded that 25% volume fraction of the snake grass fibres composite is the optimum fibre volume fraction beyond that mechanical property decreases. Table 3 show the mechanical property of the different types of grass fibre with polymer matrix composite. from table it is observed that grass fibres have potential to mix with different polymers. Its mechanical property not only depends on fibre volume fraction but also fibre length. Long napier fibre shows the best mechanical property among all the grass fibre composite which can be comparable with glass fibre and other natural fibre.

Figure 2. Napier grass fibre composite from its the extraction to its composite fabrication [16]

Table 3. Mechanical property of treated grass fibre polymer composite

Grass Fibre (wt.%)	Matrix	Tensile strength (MPa)	Tensile Modulus (GPa)	Flexural strength (MPa)	Flexural Modulus (GPA)	Impact Strength (J.m)	Ref
Napier grass fibre(25vol.%)- Short fibre (1-3 cm)	Polyester	15.64	3.02	47.9	1.89		16
Napier grass fibre(25vol.%)- long fibre (5-8 cm)	Polyester	18.17	2.00	39.5	1.80		
Napier (20 wt.%) short fibre (10mm)	Epoxy	39.53	2.95	76.21	3.57	106.29	10
Napier (20 wt.%) long fibre(185mm)	Epoxy	142.14	10.90	111.3	6.78	237.59	
Broom grass (10 length)	Polyester	82.39	1.05	78	5.3		11
Kusha grass (2mm)	Polypropylene	34	-	-	-	-	21
Snake grass (30mm, 25Vol%)	Unsaturated polyester	37	4.80	67	1.58	-	13

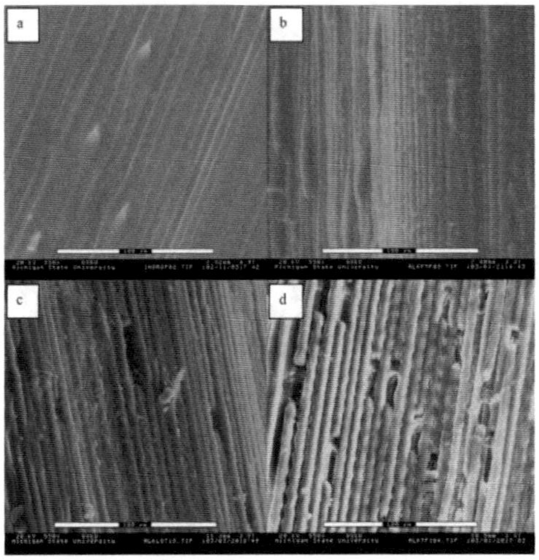

Figure. 3. ESEM micrographs of grass fibres raw Indian grass fibres and treated grass fibre (a) untreated fibre, (b) treated with 5% alkali solution for 2 hr (c) treated with 10% alkali solution for 2 hr and (d) treated with 10% alkali solution for 4 hr [31]

4. Conclusions

In present piece of paper helps to understand the different grass fibre and its property. Followings are the conclusion of the paper:

i. Simple extraction process of grass fibres with low cost can easily fabricate the composite in simple way.

ii. Fibre content and fibre length play important role in deciding the mechanical property of fibre polymer matrix composite. Mechanical property of the composite is maximum mostly at 25vol% of fibre and 10-30cm fibre length

iii. There may be the possibility that these strong and biodegradable natural materials may dominate the widely popular materials like glass fibre.

5. Scope for future work

The grass fibre is an eco-friendly fibre which may be used in the place of other wood fibre or synthetic fibre if it is fully explored in different property. It is a new trend fibre and researchers have tried to explore with different new grass fibre and try to explore its mechanical property. Though, mechanical property is the basic knowledge of any composite material for its structural application, tribological, thermal and damping behaviour can be experimented which are less explored for this fibre. Again, hybridisation of grass fibre with particulate filler is also not studied adequately. Study can be done on different conventional and non-conventional filler which have potential to modify the mechanical, tribological and other property. There is wide scope of review on water absorption behaviour of different grass fibre polymer composite and also different treating agent can be explore to improve the adhesion between fibre and matrix.

Reference

1. Jena H, Pradhan AK, Pandit MK 2014 *J Reinf Plast Compos* **33**(11) 1073.
2. Ganesamoorthy R, Meenakshi Reddy R, Raja T. et al. 2021 *Adv Mat Sci* Eng **2021**.
3. Khan A, Vijay R, Lenin Singaravelu D, Arpitha GR, Sanjay MR, Siengchin S, Jawaid M, Alamry K, M Asiri A 2020 *J Mat Res Tech* **9**(1)773-778.
4. Jena H, Panigrahi A (2021) World J Eng.doi:10.1108/WJE-08-2020-0334.,
5. Nath S, Jena H, Priyanka, Sahini S (2019 Silicon. 11: 659–671.
6. Jena H, Pradhan AK, Pandit MK 2016 Adv PolymTechnol. Doi: 10.1002/adv.21718.

7. Mohanty AK, Misra M, Drzal TL 2005 *Natural fibres, biopolymers, and bio-composites*. Boca Raton: CRC Press, Technology & Engineering.
8. Karthi S, Sasikumar KSK, Sangeetha RK, Saravanan N 2020 *Mater Today Proc* **33** 2929.
9. Kumar Gudapati SP, Chidambaranathan S, Ratna Prasad AV 2020 *Mater Today Proc* **23** 632.
10. Kommula VP, Obi Reddy K, Shukla M, Marwala T, Varada Rajulu A 2014 *Int J Polym Anal Charact* **19** 693.
11. Srinivasababu N, Suresh Kumar JK, Reddy VK 2014 *Procedia Mater Sci* **6** 1006.
12. Obi Reddy K, Uma Maheswari C, Shukla M, Varada Rajulu A 2012 *Mater Lett* **67** 35.
13. Sathishkumar TP, Navaneethakrishnan P, Shankar S 2012 *Compos Sci Technol* **72**1183.
14. Balaji AN, Karthikeyan MKV, Vignesh V 2016 Int *J Polym Anal Charact*.doi:10.1080/1023666X.2016.1192324
15. Kommula VP, Reddy KO, Shukla M, Marwala T, Reddy EVS, Rajulu AV 2016 *Int J Polym Analy Charact* **21** 18.
16. Haameem JAM, Abdul Majid MS, Afendi M, Marzuki HFA, Fahmi I, Gibson AG 2016 *Compos Struct* **136** (2) 1.
17. Herlina Sari N, Wardana ING, Surya Irawan Y, Siswanto E 2017 *J Nat Fib***15**(4) 545.
18. Li Y, Mai YW, Ye L 2000 *Compos Sci Technol* **60** 2037.
19. Lokantara IP, Suardana NPG, Surata IW, Winaya INS 2020 Int J Mech Eng Technol 11(1):84-91.
20. Vijay R, Vinod A, Lenin Singaravelu D., Sanjay MR, Siengchin S 2021 *Int J Lightweight Mat Manufact* **4** 43.
21. Hariprasad K, Ravichandran K, Jayaseelan V, Muthuramalingam T 2020 *J Mater Res Technol* **9** 14029.
22. Sreenivasan VS, Somasundaram S, Ravindran D, Manikandan V 2011 *Mat Des* **32** 453-461.
23. Suryanto H, Marsyahyo E, Irawan YS, Soenoko R 2014 *J Natur Fibre* **11**:333.
24. Rao K. Murali Mohan, Ratna Prasad AV, Ranga Babu MN, Rao KM, Gupta AVSSKS 2007 *J Mater Sci* **42** 3266.
25. Sathishkumar T, Navaneethakrishnan P, Shankar S, Rajasekar R 2014 *Fibre Polym* **15** (9)1927.
26. Fiore V, Scalici T, Valenza A 2014 *Carbohy Polym* **106** 77.
27. Sathishkumar TP, Navaneethakrishnan P, Shankar S, Rajasekar R 2013 *Compos Interf* **20** 575.
28. Ramanaiah K, Ratna Prasad AV, dan Hema Chandra Reddy K 2012 *Mat Des* **40** 103.
29. Ibrahim ID, Jamiru T, Sadiku RE, Kupolati WK, dan Agwuncha SC 2016 *J Polym Enviro*. Doi: 10.1007/s10924-016-0823-2.
30. Somseemee O, Sae-Oui P, Siriwong C 2021 *Indus Crop Prod* **171** 113881.
31. Liu W, Mohanty AK, Askeland P, Drzal LT, Misra M 2004 *Polym* **45**(22) 7589.

Effect of additional heating on corrosion of dissimilar friction stir welded aluminum alloys

M Raturi[1] and A Bhattacharya[1]

Measurement and Process Analysis Lab, Department of Mechanical Engineering, Indian Institute of Technology Patna, Bihar, 801106, India

madhav_1821me04@iitp.ac.in

Abstract. In the present work, aluminum alloys AA2014-T6, AA6061-T6 and AA7075-T651 are used to fabricate two different dissimilar friction stir welds of AA6061-AA7075, and AA6061-AA2014 using a threaded tool pin with three flat faces at tool rotational speed of 1200 rpm, weld speed 98 mm/min and employing additional secondary heating. Additional heating is provided using butane gas torch which principally expedites better material mixing supplementarily resulted in increased corrosion resistance. The higher working temperature provides steeper cooling rates necessary for grain refinement and additional breakdown or dissolution of corrosion susceptible particulates. The joints fabricated sustained corrosive environments for adequate duration and ultimate tensile strength of the joints are retained after 14 days of immersion in corrosive saline solution. AA6061 experienced pitting as well as intergranular corrosion, AA2014 shows pitting and AA7075 exhibited exfoliation corrosion. The fracture morphology shows substantial size dimples indicating the loss in ductility for all the joints due to the after effects of heating.

Keywords: Aluminum alloys, Friction stir welding, Corrosion, Strength, Fractographs

1. Introduction

Joining of aluminum alloys using solid state friction stir welding (FSW) has become quite common for aerospace, shipbuilding and automobile industries [1].The joining can take place in different configurations e.g. joint type, varied thickness, different shapes and most importantly dissimilar alloys. The dissimilar joining very often required in different engineering components with reliable quality for varied manufacturing sectors. Consequently, dissimilar joining of aluminum alloys becomes an important concern for researchers along with evaluating their mechanical and chemical performances. In FSW the heat generated due to friction between tool and workpiece, makes the material to flow around the rotating tool pin. The judicial selection of process parameters like tool rotational speed, weld speed, pin profile etc. are needed as it governs degree of heating and material mixing. The use of appropriate parameters along with additional heat supply can intensify the material flow resulting in positive mechanical response [2]. Along with the mechanical responses, corrosion behavior becomes equally important as sometimes the fabricated joints are likely to experience saline environments during their applications [3]. AA2014 owing high strength-to-weight ratios and permissible ductility lacks in corrosion resistance. It generally undergoes pitting corrosion due to its non-uniform microstructure with Al_2Cu precipitates populated within the matrix and at grain boundaries [4]. AA 6XXX series alloys are aimed to offer corrosion resistive properties compared to other series. However, during FSW the alloy might experiences thermomechanical disturbances affecting its corrosive behavior leading to pitting and intergranular corrosion (IGC) while operational in saline environments [5]. AA7075 alloys reportedly carry intermetallic particulates like Al_7Cu_2Fe, $MgZn_2$ in substantial amount [6]. These particulates act as the site of corrosion initiation and further expedites leading to pitting corrosion and stress corrosion cracking. The presence of alloying elements like Cu tends to increase the corrosion resistivity although elements like Mg and Zn encourages the corrosion process [7]. Fine grains can also deaccelerate the corrosion rate as grain refinement led to increase in grain boundaries the leads to breakdown of intermetallic particles responsible for corrosion initiation [8].

The present experimental work is intended to comprehend the effect of immersion duration in saline solution on the corrosion behavior of obtained dissimilar FSWed joints all prepared with the application of additional heating. Two different dissimilar aluminum joints are obtained namely AA6061-T6 to AA7075-T651, AA6061-T6 to AA2014-T6 by FSW employing additional secondary heat (moreover the primary heat generated by the process itself). The joints are then exposed to corrosive environment (dipped in saline solution) for different time durations to assess the extent of corrosion and appraise the corrosion morphology, amount of deterioration in tensile strength, fracture behavior of these preheating assisted dissimilar AA6061-AA7075 and AA6061-AA2014 FSW joints.

2. Materials and methods

Two different dissimilar joints of AA6061-AA7075 and AA6061-AA2014 are obtained using 6.1 mm thick plates of the alloys by FSW with the application of additional secondary heating. The dissimilar joints are termed (e.g., AA6061-AA7075) in such a manner that first mentioned alloy (e.g., AA6061) is kept on advancing side (AS) while the later (e.g., AA7075) on retreating side (RS) as shown in figure 1(a). The joint fabrication is carried out using a H13 die steel tool having threaded pin profile with three flat faces (TIF) as shown in figure 1(b). The tool rotating at 1200 rpm is plunged along the abutting surfaces of dissimilar alloy plates held rigidly together and welding along the length is completed maintaining traverse speed of 98 mm/min. Before tool plunging, the work surface is heated with a butane gas torch near the seam region upto ~ 150 ± 5 °C and heating is also continued during welding by moving the torch ahead of the tool (see figure 1(a)). Afterwards, transverse tensile samples (Figure 1 (c)) are obtained from the joints using wire-cut electric discharge machining. Immersion test (as per ASTM G110) is performed over the gage length of the tensile samples of fabricated joints. Before immersion the samples are cleaned using etch cleaner (189 mL H_2O, 10 mL HNO_3, 1 mL HF) and concentrated HNO_3 to get rid of any contaminants. Post cleaning the samples are masked using enamel paint on the areas need not to be corroded and the gauge length region is left exposed as shown in figure 1(d). The immersion is carried out in saline corrosive solution (NaCl (57 g) + H_2O (1 L) + H_2O_2 (10 mL)) for durations of 7 and 14 days as shown in figure 1(e). The saline solution is changed every 2 days to maintain the efficacy of the progressive corrosion. The samples removed after specific time duration are cleaned as per ASTM G1 for further examination. The samples are immersed (Figure 1(f)) in heated (upto 90 °C) cleaning reagent (H_3PO_4 (50 mL) + CrO_3 (20 g) + H_2O (1 L)) for upto 10 minutes followed by dipping them in heated HNO_3 for 1 minute to get rid of any corrosion residue.

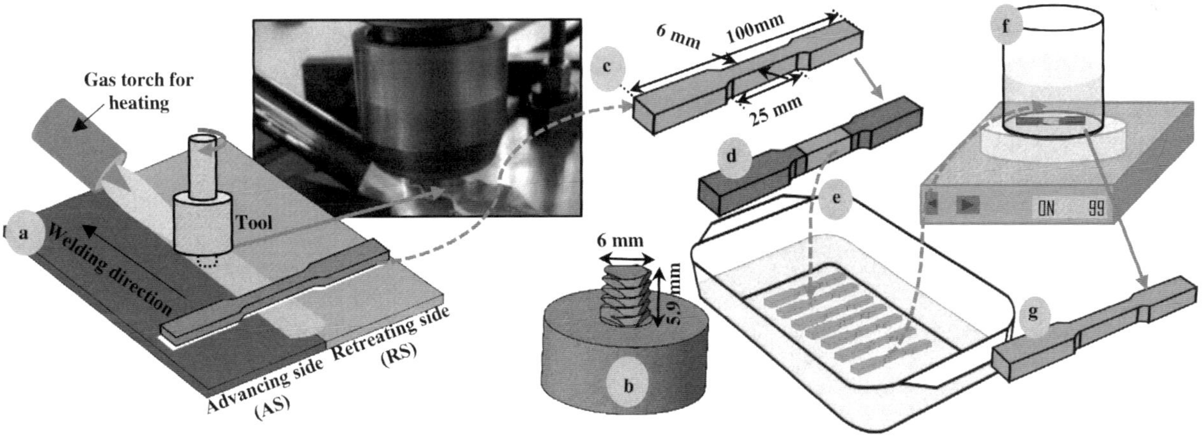

Figure 1. Schematic illustration of (a) dissimilar welded joint with additional heating, (b) pin profile, (c) tensile sample obtained from the joint, (d) tensile sample masked (red) before dip corrosion, (e) samples placed in corrosive solution, (f) cleansing of tensile sample post corrosion and (g) tensile sample post corrosion and cleansing

To evaluate the influence of corrosion on the mechanical performance of the welded joint, the tensile strength and elongation are measured for all corroded and non-corroded samples. Tensile tests of all samples are performed on 50 kN (Zwick/Roell Z50) universal testing machine maintaining cross head speed of 0.5 mm/min accompanied with contact type extensometer (25 mm gauge length). Corroded morphology and fractography is studied on field emission scanning electron microscope (FESEM) (GeminiSEM 500 of Zeiss). Optical micrographic image (Axio Imager.M2m, Zeiss) is used to exhibit the fracture location in corroded weld zone before and after the tensile test.

3. Results and discussion

The corrosion behavior, FESEM images of selected regions of additional heating assisted AA6061-AA7075 FSW joints are presented in figure 2. Figure 2(a) shows the optical micrograph of the joint after corrosion duration of 7 and 14 days respectively but before the tensile test. The aluminum matrix of AA6061 generally carries intermetallics like Al_2Cu, $FeAl_3$ and alloying elements like Si, Cu setting up electrochemical cells leading to corrosion while the required conditions are reached. Depending upon the corrosive environment (acidic or alkaline) the alloy can experience pitting or general corrosion [9].

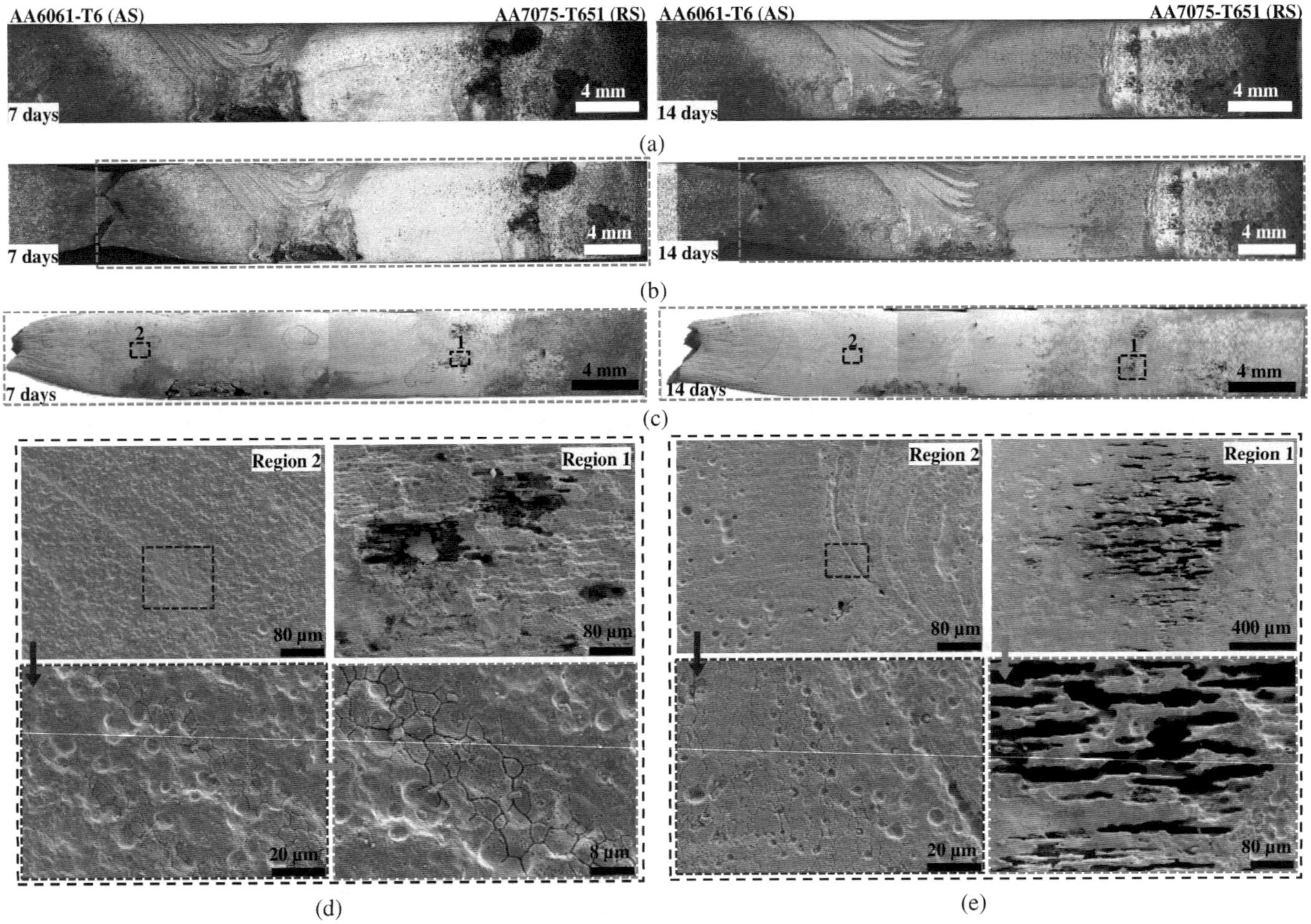

Figure 2. Optical microscopic image of additional heating assisted AA6061-AA7075 FSW joints (a) before and (b) after tensile test and corroded for 7 and 14 days, (c) FESEM image of highlighted section in (b). High magnification FESEM images of corroded regions (highlighted in (c)) for (d) 7 days and (e) 14 days respectively.

The hemispherical pits can be usually observed in corroded AA6061 and their size is subjected to the degree of heat and deformation alloy can experience during the FSW process. Thus, the use of additional heating is directly associated to the corrosion resistance of the joint. In the SZ of the nugget the AA6061 undergoes grain refinement and thus pitting due to presence of large particulate is uncertain, though IGC can take place but at minuscule level.

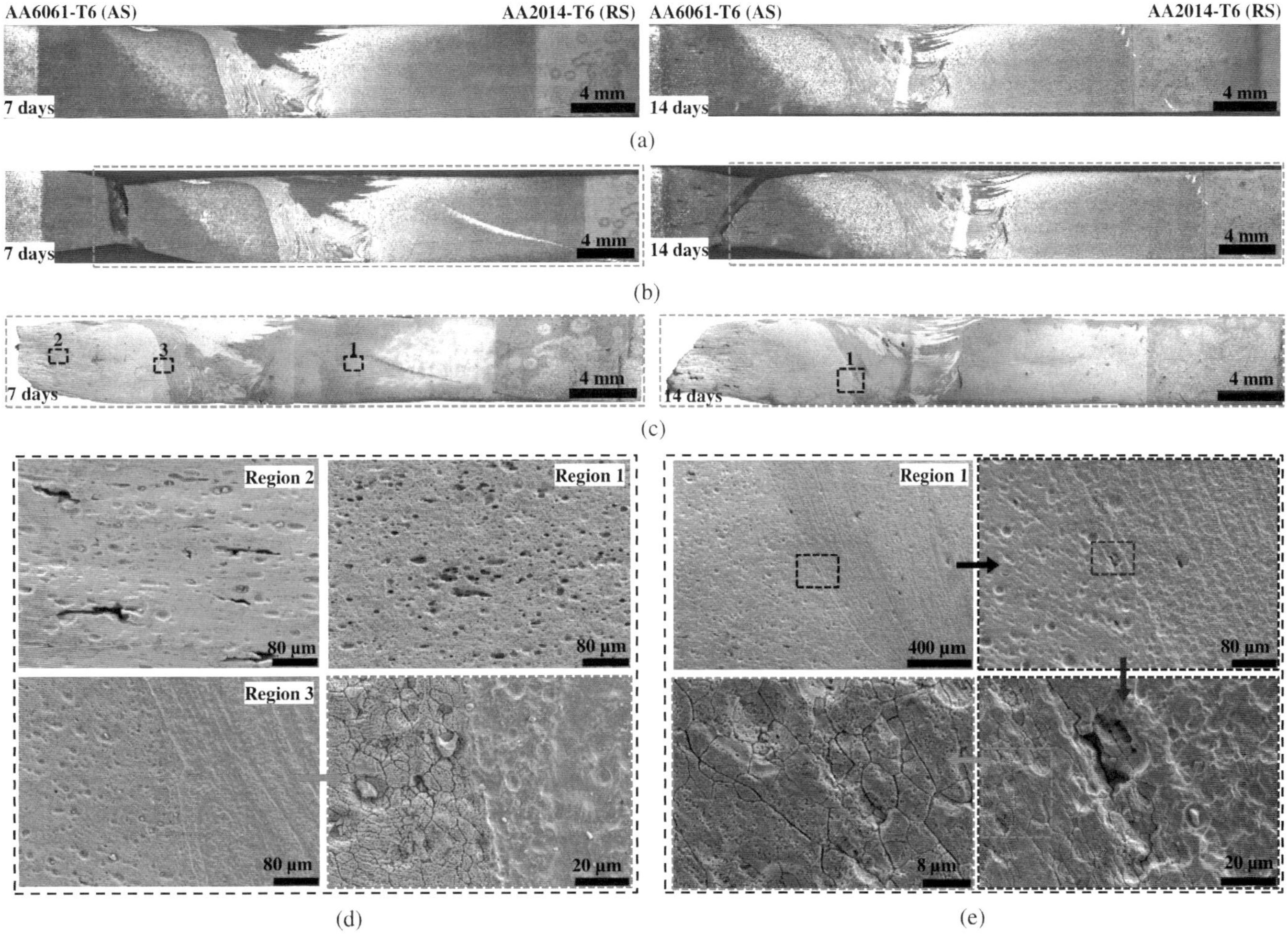

Figure 3. Optical microscopic image of additional heating assisted AA6061-AA2014 FSW joints (a) before and (b) after tensile test and corroded for 7 and 14 days; (c) FESEM image of region marked in (b); high magnification FESEM images of corroded regions (highlighted in (c)) for (d) 7 days and (e) 14 days respectively.

Contrarily the HAZ of AA6061 undergoes pitting corrosion along with IGC and the extent will be dictated by the amount of heat experienced by the zone. The extent of corrosion in HAZ observed to be reduced with the use of additional heating. The exfoliation corrosion has been observed over AA7075 in RS. The extent of corrosion for AA7075 in joints obtained with additional heating is recorded significantly low and has been fairly shifted away from SZ. Figure 2(c) shows the FESEM low magnification image of the highlighted section in figure 2(b) with marked regions for subsequent detailing. As AA7075 surface comes in contact with the corrosive solution, the hydrogen bubble formation on the surface takes place. Eventually the AA7075 surface starts experiencing blistering and pit formation. The Fe-rich particles like Al_7Cu_2Fe experience trenching around them knocking out the particles leaving larger pits. Along with pitting corrosion, IGC takes place and grain boundaries gets shallower with time. The excessive IGC causes mud crack like morphology as shown Region 1 of in figure 2 (d) and (e) to form known as exfoliation corrosion. The extent of corrosion seems to be increased with increase in immersion time from 7 to 14 days. The other particle like Mg_2Si preferentially undergoes dissolution leaving SiO_2 behind. Region 2 of figure 2(e) demonstrates the corrosion morphology at the confluence of stir zone and thermo-mechanically affected zone on AS. The high magnification image shows two different morphologies one within the SZ and the other adjacent to it. The available alloy generally carries Si rich intermetallics which are near-hemispherical in shape [5] and the presence can be justified by the depleted particles leaving craters. Fe rich intermetallics are generally irregular in shape and inhomogeneously distributed over the aluminum matrix. Si dominated particles persist along the grain boundaries making them prone to IGC and thus half of the region can be seen with decomposed grain boundaries. Region 2 in figure 2(d) represents the corrosion morphology within the SZ in layered format of AA6061-AA7075 FSW joint. This is due to the laminar mixing of two dissimilar alloys and the heterogeneity can be expected at the extremes of the SZ. Figure 3(a) and (b) represents the 7- and 14-days corroded specimens of additional heating assisted AA6061-AA2014 FSW joint before and after tensile test. The fracture location still exists on AA6061 side being weaker than AA2014. The extent of corrosion doesn't largely get affected due to the increase of corrosion duration. The extent of corrosion for AA6061 get much effected with the use of additional heating and showed less aggressive pitting corrosion and IGC. Region 1 of figure 3(d) represents the corroded morphology of AA2014. AA2014 is constituted with Al_2Cu particles within the aluminum matrix and also at the grain boundaries. During FSW the microstructural changes occurs causing variability in grain size and inhomogeneous distribution of precipitates. Al_2Cu particulates act cathodic with respect to aluminum matrix causing its dissolution.

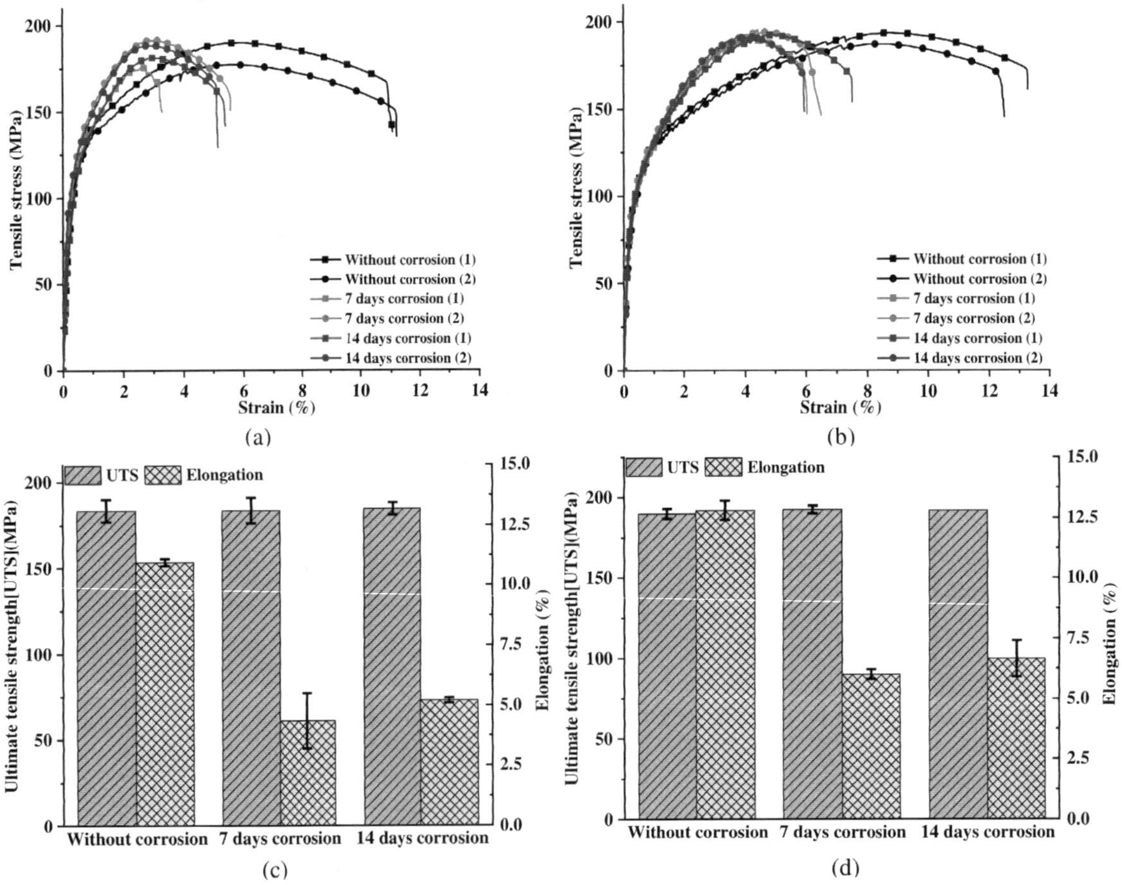

Figure 4. Tensile stress-v-s strain diagram for (a) AA6061-AA7075 (b) AA6061-AA2014 additional heating assisted FSW joints; comparison of UTS and elongation of the joints after corrosion of 7, 14 days for (c) AA6061-AA7075 (d) AA6061-AA2014 joints

The trench is created around these small rounded particles which get removed from the surface leaving pits. The breakdown of Al_2Cu leading to the deposition of inactive Cu resulting in the creation of passive layer at the bottom of craters hindering further pitting. Region 2 of figure 3(d) represents the HAZ in AS (AA6061) with little elongated pits. The elongation is due to the application of tensile loading along the direction of elongation. The HAZ of AA6061 experiences only heating in the absence of any deformation during FSW. Thus, the process induced and additional heating followed by slower cooling rate significantly changes the grain size leading to coarsening. Although the SZ experiences high temperature, intense plastic deformation that allows fine grain formation, which in return increases the corrosion resistance. Moderately occurring IGC can be observed in Region 1 of Figure 3(e) which is AA6061 dominated and in closure vicinity to the SZ.

The tensile stress-vs-strain response of the preheating assisted AA6061-AA7075 and AA6061-AA2014 joints before corrosion and after 7, 14 days of dip corrosion are presented in figure 4. The figure 4(a), (b) represents the stress-strain plot of two samples at each condition for AA6061-AA7075 and AA6061-AA2014 joints respectively. The comparison of ultimate tensile strength (UTS) and % elongation for these joints are also presented in figure 4(c), (d). Among AA6061-AA7075 and AA6061-AA2014 preheating assisted FSW joints, the later joint shows the slightly more UTS than the earlier joint. The strength of the dissimilar joint is generally governed by the strength of the lower strength material therefore the failure occurred from AA6061 side which has lower mechanical strength compared to AA7075 or AA2014. For these both joints AA6061 is placed on the advancing side and failure occurred in the vicinity of the HAZ on AS i.e. AA6061. The strain localization occurs at HAZ of the FSW joints during tensile loading due to thermal softening [10]. The dissolution of Mg_2Si strengthening precipitates in the HAZ is the reason for severe softening and thus it becomes vulnerable to fracture undergoing sufficient elongation [11]. It may be also noted from figure 4 that for the AA6061-AA7075 and AA6061-A2014 FSW joints prepared employing the additional heating the UTS of 183±6.5 MPa and 190±3 MPa respectively before dip corrosion remained almost similar after 7 or 14 days of corrosion indicating no significant loss in UTS however, authors in previous study observed a mild loss of strength with increase in corrosion period especially for the AA6061-AA2014 FSW joints fabricated under the same configuration but without the use additional heating [12]. The UTS for the joints prepared with additional heating is retained even after 14 days of immersion corrosion. Additional heating helps in lowering the aggressive corrosion by bringing the morphological changes. Changes like reduction in size of any elongated intermetallic particulate that could lead to formation of deep pits. The depth, size and alignment of these pits can lead to the faster propagation of cracks and thus failing a joint at lower loads. Thus, additional heating as a process parameter can be considered important especially when the joints are meant to be used in saline environments. A reasonably high elongation i.e. 11% and 12.5% strain is recorded respectively for AA6061-AA7075 and AA6061-AA2014 FSW joints prepared with additional heating as can be noted from figure 4. The grain size distribution specially with low angle grain boundaries as a result of additional heating helps in easy dislocation glide thereby increasing tensile ductility of the joint [13].

From figure 4 it can be also inferred that the joint elongation is higher for AA6061-AA2014 for the joints intact from corrosion. This is due to the parent alloy nature of AA6061 and AA2014 which are more ductile as compared to the AA7075 alloy. AA7075 alloy comparatively has low ductility as it is Zn based alloy with higher wt. % of alloying elements. AA2014 on other hand is Cu based alloy and thus ductility is expected to be little higher than the AA7075. Although there has been a loss in ductility of both the joints after they are immersed in corrosion saline solution for 7 or 14 days, elongation of 3.2 - 7.4% is still observed for the AA6061-AA7075, AA6061-AA2014 FSW joints prepared with employing additional heating. Pit formation due to corrosion act as the stress concentrating sites and ease the crack propagation during tensile loading and causes loss of ductility.

FESEM micrographs of the fractured tensile samples for the FSW joints experienced corrosion is represented in figure 6 with high magnification fractographs at two locations. Figure 5(a) and (b) shows the fracture morphology of the AA6061-AA7075 joint fractured from the AA6061 side. The fracture morphology is mostly dictated by the heating effects experienced by the AA6061 alloy during welding. The AA6061 alloy also experienced the fracture when in combination with AA2014 and the fractographs are presented in figure 6(c) and (d). For all four fractures clean conoidal shape pull outs formed at the top of the joints while the fracture morphologies at the bottom indicate slight shearing. This is due to the difference in heat supply which is maximum at the top due to application of flame heating from the work plate top surface by butane gas torch and reduced with the thickness of the plate.

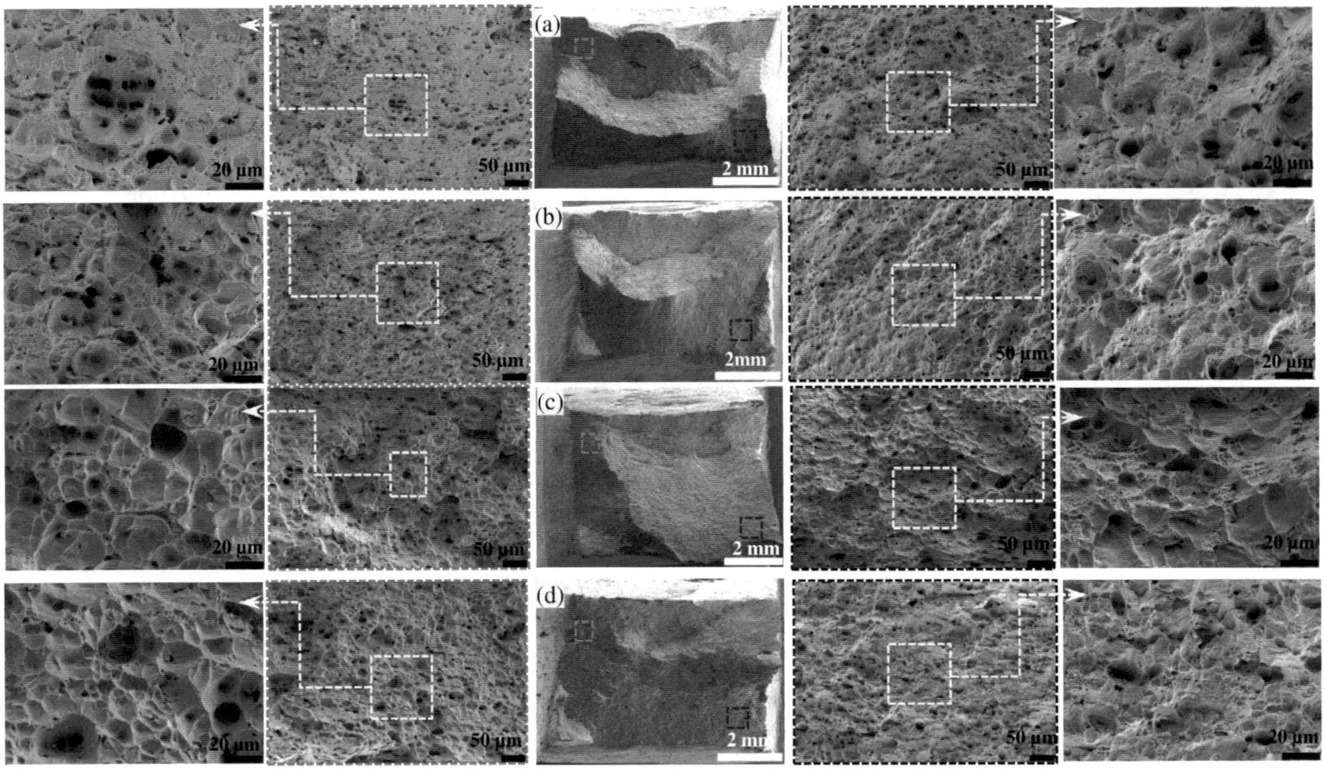

Figure 5. Fractured surfaces of tensile tested samples of additional heating assisted friction stir welded (a), (b) AA6061-AA7075; (c), (d) AA6061-AA2014 joints post 7- and 14-days corrosion respectively.

4. Conclusions

From the present experimental study on corrosion behavior assessment for AA6061-AA7075 and AA6061-AA2014 dissimilar FSW welded joints prepared using additional secondary heating, the following conclusions are drawn:

- The fabricated friction stir welds did not show any loss in joints UTS after 14 days of immersion in corrosion saline solution and even after loss of ductility for corroded samples since the pits act as crack initiation sites, a reasonably good elongation of 3.2 – 7.4% is recorded for the joints.
- AA6061 undergoes pitting corrosion and experienced IGC at a very low scale while AA2014 majorly shows hemispherical pitting. On the other hand, AA7075 initiates with deep pitting and extensive IGC later exfoliating and representing mud crack morphology.
- Additional heating with other of its advantages helped in reducing the extent of corrosion and thus can be an option to use while fabricating joints intended to be used in saline environments.
- AA7075 despite getting corroded severely doesn't fail and can be opted as a compatible alloy during dissimilar joint applications specially in corrosive environments.
- The sufficiently large sized dimples from the fracture surfaces approves the reduction in elongation for corroded joints.

Declaration of competing interest

The authors declare that they have no known competing financial interests or personal relationships that could have appeared to influence the work reported in this paper.

References

[1] Lohwasser D and Chen Z 2009 *Friction stir welding: From basics to applications* (Elsevier)
[2] Raturi M, Garg A and Bhattacharya A 2019 Joint strength and failure studies of dissimilar AA6061-AA7075 friction stir welds: Effects of tool pin, process parameters and preheating *Eng. Fail. Anal.* **96** 570–88
[3] Rosliza R and Nik W B W 2010 Improvement of corrosion resistance of AA6061 alloy by tapioca starch in seawater *Curr. Appl. Phys.* **10** 221–9
[4] Sinhmar S and Dwivedi D K 2018 A study on corrosion behavior of friction stir welded and tungsten inert gas welded AA2014 aluminium alloy *Corros. Sci.* **133** 25–35
[5] Gharavi F, Matori K A, Yunus R and Othman N K 2014 Corrosion behavior of friction stir welded lap joints of AA6061-T6 aluminum alloy *Mater. Res.* **17** 672–81
[6] Na K-H and Pyun S-I 2008 Comparison of susceptibility to pitting corrosion of AA2024-T4, AA7075-T651 and AA7475-T761 aluminium alloys in neutral chloride solutions using electrochemical noise analysis *Corros. Sci.* **50** 248–58
[7] Khoshnaw F M and Gardi R H 2007 Effect of aging time and temperature on exfoliation corrosion of aluminum alloys 2024-T3 and 7075-T6 *Mater. Corros.* **58** 345–7
[8] Ralston K D, Birbilis N and Davies C H J 2010 Revealing the relationship between grain size and corrosion rate of metals *Scr. Mater.* **63** 1201–4
[9] Zaid B, Saidi D, Benzaid A and Hadji S 2008 Effects of pH and chloride concentration on pitting corrosion of AA6061 aluminum alloy *Corros. Sci.* **50** 1841–7
[10] Suthar H, Bhattacharya A and Paul S K 2020 Local deformation response and failure behavior of AA6061-AA6061 and AA6061-AA7075 friction stir welds *CIRP J. Manuf. Sci. Technol.* **30** 12–24
[11] Rajakumar S, Muralidharan C and Balasubramanian V 2011 Predicting tensile strength, hardness and corrosion rate of friction stir welded AA6061-T6 aluminium alloy joints *Mater. Des.* **32** 2878–90
[12] Raturi M and Bhattacharya A 2020 Post corrosion tensile strength and failure of dissimilar friction stir welded aluminium alloys *MATEC Web of Conferences* vol 326 (EDP Sciences)
[13] Raturi M and Bhattacharya A 2021 Microstructure and texture correlation of secondary heating assisted dissimilar friction stir welds of aluminum alloys *Mater. Sci. Eng. A* **825** 141891

Manufacturing process planning in aerospace systems

N Subramanya[1], A R Satheesh Kumar[1], Vikas Yadav[1], R K Venkatesh[2]

[1] Scientist, Aeronautical Development Establishment (ADE), Defence R&D Organisation, Bangalore, India
[2] Technical Officer, Aeronautical Development Establishment (ADE), Defence R&D Organisation, Bangalore, India
subramanya-n.ade@gov.in

Abstract. Process planning is a task comprising a broad range of activities to design and develop an appropriate manufacturing process for producing a part. Interpretation of the part design, selection of manufacturing processes, definition of operations, operation sequences, machining datums, geometrical dimensions and tolerances are some common activities associated with the task. Manufacturing process planning can be defined as systematic determination of detailed methods by which work pieces or parts can be manufactured economically and competitively from raw material stage to final finished size. Geometrical features, dimensional sizes, tolerances, materials, and finishes, are analyzed evaluated to determine appropriate sequence of processing operations, which are based on specific, available resources viz machinery and manpower. Inputs to process planning are designing data, raw material data, resource data (machining data, tooling data, fixture data etc), quality requirement data, production type data etc. The output of a process planning is a process document which is an important document in production management. The paper mainly discusses about production planning and control of metallic manufacturing activities in an aerospace Research and Development (R&D) industry, main emphasis on manufacturing process planning. The general process planning methodology adopted in an aircraft industry is discussed. Elements and Functions of process planning, model of process planning, main mission of process planning, the interaction of processes and its process performance indices are discussed and presented. Importance of manufacturing process planning in an aerospace R&D industry is debated. Also, a case study of process planning adopted for manufacturing of a critical Aircraft system component is discussed.

Keywords: Process Planning; Aerospace; Unmanned Aerial Vehicle (UAV) ; Manufacturing

1. Introduction

Production refers to the transformation of inputs into finished goods/ or creation of services in order to satisfy the customer needs. This uses different inputs mainly including 6M's namely, man, material, machine, money, method and management. Production involves application of processes by which the inputs can be transformed into desired product (output) of potential utility while improving properties and adding economic values through the best method without compromising on quality. In any manufacturing enterprise, the main objective of production department is to produce the things in desired quantity at desired time so that they may be made available to end users when they demand it [1]. Production, being a very complex process is very difficult to manage for the people. This includes a large number of activities and operations which need to be planned appropriately and in turn controlled for the effective production of the output. The main purpose of production planning and control (PPC) is to establish routes and schedules for the work that will ensure the optimum utilization of materials, workers, and machines and to provide the means for ensuring the operation of the plant in accordance with these plans. The paper discusses about various aspects considered during a production planning in an aerospace R&D industry, main focus being on manufacturing process planning. Manual and computer based production planning are presented. Work flow chart from raw material stage to final product is explicated. Lastly, few case studies of process planning of some critical aerospace components are discussed.

2. Functions of Production Planning

In general, the engineering components are fabricated by processing through various work centers of a production firm namely raw material Stores, Machine Shop, CNC Machine shop, Welding Shop, Electro-Plating Section, and Painting Shop to meet the stringent needs of the customer or project groups. At every stage, quality is ensured by critical inspection checks carried out at Reliability, Quality and Airworthiness Division. The special processes (like Welding and Protective treatment processes) are having well defined work instructions. All the activities are meticulously planned, co-ordinated, monitored, and controlled by Production Planning and Control (PPC) section which is the central hub among the various divisions of a manufacturing enterprise. The choice of production system depends upon the nature of products, variety of products and volume of products [1]. Production system requires the optimal utilization of natural resources like men, money, machine, materials and time. Production Planning and Control (PPC) can be defined as the process of planning the production in advance, setting the exact route of each item, fixing the starting and finishing dates for each, rout them, to give production orders to shops and lastly to fallow up the progress of products according to orders. PPC is necessarily concerned with implementing the plans, i.e. the detailed scheduling of jobs, assigning of workloads to machines (and people), the actual flow of work through the system. PPC coordinate with different departments: such as production, marketing, logistics, warehouse and other departments depending upon the nature of organization. The aim of production control is to produce the product of right quality, in right quantity, at the right time by using the best and least expensive methods. In short, the principles of PPC lie in the statement 'First plan your work and then work your plan'. Key elements of PPC are shown in figure 1.

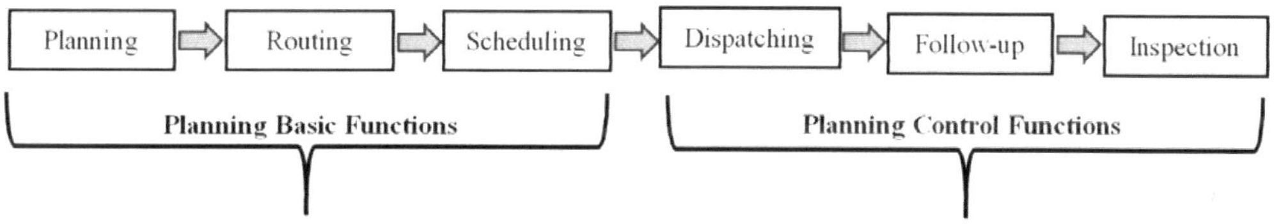

Figure 1. Key Elements of Production Planning and Control

Before starting any work, it is necessary to plan the things properly for getting best results. Similarly for production planning in advance is necessary and if there is no planning then how shall we control? It is the planning department, which decides about each element of the job in anticipation that what work shall be done, where, how and when it shall be done. Routing may be defined "as the selection of path which each part of the product will follow which while being transformed from raw material to finished product". Scheduling determines the programme for operations. In scheduling, order of sequence of each operation and their starting and finishing time is decided so that required materials, machines etc, may be kept as per schedule. Thus, Scheduling may be defined as "the fixation of time and date for each operation "as it determines the sequence of operations to be followed. This programme is prepared by planning department and assigns timing for various operations and processes. It is concerned with starting the process. It gives necessary authority to start a particular work, which has already been planned under "routing "and "scheduling." For starting the work essential orders and instructions are given. After dispatching the production orders to various shops, it is necessary to regulate the progress of jobs through various processes. For this purpose a follow-up section is formed. The function of follow-up section is to report daily the progress of work in each shop in a prescribed format and to investigate the causes of deviation from the planning performance.

Inspection is carried out for confirming that quality is of predetermined standard. During inspection defective parts are rejected so that only products of good quality should reach to the user. To reduce rejection, inspection is done at various stages starting from raw material to finished shape. Different industries use different techniques and tools for production planning and control for planning, scheduling, and controlling. To name a few are Bar charts [11], milestone charts, network charts, Program Evaluation and Review Technique (PERT) [9, 10] and CPM [9, 10].

3. Mission of Process Planning

There are three types of production systems, namely continuous production, Job production and batch production. R&D industry usually adopt job type or intermittent production model for all prototype fabrication activities. As discussed, this paper mainly focuses on aerospace R&D manufacturing systems, topics discussed here are limited to jobbing type production schemes. In this type of production, manufacturing of items will be as per customer's specifications. This ensures the simultaneous production of large number of batches/orders. Each batch or order comprises of a small lot of identical products and is different from other batches. It requires comparatively smaller investment in machines and equipment. It is flexible and can be adapted to changes in product design and order size without much inconvenience. This system is most suitable where heterogeneous products are produced against specific orders. In this system products are made to satisfy a specific order. Usually, a person who does the process planning activity is called as Process Planner, who follows the process planning model (as shown in figure 2) to arrive at logical sequences of a process to realise item [2]. A typical work flow for a production activity is detailed shown in figure 3, clearly showing the stage where process sheet is prepared. Input from user or customer will be in the form of Request For Action (RFA) along with legible, approved and controlled engineering drawings and raw materials / components as applicable. Some industry have a practice of using the word "Work Order (WO)" synonym to RFA. Once the inputs are received at PPC group, process team will carry out feasibility study of the components about the feasibility of realisation of item in-house with the available resources namely man, machine, tools, special fixtures, materials etc. Parallelly, Process engineer categorizes the items as Category-A (Cat-A), Category-B (Cat-B) or Category-C (Cat-C). The main requirements and guidelines for categorization of items are tabulated in Table 1. Eligibility for categorization is done based on the guidelines provided in table no. 2. During feasibility study a detailed study of user requirements are gathered and necessary clarifications are sought from the customer if needed. Once the feasibility study is carried out, process planning is carried out. Also detailed routing document called Job card (JC) is evolved. JC is often called as "Route Card" in some manufacturing enterprise. Once the process planning is done, process sheet, JC, drawing and material is routed for fabrication activities through various shops / work-centers. Independent inspection agency will inspect the manufactured components / parts against drawing specifications.

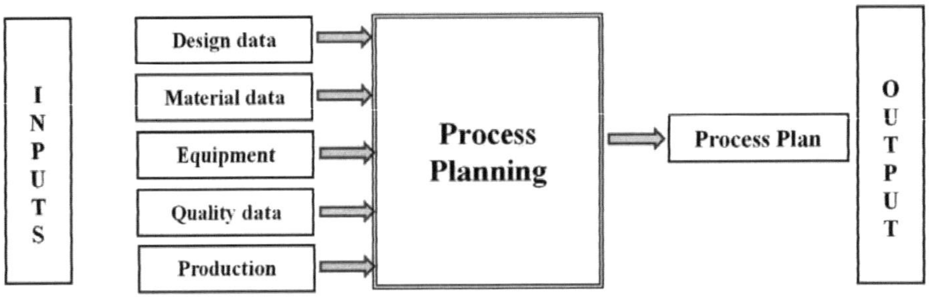

Figure 2. Process Planning Model

Table 1. Requirement Matrix for Categorization of items.

Cat	Legible Dwg ?	Approved Dwg?	Controlled Dwg?	3D CAD Model Reqd.?	COC of Raw Material	Process Plan	Stage Inspection	Final Inspection
Cat - A	Yes	Yes	Yes	Yes	Yes	Yes[b]	Yes[c]	Yes
Cat - B	Yes	Yes	No	Yes[d]	No	No[e]	Yes[c]	No[a]
Cat - C	Yes	No	No	No[d]	No	No[e]	No	No[a]

[a] Based on user requirements.
[b] Prepared by PPC.
[c] Will be carried out by Process Planning Team.
[d] Subject to complexity of the component.
[e] Prepared by Process Planning Team & Based on user requirements.

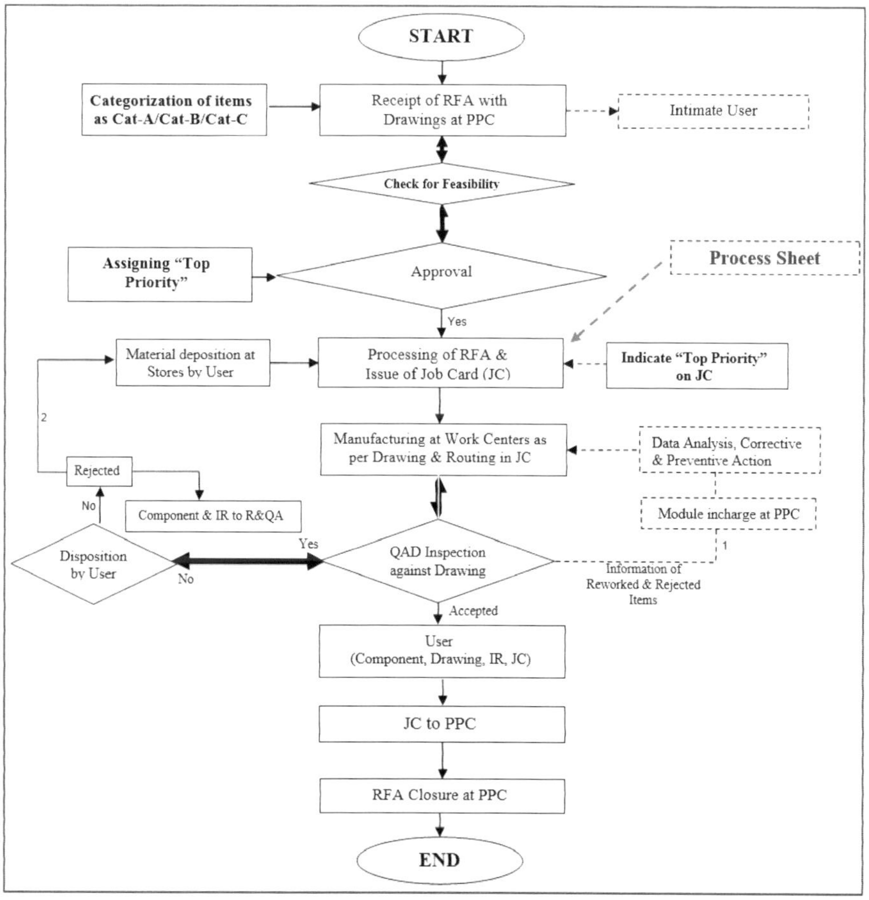

Figure 3. Work Flow Diagram for Prototype Fabrication

Table 2. Eligibility Criteria for Categorization of items at Prototype Fabrication Centre.

Cat	Eligibility Criteria
A	• Applicable for completely Flight worthy items • Items / parts / components which are Flight critical • User should provide Legible, Approved and Controlled Drawing • User should provide Raw material Certificate (Mechanical analysis, chemical analysis, hardness details) • Complete routing at fabrication group will be considered for these items • Final inspection will be carried out • Priority Processing will be considered by manufacturing team based on available slot and user requirements.
B	• Applicable for engineering mock up units, ground testing items / rigs / Test fixtures / Jigs etc • User should provide Legible and Approved Drawing • Complete routing at Prototype Engineering Centre will be considered for these items • Stage Inspection is not mandatory • Final Inspection is carried out as per user requirements. Priority Processing will be considered based on available slot and user requirements.
C	• Mostly applicable for rework items, repair work, name board and any other services • User should provide Legible Drawing or sketch • If the job duration is 2 hours or less, the RFA will be converted to RFA : Urgent Jobs

4. Process Planning of Critical Aircraft Components : A Case Study

Digital Flight Control Computer (DFCC) [7] main block for a Manned Combat Aircraft :

Digital Flight Control Computer (DFCC) is a quad redundant flight computer which is the core of any modern aircraft, both manned and unmanned [4,8]. It actuates the primary flight control surfaces, to drive the flight path of the aircraft (figure 4). It also provide finer control for stability. The blank size of the component to be manufactured was about 514mm x 278 mm x 216mm. Material was high strength aluminum alloy under temper grade of T6 condition. Important challenges were the handling of block (initial weight of the black was about 90 kgs), evolution of number of set ups to be used, innovative clamping methodology adopted to minimize clamping stresses and distortion, programming skills required to minimize the manufacturing time, handling of job, machining of heat transfer fins (which are of 400 microns thick), removal of chips etc. The location of DFCC box on a manned combat aircraft is shown in figure 4. A systematic machining strategies were evolved and a detailed process planning was generated (as given in table 3(a) to table 3(c)). CNC programming was carried out as per process plan and 3D CAD model provided by customer. Manufacturing job setting 1 to 6 utilized during the CNC programming are shown in figure 5. Figure 6 portrays various critical machining operations carried out during manufacturing of DFCC main block.

Figure 4. Location of DFCC box on a manned combat aircraft (aircraft image courtesy : web)

IOP Conf. Series: Materials Science and Engineering 1258 (2022) 012027 doi:10.1088/1757-899X/1258/1/012027

Table 3(a). Process plan sheet for Manufacturing of Digital Flight Control Computer Main box.

SL. NO.	Operation/ Process Description	Work Centre. Machine/ Equipment	Information/Key parameters/Tooling/ & Instruments	Stage Inspection		Remarks/ Signature.
				Specified Value	Measured value	
1	Raw Material Cutting	Hacksaw (HP-300)	Maintain 600x340x250 mm Tool: Carbon steel blades	600x340x250	650x360x280	02 days/12 hour
2	Blanking (All Side Machining)	FN-2V Milling machine	Maintain 513.3x278.2x216.3 mm Tool-50 Ø Face Mill DoC-2 mm, Feed-100 mm/min	513.3x278.2x 216.3mm	513.35x278.2x 216.32	02 days/12 hour
3	Stage Inspection	CMM/ Measuring Instruments	a. Use Varnier Caliper of suitable dimension b. Record the dimensions in JC	513.35 x 278.2 x 216.32 Verified the dimensions & recorded in JC		01 day
4	**Set-1-Roughing** a. Set the job Vertically & true within 0.05 b. Hold 278.2x216.3 mm the Job in Jaw & parallel	Boko-4-axis Milling machine	a. Use 16 Ø for roughing & 10 Ø for finishing b. Keep 1mm stock on wall & floor c. machine the remaining stock d. Measure with Varnier 0-300mm length	Job setting Pre finishing operation is carried out		01 days/12 hour
5	**Finishing** a. True again the job within 0.05	Boko-4-axis Milling machine	a. Use 16 Ø for roughing & 10 Ø for finishing b. machine the remaining stock	Finishing operation is carried out		01 days/12 hour
6	Stage Inspection	CMM/ Measuring Instruments	Use Varnier Caliper of suitable dimension	220.2x190.2x 17.0mm	220.25x190.22x 16.99 mm	01 day
7	**Set -2- Roughing** a. Reverse the job & true within 0.05mm	Boko-4-axis Milling machine	a. Use 16,10 Ø for roughing & 6,4 Ø for finishing b. Machine the fillet with 8 BNS	Reverse setting done Job setting Pre finishing operation is completed		02 days/24 hour
8	**Finishing** a. True again the job within 0.05	Boko-4-axis Milling machine	a. Use 6,4 Ø for finishing b. Finish the remaining stock with high speed & feed c. use the finishing strategy to machine	Finishing operation is completed		02 days/24 hour
9	Stage Inspection	CMM/ Measuring Instruments	a. Use Varnier Caliper of suitable dimension b. Record the dimensions in JC	183.2x238.0x 22.25	183.22x238.1x 22.24	01 day
10	**Set -3 Roughing OP.** a. Hold the job in Vice 513.3x278.2side	Boko-4-axis Milling machine	a. Use 16Ø for roughing & 10 Ø for finishing b. Machine the fillet with 6BNS c. Create 13.2 mm step for	Work origin defined Roughing operation is carried out		04 days/24 hour

Table 3(b). Process plan sheet for Manufacturing of Digital Flight Control Computer Main box.

SL. NO.	Operation/ Process Description	Work Centre. Machine/ Equipment	Information/Key parameters/Tooling/ & Instruments	Stage Inspection		Remarks/ Signature.
				Specified Value	Measured value	
11	**Finishing Op.** a. Reverse the job & true within 0.03mm b. Set the job in vice & hold 30mm inside	Boko-4-axis Milling machine	a. Use 16Ø for roughing & 10 Ø for finishing c. Create 33 mm step for wire cut operation up to the vertical wall	Roughing operation		04 days/24 hour
12	Stage Inspection	CMM/ Measuring Instruments	a. Use Varnier Caliper of suitable dimension	506.8x269.2x 2.5mm	506.85x269.22x 2.49mm	01 day
13	**Set -4 Roughing Op.** a. Hold the job in Vice & true within 0.03mm b. Use the parallel plates & strap clamps	Boko-4-axis Milling machine	a. Use 16,10 Ø for roughing & 6 Ø for finishing b machine the 2.5mm step c. Clear both side of corner area by using 6FEM	Rough Machining (RM) completed		02 Days
14	**Finishing Op.** a. Reverse the job in Vice & true within 0.03mm b. Use the parallel plates & strap clamps	Boko-4-axis Milling machine	a. Use 16,10 Ø for roughing & 6 Ø for finishing b. Machine the 2.5mm step c. Clear the corner side material by using 6FEM	Finish Machining (FM) completed		02 Days
15	Stage Inspection	CMM/ Measuring Instruments	a. Use Varnier Caliper of suitable dimension	504.8x267.2x 19.0mm	504.85x269.22x 18.99mm	01 day
16	**Set -5 Roughing/Finishing** a. Set the job on bed & true within 0.03mm b.Use the parallel plates & strap clamps, studs to clamp	Mikron-3-axis Milling machine	a. Use 4 Ø for machining the inside slot b. use 2 Ø FEM to machine the fin c. Maintain the Spindle rpm 2500 d. DOC should be 0.15 to 0.2 mm	Fin machining started		7 days/90 hour
17	Stage Inspection	CMM/ Measuring Instruments/Laser scanning	Use laser scanning to measure the fin thickness & gap	Fins: 2.0mm Step 2.5mm	Fins: 2.05mm Step 2.51mm	01 day
18	**Set -6 Roughing/Finishing**	Mikron-3-axis Milling machine	a. Use 1.9 Ø end mill & observe the fin waviness b. Through spindle coolant to carry the machined	Fin machining competed		7 days/90 hour

Table 3(c). Process plan sheet for Manufacturing of Digital Flight Control Computer Main box.

SL. NO.	Operation/ Process Description	Work Centre. Machine/ Equipment	Information/Key parameters/Tooling/ & Instruments	Stage Inspection		Remarks/ Signature.
				Specified Value	Measured value	
19	Stage Inspection	CMM/ Measuring Instruments/Laser scanning	a. Record the dimensions in route card b. use laser scanning to measure the fin thickness & gap	Fins: 2.0mm Step 2.5mm	Fins: 2.05mm Step 2.51mm	01 day
20	**Set -7** a. Set the job on bed & true within 0.03mm b. Center the job c. scoop out the inside material	Electronic U843 Wire Cut machine	a. Use 0.2mm brass wire d. Maintain the spark gap 0.2mm c. Maintain the appropriate current TON & TOFF d.Measure with Varnier 0-300mm	Wire cut machining		8 days/48 hour
21	Stage Inspection	CMM/ Measuring Instruments/Laser scanning	a. Record the dimensions in route card b. use laser scanning to measure the fin thickness & gap	109.85 & Rib:1.6mm	109.85 & Rib:1.62mm	01 day
22	**Set -8** a. prepare the copper electrode b. Set the job on the spark erosion machine	AG Charmilles Form 20ZNC	a. Use copper electrode and plunge to machine the vertical steps c. Maintain the 15 amps current TON d. Measure with Varnier 0-300mm	Cu electrode prepared & EDM started	14 days/84 hour	
23	Stage Inspection	CMM/ Measuring Instruments/Laser scanning	a. Record the dimensions in route card b. use laser scanning to measure the fin thickness & gap	3 mm V Steps & 137.5 mm length	3.05 mm V Steps & 137.55 mm length	01 day
24	Remove all the burrs black spot	Fitting Bench	Use the various types of hand tools available	Deburring tool used		01 days/6 hour
25	Stage Inspection	CMM/ Measuring Instruments/Laser scanning	a. Record the dimensions in route card b. use laser scanning to measure the fin thickness & gap	Recorded in JC		1/2 day
26	Final inspection	CMM/Manual inspection	As per drawing	Carried out as per drawing & values recorded in IR		03 days/18 hour
27	Use the appropriate surface	Chromate bath	As per standard IS: 11232 : 1965 [5,6]	Carried out as per drawing & values recorded in IR		03 days/18
28	Handing over to the User/Project office	Production Planning and Control (PPC) Group	Job, Job card, Drawing, IR to be received and stored at PPC for delivery to user reps	- Job, Drawing, IR handed over to User - Job Card Closed - User Feedback taken		01 Hour

IOP Conf. Series: Materials Science and Engineering 1258 (2022) 012027 doi:10.1088/1757-899X/1258/1/012027

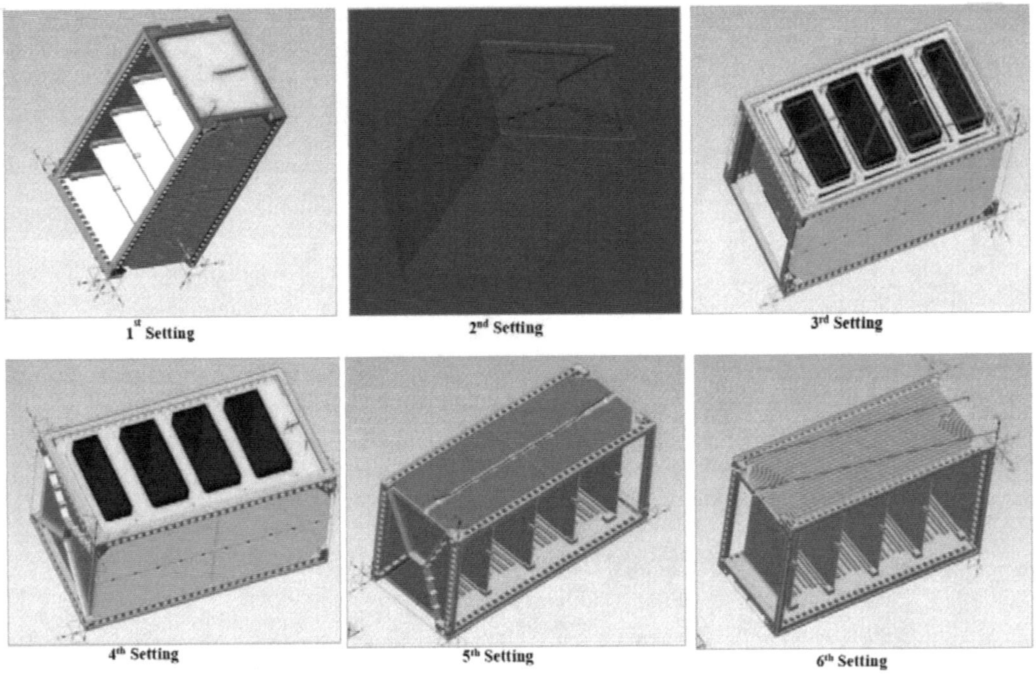

Figure 5. Different settings used in CNC programming for manufacturing of DFCC main block

Figure 6. Machining operations carried out during manufacturing of DFCC main block

5. Conclusions

This research paper discusses broadly about the importance of manufacturing process planning in an aircraft industry. Even though, it is understood that process planning starts before the manufacturing of components, the basic process planning should start from the beginning of a product life cycle where selection of materials, selection of manufacturing methods take place [3]. Manufacturing process planning takes over once the engineering drawings are released clearly defining the specifications of the components to be realized. For generating manufacturing process planning clear technical knowledge of processes, equipment, resources capability in the company and production economics etc is essential. Case studies presented also explicates the procedure in generating detailed process sheet for manufacturing. Even though the process planning concepts discussed in this paper are general in nature, they are limited to aerospace R&D industry where prototype manufacturing is of most important.

6. Acknowledgements

Authors would like to thank the Director of their establishment for providing facility to carry out the research work and for giving permission to present this technical paper. Authors are also thankful to all the internal customers who have given permission to use their design components to present as case studies in this paper. Authors would like to acknowledge the contributions of members of divisional Process planning team namely Subramani M, Manoj T V, Johnson N and Jomon Joseph in generating the case study presented in this paper. Authors are also grateful to their Technical Paper Review Committee for critically reviewing the paper and providing valuable suggestions.

7. References

[1] Dr Hemant Sharma, Professor & Area Chair- General & Operations Management, Amity Business School, Amity University Haryana, Gurgaon- 122413 *Hand out notes*

[2] Hong-Chao Zhang, Texas Tech University *Manufacturing process planning* Jan 1994

[3] Cristian Tarba and Petre Tiriplica *Process Planning in Manufacturing System* Conference Paper September 2014, DOI: 10.4028/www.scientific.net/AMM.760.745

[4] M Arjomandi, S Agostino, M Mammone, M Nelson et. al. *Classification of unmanned aerial vehicles*, Mech Engg. vol. **3016**, 2006.

[5] IS 11232 : 1985 *Code of practice for Chromate conversion coatings on Aluminum*

[6] MIL-DTL-5541 *Chemical conversion coatings on Aluminum and Aluminum alloys*

[7] https://www.tejas.gov.in/technology/fly_by_wire.html

[8] L Brooke-Holland *Unmanned Aerial Vehicles (drones): an introduction*, UK House of Commons Library Report, Standard Note **SN06493** (25 April 2013), 2012

[9] P Ramamurthy *Operations Research* **2007**, New Age Intl (P) Ltd Publishers, New Delhi

[10] https://www.investopedia.com/terms/p/pert-chart.asp

[11] https://www.gantt.com

Review on Corrosion studies of Heat Treated Al-Si Alloy

M N Sanath[1], Nihal C L[1], Prabhuling[1], Shivaprasad P M[1], Puneeth.H.V[2], Srinath M K[3]

[1]Mechanical Engineering Department, New Horizon College of Engineering, Bangalore-560103.
[2]Assistant Professor, Mechanical Engineering Department, New Horizon College of Engineering, Bangalore-560103.
[3]Associate Professor, Mechanical Engineering Department, New Horizon College of Engineering, Bangalore-560103.

Abstract. Heat-treated Al-Si alloys are widely used in the automotive, military, marine, electrical, food and chemical industries. This alloy exhibits high-strength properties used in corrosion-resistant homes. However, this alloy has low mechanical properties and a large granular structure when cast. The properties of aluminium alloys depend primarily on the microstructure. It is very important to achieve a smooth structure. The formation of fine and equivalent grains depends primarily on the amount of hardening, the addition of basic alloys (grain cleaners), the mixing and processing of alloys. Aluminium alloys are an important component of light metals used in industry. Al-Si alloys are essential for automotive, aerospace, marine and engineering applications. Al-Si alloys have excellent physical and mechanical properties. These alloys offer low weight, excellent corrosion resistance, easy machining, heat treatment, excellent casting ability and excellent machining performance. The mechanical properties of these alloys depend primarily on the size, shape and distribution of Si and Al particles. Al-Si alloy produces coarse α-Al dendrites and shark eutectic silicon. Fine structures are known to provide good mechanical properties and reduce casting defects. During the casting process, a fine-grained structure can be obtained by adding a lower alloy to the melt.

Keywords: Aluminium silicon alloys; heat treated alloys; corrosion studies; SEM and XRD;alloy microstructure.

1. Introduction

Due to the availability of various materials (metals, sand, polymer) and additional processing methods, coating technology has grown rapidly over the last 20 years [1]. This technology is developed to create models during the product development cycle, due to the flexibility of model form and physical validation, which reduces lead time [2,3]. Still, many industries (aeronautics, automotive, foundry) have introduced this new technology into the industrial sectors of manufacturing and development [4]. Construction companies in particular are still aware of the benefits of other steel construction industries [5]. The ability to create layer after layer allows designers to create a mask that combines a rare, gym, harbour and food in one mask design and is hard to find. This unique design with intricate design and traditional design. Production time is reduced by avoiding design, equipment and sample processing. As more than 70% of all metal alloys are produced by sand casting, these components are very effective. In addition, the use of sand can be reduced by the production of other products, which allows the selective absorption of acidic compounds (e.g. furans), which helps to recover energy and reduce pollution of the environment. Seals are one of the best manufacturing technologies for mould. Several studies have used different sources to improve this process. Tangram [7] the jet bonding process has been optimized with refined materials for inkjet printing of Ti_6Al_4V components. Marianet al [8] demonstrated the use of inkjet bonding technology to seal WC-Co (cemented tungsten carbon) components. This study demonstrates the importance of influencing the primary properties of powder and the mechanical properties of the printing phase. Chen Nar [9] studied wave process of a piezoelectric printer with several valves. The developed method makes it possible to reduce the number of tests to find compatible parameters of the waveforms of the 3D machine. Other researchers [10] have focused on improving advanced coating methods to attain the desired layer thickness, smoothness, and colour quality. La et al [11] solved his torque problem and did research to improve the accuracy of jet technology. To this end, various types of water pipes and composite gypsum powder were prepared, followed by element embedding tests and attribute tests to determine the optimal model of the attached powder. There are several lists of books that support the benefits of using other production technologies in metalworking.

1.1. Design and modelling

Cheese day [12] enjoyed using 3D printing designs to eliminate complex building components, suchas movable structures and sandwich panels. Thus, they close the switch at the last point, which has the function of increasing the pressure. Srinath [13] used bond injection technology to create moulds and silencers that could control the casting process by documenting various parameters (temperature, pressure, chemical gas) under different mould conditions. Although it is advantageous to use other methods in the construction industry, it is not ready to replace conventional methods. A deeper understanding of the nature of yeast and its impact on yeast quality is still needed. The standard of the cast iron depends on several factors or components related to the manufacture of the metal and the type of mould used. Depending on the treatment of the metal, the rate of heating of the metal in the mould, the temperature of the metal and the chemical composition of the initial alloy must be considered. Of course, it is mandatory to understand the mechanical properties of mass, permeability, and surface area. Numerous studies have been conducted to investigate the properties of gypsum dust and sandpaper from light alloys, such as aluminium, to produce metallic materials. Makkena Nar [14] analysed the effect of temperature and on the mechanical properties of the mould. The writer says it is a sad loss combined with increasing treatment time, increased permeability. Wagey et al [15] investigated the effect of layered thickness on fungal substances. They found that by increasing the thickness of the cast applied between the joints and reducing the thickness of the layer, an increase in tensile strength was obtained, but the effect of the mould shape was reduced.

Edwino et al [16] investigated the possibility of moulding plaster for the production of aluminium alloys. Finishing work, such as insertion, is carried out to improve the structure and thermal conductivity of the mould. The missing elements may not work to withstand the pressure of the cast iron at all times, but good heat resistance. There are many studies examining the degree of melting obtained by 3D moulding, but the yield is limited. Snelling et al [17, 18] produced 3DP plastics (Z Corps gypsum and Ex Ones silica) using two different materials and modelling to obtain A356 aluminium alloys.

1.2. Composition and microstructure

This study focused on examining various solubility factors such as hardness, porosity, density and microstructure. They culminated that moving material had a significant effect on the quality of the final part. Various details, such as the size of the cast, the particle size and the composition of the composite material affect the properties of the cast material. These functions affect the final load matrix (secondary dendritic division, size), [20] secondary intermetallic dimensions (distribution, size, morphology and dimension) and microstructure porosity. In particular, the microstructure of Al-Si alloys examined in this article makes the assembly of end-aluminium dendrites as the main and several steps in the microstructure. Current systems include eutectic Si particles, some parts of secondary intermetallic compounds such as Al_2Cu and Mg_2Si, as well as some containing Fe (AlFeSi, AlFeSiMn). The secondary systems (eutectic Si, Al_2Cu, Mg_2Si, AlFeSi) provide unique properties for the alloy and can affect the structure and corrosion of the material. Although Al-Si-Cu alloy is rare, itis highly sensitive to corrosion and chloride solutions. This is related to the presence of a second phase, since most of the material is better than the Al matrix. These micro galvanic cells form a soluble aluminium matrix [21]. The second minute is part Si. If the solubility of silicon and aluminiummatrix is low, the contents increase [22].

The size, morphology and distribution of these elements play an important role in the behaviour and degradation of Al-Si-Cu alloys [23]. The action of Si cathode on the matrix as an aluminium that acts as an anode leads to the formation of two micro galvanic pathways causing local damage. Impact of silicon morphology and its impact on the size of the galvanic corrosion zone and that of Tahmpton et al. It has been examined. [23]. Large cloud particles have a sharp edge, which increases the cathode / anode size to reduce density [24]. In addition, the electrochemical properties of Al-Si-Cu alloys affect the morphology, shape and distribution of the intermetallic compound.

It is the function of the mixture at cold rates [25, 26]. Intermetallic composites such as Al_2CuMg and Mg_2Si act as anodes in matrix A and degrade surrounding matrix A. However, parts of the intermetallic compound, such as Al_2 Cu, especially the metal-like structures (β-Al-Fe-Si and α-Al-Fe- Mn-Si), are usually close to Al cathodes. Eutectics al [27]the matrix is located near the mouth of this intermetallic compound [28]. Machine content determines the type and processing structure of the intermetallic mixture containing the machine. The flat system as $5Al_5FeSi$ is rich and has a large steel, long and pinned. This time is hard and soft, making a high impact on local corrosion. The presence of manganese supports the "Chinese grade" type dịAl15 (Fe, Mn) $3Si_2$, which improves corrosion resistance compared to non-manganese [29-31]. Studies on microstructure levels, mechanical strength and corrosion behaviour of Al-Si alloys developed by the blasting system. But gradually Information about this part of the container was analysed using a template using the embedded slash method. This work may focus on this space, due to the microstructure Al corrosion behaviour of AlSiCu alloys. For this purpose, composite materials are usually used in the form of sand-composite materials.

2. In Periodic Boundary Conditions

Aluminium is the most common metal on Earth. As a result, these alloys have become the second most popular material in the automotive industry since 2006. As a result, the automotive industry has become very popular. Large brand of aluminium wheels. More than half of them are Al. These differences are mainly due to two international problems [32] (i) the use of global resources that require lighter and richer materials. Title: For example, in the European Directive issued in 2000 (2000/53 / EC), 85% of cars completed in 2006 and reused in 2006 will be 95% in 2015. The tertiary, mechanical and corrosive properties of each compound are determined by the solubility of its resistant components. The properties of the level of IMP achieved control the mechanical and electrochemical properties of the material depending on its morphology, resistance and diffusion chemistry.

Alis Abrasion Formulation Alis is based on high-strength silicon alloys in ant diamond blocks (2.6 g cm 23) .It has good thermal expansion, corrosion resistance and mechanical properties. Zone 46 can reduce the stability of the mixture by more than 11%. Mould hardening is a complex process. Elements such as copper, magnesium, manganese and iron causes complex compounds. The first models of microstructural development, such as the isothermal synthesis model 93, were based on independent growth criteria that showed that the determining step in grain formation is the passage of the energy barrier, not the formation of the nucleus. Fixed / floating interface curves (Gibbs-Thomson effect). This model is very simple. He noted that modern hunting methods do not allow the production of aluminium alloys with sufficient wear and / or corrosion resistance to be used in all comfortable operating conditions.

3. Effects on alloying on aluminium-silicon alloys.

The factors that determine the mechanical, chemical and physical properties of a mixture are its composition and microstructure. The addition of certain alloying elements to aluminium in its pure form has been shown to improve its properties. Aluminium and iron provide the best combination of properties, making metallic materials the most suitable, economical and satisfying for a wide range of applications. Due to its mechanical properties, aluminium alloys are used as metal plates in transportation, marine and chemical applications. We also use aluminium alloys in aircraft manufacturing and the airframe industry[33].The properties of aluminium, corrosion resistance, strength, heat dissipation and fluidity make it very suitable for construction. Al-Si iron is a technical material that, due to its eutectic composition, can be used in many applications.

Silicon, a mixture of elements and impurities with this effect, is the most common impurity in commercial aluminium. Mg_2Si is made from annealed silicon alloy containing up to 1.5% magnesium. Heat treatment can be performed by the addition of small magnesium particles to silicon -containing aluminium alloy, but Mg_2S is required to produce Mg_2S.

The effect of alloys on the negative corrosion behaviour for silicon solid aluminium matrices. However, the effect of silicon on the corrosion resistance of this iron is small due to the high concentration of corrosion current due to the high polarization of the silicon particles.

The melting process is as follows.

- Forming
- Forging
- Machining
- Welding

Applications where aluminium has been completely replaced by an alloy containing one or more alloying elements or impurities. Pure aluminium and iron still experience unequal mechanical properties and low strength barriers. The microstructure and mechanical properties can be improved through heat treatment, cold treatment and alloys.

4. Wear and Corrosion Behaviour of Al-Si.

There, he explains that the matrix formed by Al-Si metal is reinforced by regions with different weights of Al_2O_3 ore up to 25 wt. Design by mixing method. Part of Al_2O_3 weight and heat treatment effect on Al-Si user behaviour mixed with the product. Material Matrix (MMC) represents a new generation of materials that have been invested in the construction of ceramic structures in a steel matrix for real estate development, including energy, special stress, special anti-corruption, anti- corruption and high technology. Studies focusing on the use of enhanced Al_2O_3 shear during heat release and heat treatment methods are different and often contradictory. Das and others estimated after heat treatment, the compactor contains anti-wear chemicals compared to die alloy wheels. The main consequence of using MMC is enhanced anti-corruption results. This is especially true of aluminium where anti-corrosion film is used [34]. At this point, the level of protection does not increase because it increases the number of places where corruption can be initiated and the number of attacks is higher. The main causes of corruption in MMC are (1) galvanic bonding between uterus and uterus, (2) uterine classification, and (3) changes in microstructural treatment of pollution resulting in MMC. The consequences of strengthening the attitude of corruption are unclear. It has been proven that the intensity of corruption increases or

decreases or does not affect its existence. In addition, amplification has been shown to increase or decrease peripheral capacity (OCP) without side effects.
Al-Si substrates with a density of 2.68 g / cm^3 and Al_2O_3 with a density of 3.9 g / cm^3 and a diameter of 75 mm were used as substrates. The chemical analysis of the Al-Si matrix used in this study was 7.1% Si, 0.3% Mg, 0.01% Mn, 0.02% Cu, 0.01% Ni. MMCs containing 0, 10, 15, 20 and 25% wt were formed. For $2O_3$ parts. Al-Si MMC was applied with Al_2O_3 pellets produced by toy method. The compound is made into a cylinder with a diameter of 10 mm and a length of 100 mm, alkaline oil is placed in a container and heated to 750 degrees Celsius. The melt is heated to 700 rpm. The mixing time was 5 minutes after eating the seeds. The temperature was raised to Al_2O_3 at 900 °C before melting and 1% Mg was added. Examples of heating are as follows: heat treatment with solution at 540 °C for 8 hours, water cooling at 60 °C and isothermal aging at 155 °C. Frequent cooling (if not used) to prevent normal aging. The results showed that the wear resistance of the test was improved by heat treatment with the addition of Al_2O_3 compounds. The effects of heat therapy on computer- damaged behaviour were also investigated using a 600rpm polymer electrolyte test. Compared to matrix with 3.5% NaCl. The anti-corrosion properties of Al_2O_3 are 10, 15 and 20% higher than those of matrix. Al-Si wheels and iron treatment have a significant improvement in anti-corruption compared to this modern equipment.

5. Microstructure and mechanical properties of Al Alloys.

In this study, we examined the microstructure and mechanical properties of LM 25 alloys under heat treatment of acetate and T6 in liquid metal. Al-C steel is mainly used in the automotive and aerospace industry for pistons, cylinder heads, cylinder blocks and lift valves with high wear resistance, good melting during casting, good corrosion resistance, and high strength and weight ratio, low temperature. Coordination of extension, heat treatment and mechanical properties above different temperatures. A356 Edge-Cast Iron is made of α-Al primary dendrites made from waste and eutectic aqueous silicone, which reduces mechanical properties and limits industrial use. Mechanical properties can be determined by controlling the size of the mixing particles. Heat treatment determines the mechanical and mechanical properties of the coating. Heat therapy is usually designed to achieve a smooth and strong combination of Al-C-Q-MG alloys. The below table shows the properties of the alloy after mixing with different elements.

Table 1. Alloys and their properties

Alloy	Properties
Al-Ti	Better stability, low corrosive properties at highly aggressive conditions.
Al-Si	Improves hardness, tensile strength, wear resistance and corrosive properties.
Al-Mg	High strength to weight ratio, better corrosion resistant when compared to Mn.
Al-Zn	High corrosive resistant property at high temperature conditions.
Al-Ni	Improves strength and hardness at elevated temperatures and reduces the coefficient of expansion.
Al-Fe	Improves resistance to oxidation, high yield and ultimate tensile strengths.
Al-Cu	Increases strength and precipitation hardening.

Therapeutic measures include treatment, elimination, and artificial aging [35]. The procedure is as follows. Heat treatment: The alloy matrix was incubated at 495 ° C for 8 h, then immersed in oil (35 ° C) for 6 hours at 175° C. Microstructure studies: LM 25 to 15 mm thick and 20 mm diameter LM alloy samples were cut and kept cold for microscopic analysis under conditions such as smelting and heat treatment. These specimens were filtered through a standard metallurgical process and recorded with Keller regiments (2.5% HNO_3, 1.5% HCl, 1% HF, and sewer). Load capacity, final load capacity,LM25 gene modules and compounds. With a stress level of 10-2 /h.

The three main stages of the MLM 25 alloy are made of the main pathway α-Al, eutectic silicone and liquid Mg_2Si. Eutectic silicone particles become round after T6 heat treatment and are evenly distributed throughout the grain. After heat treatment, the belts in the eutectic phase of Al-C are stronger and heavier than those in the cast. The strength of the LM25 alloy reached 65.98 Hv in the T6 heat treatment requirements and increased compared to the casting conditions and is 14.8% higher than the casting conditions. The LM25 alloy has a final load capacity, load capacity and load capacity of 229.2238 MPa, 196.269335 MPa and 3.26965 in thermal conditions T6 and is 49%, 73% and 18% better than molten.

6. Microstructure and hot corrosion behaviour of Al-Si-Hf

Al-C-HF coating uses cement coating on new γ hardened cobalt based concrete. He studied the corrosion resistance of Al-CHF coatings after 100 hours of exposure. The increase in total mass was only 0.29 mg / cm^2.

(1) Al-C-HF coating has succeeded in supplying a new γ-hard cobalt-based cobalt to a cement pack at 10 ° C for 10 hours.

(2) The presence of Si prevents the diffusion of hard metal parts, contributes to the stable growth of SiO_2 oxides, promotes the formation of Al_2O_3, and prevents the rapid corrosion of aluminium coatings.

(3) Al-CHF coating has better corrosion resistance than Al-C coating. HF-added aluminium balances limit the external distribution during oxygenation and suppress the growth of aluminium balances. Theaddition of HF coating is also enhanced by creating a "rust plug" on the rust-coated joint.
[36]As a result, it is expected to improve corrosion resistance at high temperatures. Data-Application for the improvement of extremely liquid cobalt Increased HF and C. modified aluminium coatings on the other hand there are several similar studies in the literature. In this study, H.F.A new modified aluminium cover has been added to the cloud.

7. Corrosion behaviour of Al-Si-Mg.

This study demonstrated the corrosion resistance of Al-C-MG $22MnB_5$ coated steel. The structural differences between the aluminium forming process and the hot-pressed MG coating include a thick, light-free layer that separates the pressed surface and the MG. Electrochemical anti-corrosion tests have shown higher resistance to polarization, especially in the first baptism compared to the Al-C-MG model. This is due to the structural changes and anti-corrosion treatment above the good corrosion protection. This study suggests that AL-C-MG coatings can be considered as an alternative to heat- treated steels with a wide protective layer on A-C skins [37]. This study determined the wear resistance of HPF limestone coated Al-CG steels. When MG was added, even after the HDDA method, there was a dense dispersion of the layer, which indicated the release of Mg during the MDA process. Heat treatment with HPF for 300 s at 930°C accelerated the dissolution of Mg, resulting in theformation of a horizontal protective layer.

IOP Conf. Series: Materials Science and Engineering 1258 (2022) 012028 doi:10.1088/1757-899X/1258/1/012028

In the case of the thickening of the metal layer of the Al-CF interlayer, the shear surface profile represents a solid surface layer rather than a large porous surface of the sample added to Mg, especially in the 0.5 mg Si sample. C Reference cover. Surface structure. The depth of the element obtained by GIDOES test showed that one Mg surface was detected and the thickening area was free from cracks. XPS and XRD studies have shown that Mg neutralization is confirmed by MgO and Mg_2Si. At the time of the first baptism, EIS results showed that each sample had a distribution system that monitored the concentration of iron and oxygen ions. High affinity values were observed especially in samples containing Mg in the first stage of baptism.

This can lead to the formation of hard surface and a very thick crack layer with Mg element. Electrochemical results showed that samples prepared with the addition of Mg produced a more protective coating, especially Al-Si-0.5 Mg, which showed the recommended corrosion resistance withhigher polarization resistance than other types. In addition, prolonged immersion in the sample eventually results in disintegration of the sample. Conclusion Shem points out that basic wear can be caused by small cracks or frontal defects associated with micro-contraction, and Al-Si-Fe shows wear on the continental layer (zone B) and substrate (zone A). Ham. Corrosion products after XRD were γ- FeO (OH) and α-FeO (OH) in all samples, an increase in Mg indicating that does not alter the electrochemical process. However, structural changes caused by the addition of MG elements increase barrier protection, resulting in a dense MG shield surface which leads to a reduction in corrosion products but does not provide long-term protection. It forms a thick layer without small cracks.

Kapil and Himanshu [39] studied The effect of increasing Si and SiC on the microstructure, corrosion properties and properties of matrix A. SiC was prepared with Al-Si crystalline solution (2% by weight)by mixing fermentation with silicon carbide. construction material from 0 to 20% by mass. A similar dispersion of SiC particles in matrix A was observed under microscopic examination. As a mixing element, 60 mesh of 99% pure SiC leather at 300 nets of 99% pure Si and SiC leather with abrasives such as aluminium (A-1100) ignots 99% pure as matrix items were used. During the melting of Al, metallic silicon powder was added and stirred at 300 r / min for 30 minutes. The process was repeated for 15 minutes at a speed of 200 r / min to ensure proper mixing of the aluminium element. SiC mounting added into molten metal and spun at 300 rpm for 45 minutes. The prepared components were analysed by SEM to study the distribution of amplified components in the matrix section. The whole pure morphology was observed without improvement. The uniform distribution of Si particles in the A-matrix was studied. In the process of oxygen production. Specific graphs are analysed for different factors. Silicon Carbide composites reinforced with silicon elements are inferior to composites made of pure aluminium. Biomarkers were rigorously tested on iron to determine root strength and flow. Taffell additive methods have been used to understand the additional effect of anode augmentation ιι cathode hydrogen reduction anode degradation. Root tests were performed at room temperature, and the samples were immersed in 3.5% NaCl solution for 1 hour. The prepared components were analysed by SEM to study the distribution of amplified components in the matrix section. The whole pure morphology was observed without improvement. The uniform distribution of Si particles in the A-matrix was studied. In the process of oxygen production. Specific graphs are analysed for different factors. Silicon Carbide composites reinforced with silicon elements are inferior to composites made of pure aluminium. Biomarkers were rigorously tested on iron to determine root strength and flow.

8. Microstructure characterization

[39] In this article is to focus on the microstructure of the Al_4Cu alloy matrix reinforced with SiC. They used conventional powder metallurgy to produce a matrix of Al-SiC and Al_4Cu alloy compoundscontaining 2.5.5, 7.5 ÷ 10% SiC particles. Two reinforcing particles of different sizes were used to determine the effect on the synthetic microstructure. The turbidity process was carried out in a nitrogen atmosphere at 600 °C and aluminium nitride was present in the composite

microstructure as an additional thickening step. The compound was then deleted and resynthesized under the same conditions. It has an average amount of argon atomic gas is less than 63 m. The electrolytic copper powder was less than 63 μm. Reinforcing materials are used sparingly and are added to the alloy matrix at weights of 2.5, 5, 7.5 and 10%. The pure aluminium powder and the aluminium alloy powderwere mixed by DC curing in a Turbula T2F mixer for 30 minutes to obtain a homogeneous mixture.Press the mixed powder in a solid press at room temperature at 300 MPa. To reduce the frictionbetween the powder and the wall, the powder is coated with glycerin, which allows the greens to beremoved from the well after compaction is complete. The green moulding product was stored in anitrogen atmosphere for 60 minutes at 600 ° C. The main purpose of the connection was to ensure anadequate distribution of the beams between the beams, indicating the size required for the connection. Effect of double compression L double concentration on the relative density of Al_4Cu alloys containing aluminium matrix composites containing SiC particles of different sizes depending on theweight of these particles. The heavy phase addition of silicon carbide reduces the relative density of the two compounds based on aluminium and aluminium alloys. Increasing the relative density makesthe hole smaller. High density values were obtained without sampling. Their relative density wasabout 95%. In all models, the additional stress and welding significantly improved the relative densityand strength compared to the first concentration action, mainly due to the strong bonds formedbetween the composite.

[40] Al-C alloys are widely used in construction due to their high strength, weight ratio, good shape and good corrosion resistance. Corrosion sink test, Corrosion fatigue test was performed to determine the effect of precipitation on the grain structure due to the corrosion fatigue properties. Corrosion sink test, corrosion fatigue test mechanical properties were performed to study the effect of precipitation onthe structure of the nuts on the corrosion resistance of the experimental trench. And corrosion fatigue properties. In addition, electron diffraction diffusion (EBSD), electron transmission microscopy (TEM), and scanning electron microscopy (SEM) are used to explain interconnection mechanisms. Corrosion fatigue was tested at room temperature using a 10ton capacity fatigue apparatus with a stress factor of R 0.1 and a frequency of 0.33 Hz at a pressure of 75% of the output power of each sample. The samples are placed in a closed vessel filled with a NaCl solution of corrosive acid. Corrosion fatigue range: The air fatigue life ratio (then RL; decomposition RL = N / air Nin) was used to determine corrosion resistance. The difference in corrosion behaviour of the three samples is attributed to the microstructure, in particular the nut structure associated with precipitation. TEM results show that the FRE sample is not stored in your HAGB, leaving PFZ on both sides of your HAGB. That is, the HA RE HAGB model has a homogeneous region for the application of the nuclear solution. Therefore, microwave uniformity cannot be achieved if the sample is immersed in a corrosive solution. As a result, the FRE sample showed higher corrosion resistance in three experimental samples.

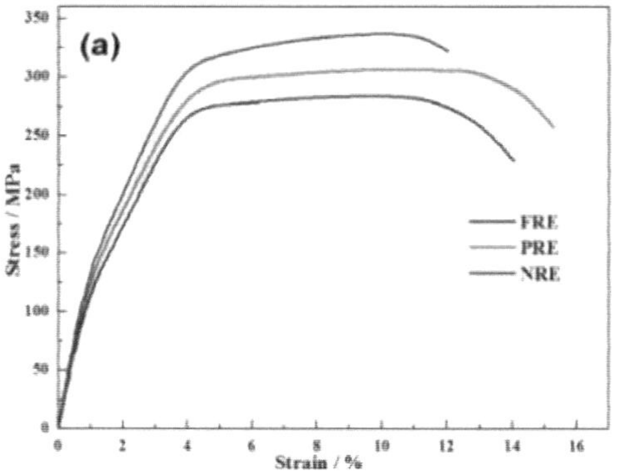

Figure 1(a). Stress vs strain curve for three models.

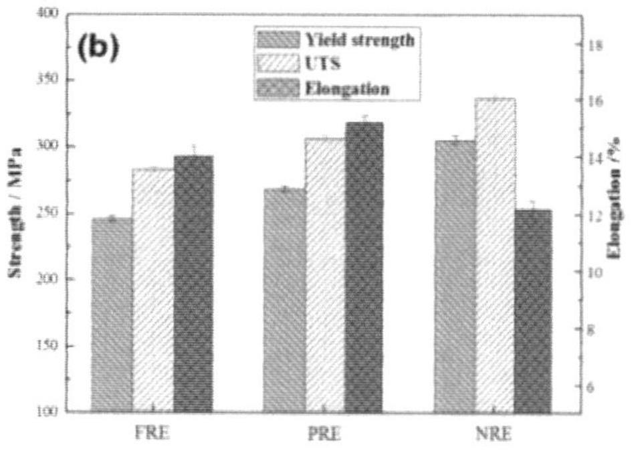

Figure 1(b). Strength vs Elongation graph for three models.

Corrosion behaviour is a practical alloy that has a strong dependence on the composition of the rain grain. The FRE model (fully equipped) has a high corrosion resistance; The PRE (Cloud Response) model goes through a little more surface corrosion, but penetrates much deeper than the NRE (non-refractory) model.

9. Conclusion

Current recycling technologies produce fragile and / or solid aluminium ingots which are not suitable for use in another work environment, but which are environmentally friendly. Research aimed at improving tertiary and / or corrosive structures could lead to better integration, adaptation to a larger industrial environment and increased use in new environments. These include applications in emerging sectors such as the electrical and marine industries, and increased limited use of existing transport sectors (especially automotive and aviation industries). The tribological, mechanical and abrasive properties of the alloy are mainly determined by the presence of various alloys such as silicon, magnesium, copper and molten alloys. The resulting mineral layers can control performance based on shape, hardness, dispersion, and chemical properties. Therefore, such a rainwater system requires careful monitoring. Typically, devices that are small (e.g., less than a micro meter) can be used for both the device and the device. Conversely, larger, more uniform bundles unite at the start of the fracture, decreasing elasticity. Although heat treatment is most often used to disperse and fill IMPs, these treatments can result in a high percentage and at the same time reduce corrosive alloys. A solution can be added to increase seed production and reduce germination time. However, it does not damage the binding properties due to its incompatible crystal structure. The new method could be quality control based FCC dispersants which can provide electrical and corrosive properties. These new ingredients generally have higher heat resistance and improved corrosion resistance than some perishable drinks. Therefore, these components are not considered to be dispersed in the classification.It has been suggested that these new technologies could play a role in diffusion, for example in recycling technologies where degradation of machinery has not yet occurred. Gravity pulls the sand and throws it away. In addition, the thermal stability of the dispersions indicates that the stable submicron IMP should remain constant during subsequent heat treatment. Although it is not possible to completely remove harmful precipitating agents by increasing the melting point and the cooling rate, the levels of these marine agents can be carefully monitored and this can be achieved by including these soluble components. This brief overview highlights the important role the IMP plays in controlling Cast Al Missy and suggests possible ways to address some of its shortcomings. Several electrical engineering methods, including weight loss and depolarization prevention, have been used to assess the effects of efficient dynamics, force efficiency, and destruction of commonly studied components.

Declarations:
The authors proclaimed that there are no conflicting interests with any single person or party.

Funding:
The authors proclaimed that there are no sources of external funding

References

[1] Yılmaz, M.S.; Özer, G.; Öter, Z.C.; Ertugrul, O. Effects of hot isostatic pressing and heat treatments on structural and corrosion properties of direct metal laser sintered parts. RapidPrototyp. J. 2021, 27, 1059–1067.
[2] Cohen, A.L. From rapid prototyping to the second industrial revolution. In Proceedings of the Third International Conference on Rapid Prototyping, Dayton, OH, USA, 23–25 June 1992; pp.189–192.

[3] Low, Z.X.; Chua, Y.T.; Ray, B.M.; Mattia, D.; Metcalfe, I.S.; Patterson, D.A. Perspective on 3D printing of separation membranes and comparison to related unconventional fabrication techniques. J. Membr. Sci. 2017, 523, 596–613.
[4] Attar, M.; Aydin, S.; Arabaci, A.; Mutlu, I. Mechanical meta-material-based polymer skin graft production by rapid prototyping and replica method. Rapid Prototyp. J. 2021, 27, 278–287.
[5] Wang, J.; Sama, S.R.; Manogharan, G. Re-Thinking Design Methodology for Castings: 3D Sand- Printing and Topology Optimization. Int. J. Met. 2019, 13, 2–17.
[6] EPA (Environmental Protection Agency). Profile of the Metal Casting Industry; Office of Compilance Sector Notebook Project; EPA (Environmental Protection Agency): Washington, DC,USA, 1997.
[7] Tang, Y.; Huang, Z.; Yang, J.; Xie, Y. Enhancing the Capillary Force of Binder-Jetting Printing Ti6Al4V and Mechanical Properties under High Temperature Sintering by Mixing Fine Powder. Metals 2020, 10, 1354.
[8] Mariani, M.; Goncharov, I.; Mariani, D.; De Gaudenzi, P.G.; Popovich, A.; Lecis, N.; Vedani, M. Mechanical and microstructural characterization of WC-Co consolidated by binder jetting additive manufacturing. Int. J. Refract. Met.
[9] Hard Mater. 2021, 100, 105639.
[10] Cheng, Y.-L.; Tseng, T.-W. Study on driving waveform design process for multi-nozzle piezoelectric printhead in material-jetting 3D printing. Rapid Prototyp. J. 2021, 27, 1172– 1180.
[11] Cheng, Y.-L.; Chang, C.-H.; Kuo, C. Experimental study on leveling mechanism for material-jetting-type colour 3D printing. Rapid Prototyp. J. 2020, 26, 11–20.
[12] Li, J.; Yan, R.; Yang, Y.; Xie, F. Water-based binder preparation and full-color printing implementation of a self-developed 3D printer. Rapid Prototype. J. 2021, 27, 530–536.
[13] Srinath M.K., Prasad M.S.G. Wear and corrosion resistance of titanium carbo-nitride coated Al-7075 produced throughPVD.2020
[14] Walker, J.; Harris, E.; Lynagh, C.; Beck, A.; Lonardo, R.; Vuksanovich, B.; Thiel, J.; Rogers, K.; Conner, B.; MacDonald, E.; et al. 3D Printed Smart Molds for Sand Casting. Int. J. Met. 2018, 12, 785–796.
[15] George, S.; Roy, A.; McKenna, N.; Singamneni, S.; Diegel, O.; Singh, D.; Neitzert, T.; St George, J.; Roy Choudhury, A.; Yarlagadda, P. Direct Metal casting through 3D printing: A critical analysis of the mould characteristics. In Proceedings of the 9th Global Congress on Manufacturing and Management (GCMM2008), Gold Coast, QLD, Australia, 12–14 November 2008; pp. 1–5.
[16] Vaezi, M.; Chua, C.K. Effects of layer thickness and binder saturation level parameters on 3D printing process. Int. J. Adv. Manuf. Technol. 2011, 53, 275–284.
[17] Garzón, E.O.; Alves, J.L.; Neto, R.J. Study of the viability of manufacturing ceramic moulds by additive manufacturing for rapid casting. Ciência Tecnologia Materiais 2017, 29, e275–e280.
[18] Snelling, D.A.; Williams, C.B.; Druschitz, A.P. Mechanical and material properties of castings produced via 3D printed molds. Addit. Manuf. 2019, 27, 199–207.
[19] Snelling, D.; Blount, H.; Forman, C.; Ramsburg, K.; Wetzel, A.; Williams, C.; Druschitz, A. The Effects of 3D Printed Molds on Metal Castings. In Proceedings of the Solid Freeform Fabrication Symposium, Austin, TX, USA, 12–14 August 2013; pp. 827–845.
[20] Srinath M.K., Ganesha Prasad M.S. Corrosion Analysis of TiCN Coated Al-7075 Alloy for Marine Applications. 2019

[21] Szklarska-Smialowska, Z. Pitting corrosion of aluminum. Corros. Sci. 1999, 41, 1743–1767.
[22] Guo, W.; Li, J.; Qi, M.; Xu, Y.; Ezatpour, H.R. Effects of heat treatment on the microstructure, mechanical properties and corrosion resistance of AlCoCrFeNiTi0.5 high-entropyalloy. J. Alloys Compd. 2021, 884, 161026.
[23] Davis, J.R. Alloying: Understanding the Basics; ASM International: Metals Park, OH, USA, 2001; p. 647.
[24] Fadavi Boostani, A.; Tahamtan, S. Fracture behavior of thixoformed A356 alloy producedby SIMA process. J. Alloys Compd. 2009, 481, 220–227.
[25] Tahamtan, S.; Fadavi Boostani, A. Evaluation of pitting corrosion of thixoformed A356 alloy using a simulation model. Trans. Nonferrous Met. Soc. China 2010, 20, 1702–1706.
[26] Ye, H. An overview of the development of Al-Si-alloy based material for engine applications. J. Mater. Eng. Perform. 2003, 12, 288–297.
[27] Chen, C.L.; West, G.D.; Thomson, R.C. Characterisation of Intermetallic Phases in Multicomponent Al-Si Casting Alloys for Engineering Applications. Mater. Sci. Forum. 2006, 519, 359–364.
[28] Donatus, U.; Thompson, G.E.; Omotoyinbo, J.A.; Alaneme, K.K.; Aribo, S.; Agbabiaka, O.G. Corrosion pathways in aluminium alloys. Trans. Nonferrous Met. Soc. China 2017, 27, 55– 62.
[29] Li, J.; Dang, J. A summary of corrosion properties of Al-Rich solid solution and secondary phase particles in al alloys. Metals 2017, 7, 84.
[30] Couture, A. Iron in aluminium casting alloys—a literature survey. AFS Int. Cast Met. J. 1981, 6, 9–17
[31] Crepeau, P.N. Effect of Iron in Al-Si Casting Alloys A Critical Review. AFS Trans. 1995, 103, 361–366
[32] Cao, X.; Campbell, J. Morphology of β-Al5FeSi Phase in Al-Si Cast Alloys. Mater. Trans. 2006, 47, 1303–1312.
[33] Impact of intermetallic precipitates on the tribological and/or corrosion performance of cast aluminium alloys: a short review. D. Culliton*1 , A. J. Betts2 and D. Kennedy 2013
[34] O.O.Joseph,P.A.Olubambi,Olakunle O. Joseph, B.M. Edun, J.O.Okeinyi,O.P.Abioye, EffectsOf Alloying On Aluminium-Silicon Alloys, 2020
[35] Alumina.Khalid Abd El-Aziz • Dalia Saber • Hossam El-Din M. Sallam, Wear andCorrosion Behaviour of Al–Si Matrix Composite Reinforced, 2015
[36] Raj Kumar Singh, Amit Telang, S. Das, Microstructure and Mechanical Properties of Al-Si Alloy in Ascast and heat treated Condition. 2016
[37] Xin Qi, Qingyan Hou, Mingyue Chen, Simin Zhu, Meifeng Li, Chungen Zhou, Microstructure and hot corrosion behavior of Al-Si-Hf coating on new γ′ -strengthened cobalt- based superalloy, 2020.
[38] Changkyu Kim, Seongkoo Cho, Wonseog Yang, Ahmad Ivan Karayan , Homero Castaneda, Corrosion behavior of Al-Si-Mg coated hot-press-forming steel. 2021.
[39] B.leszczyńska-madej, A.wąsik, M.madej, microstructure characterization of sic reinforced aluminium and Al_4Cu alloy matrix composites microstructure characterization of sic reinforcedaluminium and Al_4Cu alloy matrix composites, 2017.
[40] YuWang, YunlaiDeng, JiqiangChen, QingsongDai, XiaobinGuo, Effects of grain structure related precipitation on corrosion behavior and corrosion fatigue property of Al–Mg–Si alloy, 202

Predicting Melt-Crystal Interface Position and Shape during the Manufacturing Process of Single Crystal via Explainable Machine Learning Models

Chang Zhang[1], **Qisen Cheng**[2]
[1]Department of Chemical Engineering and Materials Science, University of Minnesota, Minneapolis, MN 55455, USA
[2]Department of Electrical Engineering and Computer Science, University of Michigan, Ann Arbor, MI 48109, USA

E-mail: zhan3709@umn.edu

Abstract. In recent years, employing machine learning models to predict the process parameters during the manufacturing process of single crystals has gained wide attention as it's reliable and much faster than traditional numerical simulation approaches. However, most machine learning models used in previous studies are black box models, which don't provide explainable results. In this paper, we present a feasibility study of applying explainable machine learning models to predict steady-state melt-crystal interface position and deflection with the set-point temperature of 5 heaters in a vertical Bridgman furnace. The dataset used to train and evaluate the machine learning models was generated by 2-D numerical simulation. We experimented with linear regression and random forest algorithms, and then used linear regression coefficient and SHAP value to quantify the impact of each input on the output, from which we inferred a heater control strategy that could potentially improve the crystal growth process. Our encouraging results show that explainable machine learning models can be applied to predict crystal growth process parameters in real-time and generate actionable insights to guide crystal manufacturing practice.

1. Introduction

The vertical Bridgman (VB) method is a popular method to manufacture bulk single crystals from the melt [1]. During a VB process, initially, a charge of raw material is placed in a crucible known as the ampoule. Inside the furnace, the temperature near the top is hot enough to melt the material, while that near the bottom is cold enough for the material to solidify. The ampoule is initially placed in the hot zone and will be translated slowly downwards towards, while the melt crystallizes from the bottom towards the top. During the process, the interface separating the melt and crystalline phases is known as the melt-crystal interface. The macroscopic melt-crystal interface shape is a crucial process parameter determining many factors associated with the final quality of the crystal, such as crystallinity, cracking, and impurity distribution [2, 3]. An interface moderately convex to the melt is ideal to minimize the thermoelastic stress near the interface and avoid the formation of internal defects [4]. If the interface shape deviates from the ideal state, then control actions, usually exerted via adjusting the power of the heaters, need to be performed to tune the interface shape back to the desired state [5]. However, adjusting heater power changes the thermal environment inside the furnace, which in turn might lead to increase in temporary thermoelastic stress level and defect density in the grown crystal [6]. Therefore, the stability of interface position also needs to be considered while controlling interface shape.

Due to the high temperature and slow process, it's technically difficult to observe or control interface shape [7]. Therefore, it's strongly desired to predict interface shape and position via computer programs. Since 1970s, physics based numerical simulations have been employed to simulate VB process and their results have proven to be in good agreement with experimental results [5]. A number of numerical simulation works explored tuning interface shape by designing furnace temperature profiles [4, 8, 9]. However, these approaches were developed from more of a theoretical viewpoint (e.g., how to design the furnace temperature profile) instead of from a practical viewpoint (e.g., how to operate the heaters), lacking straightforward guidance on crystal growth practice. Also, numerical models are usually very slow [10]. In application scenarios, real-time prediction of growth outcome is necessary. Therefore, it is necessary to develop a new approach to mitigate these disadvantages.

In this era of artificial intelligence (AI), applying machine learning models to predict crystal growth outcomes has gain popularity as they are fast and reliable [11]. Using numerical simulation results to train machine learning models is of particular interest as it allows generating massive amount of data close to actual experimental results at low cost. This approach has proved to be successful in predicting a variety of parameters in crystal growth process such as melt-crystal interface shape/position [10, 12], growth rate [13, 14], furnace temperature [10], melt flow field [15], etc. However, most recent relevant works primarily employed neural networks [11]. A major disadvantage of neural network based models is associated with the difficulties in explaining their results, which makes them "black box models" [16]. Model explainability means that a model can reveal the relationship between its input and output in a way that makes sense to human expertise. There has been ongoing interest in applying explainable methods to traditionally black box models [17, 18]. Previous works in the crystal growth field have tried to explain the results of neural networks via multi-dimensional plot [19], which can be very unintuitive to read when there are many input variables as it's difficult to visualize a space with four or more dimensions.

Linear regression has been used to predict crystal growth outcomes [20]. Linear regression is simple and it has global explainability as the regression coefficients directly quantifies the linear dependence of the output on the input. However, it makes certain assumptions with the characteristics of data, making its applications are limited. Ensemble of trees models have also been applied to predict crystal growth parameters and achieved satisfactory outcomes [21, 22]. Ensemble of trees models can be locally explained via SHAP method. SHAP is a code package released in 2020 and it enables the detailed explanation of ensemble of trees models by quantifying the dependence of output on each input variable at each data point using game theory [24, 25]. Yet, to our knowledge, very few studies have reported the utilization of explainable machine learning techniques to predict and understand crystal growth process.

In this paper, we report applying linear regression and random forest models to predict steady-state melt-crystal interface shape and position (2 output variables) with set-point temperature of the 5 heaters (5 input variables) in a VB furnace. The dataset, consisting of 96 data points, was generated for BaBrCl crystal in a 5-heater VB furnace with a 2-D finite volume method based model with CrysMAS code. Linear regression coefficients and SHAP value from the random forest models were explained to study how different models recognize the relation between input and output and show how explainable machine learning techniques could help with experimental design. Our work provides a novel approach of employing explainable machine learning models to fast predict crystal growth process parameters and generate actionable insights to guide manufacturing practice.

Table 1. Thermal physical properties of materials used in the simulation, adapted from [7].

Symbol	Description	Value	Unit
k_m	melt thermal conductivity	0.7	$Wm^{-1}K^{-1}$
k_c	crystal thermal conductivity	1.1	$Wm^{-1}K^{-1}$
k_q	quartz thermal conductivity	6.0	$Wm^{-1}K^{-1}$
T_{mp}	melting point	1168	K

2. Methodology

2.1. Vertical Bridgman System

We choose BaBrCl crystal as the candidate material. BaBrCl crystal is an alkaline-halide scintillator material which has a great potential to be commercialized for medical and security applications [26], therefore, improving its growth outcome via controlling the interface shape is of great interest to many researchers [27]. Our previous numerical simulation results for the interface position and shape of BaBrCl:Eu crystal during a VB growth process were in good agreement with neutron imaging results [7], a novel and efficient way to observe high temperature VB crystal growth process [27]. Thus we are confident that we have a good understanding about its properties and growth behaviors. Readers can refer to [7] for the properties of BaBrCl.

We choose a 5-heater VB furnace at Lawrence Berkeley National Laboratory as the candidate furnace, reported in [28]. Figure 1(a) shows the geometry of the furnace. Its heaters are named top, upper, middle, lower, and bottom heater from top to bottom, and each of them has a thermal couple inserted at its half-height along the internal surface. The controllers of the heaters control their power by minimizing the difference between the pre-specified set-point temperature and the actual temperature at the thermal couples. The position of the thermal couple of each heater, where the set-point temperature for each heater is specified, is denoted by a pink dot in the right half of Figure 1(b). The inner radius of the ampoule is 13mm.

2.2. Data Generation

2.2.1. Numerical Model We employ CrysMAS [29], a crystal growth simulator based on finite volume method, to simulate the temperature field in the furnace. The material properties used for simulation are summarized in Table 1. Some assumptions are made to simplify the model: 1) The system is assumed to be axis-symmetric, so a 2-D model is developed to simulate the 3-D furnace. 2) The system is assumed to be in steady-state, so all the transient effects are neglected. This assumption holds when growth rate is very small. 3) The effect of melt flow convection is neglected, and only radiative and conductive heat transfer modes are considered.

Based on the above assumptions, the governing equations and boundary conditions of the model are as follows. Heat diffusion equation is applied to all non-gaseous phases:

$$0 = k_a \nabla^2 T + \dot{R}. \tag{1}$$

Here T is the temperature. k_a is the thermal conductivity of phase a. $\dot{R}$ is the heat generation rate. The energy conservation equations are solved by finite volume method based on an unstructured, fixed mesh. Radiative heat flux between all surfaces is calculated using view factor method:

$$\dot{Q}_{i-j} = \sigma T_i^4 F_{i-j} A_i. \tag{2}$$

Here $\dot{Q}_{i-j}$ is the rate of radiative energy leaving surface i for surface j. σ is the Stefan–Boltzmann constant. T_i is the temperature at surface i. F_{i-j} is the view factor between the two surfaces. A_i is the area of surface i. The solid-liquid interface lies along the melt-point isotherm.

$$T = T_{mp}. \tag{3}$$

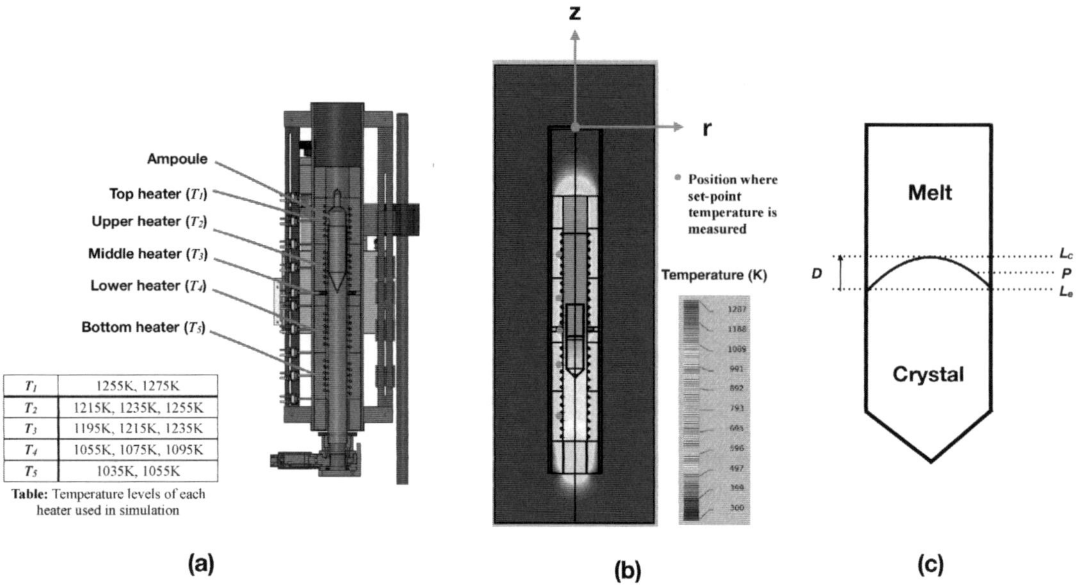

T_1	1255K, 1275K
T_2	1215K, 1235K, 1255K
T_3	1195K, 1215K, 1235K
T_4	1055K, 1075K, 1095K
T_5	1035K, 1055K

Table: Temperature levels of each heater used in simulation

Figure 1. (a) Geometry and structure of the 5-heater vertical Bridgman furnace, reproduced from [28]. The table shows the levels of the set-point temperature of each heater used in the data generation process. (b) The temperature field in the furnace generated by CrysMAS code. The origin and the directions of the horizontal axis r and vertical axis z are specified in the plot. The pink dot in each heater represents the position of the thermal couple, where the set-point temperature is measured. (c) Geometric representation of the ampoule, where crystal growth takes place. L_c, L_e are the vertical position of the center and edge of the melt-crystal interface with respect to the reference level. $\mathcal{P}$ is the vertical position of the midpoint between L_c and L_e. $\mathcal{D}$ is the difference between L_c and L_e. A positive value indicates that the center is higher than the edge and that the interface is convex. A negative value indicates that the center is lower than the edge and that the interface is concave.

Here T_{mp} is the melting point of BaBrCl. CrysMAS uses the enthalpy method to locate the interface. When the furnace is under operation, pipes with running water are wrapped around the external surface for cooling purpose. Newton's law of cooling is applied along the external surface of the furnace to approximate convective heat transfer.

$$-k_a \mathbf{n} \cdot \nabla T = h(T - T_{env}). \tag{4}$$

Here $\mathbf{n}$ is the the unit normal vector of the interface pointing towards the inside of the furnace. T_{env} is environmental temperature, which is set to be 298K. h is the heat transfer coefficient between furnace wall and ambient environment, which is set to be 300 Wm^{-2}.

We use the inverse calculation approach to calculate the temperature field in the furnace. Firstly, the set-point temperature at the controllers of the top, upper, middle, lower, bottom heater, denoted as T_1, T_2, T_3, T_4, T_5 respectively, is specified. Then, the power of the heaters is calculated and optimized such that the difference between the simulated temperature and the set-point temperature is minimized. After the solution converges, we record the shape and position of the interface. We use the average position of the center and edge of the interface, $\mathcal{P} = (L_c + L_e)/2$, to represent its position. Here L_c is the interface position at the centerline of

Table 2. Details of the Machine Learning Models

Model	Input	Output	Algorithm	Interpretation Method
1	T_1, T_2, T_3, T_4, T_5	Interface position $\mathcal{P}$	Linear Regression	Linear regression coefficients
2	T_1, T_2, T_3, T_4, T_5	Interface position $\mathcal{P}$	Random Forest	SHAP summary plot
3	T_1, T_2, T_3, T_4, T_5	Interface deflection $\mathcal{D}$	Linear Regression	Linear egression coefficients
4	T_1, T_2, T_3, T_4, T_5	Interface deflection $\mathcal{D}$	Random Forest	SHAP summary plot

the furnace, L_e is the interface position at the edge where it meets the ampoule wall. We use the interface deflection, $\mathcal{D} = L_c - L_e$, to represent interface shape. $\mathcal{D} > 0$ indicates a convex shape, and $\mathcal{D} < 0$ indicates a concave shape. L_c, L_e, $\mathcal{P}$, and $\mathcal{D}$ are denoted in Figure 1(c),

It is extremely difficult to observe interface in a high temperature vertical Bridgman process by experimental method. In a previous study, we used neutron imaging experiment to validate that the heat transfer model built with CrysMAS can accurately simulate the transient interface position and deflection of BaBrCl in a very similar vertical Bridgman furnace [7]. Therefore, we believe that the results presented in this dataset accurately represents the actual growth situation.

2.2.2. Design of Experiment We used design of experiment to choose set-point temperature of the heaters. The levels of the set-point temperature of each heater are listed in the table in Figure 1(a). The levels were chosen in a way such that: 1) The melting point isotherm, where the interface lies along, is within the region between the middle zone and lower zone; 2) The axial temperature gradient near the interface is within the range of 20K/cm - 50K/cm, which is used in real manufacturing experiments [30]. To conduct a full factorial experiment, the model needs to be run $3^3 \times 2^2 = 108$ times. We required that $T_1 \geq T_2 \geq T_3 \geq T_4 \geq T_5$ as positive axial temperature gradients are necessary to allow heat to flow downwards and drive directional solidification. This rules out 12 cases and only 96 runs are necessary. The final dataset has 96 data points, each with 5 inputs and 2 outputs.

2.3. Machine Learning Models

We chose the explainable linear regression (LR) and random forest (RF) algorithms to build the machine learning models. We used 5-fold cross validation to evaluate the performance of the 2 models. To perform 5-fold cross validation, the dataset is randomly split into 5 groups of similar size. In Step 1, one group is used as the hold-out set and the machine learning model will be fitted on the data from the other 4 groups to identify the relationship between the inputs and the output. Then the models will make predictions for the output in the hold-out set based on the corresponding inputs. This procedure will be repeated for 5 times, with one group being the hold-out set in each time. The model predicted values for the 5 groups will be compared with the corresponding true output values for validation. The mean absolute error, MAE, and R-squared statistic, R^2 will be reported. Table 2 lists the details of the models employed in this work.

After model evaluation, we proceed with model explanation. We aim to identify how a model recognizes the relationship between the set-point temperature, interface position, and interface deflection. In cross validation, the parameters of a model change at every step as the data which it is fitted on is different. For convenience and consistency, we choose to explain the linear regression and random forest models fitted at Step 1 of the cross validation procedure (Group 1 is chosen as the test-set and the other groups are chosen as the training set). A LR model can be easily explained via its regression coefficients. Each regression coefficient quantifies the linear dependence of the output on each input variable. We employed SHAP

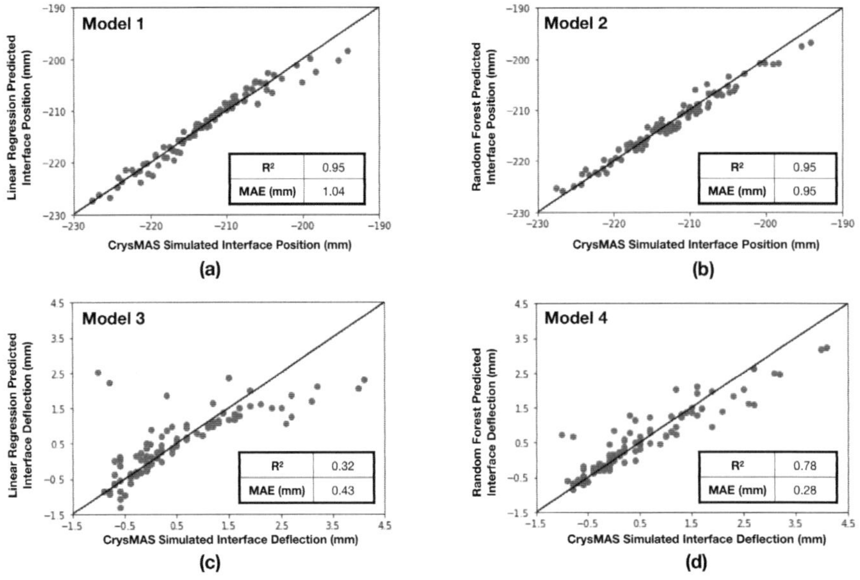

Figure 2. Comparison on $\mathcal{P}$ and $\mathcal{D}$ between CrysMAS simulation and machine learning prediction based on cross validation results. The R^2 and mean absolute error (MAE) are presented in each plot. (a) Model 1. (b) Model 2. (c) Model 3. (d) Model 4.

(Shapley Additive exPlanations) value [23], a novel quantity derived from game theory and local explanations, to interpret RF model results. SHAP value for all input variables and data points is usually visualized in a SHAP summary plot, which reveals the magnitude as well as the directionality of the dependence of model prediction on each input variable at each data point. The mathematical derivation of SHAP is prolonged and readers can refer to Lunberg's works for mode details [24, 25]. All machine learning models were built by scikit-learn package (version 0.24.2) in Python. SHAP analysis was implemented by the SHAP package (version 0.39.0) in python.

3. Results and Discussion

Our objectives are to 1) evaluate the predictive performance of LR and RF models on interface deflection $\mathcal{D}$ and position $\mathcal{P}$ with set-point temperature T_i ($i = 1, 2, 3, 4, 5$) at the 5 heaters, 2) quantify the impact of set-point temperature of each heater via explainable machine learning techniques, and 3) Use the results to compare LR and RD models and demonstrate how explainable machine learning technique can help guide crystal growth practice.

3.1. Model Evaluation

Figure 2(a) and (b) show the performance of Model 1 and 2 in predicting interface position $\mathcal{P}$. The output value is predicted when this data point is in the test set (the part of the dataset that the model isn't fitted on) during the cross validation process, as described in Section 2.3.3. As each data point in the dataset has been in the test set for exactly one time in the cross validation process, each plot contains the comparison between model prediction $\bar{y}_i$ and true value (CrysMAS simulation) y_i for all 96 data points in our dataset. A point being close to the diagonal line indicates that the machine learning prediction is close to CrysMAS simulation.

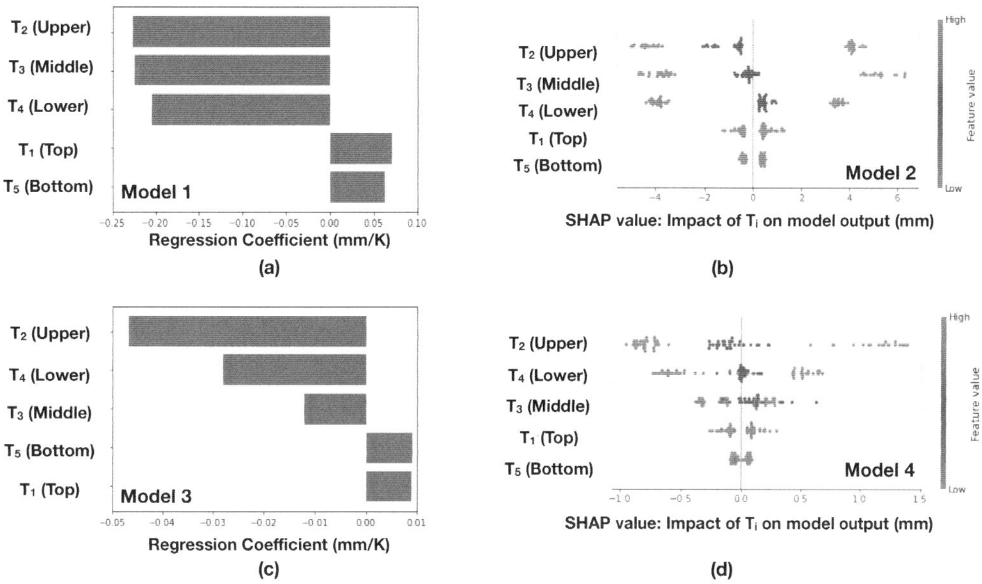

Figure 3. (a) The impact of each set-point temperature T_i on interface position $\mathcal{P}$, quantified by linear regression coefficients from Model 1. (b) The impact of each set-point temperature T_i on interface position $\mathcal{P}$, quantified by SHAP value from Model 2. The color of each dot represents the value of T_i in data point i within the range of T_i. The position of a dot along the horizontal axis represents the SHAP value of this input variable and this data point, which quantifies how much the value of T_i pushes $\mathcal{P}$ away from its base value (the average of $\mathcal{P}$ in the training dataset). (c) The impact of each set-point temperature T_i on interface deflection $\mathcal{D}$, quantified by linear regression coefficients from Model 3. (d) The impact of each set-point temperature T_i on interface deflection $\mathcal{D}$, quantified by SHAP value from Model 4.

Both algorithms perform well, with R^2 being higher than 0.95. Model 1, with an R^2 of 0.95 and a MAE of 1.0 x 10^{-3}, achieves very close performance to that of Model 2, which has an R^2 of 0.96 and a MAE of 9.5 x 10^{-4}. This demonstrates the feasibility of using LR or RF model to predict interface position. This also suggests that the impact of the set-point temperature of each heater on interface position is more linear and independent as a simple LR can well predict interface position at different set-point temperature. Figure 2(c) and (d) show the performance of Model 3 and 4 in predicting interface deflection $\mathcal{D}$. Model 3 performs poorly, with an R^2 of 0.32 and a MAE of 4.3 x 10^{-4}, while Model 4 performs significantly better, with an R^2 of 0.78 and a MAE of 2.9 x 10^{-4}. $R^2 > 0.7$ is considered statistically significant. This reveals that RF model should be preferred to predict interface deflection. As RF algorithm can capture nonlinear relationship between data while LR algorithm doesn't, this suggests that the dependence of interface deflection on the set-point temperature of the heaters is highly nonlinear. The above results suggest the feasibility of applying machine learning models to predict interface position and deflection with set-point temperature and that the RF algorithm should be preferred over LR for prediction.

3.2. Model Explanation

In this section, we will explain how the machine learning models recognize the relationship between input and output via linear regression coefficients and SHAP value from random forest models. We used the models fitted at Step 1 of the cross validation process for explanation to assure that the LR and RF models are fitted on the same data. In Step 1, 76 data points randomly selected from the dataset were used as the training set, and the rest 20 data points were used as the test set.

Figure 3(a) plots the regression coefficients of Model 1, ranked by their absolute values from high to low, which quantifies how the LR model predicts the impact of T_i on $\mathcal{P}$. According to the plot, T_2 and T_3 have the strongest impact, followed by T_4, while the impact of T_1 and T_5 is much weaker. Increasing T_2, T_3, or T_4, or reducing T_1 or T_5 raises interface position. It can be read from the plot that the LR model predicts that increasing T_2 by 1K can lead to $\sim$0.23 mm drop in interface position, while increasing T_5 by 1K can lead to $\sim$0.05 mm rise. Figure 3(b) is the SHAP summary plot of Model 2. According to the plot, for T_2, the red dots ($T_2 = 1255$K) are clustered around -4.5mm, while the blue dots ($T_2 = 1215$K) are clustered around +4.5mm. This indicates that the RF model recognizes that in general, increasing T_2 from 1215K to 1255K can lead to about 9 mm drop in interface position. The average change in interface position with per degree increase in T_2 predicted by the RF model is about -0.225 mm/K, pretty close to the estimation of the LR model. Similarly, for T_5, the red dots ($T_5 = 1055$K) are clustered around +0.5mm, while the blue dots ($T_5 = 1035$K) are clustered around -0.5mm, which indicates that increasing T_5 from 1035K to 1055K can lead to about 1mm rise in interface position. The plot shows that T_2 has the strongest impact on interface position, followed by T_3 and T_4, and that T_1 and T_5 have a much weaker impact. The SHAP summary plot also shows that increasing T_2, T_3, or T_4, or reducing T_1 or T_5 lowers interface position. Although the RF algorithm is essentially different from the LR algorithm, Model 1 and 2 yield qualitatively consistent results in terms of the dependence of interface position on the 5 set-point temperature.

Figure 3(c) plots the regression coefficients of Model 3, which quantifies how the LR model predicts the impact of T_i on interface deflection $\mathcal{D}$. According to the plot, T_2 has the strongest impact, followed by T_4, while the impact of T_5, T_3, or T_1 is much weaker. Increasing T_2, T_3, or T_4 or reducing T_1 or T_5 reduces interface deflection. Figure 3(d) is the SHAP summary plot of Model 4. The plot shows that the RF model predicts that T_2 has the strongest impact on interface deflection, followed by T_4, T_3, T_1, and T_5. Figure 3(d) also shows that the relationship between set-point temperature and interface deflection is more complicated. Interface deflection $\mathcal{D}$ increases by decreasing T_2 from 1255K to 1215K. However, the clusters are more "spread" than in those in Figure 3(b), indicating that the impact of T_2 on interface deflection is more sensitive to the set-point temperature of other heaters, a result from a stronger interactive effect between different heaters on interface deflection. The impact of T_3 on interface deflection is even more complicated. For T_3, there is a nearly full overlap between the blue dots ($T_3 = 1195$K) and the purple dots ($T_3 = 1215$K), indicating that increasing T_3 from 1195K to 1215K won't lead to significant change in interface deflection. However, the red dots ($T_3 = 1135$K) is left to the purple cluster, indicating that further increasing T_3 to 1235K will lead to reduction in interface deflection (less convex or more concave). It is noteworthy to point out that this nonlinear relationship between T_3 and interface deflection can't be recognized by the LR model as it assumes that the dependence of interface deflection on any set-point temperature is linear, which is why the RF model outperforms the LR model in predicting interface deflection.

3.3. Applications to Crystal Growth Practice

In this section, we will illustrate how those insights generated by the explainable machine learning models can be applied to guide crystal growth practice. As we have concluded in Section 3.1 that RF is more accurate than LR, we will focus on the results of the RF models shown in

Figure 3(b) and (d). As a common objective in crystal manufacturing process is to increase interface deflection while stabilizing interface position. We will explore how to tune set-point temperature in our furnace to achieve this objective. As shown in Figure 3(b) and (d), both interface deflection and position have strong dependence on T_2, T_3, and T_4. Therefore, it is not feasible to tune interface shape without much change in interface position by adjusting the set-point temperature of one single heater. It has been discussed in Section 3.2 that reducing T_2 from 1255K to 1215K (red $\rightarrow$ purple $\rightarrow$ blue) leads to increase in interface deflection (increase in convexity or reduction in concavity). For T_3, increasing T_3 from 1195K to 1215K (blue $\rightarrow$ purple) minor impact on interface deflection; however, further increasing T_3 to 1235K (purple $\rightarrow$ red) will reduce interface deflection. As for interface position, reducing T_2 leads to rise in interface position while increasing T_3 at the same time can act as a counter effect and avoid a big shift in interface position. In summary, it can be inferred from SHAP summary plots that reducing T_2 from 1255K to 1215K and increasing T_3 from 1195K to 1215K at the same time can increase interface convexity and stabilize interface position, given that the other set-point temperature is within the range used in our study. To verify this statement, we selected 12 scenarios in our dataset from CrysMAS simulation results where $T_2 = 1255\text{K}$ and $T_3 = 1195\text{K}$ as the initial states. We propose 4 different controls actions on heaters. Action 1: Only reduce T_2 from 1255K to 1215K. Action 2: Only increase T_3 from 1195K to 1215K. Action 3: Reduce T_2 from 1255K to 1215K and increase T_3 from 1195K to 1235K. Action 4: Reduce T_2 from 1255K to 1215K and increase T_3 from 1195K to 1215K, which is inferred from SHAP summary plots in Figure 3(b) and (d). We are interested in comparing how interface deflection $\mathcal{D}$ and position $\mathcal{P}$ will change from the 12 initial states with those 4 different heater actions. As can be observed from Figure 4, Action 4) in general leads to a higher increase in interface deflection and lower change in interface position as compared with the other 3 actions. It is noteworthy to point out scenarios 1, 10, 11, 12 are in the test set at Step 1 of cross validation process and are not used to generate Figure 3(b) and 3(d). Surprisingly, the insights inferred from SHAP summary plots successfully predict the change in interface position and deflection in those 4 scenarios. This demonstrates the feasibility of using explainable machine learning results to guide and improve crystal manufacturing practice.

4. Conclusions

In this study, we applied machine learning models, trained with data from the numerical simulation results from the physics-based CrysMAS model, to predict the interface position and shape during a vertical Bridgman process. Then we explained the linear regression and random forest models via linear regression coefficients and SHAP value. Both RF ($R^2 = 0.95$) and LR ($R^2 = 0.95$) models performed well in predicting interface position, while RF model ($R^2 = 0.78$) significantly outperformed LR model ($R^2 = 0.32$) in predicting interface deflection. This is because RF is better at recognizing the nonlinear relationship between set-point temperature and interface deflection. We inferred a novel heater control strategy from SHAP summary plot to optimize the growth process by increasing interface deflection and stabilizing interface position, which was then validated by CrysMAS simulation. We believe this paper is one of the first studies to apply explainable machine learning concept to optimize crystal manufacturing process.This work can serve as a framework to be applied to predict, understand, and optimize other crystal manufacturing parameters of interest.

Acknowledgments

We would like to thank Dr. Edith Bourret, Dr. Didier Perrodin, and Dr. Jeffery Peterson for their inputs on furnace design and operation. We would also like to thank Dr. Jeffery Derby for letting us use CrysMAS code.

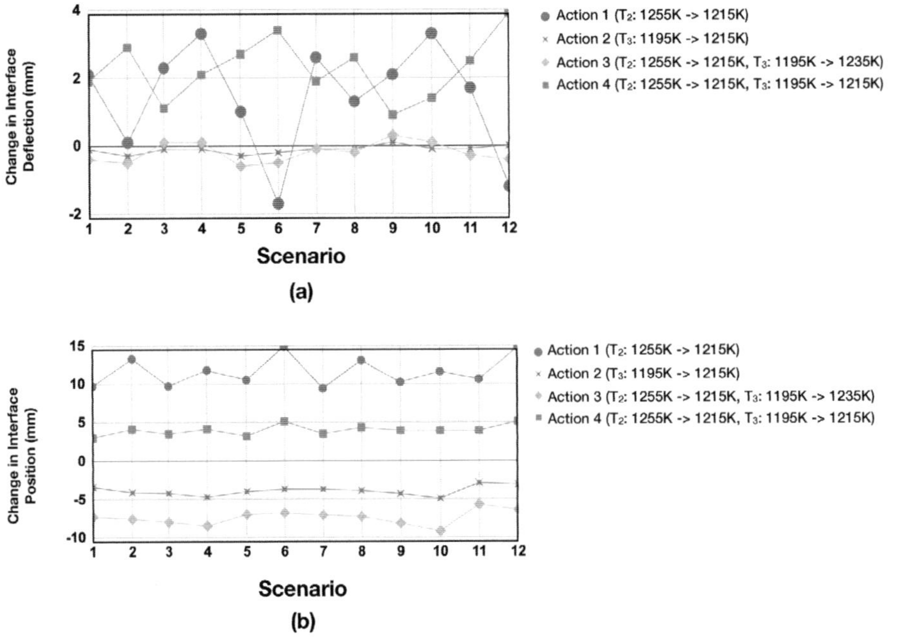

Figure 4. Change in interface deflection $\mathcal{D}$ and position $\mathcal{P}$ from the initial states in 12 scenarios with 4 different types of heater actions.

References

[1] W. Bridgman. Nobellecture: General survey of certain results in the field of high-pressure physics, Nobelprize.org.

[2] W.A. Tiller, Principles of solidification, in: J.J. Gilman (Ed.), The Art and Science of Growing Crystals, John Wiley Sons Inc, 1963, p. 276.

[3] Chang Zhang, Bing Gao, Anton S. Tremsin, Didier Perrodin, Tetiana Shalapska, Edith D. Bourret, Drew R. Onken, Sven C. Vogel, Jeffrey J. Derby, Analysis of chemical stress and the propensity for cracking during the vertical Bridgman growth of BaBrCl:Eu, Journal of Crystal Growth, Volume 546, 2020, 125794, ISSN 0022-0248, https://doi.org/10.1016/j.jcrysgro.2020.125794.

[4] Lisa Lun, Andrew Yeckel, Michael Reed, Csaba Szeles, Prodromos Daoutidis, Jeffrey J. Derby, On the effects of furnace gradients on interface shape during the growth of cadmium zinc telluride in EDG furnaces, Journal of Crystal Growth, Volume 290, Issue 1, 2006, Pages 35-43, ISSN 0022-0248, https://doi.org/10.1016/j.jcrysgro.2006.01.006.

[5] C.W. Lan, Recent progress of crystal growth modeling and growth control, Chemical Engineering Science, Volume 59, Issue 7, 2004, Pages 1437-1457, ISSN 0009-2509, https://doi.org/10.1016/j.ces.2004.01.010.

[6] C.S. Bohun, I. Frigaard, H. Huang, S. Liang, A semianalytical thermal stress model for the Czochralski growth of type III–V compounds, SIAM J. Appl. Math. 66 (5) (2006) 1533.

[7] Jeffrey J. Derby, Chang Zhang, Jan Seebeck, Jeffrey H. Peterson, Anton S. Tremsin, Didier Perrodin, Gregory A. Bizarri, Edith D. Bourret, Adrian S. Losko, Sven C. Vogel, Computational modeling and neutron imaging to understand interface shape and solute segregation during the vertical gradient freeze growth of BaBrCl:Eu, Journal of Crystal Growth, Volume 536, 2020, 125572, ISSN 0022-0248, https://doi.org/10.1016/j.jcrysgro.2020.125572.

[8] Nan Zhang, Andrew Yeckel, Jeffrey J. Derby, Maintaining convex interface shapes during electrodynamic gradient freeze growth of cadmium zinc telluride using a dynamic, bell-curve furnace profile, Journal of Crystal Growth, Volume 355, Issue 1, 2012, Pages 113-121, ISSN 0022-0248, https://doi.org/10.1016/j.jcrysgro.2012.06.042.

[9] Jeffrey H. Peterson, Jeffrey J. Derby, An axial temperature profile curvature criterion for the engineering of convex crystal growth interfaces in Bridgman systems, Journal of Crystal Growth, Volume 468, 2017, Pages 899-904, ISSN 0022-0248, https://doi.org/10.1016/j.jcrysgro.2016.09.064.

[10] Natasha Dropka, Martin Holena, Stefan Ecklebe, Christiane Frank-Rotsch, Jan Winkler, Fast forecasting of VGF crystal growth process by dynamic neural networks, Journal of Crystal Growth, Volume 521, 2019, Pages 9-14, ISSN 0022-0248, https://doi.org/10.1016/j.jcrysgro.2019.05.022.
[11] Dropka, N.; Holena, M. Application of Artificial Neural Networks in Crystal Growth of Electronic and Opto-Electronic Materials. Crystals 2020, 10, 663. https://doi.org/10.3390/cryst10080663.
[12] Yifan Dang, Lijun Liu, Zaoyang Li, Optimization of the controlling recipe in quasi-single crystalline silicon growth using artificial neural network and genetic algorithm, Journal of Crystal Growth, Volume 522, 2019, Pages 195-203, ISSN 0022-0248, https://doi.org/10.1016/j.jcrysgro.2019.06.033.
[13] Yuto TAKEHARA, Atsushi SEKIMOTO, Yasunori OKANO, Toru UJIHARA, Sadik DOST, Explainable machine learning for the analysis of transport phenomena in top-seeded solution growth of SiC single crystal, Journal of Thermal Science and Technology, 2021, Volume 16, No. 1, p. JTST0009, Online ISSN 1880-5566, https://doi.org/10.1299/jtst.2021jtst0009.
[14] Wancheng Yu, Can Zhu, Yosuke Tsunooka, Wei Huang, Yifan Dang, Kentaro Kutsukake, Shunta Harada, Miho Tagawa, Toru Ujihara, Geometrical design of a crystal growth system guided by a machine learning algorithm, CrystEngComm, 2021, 23, 2695-2702, https://doi.org/10.1039/D1CE00106J.
[15] Yosuke Tsunooka, Nobuhiko Kokubo, Goki Hatasa, Shunta Harada, Miho Tagawaab, Toru Ujiharaab, High-speed prediction of computational fluid dynamics simulation in crystal growth, CrystEngComm, 2018, 20, 6546, https://doi.org/10.1039/C8CE00977E.
[16] Jack V. Tu, Advantages and disadvantages of using artificial neural networks versus logistic regression for predicting medical outcomes, Journal of Clinical Epidemiology, Volume 49, Issue 11, 1996, Pages 1225-1231, ISSN 0895-4356, https://doi.org/10.1016/S0895-4356(96)00002-9.
[17] X. Qu, P. Liu, Z. Li and T. Hickey, "Multi-class Time Continuity Voting for EEG Classification" in Brain Function Assessment in Learning. BFAL 2020. Lecture Notes in Computer Science, Cham:Springer, vol. 12462, 2020.
[18] Ziyuan Zhao, Kartik Chopra, Zeng Zeng and Xiaoli Li, "Sea-net: Squeeze-and-excitation attention net for diabetic retinopathy grading", 2020 IEEE International Conference on Image Processing (ICIP), pp. 2496-2500, 2020.
[19] Natasha Dropka, Martin Holena, Optimization of magnetically driven directional solidification of silicon using artificial neural networks and Gaussian process models, Journal of Crystal Growth, Volume 471, 2017, Pages 53-61, ISSN 0022-0248, https://doi.org/10.1016/j.jcrysgro.2017.05.007.
[20] J. Ding, L. Liu, Real-time prediction of crystal/melt interface shape during Czochralski crystal growth, CrystEngComm, 2018, 20, 6925.
[21] Schimmel, S.; Tomida, D.; Saito, M.; Bao, Q.; Ishiguro, T.; Honda, Y.; Chichibu, S.; Amano, H. Boundary Conditions for Simulations of Fluid Flow and Temperature Field during Ammonothermal Crystal Growth—A Machine-Learning Assisted Study of Autoclave Wall Temperature Distribution. Crystals 2021, 11, 254. https://doi.org/10.3390/cryst11030254.
[22] Yao T S, Tang C Y, Yang M, Zhu K J, Yan D Y, Yi C J, Feng Z L, Lei H C, Li C H, Wang L, Wang L, Shi Y G, Sun Y J and Ding H, Machine Learning to Instruct Single Crystal Growth by Flux Method, Chin. Phys. Lett., 2019, 36, 068101.
[23] Hart, S. (1989). Shapley Value. In: Eatwell, J., Milgate, M., Newman, P. (eds) Game Theory. The New Palgrave. Palgrave Macmillan, London.
[24] Scott Lundberg, Su-In Lee, A unified approach to interpreting model predictions, arXiv preprint, arXiv:1705.07874.
[25] Lundberg, S.M., Erion, G., Chen, H. et al. From local explanations to global understanding with explainable AI for trees. Nat Mach Intell 2, 56–67 (2020). https://doi.org/10.1038/s42256-019-0138-9
[26] Z. Yan, T. Shalapska, E.D. Bourret, Czochralski growth of the mixed halides BaBrCl and BaBrCl:Eu, Journal of Crystal Growth, Volume 435, 2016, Pages 42-45, https://doi.org/10.1016/j.jcrysgro.2015.11.032.
[27] Anton S. Tremsin, Didier Perrodin, Adrian S. Losko, Sven C. Vogel, Takenao Shinohara, Kenichi Oikawa, Jeff H. Peterson, Chang Zhang, Jeffrey J. Derby, Alexander M. Zlokapa, Gregory A. Bizarri, and Edith D. Bourret, In-Situ Observation of Phase Separation During Growth of Cs2LiLaBr6:Ce Crystals Using Energy-Resolved Neutron Imaging, Crystal Growth & Design 2017 17 (12), 6372-6381, DOI: 10.1021/acs.cgd.7b01048.
[28] C. Zhang, Computational Modeling for the Vertical Bridgman Growth of BaBrCl: Eu Crystal, Ph.D. thesis, University of Minnesota (2020).
[29] CrysMAS user manual and tutorial [online] (2013). http://www.cglerlangen.com/downloads/Manual.pdf.
[30] Tremsin, A., Perrodin, D., Losko, A. et al. Real-time Crystal Growth Visualization and Quantification by Energy-Resolved Neutron Imaging. Sci Rep 7, 46275 (2017). https://doi.org/10.1038/srep46275.
[31] Lundberg, S.M., Nair, B., Vavilala, M.S. et al. Explainable machine-learning predictions for the prevention of hypoxaemia during surgery. Nat Biomed Eng 2, 749–760 (2018). https://doi.org/10.1038/s41551-018-0304-0

[32] R.J. Naumann, An analytical approach to thermal modeling of Bridgman-type crystal growth: I. One-dimensional analysis, Journal of Crystal Growth, Volume 58, Issue 3, 1982, Pages 554-568, https://doi.org/10.1016/0022-0248(82)90143-9.

Experimental study and analysis on single slope single basin solar still with zeolite crystals as storage materials

K.Senthil Kumar[1] A.S.Manirathnam[1], Anish. L[2], Ajay. A[2], Jaseem Ahmed.S[2], Kaleeswaran.B[2]

Sri Krishna College of Technology, Coimbatore

senthil181189@gmail.com

Abstract. We humans find solutions for every problem around us. Taking an environmental problem for water scarcity we have come up with a traditional idea but a bit modified. We have used solar distillation as the basic concept and developed a single slope single basin solar still (SSSBSS) where we have used a storage material that is rarely used for this process and have carried out the research work in it. The experimental study is carried for a single slope single basin solar still (SSSBSS) under two cases (i) SSSBSS (ii) SSSBSS with zeolite crystals as storage materials. The still was made of galvanized iron (GI) and the sides were insulated with polystyrene. It has a basin area of 1 m^2. Bore water of Coimbatore which is a bit salty was taken as the feed water. The test was carried out at Kuniyamuthur, Coimbatore during 10th, 11th, 12th of February 2021. Even though it was winter, the output was average. By the comparison using zeolite crystals, the efficiency was doubled compared to the first case.
Keywords: Solar distillation, Single slope single basin solar still (SSSBSS), Zeolite crystals, Galvanized iron (GI), Bore water, Efficiency

1. Introduction

Though our world has 71% water compared to land, only 1% of it is pure water that is available for humans. In a country like India, we have different landscapes ranging from high peaks to the seashore and mangroves. In some places which are so arid, people walk about 40 kilometers to fetch fresh water for their daily needs. If we take a deep look into this, these areas have a lot of unutilized solar energy [16], so there is a technical solution for them that is very cost-effective and has a higher rate of an outcome. Frankly speaking, desalination and distillation are easy methods to convert any available water for drinking purposes and other pure water needs. Taking the best two processes, solar distillation is the process that can yield a perfect solution to such situations.

The higher the intensity of solar radiation, the efficiency of the solar still is increased [17]. It is a basic process similar to that of the water cycle or so-called rain cycle. The saline or impure water inside the still is heated and evaporated to the top glass cover in which the vapour gets attached. After evaporation, the water molecules in the glass are condensed and it flows down to the output collector. The salt or impurities settle down at the bottom of the basin where the output water is free from impurities and is pure. The amount of water that we get in the output shows the efficiency of the still. An average solar still with a basin area of 0.54 m^2 gives 1.74litre/day of output water [10].

The modelled single slope single basin solar still has corrugated sheets (CS) as absorbent material. It is used to increase the surface area in which the solar rays fall and also to increase the distillate output. GI is used to make the still, as the saltwater is used there is a chance for corrosion. The main modification is made to the storage material, which is used to retain the heat in the basin. In this project, zeolite crystals are taken. The thermal property of zeolite states that it can retain heat four times than that of water. So by this, we can clearly say that the performance of the still will surely increase. For further proof experiment is carried out in the solar still with and without zeolite crystals and the conclusions are made.

A single slope single basin solar still can remove salt from the water at the efficiency of 99% [1]. Also using copper tubes to the conventional still can increase its efficiency of output. For an average single basin solar still, the efficiency throughout the year can be approximated to 27% [2]. [Bilal A et al., 1998] Using different absorber materials in a single basin solar still can vary the efficiency of the still. Further using black colour to the material can have a better yield than any other colours available. [H.S. Deshmukh et al., 2017] Usage of sensible heat storage materials can also increase the efficiency of the still. Also, the night productivity of the still is enhanced using storage materials. [Vimal Dimria et al., 2007] The properties of the condensing surface like the thickness of the glass cover and the area of the condensing surface can also affect the distillation process.

[A.A. El-Sebaii et al., 2009] Comparing the single basin solar still with and without sensible heat storage materials, annually the still with storage materials have an efficiency 23.8% more than the still without the storage materials. [A.A. El-Sebaii et al., 2008] PCM (Phase Change Materials) is used as the heat storage materials and notably it increases the efficiency 85.3% more than the still used without PCM. [A. Hanson et al., 2003] The stills are very successful in removing inorganic contaminants, where the organic pollutants have a mixed success. The results did not have much difference in the lab and the field for the removal of contaminant compounds. [Muhammad Ali Samee, et al., 2005] Transient mathematical models were modelled for single basin solar still and the simulations were done. The results show that the still with storage materials have efficiency of 60% than the still without storage materials. [A.E. Kabeel et al., 2018] A single basin solar still with and without sensible heat storage materials, which are considered graphite here. The efficiency of the still with graphite was 59.9-60.54 % and without graphite were 33.41- 34.6 %. [K. Kalidasa Murugavel et al., 2007] Many materials and compounds are used in the basin along with the water and the best properties to the components were identified. Some important conclusions were rubber for the basin, black dye for the deep basin, mica sheet as the absorber, and the sidewalls of the basin are fixed with reflecting mirrors. [S. A. Lawrence et al., 2007] Both experimental and theoretical analysis was done for single basin solar still with dye. Both analyses had a similar result 0.02m water depth and also for deeper basins the effect of dye was significant.

Zeolite crystals can be considered sensible heat storage materials that can be used for solar stills due to their thermal properties. Zeolite can store up to 4 times the heat stored by water. [Vladimir V. Murashov et al., 2002] The study was determined to find the effective thermal conductivity of powdered zeolite 4A. The upper estimate of heat transfer through radiation was found to be 0.2% of effective thermal conductivity based on quartz. [Yazan Taamneh & Ahmad Al-Shyyab et al., 2016] The absorption heat storage material in solar still was taken as Jordanian natural zeolite. The result clearly shows that the amount of zeolite crystals increases the volume of freshwater, i.e. the higher the mass of zeolite the more the freshwater is obtained. Also, the efficiency was increased up to 43%. Hence, from the above references, we can clearly say that the usage of zeolite crystals as storage materials can drastically improve the efficiency of the single-slope single basin solar still. So to practically examine the efficiency and characteristics of zeolite as storage material in solar still, the study was carried out and notable outcomes were obtained.

2. Proposed model and Experimental setup

The schematic diagram of solar distillation using single slope single basin solar still (SSSBSS) is shown in figure 1. Figure 2 shows the dimensions of the single slope single basin solar still (SSSBSS). Figure 3 is the pictorial view of a single slope single basin solar still (SSSBSS) model used for the present experiment. The experimental setup consisted of a single slope single basin solar still (SSSBSS) with corrugated sheets and energy storage materials.

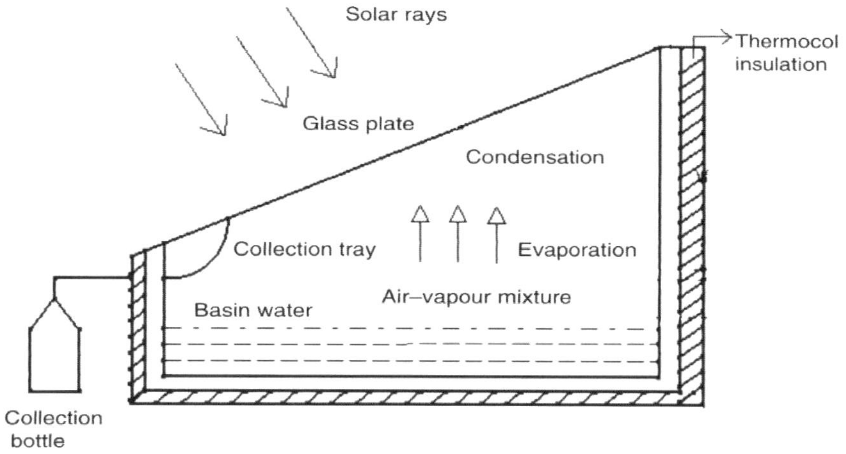

Figure 1. Schematic diagram of solar distillation using single slope single basin solar still (SSSBSS)

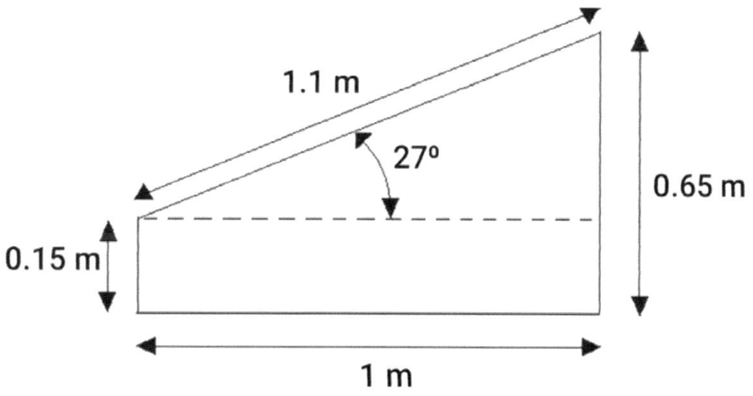

Figure 2. Dimensions of single slope single basin solar still (SSSBSS)

Figure 3. Proposed model of single slope single basin solar still (SSSBSS) used for the present experiment.

2.1. *Single slope single basin solar still (SSSBSS)*

The single slope single basin solar still consists of a single basin and the type is of single slope along with evaporation and condensation in it. The basin material is made of galvanized iron (GI). The basin area of the still is 1 m^2 with an angle of the glass cover equal to 27^0. The basin of solar still is coated with black paint to absorb maximum amount of solar energy. The casing of the solar still is also made of galvanized iron (GI) of 2.5 mm thickness. A plain glass cover of area 1.1 m^2 and 3 mm thickness is mounted on the top of the casing to receive high solar radiation in the solar still [19]. The angle of the glass cover is equal to 27^0. To avoid water leakage, the edges are sealed with M sealant gel. In order to reduce the heat dissipation to the atmosphere, the solar still is covered with a thermocol box of 16 mm thickness.

2.2. *Corrugated Sheet*

Corrugated sheets are used to increase the surface area. Inorder to improve the distillate output corrugated sheets are necessary in the solar still. For this experiment locally available galvanised iron corrugated sheets are used as absorbent material. Two corrugated sheets each of area 1 m^2 and thickness 1 mm are kept inside the still [18].

2.3. *Storage materials*

The purpose of the storage materials used in the solar still is to store excess solar heat and reuse it. In the present experiment, zeolite crystals are used as storage materials. Zeolite crystals have higher heat absorbing capacity than water thus increasing the distillate output. Six kilograms of zeolite crystals of size about 10 mm to 50 mm in diameter are used. Figure 4 shows the picture of single slope single basin solar still with zeolite crystals and corrugated sheets in it. Figure 5 is the experimental setup of single slope single basin solar still (SSSBSS) with zeolite crystals and corrugated sheets in it.

Figure 4. Picture of single slope single basin solar still with zeolite crystals and corrugated sheets in it.

Figure 5. Experimental setup of single slope single basin solar still (SSSBSS) with zeolite crystals and corrugated sheets in it.

3. Instruments

Various instruments like pyranometer, anemometer, thermocouples, temperature indicator, measuring jar, electronic weighing machine are used to determine the performance of the present experiment. The solar radiation incident on the solar still is measured by a pyranometer [16] [17]. Vane type anemometer is used to measure the velocity of wind. The temperature of the basin water and glass cover temperature of the are measured using K type thermocouples. The thermocouples are connected to the temperature indicator which shows the temperature readings. The distillate output is measured using a measuring jar.

4. Experimental Procedure

The goal of the experimentation is to find and compare the distillate output and thermal efficiency of conventional SSSBSS and SSSBSS with zeolite crystals as heat storage materials. The experiment was carried out on 10th, 11th and 12th of February 2021 from 8:30 a.m. to 6:30 p.m. Every morning at 8:30
a.m. the solar still is filled with water. About 10 litres of water is taken and filled inside the still through the inlet pipe. Bore water is used as the feed water. The still is kept facing eastwards on a sunny day. As the day progresses, the water, still, corrugated sheets and zeolite crystals gets heated by the sun's radiation and the water inside the still starts to evaporate and touches the inner glass surface. Then the water vapour condenses and slides down to the lower end of the glass cover. Finally the distillate water is collected in a bottle through the outlet pipe. All experiments had been carried out at Kuniyamuthur (10.96^0 N, 76.95^0 E), Coimbatore, Tamilnadu. The glass temperature, basin water temperature, ambient temperature, solar intensity, wind velocity and distillate output are taken on an hourly basis from 8:30 a.m. to 6:30 p.m.

5. Result and Discussion

Experiments were conducted in two cases, case (i) SSSBSS and case (ii) SSSBSS with zeolite crystals. Both the cases experiments were conducted for three days and we have measured for various parameters like temperature of glass and basin water, ambient temperature, solar intensity, wind velocity and distillate output on an hourly basis for both cases. We have tabulated and plotted the average readings obtained from three days of experimentation for all the parameters for both cases. The efficiency of the solar still is calculated using the following formula:

$$\eta = \frac{m \times}{h_{fg}\ A \times Ig} \times 100$$

where η = Still efficiency (%); m = Daily distillate output (l/day); h_{fg} = Latent heat of water (MJ/kg); A = Basin area of still (m^2); I_g = Daily global solar radiation (kWh/m^2/day).

Table 1 represents the experimental average readings of various parameters conducted for case (i) SSSBSS and case (ii) SSSBSS with zeolite crystals. Table 2 represents the experimental average readings of ambient temperature, solar intensity and wind velocity which is common for both the cases. The figuration of solar intensity and wind velocity with respect to time is shown in figure 6. From the graph we can note that the solar intensity gradually increases from morning 8.30 a.m. and reaches a maximum of 860 W/m^2 at 1.30 p.m. and it gradually decreases and reaches 220 W/m^2 at 6.30 p.m. while the velocity of wind had some fluctuations remaining within 8 to 12 km/h.

Table 1. Experimental average results of various parameters conducted for case (i) SSSBSS and case (ii) SSSBSS with zeolite crystals on February 10, 11, 12, 2021.

Time (hr)	Basin water Temperature (°C)		Basin water Temperature (°C)		Hourly distillate Output (kg/hr)		Cumulative distillate Output (kg/hr)		Efficiency (%)	
	Case (i)	Case (ii)	Case (i)	Case (ii)	Case (i)	Case (ii)	Case (i)	Case (ii)	Case (i)	Case (ii)
8.30	25	25	23	23	0	0	0	0	0	0
9.30	32	36	33	36	0.04	0.07	0.04	0.07	11.73	20.54
10.30	35	47	39	48	0.07	0.11	0.11	0.18	9.45	14.85
11.30	57	58	47	47	0.13	0.19	0.24	0.37	12.18	17.81
12.30	61	69	57	65	0.19	0.3	0.43	0.67	15.17	23.96
13.30	64	71	60	67	0.24	0.45	0.67	1.12	18.83	35.31
14.30	63	70	61	61	0.22	0.4	0.89	1.52	18.33	33.33
15.30	59	69	57	57	0.18	0.33	1.07	1.85	17.35	31.82
16.30	48	60	51	51	0.11	0.2	1.18	2.05	13.14	23.89
17.30	38	56	36	48	0.10	0.17	1.28	2.22	16.46	27.98
18.30	34	48	30	45	0.05	0.12	1.33	2.34	15.34	36.81

Table 2. Experimental average results obtained for case (i) SSSBSS and case (ii) SSSBSS with zeolite crystals on February 10, 11, 12, 2021.

Time (hr)	Ambient Temperature (°C)	Solar Intensity (W/m^2)	Wind Velocity (km/hr)
8.30	27	170	9
9.30	29	230	10
10.30	30	500	10
11.30	31	720	11
12.30	32	845	10
13.30	33	860	11
14.30	32	810	12
15.30	32	700	9
16.30	32	565	9
17.30	31	410	8
18.30	29	220	9

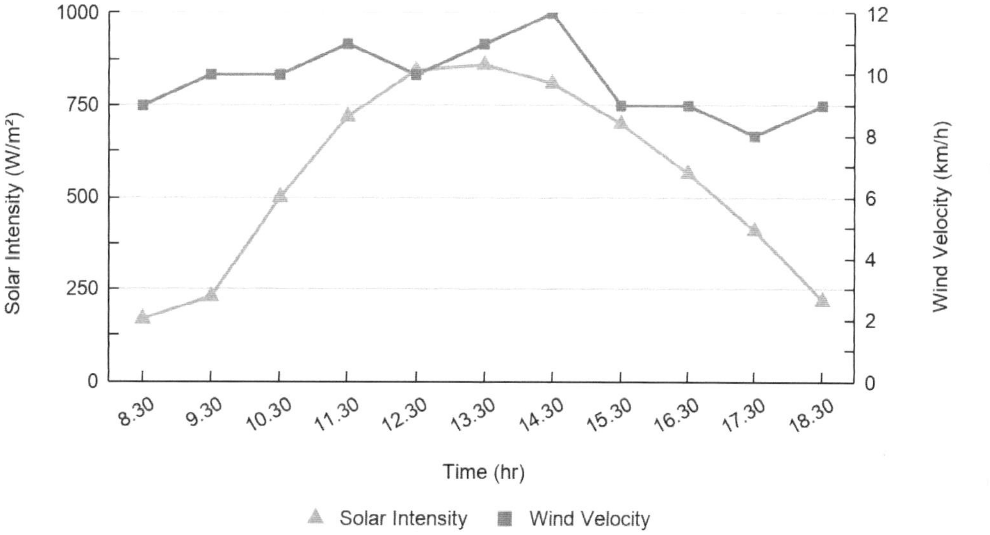

Figure 6. Hourly variations of solar intensity and wind velocity on February 10, 11, 12, 2021

Figure 7 shows the change in the basin water temperature and glass temperature of both stills and ambient temperature with respect to time. It is noted that as the time moves on, the basin water temperature of SSSBSS with zeolite crystals always remains higher than the basin water temperature of conventional SSSBSS. It is due to the high heat absorbing and heat storing capacity of zeolite crystals. The basin water temperature of SSSBSS with zeolite crystals reached a maximum value of 71 ^{0}C at 1.30 p.m. whereas the basin water temperature of conventional SSSBSS was 64 ^{0}C at the same time. As the basin water temperature of both the stills and solar intensity increases, it reflects in the increment of the glass temperature also. The glass temperatures of conventional SSSBSS and SSSBSS with zeolite crystals were maximum at 1.30 p.m. The ambient temperature gradually increased from morning and it was maximum during noon time and gradually decreased in the evening time.

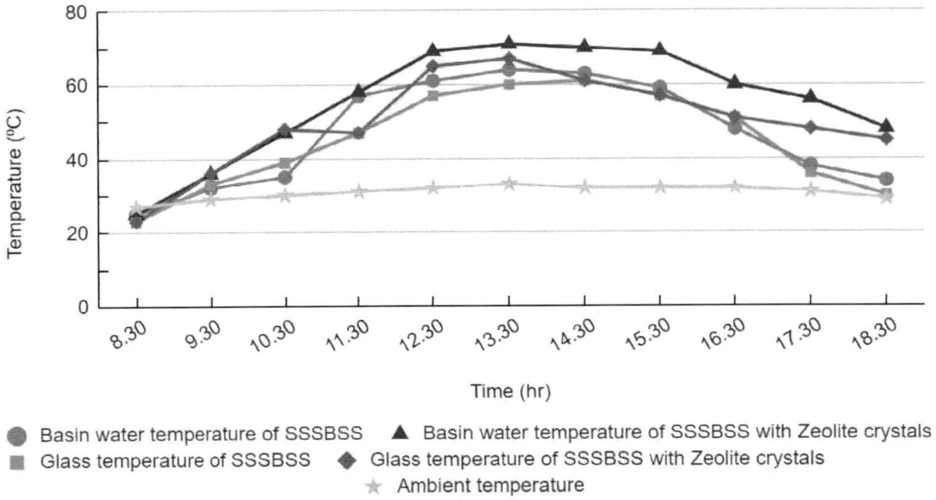

Figure 7. Average hourly variations of ambient temperature and glass and basin water temperatures of SSSBSS and SSSBSS with zeolite crystals on February 10,11, 12, 2021

The hourly distillate output obtained from conventional SSSBSS and SSSBSS with zeolite crystals is shown in figure 8. It is found that more distillate output is obtained from SSSBSS with zeolite crystals than conventional SSSBSS which is due to the high thermal storage capacity of zeolite crystals. A maximum of 0.24 kg/h and 0.45 kg/h is obtained at 1:30 p.m. for conventional SSSBSS and SSSBSS with zeolite crystals respectively. Figure 9 shows that the cumulative distillate output collected from SSSBSS with zeolite crystals is more than the conventional SSSBSS.

The productivity of SSSBSS with zeolite crystals is 2.34 kg/h/day whereas it is 1.33 kg/h/day for conventional SSSBSS. There is an increase of 76% of cumulative distillate output in SSSBSS with zeolite crystals compared to conventional SSSBSS. The reason for this high productivity rate in the climatic conditions of Coimbatore during the month of february is due to the impact of zeolite crystals high thermal storage capacity. The hourly still efficiency of conventional SSSBSS and SSSBSS with zeolite crystals can be seen in figure 10. The hourly still efficiency of SSSBSS with zeolite crystals varies from 14.85% to 36.81%, which verified that zeolite crystals thermal storage medium has better efficiency. The maximum still efficiency obtained for SSSBSS with zeolite crystals is 36.81 at 6:30 p.m., which shows the heat storing capacity of zeolite crystals, whereas the conventional SSSBSS hourly still efficiency is 18.83%.

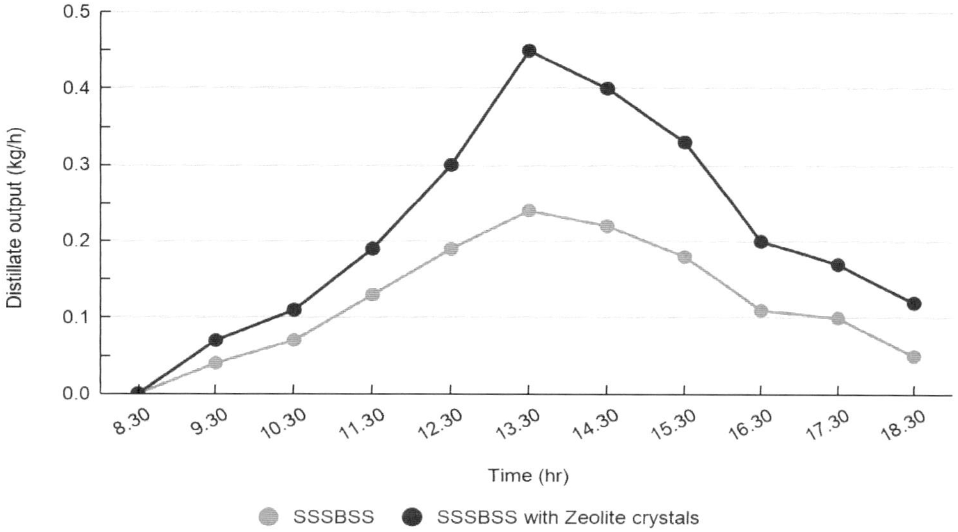

Figure 8. Average hourly distillate output obtained from SSSBSS and SSSBSS with zeolite crystals on February 10, 11, 12, 2021

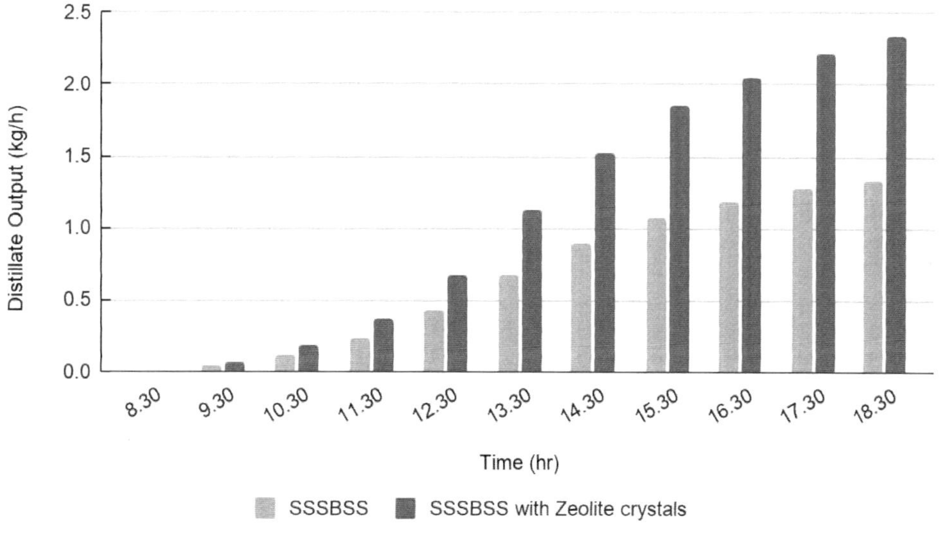

Figure 9. Average cumulative distillate output obtained from SSSBSS and SSSBSS with zeolite crystals on February 10, 11, 12, 2021

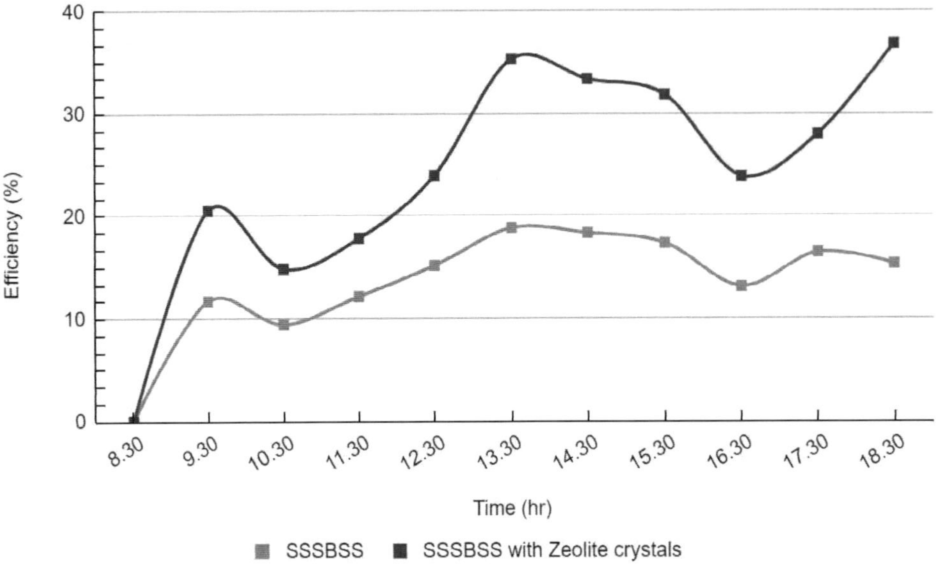

Figure 10. Average hourly still efficiency of SSSBSS and SSSBSS with zeolite crystals on February 10, 11, 12, 2021

6. Conclusion

From this current experimental work of solar distillation using SSSBSS and zeolite crystals as heat storage materials, the following conclusions have been made:

- SSSBSS with zeolite crystals showed profound performance in maintaining the basin water temperature as compared to conventional SSSBSS of the same area.
- The distillate output and still efficiency of SSSBSS with zeolite crystals was pretty good as compared with the conventional SSSBSS.
- About 2.34 kg/h/day of distillate output was obtained from SSSBSS with zeolite crystals as storage medium compared to 1.33 kg/h/day of distillate output from conventional SSSBSS.
- An increment of 75.93% of distillate output was achieved from SSSBSS with zeolite crystals in comparison with conventional SSSBSS.
- Still efficiency of SSSBSS with zeolite crystals as storage medium had an increment of 79.92% than conventional SSSBSS.
- Thus zeolite crystals are a good choice for using as sensible heat storage materials.

References

[1] Dhivagar R and Subramanian Sundararaj 2018 Thermodynamic and water analysis on augmentation of a solar still with copper tube heat exchanger in coarse aggregate *Journal of Thermal Analysis and Calorimetry* **136** 89-99

[2] Al-hassan Ghassan and Algarni Salem 2013 Exploring of Water Distillation by Single Solar Still Basins *American Journal of Climate Change* **2** 57-61

[3] Bilal A A, Mousa S M, Omar Osta and Yaser Elayan 1998 Experimental evaluation of a single-basin solar still using different absorbing materials *Renewable Energy* **14** 307-10

[4] Deshmukh H S and Thombre S B 2017 Solar distillation with single basin solar still using sensible heat storage materials *Desalination* **410** 91-8

[5] Vimal Dimri, Bikash Sarkar, Usha Singh and Tiwari G N 2008 Effect of condensing cover material on yield of an active solar still: an experimental validation *Desalination* **227** 178-89

[6] El-Sebaii A A, Yaghmour S J, Al-Hazmi F S, Adel S F, Al-Marzouki F M and Al-Ghamdi A A 2009 Active single basin solar still with a sensible storage medium *Desalination* **249** 699-706
[7] El-Sebaii A A, Al-Ghamdi A A, Al-Hazmi F S and Adel S F 2009 Thermal performance of a single basin solar still with PCM as a storage medium. *Applied Energy* **86** 1187-95
[8] Hanson A, Zachritz W, Stevens K, Mimbela L, Polka R and Cisneros L 2004 Distillate water quality of a single-basin solar still: Laboratory and field studies *Solar Energy* **76** 635-45
[9] Pamesh K L and Prayagi S V 2016 Mathematical analysis of solar still with storage : A review *International Research Journal of Engineering and Technology (IRJET)* **3** 2820-22
[10] Muhammad Ali Samee, Umar K M, Tariq Majeed and Nasir Ahmad 2007 Design and performance of a simple single basin solar still *Renewable and Sustainable Energy Reviews* **11** 543-9.
[11] Kabeel A E, Mohamed Abdelgaied and Amr Essa 2018 Enhancing the performance of single basin solar still using high thermal conductivity sensible storage materials *Journal of Cleaner Production* **183** 20-5
[12] Kalidasa Murugavel K, Chockalingam Kn and Srithar K 2008 Progresses in improving the effectiveness of the single basin passive solar still *Desalination* **220** 677-86
[13] Lawrence S A, Gupta S P and Tiwari G N 1988 Experimental Validation of Thermal Analysis of Solar Still with Dye *International Journal of Solar Energy* **6** 291-305
[14] Vladimir V M and Mary Anne White 2002 Thermal properties of zeolites: effective thermal
Conductivity of dehydrated powdered zeolite 4A *Mater. Chem. Phys.* **75** 178-80
[15] Yazan Taamneh and Ahmad Al-Shyyab 2016 Improvement of the performance of a solar still by utilizing the sorption thermal storage of natural zeolite *Desalination and Water Treatment* **57** 27450-57
[16] Manirathnam, A.S., Manikandan, M.D., Prakash, R.H., Kumar, B.K. and Amarnath, M.D., 2021. Experimental analysis on solar water heater integrated with Nano composite phase change material (SCi and CuO). Materials Today: Proceedings, 37, pp.232-240.
[17] K Senthil Kumar, H Ashwin Kumar, P Gowtham, S Hari Selva Kumar, R Hari Sudhan., 2021 Experimental analysis and increasing the energy efficiency of PV cell with nano-PCM (calcium carbonate, silicon carbide, copper) Materials Today: Proceedings, 37, 1221-1225
[18] K Senthil Kumar, S Revanth, D Sanjeev, P Sabesh Kumar, P Surya., 2021, Experimental investigation of improving the energy conversion efficiency of PV cell by integrating with PCM, Materials Today: Proceedings,37, 712-716
[19] AS Manirathnam, K Senthil Kumar, S Sudhakar, S Vasanth Balaji, S Sasi Kumar, R Vibishan., 2021, Improving the efficiency of solar drying unit with PCM (Paraffin wax), Journal of Physics: Conference Series, 2070, 012213
[20] Abdelgaied, M., Seleem, M.F. and Bassuoni, M.M., 2022. Recent technological advancements in membrane distillation and solar stills: preheating techniques, heat storage materials, and nanomaterials—a detailed review. Environmental Science and Pollution Research, pp.1-20.
[21] Das, D., Bordoloi, U., Kalita, P., Boehm, R.F. and Kamble, A.D., 2020. Solar still distillate enhancement techniques and recent developments. Groundwater for Sustainable Development, 10, p.100360.
[22] Rufuss, D.D.W., Arulvel, S., Kumar, V.A., Davies, P.A., Arunkumar, T., Sathyamurthy, R., Kabeel, A.E., Vishwanath, M.A., Reddy, D.S.C., Dutta, A. and Agrawal, M., 2022. Combined effects of composite thermal energy storage and magnetic field to enhance productivity in solar desalination. Renewable Energy, 181, pp.219-234.

Optomechanical design and analysis of lens assembly for space application

M V Shah[1] **and A V Patel**[1]

[1]*CHARUSAT Space Research and Technology Center, Chandubhai S. Patel Institute of Technology, Charotar University of Science and Technology, CHARUSAT Campus, Changa-388421, Gujarat, India.*

axatpatel.me@charusat.ac.in

Abstract. This study illustrates how commercial off-the-shelf optics may be employed in space exploration cameras. The most frequent loads on optical systems are self-weight, which is caused by gravitational orientations, vibration, and temperature ranges, which is caused by exposure to quickly changing temperatures ranging from extremely cold to very hot and during launch. This study describes the modulation transfer function (MTF) performance of the selected lens assembly, followed by a top-down method to disassemble the optical parts of the commercial lens housing. The lens enclosure was designed and modelled by retaining the interelement lengths determined from physical measurement. The performance of the lens barrel and optical elements have also been validated in ANSYS using a quasi-static loading and thermoelastic condition. As a post-performance assessment owing to stress and vibrations during satellite launch, a tilt-separation simulation was conducted. As a result of this research, it has been demonstrated that optomechanical design of commercial lens assembly is feasible for use in space application camera systems.

Keywords: Optomechanical Design, Quasi-static analysis, Disassemble of optics, Sensitivity analysis, Thermo-elastic analysis.

1. Introduction

The camera's imaging optics, detector, and processing circuitry all contribute significantly to the quality of the picture frame. The basic objective of imaging optics is to transmit the reflected light from object space to the sensor's image plane. The spectral range to be covered, the overall field of view (FOV), and the aperture size all affect the choice of imaging optics for a satellite mission. Optical instrument design is a highly integrated process that embraces several technical domains. The optical system used in satellite technology should be able to withstand a broad range of environmental pressures experienced during component manufacturing, launch vibrations, and space operation, including mechanical and thermal stresses as well as environmental radiation. As a consequence, it is necessary to pay special attention to opto-mechanical design in order to ensure that these pressures have no effect on the shape or position of the system's optical components and that the system meets its performance criteria [1-5].

Complete optomechanical research is needed to anticipate high-performance lens assembly. This necessitates the results of finite element analysis (FEA) being processed suitably using optical analysis software. Extensive research studies on optomechanical design have been conducted, starting with the optical designer's lens design and concluding with a framework that keeps the lens's numerous components in appropriate axial and radial alignment. As a response, the lens must be installed in such a manner that it maintains its degree of perfection while being used. Additionally, when extremely tight tolerances and a wide thermal temperature range exist, the mechanical designer endorses an extremely low residual distortion using a sub-cell approach with elastomeric bond material to reduce axial and radial stresses [6-9]. The founding of (Pawar et al., 2013) may be utilised to analyse multi-element lens assemblies with optical performance for spaceborne applications, as well as to construct very rigid, thermally and mechanically stable optical mounts [10]. The Micro Yield Stress (MYS) requirements were fulfilled in the specifics of the lens assembly's finite element analysis and in the lenses' performance stress summary under static loading and temperature fluctuation boundary conditions. While there are various research studies on lens installation with the barrel, this field is always evolving in order to uncover more effective design solutions. When it comes to gluing a lens to a barrel, there are two important aspects to take into account. First, its position must stay consistent amid external conditions such as shock, vibration, or temperature fluctuations, and must first be securely and accurately kept in place. Second, it must be held in such a manner that it does not break and that the stresses on the components do not produce distortions or deformations in the components themselves [11].

Dismantling a commercial lens assembly is one of the major parts of the study, which describes the technique to be followed. This study also describes the optical analysis before and after the assembly, as well as the overall performance of the assembly. To meet the requirements of comprehending the camera in a short period of time while also achieving the greatest possible reduction in size and weight, it was appropriate to use readily available off-the-shelf optics with the understanding that these optical elements would be used in earth observation rather than developing custom optics. In order to achieve the original performance, it is possible to reduce the size of this lens assembly by detaching its lens parts from their existing housing and reconstituting them in a newly built compact housing. Aside from that, the paper proposed the use of the ANSYS software for the design modelling of optical lenses as well as the survivability study of optical lenses in a space environment. The manuscript highlights the major issues like stress on glass, glass strength, thermoelastic effects, and distortion associated with mounting lenses.

1.1. Product description
Figure 1 shows the tiny 20 mm length of the f/2.8 Alpha E-mount Lens. It has a 20 mm focal length, which categorises it as a wide-angle lens. Its maximum aperture of f/2.8 enables low-light capture and shallow depth of focus shooting. The lens assembly's physical measurements are approximately 20 mm in length and 63 mm in diameter. It's small and sturdy, measuring less than an inch long and composed of an aluminium alloy. The lens assembly weighs approximately 70 grams with the mechanical housing [1].

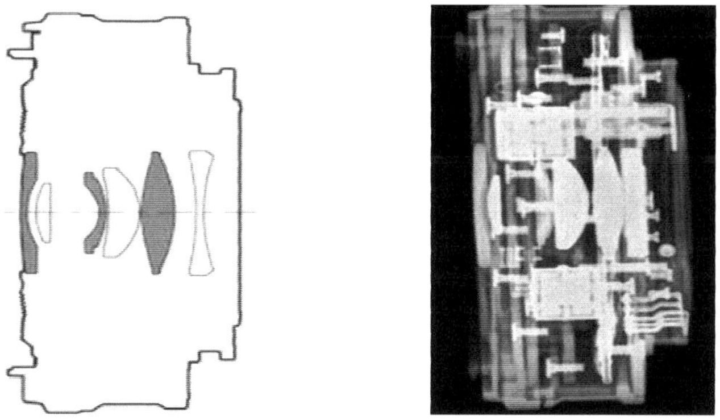

Figure 1. Commercially procured lens assembly with its x-ray image.

2. Dismantling procedure

2.1. Initial tests

Initial tests were performed to establish a baseline for comparisons after dismantling and reassembling the system. Initially, the modulation Transfer Function (MTF) for f/2.8 and infinity focus mode was measured in order to assess the lens assembly's performance. The MTF was calculated by attaching the lens assembly to a ILCE Alpha 3500J camera and using a white LED panel with Knife edge Blades as a target as shown in figure 2(a). The back focal length was also measured using an autocollimator, which used an identical combination of target, USAF resolution chart, and objective lenses. The test results are obtained by capturing image from an infinite distance of 380 mm which is technically considered the twenty times of the focal length and aperture stop set as f/2.8. The captured image has been taken to the software for computing MTF by selecting slant edge as area of region which gives the result as 63.1 cy/mm with Imatest SFR plus Test Chart as shown in figure 2(b).

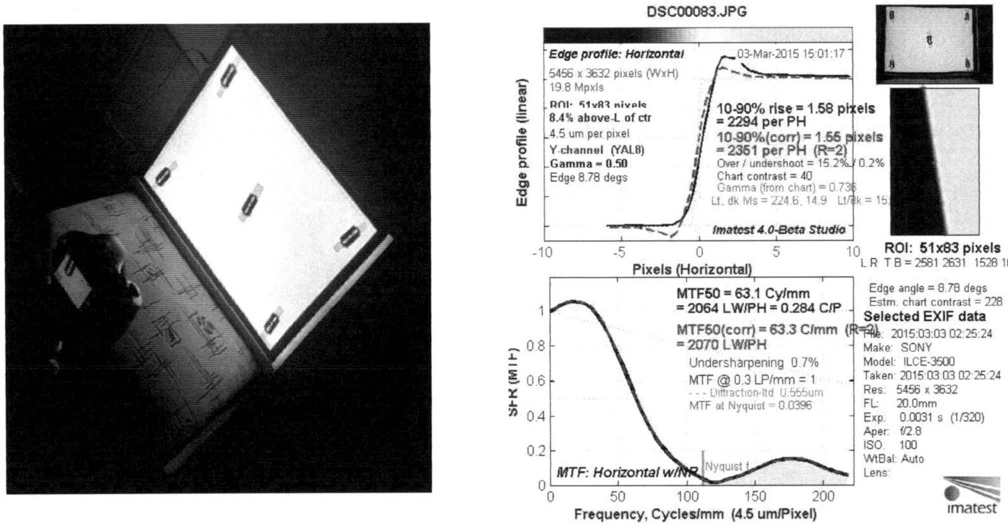

(a) (b)

Figure 2. (a) MTF test setup, (b) Imatest result for MTF.

2.2. Sequential Procedure

Tools and accessories required for dismantling the lens assembly are as follows;

- Micro Screwdriver set
- Nose head and flat head pliers
- Paper tape
- Adjustable Spanner with two legs for lock ring opening
- Solvent Zorrik-88
- Alcohol and clean rust-free cloth
- Digital camera

To begin, calibrate height gauge using a surface plate and dial gauge to obtain the separation distance between consecutive lens elements as illustrated in figure 3. Apply a tiny amount of solvent on the collar face from the outside for 4 screws. After the solvent has been administered, leave the assembly for 15 minutes. Optical components should not be sprayed with solvent. Do not spray directly on the skin. With the help of a thin stick, apply the solvent. Wear gloves when handling any optical components. Remove the three PCB mounting screws after opening the flange. Figure 4 depicts the technique.

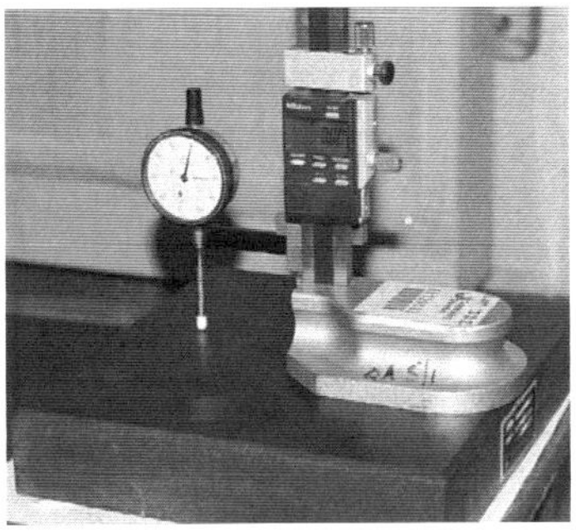

Figure 3. Calibration of dial gauge indicator on the surface plate.

Next, remove screws for takeout the 6th lens element housed with plastic mount from rear end of the lens assembly. There is no access to screw removal space along the diametrical side of the 6th element mounting so it is essential to drill into the barrel and remove 2 no. of screws. The process is shown in figure 5. By doing so, it has a clear passage to remove the lens 6 from the dowel pin. At this stage extra precaution has to be ensured that tools can damage optical components if not gripped properly. Use optimum force for dismantling mechanical components. The height measurement of this lens assembly with the reference set at the front-end hood with the use of surface plate and dial gauge indicator, as shown in figure 6, was made to get separation distance between consecutive lens elements. The total height from the 6th element to the front-end reference is 19.59 mm. After removing the 6th element, the total height measurement is 18.65 mm.

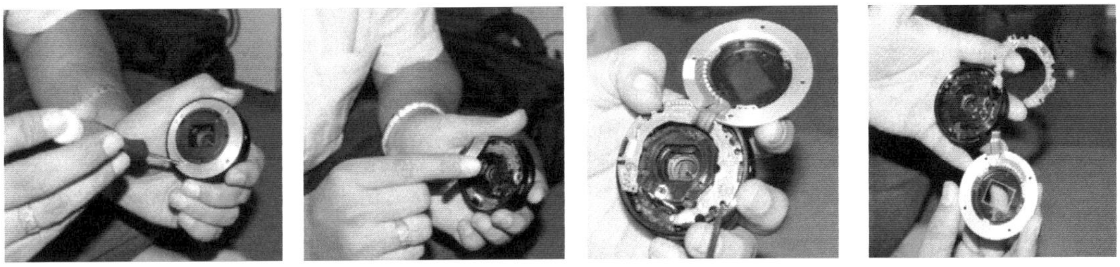

Figure 4. Application of solvent on the screws and Removal of Rear glass window with its mount.

Figure 5. Removal of screws and making the access for screw removing.

Figure 6. Instrument setup for the measurement of height as well as consecutive elemental separation

There is no access provided to take out the central lens group from the housing, it is necessary to remove the casing with the help of hand drill. Figure 7 shows the central lens group is taken out by machining and the remaining portion of the barrel. Remove the three screws from the central lens group after applying solvent on them and leaving for 10 minutes. The 1st and 2nd lens group separated from the central lens group and the remaining central lens group has 3rd, 4th and 5th elements. After Removal of two more screws located at the central lens group, separate the aperture stop attached with diaphragm blade with stepper motor as depicted in figure 8 and figure 9. There are more than three lenses remaining in the central lens group in which one of the front lenses is mounted by three-point RTV glue. Remove the RTV from all three sides with the help of a sharp tip tool to have the 3rd lens element from the central lens group as shown in figure 10. Here there is no access to remove lens 4 from central lens housing, moreover lens 5 is burnished from the edge of the lens. So, a decision has to be made to remove the burnishing of the 5th lens element with the help of a bench top grinder.

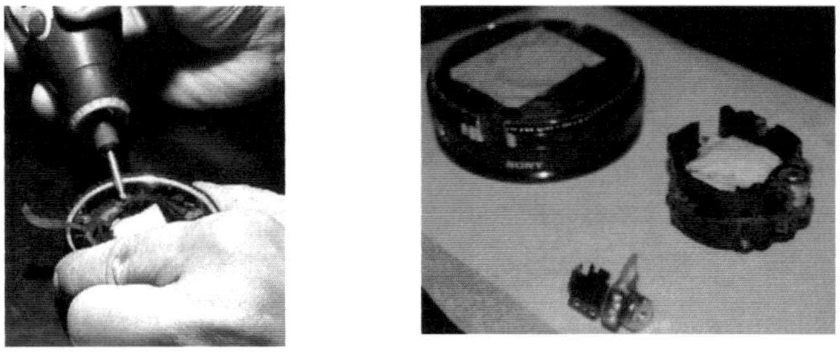

Figure 7. Removal of central Lens group from the barrel

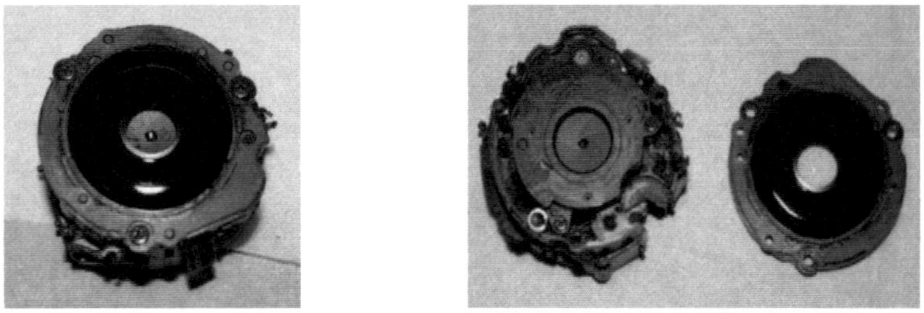

Figure 8. Removal of 1st and 2nd Lens elements from the central lens group

Figure 9. Separation of aperture stop from the lens housing

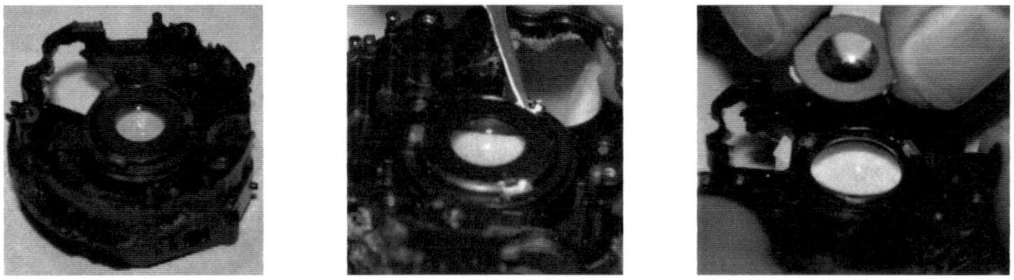

Figure 10. Separation of 3rd lens from the lens group.

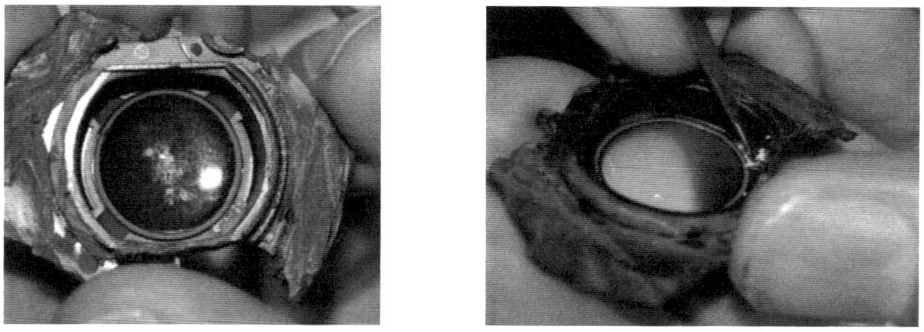

Figure 11. After deburring, the 4th and 5th lens elements are taken out

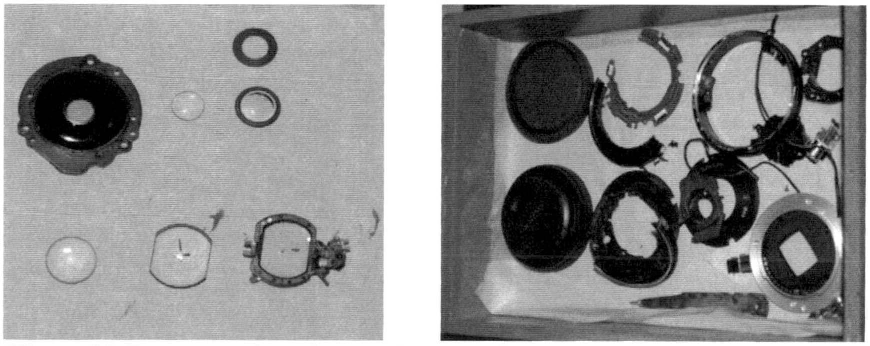

Figure 12. Demonstration of Lens elements after removal from the entire barrel and remaining components obtained from the dismantled lens assembly.

After removing the burnishing, the 5th lens element is taken out and simultaneously finding the access of removing the 4th lens element from the central lens group. Here now repeat the process described above to remove the 4th lens element from the housing as depicted in figure 11. Here there is no access to remove lens 1 and lens 2 from front lens housing, moreover lens 1 is burnished throughout the periphery of the lens. So, a decision has to be made to remove the burnishing of the 1st lens element with the help of a bench top grinder. Figure 12 depicts the dismantled optical lenses from the housing of the assembly along with the remaining parts. At last, with the help of a calibrated weighing machine, accuracy up to 1 gms, individual lens element's weight is measured which is shown in Table 1.

Table 1. Individual optical elements weight.

Sr. No.	Optical Elements	Weight (grams)
1	Lens 1	1.03
2	Lens 2	0.35
3	Lens 3	0.45
4	Lens 4	1.6
5	Lens 5	2.75
6	Lens 6	1.55
7	Thermal Filter	1.5012
	Total Weight	9.2312

3. Optomechanical Design

A popular mounting method for lenses is to install them into metallic housing that have precisely-made surfaces that make connection with the polished optical surfaces of the lenses. To keep the lens in position, additional spacers and retainers come into contact with the lenses as well. This document outlines both the art and the science that are employed in the design of glass-to-metal interfaces, as well as the results of such efforts [12].

3.1 Material Selection
Stress and strain are induced in the lens elements in a number of ways when a lens barrel is added to the lens. Due to the fact that the lens barrel and optical element are not constructed of the same material, a change in temperature will cause the components to expand and contract in distinctly different ways. It is possible to choose from a number of materials in order to minimise thermal expansion; however, sacrifices must be made between performance, weight, cost, and manufacturability. The selection of a lens barrel material is a critical step in ensuring that the optical and mechanical integrity of the lens system is maintained under all operating and non-operating situations. It is also necessary to consider the material's reactivity to stress and vibration in the optical system. It will be necessary to apply a force from the lens barrel in order to reduce the axial and radial strains that occur during temperature and vibration extremes. It is possible that this force will need to be significant in order to maintain proper axial and radial alignment during stress and vibration; thus, the ultimate strength of the glass must also be determined in order to avoid a fracture in the lens element [13].

Alloys such as 6061-T6 and 2024-T4 are the most commonly seen. Although aluminium has a high thermal expansion coefficient when compared to glass, which can result in image degradation and breakage. Despite this, its low cost, ease of machining, high conductivity, and light weight make it a popular choice among optomechanical engineers due to its low cost, ease of machining, high conductivity, and lightweight. For example, the qualities of different materials, such as stiffness, density, yield strength, thermal conductivity, and coefficient of thermal expansion, are taken into consideration while constructing the barrel. Lens assembly consists of the following major components i.e.; barrel, lens elements, spacers, aperture stop, adhesives. The lens barrel, aperture stop, optical elements, and thermal filter are made of Al6061-T6, NBK-7, and fused silica respectively. Table 2 shows the different properties of the materials used in this study [14].

Table 2. Mechanical properties of various materials.

Materials	Young's Modulus (MPa)	Poisson-Ratio	Density (Tonne/ mm^3)	Yield Strength (MPa)	Specific Heat (J/Kg K)	Thermal Conductivity (W/mK.)	Coefficient of Thermal Expansion (x 10^{-6}/K)
AL6061T6	0.7E+5	0.33	2.70E-09	275	896	167	23.6
NBK-7	0.82E+5	0.206	-	82	858	1.114	7.1
EC2216B/A	0.03E+5	0.48	1.25E-09	5	-	0.3946076	100
Fused silica	0.73E+5	0.17	2.2E-09	50	-	1.3	0.55

3.2 Design criteria
For the optomechanical design, a combination of static and dynamic loads must be given as an equivalent static load, known as "Quasi-static" load. The design will be approved based on simulations of quasi-static and thermo-elastic analyses. 18.3g Grms would be the design's qualification criteria. In optomechanical vibration engineering, it was usual practice to presume that the structure damage was caused by 3-sigma acceleration. As a result, the proposed configurations were assessed using a Quasi-Static load of 55g Grms applied in the x, y, and z directions, respectively. The results are observed along the x-axis, considered as an optical axis. In addition to quasi-static loads, thermal in-equilibrium conditions generated by operating in a zero-gravity environment will be dominant. The assembly must be examined using thermo-elastic simulation, which simulates the influence of a +10°C heat gradient on the structure, which can cause thermal strains [14].

3.3 Mechanical Design
The optomechanical assembly that is included in the proposed simulation model comprises lens elements and a barrel, and it is described below. In CREO 3.0 parametric modelling software, the lens elements were created from scratch. Using the geometrical dimension and separation readings obtained from the parametric measurement activity for the lens elements, the modelling process was carried out in order to create the final product. Due to the fact that each component differed in shape and physical

measurements, it was a difficult process to assemble all of them. CAD model arrangement of each lens, with the thermal filter in the foreground, is shown in figure 13. All of the system's parts' measurements and placements were maintained throughout the enclosure design process, by lockrings and spacers, respectively, to prevent movement. The single barrel construction was chosen, with a three-point flange mounting method, to allow the system to be suspended to the next connecting block. Because of its small physical dimensions of the lens assembly, the single-side flange mechanism. should be the most straightforward solution for housing the lenses. Figure 14 depicts a 3D cross section of the camera lens assembly, as well as the enclosed barrel construction of the lens assembly. Figure 15 depicts an expanded image of the optical assembly that showcase better understanding of the lens element, lockrings, and spacers.

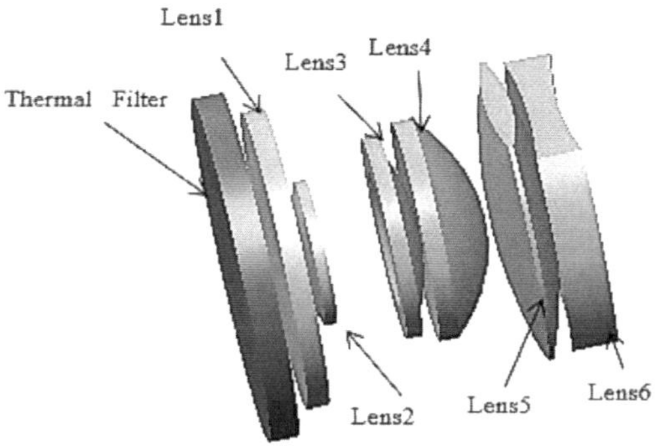

Figure 13. CAD model of optical elements

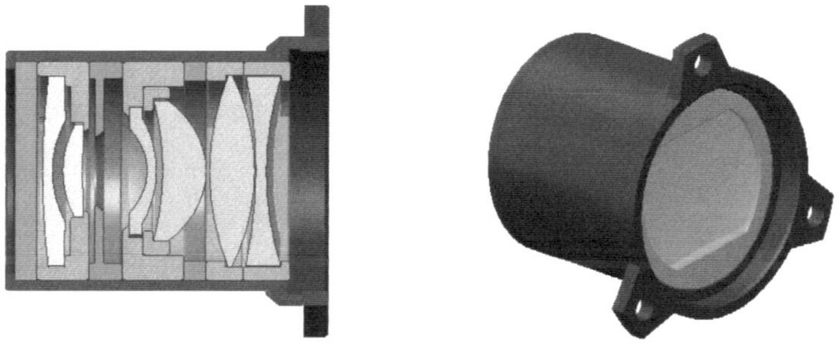

(a) (b)

Figure 14. (a) 3D cross section view of optomechanical assembly and (b) 3D model of optomechanical assembly

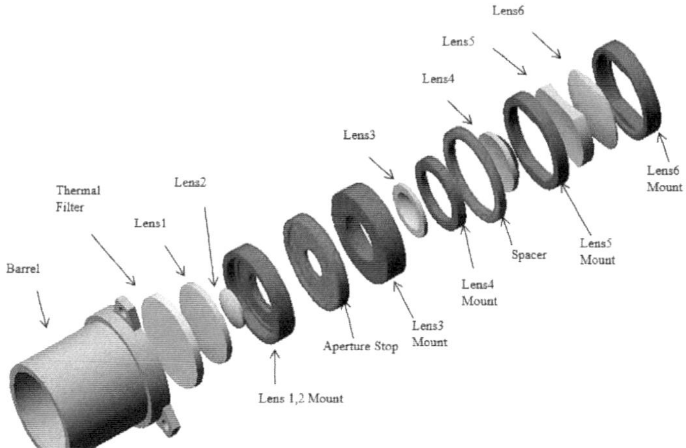

Figure 15. Exploded view of ruggedized lens assembly in sequence

4. Results and Discussions

4.1 Quasi-static stress analysis

ANSYS software was used to build simulation model of the lens assembly. The flange of the lens barrel has three holes, are fixed with six degrees of freedom (DOF). The DOF in solids are the X, Y, and Z translations and rotations. The lens assembly was examined for the quasi-Static load of 55g grms in the x, y, and z direction. Failure of optical elements is observed by maximum principal Stress theory having brittle characteristics subsequently von mises theory for lens barrel having ductile nature. The results of von mises stresses and maximum principle stresses are shown in table 3 and 4 respectively. The highest stress is obtained in lens 5 and was also under permissible range. Figure 16 depicts the stress distribution contours of optical elements and lens barrel in axial direction.

Table 3. Von mises stresses on lens assembly components.

Components	Von mises stress (MPa)		
	X - axis	Y - axis	Z - axis
Barrel	1.686	2.6976	2.4586
Spacer	0.1383	0.2748	0.2300
Aperture Stop	0.0820	0.1550	0.6110

Table 4. Maximum principle stresses on lens elements.

Components	Maximum principle stress (MPa)		
	X - axis	Y - axis	Z - axis
L1	0.037	0.0367	0.0437
L2	0.048	0.0307	0.0193
L3	0.036	0.3112	0.3187
L4	0.028	0.1442	0.1101
L5	0.4149	0.6549	0.6938
L6	0.1268	0.2704	0.2442
Thermal Filter	0.034	0.0203	0.0280

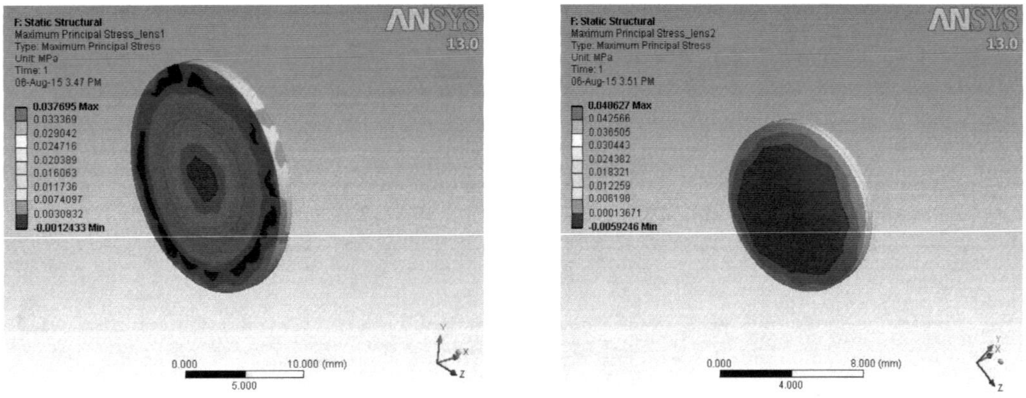

(a) (b)

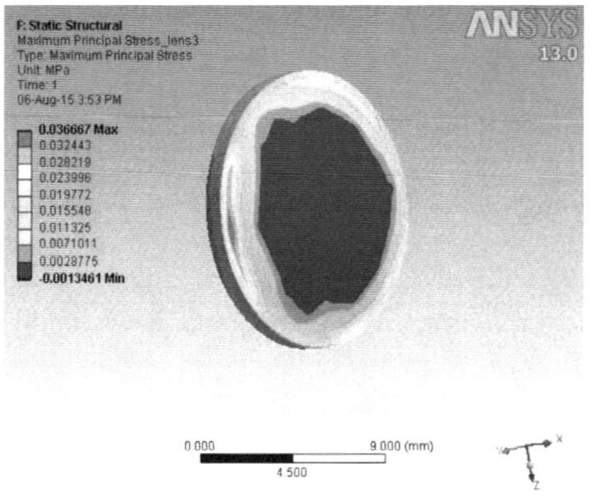

(c)

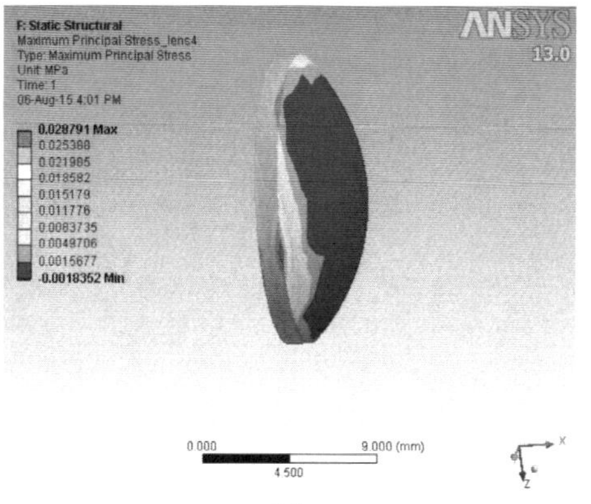

(d)

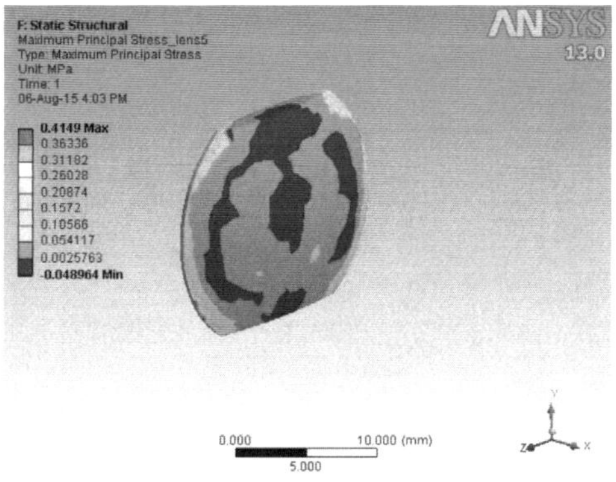

(e)

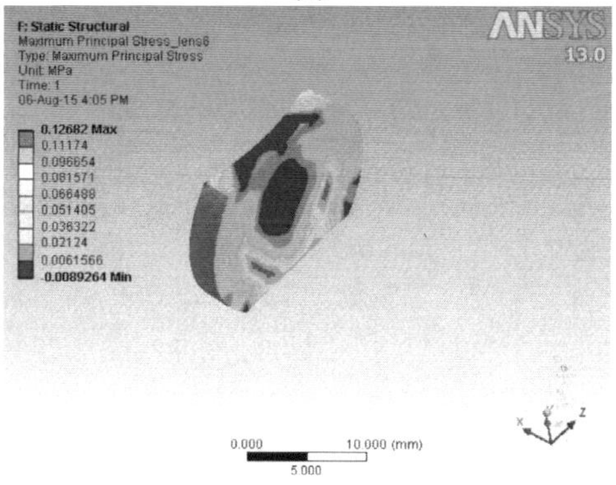

(f)

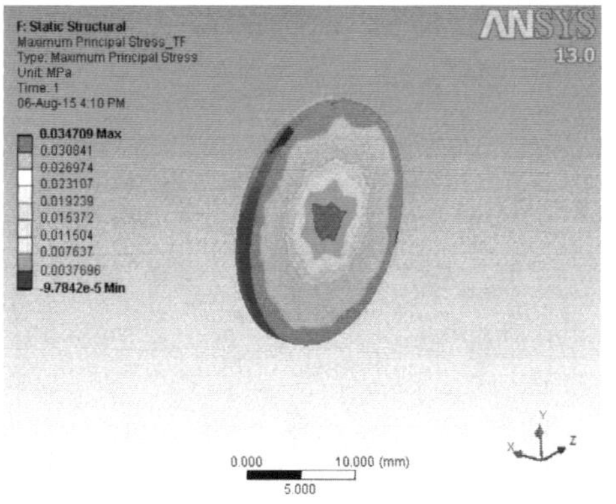

(g)

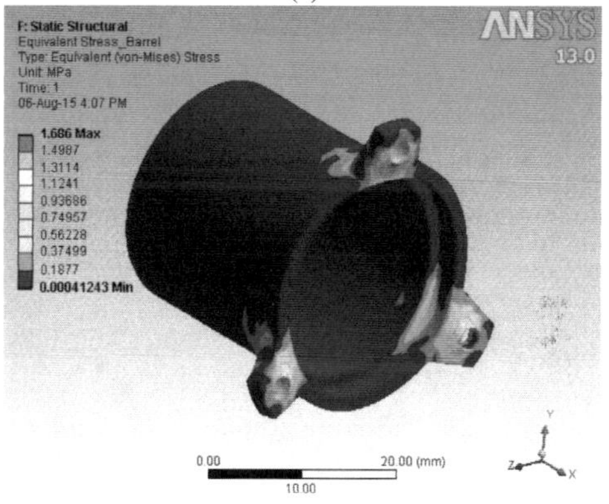

(h)

Figure 16. maximum principle stress distribution in (a) lens 1, (b) lens 2, (c) lens 3, (d) lens 4, (e) lens 5, (f) lens 6, (g) thermal filter and (h) von mises stress distribution in lens barrel during quasi-static loading.

4.2 Thermo-elastic stress analysis

The stress of the optical elements due to thermal loading was assessed using thermo-elastic stress analysis. A structure's internal thermal stresses occur due to the coefficient of thermal expansion when it is exposed to a temperature environment that is different from that of a stress-free structure. Thermo-elastic stresses were estimated at +10°C, which accounted for the lens assembly's non-operational testing conditions. Figure 17 (a) to (g) give the stress distribution of optical elements and (h) show von mises stress distribution of lens barrel in axial direction under thermo-elastic loading. The highest stress is obtained in lens 5 and was also under permissible range.

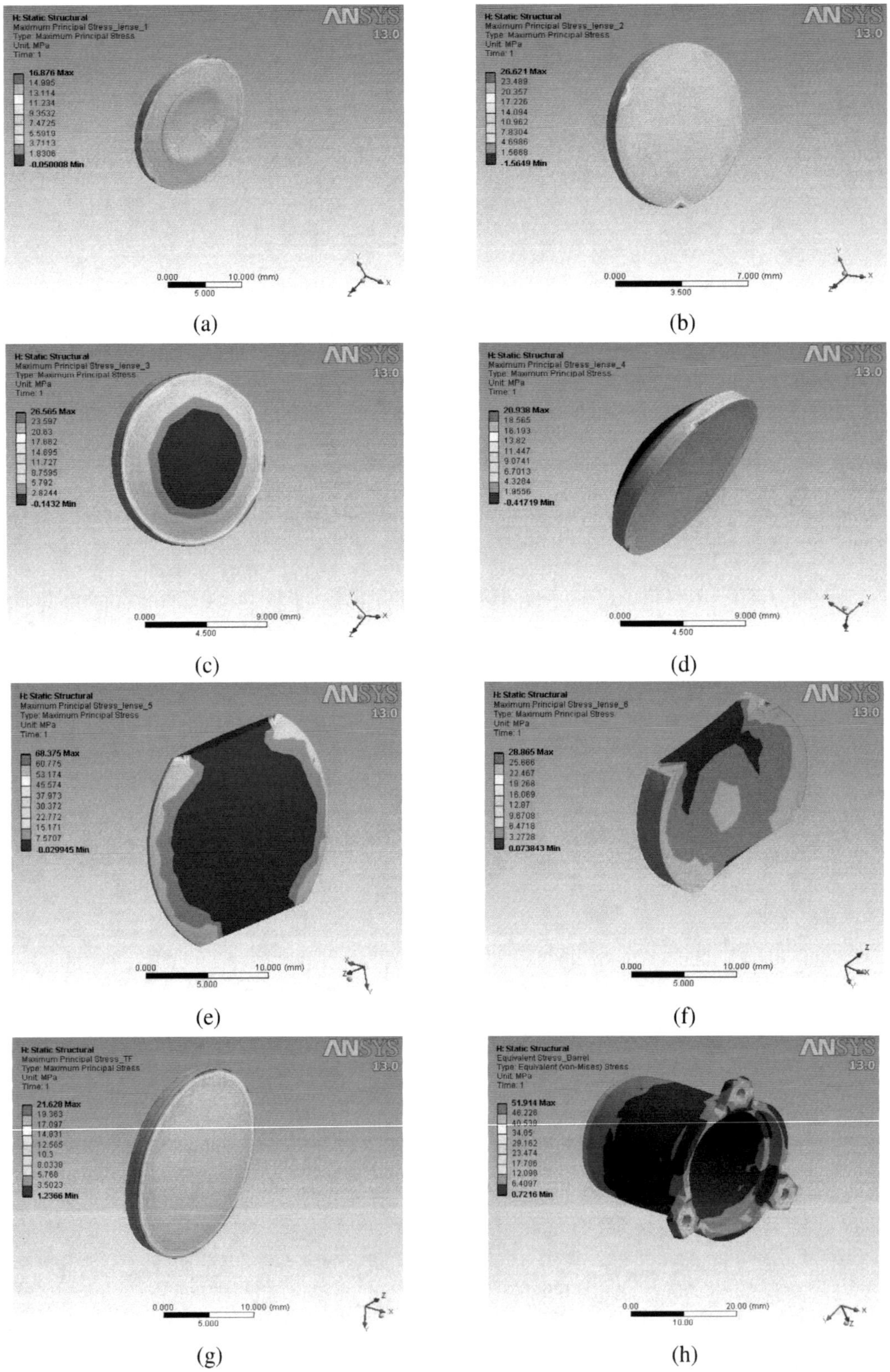

Figure 17. maximum principle stress distribution in (a) lens 1, (b) lens 2, (c) lens 3, (d) lens 4, (e) lens 5, (f) lens 6, (g) thermal filter and (h) von mises stress distribution in lens barrel during thermoelastic analysis.

4.3 Alignment sensitivity

The optical defects that can occur in the alignment of the optical elements will be highlighted in this section. Optical defects such as tilt, decentre and separation generated by lens manufacturing processes are reasonably controllable and measurable. The lens elements in most optical systems are of varying sizes. As a result, each lens cell in the barrel can be inclined and decentred with respect to the mechanical mount's datum axis. The barrel mechanical and mounting tolerances were necessary for the lens element's positioning, tilt, and decentre. A decentration or lateral misalignment of the lens element is indicated by a decentring or tilt of the element axis with respect to the optical axis [15].

All these lens elements have very low induced stress under the static 55 g loading condition. Hence, these lens elements are safe. However, these lens elements may deform under loading condition from their particular position which may cause either interference of two consecutive lens elements or tilting effect from the central optical axis position. These axial and lateral deformation of the lens elements may cause problems in acquiring the perfect image. To avoid these problems, one needs to check the effect of loading condition in individual direction over the central vertex points either side of the lens elements. These can be possible by checking tilt, separation and decentre of individual lens elements with the reference of ideal vertex point or axis. Table 5 indicates the pre analysis vertex positions of each lens element considering the first vertex point of the first lens as an absolute zero.

Table 5. Vertex point of individual Lens elements.

Lens Elements	Vertex Points	Pre_X	Pre_Y	Pre_Z
1	0	0	0	0
	1	0.8	0	0
2	2	1.682768	0	0
	3	3.302768	0	0
3	4	8.829998	0	0
	5	9.579998	0	0
4	6	9.870741	0	0
	7	13.38074	0	0
5	8	13.64805	0	0
	9	17.36805	0	0
6	10	17.45266	0	0
	11	18.30266	0	0

After completion of quasi-static analysis, the deformation at each vertex point of the lens elements will be derived in terms of x, y, and z coordinates and tabulated in table 6.

Table 6. Separation, Decentre and tilt analysis for Vertex point of individual Lens elements.

Lens Elements	Vertex Points	Separation (mm)	Decentre (mm)	Tilt (rad)
1	0	0.00011787	0.000328932	0.000575836
	1	0.00011794		
2	2	0.00012316	0.000228703	0.00229855
	3	0.00012324		
3	4	0.00010922	0.000228671	0.000494282
	5	0.00010919		
4	6	0.00010914	0.000216871	0.000546642

	7	0.00010898		
5	8	0.00011731	0.00016246	0.000577747
	9	0.00011759		
6	10	0.00011759	0.00012494	0.000568247
	11	0.00011753		

The tilt, separation and decentre of individual lens element vertex position are very less. i.e., under 1 micron. These results are very less significant and will not affect the optical performance of lens assembly. In this publication, the unique approach of optomechanical design was employed to qualify the quasi-static and thermal environments that impart component manufacturing processing. The complete optomechanical lens assembly components with optical elements are depicted in Figure 18.

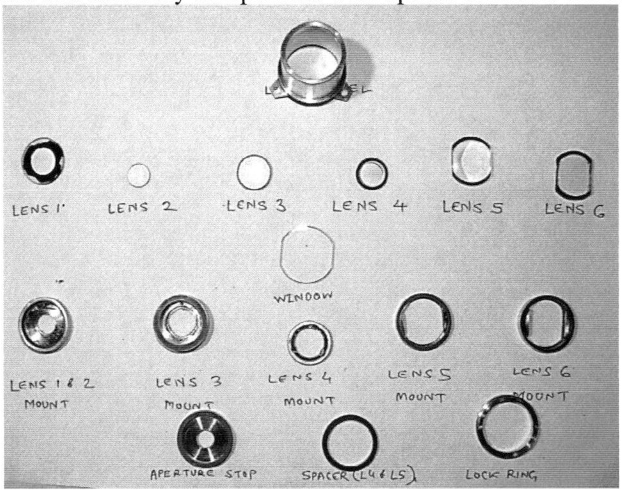

Figure 18. Actual optomechanical lens assembly components.

5. Conclusion

This article mainly envisaged the feasibility towards the space application camera through the ruggedization process of commercial optics. As a part of the study, identification of commercial optics based on the system parameter alongside its optical performance was also verified. A Sequential dismantling of optical elements was carried out successfully. Optomechanical lens assembly design has been developed using physical parameters of the elements. This optomechanical design was also verified to vibration and environmental conditions. The result obtained from the quasi-static analysis and thermoelastic analysis showcases the capability to withstand the loading condition alongside reduced effect of stresses on optical elements. Alignment sensitivity analysis has also been performed to check its physical parameters post analysis part. Manufacturing of optical mounts has been carried out as a part of successful results of optomechanical analysis. Through this research, it is quite evident that optomechanical design ensured the feasibility towards the space application camera.

Reference

[1] Shah M, Patel A, Patel K and Prajapati C 2018 *International Journal of Applied Engineering Research*, ***13***(12) 10276.
[2] Joseph G 2015 *Building earth observation cameras*. CRC press.
[3] Ahmad A ed. 2017 *Handbook of optomechanical engineering*. CRC Press.
[4] Yoder P and Vukobratovich D eds. 2017 *Opto-Mechanical Systems Design, Volume 2: Design and Analysis of Large Mirrors and Structures*. CRC Press.
[5] Schwertz K and Burge J 2012 *Field guide to optomechanical design and analysis*. Society of Photo Optical Instrumentation Engineers (SPIE).
[6] Bayar M 1981 *Optical Engineering*, **20** 181.
[7] Fischer R 1991 *Optomechanics and dimensional stability*. In: SPIE **1533** 27.

[8] Valente T and Richard R 1994 *Opt. Eng.* **33** (4) 1223–28.
[9] Jones G 1975 In: Baker L (Ed.) *Proc. of SPIE, Quality Assurance in Optical and Electrooptical Engineering* **0073**
[10] Pawar M, Raha B, Chandru B, Venkateswaran R and Krishna M 2013 *Int. J. Enhanced Res. Sci. Technol. Eng* **2** (7) 77–82.
[11] Maamar F and Boudjemai A 2020 *Advances in Space Research,* ***65***(10) 2263-75.
[12] Chase K and Burge J 2011 *Optomechanics: Innovations and Solutions.* International Society for Optics and Photonics **8125** 81250F.
[13] Harwell J 2010 *Lens barrel optomechanical design principles.*
[14] Shah M and Patel A 2020 *International Journal of Innovative Technology and Exploring Engineering (IJITEE)* **9**(4) 972-76
[15] Chang C 2007 *Optical Manufacturing and Testing VII.* In SPIE **6671** 307-14.

Investigation of surface roughness in multi point incremental sheet metal forming

Shafeek M[1], Narayanan Namboothiri V N[2], C Raju[3]
[1]Research Scholar, SoE, CUSAT, Kochi, Kerala, India
[2] Professor in Mechanical Engineering, SoE, CUSAT, Kochi, Kerala, India
[3]Associate Professor in Mechanical Engineering, CET, Thiruvananthapuram, Kerala, India

shafeek@cet..ac.in

Abstract. Surface quality is one of the key parameters in Multipoint Incremental Sheet forming (MPIF). This work focuses on finding the factors influencing surface quality in MPIF. Three different parameters namely speed, feed and step depth are considered for the study. The effects of the above input variables on surface quality was investigated by Taguchi's design of experiments and finally the optimum input parameters combination for minimum surface roughness is found. The most influencing parameter is found to be step depth followed by speed and feed. The findings were validated by a confirmation test.

1. Introduction

Conventional metal forming methods are most appropriate for high volume production due to the large cost associated with die designing and fabrication. ISF is a die-less forming method in which the sheet metal is firmly fixed on a fixture and a spherical or ball-ended tool is moved over the sheet metal part based on the program. The shape of the final part can be changed by changing the program which enhances the applications of ISF. ISF is mainly suitable for low volume and rapid prototyping. ISF is classified into two i) Single Point Incremental Forming (SPIF) and ii) Two-point incremental forming (TPIF). A spherical or ball-ended tool is used in SPIF and for TPIF uses a partial die for forming the sheet metal to the required shape. The time required to complete the process is a major disadvantage of these methods. So in this work, a multipoint tool with six balls of 10mm diameter is used to shape the sheet metal to the required shape. The objective of this work is to find out the most significant factors influencing the surface finish of the formed part.

The effect of input variables on surface roughness during SPIF was first investigated by Ham and Jeswiet [1]. Taguchi DOE was used and found that material thickness, tool size and their interaction significantly affect the surface roughness and formability. Various reviews are available for tracking the history and development of ISF [2,3]. DoE is a powerful tool for assessing the influence of process parameters on ISF. Input parameters like step depth, sheet thickness, speed, feed and diameter of the tool significantly affect the surface finish of the formed parts[4,5].

Durante et al.[6]compared experimental and analytical results of surface roughness in ISF. The initial model created for the spacing parameters and modified model for the amplitude parameters are found to have accurately predicting the surface roughness. Salah et al.[7]studied the influence of input variables on surface roughness in ISF. The major input parameters selected for study include vertical step depth, diameter of tool, speed and feed rate. Tool diameter and vertical step depth are found to have major impact on surface roughness. Mulay et al.[8]conducted high speed ISF on aluminium and steel sheets with the main objective of reducing the processing time and revealed that the surface quality is largely influenced by vertical step depth followed by feed rate. Ajay Kumar and Vishal Gulati [9] investigated the influence of input variables on surface finish in SPIF. The most dominant factor was found to be the tool diameter. Praveen et al.[10]used image processing techniques to evaluate the surface roughness (Ra) in ISF. The proposed method shows excellent classification accuracy. Yoganjaneyulu et al.[11]conducted multi sheet incremental forming using titanium alloy sheets. They studied the fracture behaviour and formability in multi sheet forming.

Dabwan et al.[12]conducted multiobjective optimization in SPIF using GRA coupled with entropy analysis. Ajay et al.[13]studies SPIF of titanium alloy sheets. The optimum conditions for minimum surface roughness, maximum wall angle and wall thickness are obtained when the feed rate is 1000mm/min, speed 150 rpm and step depth 0.1mm. Mohanraj et al.[14] using Taguchi method studied the input factors influencing surface roughness in ISF. AA3003-H18 sheets were selected for the study. The vertical step over is found to be the most influencing factor on surface roughness followed by feed rate and diameter of tool.

Few research works are available in MPIF[15–17]. In this study, three input parameters namely feed, speed and step depth are used to experimentally study their influence on surface roughness. Each input variable has three levels. Finally, the optimum set of process parameters for minimum surface roughness are found out and experimentally validated.

2. Materials and methods

Titanium grade 2 sheets of 0.5mm thickness and 150mmx150mm size was used for the experimental study. The experimental setup consists of a blank holder, a backplate and a multi point tool. The fixture is firmly fixed on the BFW BMV 45 CNC milling machine worktable. The multipoint tool having six 10mm diameter hardened steel balls is mounted on the machine spindle using a tool holder. Schematic diagram of multipoint incremental forming (MPIF) is shown in Figure 1. Figure 2 shows the experimental setup. A CNC program in order to form the sheet metal into conical shape was created by using CAM software and transferred to the machine. The tool follows the tool path based on the CNC program and deforms the sheet metal. The forming operation is carried out up to the fracture point of the sheet metal part. The input parameters selected based on previous works[17] are selected and their levels are shown in Table 1.

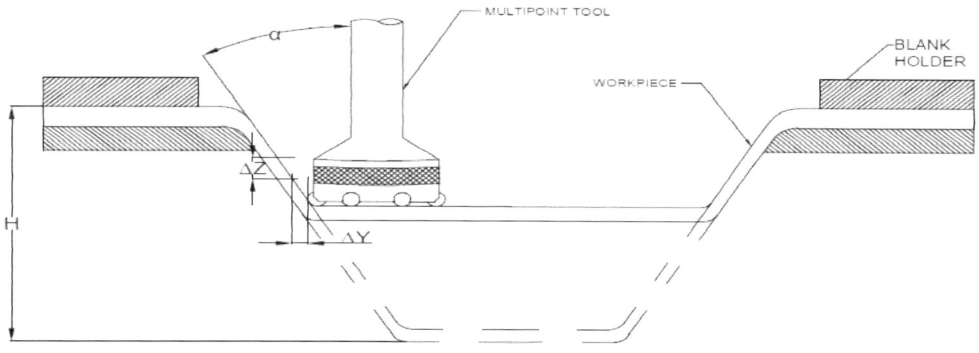

Figure 1. MPIF schematic diagram

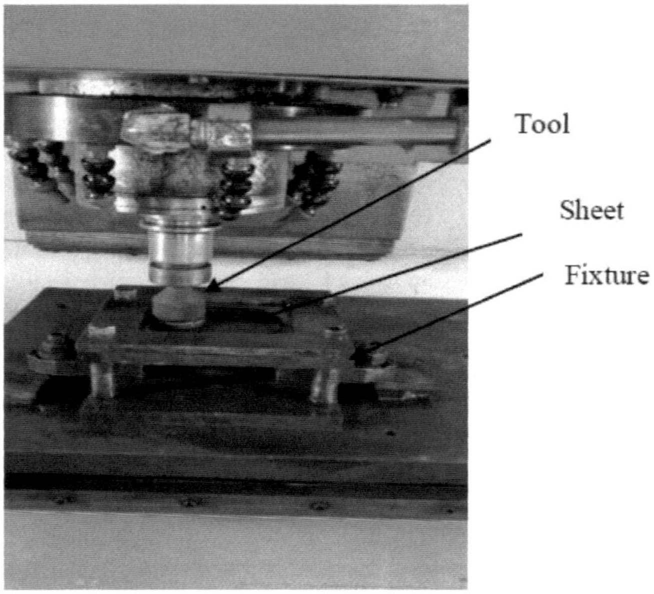

Figure 2. MPIF experimental setup

Table 1. Input variables and levels

Sl No	Factors	Factor Notation	Levels 1	2	3	Unit
1	Feed	A	300	450	600	mm/min
2	Speed	B	300	450	600	rpm
3	Step depth	C	0.2	0.4	0.6	mm

Total 9 experiments were conducted and coconut oil is used as the lubricant for all the experiments. The average surface roughness(Ra) of the formed part was measured by using Mitutoyo surface roughness tester SJ-210. The data collected are represented in Table 2.

Table 2. Experimental Data

Feed in mm/min	Speed in rpm	Step depth in mm	Ra in µm
300	300	0.2	0.5792
300	450	0.4	0.964
300	600	0.6	1.4194
450	300	0.4	0.9904
450	450	0.6	1.4665
450	600	0.2	1.0169
600	300	0.6	1.4686
600	450	0.2	0.7751
600	600	0.4	1.6108

3. Results and discussion

Analysis of Variance (ANOVA) helps to determine the significance of the factors affecting surface roughness. P-value less than 0.05 determine the significance of the factors at 95% confidence level. Tables 3and 4 show ANOVA for S/N ratio and means and it is clear that step depth is the most influencing factor on surface roughness followed by speed and feed rate. Increase in step size increases the surface roughness of the formed part. This occurs mainly due to increase forming force which leads to large waviness formed on the sheet metal part[18]. Increase in speed enhances proper lubrication and reduces the friction at the tool work piece contact zone which increases the finish of the formed part[19]. Feed rate only speed up the process and has least influence on the surface finish[5].

Table 3. ANOVA for S/N ratios

Source	DF	Seq SS	Adj SS	Adj MS	F	P
Feed	2	13.3366	13.3366	6.6683	637.73	0.002
Speed	2	15.9239	15.9239	7.9620	761.45	0.001
Step depth	2	56.7060	56.7060	28.3530	2711.56	0.000
Residual Error	2	0.0209	0.0209	0.0105		
Total	8	85.9874				

.

Table 4. ANOVA for Means

Source	DF	Seq SS	Adj SS	Adj MS	F	P
Feed	2	0.16352	0.163523	0.081762	21.31	0.045
Speed	2	0.23955	0.239547	0.119774	31.22	0.031
Step depth	2	0.85748	0.857479	0.428740	111.77	0.009
Residual Error	2	0.00767	0.007672	0.003836		
Total	8	1.26822				

Both response plots as shown in Figure 3 and response tables as shown in Tables 5 and 6 are employed to analyze the experimental results. The ranking and impact of each parameter can be easily understood from the response table. From Figure 3 and tables (Table 2 and 3) step depth (rank 1) is mostly influencing surface roughness followed by speed (rank2) and speed (rank 3). The best surface roughness values are obtained when the step depth is kept at level 1, the speed at level 2 and the feed at level 1.

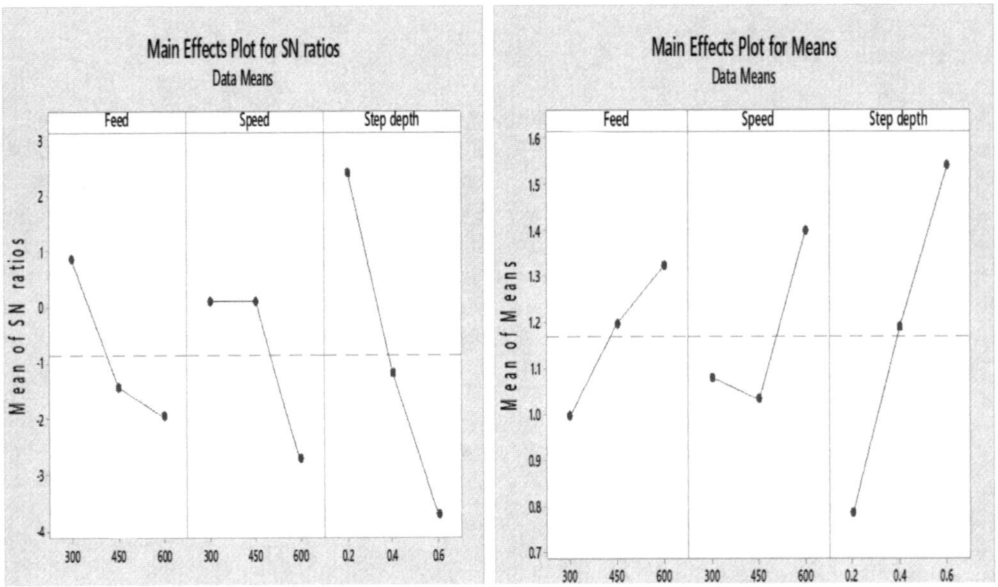

Figure 3. Main effect plot for S/N ratio and mean

Table 5. Response Table for Signal to Noise Ratios

Level	Feed	Speed	Step depth
1	0.84003	0.08763	2.38079
2	-1.43933	0.08432	-1.20729
3	-1.96448	-2.73572	-3.73728
Delta	2.80451	2.82334	6.11808
Rank	3	2	1

Table 6. Response Table for Means

Level	Feed	Speed	Step depth
1	0.9942	1.0795	0.7832
2	1.1953	1.0320	1.1892
3	1.3215	1.3994	1.5385
Delta	0.3273	0.3674	0.7554
Rank	3	2	1

4. Confirmation test

Comparisons between predicted and experimental results were conducted and Table 7shows the results. The percentage error between experimental and predicted surface roughness is 8.5% which shows good agreement between predicted and experimental results.

Table 7. Confirmation test

Response	Optimum set of parameters	Predicted optimal value	Actual value	% error
Surface Roughness(Ra)	A1B2C1	0.4687	0.5125	8.546341

5. Conclusion

Through this work, an effort has been made for finding the most influencing parameter in the MPIF process.

- Vertical step depth is found to be the most significant factor affecting surface finish followed by speed.
- Feed rate gives the least significance compared to step depth and speed.
- Step depth 0.2mm, speed 450 rpm and feed rate 300mm/min are found to be the optimum parameter combination for minimum surface roughness.
- The confirmation test with optimum input parameters were performed. The predicted and actual surface roughness values also are found to have in good agreement.

6. References

[1] M. Ham, J. Jeswiet, Single point incremental forming and the forming criteria for AA3003, CIRP Ann. - Manuf. Technol. 55 (2006) 241–244. https://doi.org/10.1016/S0007-8506(07)60407-7.

[2] C.V. Ajay, C. Boopathi, P. Kavin, Incremental sheet metal forming (ISMF): A literature review, Int. Conf. Mater. Manuf. Mach. 2019. 2128 (2019) 030012. https://doi.org/10.1063/1.5117955.

[3] S. Gatea, H. Ou, G. McCartney, Review on the influence of process parameters in incremental sheet forming, Int. J. Adv. Manuf. Technol. 87 (2016) 479–499. https://doi.org/10.1007/s00170-016-8426-6.

[4] C. Raju, C. Sathiya Narayanan, Application of a hybrid optimization technique in a multiple sheet single point incremental forming process, Meas. J. Int. Meas. Confed. 78 (2016) 296–308. https://doi.org/10.1016/j.measurement.2015.10.025.

[5] A. Baruah, C. Pandivelan, A.K. Jeevanantham, Optimization of AA5052 in incremental sheet forming using grey relational analysis, Meas. J. Int. Meas. Confed. 106 (2017) 95–100.

https://doi.org/10.1016/j.measurement.2017.04.029.

[6] M. Durante, A. Formisano, A. Langella, Comparison between analytical and experimental roughness values of components created by incremental forming, J. Mater. Process. Technol. 210 (2010) 1934–1941. https://doi.org/10.1016/j.jmatprotec.2010.07.006.

[7] S.B.M. Echrif, M. Hrairi, Significant parameters for the surface roughness in incremental forming process, Mater. Manuf. Process. 29 (2014) 697–703. https://doi.org/10.1080/10426914.2014.901519.

[8] A. Mulay, B.S. Ben, S. Ismail, A. Kocanda, C. Jasiński, Performance evaluation of high-speed incremental sheet forming technology for AA5754 H22 aluminum and DC04 steel sheets, Arch. Civ. Mech. Eng. 18 (2018) 1275–1287. https://doi.org/10.1016/j.acme.2018.03.004.

[9] A. Kumar, V. Gulati, Experimental investigation and optimization of surface roughness in negative incremental forming, Meas. J. Int. Meas. Confed. 131 (2019) 419–430. https://doi.org/10.1016/j.measurement.2018.08.078.

[10] P.K. Gandla, V. Inturi, S. Kurra, S. Radhika, Evaluation of surface roughness in incremental forming using image processing based methods, Meas. J. Int. Meas. Confed. 164 (2020) 108055. https://doi.org/10.1016/j.measurement.2020.108055.

[11] G. Yoganjaneyulu, C. Raju, N. Manikandan, C. Sathiya Narayanan, Investigations on multi-sheets single point incremental forming of commercial pure titanium alloys, Mater. Manuf. Process. 00 (2020) 1–8. https://doi.org/10.1080/10426914.2020.1753882.

[12] A. Dabwan, A.E. Ragab, M.A. Saleh, A.M. Ghaleb, M.Z. Ramadan, S.H. Mian, T.M. Khalaf, Multiobjective optimization of process variables in single-point incremental forming using grey relational analysis coupled with entropy weights, Proc. Inst. Mech. Eng. Part L J. Mater. Des. Appl. 235 (2021) 2056–2070. https://doi.org/10.1177/14644207211020401.

[13] C. Veera Ajay, S. Elangovan, C.T. Justus Panicker, Studies on single point incremental forming of Ti-6Al-4V alloy for aerospace applications, Mater. Today Proc. 22 (2019) 1967–1976. https://doi.org/10.1016/j.matpr.2020.03.162.

[14] M. Murugesan, J.H. Yu, K.S. Jung, S.M. Cho, K.S. Bhandari, C.W. Lee, Optimization of Forming Parameters in Incremental Sheet Forming of AA3003-H18 Sheets Using Taguchi Method, Materials (Basel). 15 (2022) 1–16. https://doi.org/10.3390/ma15041458.

[15] K. Ramkumar, G. Paulraj, K. Elangovan, C. Sathiya Narayanan, Forming limit diagram, void analysis, strain distribution and surface roughness for SS430 sheets during multipoint incremental forming, Arch. Metall. Mater. 63 (2018) 1709–1714. https://doi.org/10.24425/amm.2018.125096.

[16] K. Ramkumar, N. Baskar, K. Elangovan, C.S. Narayanan, K.A. Selvarajan, C.P. Jesuthanam, Comparison of Multi Point Incremental Forming Tool with Single Point Incremental Forming Tool Using FLD, Fractography and 3D-Surface Roughness Analysis on Cr/Mn/Ni/Si Based Stainless Steel, Silicon. (2020). https://doi.org/10.1007/s12633-020-00468-4.

[17] R. Kathalingam, B. Neelakandan, E. Krishnan, S.N. Chinnayan, S. Arangulavan, V. Muthukrishnan, Comparison of incremental sheet metal forming process using newly designed multipoint tool with the existing single point tool by optimization and characterization methods on austenite stainless steel 202, Proc. Inst. Mech. Eng. Part B J. Eng. Manuf. 235 (2021) 1298–

1308. https://doi.org/10.1177/0954405420986098.

[18] X. Li, K. Han, Z. Li, D. Li, L. Zhang, One novel method to improve surface quality in incremental sheet forming, Procedia Eng. 207 (2017) 842–847. https://doi.org/10.1016/j.proeng.2017.10.839.

[19] A. Kumar, V. Gulati, P. Kumar, Investigation of Surface Roughness in Incremental Sheet Forming, Procedia Comput. Sci. 133 (2018) 1014–1020. https://doi.org/10.1016/j.procs.2018.07.074.

A review on classification of trains for evaluate the railway network capacity of various speed trains

Manoj Mittal and Anand Kishore Chaturvedi

Department of Mechanical Engineering, Rajasthan Technical University, Kota (Rajasthan)-324001, India

mmittalgecj@gmail.com, chaturvedi101@gmail.com

Abstract. Railway is a very complicated decision making sector in a realistic way with numbers of constraints. Several thousands of trains are running over thousands of kilometers. The aim of this paper is to measure the rail network capacity by efficient utilization of existing resources of railway network by analytical model. In this model headway is reduced for variable speed trains so that track blocking time and track running time will be reduced. A case study of 14 sections with 5 corridors have analyzed by CIPLEX solver to develop the generic solution for optimizing rail network capacity to solve real problem. These models are useful in planning phase and can evaluate the entire network of railway.

Keywords: Analytical Model; Capacity analysis; Track blocking time; Track running time; Rail network capacity; Headway

1. Introduction

The railway consists both freight and passengers moving approach in various countries. The railway sector is a large investment industry in equipment and employees. Railway operation are very complicated decision making processes due to the need to schedule several hundred trains over thousands of kilometers distances. The railway business in many countries demands more new features in the planning and development process and is intense to execute better modeling and result. The present circumstances provide great incentives for pursuing better optimization and control strategies for the operation of the rail transportation system. A object consists three main steps: making the model, solving the problem and implementing the result. Dirneberger (2002) gave model for improving the service of the railroad industry [7]. Determining by the train scheduling, how many trains to run from origin to destination, and route of each train, the train arrival and departure times for each station at which it stops [14]. The train schedule must satisfy several realistic constrains and achieve the minimum cost of transportation [17]. The railway problem is a very complicated large scale, various dimensional problems which contains trillions of decision variables for finding optimum results.

2. Literature review of existing approaches on train network capacity

According to A. Jaimili (2018), The method of computation of capacity in single-track railway lines is different from double or more track ones [2]. Despite the normally larger number of tracks, stations and larger nodes may represent capacity bottlenecks of the networks as per Armstrong J., Preston J., 2017. Atieh Kianinejadoshah and Ricci Stefano (2021) focused on identification of parameters affecting the capacity of both lines and nodes and the quantification of these effects by sensibility analysis on the identified fundamental parameters[1]. Pengling Wang et al. gave many different things including locomotive utilization, yard capacity, crew availability, track structure, and traffic levels all play a role and combine to determine railing efficiency [7].

The details of past research on train scheduling and capacity analysis are given below.

Table 1. Existing approaches on train scheduling.

Sr. No.	Approaches/ Methods / Techniques	Description
1.	Capacity Analysis	The high density and long route rail services of the same track was done by humans not computers in 1970s.
2.	Exact solution Techniques	The public railway sector has improved the awareness of the importance of service for passengers.
3.	Heuristic solution Techniques	This feature provides extensive computational result on real world.

2.1 Researcher's contribution for capacity analysis

Table 2. Capacity Analysis.

Techniques			**Capacity Analysis**	
S. No	Researcher	Year	Description	Applications
1	Abdekhodaee et al.	2004	Integrated the schedule of rail network by studied mixed integer programming models.	Utilised the operations in a coal terminal system with limited capacity.
2	Kozan and Burdett	2005	capability and pricing are two major inputs.	Train access charging tactic developed for getting railway capacity.
3	Burdett and Kozan	2006	Developed a model of capacity analysis techniques and methodologies	These approaches can be used to enumerate the possible interaction affects in interrelated corridors.

2.2 Researcher's contribution for Exact Solution Techniques

Table 3 Exact solution Techniques.

Techniques			**Exact solution Techniques**	
S. No	Researcher	Year	Description	Applications
1	Ingolotti et al.	2005	Made a model of the Decision Support System (DSS).	Quickly and efficiently solve single track railway scheduling problem.
2	Linder and Zimmermann	2005	Gave a model by using Mixed integer programming for train scheduling.	Optimized the railway networks in Germany and Netherland for minimizing the running cost.
3	Zhou and Zhong	2004	Focused on high speed train for minimizing the waiting time and travel time.	Uses for high speed railway in China for double track by multi objective criteria.

2.3 Researcher's contribution for Heuristic Solution Techniques.

Table 4. Heuristic solution Techniques.

Techniques			**Heuristic Solution Techniques**	
S.No.	Researcher	Year and Ref. No.	Description	Applications
1	Yuri N. Sotskov et.al	2012 [18]	Developed a model for single track railway.	Bottleneck shifted for optimizing the train scheduling for single track railway.
2	Pengling Wang et.al	2016 [7]	Analysed a heuristic model for rescheduling trains by Petri nets.	In this study work focused on rescheduling the trains.
3	Peng-Sheng You et.al	2013 [8]	Developed heuristic algorithm	Worked on train scheduling with parallel track.

2.4 Researcher's contribution for various issues

Table5: Based on various issues.

Techniques				**Based on various issues [30,32,33]**	
S. No	Researcher	Year	Ref. No.	Description	Applications
1	T wan Dollevoet et. al.	2014	[14]	Developed a framework for optimizing the train scheduling.	Maximum allowable delay provided to measure the optimum result.

2	Xiang Li et. al	2013	[16]	Multi objective train scheduling model gave.	Focused on the lowering the fuel consumption cost.
3	Xiaoming Xu et. al	2014	[17]	Studied for balanced train time tabling with optimized velocity for single line.	This model is used for non constant train velocity.

3. Classification:

Train scheduling is a major constraint of rail network capacity. It can be classified on the basis of past study in following parameters:

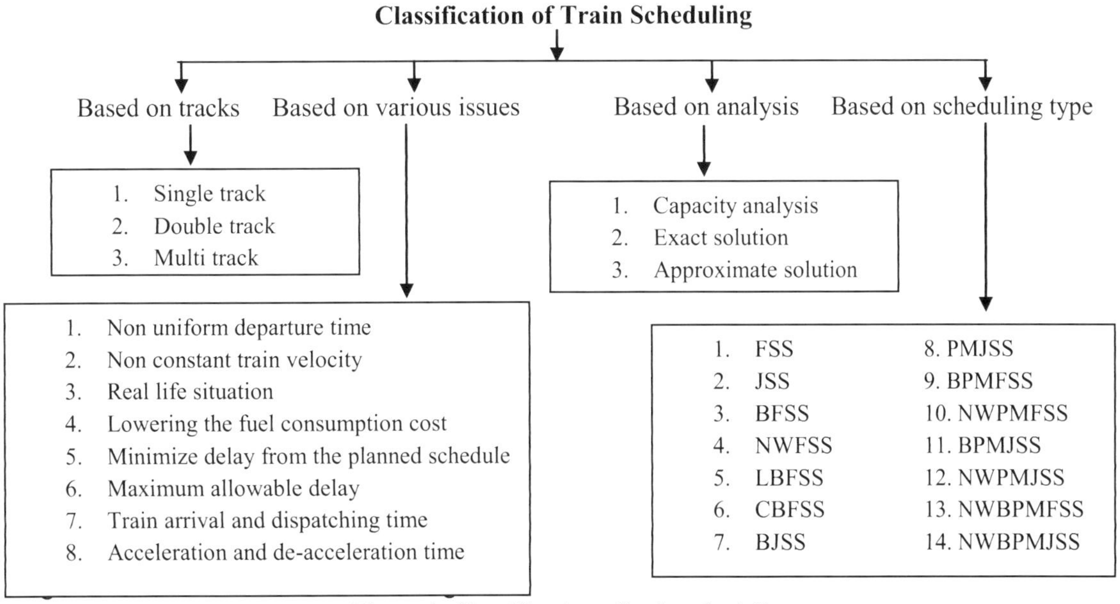

Figure 1. Classification of train scheduling

3.1. Based on tracks

Train scheduling classified on the basis of number of tracks as like one way track, two way track or multi way track.

3.1.1. Single Track. Single track means when train move to and fro in same track, If a track is to be employed by two or more trains at a time, it must have passing sidings or passing loops or cross over at interval along the road to allow trains running in several or same directions to pass one another [21]

3.1.2. Double Track. A double track railway means a pair of single line, one track in each direction left and right [21].

3.1.3. Multi Track. A railway line consists of two or more than two track of single lines is known as multi track.

3.2. On the basis of various issues

There are various issues found in the railway system as like speed of the train, arrival and departure time

of the train etc. It is discussed in the following subdivision.
3.2.1. Non uniform time of departure. Variable departure time of the train is based on the availability of track if planned schedule is not meet.
3.2.2. Non constant train velocity [17]. The train velocity varies with respect to time.
3.2.3. Reality situation. Actual live status of train at particular instant of time.
3.2.4. Lowering the fuel consumption cost [16]. How to minimise the fuel consumption cost.
3.2.5. Minimize delay from the planned schedule. Delay should be zero for meeting the planned schedule.
3.2.6. Maximum allowable delay [14]. Some buffer time is provided for avoiding the delay.
3.2.7. Train arrival and dispatching time. Reaching at the platform is arrival time and leaving the platform is dispatching or departure time.
3.2.8. Acceleration and de-acceleration time. Increasing the train speed is acceleration and slowness the speed of train is de-acceleration.

3.3. Based on analysis
The railway system has categorised on the basis of various analysis as like services, operation etc.
3.3.1. Capacity analysis. The high density and long route rail services of the same track was done by humans not computers in 1970's, [6], [15].
3.3.2. Exact solution. The public railway sector has improved the awareness of the importance of service for passengers, [11], [12].
3.3.3. Approximate solution. This feature provides extensive computational result on real world, [3], [7], [9], [10], [11], [18].

3.4. Based on scheduling type.
Various kinds of train scheduling are given below.
3.4.1. FSS. FSS means Flow shop scheduling in this each job is processed in particular sequence on various machine in single direction.
3.4.2. JSS. JSS represents job shop scheduling, arranging routines (in order) through each job feature machines, but this machine sequence can also be different for different job.
3.4.3. BFSS. It means blocking flow shop scheduling. In this a track section cannot release and must hold a train until the subsequent section on the routing becomes available.
3.4.4. NWFSS. No wait flow shop scheduling means no wait constraints restrict the operation of every train should be nonstop from the beginning to the end point without interruptions.
3.4.5. LBFSS. Limited buffer flow shop scheduling is a particular track segment can departure and must hold a train until the succeeding section on the routing becomes available.
3.4.6. CBFSS. Combined buffer flow shop scheduling consist FSS with infinite buffer, BFSS with no buffer, NWFSS and LBFSS.
3.4.7. BJSS. Blocking job shop scheduling represent train cannot leave and must hold a track until the subsequently track on the routing becomes available.
3.4.8. PMJSS. Parallel machine job shop scheduling consists issues without considering blocking conditions.
3.4.9. BPMFSS. Blocking parallel machine flow shop scheduling represents PMFSS problems with the blocking conditions.
3.4.10. NWPMFSS. No wait parallel machine flow shop scheduling shows PMFSS problems with no wait constraint.
3.4.11. BPMJSS. Blocking parallel machine job shop scheduling model for train scheduling, train and section are synonymous with jobs and machines respectively and an operation is considered as a traversal of a train in a section. Due to buffer space constraints, real-life train scheduling must consider blocking

during wait interruptions, meaning the track section may not be released and the train must be on hold until the next section on the routing is available. As a result, the problem is treated as BPMJSS with blocking conditions

3.4.12. NWPMJSS. No wait parallel machine job shop scheduling shows PMJSS problems with no wait constraints.

3.4.13. NWBPMFSS. No wait blocking parallel machine flow shop scheduling based on PMFSS problems with blocking and no wait constraints.

3.4.14. NWBPMJSS. No wait blocking parallel machine job shop scheduling, this is a generalized case of BPMJSS problem in real environment, no wait constraints are more restrictive than blocking interrupts.

4. Analytical Approach of Rail Line Capacity

These models are easy to use and have less mathematical calculations so, It may be provided solutions quickly. These models are suitable for planning phase. Poole gave a mathematical method to evaluate maximum capacity of the one way track line. Two distinguished capacity analysis methods Capacity Utilization Index (CUI) and Union of Railways proposed UIC-406 method for the capacity calculation studied in this criterion.

Scott's formula for Line Capacity:

$$C = T / (T_1+T_2) \times \eta$$

Where, T = Total duration in minutes in a day (24 hours) = 1440 minutes

T_1 = Over the critical block section running time of slowest train

T_2 = Block operation time

η = Factor of efficiency

Capacity is determine of the ability to travel a definite quantity of transfer over a defined rail line with a given set of assets under a definite service plan (Krueger, 1999). The railway network transported over 22 million of passengers every day. Over one billion metric tons of goods were passed across the India in 2020.

4.1. Capacity Utilization Index (CUI) Method

This technology is based on the timetable compression rule. The timeline is squeezed according to least headway distance as shown in fig. 2. According to Gibson the Capacity Utilization Index is the ratio of compressed time to actual time of the timetable graph [8]. CUI method is applied at route section level.

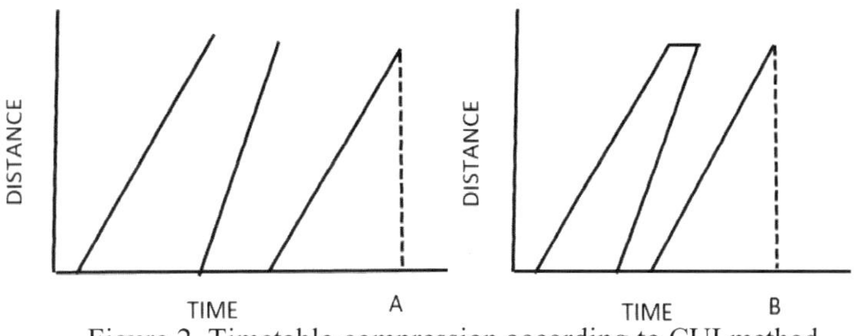

Figure 2. Timetable compression according to CUI method

4.2. UIC-406 Method
The International Union of Railways proposed UIC-406 method based on timetable compression process as shown in figure 3.UIC-406 method is applied at signal block level.

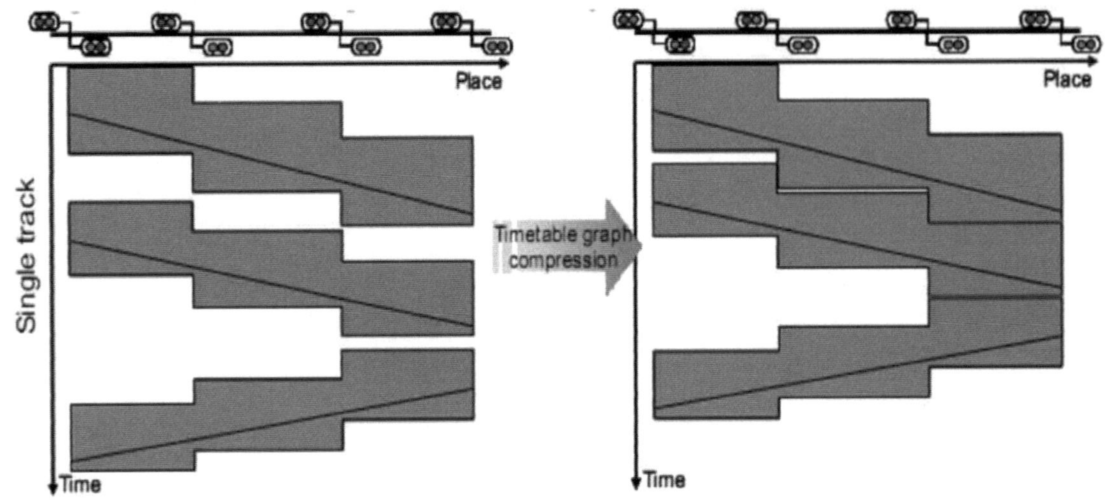

Figure 3. Time-table compression according to UIC-406 method

5. Work on rail network capacity: Related case study
In this case study a capacity model is demonstrated. This consists of single and double track and total 14 numbers of sections, each section has definite distance. This network has 5 corridors. All data is given in the following tables.

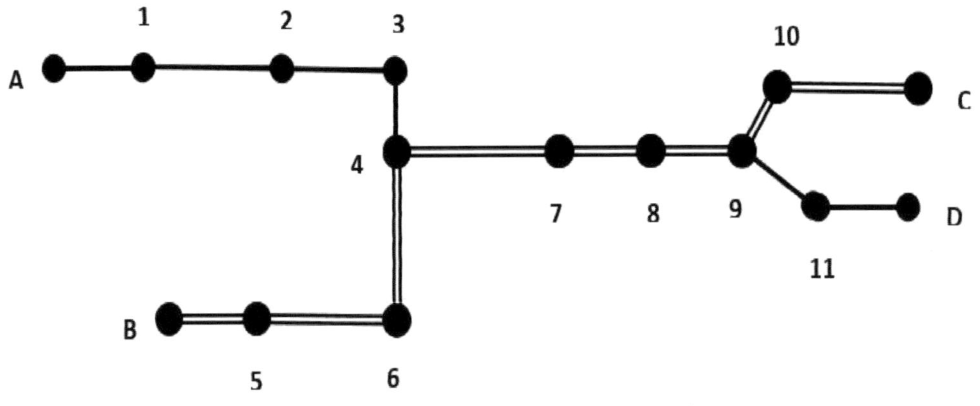

Figure 4. Network Diagram

Variable speed of trains is 50, 70, 90.KM per Hour for both directions. This model solved by CPLEX and Corridors capacity is shown in Table 8. The total network capacity is 532.935.

Table 6. Route Attributes.

Corridor No	Corridor	Section. No	Section	Length	Line
1	A-D	1	A-1	5	1
2	A-C	2	1-2	8	1
3	B-C	3	2-3	6	1
4	B-D	4	3-4	4	1
5	A-B	5	B-5	5	2
		6	5-6	8	2
		7	6-4	9	2
		8	4-7	7	2
		9	7-8	5	2
		10	8-9	4	2
		11	9-10	5	2
		12	10-C	5	2
		13	9-11	4	1
		14	11-D	7	1

Table 7. Proportional and Directional distributions.

Proportional distributions				Directional distributions (forward)			
Corridor	Train types			Corridor	Train types		
	1	2	3		1	2	3
A-D	0.35	0.45	0.20	**A-D**	0.45	0.52	0.46
A-C	0.40	0.50	0.10	**A-C**	0.40	0.55	0.50
B-C	0.35	0.40	0.25	**B-C**	0.45	0.55	0.65
B-D	0.40	0.30	0.35	**B-D**	0.45	0.60	0.40
A-B	0.25	0.60	0.15	**A-B**	0.50	0.40	0.65

Table 8. Absolute Capacity of Corridors and Network.

Corridor	Absolute Capacity
A-D	53.016
A-C	125.29
B-C	192.15
B-D	97.422
A-B	65.057
	532.935
Total Network Capacity	

6. Conclusion

In past, most of the previous research was specific and suitable for some particular scenario. In this paper capacity of railway network has analyzed for 14 sections with 5 corridors by optimization software CIPLEX solver for single and double line railway track in both direction. It is found that the maximum 532.935 trains can run on this network. The main objective of this paper is to achieve the efficient capacity analysis of the Indian railway network for various speed trains on the basis of development of an algorithm and analysis of the model. The railway network capacity is analysed by analytical methods using timetable compression techniques for various speed trains and can increase the capacity of the railway network by increasing the speed of the train.

7. Future Research

More tests and models are needed in day to day real situations to efficiently apply the results to the actual situations of railway industries. Railway industry is a very important decision-making system. For railway capacity analysis, maximum number of tardy trains or maximum tardiness can be considered. In this paper, tardiness is not considered for analysis of train capacity.

References

[1] Atieh Kianinejadoshah and Ricci Stefano 2021 Comparative Application of Methods for Nodes Capacity Assessment *Transportation Research Procedia* **53** pp 72-9

[2] A. Jamili 2018 Computation of practical capacity in single-track railway lines based on computing the minimum buffer times *Journal of Railway Transport Planning & Management* **8** pp 91-102

[3] Clive Roberts, Shaofeng Lu, Paul Weston, Stuart Hillmansen and Hoay Beng Gooi 2014 Increasing the Regenerative Braking Energy for Railway Vehicles, *IEEE Transaction on Intelligent Transportation Systems* **15** 6 pp 2506-15

[4] Farzaad Soleymaani, Mohammad Ali Sandidzadeh, and Tofigh Faizy 2019 Controlling Train Power Consumption with Energy Storage Based on Fuzzy Control Case Study on Line 3 of Tehran Metro *International Journal of Railway Research* **6** 1 pp 39-49

[5] Gao Xing, Jia Feng , Xiamiao Li, Haidong and Liu Baohua Mao 2017 Optimizing the Energy Efficient Metro Train Timetable and Control Strategy in Off Peak hours with uncertain Passenger Demands, *Energies* **10** 4 p 436

[6] Keping Li, Hangfei Huang and Paul Schonfeld 2018 Real time energy saving metro train rescheduling with primary delay identification *PLOS – one* **13** 2

[7] Pengling Wang, L ei Ma, Rob M P Goverde and Qingyuan Wang 2016 Rescheduling trains using Petri nets and heuristic search *IEEE Transactions on intelligent transportation systems* **17** pp 726-35

[8] Peng Sheng You, Yi Chin Hsieh,Ta Cheng Chen and Yung Cheng Lee 2013 A Heuristic Approach For Assembly Scheduling and Transportation Problems with Parallel Machines *Scientific Research iBusiness* **5** pp 27-30

[9] Ralf Ruckert, Martin Lemnian, Christoph Blendiger, Steffen Rechner and Matthias Muller Hannemann 2016 PANDA a software tool for improved train dispatching with focus on passenger flows *Springerlink Integrated planning in Public Transport*

[10] R. Venkata Rao 2016 Jaya A simple and new optimization algorithm for solving constrained and unconstrained optimization problems *International Journal of Industrial Engineering Computations* **7** pp 19-34

[11] Singh Varsha IRSEE (P) Ministry of Railways Government of India 2017 Efficient Utilization of Regenerative Braking Energy in Railway operations *International Research Journal of Engineering and Technology (IRJET)* **04** 12 pp 1421-28

[12] Selim Dundar, Ismail Sahin 2013 Train rescheduling with genetic algorithms and artificial neural network for single track railways *Transportation Research Part C (Elsevier)* **27** pp 1-15
[13] Sundaravalli Narayanaswami and Narayan Rangaraj 2013 modelling disruptions and resolving conflicts optimally in a railway schedule *Computers & Industrial Engineering (Elsevier)* **64** pp 469–81
[14] T wan Dollevoet, Francesco Corman, Andrea D Ariano, Dennis Huisman 2014 An iterative optimization framework for delay management and train scheduling *Springer Flex Serv Manuf Journal* ***26*** pp 490—515 doi 10.1007/S 10696-013-9187-2
[15] Werner Brandauer, Mahdiyeh Khodaparastan and Ahmed A Mohamed 2019 Recuperation of Regenerative Braking Energy in Electric Rail Transit Systems *IEEE Transactions on Intelligent Transportation Systems* **20** 8 pp 2831 – 47
[16] Xiang Li, Dechun Wang, Keping Li and Ziyou Gao 2013 A green train scheduling model and fuzzy Multi objective optimization algorithm*Applied Mathematical Modelling (Elsevier)***37**pp2063-73
[17] Xiaoming Xu, Keping Li, Lixing Yang and Jingjing Ye 2014 Balanced train time tabling on single line railway with optimized velocity *Applied Mathematical Modelling(Elsevier)***38** pp 894-909
[18] Yuri N Sotskov and Omid Gholami 2012 Shifting Bottleneck algorithm for train scheduling in Single track railway *Proceedings of the 14th IFAC Symposium on Information Control Problems in Manufacturing Bucharest Romania* pp 23-25
[19] Zhang Huiru, Limin Jia, Li Wang and Xinyue Xu 2019 Energy consumption optimization of train operation for railway systems Algorithm development and real world case study *Journal of Cleaner Production (Elsevier)* **214** pp 1024-37
[20] Zhou, X, and Zhong, M 2004 Bi criteria train scheduling for high speed passenger railway planning applications *European Journal of Operational Research* **167** pp 752-71
[21] Mittal Manoj, Prof. (Dr.) Chaturvedi A. K. 2020 Utilisation of Regenerative Energy Of Indian Trains in The Present Scenario. *Proceedings of APGRES International Conference on Advances in Power Generation* **9 1** pp 166-70

Machining of high temperature heat resistant super alloys- a concise review on cooling aspects

Ramai Ranjan Panigrahi, Ramanuj Kumar, Ashok Kumar Sahoo andAmlana Panda

School of Mechanical Engineering, Kalinga Institute of Industrial Technology (KIIT),Deemed to be University, Bhubaneshwar-751024, Odisha, India

**ramanujkumar22@gmail.com*

Abstract. Analyzing and comprehending cutting process mechanisms is a crucial step in creating a cost-effective, long-lasting, and safe machining process.In modern manufacturing, it is very challenging to achieve a high surface finish product with high dimensional accuracy. To reduce the high production cost as well as for the sustainable manufacturing selection of both cutting tool materials and cutting environments is necessary. Nowadays high-temperature alloys are in very demand in industries like power generation, gas turbines, and chemical processing. Superalloys are recognised as a significant problem in a high-temperature work environment with the sophisticated material technology in space and aviation industries. At the time of machining of these superalloys, a high amount of heat developed at the cutting area which affects the microstructure of the work specimen and also the cutting inserts. To minimize that excess heat, researchers and manufacturing industries are adopting different cooling lubrication techniques. When the lubricant penetrated the cutting region, the temperature is reduced since cooling effect. Several studies focus on the optimization of individual performance features in the machining processes and in recent past, Super-alloys and cutting materials like ceramic, carbide, and hybrid tools are gaining popularity. Since the application of the superalloy in various sectors is ever- increasing, therefore it is necessary to evaluate a machining behavior during the experimental investigation. In this contemporary review, the effects of various cooling and lubrication strategy on machining outputs will be discussed considering the sustainable and greenway of manufacturing.

1. Introduction

With the superior material technology in space and aviation enterprises, superalloys are recognized as a foremost challenge in a high-temperature heat-resistant working environment. Heat resistant super alloy have a high temperature; strength, creep, and corrosion-resistant material finds their application in the industrial turbine, atomic and submarine aircraft sectors. In addition, the important properties are needed generally utilized in heat exchangers, atomic reactors, and turbine blades. Because of its high weight to strength proportion, its exploitation in the aerospace and aeronautical industry is widespread. The current article presents a comprehensive overview of various machining processes (turning and milling) during machining of superalloys particularly, Inconel 800, Incoloy 800, Inconel 706, Inconel 825, and Inconel 718, titanium alloys(Ti-6Al-4V titanium alloy), etc. The machining attributes exploration of Nickel-based superalloys grades have become a promising topic for research advanced technology-based thermal barrier insulating materials are used to progress service performance of parts. Moreover, comparisons are reviewed between various cutting systems like cryogenic, MQL, and dry environments. In total, experimental and theoretical perceptions for surface quality, temperature, and wear of the tool exerting on the various types of cooling conditions and cutting inserts were summarized. This overview will definitely be helpful for the practical perspective of manufacturing, academic researchers as well as cutting tool manufacturers.

Machining manufacturers, that are continuously striving to raise the component's quality, diminish its cost, reduced environmentally impact, and operator health hazards for cleaner process technology is escalating rapidly. The high resistant superalloy has been associated with the group of Ni-Fe-Cr superalloy. It provides higher yield strength, and tensile strength, and is better performed even up to a temperature of 1300F. This combination shows extreme resistance to erosion and oxidation, so this is appropriate for service in the most intense atmosphere with high temperatures. This high-temperature alloy is utilized owing to its excellent corrosive resistance and physical properties at higher temperatures [1]. Nevertheless, few attributes like high durability, combined with the heat transfer coefficient and the limited scope of metallic environments make the metal highly susceptible and should not be viewed concerning machining performance [2–5]. Alternatively, to achieve higher productivity, the minimal cost of production and better surface finish is the main reason behind the machining of superalloys [6-9]. New exploration in the aerospace sector views for the production process regarding the yearly increment on cutting fluids suitable to a minimum proportion rate of (i.e., 20%), signifying the intact cutting lubrication and cutting tool process [10,11]. In this manner, to prevail a sustainable, green, and clean environment, this might be crucial to utilize dry conditions as well as near dry machining that may be frequently viewed as the substitute effective strategy [12,13]. A process known as dry machining produces higher temperatures during the machining of nickel superalloys [14]. Many researchers explored different machining measures like surface quality, tool wear and temperature on Inconel alloy under MQL and dry conditions [15].

As Titanium alloys have superior properties, it leads to widespread popularity among aerospace, biomedical, and other engineering sectors. Yet, the hard turning of titanium alloys has always been confronted because of its high tool wear rates and high-temperature generation. Machining in a dry environment has an inadequate variety of acceptable cutting environments and is consequently not proper for industrial manufacture. So that, flood cooling by utilizing cutting fluids is predictably utilized to decrease the cutting temperatures can be a good solution. Still, it is frequently depressed in light of the allied health and environmental problems. As a result, to accomplish sustainable machining practice, various methods for implementing the cutting fluids are developed. Few of the well-known strategies consist of cryogenic cooling, minimum quantity lubrication (MQL), quantity cooled lubrication (MQCL). Usual trouble that can be related with machining operations varies from the high tool cost and setup time for huge quantity fabrication to machinery often requiring various machining attributes, As a result, managing the machine shop effectively becomes more difficult, which leads to an increase in work that is in progress. These, together with the enormous amount of scrap created, often serve as the foundation for ongoing investigation in this area of manufacturing technology.

Consequently, the overview generates the idea regarding the various machining practice. The overview of the machining practices and their requirements are disclosed in Fig. 1.

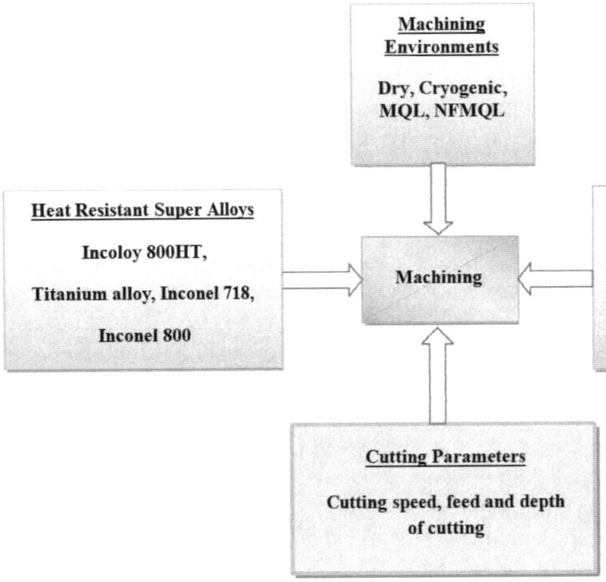

Figure 1. Schematic overview of machining requirements

2. Machinability performance of superalloys in various cooling conditions
2.1 Dry machining
2.1.1 Coated carbide tool

Dudzinski et al.[16] explored that the Al_2O_3 CVD coated cemented carbide displayed additional notch wear at each of the three angles tried and probably won't reasonable for turning of Inconel 718. Devillez et al. [17] obtained that the lower roughness Ra value was obtained while machining of Inconel 718 at the same cutting speed in accordance with wet. Due to the higher temperatures produced, residual built-up edges are deposited on the machined surface, altering the surface quality. Sharma et al.[18] experimented in both TiAlN and CrN coated tungsten carbide and obtained that, Because of greater hardenibility , TiAlN coated tools accomplished superior as compare to CrN coated tools. Liao and Shiue [19] noticed a extremely thin layer next to the cutting edge during machining of inconel 718.While comparing TiCN and TiN coatings, it is revealed that the PVD-TiAlN coating was most appropriate for dry machining of super alloy like Inconel 718 as least wear was shown with TiAlN tool (Fig. 2).

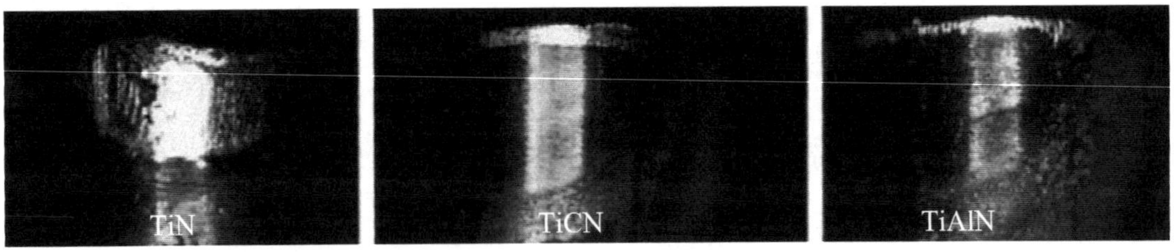

Figure 2. Comparison of nose wears after 5 minutes machining with 46 m/min [20]

2.1.2 Uncoated carbide tool

Palanisamy et al. [21] noticed a large abrasive marks onto the tool-tip due to enhancement in MRR at a particular speed. The surface quality and chip-tool interface temperature was severely affected and increased when feed-rate was increased from 0.05 to 0.1 mm/rev. Alaudin et al. [22] performed extensive research on the full immersion and half immersion (both in up cut and down cut) by using end milling of Inconel 718 and tungsten carbide inserts (K20 grade). The results showed that down cut provided greater performance while full immersion extended tool life in compared to half immersion. Thakur et al. [23] witnessed that, at prominent speed, tool wear was mostly influenced by tool-chip contact length. According to Choudhury and El-Baradie [24], uncoated carbide tool exhibited the superior surface quality relative to coated carbide tools at inferior cutting speed (50 m/min), while using coated tool, 39.79% reduction in cutting force was observed at the maximum cutting speed [25].

2.1.3 Other cutting tools

Arunachalam et al. [26] examined that the CBN (type C1) cutting tool produced the compressive residual stresses up to 15 mm depth in workpiece from its periphery, followed by tensile stresses near the centre of the workpiec. Zhou et al. [27] experimentally found that the TiN coated PCBN cutting tool life was about 20% superior to equivalent cutting insert in machining Inconel 718 alloy [28, 29]. Coelho et al. [30] observed the notch wear in finish turning of Inconel 718 at elevated cutting depth.

2.2 Cryogenic cooling assisted machining

2.2.1 Coated carbide tool

Ross and Manimaran [31] found that the implementation of cryogenics coolant in machining exhibited a great minimization in cutting force when speed was enhanced from 30 to 75 m/min. The surface quality was also improved and greatly affected by cryogenic delivery nozzle angle. Over wet and MQL cutting, LCO_2 imparted a great enhancement in surface finish [32]. In turning Ti64, Dhananchezian and Kumar [33] reported that the cryogenic cooling conditions bring considerable improvements in the life span of the tool, surface finish, and decreased the cutting temperature. Venugopal et al. [34] compared the turning performances in dry, wet and cryogenic cooling turning by considering the values of tool wear and tool life. Cryogenic was the best cooling strategy found in comparison to others.

2.2.2 Uncoated carbide tool

Biermann et al. [35] revealed that, as compared to a flood emulsion cooling, cryogenic (CO_2) cooling process exhibited an increased tool life (25% greater). At lower speed (50m/min) with moderate feeds, the uncoated WC tool performed the best [36].

2.2.3 Other cutting tool

Costes et al. [37] confirmed the occurrence of adhesion and diffusion wear mechanism in machining Inconel 718 alloy. Persson et al. [38] experimentally found the worst turning ability as undesirable wear and cutting forces were seen when high grade CBN tool was employed in machining AD370 superalloy. Minimum Quantity Lubrication assisted machining.

2.2.4 Coated carbide tool

Ramana et al. [39] experimented with machining investigations on Ti–6Al–4V by implementing different cooling environments and stated that, MQL causes inferior significance of wear of tool as compared to the wet and dry cooling technique. Both Kamata and Obikawa [40] used three types of coated carbide tools during MQL assisted turning on Inconel 718 superalloys. The evaluation criteria were tool life and surface finish. It was observed that tool life of CVD coating $TiCN/Al_2O_3/TiN$ layers was very less. Yazid et al. [41] used three machining conditions (dry, MQL 50 mL/h and MQL 100 mL/h) during the finish turning of Inconel 718 nickel-based super alloy. The results revealed that that MQL may possibly enhance surface integrity features (surface roughness, surface texture, microstructure alteration). The severity of changes in microstructure on dry was converted into more hardness signifying more work hardening when compared with MQL assisted samples is another key finding. Pusavec et al. [42] performed CNC enabled turning experiment using dry MQL, cryogenic and cryoMQL machining condition with carbide insert. Under the same cutting parameters cryogenic machining provides minimum surface roughness. The Cryogenic machining influences slightly the final product microstructure and produces less plastic deformation on the turned surface. Usung coated carbide tool and Inconel 718 alloy, Behera et al. [43] adopted high-pressure jet machining, MQL, Cryogenic, and nano fluid assisted MQL condition in turning process. Because of the improved cooling and lubrication effect nMQL cutting environment improves the machining characteristics of Inconel 718 by minimizing tool wear and cutting force.

2.2.5 Uncoated carbide tool

Sulaiman et al.[44] found that the most significant cutting parameter towards tool life was cutting speed succeeded by feed rate and depth of cut. Joshi et al. [45] examined that the minimal roughness value of 3.70 µm and lower flank wear were noticed in MQL2 machining conditions as disclosed in Fig. 3. The outlines of literatures are reported in Table 1.

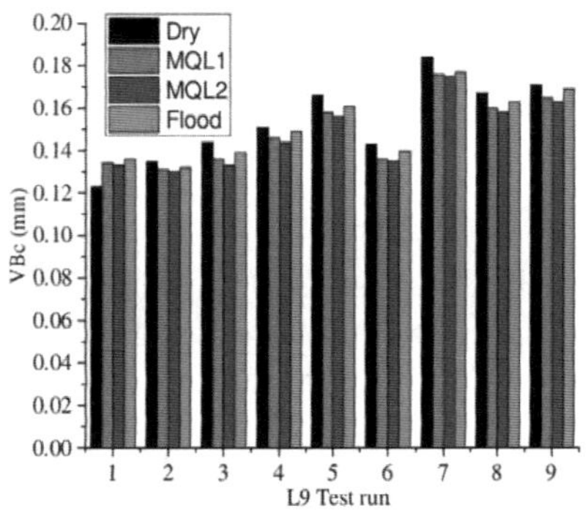

(a)

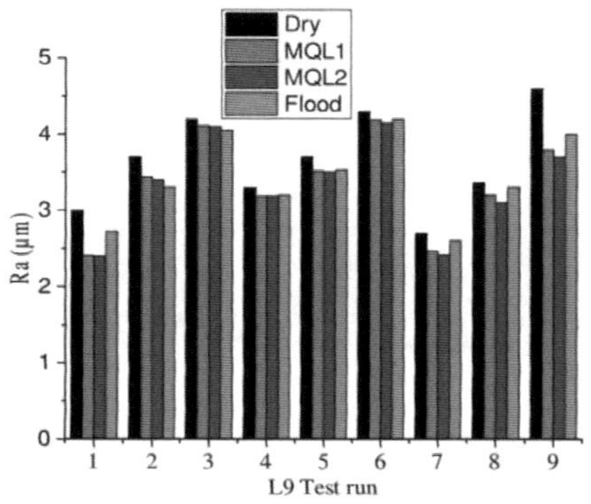

(b)

Figure 3. Comparison of (a) flank wear and (b) surface roughness under different cooling conditions [45].

Table 1. Outlines of Literatures based on various high temperature superalloys grades

Year, Author, Reference	Cutting Tool / Work Material	Cooling environment	Responses	Remarks
2011 Devillez et al.[16]	coated carbide inserts, Inconel- 718	Dry cooling	Tool wear, chip thickness ratio, Chip morphology, sub-surface integrity	Coated inserts were found to be better compared to uncoated one
2012, Zhang et al.[47]	Coated Carbide Inserts, Inconel- 718	MQLand Dry cooling	Cutting force and Tool life	Prolonged tool -life with lower cutting forces
2014, Pusavec et al.[42]	Coated Carbide Inserts, Inconel-718	Cryo- lubrication, MQL, Dry	Tool-wear, chip breakability measurements, cutting forces, surface roughness,	Cryo- lubrication perform beneficial efficacy due to low temperature of liquid nitrogen
2018, Palanisamy et al.[21]	Uncoated tungsten carbide, Incoloy 800HT	Dry	Surface roughness, Material removal rate	Optimal controllable factors by Taguchi , Signal to noise ratio
2018, Gupta et al.[46]	Cubic Boron Nitride insert Incoloy 800HT	Near Dry Machining (NDM)	Cutting force, surface roughness, Tool wear, Chip control	NDM Application lower cutting force, machined surface shows a better quality and excellent performance in tool wear
2015, Marques et al.[48]	Coated Carbide Inserts, Inconel 718	MQSL, MQL	Tool-wear, surface roughness	Wet cutting forces results lead by
2007, Satyanarayana et al.[49]	Carbide inserts, Ti-6Al-4V alloy	Dry	Tool wear	depth of cut increases the interface temperature also increases

3. Conclusion

In conclusion, for addressing the challenge of lesser tooling life, the appropriate use of various cooling methods (flood and MQL) has been implemented to control the raised temperature during machining practice. The introduction of advanced tool materials has to stand a greater challenge to production, design, and industrial engineers owing to the generation of high temperature and stress. When compared to dry machining, nanofluid MQL extensively reduces cutting force and tool-tip temperature. Investigations revealed that cryogenic cooling technique has shown a perspective to enrich the machinability aspects of Ti-6Al-4V superalloy as compared to conventional machining techniques by producing relatively lower surface quality and tool wear besides major impacts on the power consumption of the machine tool. The purpose of recently developed methodology and cutting tools in the machining practice of titanium and of Ni-based alloys leads to a number of fold increases in tool life by not eliminating the surface integrity of the machined materials. Alloys which are tough to cut at room temperature may effortlessly capable of acquiring optimum temperature, by adopting the hot machining concept.

References

[1] Singh G, Gupta M K, Mia M and Sharma V S 2018 *International Journal of Advanced Manufacturing Technology* **97** pp 481-494
[2] Davoodi B and Hosseini A and Tazehkandi 2016 *Proceedings of the Institution of Mechanical Engineers, Part B: Journal of Engineering Manufacture* **230** pp 215-226
[3] Gupta M K and Sood P K 2017 *Proceedings of the Institution of Mechanical Engineers, Part C: Journal of Mechanical Engineering Science* **231** pp 1445-1464
[4] Yılmaz B, Arabulut S K and Güllü A 2018 *Journal of Manufacturing Processes* 32 pp 553-563
[5] Güllü A and Arabulut S K 2008 *Materials and Manufacturing Processes* **23** pp 852-857
[6] Yazid M Z A, CheHaron C H, Ghani J A, Ibrahim G A and Said A Y M 2011 *ProcediaEngineering* **19** pp 396-401
[7] Abbas A, Pimenov D, Erdakov I, Taha M, Rayes M EI and Soliman M 2018 Metals 8 pp 394
[8] Wojciechowski S, Maruda R W, Barrans S, Nieslony P and Krolczyk G M 2017 *Measurement* **111** pp 18-28
[9] Maruda R W, Krolczyk G M, Nieslony P, Wojciechowski S, Michalski M and Legutko S T 2016 *Journal of Manufacturing Processes* **24** pp 107-115
[10] Gupta M, Pruncu C, Mia M, Singh G, Singh S, Prakash C and Gill H 2018 *Materials* **11** pp 2088
[11] Gupta K, and Laubscher R F 2017 *Proceedings of the Institution of Mechanical Engineers, Part B: Journal of Engineering Manufacture* **231** pp 2543-2560
[12] Sharma V S, Singh G and Sorby K 2015 *Materials and manufacturing processes* **30** pp 935-953
[13] Gupta M K and Sood P K 2017 *Friction* **5** pp 155-170
[14] Dureja J S, Singh R, Singh T, Singh P, Dogra M, and Bhatti M S 2015 *International Journal of Precision Engineering and Manufacturing-Green Technology* **2** pp 123-129
[15] Sharma V S, Manu D., and Suri N. M. 2009 *2007 International Journal of Machine Tools and Manufacture* **49** pp 435-453
[16] Dudzinski D, Devillez A, Moufki A, Larrouquere D, Zerrouki V, Vigneau J A 2014 *International Journal of Machine Tools and Manufacture* **44** pp 439-456
[17] Devillez A, Le Coz G, Dominiak S and Dudzinski D 2011 *Journal of Material Processing Technology* **211** pp 1590-1598
[18] Sharman A, Dewes RC, Aspinwall DK 2001 *Journal of Materials Processing Technology* **118** pp 29–35.
[19] Liao YS and Shiue 1996 *Wear* **193** pp 16–24

[20] Jindal PC, Santhanam AT, Schleinkofer U and Shuster AF 1999 *International Journal of Refractory Metals and Hard Materials* **17** pp 163–170
[21] Palanisamy A and Selvaraj T 2018 *Materials Today: Proceedings* **5** 7708-7715Costes JP, Guillet Y, Poulachon G and Dessoly M 2007 *International Journal of Machine Tools and Manufacture*, **47** pp 1081-108
[22] Alaudin M, Baradie MA and Hashmi MSJ 1995 *Journal of Materials Processing Technology* **55** pp 321–330
[23] Akıncıoğlu G 2021 *Journal of Materials Engineering and Performance* **30** pp1280-1288
[24] Thakur DG, Ramamoorthy B, Vijayaraghavan L, 2009 *Materials and Manufacturing Processes* **24** pp 497–503.
[25] Choudhury IA, El-Baradie MA, 1998 *J. Mater. Proc. Technol. 77 278– 284*
[26] Behera GC, Thrinad J, & Datta S. 2021 *Materials Today: Proceedings* **38** 2664-2670
[27] Arunachalam RM, Mannan MA Spowage AC 2004 *International Journal of Machine Tools and Manufacture* **44** pp 879-88
[28] Zhou J, Bushlya V, Avdovic P, Stahl JE 2012 *Int J Adv Manuf Technol* ***58*** *141-151.*
[29] Uhlmann E, Ederer G. 2001 *Ind Diamond Rev* **61** pp 169-74
[30] Pawade RS, Joshi SS, Brahmankar PK, Rahman M 2007 *J Mater Process Technol* **192** pp 139-46
[31] Coelho RT, Silva LR, Braghini A, Bezerra AA. *2004 Journal of Materials Processing Technology* **148** pp 147–53
[32] Ross K and Manimaran G 2020 *Arabian Journal for Science and Engineering* **45 pp** 9267-9281
[33] Dhananchezian M, Kumar MP 2011 *Cryogenics* **51** pp 34-40
[34] Venugopal KA, Paul S, Chattopadhyay 2007 AB *Wear* **262** pp 1071–1078
[35] Biermann D, Abrahams H and Metzger M 2015 *Advances in Manufacturing* **3** pp 292-299
[36] Bhatt A, Attia H, Vargas R, Thomson V 2010 *Tribology International* **43** pp 1113-1121.
[37] Costes JP, Guillet Y, Poulachon G and Dessoly M 2007 *International Journal of Machine Tools and Manufacture* **47** *pp* 1081-1087
[38] Persson H, Bushlya V, Franca L, Zhoua J, Ståhl J-E, Lenrick F 2022 *Ceramics International*, 48(16) 22733-22742
[39] Venkata Ramana M, Krishna Mohan Rao G and Hanumantha Rao D 2014 *International Journal of Materials, Mechanics and Manufacturing* **2** pp 272–277
[40] KamataY and Obikawa T 2008 *Journal of Materials Processing Technology* **192-193** pp 281-286
[41] Yazid M Z A, CheHaron C H, Ghani J A, Ibrahim G A and Said A Y M 2011 *Procedia Engineering* **19** pp 396-401
[42] Pusavec F, Hamdi H, Kopac J, Jawahir I S 2011 *Journal of Materials Processing Technology* **211** pp 773–783
[43] Behera B C, Alemayehu H, Ghosh S & Rao P V 2017 *Journal of Manufacturing Processes* **30** pp 541-552
[44] Sulaiman M A, Asiyah M S, Shahmi R., Mohama E, Mohamad N A, Ali M M and Ito T 2018 *Journal of Advanced Manufacturing Technology* **12** pp 63-72\
[45] Joshi, KK , Kumar, R, Anurag (2018). An experimental investigations in turning of Incoloy 800 in dry, MQL and flood cooling conditions. *Procedia Manufacturing*, **20** pp 350-357
[46] Gupta M K, Sood P K, Singh G and Sharma V S 2017 *Advanced Manufacturing Technologies* 237-248
[47] Zhang S, Li J F and Wang Y W 2012 *Journal of Cleaner Production* **32** 81-87
[48] Marques A, Narala S K R, Machado A R, Gunda R K, Josyula S K, Da Silva R B and Da Silva M B 2017 *Part B: Journal of Engineering Manufacture* **231** pp 1144-1159
[49] Satyanarayana B, Reddy M D and Nitin P R 2017 *Materials Today: Proceedings* **4** pp 2203-2212.

Corrosion Inhibition in Magnesium by using Green Inhibitor

D.Jones Joseph Jebaraj[a], N. Rajesh Jesudoss Hynes[b*] , Basanth Navin K[c]

[a,b,c]*Department of Mechanical Engineering, Mepco Schlenk Engineering College, Sivakasi - 626005, India*
[d]*Associate Professor (Sr. Grade), Department of Mechanical Engineering, Mepco Schlenk Engineering College, Sivakasi - 626005, India*

[d] Corresponding Author: findhynes@yahoo.co.in

Abstract. Though magnesium has attractive properties in many transport applications, its use is limited because of the issue of corrosion. Corrosion rate increases under influence of pressure due to stresses acting over the metal surface. To avoid this stress induced corrosion, there is a need of an inhibitor to inhibit the corrosion. In the present work, a natural inhibitor is used as the inhibitor for preventing corrosion. The plant extract consist of active materials of plant material which has the antioxidant properties. The corrosion property of magnesium alloy in 3.5 wt% NaCl solution is analyzed. As per the study, the Moringa Oleifera leaf extract is chosen as green inhibitor. The effective inhibition of plant extract is analyzed in electrochemical workstation with varying concentration of extract.

Keywords: inhibition performance, hydrostatic corrosion test under pressure, corrosion protection, electrochemical technique.

1. INTRODUCTION
1.1 Magnesium alloy

The magnesium alloy is lightest metal which has the density of 1.738g/cm^2. It considered to be stronger than aluminum alloys [1]. Due its lighter weight properties it is used in automobile application [2]. Its strength is comparatively equal to titanium. It also has damping properties, which is used in sports car engine manufacturing [3,4]. Many researchers research for lighter metal for automotive and aerospace application, which the magnesium alloy show lighter in weight and high strength. The only drawback in magnesium alloy is highly corrosive in corrosive medium.

Magnesium composites are combinations of magnesium (the lightest underlying metal) with different metals, frequently aluminum, zinc, manganese, silicon, copper, uncommon earths and zirconium. Magnesium composites have a hexagonal grid structure, which influences the key properties of these combinations [5]. The upside of magnesium amalgams to be utilized in such applications is their light weight, high solidarity to-weight proportion, high firmness to-weight proportion, castability, machinability, and incredible damping [6].

The only problem in this is reactivity of magnesium in saline atmosphere, which means the magnesium get corroded. The salt in the sea water acts as a catalyst in the corrosion of Magnesium [7,8]. Magnesium reacts in these conditions to form a porous magnesium hydroxide layer. The magnesium hydroxide layer thus formed does not passivate the corrosion, which explained in the paper of hydrogen evolution in magnesium at corrosion reaction.

$$Mg + 2H_2O \rightarrow Mg\,(OH)_2 + H_2$$

The magnesium alloy can be used in underwater vehicle application, such as under water exploring device and vehicle. To study the corrosion rate under hydrostatic pressure effect on magnesium alloy in saline environment, a setup is to be made. The setup is made to inbuilt pressure inside it as pressure vessel. The compressed air is used to pressurize the vessel. The sample metal piece is to be kept inside the setup during test. [9]

Underwater steel structure such as remote operated vehicle (ROV) corrosion was prevented by coating. The corrosion efficiency is tested by testing the sample in the pressure vessel under hydrostatic pressure. As per the test the corrosion rate increases with increasing rate of pressure. By analysis corrosion rate is affected by pressure. The increase of corrosion rate is caused by the stress induced corrosion on the surface of metal. The stress induced corrosion can cause pit formation over the metal surface.[10]

1.2 Study of Corrosion in Magnesium Alloy

Uniform corrosion is the type of corrosion that affect entire surface of magnesium metal that expose to corrosion environment. This type of corrosion lead to weight loss of metal surface at equal and uniform rate throughout the exposed metal surfaces. In general uniform corrosion is caused by continues interaction of corrosion medium with metal or alloy surface. In magnesium alloy uniform corrosion causes magnesium hydroxide layer in uniform manner throughout metal surface. Due to this corrosion the thickness of metal gets reduced by continues exposure of corrosive medium with metal surface. This type of corrosion can be reduced by coating the metal surface with corrosion resistance coating [11].

Galvanic corrosion occurs at place where two dissimilar metal get contact with each other. In this corrosion one of the metal act as anode and other as cathode, In which the potential difference between the metal cause corrosion on anodic metal surface. This type of corrosion can be reduced by insulating two dissimilar metal to avoid contact between them. Galvanic corrosion also can be reduced by corrosion resistive coating [12,13].

Pitting corrosion caused by localized corrosion which forms cavity like structure over the metal surface. It can also form hole on thin metal surface. Pitting corrosion can weaken the metal structure. In magnesium alloy pitting corrosion can be caused by some stress induced corrosion. Mostly this type of corrosion caused due to interaction of corrosion medium in same place of metal surface. For example, if the salt water droplet is dropped over the metal surface causes localized corrosion, which leads to formation of pitting over the metal surface.

Filiform corrosion is caused in coated metal surface. It occurs inside the coated layer by penetration of water into the coated surface. Filiform corrosion caused in moisture environmental condition such as coastal area. This type of corrosion can be prevented by coating the metal surface by removing the old coating. Filiform corrosion is common in magnesium and aluminium based metals.

1.3 Green Inhibitors

Green Inhibitor are organic material that are used as corrosion protection material [14 -17]. The active materials found in plant extract has the anti-oxidant properties, which help in inhibiting the corrosion [18,19]. The efficiency and nature of corrosion resistance of plant extract can be identified by electrochemical workstation and weight loss method [20,21]. The impact of erosion restraint by green inhibitor on magnesium composite under hydrostatic pressure can be contemplated through test arrangement. The effect of plant extract in corrosion medium is evaluated by altering their concentration in corrosion solution and metal is analysed in the solution.

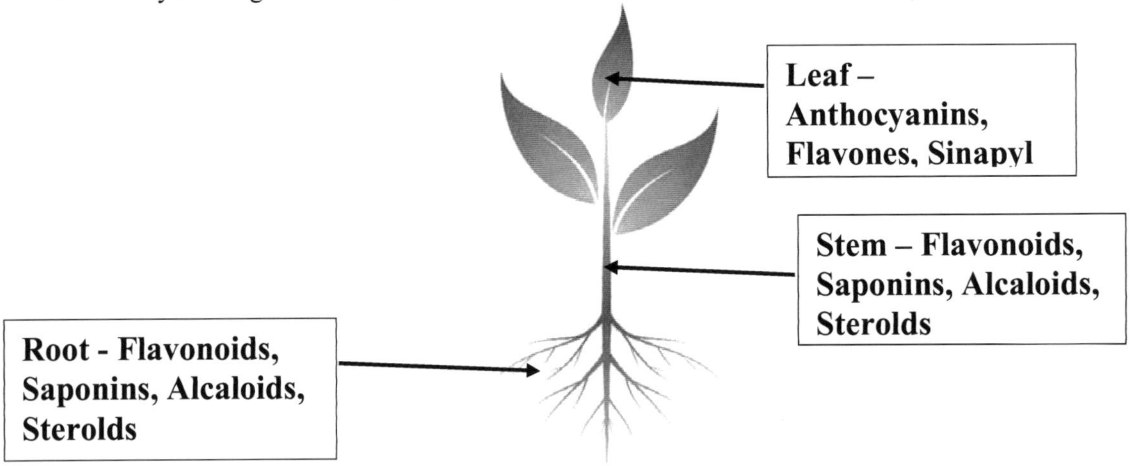

Figure 1. Active Component in Plant

Corrosion inhibitors are the material which prevent corrosion on metal surface by forming the protective surface. For magnesium commonly used corrosion inhibitor are chromium. The chromium is highly hazardous to environment and human. For chromium alternative organic based inhibitors can be used, which is an ecofriendly product.

Plant material consist of some phytochemicals which is found in plant cells, which are called as active materials. These active materials are called as extract component that we get from extraction process. The extraction process is done in different methods and they are Soxhlet and maceration. The Soxhlet is apparatus in which large amount of plant extract can be yielded. In this method the plant material is heated and circulated through the filter paper. At final the separate extract materials are collected at bottom of the apparatus. This apparatus is available in market, which can be used in laboratory over the hot plate apparatus. The collected extract material will be in liquid form, which can dry to get powdered form.

Maceration is conventional method which is followed in ayurvedic medicine process to extract plant active materials [22,23]. Maceration is otherwise called as digestion of plant material in extracting solution such as water, ethanol, methanol. During maceration the active material from the plant material get dissolved with the extracting solution. The remaining waste material settle at bottom as sediment. Then the extract solution is separated by filtration process. This process takes three to four days of maceration process.

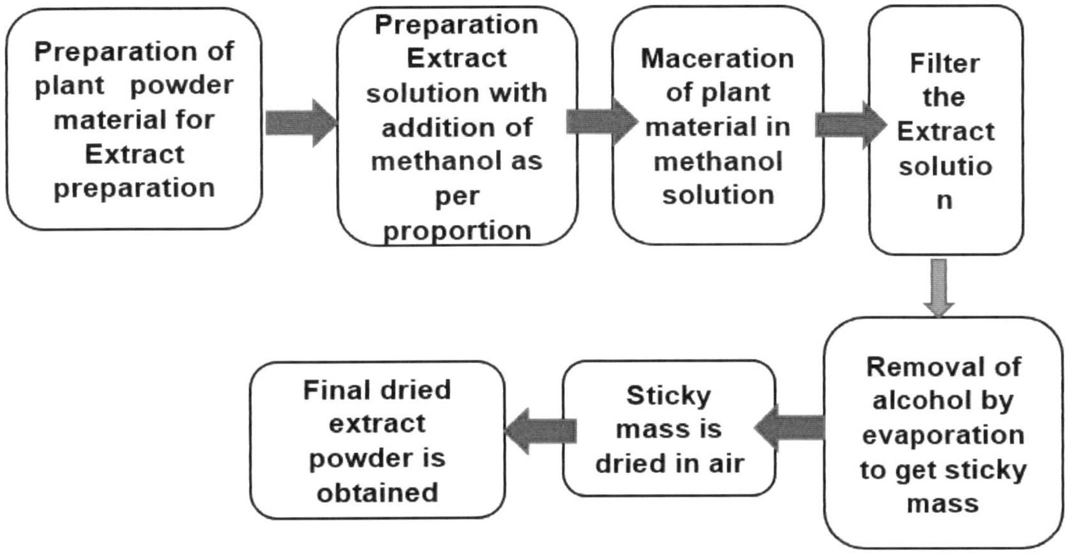

Figure 2. Extraction Process.

The corrosion restraint can be accomplished by inhibitor that acquired regular plant Extract. The Moringa Oleifera Extract is a non-harmful and best inhibitor for corrosion which has been tried with zinc metal. The presence of sweet-smelling compounds in the Extract can shape metallic complex that give high erosion restraint execution. The moringa extricate explored in this paper goes about as great erosion inhibitor for zinc in corrosion medium. The substance and electrochemical estimations affirm the inhibitive idea of moringa extract. Moringa Oliefera has the active material that are highly effective in corrosion inhibition. The moringa shows certain corrosion inhibition for respective concentration of extract in corrosive medium. By this idea it also can be used in magnesium-based metal. Moringa Oliefera extract has the properties of forming coating layer over metal layer by absorption.

The Orange Extract [14] are best in corrosion resistance for magnesium. By analysing the extract in FTIR, the result show that the functional groups found in extract such as OH,

Aromatic are best suited for corrosion inhibition. Orange Extract give better corrosion inhibition for magnesium metal. From this it shows that the magnesium alloy or metal corrosion can prevented with plant extract. The extract is effectively absorbed by the metal surface to form the protective layer.

The corrosion prevention is to be done by forming passive layers by some green inhibitor by formation of passive coating layer which inhibit corrosion. From the study best and effective green inhibitor is chosen for corrosion inhibition in application. The moringa extract also can be used as corrosion inhibitor for magnesium metals.

The current research is focused on green inhibitor that used to inhibit the corrosion in magnesium metal. The magnesium metal corrosion can be inhibited using plant extract which is has researched in a paper and given better result. In this research the Moringa Oleifera extract is used as green inhibitor for inhibiting corrosion over magnesium metal and its alloy, which is a novel method for inhibiting corrosion over magnesium alloy.

2 Experimentation.

The experimental research was done in 3.5 wt% of NaCl solution to stimulate the sea water as corrosion medium. The extract was prepared from the Moringa Oleifera leaf with influence of methyl alcohol, which is efficient in extracting extract from moringa oleifera extract. The corrosion medium of 3.5 wt% of NaCl solution is prepared in 100 ml of water and the 600ppm concentration of dried extract powder is mixed by dissolving it into the corrosion medium as the testing solution. Extract powder is mixed to corrosion medium to analyze the inhibition efficiency. The magnesium alloy sample is immersed in the testing solution and analyzed in electrochemical workstation to identify the inhibition efficiency.

$$Inhibition\ Efficiency\% = \frac{R_{p(inh)} - R_p}{R_{p(inh)}} \times 100$$

Table. 1 Equivalent circuit used for modelling the impedance data obtained for Magnesium alloy in different testing medium

Concentration	R_s	R_p	C_{dl}	θ	%IE
ppm	Ωcm^{-2}	Ωcm^{-2}	$\mu\Omega^{-1}S^n cm^{-2}$		
Bare	4.605	9.67	104.5	-	-
600	4.99	15.42	257.5	0.34565	34.56

$$\text{Protection Efficiency\%} = \frac{1 - I_{corr}}{I''_{corr}} \times 100$$

I_{corr} = Corrosion current of the inhibited specimen
I''_{corr} = Corrosion current on bare metal

Table 2. Values of polarization parameters for magnesium alloy in 3.5 wt% NaCl uninhibited and inhibited with various concentrations of Moringa oliefra extract at 25 °C

Concentration	$-E_{corr}$	I_{corr}	Tafel slope		Linear polarization	θ	%IE
ppm	Volts	Amp	Anodic slope	Cathodic slope	Ω cm^2		
Blank	1.584	0.007466	4.897	3.277	7.1	-	-
600	1.573	0.004732	4.955	3.350	11.1	0.3662	36.62

3. Conclusion

According to the study, the corrosion of magnesium alloy can be prevented by the influence of plant extract. Plant extract will be cheaper, easily available and act as alternatives for heavy metal-based inhibitors. Plant extract as an inhibitor for corrosion minimize the usage of hazardous inhibitor such as chromium oxide. From the table 1 and 2 the values shows that the inhibitor shows inhibition efficiency of extract solution.

Table 1 shows the impedance spectroscopy equivalent circuit value which shows that the extract has inhibition efficiency of 34.56%. Table 2 shows the Tafel plot result of extract in corrosion medium shows inhibition efficiency of 36.62%.

References

[1] J.A. Yuwono, C.D. Taylorc, G.S. Frankel, N. Birbilis, S. Fajardo - Understanding the enhanced rates of hydrogen evolution on dissolving magnesium –Electrochemistry Communications – 2019.

[2] Shenbaga Velu, P., Vignesh, N.J., Rajesh Jesudoss Hynes, N. (2020). Surface Characteristics When Solid State Welding of Aluminium Alloys to Magnesium Alloys. In: Gupta, K. (eds) Surface Engineering of Modern Materials. Engineering Materials. Springer, Cham. https://doi.org/10.1007/978-3-030-43232-4_4

[3] Pitchumani, Shenbaga Velu. Hynes, Rajesh Jesudoss. (2015). Simulation on Friction Welding Of MgAZ31 / AA 6061 T6 Joints. Recent Advances in Computer Science.

[4] Rajesh Jesudoss Hynes N, Vivek Prabhu M, Shenbaga Velu P, et al. An experimental insight of friction stir welding of dissimilar AA 6061/Mg AZ 31 B joints. Proceedings of the Institution of Mechanical Engineers, Part B: Journal of Engineering Manufacture. 2022;236(6-7):787-797. doi:10.1177/09544054211043474

[5] Cristina R. Weber, Gerhard Knörnschild and Luís F. P. Dick - The Negative-Difference Effect during the Localized Corrosion of Magnesium and of the AZ91HP Alloy –Sociedade Brasileira de Química – 2003.

[6] Pavel Predko, Dragan Rajnovic, Maria Luisa Grilli, Bogdan O. Postolnyi, Vjaceslavs Zemcenkovs, Gints Rijkuris, Eleonora Pole, and Marks Lisnanskis - Promising Methods

for Corrosion Protection of Magnesium Alloys in the Case of Mg-Al, Mg-Mn-Ce and Mg-Zn-Zr - Metals – 2021.
[7] R. M. Katona, J. C. Carpenter A. W. Knight, C. R. Bryan, R. F. Schaller , R. G. Kelly, E. J. Schindelholz - Importance of the hydrogen evolution reaction in magnesium chloride solutions on stainless steel - Elsevier – 2020.
[8] Yuhong Yao, Wei Yang, Dongjie Liu, Wei Gao and Jian Chen - Preparation and Corrosion Behavior in Marine Environment of MAO Coatings on Magnesium Alloy - Materials - 2020.
[9] H. Gao, Q. Li , F.N. Chen, Y. Dai, F. Luo, L.Q. Li - Study of the corrosion inhibition effect of sodium silicate on AZ91D magnesium alloy - Elsevier – 2011.
[10] T. Hyakudome, S. Ishibashi, Y. Watanabe, H. Yoshida, S. Tsukioka, T.Aoki - Application to Pressure Vessels for Underwater Vehicle of Magnesium Alloys – IEEE – 2008.
[11] A.S.Fouda , Y.Aggour, G.Bekheit, M.A.Ismail – Moringa Oleifera Extract as Green Corrosion Inhibitor for Zinc In Polluted Sodium Chloride Solution – International Journal Of Advanced Research – 2014.
[12] Nurulnatisya Ahmad etal., - The Effect of Hydrostatic Pressure on Corrosion of Coated Stainless Steel – ICAME – 2019.
[13] A.S.Fouda , Y.Aggour, G.Bekheit, M.A.Ismail - Moringa Oleifera Extract as Green Corrosion Inhibitor for Zinc In Polluted Sodium Chloride Solution. - International Journal of Advanced Research - 2014.
[14] Yaxu Wu, Yuanmi Zhang, Yumiao Jiang, Yafeng Qiand , Xugeng Guo, Li Wang, Jinglai Zhang - Orange peel extracts as biodegradable corrosion inhibitor for magnesium alloy in NaCl solution: Experimental and theoretical studies - Journal of the Taiwan Institute of Chemical Engineers - 2020.
[15] Mukesh, A. and Hynes, Rajesh. (2019). Corrosion behaviour of fibre metal laminates and control by inhibitors. AIP Conference Proceedings. 2142. 070008. 10.1063/1.5122400.
[16] Hynes, N.R.J., Selvaraj, R.M., Mohamed, T. et al. Aerva lanata flowers extract as green corrosion inhibitor of low-carbon steel in HCl solution: an in vitro study. Chem. Pap. 75, 1165–1174 (2021).
[17] N. Rajesh Jesudoss Hynes; Vignesh, N.J.; et al., A. Green Corrosion Inhibition on Carbon-Fibre-Reinforced Aluminium Laminate in NaCl Using Aerva Lanata Flower Extract. Polymers 2022, 14, 1700. https://doi.org/10.3390/polym14091700
[18] G. Sundar, N. Rajesh Jesudoss Hynes, Corrosion issues in metal matrix composites & Bi-metals, AIP Conference Proceedings 2142, 070007 (2019); https://doi.org/10.1063/1.5122399
[19] A. M. Mukesh and N. Rajesh Jesudoss Hynes, Corrosion behaviour of fibre metal laminates and control by inhibitors, AIP Conference Proceedings 2142, 070008 (2019). https://doi.org/10.1063/1.5122400
[20] Winston, Arockia & Hynes, Rajesh & Sankaranarayanan, R., (2020). Risk assessment of corrosion inhibitors of magnesium and its alloys. AIP Conference Proceedings. 2220. 140034. 10.1063/5.0001241.
[21] Afia, L., Salghi, R., Bazzi, E.H., Zarrouk, A., Hammouti, B., Bouri, M., Zarrouk, H., Bazzi, L., Bammou, L., 2012. Argan hulls extract: green inhibitor of mild steel corrosion in 1 M

HCl solution. Res. Chem. Intermediat. 38, 1707 -1717. DOI: 10.1007/s11164-012-0496-y

[22] Bouraoui, M.M., Chettouh, S., Chouchane, T., Khellaf, N., 2019. Inhibition Efficiency of Cinnamon Oil as a Green Corrosion Inhibitor. J. Bio Tribo. Corros. 5, 28. https://doi.org/10.1007/s40735-019-0221-0

[23] Appapalam, S.T., Panchamoorthy, R., 2017. Aerva lanata mediated phytofabrication of silver nanoparticles and evaluation of their antibacterial activity against wound associated bacteria, J. Taiwan Inst. Chemical Engrs. 78, 539-551. https://doi.org/10.1016/j.jtice.2017.06.035

Selection of aircraft materials and aircraft air conditioning technologies: A review

Shankar Durgam*, **Sweety Rajesh Deshmukh, and Preetam Selmokar**
Department of Mechanical Engineering, College of Engineering Pune - 411005, India
E-mail: *sodiitm@gmail.com

Abstract. This article reviews the demands of innovative material and the best refrigeration cycle to dissipate heat in airplane cabin. The suitable materials of excellent mechanical properties are used for aircraft applications that makes the aircraft to live longer, stay safe, reliable, and durable. Airplane is a special type of aircraft. The helicopter, and hot air balloon are aircrafts, but airplane is special type of aircraft which is weighing its body and it can be controlled and powered. Refrigeration in airplane, does not so far involve the use of any techniques which have not already established and used in other fields but of prime importance for comfort of occupants. The analysis for aircraft materials is briefly described with its significance. The material production is going to be innovated in various countries across the globe and among all the countries Russia is on top level for better using materials for aircraft production. The material used for aircraft production must be light, and efficient. It should not be affected by any impurities. This paper also gives the merits and demerits of such material and their sustainability. The motive of this review is to describe, briefly properties of materials for aircraft, and air conditioning technologies that best suited system for aircraft to dissipate heat.

Keywords: Materials for Aircraft, Comparative analysis of material, Reinforcement Material, Simple air refrigeration system, bootstrap air refrigeration system, reduced ambient air refrigeration cycle, Composite structural materials.

1. Introduction

Due to common or any resistance the materials which are useful in aircraft will get affected by humidity, salty rain then UV radiation for that material should be selected in proper manner by considering all the parameters should not get effected and having long lasting life [1]. The new development and the recent technologies find their best innovative ways for aircraft and space vehicle, Besides the Engineers are challenged to develop engines and other functions in aircrafts by suing various thermodynamic properties and reducing the weight and effective material for the aircraft structure [2]. The designed and development of new material and various manufacturing processes for aircraft uses having defined procedures that is followed and the long-life span [3]. For selection of material the weight ratio is major factor and the World Civil Transport Aviation have also described the overall efficiency and production of aircraft and they also conclude it is to be improved by 60 to 70% [4, 5]. In respect to achieve this all-strategies various product design methods the manufacturing process and the thermodynamics properties are essential [6]. The materials selected should have significant characteristics like maximum possible stress and fatigue, resistance to heat and corrosion, behaviour of crack growth, and ease of manufacturing. The structure of the materials used in aircraft is shown in Fig. 1. It consists of three types metallic, composite and non - metallic structural materials. This metallic material include elemental metal and compound or alloy. Non metallic seems to be used in construction industry and composites are versatile and used for both structural application and components, in all aircraft and spacecraft.

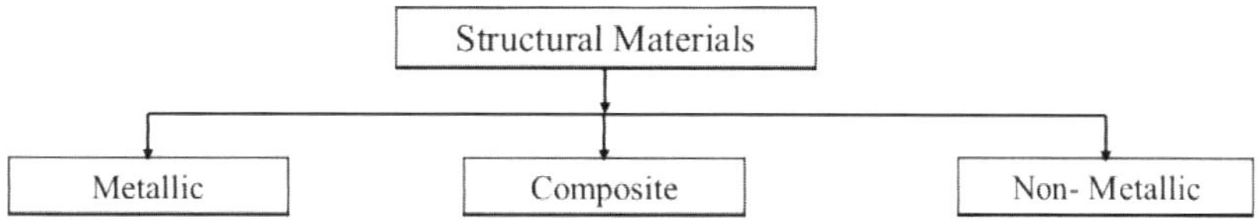

Figure 1. Structure of Materials

1.1. Equipment Heat and Composite Structured Materials
Most airplanes have quite heat loads from equipment. These may be of the order of two or three kilowatts for radio equipment or as much as fifteen kW of galley load. There are some cases of control which require nearly 100 kW of cooling for short periods although such cooling can usually be done at a high temperature level. Fig. 2 show the composite structured materials of three types which includes fiber glass plastics used in manufacturing of construction materials, ship building, sports gear. The carbon fiber plastics used in manufacturing of rocket engines and supersonic aircrafts. The reinforced plastic used in manufacturing of rocket socket parts, nuclear reactor construction materials. The detailed explaination of materials, material requirements for aerospace structures and engines, their importance and future advances in aerospace mateials for the global aerospace industry are appeared in [7].

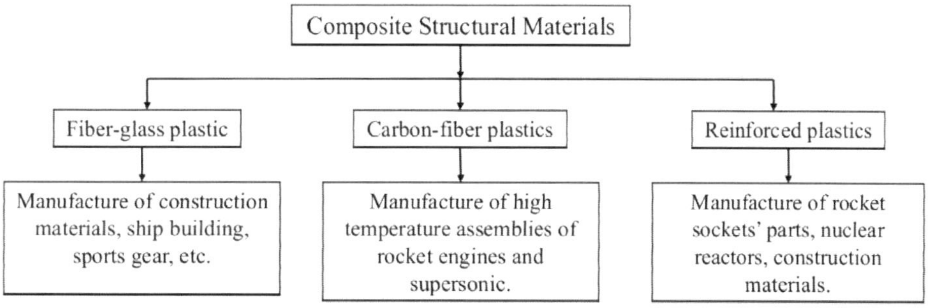

Figure 2. Classification of composite structural materials

2. Materials for airplane: solid monolithic composite structures (SMCS)
In solid monolithic structures, the mail beams stringers and the mainframes and ribs are used and rigidity is accomplished by maintaining the costs and also by engaging the longitudinal and transverse stiffeners it is shown in Fig. 3.

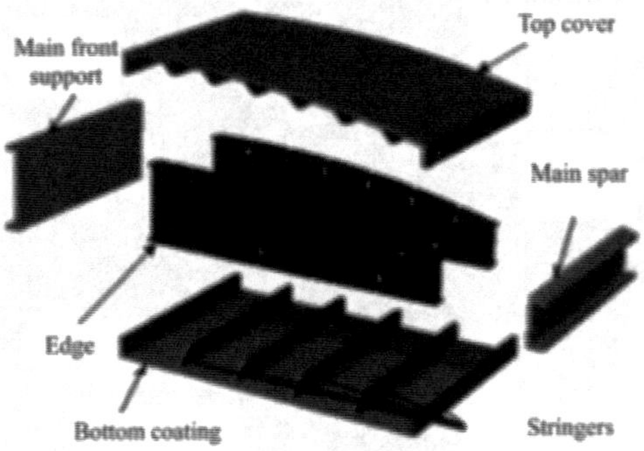

Figure 3. composite structure of solid monolithic [1]

2.1. Merits and demerits of Solid Monolithic Composite Structures (SMCS)
Merits of SMCS:

(i) It helps in reducing the maintenance costs.
(ii) It reduces any damages and resistance to shock.
(iii) Decrease damages due to humidity.

Demerits of SMCS:

(i) It has low weight ratio as compared to multi-layer structure.
(ii) it is used in torsion rigidity and bending.

2.2. Multi-layered Composite Materials (MLCM)
In this the plywood layer is most used and the cell shaped materials have no supplementary stiffness and are more frequently used in this layer structure, it is intended good interior and the exterior properties to which the materials are attached in this layer. Fig. 4 shows a composite structure of MLCM.

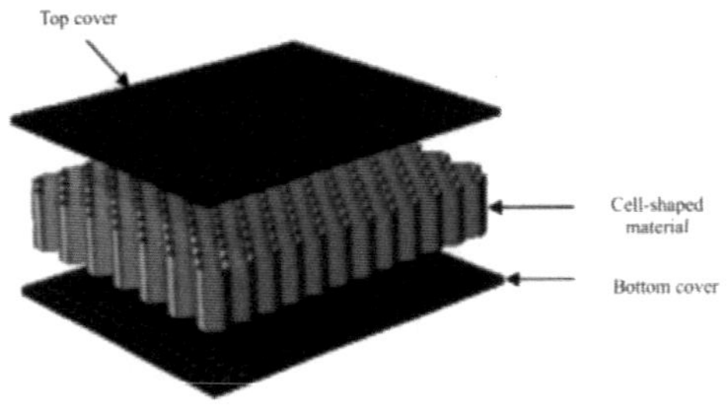

Figure 4. Composite Structure of Multi-Layer Composite Materials [1]

2.3. Merits and demerits of MLCM
Merits:

(i) The structural rigidity is high at the lowest weight.
(ii) The torsional rigidity is high.
(iii) It has excellent insulation to heat.
(iv) The damping properties are good.

Demerits:

(i) The moisture absorption property leads to destruction.
(ii) It has low resistance as compared to solid structures and also have high maintenance costs.

3. Requirements of cooling in airplane

The total heat flow into an airplane cabin is calculated by adding the heat flow through the walls by convection, the heat flow through the transparencies by solar radiation, the heat produced by the passengers and the equipment load. In the past it has been considered sufficient to assume that temperature of cabin is the cooling air temperature when it has taken up the heat load. This method has been shown to be inadequate in practice and a more sophisticated method of calculating the cooling required is now used. This new method, gives much better results and is based on calculating the conditions which must prevail to produce a mean skin temperature of 33 $°C$. For any passenger of crew member, it takes into account solar radiation, shade dry bulb temperature, cabin wall temperature, the metabolic heat production, the insulation of cabin air was as it effects the man, the insulation of clothing and the presence of heating or cooling by specialized garments. One of the most significant results of applying this technique is comparatively low air temperatures requirement. In the past, when a cabin was too hot, it has often been the practice to increase the cold air supply, but this is now shown to be of limited value. Airplane cruise at 10000 m altitude where ambient temperature is -50 $°C$ and pressure at 0.15 bar. Higher we go, cooler we find, is factual true.

4. Comparative analysis of innovative materials for aircraft

Fig. 5 shows the comparative analysis of innovative materials which are proved to be having great popularity in aircraft application. Conventional aluminum and alloys get replaced by titanium. As the demand for titanium increases due to this aircraft industry is in pursuit of solution to the problem of reduction in structural weight. Titanium is considered as main metal and mostly used metal in aircraft application. Nowadays the demands for titanium metals are 50% for Aircraft Industry, so that 50% of titanium is used in Aircraft Application. All the applications which are associated with composite material for aircraft are not possible without titanium metal. So that titanium plays major role in aircraft industry.

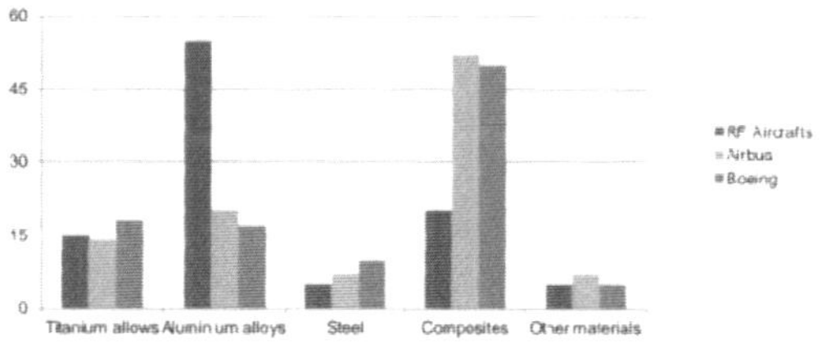

Figure 5. Comparative analysis of innovative material [1]

5. Reinforcement material

It is observed that the main component of composite material is reinforcement material which includes their mechanical strength, toughness, rigidity, reliability, and sustainability. The reinforcing materials namely long continuous filament and staple fiber threads or needles are commonly used for construction purposes. Mostly with having the filler material or without having it does not matter too much in nano technologies or if it is used then it is only using in both combinations. Examples of reinforcement material are carbon fiber graphite, glass fiber aramid, polyamide fiber, ceramic fiber mineral-based reinforcement fiber.

(a) Sukhoi superjet 100

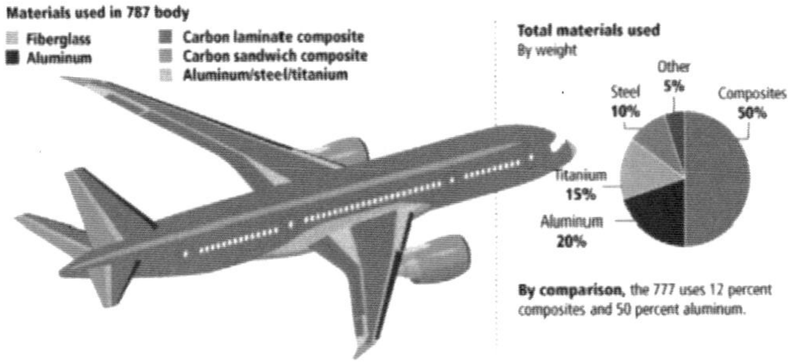

(b) 787 body materials

Figure 6. Body materials used in Sukhoi superjet and 787 [1]

6. Types of airplane refrigeration system

The comfort conditions need to be maintained in the airplane for the comfort of all occupants. There are four popular refrigeration systems which are commonly used in aircraft applications.

(i) Simple Refrigeration System.
(ii) Boot Strap Refrigeration System.
(iii) Regenerative Refrigeration System.
(iv) Reduced Ambient Refrigeration System.

The operation of expenditure is used for further compression of air to get maximum refrigerating effect. Simple and Boot strap have arrangement for evaporative cooling. Muller et al. [8] presented detailed simulation study of innovative aircraft air conditioning for maintaining aircraft cabin temperature and ventilation. Jianzhong et al. [9] studied aircraft air conditioning of airplane conditioning monitoring system for health state estimation and predictive maintenance to proactively plan futre maintenance. The airconditioning systems commonly used in aircraft are explained in the following sections.

6.1. Simple air refrigeration cycle

Fig. 7 shows the schematic and T-S diagram of simple air refrigeration system. Assume point 0 below the point 1 and the air is coming at certain velocity. Stagnation of air takes place due to this stagnation the pressure of the air rises we get state 1. This action is known as ramming action, so the kinetic energy of air is converted into pressure energy. And in this process the pressure of air also rises. This state 1 is also shown in Fig. 7 of simple aircraft refrigeration system. The compressor which is used for gas turbine in the T-S diagram at state 1 stagnated air is attain this air gets compressed in compressor and we get state 2 as shown in T-S diagram. This at state 2 air is at very high temperature as well as at high pressure. Now this air is cooled with air available at state 1 and cooling of air takes place in process (2-3). Now at state 3 there is expander, and it is used for expansion of air as we have pressure up to 0.15 bar, so air expands to 1 bar and get supplied to cabin. The power developed in the expander is used to run the fan this air is also used for cooling purpose.

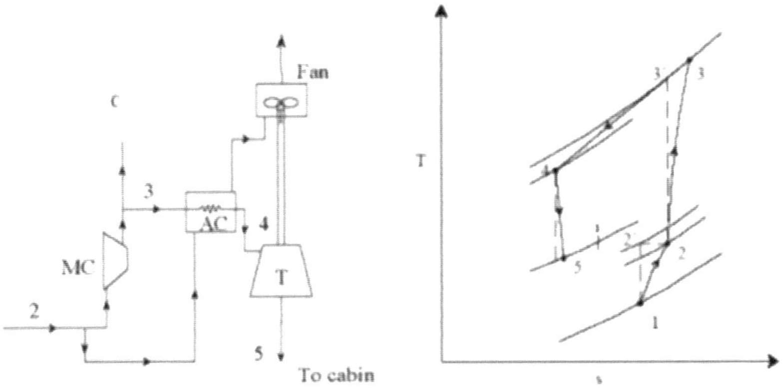

(a) Schematic diagram (b) T-S diagram

Figure 7. Schematic and T-S diagram of simple air refrigeration cycle [10].

6.2. Boot strap air refrigeration system

The process 0 to 1 is a ramming action takes place and compression takes place in compressor and after the compression cooling is done up to state 3 In a heat exchanger and this cooling is again done by the air available at state 1. After state 3 either compressor is provided powered by turbine which is also an expander. So, in T-S diagram the state 3 air is further compressed with another compressor and to get the state 4. After state 4 again cooling takes place and it reaches state 5, the cooling is also done with the air available at the entry of compressor, and after state 5 expansion takes place in expander and air is supplied to the cabin. Now here in this case expansion is taking place at higher pressure, so P4 is greater than P2 and if we expanded air from state 6 to state 3 have got higher temperature. Since the compressed cooled and it attend state 5 to low temperature. So, the benefit of boot strap type of refrigeration is that it is used for aircraft with readily high speed than in case of simple refrigeration cycle. Hence in boot strap cycle air fan for the purpose of cooling is not required. The fan can eliminate instead of this fan and power going to the expander of this fan it goes to compressor. The compressor is driven by expander, or the turbine and this compressor gives additional pressure rise in air. Results in lower temperature after expansion at state 6.

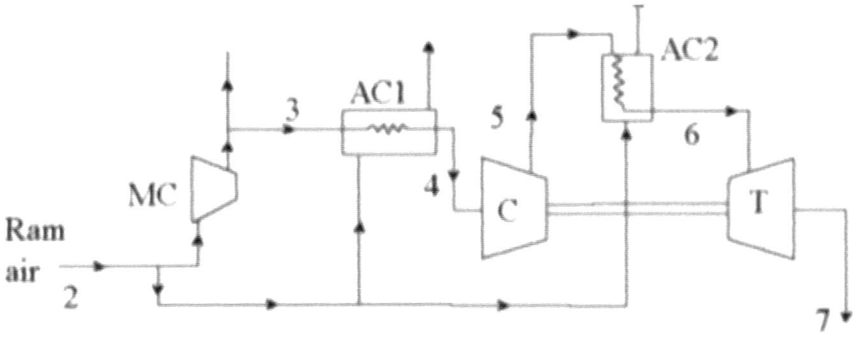

Figure 8. Schematic of boot strap air refrigeration system [11]

6.3. Boot strap air refrigeration cycle with evaporative cooling

Fig. 9shows the arrangement of boot strap refrigeration cycle with evaporative cooling. Some arrangement is made in boot strap evaporating cycle. In boot strap cycle there two types of compression first is state 2 and second compression is state 3 then the cooling state 4. From state 4-5 expansion takes place but, in this case, gas will get further cooled to 6 and expansion will take place up to state 7. Now evaporative cooler they provide addition cooling to gas which is entering the expander. By virtue of which we get cooler air at the exit of expander or turbine. Now evaporative cooling is attaining either with the help of refrigerant or liquid nitrogen. Now this evaporative cooling arrangement is very useful when aircraft is flying with very high velocity or supersonic velocity. Airplane cabin environment or temperature conditions and airflows are very important in view of comfort conditions for the occupants [12, 13, 14].

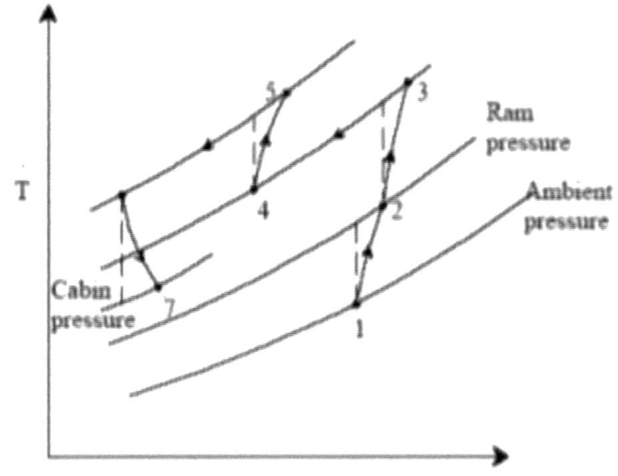

Figure 9. Boot strap air refrigeration cycle with evaporative cooling [15]

6.4. Material properties

The material properties play a significant role in performance, safety and life of a aircraft. Table 1 shows the mechanical properties of materials that ensures the strength and the durability of materials [16]. The material selected for a specific application of airplane should be such that it meets the requirements of strength, safety, durability, and life. Fig. 10 shows tensile yield strength for magnesium casting alloys from 2000 - 2009 and wrought alloys from 2009 - 2016. The values of yield tensile strength for wrought alloys are higher compared with casting alloys in both aluminium and magnesium. The yields strength for magnesium wrought alloy is 500 Mpa whereas that in case of aluminium wrought alloy is 700 Mpa.

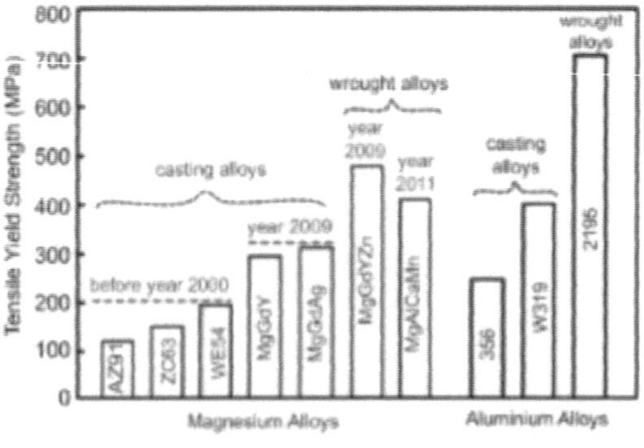

Figure 10. Tensile yield strength for magnesium and aluminum alloys [16]

Table 1. Material Properties [7]

Material	Specific gravity g/cm^3	Young's Modulus GPa	Specific modulus MPa m^3/kg	Yield strength Mpa	Cost per kg, in Rs.
Aluminum Alloy (7075-T76)	2.7	70	25.9	470	1250
Carbon/ epoxy composite	1.7	50	29.4	760	1600
Magnesium Mg	1.7	45	26.5	200	250
Polymer (epoxy)	1.2	3	2.5	100	1570
Titanium alloy Ti-6Al-4V*	4.6	110	23.9	1000	2150

Table 2 and Table 3 shows the advantages and limitations in terms of mechanical properties of thermoplastic, thermoset, and elastomer.

Table 2. Advantages of thermoplastic, thermoset and elastomer [17]

Thermoplastic	Thermoset	Elastomer
Non reacting, No cure required	Low processing temperature	low processing temperature
Rapid processing	low viscosity	High ductility
High ductility	Good compression properties	High fracture roughness
High impact resistance	Good fatigue resistance	High impact resistance
Absorb little moisture	Good creep resistance	—-
Recycling possible	high resistant to solvents, good fiber wetting for composites	—-

Table 3. Advantages of thermoplastic, thermoset and elastomer [17]

Thermoplastic	Thermoset	Elastomer
Non reacting, No cure required	Low processing temperature	low processing temperature
Rapid processing	low viscosity	High ductility
High ductility	Good compression properties	High fracture roughness
High impact resistance	Good fatigue resistance	High impact resistance
Absorb little moisture	Good creep resistance	—-
Recycling possible	high resistant to solvents, good fiber wetting for composites	—-

7. Results and discussion

To pressurize the aircraft cabin, air cycle air conditioning system prepares engine bleed air. At ground and at all altitudes the temperature and air quantity must be controlled to maintain a comfortable cabin environment. From the previous literature it is found that at ground simple air cooling system gives highest cooling effect but decreases at increased airplane speed. On the other hand, boot strap system requires the airplane to be in flight thereby using ram air for cooling in heat exchanger. For high-speed airplanes the boot stap system is preferred as it maintains a high pressure of air after ramming. As the ram air automatically flow over the heat exchanger tubes hence cooling fan is not required. It is found from this study that boot strap is most preferable over all the system for cooling purpose in airplane cabin. As the air is easily available non-toxic and non-inflammable and the refrigerant (air) in this system is free of cost. Also, in boot strap system the leakage of air in small amount that is tolerable, the chilled air is directly used for cooling thereby eliminating the cost of separate evaporator, in this system pressure in the whole system is comparatively low hence the piping, ducting is quite simple to design fabricate and maintain. By having the comparative analysis of material, the use of proper material is mandatory for aircraft like polyester and aluminum used on their contents by considering all the parameters.

8. Concluding remarks

From the review of selection of materials for aircraft structures and air conditioning technologies it is concluded that materials should be innovative light weight and should have good quality to ensure safety of aircraft and long life. It does not matter the material is metallic or nonmetallic it should have good mechanical properties, high strength and durability. Different materials have their own merits and demerits. Hence it is imperative to choose a suitable material for specific application depending on properties of the material. The refrigerator consists of two compartments one for frozen items and other for items requiring refrigeration but not freezing. It can remove all the heat from the loads inside the compartments and make it cooled for longer time. It is also suitable for applications in food preservation and medical industry. From this study it is concluded that for cabin cooling purpose boot strap air refrigeration cycle is most suitable. This contribution introduces that using the boot strap system it become very convenient to dissipate heat in the airplane cabin. It is possible to maintain comfort air conditioning under hot and humid weather conditions using desiccant-augmented evaporative cooling systems. However, the size required for this unit is much larger than that of vapor compression refrigeration systems used for this purpose. The COP values obtained under high humidity are only 0.4 that are quite low compared with other system such as absorption refrigeration based air conditioning machines.

References

[1] M. Smurov, A. Gubenko, T. Ksenofontova, V. Staroselets, Comparative analysis of innovative materials application in aircraft building of different countries, International Journal of Applied Engineering Research 12 (3) (2017) 394–401.
[2] K. Yıllıkçı, F. Findik, A survey of aircraft materials: Design for airworthiness and sustainability, Periodicals of Engineering and Natural Sciences 1 (1) (2013).
[3] S. Elbel, Historical and present developments of ejector refrigeration systems with emphasis on transcritical carbon dioxide air-conditioning applications, International Journal of Refrigeration 34 (7) (2011) 1545–1561.
[4] M. Merzvinskas, C. Bringhenti, J. Tomita, C. de Andrade, Air conditioning systems for aeronautical applications: a review, The Aeronautical Journal 124 (1274) (2020) 499–532.
[5] S. Jafari, T. Nikolaidis, Thermal management systems for civil aircraft engines: Review, challenges and exploring the future, Applied Sciences 8 (11) (2018) 2044.
[6] J. L. Warner, Multi-zone temperature control in air cycle refrigeration systems, uS Patent 4,445,342 (May 1 1984).
[7] A. Mouritz, Introduction to Aerospace Materials, Woodhead Publishing in Materials, Elsevier Science, 2012.
[8] C. Müller, D. Scholz, T. Giese, Dynamic simulation of innovative aircraft air conditioning, DGLR: Deutscher Luft-und Raumfahrtskongress. Bonn: Deutsche Gesellschaft für Luft-und Raumfahrt (2007).
[9] S. Jianzhong, W. Fangyuan, N. Shungang, Aircraft air conditioning system health state estimation and prediction for predictive maintenance, Chinese Journal of Aeronautics 33 (3) (2020) 947–955.

[10] S. Wright, G. Andrews, H. Sabir, A review of heat exchanger fouling in the context of aircraft air-conditioning systems, and the potential for electrostatic filtering, Applied Thermal Engineering 29 (13) (2009) 2596–2609.
[11] D. Leech, Aircraft refrigeration systems, Aircraft Engineering and Aerospace Technology (1961).
[12] E. H. Hunt, D. R. Space, The airplane cabin environment, TSP 3 (2) (1994) 1–000.
[13] J.-P. Emond, F. Mercier, M. Nunes, In-flight temperature conditions in the hold of a widebody aircraft, Proc. of 20th IIR/IIF. Sydney, Australia (1999) 19–24.
[14] Y. Zhang, Y. Sun, A. Wang, J. L. Topmiller, J. S. Bennett, Experimental characterization of airflows in aircraft cabins, part ii: results and research recommendations, ASHRAE transactions 111 (2005) 53.
[15] A. Ebeling, Fundamentals of Aircraft Environmental Control, Hayden Book Company, Madison, USA, 1968.
[16] V. Izquierdo, 8 - testing fabrics for flammability and fire safety, in: P. Dolez, O. Vermeersch, V. Izquierdo (Eds.), Advanced Characterization and Testing of Textiles, The Textile Institute Book Series, Woodhead Publishing, 2018, pp. 189–209.
[17] N. Barrington, M. Black, Aerospace materials and manufacturing processes at the millennium, Aerospace Materials 1 (2001) 3–14.

Modeling and simulation of SI engine muffler for prediction of backpressure by optimizing porosity and muffler diameter

Shankar Durgam*, **Snehal Thanekar, Sayali Hajare, Saharsha Bartakke, Chaitra Shet**

Department of Mechanical Engineering, College of Engineering Pune - 411005, India
*Corresponding Author's Email: sodiitm@gmail.com

Abstract. Air Pollution produced from automobiles has become one of the most critical problems of environmental concern and likely to increase as vehicle population is assumed to grow approximately by 1300 million by the year 2030. In India, automobiles will have to meet the BS-VI emission standards to reduce vehicular pollution that impacts on environment and human health. These norms impose various restrictions on emissions of exhaust gases like NOx, HCs, particulate matter. Many recent exhaust control devices are being developed to meet the required needs. Therefore, the present study aims to reduce CO_2 emissions generated in SI engine automobiles using activated carbon/calcite in a muffler. The backpressure parameter of the muffler affects the efficiency of an internal combustion engine. The porosity and diameter of the inner tube of the muffler has a prominent effect on backpressure. Therefore, the muffler with 48, and 60 holes, respectively and 1mm and 2mm diameter, respectively on a perforated tube have been investigated. The result shows that the concentration of CO_2 and back pressure is greatly reduced by increasing porosity and diameter, respectively of the tube. Thus, an environmentally friendly and efficient muffler is modeled and simulated.

Keywords: Muffler; Insertion loss; Flow; Backpressure; CFD.

1 Introduction

In recent times the automobile population is 1200 million and will increase by 2000 million in the next fifteen years. The forecast is, 2.5% of automobiles are electric vehicles, hybrid and fuel cell operated [1]. These harmful emissions thereby pose challenges to curb them. In view of this fact the reduction of harmful pollutants from conventional internal combustion engines are of prime importance. Beken et al. [2] have claimed carbon dioxide produced by emissions adds to global warming. De Lima et al. [3] have optimized various parameters in the field of silencers used to optimize the design. The authors found that the optimum efficiency of the inlet pipe profile without shape optimization in frequency ranges with gains up to 20 dB with respect to silencers. Blanco et al. [4] investigated the legal noise limitations and to meet this requirement it is necessary that BEM methodology is used to decrease noise radiation in which the acoustic variables in the outer fluid of silencer and interior fluid of silencer are calculated. Kant [5] investigated the design modification and CFD of two different types of car muffler, they found turbulence optimization was the main reason to decrease noise radiation caused by vehicles. Sherekar and Dhamangaonkar [6] found that adequate insertion loss, minimal backpressure, etc. are parameters required to design an automatic muffler. Hield [7] explained that the increase of pumping loop is directly proportional to the back pressure in a four-stroke internal combustion engine. This is because of the excess piston work to pump the gas outside of the cylinder during the final stroke along with constant load demand as shown in the volume ratio versus pressure plot, Fig. [1]. On the pressure axis the values shown are the logarithmic scale. Arrows show direction of change with increasing back pressure. The pressure ratio remains constant throughout the cycle and maximum pressure of the cycle reduces due to the reduction in the compressor pressure ratio. With the increase in back pressure, they found a decrease in gradient of the power stroke curve.

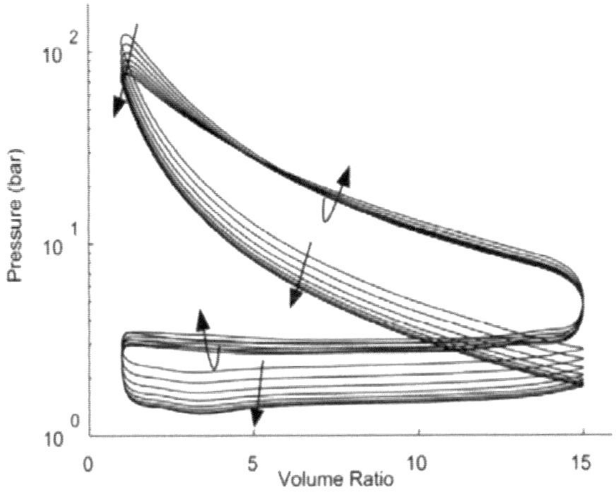

Figure 1. The pressure-volume diagram for various back pressures [7].

1.1 Exhaust chamber element having impact on backpressure

a. Head pipe: A strong peak power is a result of a small head pipe. Generally, large head pipes bring large bottom-end power at the cost of peak power. In conventional automobiles, tapered head pipes are performance boosters, but their drawback is that it is costlier and hard to manufacture.

b. Converging and diverging cones: Peak power is greatly affected by taper volume and length of divergent cones. The function of a divergent cone is to overrun i.e., to control power after peak. A relatively small, steep cone gives less overrun whereas a large, tapered cone gives higher overrun

c. Straight pipe: Dimensional adjustment makes amends for ideal head pipe, outlet pipe, convergent and divergent cone at mid-section.

d. Tailpiece: Perkins et al. [8] explained that tailpiece/outlet pipe dimensions affect peak power, bottom-end, and engine piston resistance. Outlet pipe diameter is inversely proportional to horsepower value and its thermal breakdown. It performs as a resonant shell which bounds with the muffler cavity. To control noise reduction in a muffler, a tailpiece is a must.

e. Perforations on inner pipe: Panigrahi et al. [9] found that higher porosity with small hole diameter implies large peak power along with legal backpressure value. Whereas higher porosity with larger hole diameters reduced the performance because of inadequate back pressure. When sound energy passes through the perforated tubes it gets dissipated considerably also it adds to the total attenuation. Kang et al. [10] investigated the numerical calculation of an acoustic perforated tube and they recorded its influence on elements. The researchers developed, (3-D FEM) to calculate the performance of a straight-through perforated muffler with gas flowing through it. In an internal combustion engine, a conventional muffler is manufactured with the help of perforated ducts, perforated baffle, and expansion chamber. Low fuel efficiency was a result of higher back pressure. A silencer based on counter-phase counteract as well as split-gas rushing was developed to solve the problem of silencer possessing bad characteristics of noise reduction in high exhaust resistance. Shao [11] have compared the aftereffects of new mufflers to those not having mufflers in the single-cylinder diesel engine. Mehdizadeh and Paraschivoiu [12] used a 3-D FEM approach to predict the transmission loss in mufflers as well as silencers with zero mean flow. A closed muffler and a parallel baffle silencer show the ability of FEM in the field of acoustic wave propagation modeling, a non-homogenous domain. These results also developed the Spectral Element Method (SEM) and more effective iterative solvers. By this, they figured out how to address large frequency ranges, as this 3-D FEM is costly, in computation for solving very high frequency acoustic fields. Much research was carried out in this field of regulation of post combustion by means of simulation or practical experimentations. Subramanian et al. [13] have done tests using a 2-cylinder compression ignition engine and checked CO_2 emission, and brake specific energy consumption by using calcite and activated carbon also without using these two materials. For CO_2 emissions, calcite showed better results compared with activated carbon. Sudarshan et al. [14] investigated mufflers and concluded that they form an important part of the exhaust system and are located near the rear end of your vehicle. They concluded that muffler dampening noises produced and reduced emissions. The opening of the exhaust valve brings a sudden mixture of high temperature exhaust gases. Muffler has multiple baffles in its design aimed at inducing resonance. They found that this reduction of flow of exhaust in the muffler induces a back pressure. Results show that back pressure is reduced by increasing the hole diameter. From the above literature it is found that though some studies are available on the SI engine muffler but the focus on computation of reduction in back pressure for efficiency is scarce. Therefore, this paper aims to study modeling and CFD simulation to reduce back pressure and increase engine efficiency by increasing the number of holes and hole diameter on perforated inner pipe. More exhaust in the system means higher back pressure, or more importantly more resistance to flow of gases from the cylinder, which means that lose valuable engine power to it. The result shows that the concentration of CO_2 is greatly reduced.

2 Materials and Methods

2.1 Modeling and CFD simulation for muffler and optimization of back pressure

Three-dimensional model of the muffler in Ansys design modeler is shown in Fig. 2. The air passed having specific mass flow rate and the pressure variation is measured. Inside the exhaust tube system of an engine, the back pressure and pressure drop are the same. Ideal steady state time conditions are applied. Constant mass flow rate is chosen hence the flow is subsonic. Based on density and pressure change with respect to the normal speed of sound, a flow becomes compressible flow for $M \geq 0.3$ and incompressible for $M < 0.3$. Therefore, by a variety of physical conditions, computational schemes are dictated like compressibility-incompressibility of flow, viscosity. The properties of test fuel are shown in Table 1, and the muffler dimensions in Table 2.

Table 1. Test fuel properties

Properties	**Unit**	**Values**
Density	kg/m^3	1.225
Temperature	K	773
Velocity	m/s	40
Viscosity	kg/m-s	1.7894e-05
Specific Heats Ratio	-	1.4

Table 2. The muffler dimensions

Inner perforated tube diameter	30mm
Outer tube diameter	60mm
Length	100mm

Figure 2. Three-dimensional model of the muffler in Ansys design modeler.

2.2 Solver model and inputs used for CFD simulation
For incompressible flow pressure-based solvers are used to avoid oscillations in the pressure field. This occurred because of lack of prevention of incompressibility conditions and/or mass conservation as the sound speed becomes greater than components of conventional velocity. Hence, to avoid these incompressibility conditions, use pressure-based solvers having pressure correct algorithms.

Spatial Discretization

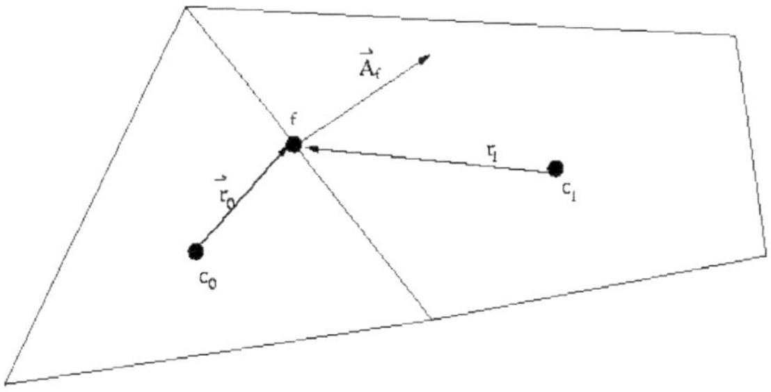

Figure 3. Volume Control to show discretization of a transport equation.

By default, Ansys Fluent at the cell centres (c_0 *and* c_1 in Fig. 3 stores discrete values of the scalar ϕ. However, face values ϕ_f is required for the convection terms in Eq. 1 and must be interpolated from the values of cell centers. This is proficient using an upwind scheme. The meaning of unwinding is that the face value ϕ_f is derived in the cell upstream, or "upwind", from quantities relative to the direction of the normal velocity v_n in Eq. 1.

$$\frac{\partial\rho\phi}{\partial t} + \sum_f^{N_{faces}} \rho_f v_f \phi_f \cdot A_f = \sum_f^{N_{faces}} \mathrm{T}_\phi \nabla_{\phi_f} \cdot A_f + S_\phi V \tag{1}$$

2.3 Pressure based solver

In a fluent solver, the variables are located at the middle of grid cells. Next step is to solve equation (2) and (3) the transport equation. To do so the values at the control volume faces and its gradient are required. The quantities required mostly are turbulent kinetic energy, momentum, and turbulent dissipation rate. In Ansys fluent this interpolation is known as upwind schemes. Thus, while interpolating the centre cell, a variety of interpolating schemes are available in fluent solver.

$$\oint \rho v \cdot dA = 0 \tag{2}$$

$$\oint \rho v \cdot A = -\oint p\mathbf{I} \cdot dA + \oint \bar{\bar{\tau}} \cdot dA + \int_V F dV \tag{3}$$

2.4 Discretization of the momentum and continuity equation

After integration of equation (2) over control volume in Figure [3], implies the following discrete equation (4) where J_f is the mass/face flux, ρv_n

$$\sum_f^{N_{faces}} J_f A_f = 0 \tag{4}$$

To move in next step it is important to compare velocity face values, $\vec{v_n}$, to the cell centres velocity value that are already stored. Linear interpolation of the velocity of the Cell-centre to the results of the face in an unphysical interrogator board of pressure. Linearly the face value of velocity is not averaged; alternatively, momentum-weighted mean with the help of these elements based upon the coefficient of Eq. 5 is performed.

$$a_p u = \sum_{nb} a_{nb} u_{nb} + \sum p_f A \cdot \vec{i} + S \tag{5}$$

By this method, the face flux J_f is depicted in Eq. (5), where v_{n,c_0} , v_{n,c_1} and p_{c_0} , p_{c_1} are normal velocities and pressures respectively. Inside these two cells on either side of the face, also Jf contains the impact of velocities in these cells (see Fig. 3). The term d_f is a function of the term $\vec{a_p}$, the mean of the momentum equation ap is a coefficient on only one side of the face f.

$$J_f = p_f \frac{a_{p,c_0} v_{n,c_0} + a_{p,c_1} v_{n,c_1}}{a_{p,c_0} + a_{p,c_1}} + d_f((p_{c_0} + (\nabla_p)_{c_0}) - (p_{c_1} \cdot \vec{r_1})) = \vec{J_f} + d_f(p_{c_0} - p_{c_1}) \tag{6}$$

2.5 Pressure-velocity coupling

Pressure velocity coupling is an algorithmic term carried out with the help of Eq. 6. It is an equation consisting of discretization of the momentum and continuity equation in it. Additional condition for pressure from continuity Eq. 4 is found. These combined equations are used for derivation work of pressure in pressure-based solvers. Few algorithms are already available in it. Therefore, SIMPLE semi-implicit splitting of operators is used. It gives necessary solutions without the need of iteration i.e., it has less computing but more time steps. Deriving the face value from the above-mentioned quantities with respect to direction of normal velocity in a cell is called unwinding. From by default schemes, a first order unwinding method is used for interpolating these quantity values to minimize computation. Gradient is required to build values of a scalar at the faces cell as well as computing secondary diffusion terms, derivatives of velocity. This gradient reduces the convection and diffusion in the flow of the conservation equation. Therefore, to find the gradient, Green Gauss node method is used. The typical mesh used for all simulations is shown in Fig. 4. Accurate value of linear function from the surrounding centre of cell values at the node on random meshes by solving a constrained minimization problem is found by this scheme. The node-based gradient is more precise as compared to other schemes especially on skewed, distorted meshes with more computing price. To find the pressure of the cell face a standard interpolation scheme and a pressure-based solver is used.

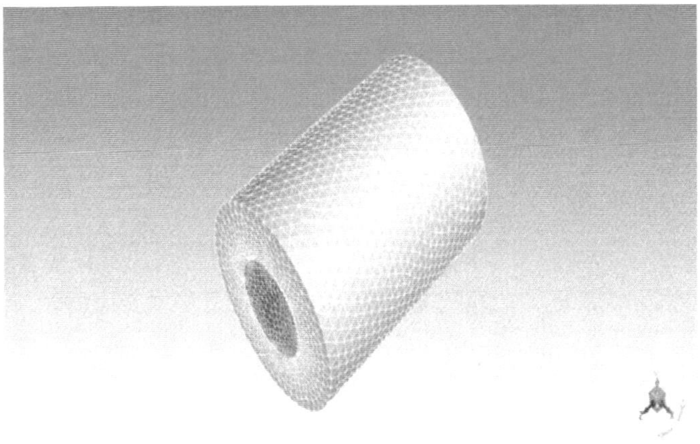

Figure 4. Typical Mesh used for turbulence models.

Mostly the fluids considered are turbulent, turbulence comes into picture when there are higher velocity gradients that result in oscillations i.e., flow area as a function of time, space. Turbulent flows are 3D in nature. These types of turbulent flow occur in contact with walls and/or within neighboring layers of variant velocities. As velocity gradient increases, the flow turns rotational in nature, resulting in a stretch of vortex lines. Small and large continuous energy scales are mixed which are directly dependent on the size of the Eddies motions in turbulent flow. Eddy motions are overlapping with larger eddies carrying the smaller ones. There is a transfer of turbulent kinetic energy from larger eddies to smaller eddies. The smaller eddies dissipate into form of heat by molecular viscosity. Hence, an approximate turbulent model is mandatory. Standard K-Epsilon (SK-ϵ) model is used to represent turbulence in various ranges of industrial applications. In isotropic turbulence, SK-ϵ represents lower speed of incompressible flow. The SK-ϵ model consists of an equation of turbulence model and it determines both equations. These two equations consist of turbulent length and time scale. It depends on transport equations for the dissipation rate equation and Kinetic energy given in the equation. The turbulent kinetic energy (k) is given by Eq. 7.

$$\frac{\partial}{\partial t}(\rho k) + \frac{\partial}{\partial x_i}(\rho k u_i) = \frac{\partial}{\partial x_i}\left[(\mu + \frac{\mu_t}{\sigma_k}\frac{\partial k}{\partial x_j})\right] + P_k + P_b - \rho\epsilon - Y_M + S_k \quad (7)$$

The turbulent dissipation rate (ϵ) is given by Eq. 8.

$$\frac{\partial}{\partial t}(\rho \epsilon) + \frac{\partial}{\partial x_i}(\rho \epsilon u_i) = \frac{\partial}{\partial x_i}\left[(\mu + \frac{\mu_t}{\sigma_\epsilon}\frac{\partial \epsilon}{\partial x_j})\right] + C_{1\epsilon}\frac{\epsilon}{k} + (P_k + C_{3\epsilon}P_b) - C_{2\epsilon}\rho\frac{\epsilon^2}{k} + S_\epsilon \quad (8)$$

From the equation, the model transport equation for K is obtained and from physical reasoning, the dissipation rate (ϵ) is obtained. While deriving the K-ϵ model, assume flow to be purely turbulent and neglect the molecular viscosity effects. Therefore, for pure turbulent flow the K-ϵ model is valid [15] [16].

3 Results and Discussion

The simulation results for four cases, one with 48 holes and another with 60 holes on a perforated tube, third with 1mm diameter keeping number of holes constant and last case with 2mm hole diameter, are presented in this section. The element size considered was 0.5 mm with tetrahedral mesh and performed 500 number of iterations to converge the problem. It is observed that as mesh size decreases the accuracy increases. So, the results will be more accurate and reduction in back pressure.

3.1 Case I: The muffler with 48 Holes on Perforated Tube

The pressure contour 2 for a one-meter length muffler is shown in Fig. 5. It has 48 holes on it and this diagram represents the pressure distribution. At the inlet of the perforated tube, the pressure is found high whereas on the outer side of the tube, it is low. The number of holes on the tube is less therefore the backpressure obtained here is very high.

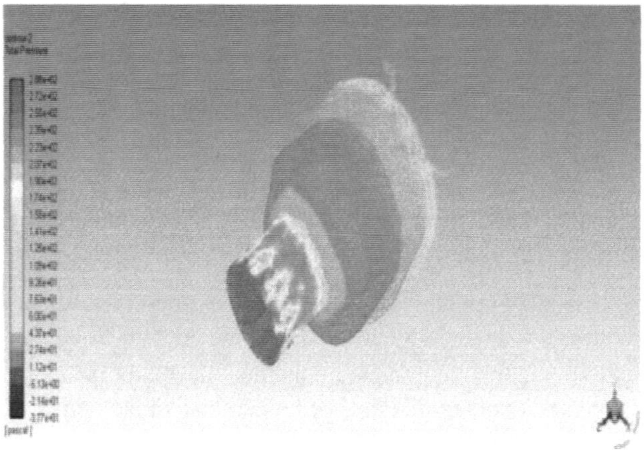

Figure 5. Pressure contour-2 for muffler with 48 holes on perforated tube.

3.2 CaseII: The muffler with 60 holes on perforated tube

The muffler considered in this case has different porosity as compared to the previously considered muffler. In this case, the number of holes is increased on the inner side of the tube. The pressure contour-1 diagram with 60 holes on it is shown in Fig. 6.

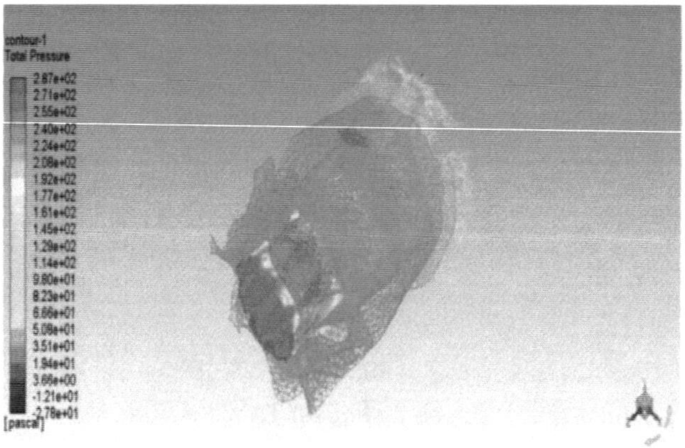

Figure 6. Pressure contour-1 for muffler with 60 holes on perforated tube.

This also shows the pressure variation across the muffler. By observing the simulated results in both cases, pressure obtained in Fig. 5 is higher than that in Fig 6. Therefore, less back pressure is obtained. Also, greater number of holes on outlet pipe implies high air flow velocity hence low back pressure that results in low obstruction and higher engine efficiency.

3.3 Case III: The muffler with 1mm diameter

The pressure contour 2 for a one-meter length muffler, 1mm diameter is shown in Fig. 7. At the inlet of the perforated tube, the pressure is found high whereas on the outer side of the tube, it is low. The size of the perforation is less therefore the backpressure obtained here is very high.

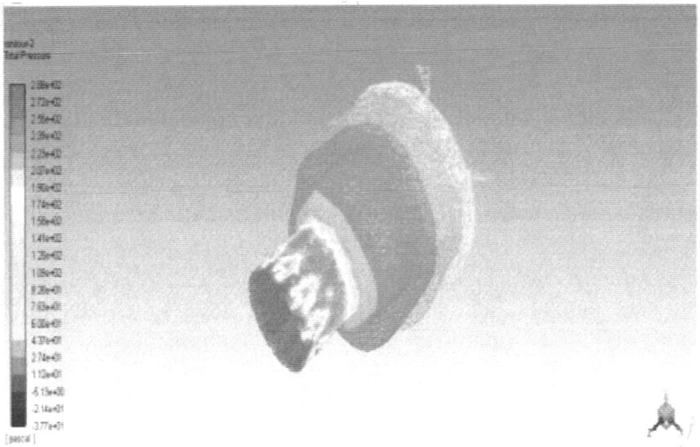

Figure 7. Pressure contour-2 for muffler with 1mm hole diameter.

3.4 Case IV: The muffler with 2mm diameter

Pressure contour 2 for a 2 mm diameter on 1m length is shown in fig8. This also shows the pressure variation across the muffler.

Figure 8. Pressure contour-2 for muffler with 2mm hole diameter.

By observing the simulated results in both cases, pressure obtained in case of 1 mm diameter is higher than that in case of 2 mm diameter. Therefore, less backpressure is seen with the increase in hole diameter.

3.5 Effect of number of holes on pressure drop

Fig. 9 show the variation of back pressure with the number of holes. To study the effect of number of holes on back pressure, different cases with 36, 42, 54 holes have been considered and the results are presented. It is found from the present study that by increasing number of holes results in a reduced back pressure significantly.

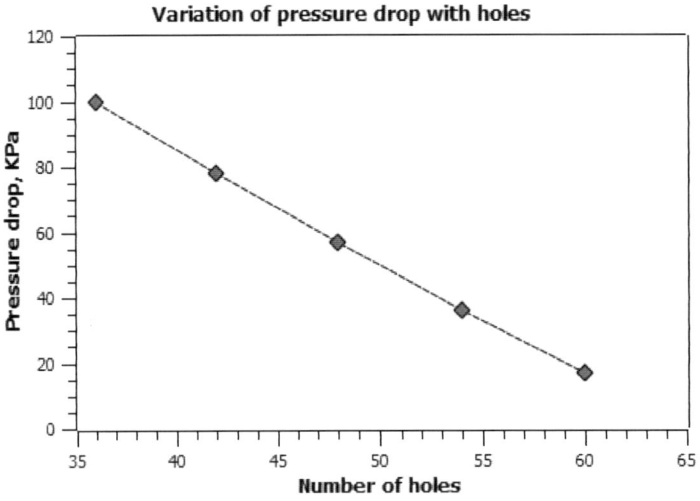

Figure 9. Variation of back pressure with number of holes.

4 Conclusions

From the study of modelling and simulation of SI engine muffler for prediction of back pressure by optimizing porosity and diameter results in the following conclusions:

1. The simulations carried out pointed towards the effect of porosity and diameter of the inner tube on back pressure produced. As the number of holes on the perforated inner pipe as well as hole diameter increases, back pressure (pressure drop) reduces significantly that helps in improving the efficiency.
2. The activated carbon/calcite which can be applied to the walls of tubes of muffler is used to reduce pollutants to a considerable extent due to their properties.
3. Thus, the modeling and simulation study of an eco-friendly muffler accomplishes the aims that reduces pollution by increasing engine efficiency due to reduced backpressure and curbs the emission of harmful pollutants

5 NOMENCLATURE

$\vec{A}$ = Surface area vector
A_f = face (f) area, |A| (= $|A_x \vec{i} + A_y \vec{j}$— in 2D)
CO_2 = Carbon dioxide
F = force vector
HCs = Hydrocarbons
NO_x = Oxides of Nitrogen

N_{faces} = number of faces in a cell
υ = volume cell
I = Identity matrix
p_{c_0}, p_{c_1} = pressures in cell
$\upsilon_{n,c_0}, \upsilon_{n,c_1}$ = normal velocities in cell
M = Mach Number
K = Turbulent kinetic energy
S = Modulus of mean rate of strain tensor

References

[1] F. Aktas, N. Dinler, S. Karaaslan, A. Turker, N. Yucel, Experimental investigation of cold start emission using preheating system on the exhaust line at the idle conditions on a spark ignition engine, Sādhanā 46 (3) (2021) 1–11.
[2] M. Beken, B. Hangun, O. Eyecioglu, Classification of turkey among european countries by years in terms of energy efficiency, total renewable energy, energy consumption, greenhouse gas emission and energy import dependency by using machine learning, in: 2019 8th International Conference on Renewable Energy Research and Applications (ICRERA), IEEE, 2019, pp. 951–956.
[3] K. F. De Lima, A. Lenzi, R. Barbieri, The study of reactive silencers by shape and parametric optimization techniques, Applied Acoustics 72 (4) (2011) 142–150.
[4] J. Blanco, A. Earnshaw, Study of the noise characteristics of motorcycle silencers, Tech. rep., SAE Technical Paper (2001).
[5] B. Shailender, A. R. Kanth, Computational fluid dynamics analysis of a resistance muffler, International Journal of Innovative Technology and Research 5 (3) (2017) 6409–15.
[6] V. Sherekar, P. Dhamangaonkar, Design principles for an automotive muffler, International Journal of Applied Engineering Research 9 (4) (2014) 483–489.
[7] P. Hield, The effect of back pressure on the operation of a diesel engine, Tech. rep., Defense Science and Technology Organization Victoria (Australia) Maritime (2011).
[8] J. Perkins, J. Robichaud, B. Ellis, Advanced two-stroke tuned exhaust system (2002).
[9] S. Panigrahi, M. Munjal, A generalized scheme for analysis of multifarious commercially used mufflers, Applied acoustics 68 (6) (2007) 660–681.
[10] Z. Kang, S. Zheng, X. Lian, Z. Ji, Study on the acoustic performance of perforated duct muffler, in: 2010 International Conference on Mechanic Automation and Control Engineering, IEEE, 2010, pp. 2346–2349.
[11] Y.-l. Shao, A study on exhaust muffler using a mixture of counter-phase counteract and split-gas rushing, Procedia Engineering 15 (2011) 4409–4413.
[12] O. Z. Mehdizadeh, M. Paraschivoiu, A three-dimensional finite element approach for predicting the transmission loss in mufflers and silencers with no mean flow, Applied acoustics 66 (8) (2005) 902–918.
[13] T. Subramanian, A. Sonthalia, E. G. Varuvel, Effect of calcite/activated carbon-based post-combustion co2 capture system in a biodiesel-fueled ci engine—an experimental study, Energy Sources, Part A: Recovery, Utilization, and Environmental Effects 41 (16) (2019) 1972–1982.
[14] S. D. Pangavhane, A. B. Ubale, V. A. Tandon, D. R. Pangavhane, Experimental and cfd analysis of a perforated inner pipe muffler for the prediction of backpressure, International Journal of Engineering & Technology 5 (5) (2013) 115–119.
[15] Ansys user's guide.
[16] Ansys theory guide.

Investigations on fracture characteristics of Aluminium 6082-T6 alloy by Using Single Edge Notch Bending

Tippeswamy M S[1, a *], Arun K V[2, b*] and Shankar Naik[3,c]

[1]Department of Mechanical Engineering, Vemana Institute of Technology, Bengaluru-560034, India
[2,3]Department of Mechanical Engineering, Government Engineering College, Haveri-58110, India

[a]*mst.aradhya@gmail.com*, [b]*bdt.arun@gmail.com*, [c]*shankarnaiksb@gmail.com*

Abstract. Aluminium 6082-T6 is a structural alloy having high strength has seen replacing 6061 alloy in engineering applications. During manufacturing process geometric discontinuities are inevitable. The presence of notch in the component is more dangerous. In this paper, the influence ofnotch length on miniature fracture characteristics of aluminium 6082-T6 alloy has been studied. Experiment was carried in accordance with ASTM standard and specimens were tested by varying the notch length within the specified constraint. Single edge notch bending test was performed on each of specimens under mode-I condition. Results showed that the specimen with minimum notch length has high crack tip initial instability, further increase in the notch length leads to decrease the load carrying capacity, extended portion of crack length and nature of fracture. On experimental observation, plastic zone size near the end of the cut is majorly dependent on notch length of the specimen. The fractured surface analysis also carried on selected specimen with the help of scanning electron microscope. Overall results shows that crack growth and crack propagation vary as a functionof notch length and is different in each of the specimens.

Keywords: Aluminium 6082-T6, Stress intensity factor, Crack tip instability, SEM images

1. Introduction

Aluminium wrought alloys have gained much attention in engineering applications because of their beneficial properties: i.e., high strength-to-weight ratio, ease of fabrication, high degree of workability, considerable ductility and attractive appearance [1,2]. Among all different series of aluminium alloys, 6000 series alloys are being used more widely in the field of automobile, aviation and ship building applications. In many cases aluminium replaces steel due to its less low mass density, cost, etc., especially in case of tubes [3]. 6082 alloy in wrought aluminum magnesium-silicon family possesses excellent corrosion resistance and high strength. The addition of manganese controls grain structure which inturn results as stronger alloy and has seen to replace 6061 in many aeronautical applications [4-6].

Aluminium beams are being used in electric passenger car and even in construction industry owing to their excellent corrosion resistance. Structural parts made up of aluminium alloys generally fails because of tensile, compression, eccentric compression, buckling, corrosion and other environmental effects [7-9]. Predominant failure mode under most of the aluminium parts fails is bending. Fatigue failure of weld joints subjected bending results in astonishing failure. Aluminium alloys used in ship structures merely subjected to vertical hogging and sagging bending moment due to impact of large waves [10]. Researchers have shown much attention on failure of the materials subjected to bending load.

Fracture tests with surface cracks, corner cracks etc., dictates the behaviour of the material in the presence of flaws. There are several methods to find crack driving force as a function of crack size. Among them single edge notch bending [SENB] test and compact tension small-scale specimens are commonly used in industries to evaluate the fracture toughness of the materials [11]. SENB test can be conducted without fixtures, which involves triaxial state of loading condition as crack growth occurs. Effect of notch sensitivity, crack growth, crack propagation etc., can be validate qualitatively by using SENB test specimens [12-14].

Microscopic images are essential to characterize the materials in the field of material science. Scanning electron microscopy (SEM), transmission electron microscopy (TEM) and scanning tunneling microscopy (STM) are the most powerful electron microscopes used to characterize the materials. Micromechanics, surface morphology, chemical composition, crystal structure orientation, phase discrimination, etc., of solid materials can be potentially discuss with the help of SEM images compare to other microscopes. SEM images are also used in the field of surface engineering, fracture mechanics etc., to witness type of bond at interface of coating, crack behaviour etc., [15-18].

From the above survey, very less work has been carried out on aluminium 6082-T6 SENB specimens and extensive work have been done using FEM. In this an attempt has been made to study the notch sensitivity of 6082-T6 SENB specimens through appropriate methods.

2. Materials and Experimentation

To investigate the effect of notch length on crack growth, quasi-static fracture test has been conducted on wrought aluminium alloy. The properties of the selected alloy, specimen preparation and experimental procedure is as follows

2.1 Material

Aluminium 6082-T6 has high strength in 6000 series and has excellent corrosion resistance among their family. The addition of manganese in large amount results in stronger alloy by controlling grain structure. It also possesses high strength to weight ratio and good wear properties when it is used as structural alloy in high stressed applications. The chemical composition and properties of aluminium 6082 alloy are given in the following Table 2.1 and Table 2.2 respectively.

Table 2.1 Elemental composition of Al6082-T6 in wt%

Element	Si	Fe	Cu	Mn	Mg	Zn	Ti	Cr	Al
Composition percentage	0.7 - 1.3	0.5	0.1	0.58	0.6 - 1.2	0.2	0.1	0.25	97.14

Table 2.2 Properties of Al6082-T6

Parameter	Value
Density	2.70 g/cm^3
Hardness (Vickers)	95 HB
Young's modulus	71 GPa
Ultimate tensile strength	330 MPa
Yield strength	290 MPa
Poisson's ratio	0.33
Thermal expansion	23.1 μm/m-k
Melting point	555 °C

2.2 Specimen preparation

Aluminium 6082 alloy procured from the market in the form of sheet and cut into required dimension for conducting single edge notch bending (SENB) test. Figure 2.1 shows schematic and pictorial representation of SENB specimen. Specimen were prepared in accordance with ASTM E399.

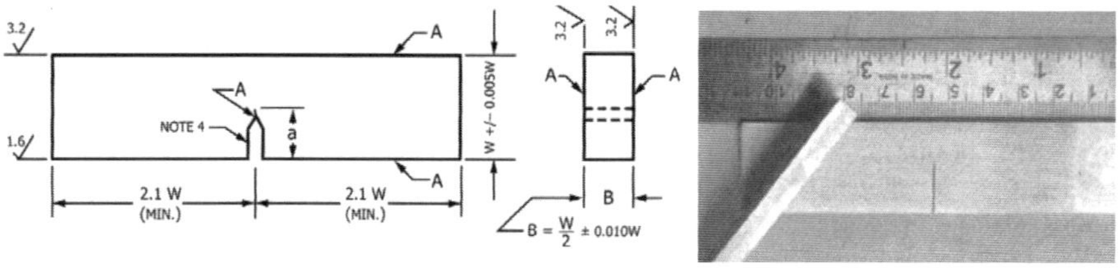

Fig 2.1 Schematic and pictorial representation of SENB specimen.

Notches were introduced into specimens by using wire EDM with a width of 0.2mm and the notch length was varied with maintaining the ratio of notch length to width of specimen within the range $0.45<\frac{a}{W}<0.55$

2.3 Experimentation

The main intention of fracture test is to find the resistance of the specimen in the presence of flaws. Three-point bending test is the one which enables us to test the specimens with mixed mode of loading a typical bending test includes the combined shear and tensile/ compressive stresses. Single Edge Notch Bending is as same as the plain bending which is used to find the fracture toughness of a specimen.

Fig 2.2 Experimental set-up, during fracture test and broken specimen.

The specimen is placed in vertical shape notch inverted position such that the application of load becomes trans laminar load. As the load applied on the face vertically opposite to the notch, notch tends to open as the load increases which results in mode-I type of failure. The test was carried on the three-point bending by using 5Ton capacity UTM. The distance between the centers were 100mm and the compressive load was acting at the mid span of the specimen. An experimental set-up, specimen during fracture test and broken specimen is as shown in the Figure 2.2.

3. Result and Discussion

The effect of notch size, variation of peak load, determination of fracture toughness and micro-mechanism of crack growth was studied by using SEM images.

3.1 Effect of notch size on the load bearing capacity of SENB specimen

SENB tests were conducted for all the varied notch length specimen. During bending test upper portion faces the compression load while lower portion undergoes tensile load. Figure 3.1 shows load-deformation curves for the tested specimen having maximum and minimum notch length. It was observed that, the trend remains same for both specimen under bending load but, the specimen with

11.75 mm notch length has high strength compared to specimen with 13.75 mm notch length. Load-deformation curve for both maximum and minimum notch length is as shown in the figure 3.1. It can clearly observed that, deformation of the notched specimens tend to increase as the load increases up to points a and b, after that initiation of the crack length takes place and propagates.

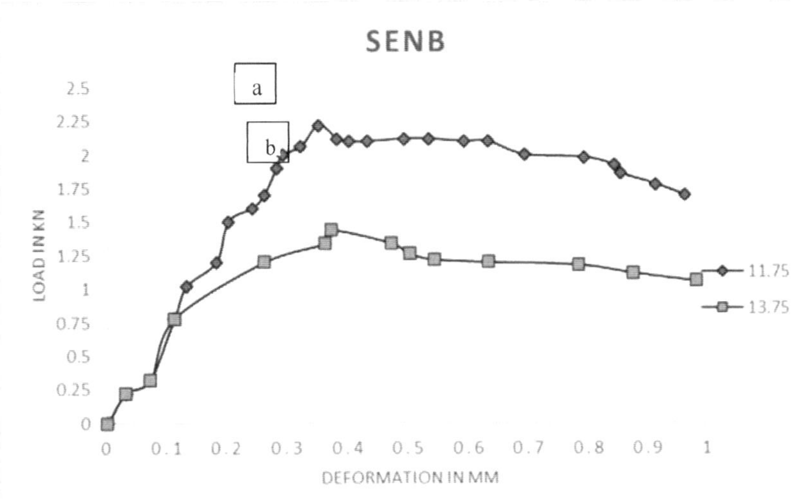

Fig 3.1 Load-deformation curve for maximum and minimum notch length

Points a & b in the Figure 3.1 represents load bearing capacity of the specimen in the presence of notch length. Slope for the specimen having 11.75 mm notch length is higher than the slope of specimen having 13.75 mm notch length. Both the specimens exhibit same nature of failure till initiation of crack length, but stiffness varies individually. Intermediate notch length specimen also exhibit same behaviour as that of 11.75 mm and 13.75 mm specimen and the load-displacement curves lies in between them. Figure 3.2 shows fractured surface of notched specimens excluding 11.75 mm notch under bending loading condition.

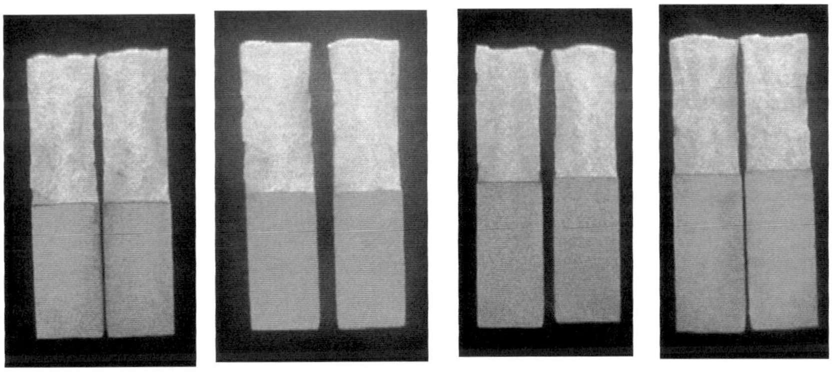

Fig 3.2 Fractured specimens showing 12,12.5,13,13.75 mm notch length

The nature of crack propagation is not same in all notched specimen after initiation of crack as the capacity to withstand load will differ as a function of notch length and can be seen from the bar chart as shown in the figure 3.3. It is also clear from the bar chart that, peak load will tend to decrease as notch length increases from 0.45W to 0.55W.

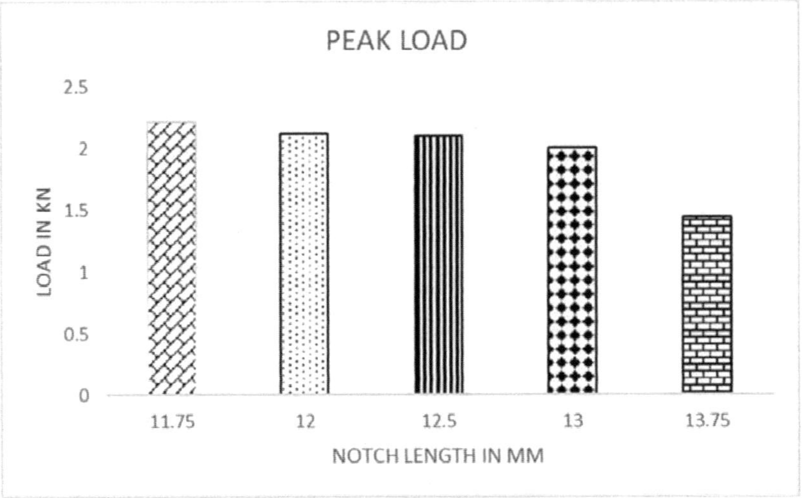

Fig 3.3 Bar chart showing variation in peak load.

3.2 Fracture toughness

SENB tests were conducted on varied notch length to find the fracture toughness of the material. Stress intensity factor is determined by using energy criteria, Energy release rate is the main factor required to calculate the fracture toughness and is defined as the energy required for crack tip propagation. Figure 3.4 shows variation of compliance with crack size.

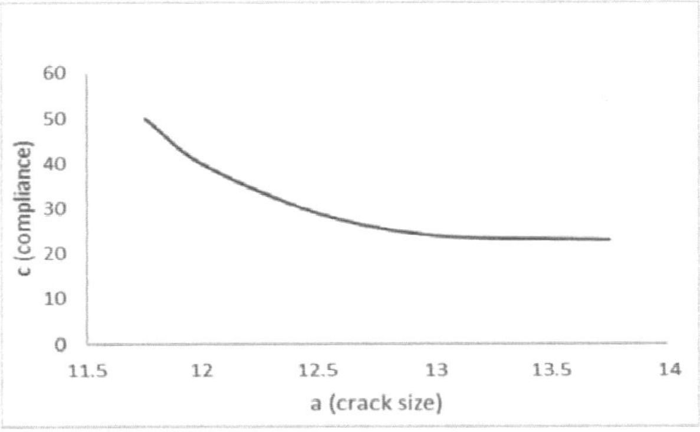

Fig 3.4 Compliance-crack size curve.

By using relation G=K2 /E the fracture toughness of aluminium 6082-T6 is found to be 17.41Mpa$\sqrt{m}$. where G is the energy release rate, E is modulus of elasticity and K is stress intensity factor under mode I.

3.3 Micromechanics of crack growth

The nature of plastic flow at the end of the crack tip will serve as basis for crack models. A plastic zone is developed at the end the end of the cut(crack) during SENB test. It is clear from the figure 3.5 that considerable strain, which appears as necking on the surface of the specimen along the direction of thickness. The shape and size of the plastic zone will depend upon the type of test, material and loading condition.

Fig 3.5 Formation of plastic zone at the end of cut.

The crack will tend to open as cohesive force decreases and the width of the crack widen. From theschematic drawing of various regions of fracture specimen i.e., figure 3.6 predominantly square fracture was observed after crack growth. SEM images were taken at interested regions for the fractured surface of 11.75 mm notch length specimen and is as shown in the figure 3.6.

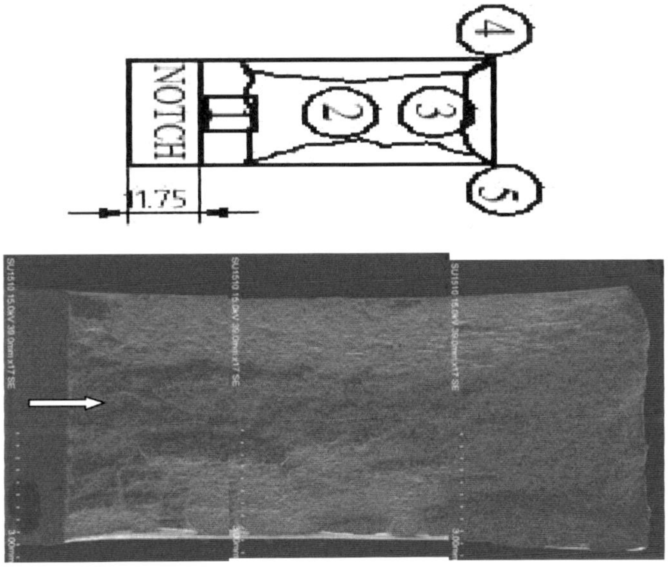

Fig 3.6 Formal drawing of various portions of fractured specimen having 11.75 mm notchand overall SEM image of 11.75 mm notch length specimen.

SEM analysis shows that, Near the crack tip an extended portion (band width indicated by white arrow) of crack length is observed due to crack tip initial instability arrest and arrow pointing in figure 3.6 also indicates direction of crack propagation. This band width decreases as notch length increases and can be seen from figure 3.1 where the load carrying capacity of 11.75mm notch specimen is high compared to 13.75 mm.

Figure 3.7 it can be noticed that crack propagates obliquely into the material and fibrous like pattern at the crack front (scan 3.7. a) indicates ductile fracture. It can be seen from scan 3.7.b that, circular dimple ensures failure at that region due to tensile stress and parabolic dimple from the scan 3.7.c indicates the failure at region is due to mixed mode and scan 3.7.d and 3.7.e shows end shear lips at different magnification. Light grey region in scan 3.7.d is the border which separates shear lip and failure zone due to triaxial state of stress. For other notch lengths, the nature of crack growth and propagation is different as the length of the notch varies.

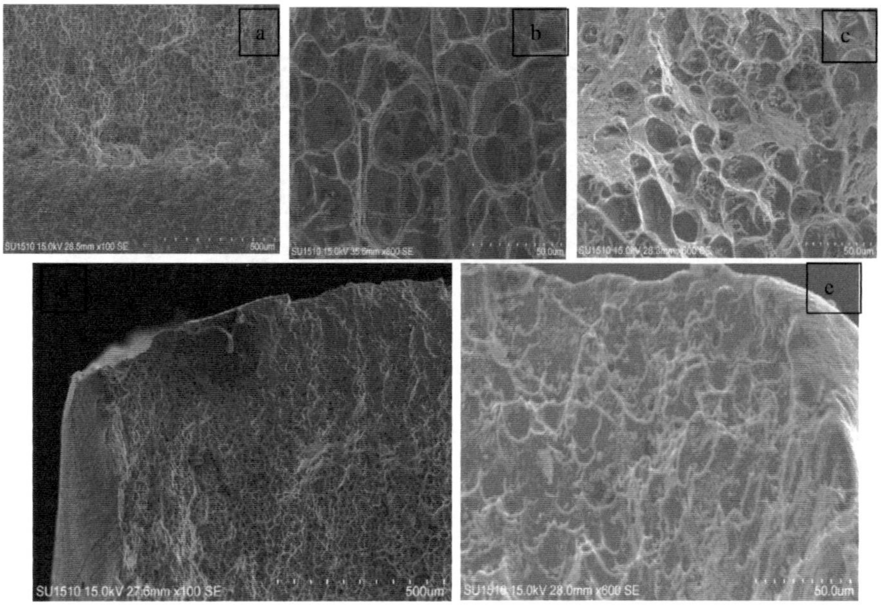

Figure 3.7 SEM images of 11.75 mm notch taken at various regions shown in figure 3.6.

4. Conclusions

From the experimentation and based on observations during test and fractography, following conclusions can be drawn.

- Load bearing capacity varies as the notch length varies and is found to be maximum forminimum notch length and decrease as the notch length increases.
- Fractured toughness of the SENB specimen was 17.41Mpa$\sqrt{m}$ and predominantly ductilefailure can be observed.
- Plastic zone size infront of the crack length not only depends upon type of load, test, andmaterial, also majorly depends on notch length.
- Band width (crack tip initial instability) decreases as the notch length.

References

[1] Evangelia Georgantzia, Michaela Gkantou, George S. Kamaris, Aluminium alloys as structural material, a review of research, Engineering Structures 227 (2021) 111372.

[2] Xuan Meng, Shanglei yang, Yubao Huang, Yu Fang, Jiaxing Gu, Qi Xiong, Chenfeng Duan, Microstructure characterization and mechanism of fatigue crack propagation of 6082 aluminum alloyjoints, Materials and Physics 257 (2021) 123734.

[3] Shafayat Bin Ali, George S. Karmis, Michaela Gkantou, Kunal D. Kansara, Concrete-filled and bare 6082-T6 aluminium alloy tubes under in-plane bending, Experiments, finite element analysis and design recommendations, Thin-Walled Structures 172 (2022) 108907.

[4] Ping Zhang, Aili Song, Yuxin Fang, Xiujie Yue, Youqiang Wang, Xiao Yu, A Study on the dynamic mechanical behavior and microtexture of 6082 aluminum alloy under different direction, vacuum 173 (2020) 109119.

[5] K.K. Saravanan, S. Mahendran, Aluminum 6082-boron carbide composite materials preparation and investigate mechanical-electrical properties with CNC turning, Materials Today: Proceedings 21(2020) 93-97.
[6] D. Morin, B.L. Kaarstad, B. Skajaa, O.S. Hopperstada, M. Langsetha; Testing and modelling of stiffened aluminium panels subjected to quasistatic and low-velocity impact loading, International Journal of Impact Engineering 000 (2017) 1-15.
[7] Qiang Liu, Jinsha Wang, Xinshen Huang, Tianming Wu, Zhijian Zong, Qing Li, In-plane and out-of-plane bending responses of aluminum mortise-tenon joints in lightweight electric vehicle inspiredby timber structures, Thin-walled Structures 127 (2018) 169-179.
[8] Yuanzheng Zhao, Ximei Zhai, Bending strength and design methods of the 6082-T6 aluminum alloy beams with circular hollow sections, Structures 26 (2020) 870-887.
[9] Jian Pu, Yali Zhang, Xiaogang Zhang, Xinlu Yuan, Pingdi Ren, Zhongmin Jin; Mapping the fretting corrosion behaviors of 6082 aluminum alloy in 3.5% NaCl solution, Wear 482-483 (2021) 203975.
[10] M.S. Syrigou, R.S. Dow, Strength of steel and aluminum alloy ship plating under combined shear and compression /tension, Engineering Structures 166 (2018) 128-141.
[11] Yifan Huang, Xin Wang, On the fracture toughness testing for single-edge notched bend specimen of orthotropic materials, composite Structures 281 (2022) 114970.
[12] Tsutomu Iwashita, koji Azuma, Effects of notch sharpness and depth on brittle fractures in single-edge notched bend specimens, Engineering Fracture Mechanics 164 (2016) 60-73.
[13] H.M Tlilan, N. Sakai, T. Majima, Effect of notch depth on strain-concentration factor of rectangular bars with a single-edge notch under pure bending, International Journals of Solids and Structures 43 (2006) 459-474.
[14] Dong-Yeob Park, Jean-PhilippeGravel, Fracture toughness measurements using two single-edge notched bend test methods in a single specimen, Engineering Fracture Mechanics 144 (2015) 78-88.
[15] HuiHuan Ma, Quanchao Hou, Yuqi Jiang, Zhiwei Yu, Mechanical performance of 6082-t6 aluminum alloy columns under eccentric compression at elevated temperatures, Thin-Walled Structures 171 (2022) 108824.
[16] Gangchang Ji, Shengru Qiao, Shuangming Du, Dong Han, Mei Li, Three-point bending fracture characteristics of three-dimensional-C/Si with single-edge notch beam specimens, Materials Scienceand Engineering A 483-484 (2008) 123-126.
[17] K.V. Arun, C.S. Venkatesha, Experimental investigations on fracture toughness and transition temperature in hard chrome coated structural materials, journal of materials processing technology 207 (2008) 336–342.
[18] Basukumar H K, Arun K V, Experimental Approach to Study the Effect of Hydrogen Embrittlement on Structural Steel under the Influence of Impact Loading and Wear Characterization,IJSTE- 2349-784X.

Study of Distribution of Magnetic Field Strength in Magnetorheological Elastomers using Artificial Neural Network

Manjeet Keshav[1] and Sujatha C[2]

[1]Department of Mechanical Engineering, SOT, Pandit Deendayal Energy University, Gandhinagar-382007, India.

[2]Department of Mechanical Engineering, Indian Institute of Technology Madras, Chennai-600036, India.

[1]Corresponding author's email id: manjeetkshv@gmail.com

Abstract. Magnetorheological elastomers (MRE) find a large number of applications in vibration control. For effective working of MRE, proper design of a magnetic circuit is needed so that it can draw the maximum amount of magnetic field through MRE region and at the same time satisfy all geometric constraints for the related application. With the availability of B-H curve data for MRE, finite element (FE) modelling of the full magnetic circuit is possible. FE modelling presents a good perspective on the distribution of the magnetic field inside the circuit. However, FE results cannot be obtained frequently for different geometries of the circuit. In the present study, the feedforward neural network (FNN) is used to predict the magnetic flux density in the MRE region of magnetic circuit setup. Training data for the FNN is generated from the results of the FE model where a log file has been coded keeping geometric parameters as variables and magnetic flux density as output. The FNN model is trained to have minimum mean square error. The FNN is further used for the optimization of dimensions of the magnetic circuit to have maximum magnetic field in the MRE region for the given geometric constraint.

Keywords: Magnetorheological Elastomers, Finite Element Analysis, Artificial Neural Networks, Optimization.

1. Introduction

A magnetorheological elastomer (MRE) is a composite material with magnetic particles suspended or arranged within a non-magnetic elastomer matrix [1]. On the application of the magnetic field, it exhibits controllable stiffness and damping properties. When compared to magnetorheological fluid, MRE has added the advantage that it is leak-proof, deposition resistant, has a wide range of controllable properties, compatibility to mechanical components, has a fast response and low power requirements. These advantages make it a very good option in various engineering applications. Changes in composition, shape, and size of the MRE has led to a wide range of applications in vibration absorber, adaptive base isolator, and sandwich beams [2].

For every application, magnetic field generated inside the MRE plays an important role in evaluating its properties based on the external load. Therefore, in the design phase, it becomes important to model the MRE and its associated appliances properly so that it can meet the application requirements by generating a controlled magnetic field inside the MRE. However, due to non-availability of B-H curve data for MRE, it becomes cumbersome to numerically model it. B-H curve data for MRE can be found experimentally at a particular composition and size of MRE [3]. If this data is available, the MRE and its magnetic circuit can be modelled in finite element (FE) software. Li and Li [4] did FE analysis of adaptive base isolators utilizing laminated base MRE layers proposed by Li *et al* [5] and studied the magnetic field distribution in MRE layers in the deformed position. Therefore, FE modelling of MRE device can be an effective way to determine its optimal geometry and shape to draw maximum magnetic field through it. However, all possible geometric combinations cannot be made every time to find out the maximum magnetic field through it. Data-driven modelling saves the day for it. In literature, most of the authors have applied the data-driven technique to model the dynamic behaviour of MRE [6][7]; however, predicting the magnetic field inside MRE using neural network data-driven technique has not been reported till now.

In this study, a feedforward neural network (FNN) is fitted for the magnetic flux data generated from FE model for different dimensions of the magnetic circuit and the study proposes an optimal dimension of it that will generate maximum magnetic flux density inside the MRE region of the magnetic circuit. The magnetic circuit design has been taken from the work from Li *et al* [5]. FE model of the circuit has been idealized in the ANSYS-APDL. A log file is generated from the software in which geometric variables and output parameters have been defined. For fitting the FNN, a method used by Keshav and Sujatha [8] has been employed. After the ANN model fitted, it is used for suggesting the optimized geometry of the magnetic circuit to maximize magnetic flux density in the MRE region. The optimization has been done using SQP method in MATLAB® 2017a optimization toolbox.

2. Magnetic Circuit and Electromagnetic FE Model

An effective magnetic circuit is very important to maintain a strong magnetic in the MRE region. Figure 1(a) shows the schematic diagram of the magnetic circuit setup for an MRE develop by Li *et al* [5]. An axisymmetric 2D plane view of the magnetic circuit is shown in figure 1(b). H_{sb} and R_{sb} are the height and radius of the steel block. H_{ay}, R_{ay} and t_{ay} are the height, radius and thickness of the annular yoke. t_{cp} and t_c are the thickness of the cover plate and coil. H and R are the overall height and radius of the setup. The block and annular yoke are made of steel AISI 1008, B-H curve of which is shown in Figure 2(a) [4] and experimentally obtained B-H curve data for the MRE is shown in Figure 2(b) [9]. These magnetic characterisation data are fed in electromagnetic module of ANSYS-APDL software as material properties and an axisymmetric FE model of the magnetic circuit is developed which is shown in figure 3(a). The current density in the coil region has been set at 3 A of current. The FE model is based on Maxwell's equation of magnetostatic model for any electromagnetic system which is given by

$$\nabla \times H = J \text{ and } \nabla \cdot B = 0 \tag{1}$$

where H is the magnetic field intensity (A/m), J is the current density (A/m^2), B is magnetic flux density (T) and μ is the magnetic permeability (H/m). Here, for the FE analysis, the magnetic field is calculated in terms of magnetic flux density generated in the mid-section of the MRE and this has been set as an output parameter for a given geometry and applied current. Nodal solution for magnetic flux density at 3 A of current is shown in figure 3(b) for the geometry given in Table 1. It can be observed that distribution of magnetic flux density in the MRE region is nearly constant; however, larger variations have been observed near top and bottom steel cover plates. This uneven distribution is due to varying geometry of steel plates and difference in magnetic permeabilities of MRE and steel. Further, after the FE model is developed, a log file is generated, which is later modified to include defined geometric variables. In the present study, four geometric variables have been defined which would affect the magnetic field in the MRE region. These geometric variables are H_{sb}, R_{sb}, t_c and t_{ay}. These geometric parameters are kept as the design variables (DVs) for the optimization problem. It is to be noted that current remains constant at 3 A.

Table 1. Values of different geometric parameters of the magnetic circuit of MRE.

Geometric parameters	Values (mm)
Height of steel block, H_{sb}	50
Radius of steel block, R_{sb}	60
Thickness of the coil, t_c	25
Thickness of annular yoke, t_{ay}	10
Overall height, H	173
Overall radius, R	130
Thickness of cover plate, t_{cp}	12
Height of annular yoke, H_{ay}	144
Radius of annular yoke, R_{ay}	100

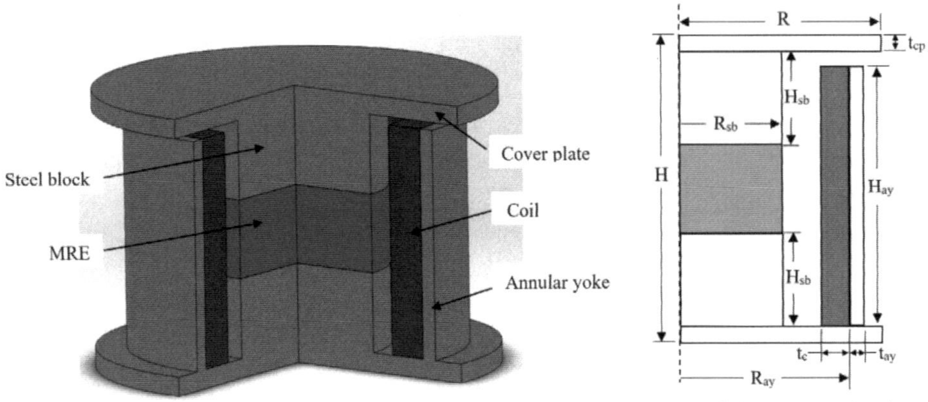

Figure 1. (a) Magnetic circuit setup of MRE, (b) Axisymmetric 2D view of the magnetic circuit setup.

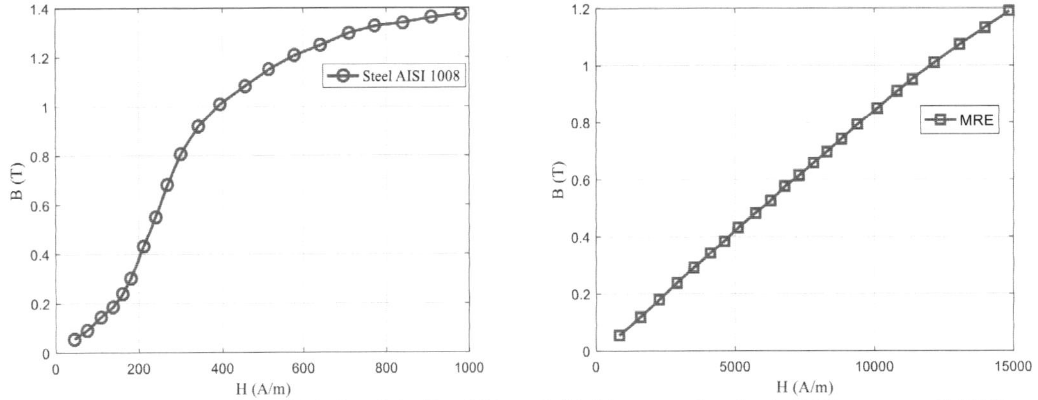

Figure 2. B-H curve of (a) Steel AISI 1008 and (b) Magnetorheological Elastomer (MRE)

3. Artificial Neural Network

In the present work, a feedforward neural network (FNN) is fitted to predict the magnetic flux density in the MRE region. Training data for the FNN are generated from the results of the FE model encoded in a log file for different values of geometric variables. For generating training data, ranges of values are decided as 40 mm < H_{sb} < 60 mm, 45 mm < R_{sb} < 65 mm, 15 mm < t_c < 30 mm and 10 mm < t_{ay} < 15 mm. The levels for each variable are decided based on three equal division for H_{sb}, R_{sb} and t_{ay}, and two equal division for t_c between the bounds. For the number of levels chosen for the variables, total of 192 full factorial combinations are generated and randomized to give a proper run. These 192 combinations of variables are fed to the log file and run 192 times to generate magnetic flux density training data. The training data have been normalized between -1 and 1 to negate the large range of values in the input data. The total training data have been divided as 80% for training, 10% for validation and 10% for testing. Figure 4(a) shows architecture adopted for the FNN where the input layer (layer 1) contains all four geometric DVs, the hidden layer (layer 2) contains activation units (neurons or hidden nodes) and the output layer (layer 3) contains the magnetic flux density. Here, $a_i^{(l)}$ is the ith activation unit (or hidden node) of the lth layer. The activation function for the hidden layer has been chosen '*Hyperbolic tangent sigmoid transfer function*', whereas the activation function for the output layer has been chosen as a linear function; formulations for the same are given as follows

$$f(x) = \frac{2}{1+e^{-2x}} - 1 \tag{2}$$

$$g(x) = px \tag{3}$$

where x is any input value given to the functions and p is a constant to be determined during the fitting of FNN. Here, the input vector $\underset{\sim}{x}_{5\times1}$ is given as

$$\underset{\sim}{x}_{5\times1} = \begin{bmatrix} x_0 \\ w_c \\ l_a \\ I \end{bmatrix} \tag{4}$$

The normalized data to be fed to activation function of hidden layer is $\underset{\sim}{z}^{(2)} = W_{m\times5}^{(2)} \cdot \underset{\sim}{x}_{5\times1}$ and further the vector of activation (hidden) units is given by

$$\left[a^{(2)}\right]_{m\times1} = f(\underset{\sim}{z}^{(2)}) \tag{5}$$

where m is the number of activation units in the hidden layer to be determined from error analysis and $W_{m\times5}^{(2)}$ is the weight matrix of the features in hidden layer to be determined at final fitting of FNN. Therefore, the mapped value of the activation unit in output layer is expressed as

$$a^{(3)} = g(\underset{\sim}{z}^{(3)}) \tag{6}$$

$$a^{(3)} = g\left(W_{1\times(m+1)}^{(3)} \cdot \begin{bmatrix} a_0^{(2)} \\ \left[a_i^{(2)}\right]_{m\times1} \end{bmatrix}\right) \tag{7}$$

Here, $W_{1\times(m+1)}^{(3)}$ is the weight matrix for the values of the activation units of hidden layer in output layer.

The FNN is fitted to minimize the mean sum square error between the predicted and observed values. If X comprises N number of given data and Θ is the fitted model parameters, then the optimization problem is formulated as

$$\min_{\theta} \; E(X,\underset{\sim}{\theta}) = \frac{1}{2N}\sum_{d=1}^{N}\left(\hat{y}_d - y_d\right)^2 \tag{8}$$

To solve this least-squares formulations the Levenberg-Marquardt algorithm (LMA) has been applied which combines two minimization algorithms: the Gradient-Descent method and the Gauss-Newton method, i.e., it tries to interpolate the results between these two methods. The data fitting has been done using Neural Net Fitting toolbox, MATLAB® 2017a.

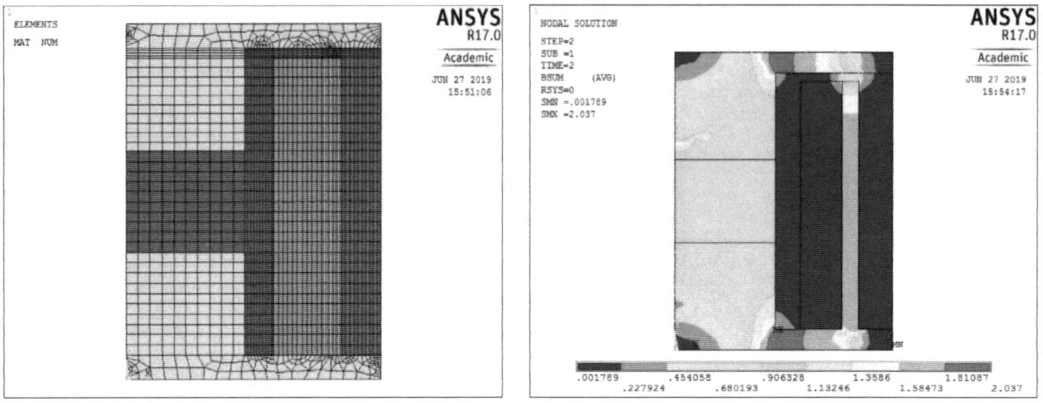

Figure 3. (a) Axisymmetric FE model and (b) nodal solution for magnetic flux density at 3 A.

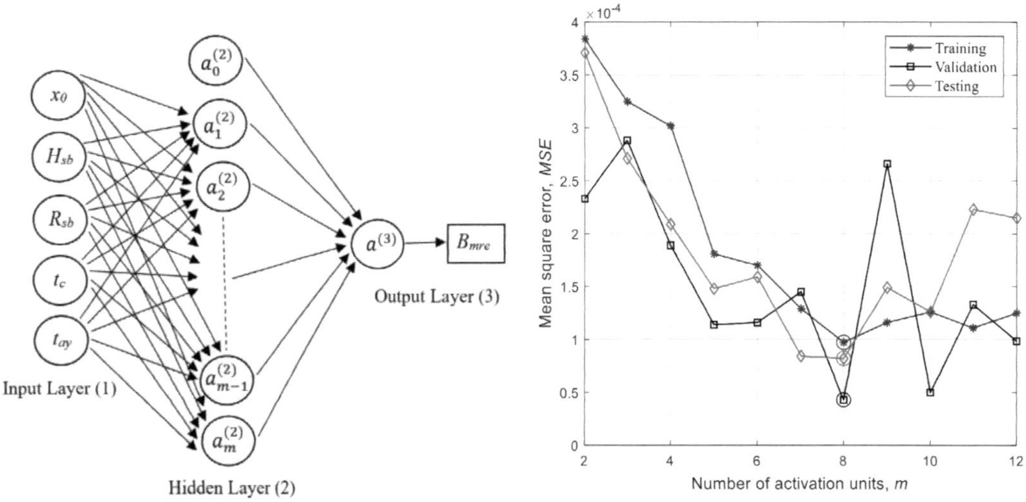

Figure 4. (a) Architecture of FNN (b) Error analysis with respect to activation units.

4. Results and Discussion

As discussed before, the number of activation units, m, in figure 4(a) is decided through error analysis. Figure 4(b) shows the mean square error (MSE) for training, validation and testing data at different activation units of the hidden layer. It can be observed that error for different set of data is decreasing with the number of activation units in the hidden layer and saturates for larger values of activation units. After analysing error for different data, it is found that 8 number of activation units would be most suitable as it gives minimal MSE of all data sets as compared to other activation units. At $m = 8$, MSE for training data, validation data and testing data are 9.73×10^{-5}, 4.3×10^{-5} and 8.21×10^{-5}, respectively, which is quite acceptable. Therefore, with this architecture of FNN, i.e., single hidden layer with 8 hidden nodes, data fitting is done and the optimal weights mapping the input layer to the hidden layer is given in the matrix shown below

$$W_{8\times5}^{(2)} = \begin{bmatrix} -2.15 & 2.01 & 0.49 & 0.04 & 2.15 & -4.64 & 1.95 & -1.93 \\ 0.53 & -1.57 & 0.61 & -0.27 & 1.73 & -1.8 & -0.01 & -0.06 \\ -0.64 & -5.6 & 0.73 & 1.77 & 5.09 & 2.74 & -0.85 & 1.43 \\ 0.56 & 0.19 & 1.42 & -1.16 & 3.13 & -2.71 & 0.76 & -0.36 \\ 0.22 & -1.84 & 0.3 & -1.81 & 0.96 & -7.87 & -0.76 & -1.29 \end{bmatrix}^T.$$

The weights from the hidden layer to the output layer are

$$W_{1\times9}^{(3)} = [0.07 \quad 0.76 \quad 0.06 \quad 0.11 \quad -0.12 \quad -0.07 \quad 0.1 \quad 0.49 \quad -0.63].$$

The optimal weights obtained can be fed in MATLAB® software for given number of inputs and activation units and magnetic flux density can be predicted for any set of geometric parameters input through input vector $\underset{\sim}{x}_{5\times1}$. Figure 5 shows the regression analysis (prediction) for the same where training data achieve the correlation coefficient value, R=0.9967, validation as R = 0.9984 and testing as R = 0.9971. It can be observed that all the correlation coefficient is more than 0.99, i.e., they are closer to 1 resulting in lesser difference between predicted and actual values.

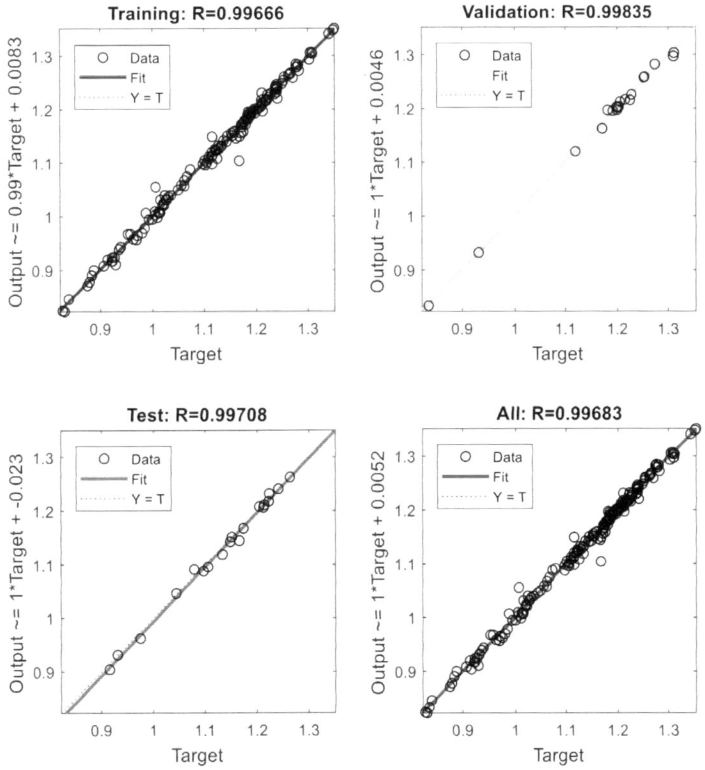

Figure 5. Regression analysis of the fitted FNN model.

4.1. Optimization

Optimization problem aims at maximising magnetic flux density in the MRE region of the magnetic circuit for the given constraints on the geometric variables of the magnetic circuit. It can be formulated as

$$\text{Minimize } f = 1/B_{mre} \tag{9}$$

subjected to

40 mm < H_{sb} < 60 mm, 45 mm < R_{sb} < 65 mm, 15 mm < t_c < 30 mm and 10 mm < t_{ay} < 15 mm.

The above-stated optimization problem is solved using sequential quadratic programming (SQP) method in optimization toolbox, MATLAB® 2017a. SQP method requires an initial point to be defined before the optimization process starts, therefore, due to lack anonymity of initial point, 1000 uniformly distributed initial points are generated. The optimization problem is run for these 1000 initial points and whichever initial points gave minimum f, the result at that initial point is considered as optimized.

When the optimization problem is solved, it is found that 877 initial points converged to the same optimal design variables and resulted in minimum f. From the optimal results, as shown in Table 1, it is observed that if the height of the steel block, H_{sb}, coil thickness, t_c and annular yoke thickness, t_{ay} tend to take on their upper bound values of 60 mm, 30 mm, and 15 mm, respectively, whereas the radius of the steel block, R_{sb} takes on the lower bound of 45 mm, the magnetic flux density in the MRE region is maximum. So, it is inferred that higher values of H_{sb}, R_{sb} and t_{ay}, and lower values of R_{sb} lead to higher magnetic flux density in the MRE.

Table 1. Optimal values of geometric design variables

Design variables (DVs)	H_{sb}	R_{sb}	t_c	t_{ay}
Optimal values (mm)	60	45	30	15

5. Conclusions

The current work presented the application of artificial neural networks to predict the magnetic flux density in MRE for a given range of geometric parameters of the magnetic circuit setup and optimize its magnetic circuit to maximize magnetic flux density. Results of electromagnetic FE model has been considered for training the ANN. FNN architecture was chosen with single hidden layer and the number of hidden nodes is calculated using error analysis. The fitted FNN was then used to optimize the magnetic circuit geometry with an aim of maximizing the magnetic flux density in MRE region. Overall, the study presented the way to analyse the distribution of magnetic field in the MRE for different geometric dimensions of the magnetic circuit using data-driven modelling technique and 30-50% of what the proposed model predicts that would be considered as approximate for real-time application. The only problem it may face is that generation of large training data is required, however, that can be solved if the log file of the FE model in ANSYS can be encoded properly.

The application of data-driven modelling in MRE can be extended to decide the optimization of the geometric configuration of the magnetic circuit with the constraint on magnetic saturation limit and to be able to predict the magnetic field inside the MRE when it is in a deformed shape. This would help in evaluating the performance of MRE at its different shapes and configurations of the magnetic circuit.

Reference

[1] Carlson J D and Jolly M R 2000 MR fluid, foam and elastomer devices *Mechatronics* **10(4-5)** 555-69

[2] Li Y, Li J, Li W and Du H 2014 A state-of-the-art review on magnetorheological elastomer devices *Smart Mater. Struct.* **23** 123001

[3] Zeng J, Guo Y, Li Y, et al. 2013 Two-dimensional magnetic property measurement for magneto-rheological elastomer *Journal of Applied Physics* **113(17)** 919

[4] Li Y and Li J 2014 Finite element design and analysis of adaptive base isolator utilizing laminated multiple magnetorheological elastomer layers *14th Int. Conf. on Electrorheological Fluids and Magnetorheological Suspensions* Granada, July 2014, **26(14)** 1861-70

[5] Li Y, Li J, Li W, et al. 2013 Development and characterization of a magnetorheological elastomer based adaptive seismic isolator *Smart Mater. Struct.* **22(3)** 035005

[6] Jie Fu, Guanyao Liao, Miao Yu, Peidong Li and Junjie Lai 2016 NARX neural network modeling and robustness analysis of magnetorheological elastomer isolator *Smart Mater. Struct.* **25** 125019

[7] Dingxin Leng, Kai Xu, Yong Ma, Guijie Liu and Lingyu Sun 2018 Modeling the behaviors of magnetorheological elastomer isolator in shear-compression mixed mode utilizing artificial neural network optimized by fuzzy algorithm (ANNOFA) *Smart Mater. Struct.* **27** 115026
[8] Manjeet Keshav and Sujatha Chandramohan 2019 Geometric optimization of magnetorheological valve using feedforward neural networks for distribution of magnetic flux density inside the valve *Smart Mater. Struct.* **28(10)** 105018
[9] Zeng, J, Guo, Y, Li, Y. 2013 Two-dimensional magnetic property measurement for magneto-rheological elastomer. Journal of Applied Physics **113(17)** 17A919.

Design of Hydraulic Biomass Briquette Machine

Hetav Kinkhabwala[1], V M Bhojawala[2*], Saumil Desai[3] and D V Patel[4]

[1] M.Tech student, Mechanical Engineering Department, Institute of Technology, Nirma University, Ahmedabad-382481, Gujarat, India.

[2, 3, 4] Assistant Professor, Mechanical Engineering Department, Institute of Technology, Nirma University, Ahmedabad-382481, Gujarat, India.

* corresponding author

Abstract. In many of our developing countries, we face many problems regarding agro-waste. If the wastes are not managed then there will be increased in the quantity of waste. In order to manage these wastes, solutions are required and one of the solution is discussed here. The present study discusses about the hydraulic briquette line which is used to make briquettes. This hydraulic briquette line basically consists of three equipment's a shredder machine, a screw conveyor and a hydraulic briquette machine. Because of good characteristics of sugar-cane bagasse such as low ash content compared to coal, lignin cells are present which acts as a binder, etc., it has been used for making briquettes. Bagasse having long tissues, it is first cut into small parts with the help of shredding machine. These cut bagasse is conveyed to the hydraulic briquette machine with the help of screw conveyor. This conveyed bagasse is compressed in hydraulic briquette machine with the help of hydraulic cylinders. This paper discusses the detailed design of all three equipment i.e. a hydraulic briquette machine, a screw conveyor and a shredding machine.

1. Introduction

Biomass solid fuels such as wood, coal, etc. are used in large quantity in households, industries, power generation unit etc. There is decreased in availability of wood because of deforestation. On the other hand, fuels are being used in large quantity to achieve energy demands without utilizing agro as well as forest waste which are getting generated every year. In India we face many problems regarding agro-waste. Large tons of agricultural wastes are generated every year in India. If these wastes are not managed then there will be increased in the quantity of waste. These wastes are burnt in their loose form which gives rise to air pollution. So to overcome these problems, solutions are needed and one solution is compaction of waste crop into biomass briquettes which helps in providing an alternative solid fuel. When the pressure is applied to the different sizes of loose biomass in order to produce compact solid blocks of different sizes and of high density, then this process is called biomass briquetting and the products formed are known as briquettes. These briquettes can provide a renewable source of energy. In order to produce strong briquettes transformation of heat, application of pressure and binding agents are playing the main part. In some biomass, lignin cells are present which acts as binding agent and thus no binding agents are used.

Advantages of using briquettes as solid fuel are dependency on fuel wood decreases, consumption of wood, coal, etc. decreases, easy to handle, store and transport because of its high density, size of briquettes will be uniform and good in quality, waste disposal problems will be solved with this process, deforestation will also decrease with this process and air pollution will also decrease. In order to produce briquettes, machines are required and these briquettes can be manufactured in two ways [1] (1) High pressure compaction and (2) Low pressure compaction. In high pressure compaction, briquettes are made by applying high pressure without using any binders and in low pressure compaction pressure applied is low as binders are used.

After removing the juice from the sugar-cane, the crushed material left is called bagasse. Here bagasse is used as a raw material to produce briquettes because of its good characteristics such as low-ash content than coal; lignin cells are already present which acts as binders, etc. Bagasse characteristics changes in consistency, composition, and values depends on the type of soil, amount of washing, climate, varieties of cane, efficiency of milling plant and harvesting methods. Bagasse is having a high calorific value of 4400 kcal/kg. As sugar cane bagasse having long tissues, it cannot be compressed directly in order to produce briquettes. Bagasse needs to be cut into pieces up-to 10mm so that it easily gets compress. Cutting of bagasse can be done using shredding machine.

2. Briquette line

The present study discusses the design of briquette line which includes hydraulic briquette machine, screw conveyor and shredding machine. Hydraulic machine will be used to compress the bagasse which is supplied through the screw conveyor and shredding machine will be used to cut bagasse with the help of rotating blades.

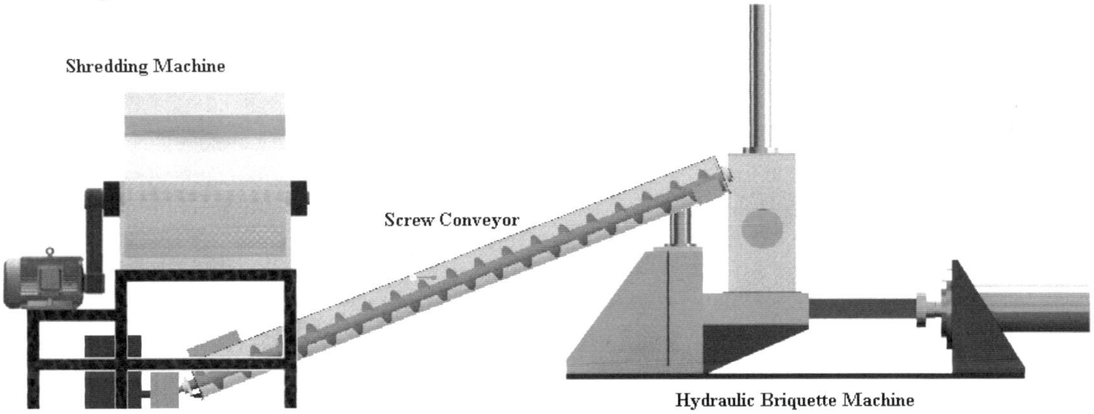

Figure 1. Layout of Briquette Line

3. Procedure of Briquette formation

Bagasse is collected from the sugar-cane mills after the removal of juice from sugar-canes. Bagasse having very high moisture content, it has to be stored under sunlight for days until the moisture content reduces to 8%. As bagasse having long tissues it has been pre-shredded i.e. cut into small tissues. After pre-shredding of bagasse is completed, pre-shredded bagasse are allowed to fall in the shredding machine. Bagasse will fall directly on the stationary blades. This bagasse will get cut into fine particles with the help of the sharp blades which are rotating at 800 rpm. These cut bagasse will fall down from the strainer into the screw conveyor, where bagasse will be conveyed to the hydraulic briquette machine. In hydraulic briquette machine, all the three cylinders i.e. stopper cylinder, shutter cylinder and striker cylinder shown in Fig. 1 will be at home position. Stopper block will move to its final position which is connected to the piston rod of stopper cylinder with the help of connector. When the stopper block comes to its final position, the screw conveyor will convey bagasse in the hopper's inlet chute. As soon as the material is filled in the hopper up-to its limit, with the help of shutter cylinder, the shutter plate which is connected to the piston rod of shutter cylinder will exert the force and levels the fed bagasse. After the bagasse gets levelled, the striker cylinder will act and the striker block will move from its home position to final position compressing the bagasse. The striker cylinder will act up-to its required pressure and force will be exerted on the bagasse. The briquette formation takes place and the desired shape of the briquette is achieved. After the formation of a briquette, the back block with the help of back cylinder will move to its home position. The striker cylinder will give pressure push and thus the briquette formed will fall down directly on the belt conveyor. After the falling of the briquette, the striker block and shutter plate will return to its home position with the help of striker cylinder and shutter cylinder respectively.

4. Design of Hydraulic Briquette Machine

4.1. Design of Hydraulic cylinders

There are three hydraulic cylinders: first striker cylinder which is used to press the bagasse, second shutter cylinder used to level the fed bagasse, third stopper cylinder used to stop bagasse until it forms briquette. Selection of hydraulic cylinder bore diameter is very important. This selection was done on the basis of the experiment done in order to decrease the volume of the briquette. In a hollow rectangular box as shown in Fig. 2 bagasse was filled up-to specific level and force was exerted to it until the volume of briquette reduces to one third of its original volume. During experiment maximum pressure noted was 170 bar and the hydraulic cylinder used was having piston bore diameter of 160 mm. This indicates that the maximum force applied was 35 tons.

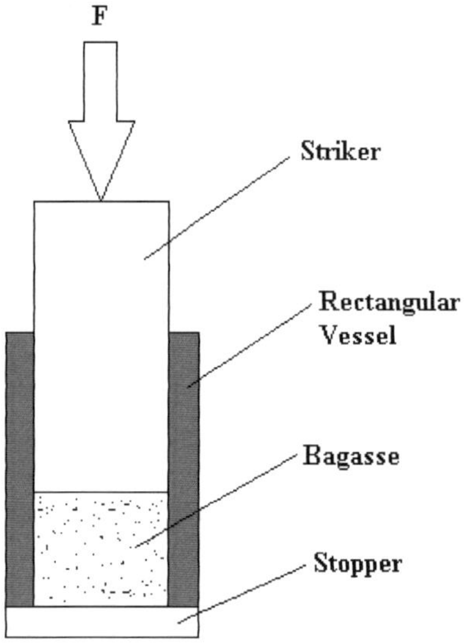

Figure 2. Experimental setup

Therefore the bore diameter of striker cylinder was taken as 160 mm. As the force developed by striker cylinder will be exerted on the sliding plate of the shutter cylinder the bore diameter of shutter cylinder was also selected as 160 mm. Material selected for hydraulic cylinders is ASTM A106 Grade B [2]. It is similar to plain carbon steel with carbon content of 0.3% and tensile strength = 415 N/mm^2. As the pressure applied by the pump is high, the wall thickness of hydraulic cylinders were found considering as thick cylinders. Cylinder wall thickness has been obtained by Barlow's equation which is used for calculation of thickness of gas pipes and high pressure oil cylinders. According to Barlow's as follows [3],

$$t_w = \frac{p_i \times d_0}{2\sigma_t} = \frac{d_0 - d_i}{2} \tag{1}$$

Thicknesses obtained using equation (1) are 8, 8 and 5 mm for striker, shutter and stopper cylinder respectively excluding corrosion allowance. The minimum diameter required for the piston rod has been obtained as 90 mm from reference [4] for both striker and shutter cylinder. Material selected for piston rod is EN8/C40 [2]. The slenderness ratios of striker cylinder and shutter cylinder have been found to be 15.55 and 8.88 respectively. As the slenderness ratios are less than 30 [3], there is no need to verify piston rod for buckling.

4.2. Design of Hollow rectangular vessel

As the cross sectional shape of the briquette is rectangular. The shape of vessel is hollow rectangular in which the raw material bagasse is being compressed. The material gets compressed when all the sides are closed, so this is considered as a closed pressure vessel. The analysis of this block can be approximated by considering cylindrical pressure vessel. The thickness was estimated as 20 mm with material as mild steel having 0.2% carbon and allowable stress of 60 N/mm^2. Plates in hollow rectangular vessel are fastened by bolts. Bolts are subjected to shear. This hollow rectangular vessel is bolted with 32 numbers of bolts. Material taken for bolts from IS 1871-2(1987) [5] is 40C8 plain carbon steel with yield strength of 380 N/mm^2. Diameters of bolts were found considering shearing failure:

$$F = 32 \times \frac{\pi \times d^2}{4} \times \tau \tag{2}$$

Diameter obtained from equation (2), is 14.61 mm. So referring IS: 4218-3(1999) [6] bolts of M16 are selected.

5. Hydraulic circuit of briquette machine

All hydraulic circuits contains four basic components [7]; a reservoir which contains the fluid; a pump which is used to force the fluid through whole system; valves which are used to control the flow; and an actuator which is used to convert fluid energy into mechanical energy.

5.1. Capacity of hydraulic reservoir

Hydraulic reservoir stores required quantity of fluid which is supplied to the whole system. When all cylinders are extended minimum fluid indicated in reservoir and maximum when all cylinders are retracted. The capacity of a reservoir is four to five times the volume of the fluid taken by the system every minute. Oil capacity for cylinder is calculated to decide the capacity of the hydraulic reservoir. Oil Capacity of a cylinder can be calculated from equation (3),

$$O_c = \frac{\pi \times {d_i}^2}{4} \times l_s \tag{3}$$

The capacity of hydraulic reservoir is selected as 4 times the total oil capacity of cylinders which is 70 litres.

5.2. Pumps

Two types of pumps are selected for this purpose:

(1) Gear Pump: - Gear Pump is normally used for low pressure. When good speed is required, these pumps are used. Gear pump was selected to make striker block move at high speed.
(2) Radial Piston Pump: - Radial Piston Pump is normally used when high pressure is required. Radial Pump was selected because of the capacity to create large amount of pressure.

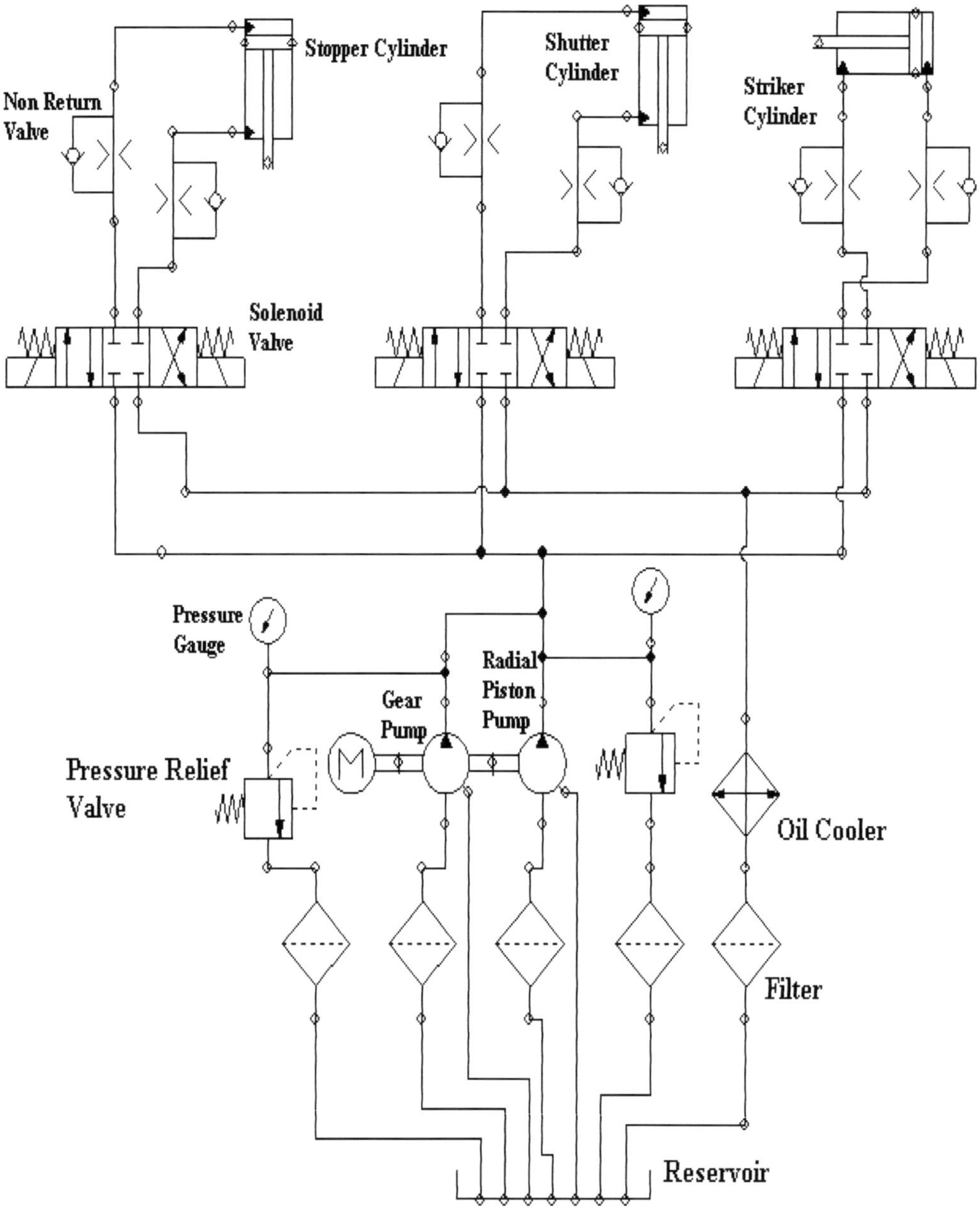

Figure 3. Hydraulic circuit diagram

The hydraulic circuit shown in Fig. 3 is prepared in Automation Studio 5.6. The home position of all the three cylinders is shown in Fig. 3 and different operating positions are explained below.
Step 1: - Movement of stopper cylinder
In the first step, the stopper cylinder will act and the other two cylinders will rest at their home position. In this step hydraulic pump supplies the fluid from reservoir through filter and sends it to the solenoid

valve. This valve acts, changes its position to parallel connection in which the fluid will enter the cylinder creating extension stroke i.e. brings stopper cylinder from its home to final position.

Step 2: - Levelling the fed bagasse with shutter cylinder

In the second step, the stopper cylinder and striker cylinder will stay at its final and home position respectively and the shutter cylinder will act. The valve of stopper cylinder will move to its center position where no fluid is allowed to flow. Hydraulic pump supplies fluid from reservoir through filter. Solenoid valve changes its position to left to bring shutter cylinder from its home to final position. During this step fed bagasse gets levelled.

Step 3:- Compressing the bagasse with striker cylinder

In the third step, the stopper cylinder and shutter cylinder will stay at its final position and the striker cylinder will act. Solenoid valve of stopper cylinder and shutter cylinder will move to its center position where no fluid is allowed to flow. Hydraulic pump supplies fluid from reservoir through filter. This valve acts and changes its position to left to bring striker cylinder from its home to final position. During this step, final compression of bagasse takes place.

Step 4: - Return of the stopper cylinder

In the fourth step, the shutter cylinder and striker cylinder will stay at its final position and the stopper cylinder will act. Hydraulic pump pulls fluid from reservoir through filter and then sends it to the solenoid valve. This valve acts and changes its position to right to bring stopper cylinder from its final to home position. During this step, the briquette formed will fall down.

Step 5: - Beginning of next cycle

In this step the shutter and striker cylinder will move back to its home position with the help of solenoid valves which moves to the right and with the flow of fluid from pump. Again the stopper cylinder moves to its final position and the process continues

6. Screw conveyor and its component design

Different materials are transferred by screw conveyors. Advantages of screw conveyor are, it can handle material having sluggish properties. Screw conveyors are considered as screw feeders when used as metering device. The disadvantage is that the capacity of conveyor decreases with the increase in the angle of inclination.

6.1. Screw conveyor design procedure

The design procedure for screw conveyor has been referred from reference. [8]

Step 1: Specifications: - Material to be handled is bagasse with the capacity required of 2.83m^3/hr and the length of conveyor should be 3 meter.

Step 2: Properties of Bagasse [9]: - The bulk density of bagasse is found to be 112 - 160 kg/m^3. Bagasse can be designated as E-45RVXY where E-Irregular stringy, fibrous, cylindrical, slabs, etc. 4–Sluggish 5-Mildly abrasive R-Gives off harmful or toxic gas or fumes V-Interlocks, mats or agglomerates X-Packs under pressure Y-Very light and fluffy.

STEP: 3 Calculation of Conveyor Speed:-

The conveyor speed can be calculated by the equation (4) [9]:

$$N = \frac{E_C}{R_m} \tag{4}$$

Where Ec will be equal to equivalent capacity 2.83m^3/hr and Rm will be equal to 0.04247 m^3/hr at 1 revolution per minute from [9]. Therefore, N (RPM) = 66.635 RPM = 67 RPM

STEP: 4 Core Diameter:-

From [9] we select the diameter of screw = 153 mm. Screw diameter according to IS: 5563(1985) [10] = 160 mm. Normally, for diameter smaller than 400 mm Pitch = 125 mm. The equation to find the core diameter of the shaft is [8]

$$C = \frac{0.7854}{1728} \times \left(D_s{}^2 - D_p{}^2\right) \times ps \times ks \times BD \, x \, N \times 60 \quad (5)$$

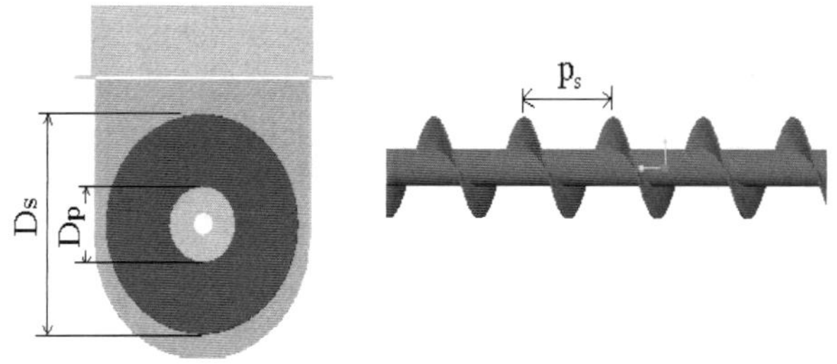

Fig. 4 Screw of the conveyor

Where, C = Capacity in kg/hr = 450, Ds = Diameter of screw = 0.160 m, Dp = Core diameter of hollow pipe in m, ps = pitch of screw = 0.125 m, ks = Trough loading = 30% = 0.3 B.D = Bulk Density = 160 kg/m^3 N = 67 RPM. Therefore outside diameter of screw pipe, Dp = 0.04123 m = 41.23 mm. According to IS: 5563(1985) [10], pipe outer (core) diameter Dp selected is 42.4 mm with 4.0 mm wall thickness. Therefore inner diameter Di = 42.4 - 2(4) = 35.2 mm.

STEP: 5 Calculation of Power Requirement

Power requirement to operate a screw conveyor is calculated [11]

Drive Power of the screw at no load $P_N = \frac{D_s \times L}{20} \; kW$

Power required for conveying the material $P_H = \frac{Q \times L \times \lambda}{367} \; kW$

Power Due to Inclination $P_{st} = \frac{Q \times H}{367} \; kW$

Total Power P = $P_N + P_H + P_{st}$

Where, Q = mass flow rate (capacity) = 0.45, L = Length of screw = 3 m, Ds = Screw Diameter = 0.160 m, H = Height = 1.524 m, λ = progressive resistance coefficient = 4, Efficiency of gear wheel = 0.85. Therefore, Total Power P = 0.04058 kW = 0.0544 hp. Now, Brake horse power (B.H.P) [8] B.H.P. = hp × p = 0.0544 × 2 = 0.1088. Actual B.H.P. = 0.1088 / efficiency of gear wheel = 0.1280 hp.

STEP: 6 Development of Flight [8]

The screw shaft with the profile of the helix of flight and its development is shown in Fig. 5. Sectional flights are selected in screw conveyor.

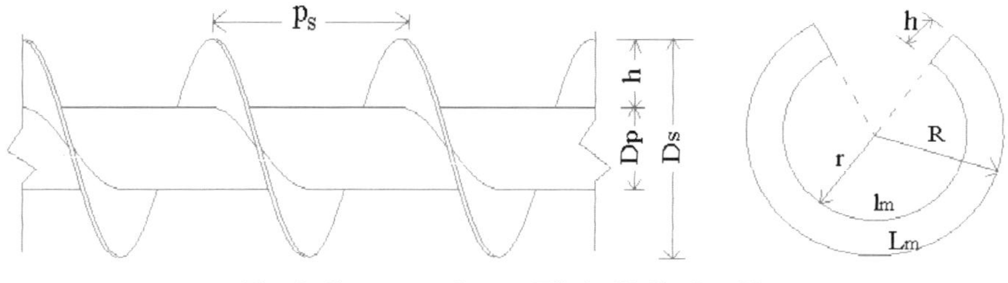

Fig 5. Conveyor Screw Flight Helix Profile

Here h = (Ds-Dp)/2 =0.0588m. Now the minor helix length (lm) and major helix length (Lm) has been calculated after the development of helix flight forms right angle triangle to be 0.1826m and 0.5179m respectively. The disc outer radius = (Lm x h)/ (Lm-lm) and disc inner radius = R-h. The disc internal diameter 0.0718 m and outer diameter 0.1895 m will be cut and then bent in the form of flight over the hollow shaft. This flight section will be butt-welded to form the complete screw. Now, No. of flights = 3000/125= 24 flights. 23 right handed flights from inlet end of trough to outlet and 1 left handed flight from the end plate of trough at outlet side to discharge has been selected. Thickness of flight selected from IS 5563(1985) [10] is 3.15 mm.

7. Shredding machine

Shredding machines are used to cut biomass into small pieces to reduce size, for better handling and for the production of solid fuel. Biomass shredding is used to transform waste to renewable source of energy. It also improves the efficiency of recycling and also helps in lowering waste-filled lands. Biomass shredding equipment is used in the process of untreated biomass. Shear test was performed on sugar cane bagasse by Amer Eissa A.H et al.[12] and shear strength obtained experimentally at bottom part, middle part, top part of sugar cane bagasse were (2.41 ± 0.095) MPa, (2.23 ± 0.095) MPa and (2.42 ± 0.095) MPa respectively and average shear strength was considered as 2.23 MPa. [12].Shear force = 2.23 MPa, Cutting length = 100 mm, Cutting thickness = 2 mm, No. of revolutions per minute = 800 rpm, Cutting force = 2.23 × 100 × 2 = 446 N Torsional moment (Mt) = Cutting Force x (PCD/2) = 44600 N-mm, Power (kW) = $(2\pi n M_t)/(60 \times 10^6)$ = 3.7 kW.

8. Conclusions

A fully automated hydraulic briquette machine has been designed that can produce briquettes automatically, with bagasse as a material and without any human effort. The hydraulic circuit for the briquetting machine has been prepared and various components of the same have been selected. The circuit diagram has been prepared in Automation studio 5.6. Design of screw conveyor has been carried out which includes different components such as screw shaft, trough, flights. Design of shredding machine has been discussed with estimation power requirement for shredding.

9. References

[1] Catalogue of *"Ashden Technology"* Biomass Briquettes and Pellets.
[2] Q.S Khan, *Design and Manufacturing of Hydraulic Cylinders.* (Tanveer Publications).
[3] V B Bhandari, *Design of Machine Elements,* (McGraw Hill Publication ltd).
[4] Catalogue of *"Parker" MMB Hydraulic Cylinders,* HY07-1215/UK pp: 15-16.
[5] IS 1871-2 (1987): *Commentary on Indian Standard schedules for wrought steels,* Part 2: Carbon steels.
[6] IS 4218-3 (1999): *ISO General Purpose Metric Screw Threads*, Part 3: Basic Dimensions.
[7] S R Majumdar, *Oil Hydraulic Systems Principles and Maintenance* (McGraw-Hill publication ltd).
[8] Shirbhate D, 2008. *Design and Fabrication of Screw Conveyor* found on http://www.scribd.com/doc/7370601/13/DESIGN-CALCULATIONS
[9] Catalogue of *"Conveyor Engineering and Manufacturing (CEMC)"* Screw conveyor components and design.
[10] IS 5563 (1985): Screw conveyors for industrial use.
[11] IS 12960 (1990): Determination of power requirements of screw feeders – General requirements.
[12] Amer Eissa A.,Gomaa A..,Baiomy M. and Ibrahim A., 2008. *Physical and mechanical characteristics for some agricultural residues* Misr J. Ag. Eng., 25(1) pp: 121-14
[13] Catalogue of ABB *"Low voltage general performance motors"* EN 02-2010

Review on Li-Ion Based Battery Chemistry: Challenges and Opportunities

A Daya[1] and S Paul Sathiyan[2]

[1]Department of Mechanical Engineering, [2]Department of Electrical and Electronics Engineering, Karunya Institute of Technology and Sciences, Coimbatore, India, sathiyanme@yahoo.co.in

Abstract. Level of demand of Li-ion battery (LIB) applications are arising, including Electric Drive Vehicles and Energy Storage devices, cell design and performance requirements are continually changing, that poses unique difficulties to existing battery manufacturers. In response, there is an inevitable requirement of high-power density and energy in the function of LIB. In this review we have discussed about the alternate battery technologies which can potentially replace the existing LIB technology. Li battery's electrical performances, their challenges the world will face in the near future as well as the upcoming battery technologies like graphene batteries, redox flow batteries which have the potential to compete in the current market have been discussed in this paper.

1. Introduction

Lithium-based batteries have great energy efficiency, specific power, density, and negligible memory effect, so dominate the battery industry and research environment, making them a potential choice among other existing batteries. Indeed, the battery revolution is hastening the transition to renewable energy sources while also increasing the adoption and commercialization of Electric Vehicles (EVs). They are used extensively in rechargeable consumer electronics too. EVs have made great use of these batteries as on-board energy storage devices. By 2028, up to 200 million EVs shall hit the road [1,2]. The electric vehicle sector in India is still in its infancy, but it is anticipated to develop on a 90 % Annual Compound Growth Rate (ACGR) between 2021 and 2030[3]. Figure. 1 depicts the rise of India's EV market [4,5].

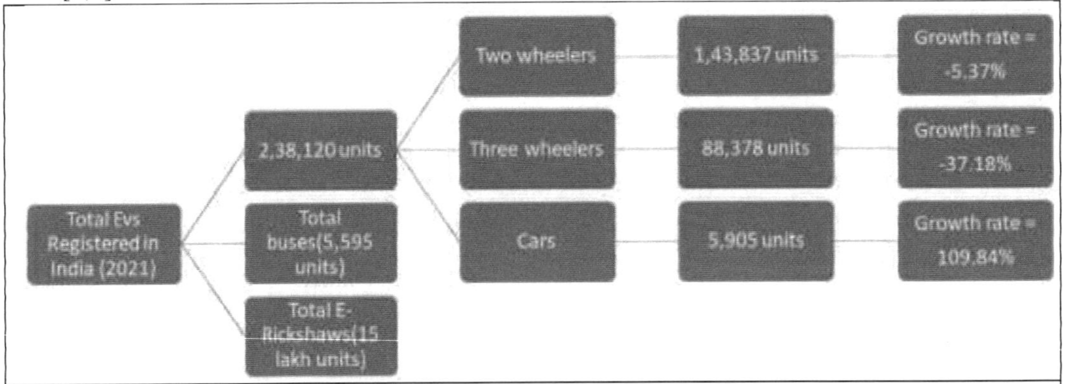

Figure 1. Indian EV market growth [6].

LIB's development to fulfil future demand, particularly in the electric vehicle arena, necessitates major improvements in specific and volumetric energy density, cyclability, rate of charging , strength, and safety features.

Because of its very reactive elemental state, lithium has been used in the form of metal oxides such as Li-Cobalt oxide ($LiCoO_2$). In cathode Li-metal oxides and in anode Li-carbon compounds are used. Structure can affect material characteristics at the microscale, and doping processes can assist in boosting conductivity. By combining and complementing materials, this is one-of-a-kind structure creates unique features. These materials allow for intercalation, which allows lithium ions to move freely in and out of the electrode structure. Traditional Lithium based ($LiCoO_2$, $LiFePO_4$, and $LiMn_2O_4$), which rely on the Li^+ intercalation process, are unable to provide the high-capacity charging (more than 200 mA h/g) necessary for unique usages. As a result, high-energy density battery technologies with transition metal oxides such as abundant in Lithium ($Li_{1.2}Mn_{0.54}Ni_{0.13}Co_{0.13}O_2$) and abundant in nickel oxides (e.g., $LiNi_{0.8}Co_{0.15}Al_{0.05}O_2$ or Li [$Ni_{0.8}Co_{0.1}Mn_{0.1}$] O_2), spinels with a higher voltage, sulphur of lithium, lithium–air, carbon-based conducting materials, lithium–carbon dioxide, and batteries in compact form have recently piqued researchers' interest [7, 8].

For energy storage applications, four potential cell chemistry have been considered: 1) $LiMn_2O_4$ / low cost materials like graphite, a naturally abundant material; 2) LiNi-X-Y_2CoXAlYO_2 / graphite with long life span with excellent energy; 3) $LiFePO_4$ / good safety features and 4) $Li_4Ti_5O_{12}$/ used as cathode with extended durability and outstanding precautionary measures [9-12]. The possible combination of battery materials and the features are obtainable from Table 1 and Figure 2.

Table 1. Component materials used in various Li-Ion based batteries

S.No	Anode	Cathode	Electrolyte	Ref
1	Lithium metal	Li-Mn-O	LiPF6	[13, 14]
2	Carbon (graphite)	LiCoO2	LiClO4	[15–17]
3	Carbon (Hard)	FeS2	LiAsF6	[18–20]
4	Titanate (Lithium)	V2O5	LiCF3SO3	[21, 22]
5	Synthetic Graphite	LiNiMnCoO2	Polyethylene Oxide (PEO)	[23–25]
6	Tin-based alloys	LiFePO4	-	[26, 27]
7	Silicon-based materials	Electronic Conducting Polymers	-	[28, 29]

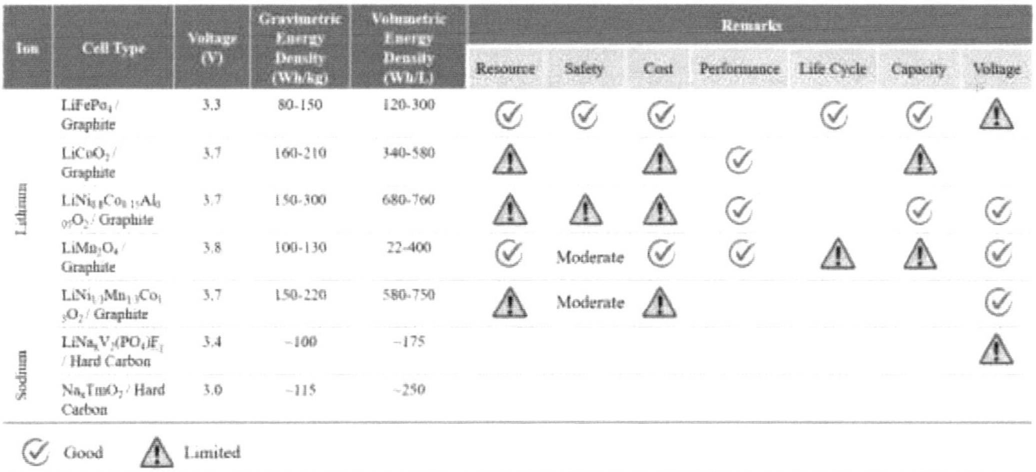

Ion	Cell Type	Voltage (V)	Gravimetric Energy Density (Wh/kg)	Volumetric Energy Density (Wh/L)	Remarks: Resource	Safety	Cost	Performance	Life Cycle	Capacity	Voltage
Lithium	$LiFePO_4$ / Graphite	3.3	80-150	120-300	Good	Good	Good		Good	Good	Limited
	$LiCoO_2$ / Graphite	3.7	160-210	340-580	Limited		Limited	Good		Limited	
	$LiNi_{0.8}Co_{0.15}Al_{0.05}O_2$ / Graphite	3.7	150-300	680-760	Limited	Limited	Limited	Good		Good	Good
	$LiMn_2O_4$ / Graphite	3.8	100-130	22-400	Good	Moderate	Good	Good	Limited	Limited	Good
	$LiNi_{1/3}Mn_{1/3}Co_{1/3}O_2$ / Graphite	3.7	150-220	580-750	Limited	Moderate	Limited				Good
Sodium	$LiNa_xV_2(PO_4)F_3$ / Hard Carbon	3.4	~100	~175							Limited
	Na_xTmO_2 / Hard Carbon	3.0	~115	~250							

Good Limited

Figure 2. Characteristics comparison of electrode materials [30].

Inability to use lithium-ion batteries at higher temperatures is a severe disadvantage. If Li- ion batteries in cell phones get overheated, they can become explosive and catch fire. Despite the fact that electronic controllers have been used under controlling temperature, these batteries are particularly hazardous when utilized in portable devices. Additional bottlenecks include recent price increases for Li-Ion batteries, rapid depletion of lithium ores, and mining-related pollution. The limitations with Li-Ion batteries prompted the development of graphene batteries and flow batteries, also known as redox vanadium flow batteries. Graphene has the advantage of recharging faster than lithium-ion batteries. Graphene is also more durable than other materials. It should, in actuality, have a longer life expectancy and consequently, fewer battery replacements are required. As Li-ion batteries have a tendency to get ignited and exploded in harsh circumstances, graphene-based batteries are found to be safer than batteries. The components of batteries made of lithium-ion types and their electrical characteristics, as well as the future possibilities of battery technology and alternatives for Li-based batteries, were examined in this study.

2. Structural analysis of electrodes of batteries made of lithium-ion

At the microscale, fundamental qualities can change material characteristics, and doping procedures can increase conductivity of such materials. By this way structural modifications provide unique properties through material integration and complementation. Sun et al. created a well-functioning negative electrode made of a transition metal oxide with Li, along with a capacity of Full Concentration Gradient (FCG) inside every atom as shown in Fig. 3. The construction makes use of nickel-rich materials' high capacity, manganese-rich materials' high thermal stability, and the high-rate ability of great infiltrated and associated morphology as nanorods. In a full-cell arrangement, this recently formed material may offer a specialised capability of up to 215 milliamp Hour per gram with exceptional stability, and of ninety % retention capacity even after thousand cycles [31]. Improved materials for positive electrodes, such as NASICON (super Na^+ conductors) or Olivine (silicate of Mg and Fe) oxyanion scaffolded materials, raised from corner-sharing MO_6 octahedral (where M is iron, titanium or vanadium) and XO_4^{n-} tetrahedral anions (where X is sulphur, molybdenum, phosphorus or tungsten) provide intriguing ideas [32].

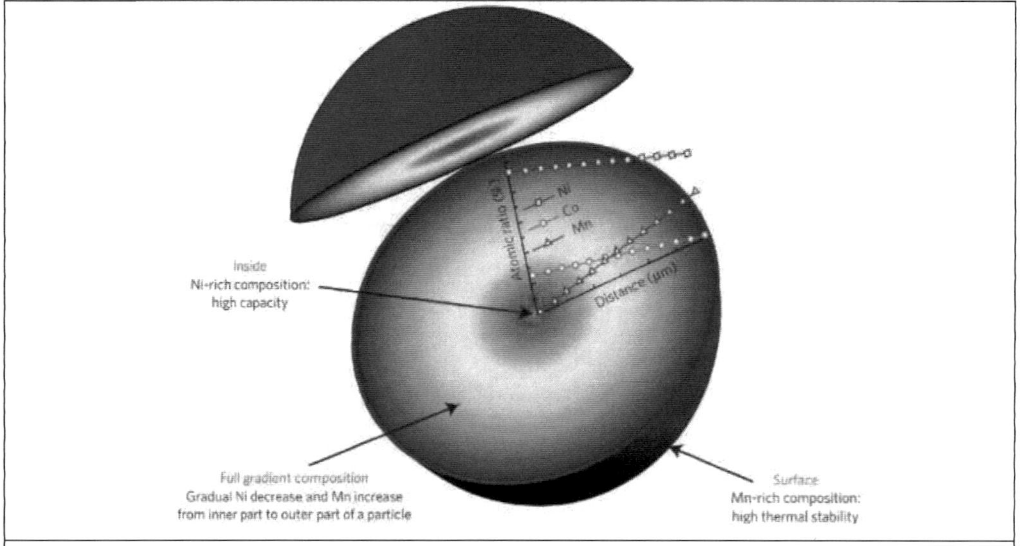

Figure 3. FCG lithium transition-metal oxide particle [33].

The electrode's potential value fluctuates marginally, and the voltage is maintained by those active drive of lithium ion, causing distribution factor to be enhanced for improved cycle output. Mg^{2+} doped structures come in four dissimilar forms. Carbon packing, on the other hand, can improve the conductivity of LFP material much further.

$LiFePO_4$ is far harmless than conventional materials for applications involving high current discharge. $LiFePO_4$ has an orthogonal olivine structure, in which the atoms of oxygen mound tightly through lithium and iron in the core to create octahedrons of iron and lithium (FeO_6, LiO_6). Each of the adjacent FeO_6 octahedrons shares one atom of oxygen and forms a layer. Li^+ can spread in one dimension because nearby LiO_6 octahedrons are linked two atoms of oxygen within layers of FeO_6. Along the crystallographic b-axis, the diffusion is one-dimensional as presented in Figure 4.

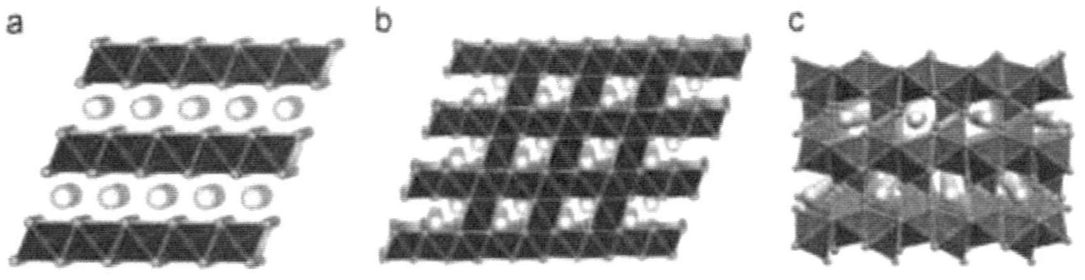

Figure 4. LIB energy system structure comparison [34]: arrangement of (a) lattice $LiCoO_2$, (b) spinel $LiMn_2O_4$ and (c) olivine $LiFePO_4$.

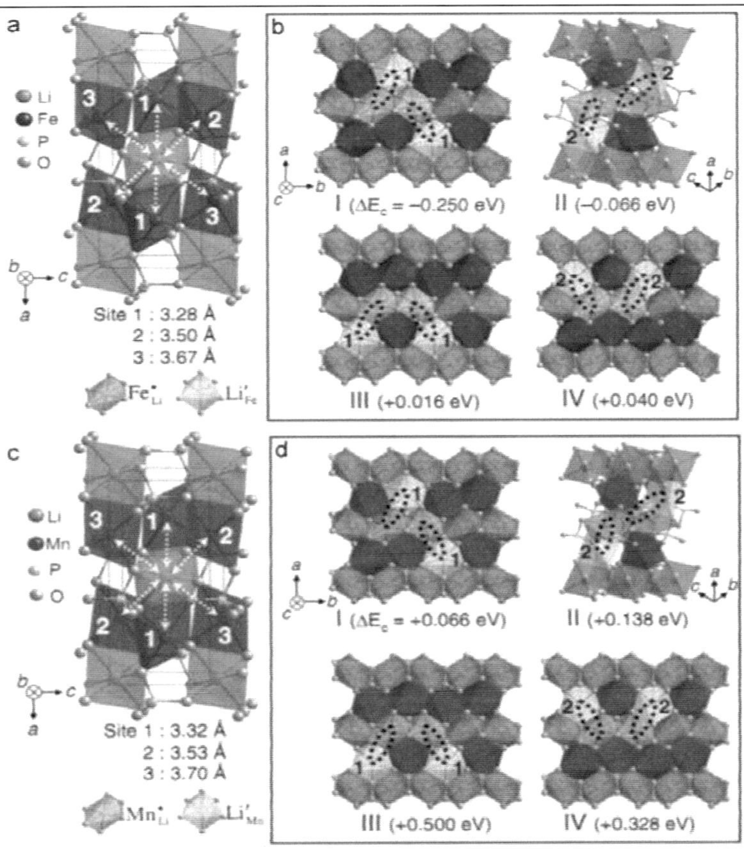

Figure 5. Model structure of $LiFePO_4$ and $LiMnPO_4$ [35]:
$LiFePO_4$ (a) sole cell assembly (b) Link models in four types $LiMnPO_4$ (c) sole cell assembly (d) Link models in four types.

Antisite exchange defects (when an atom of a different element occupies a place inside a chemical complex that would typically be occupied by that element) in $LiMnPO_4$ and $LiFePO_4$ exhibited radically different topologies, with random distribution of the defects, according to Cheng et al. (Fig. 5). In the former, exchange pairs without grouping are used, whereas in the later, grouping is used. In the latter, there is a lot of clustering. Despite their similar compositions and crystal structures, those two are not the same. Liu et al. developed a flexible thin film assembly that employs a three-dimensional gradient to absorb and retain electrolyte while conducting high-power charging and discharging [36].

2.1. Lithium-ion battery challenges

Li-ion batteries have a shorter lifespan in comparison to existing other rechargeable batteries. They must be shielded from being significantly overloaded and discharged. As a result, in order to maintain these safe limitations, every LIB requires safety, which is a significant downside. The effectiveness and performance of Li-ion batteries damage with time. Furthermore, Li-ion batteries can lose their functionality whether or not they are in use. In addition to consumption, a time-related component contributes to the capacity decline. Another significant disadvantage of lithium-ion batteries is their inability to function at higher temperatures.

The fundamental difficulty with Li-metal anodes, according to Fang et al., is poor coulombic efficiency (CE), rather than Li-dendrite generation, causing a short circuit. Metal cells of lithium suffer because of absorption by activated metal lithium or else the exhaustion due to electrolyte. The increase of residual lithium metal, relatively compared with the metaphase of solid electrolyte that is connected with the morphology of deposition of lithium, dominates the loss of CE [37]. The intrinsic capacitance (contrariwise) and the released temperate just by duration of process define the battery's power. The quantity of generated heat by Li-Ion batteries is influenced by the given main factors: a) instigation (kinematics at the interface), b) aggregation (type translocation), and c) ohmic losses (electron dispersion). The principal causes of heat emissions are the battery's irreversibility and electrochemical activity. The most abiotic depletion (fossil fuels) occurs during the production phase of the system $Li(NiCoMn)O_2$, followed by the usage period. During the process of production, it inevitably creates toxicity to the waterbodies [38]. The fabrication and usage phases of $Li(NiCoMn)O_2$ batteries have virtually comparable contributions, with a variation of up to 10% between the two. The bulk of abiotic diminution and exhaustion of ozone layer occur during the production process for the $LiFePO_4$ battery system, whereas practically all abiotic reduction (hydrocarbons), Global Warming (GWP 100), and toxicity to human occur during the usage phase. BEVs using $LiMn_2O_4$ cathodes lessen environmental difficulties (consumption of energy, air pollutants, such as GHGs), however material manufacture, construction, and reprocessing of automobile batteries of Li-Ion in hybrid electrical vehicles are still in their infancy. Hollatz S et al. [39] deliberated the consumption of energy and release of air, produced during the recovery of $LiMn_2O_4$, aluminium, and Cu using three different recovering methods (recycling by direct physically, intermediate physical recycling, and hydrometallurgical), as well as the outcome(s) of locked-loop recovering on the battery production's environmental effects. The biggest problem with Li-ion batteries is that they generate a lot of heat during the charging / discharging process, which causes battery pack temperature to increase, especially at higher working currents. Kwak W-J et al. [40] identified a number of thermal management difficulties and the several cooling solutions developed to meet them, as well as certain research gaps in the retrieval and consumption of low-grade heat produced in battery packs. Because Li-ion batteries must operate safely and reliably in such a compact size, good battery management system control and administration are essential. Lu et al. [41] gave a brief overview by examining the research and integrating it with our own personal experiences. The constituents of the management system for batteries (MSB) is discussed in the hope of providing some ideas for battery management design and research, with critical concerns like measurement of voltage in battery cell, estimation of states of battery, battery homogeneity and equalisation, diagnostics in battery error, etc. During operation, adaptive control technology is used to detect battery system characteristics on-line or offline based on data collected, estimate battery states (condition of health, charge, function, and errors), and update the vehicle regulator over the system to warrant vehicle safety and dependability. As a result, (1) battery performance studies; (2) the development of practical models of battery; and (3) the implementation of adjustable control technologies or the theories of adept scheme to the management of the battery are all aspects of the MSB that need to be investigated further [42].

2.2. Future prospects of Lithium-ion batteries

EV industry is changing at such a rapid speed, and will continue to do so, potential investors will need to make cautious decisions to ensure that their earnings are not limited to the short term. Each step toward entirely zero-emission vehicles (ZEVs) is not only a remarkable accomplishment, but also a stride toward reducing the global effect of human activities. Prior to the deployment of the solid-state batteries' first generation, the next group of better batteries of Li-Ion was planned which will be suited for applications requiring high energy, high power, and safety, also sustainable energy stowage and transportation. Liu et al. analysed the study data, resolving the high cost, energy-intensive, and time-consuming stages of LIB production, as well as the problems of LIB production, and made recommendations for future LIB manufacturing research development paths [43]. Conversion-reaction anodes and cathodes have attracted a lot of attention as their capacity has increased by an order of magnitude above insertion-reaction electrodes, but their practicality has been questioned. Although advances in membrane technique have increased the likelihood of safe batteries, the industry's primary concern is still safety. In response to the demand for more effective electrical energy storage devices, new electrode materials are being investigated. In my perspective, intercalation chemicals offer a realistic approach to high-energy density batteries. Using a unified methodology, researchers developed a framework for generating similar models of circuit that can properly mimic the multi-physics phenomena of LIB packages, including cooling arrangements with liquid. Hollatz et al. [39] investigated a predisposition for current to flow mostly near the edges of a contact by influencing the predilection and establishing a corresponding linking using a dual joint stratum with the longest plausible spacing considerably enhances linkage conductivity [44]. Kwak et al. [40] examined recent findings as well as advances in cathode and anode metal, electrocatalysis on surfaces, in the solution phase of electrolytes, and the singlet oxygen atom's degradative effect, is often generated in $Li–O_2$ cells [45]. Panwar et al. also researched at smart technologies and advanced intelligence, which included cyber-physical systems, battery swapping technology, digital-twin battery packs, non-disparaging analysis, self-reconfigurable batteries, and mechanisation-assisted recovering/reusing [46].

Recent years have seen a lot of research on graphene-based materials for a variety of applications. Nanosheets consisting of hexagonal BN (boron nitride), graphene with germanium, silicone, different types of d-block metals chalcogens, phosphor are developing as new and unique nanostructures for the future group of nanomaterial research. Solid-state batteries may be used by EV manufacturers to improve the complete performance and price of their cars. The suggested work might be expanded to assess the function of an electric vehicle's full battery management system. The structural flexibility of graphene was used to build 2-D electrodes made of graphene paper. By mixing graphene-foam with the electrodes, a high conductivity channel was constructed, and this revolutionary supple designed battery structure has an outstanding porosity structure, dependability, chemical resistance even under harsh chemical circumstances, and spectacular electronic and ionic permittivity. The promising materials for future generation energy stowage uses is 2D graphene, particularly in the devices like supercapacitor (130 GPa), because of its incredible inherent properties, which include the hight surface area (2600 m^2g^{-1}), high electrical charges mobility (23 x 10^4 cm^2/Vs), heat conductivity (3000 W/mK), and high power. Graphene's unique properties, such as its enhanced ratio of surface to volume (SA/V), resistance to chemicals, good flexibility, and grander electrical conductivity, being used to combine it with numerous oxides of transition metal nanoparticles with extremely high theoretical capacities, such as SnO_2, Fe_3O_4, Co_3O_4, and Mn_3O_4. Li-S, LiO_2, batteries of Na^+, and Li^+ will all benefit from this compound.

Redox flow batteries (RFBs) are among the supreme popular electrochemical devices especially aimed at large as well as medium purposes among the various electrochemical devices present. Due to its quick time response (in milliseconds), high cycle life (higher than $20x10^3$ cycles), and certainly eco-friendly components, RFBs are increasingly gaining market share. Redox flow batteries are slowly but surely gaining popularity in a rising number of projects throughout the world, but LIBs currently dominate the stationary power stowage business. Manufacturers of flow batteries, especially vanadium redox flow batteries (VRFBs), have spent years researching, testing, and proving the technology, with a total setting up of approximately seventy MW and 250MWh of power and energy. Manufacturers of flow batteries continue to get financing to increase production, enhance their products, and lower the cost of technology. Furthermore, substantial partnerships are taking place interconnecting flow battery manufacturers all across the world, along with industries like mining and chemical, with the shared goal of making this technology competitive on the market [47]. This technology is also highly environmentally friendly because it does not rely on the usage of relatively rare Earth elements such as vanadium or lithium. Most other commercial battery types are intended to discharge for one or two hours at a time, whereas flow batteries may discharge for up to ten hours at a time. Microgrids are little electric grids that may function independently of the main grid at any moment. For the same reasons that utility companies are interested in adopting flow batteries, they might be a good fit for microgrid architecture. In March 2020, IDTechEx conducted thorough research after a methodological and marketplace standpoint on RFBs to have a deeper understanding the present state of development and market opportunities [48]. The findings of this study, available in the paper "RFB 2020-2030: Forecast, Challenges, Opportunities," suggest a CAGR of 30% over the following ten years, driven with a surge in attention in energy stowage schemes to sustenance the grid [49]. As per Benchmark Mineral Intelligence, demands for lithium would grow by 26.1 percent, or around 10 x 10^4 tonnes of $LiCO_3$ equal, to a whole of 4.5 x 10^4 tonnes, resulting in a 10,000-tonne shortage in the market. Albemarle, a lithium miner, forecasts a tenfold increase in global EV manufacturing before 2030, and a 30 percent increase in lithium demand by 2025.

Flow batteries, which can discharge for up to 10 hours at a time, are perfect energy storage options for large-scale applications. This is a significant discharge duration, particularly when compared to other battery types, which can only drain for up to two hours at a time. Their unique design distinguishes them from other rechargeable battery types, such as lithium-ion batteries. This design divides fluid into two outside tanks, through which electrons travel through electrochemical cells and a membrane that separates the electrons. The quantity of power that can be created is determined by the size of these outside tanks; the greater the tank size, the more electricity that can be generated. Flow batteries can be used in utilities, microgrids, and electric cars (EVs). Because of the lengthy discharge capacity of flow batteries, bigger applications such as utilities or microgrids (smaller versions of electric grids that may function independently) frequently use this battery storage technology. Because of its ability to swiftly refill electrolyte liquid or "recharge," flow batteries are a great choice for EVs. Vanadium and iron are two common elements found in flow batteries. Despite its poor energy density (max. 40 W h L) as compared to Li-ion batteries (700 W h L), the all-vanadium redox flow battery (VRFB) has previously piqued commercial attention. Flow batteries make a lot of sense for the future of electric car fuel since they can be quickly "charged" by replenishing the electrolyte liquid. At a fuelling station, the expended electrolyte could potentially be readily emptied and replenished. That would be a far faster option than having to wait up to 20 minutes for your EV to suck electrons back up into its lithium ion cells. Current flow batteries, on the other hand, have a poor energy density. That means the driving range would be extremely limited. Some electrochemistry experts are investigating the use of high solubility salts in the electrolyte solution to extend range, however the use of flow batteries in EVs appears to be wishful thinking for the time being. Flow battery systems are quite safe since they do not use flammable electrolytes. While uncommon and costly, the vanadium fluid most commonly utilised in the tanks is also ecologically beneficial. The tanks are even safer since they may be housed further away from the conducting cell membrane and power stack. The life expectancy of Li-ion batteries are expected to be around 10 years whereas Vanadium flow batteries have a lifespan of up to 30 years. This is mostly due to the absence of phase-to-phase chemical interactions in flow batteries. The vanadium electrolyte may have electrons supplied and withdrawn without deterioration. This results in a possibly infinite battery cycle life. With more utility providers adopting time-of-use pricing systems, flow batteries offer an attractive alternative for microgrid operators or big manufacturing sites looking to transfer expensive peak loads to long-duration battery use. However, flow battery storage technologies capable of meeting the enormous energy demands of utility-scale applications are still prohibitively expensive. Regardless, the flow battery market is expected to develop at a moderate compound annual growth rate (CAGR) of more than 12% through 2025. The majority of demand is expected to come from Asia, notably China and India. Vanadium redox flow batteries are predicted to be the most widely used form of flow battery, owing to their ability to be charged and drained without deterioration.

3. Conclusion

The findings Future generations will benefit greatly from sustainable energy storage technologies in general, as well as electric cars in particular, paving the way for a cleaner energy future with low energy losses. Aside from the growing number of electric cars on the road, which will continue to boost the demand for Li-ion batteries, the battery's lifespan is finite. In just five years, Li-ion batteries lose 70-90 percent of their capacity and will continue to be in great demand to replace those in regularly used battery-powered goods like electric autos due to their short lifespan. As technology advances, batteries will surely get more efficient. The future potential of battery technology is essentially endless now that graphene has been added to the mix. Graphene is a brand-new material that is only now making its way into battery technology. The most difficult roadblock to overcome is the exorbitant expense of creating small graphene sheets. As production processes improve and become more cost efficient, graphene's potential applications will grow.

4. References

[1] Marinaro M, Bresser D, Beyer E, Faguy P, Hosoi K, Li H, Sakovica J, Amine K, WohlfahrtMehrens M and Passerini S 2020 Bringing forward the development of battery cells for automotive applications: Perspective of R&D activities in China, Japan, the EU and the USA *Journal of Power Sources* **459** 228073

[2] dos Reis G, Strange C, Yadav M and Li S 2021 LIB data and where to find it *Energy and AI* **5** 100081

[3] Bismah Malik 2021 India's EV market to grow by 90% to touch $150 billion by 2030: Report *Business Today*

[4] Ajit Dalvi SMEV: Cumulative EV sales down 19.41 percent in FY2021 *Autocar India*

[5] Anon Electric Bus Market Size, Share (2022 - 27) | Industry Forecast

[6] Mishra A, Mehta A, Basu S, Malode S J, Shetti N P, Shukla S S, Nadagouda M N and Aminabhavi T M 2018 Electrode materials for lithium-ion batteries *Materials Science for Energy Technologies* **1** 182–7

[7] Kim T, Song W, Son D-Y, Ono L K and Qi Y 2019 LIB: outlook on present, future, and hybridized technologies *J. Mater. Chem. A* **7** 2942–64

[8] Horiba T 2014 Lithium-Ion Battery Systems *Proceedings of the IEEE* **102** 939–50

[9] Yang Y, Shu D, Yu H, Xia X and Lin Z G 1997 Investigations of lithium manganese oxide materials for lithium-ion batteries *Journal of Power Sources* **65** 227–30

[10] Botte G G, White R E and Zhang Z 2001 Thermal stability of LiPF6–EC:EMC electrolyte for lithium ion batteries *Journal of Power Sources* **97–98** 570–5

[11] Yan Z, Hu Q, Yan G, Li H, Shih K, Yang Z, Li X, Wang Z and Wang J 2017 Co 3 O 4 /Co nanoparticles enclosed graphitic carbon as anode material for high performance Li-ion batteries *Chemical Engineering Journal* **321** 495–501

[12] Shu J, Shui M, Huang F, Ren Y, Wang Q, Xu D and Hou L 2010 A New Look at Lithium Cobalt Oxide in a Broad Voltage Range for Lithium-Ion Batteries *J. Phys. Chem. C* **114** 3323–8

[13] Gunawan I, Deswita, Sugeng B, and Sudaryanto 2017 Synthesis and characterization of PVA blended $LiClO_4$ as electrolyte material for battery Li-ion *IOP Conf. Ser.: Mater. Sci. Eng.* **223** 012039

[14] Wang K, Jin Y, Sun S, Huang Y, Peng J, Luo J, Zhang Q, Qiu Y, Fang C and Han J 2017 LowCost and High-Performance Hard Carbon Anode Materials for Sodium-Ion Batteries *ACS Omega* **2** 1687–95

[15] Tran D T, Dong H, Walck S D and Zhang S S 2015 Pyrite FeS2–C composite as a high-capacity cathode material of rechargeable lithium batteries *RSC Adv.* **5** 87847–54

[16] Li Q, Chen J, Fan L, Kong X and Lu Y 2016 Progress in electrolytes for rechargeable Li-based batteries and beyond *Green Energy & Environment* **1** 18–42

[17] Chauque S, Oliva F Y, Visintin A, Barraco D, Leiva E P M and Cámara O R 2017 Lithium titanate as anode material for lithium-ion batteries: Synthesis, post-treatment and its electrochemical response *Journal of Electroanalytical Chemistry* **799** 142–55
[18] Yang Y, Li L, Fei H, Peng Z, Ruan G and Tour J M 2014 Graphene Nanoribbon/V_2O_5 Cathodes in Lithium-Ion Batteries *ACS Appl. Mater. Interfaces* **6** 9590–4
[19] Elia G A, Park J-B, Sun Y-K, Scrosati B and Hassoun J 2014 Role of the Lithium Salt in the Performance of Lithium-Oxygen Batteries: A Comparative Study *CHEMELECTROCHEM* **1** 47–50
[20] Ma C, Zhao Y, Li J, Song Y, Shi J, Guo Q and Liu L 2013 Synthesis and electrochemical properties of artificial graphite as an anode for high-performance lithium-ion batteries *Carbon* **64** 553–6
[21] Kim Y 2012 Lithium Nickel Cobalt Manganese Oxide Synthesized Using Alkali Chloride Flux: Morphology and Performance As a Cathode Material for Lithium Ion Batteries *ACS Appl. Mater. Interfaces* **4** 2329–33
[22] Mekonnen Y, Sundararajan A and Sarwat A I 2016 A review of cathode and anode materials for lithium-ion batteries *SoutheastCon 2016* SoutheastCon 2016 (Norfolk, VA, USA: IEEE) pp 1–6
[23] Crosnier O, Brousse T and Schleich D M 1999 Tin based alloys for lithium ion batteries *Ionics* **5** 311–5
[24] Hsieh C-T, Pai C-T, Chen Y-F, Chen I-L and Chen W-Y 2014 Preparation of lithium iron phosphate cathode materials with different carbon contents using glucose additive for Li-ion batteries *Journal of the Taiwan Institute of Chemical Engineers* **45** 1501–8
[25] Casimir A, Zhang H, Ogoke O, Lu J and Wu G Silicon-based Anode for Lithium-ion Batteries: Effectiveness of Materials Synthesis and Electrode Preparation 56
[26] Kam K and Doeff M 2012 Electrode Materials for Lithium Ion Batteries *Material Matters* **7** 56–60
[27] Sun Y-K, Chen Z, Noh H-J, Lee D-J, Jung H-G, Ren Y, Wang S, Yoon C S, Myung S-T and Amine K 2012 Nanostructured high-energy cathode materials for advanced lithium batteries *Nature Mater* **11** 942–7
[28] Tarascon J-M and Armand M 2001 Issues and challenges facing rechargeable lithium batteries *Nature* **414** 359–67
[29] Chung S-Y, Choi S-Y, Lee S and Ikuhara Y 2012 Distinct Configurations of Antisite Defects in Ordered Metal Phosphates: Comparison between LiMnPO 4 and LiFePO 4 *Phys. Rev. Lett.* **108** 195501
[30] Li N, Chen Z, Ren W, Li F and Cheng H-M 2012 Flexible graphene-based lithium ion batteries with ultrafast charge and discharge rates *Proceedings of the National Academy of Sciences* **109** 17360–5
[31] Fang C, Wang X and Meng Y S 2019 Key Issues Hindering a Practical Lithium-Metal Anode *Trends in Chemistry* **1** 152–8
[32] Kim H-J, Krishna T, Zeb K, Rajangam V, Gopi C V V M, Sambasivam S, Raghavendra K V G and Obaidat I M 2020 A Comprehensive Review of LIB Materials and Their Recycling Techniques *Electronics* **9** 1161
[33] Shu X, Guo Y, Yang W, Wei K and Zhu G 2021 Life-cycle assessment of the environmental impact of the batteries used in pure electric passenger cars *Energy Reports* **7** 2302–15
[34] Dunn J B, Gaines L, Sullivan J and Wang M Q 2012 Impact of Recycling on Cradle-to-Gate Energy Consumption and Greenhouse Gas Emissions of Automotive Lithium-Ion Batteries *Environ. Sci. Technol.* **46** 12704–10
[35] Aswin Karthik C, Kalita P, Cui X and Peng X 2020 Thermal management for prevention of failures of lithium ion battery packs in electric vehicles: A review and critical future aspects *Energy Storage* **2**
[36] Liu Y, Zhang R, Wang J and Wang Y 2021 Current and future lithium-ion battery manufacturing *iScience* **24** 102332

[37] Armand M B 1980 Intercalation Electrodes *Materials for Advanced Batteries* ed D W Murphy, J Broadhead and B C H Steele (Boston, MA: Springer US) pp 145–61
[38] Yamanaka T, Kihara D, Takagishi Y and Yamaue T 2020 Multi-Physics Equivalent Circuit Models for a Cooling System of a Lithium Ion Battery Pack *Batteries* **6** 44
[39] Hollatz S, Kremer S, Ünlübayir C, Sauer D U, Olowinsky A and Gillner A 2020 Electrical Modelling and Investigation of Laser Beam Welded Joints for Lithium-Ion Batteries *Batteries* **6** 24
[40] Kwak W-J, Rosy, Sharon D, Xia C, Kim H, Johnson L R, Bruce P G, Nazar L F, Sun Y-K, Frimer A A, Noked M, Freunberger S A and Aurbach D 2020 Lithium–Oxygen Batteries and Related Systems: Potential, Status, and Future *Chem. Rev.* **120** 6626–83
[41] Lu L, Han X, Li J, Hua J and Ouyang M 2013 A review on the key issues for lithium-ion battery management in electric vehicles *Journal of Power Sources* **226** 272–88
[42] Panwar N G, Singh S, Garg A, Gupta A K and Gao L 2021 Recent Advancements in Battery Management System for Li-Ion Batteries of Electric Vehicles: Future Role of Digital Twin, CyberPhysical Systems, Battery Swapping Technology, and Nondestructive Testing *Energy Technol.* **9** 2000984
[43] Tiwari S K, Sahoo S, Wang N and Huczko A 2020 Graphene research and their outputs: Status and prospect *Journal of Science: Advanced Materials and Devices* **5** 10–29
[44] Loganathan M K, Ming Tan C, Mishra B, Msagati T A M and Snyman L W 2019 Review and selection of advanced battery technologies for post 2020 era electric vehicles *2019 IEEE Transportation Electrification Conference (ITEC-India)* 2019 IEEE Transportation Electrification Conference (ITEC-India) (Bengaluru, India: IEEE) pp 1–5
[45] Priyadarshi H, Choudhary H, Agarwal S N, Shrivastava A, Singh K and Akhtar J 2019 Graphene Based Futuristic Green Batteries For Energy Harvesting *2019 IEEE International Conference on Engineering, Technology and Education (TALE)* 2019 IEEE International Conference on Engineering, Technology and Education (TALE) (Yogyakarta, Indonesia: IEEE) pp 1–4
[46] Chou S-L, Wang J-Z, Chew S-Y, Liu H-K and Dou S-X 2008 Electrodeposition of MnO2 nanowires on carbon nanotube paper as free-standing, flexible electrode for supercapacitors *Electrochemistry Communications* **10** 1724–7
[47] Contributor G 2020 What's on the energy storage market besides lithium-ion batteries *Smart Energy International*
[48] Anon 2021 Electric vehicle sales surge in 2021 *Power Technology*
[49] Anon 2020 *Redox Flow Batteries 2020-2030: Forecasts, Challenges, Opportunities*

Study on Performance of Ramjet Intake by Changing the Cowl Angle

Levi Reuben Vr[1,a], Amrutha Rajamani[2 B]

[1]Coimbatore, Tamil Nadu, India
[2]Palakkad, Kerala, India

levireuben1997@gmail.com [a], amrutha1196@gmail.com [b]

Abstract: Most gas turbine engines require the Mach range on the engine face to be at a mild subsonic pace (around Mach 04). Therefore, for a supersonic plane with a gas turbine engine, the characteristic of the air inlet is to slow the supersonic unfastened flow to a subsonic pace and offer a matched air mass float charge to the engine. The gas turbine engine calls for the delivery of uniform excessive general stress restoration air for proper overall performance and operation because the quality of the airflow on the engine face substantially influences the overall performance of the engine, in particular, the total stress loss, which influences the engine thrust and therefore the gasoline consumption. A 2-D supersonic intake is designed for Mach No. 2.4 at an altitude of 11,000 m. The design of the mixed compression supersonic intake is done numerically, and the design with the k- turbulent module is done using CFD. Optimization of the intake is done by changing the cowl deflection in order to get maximum total pressure recovery. Results are compared to knowing the optimal design. The main advantage of mixed compression is that it gives the maximum total pressure recovery at any Mach number. The computation is done with the ANSYS software.

Keywords: Aerodynamics, Ramjet intake, combustion performance, cowl angle, shock train, CFD analysis, supersonic intake

1. INTRODUCTION

The air inlet ought to be designed such that it yields a satisfactory viable propulsive balance, which may be expressed in objectives:

- offer most thrust
- resulting in minimal drag.

Maximum thrust is acquired through designing the air inlet to convert kinetic energy (i.e., the speed of the go with the drift because it arrives in the front of the air inlet) into capacity energy (the stress after the diffuser, on the engine entrance) with satisfactory viable efficiency. Efficiency is a parameter calculated through taking the ratio of the entire stress in the front of the engine to that of the upstream go with the drift. The thrust might be greatest if the air inlet captures simply what the engine wishes for every flight configuration and presents the engine entrance with a homogeneous flow with the drift to make certain accurate engine operation, which is important for turbojets.

Minimum drag is achieved with air inlets that can be dimensioned for the essential regime. Due to the small variations in the attack and yaw attitude about the mild configurations, special attention should be paid to the facet partitions and cowls. A characteristic of combined supersonic compression is that low, in passing via the compression device produced via means of the forebody (each isentropic and discrete shocks), it grows to become outwards from the inlet axis. As a result, at the access aircraft, the waft has a tendency similar to the overall turning attitude, and in practice calls for it to grow to become lower back to the axial path in the subsonic diffuser. With green compression, the turning angles are required to be pretty large.

Whether this may be found out in a given state of affairs relies upon elements regarding both the outside flow across the cover or the inner waft at once in the duct access.

2. DESIGN

2.1 SPECIFICATIONS

- Mach no. – 2.4
- Altitude(A) – 11000 m
- Density(ρ) – 0.3403 Kg/m3
- Static temperature(T) – 216 K
- Static pressure(P) -22700 N/m2
- Gamma(γ) – 1.4
- Ideal gas constant(R) -287.05 J/Kg K

2.2 SUPERSONIC INLET DESIGN

The following equations were employed to calculate the wedge angle and the shock angles: Keeping in mind that for max pressure recovery, the normal component of Mach number over the shock is constant.

$$M_1\sin\beta_1 = M_2\sin\beta_2 = \ldots = M_{n-1}\sin\beta_{n-1} \qquad \text{Eq. 1}$$

Table 1. Properties of shock waves in the intake

M_∞	θ	β	M_n	P	ρ
2.4	15°	38.21	1.79	789KPa	3.202Kg/m³
1.79	10°	44.05	1.46	1312KPa	4.524 Kg/m³

The oblique shock pressure recovery was calculated using the formula

$$\frac{P_{02}}{P_{01}} = \left[1 + \frac{2\gamma}{\gamma+1}(M_1^2\sin^2\beta - 1)\right]^{\frac{-1}{(\gamma-1)}} \left[\frac{(\gamma+1)M_1^2\sin^2\beta}{(\gamma-1)M_1^2\sin^2\beta+2}\right]^{\frac{\gamma}{(\gamma-1)}} \qquad \text{Eq. 2}$$

Similarly, for normal shock pressure recovery this equation was used

$$\frac{P_{02}}{P_{01}} = \left[1 + \frac{2\gamma}{\gamma+1}(M_1^2 - 1)\right]^{\frac{-1}{(\gamma-1)}} \left[\frac{(\gamma+1)M_1^2}{(\gamma-1)M_1^2+2}\right]^{\frac{\gamma}{(\gamma-1)}} \qquad \text{Eq. 3}$$

2.3 INLET GEOMETRY

Now we know the optimum value of shocks and the deflection ability of wedges. As a result, we can now use the following formula to calculate the geometry of the inlet wedge based on these angles.

$$\frac{L_W}{L_N} = \frac{L_W}{h_c} tan\beta_{1D} = \frac{\tan(\beta_{2D}+\delta_1)-tan\beta_{1D}}{\tan(\beta_{2D}+\delta_1)-tan\delta_1} \quad \text{Eq. 4}$$

Where,

L_W is the length of the first wedge (in mm)
L_N is the total length of the wedge (in mm)
$\beta_{1D,2D}$ are the required angle of shocks at critical condition
$\delta_{1,2}$ are the deflection angle of wedges

Now, substituting the values of β and δ in the above equation we get the resultant ratio of L_W and L_N as follows:

$$\frac{L_{W1}}{L_N} = 0.62656,\ L_N = 31.92 \quad \text{Eq. 5}$$

Capture height for the intake is calculated for two wedges and added at the end,

$h_{c1} = 15.05$
$h_{c2} = 11.05$
$h_c = h_{c1} + h_{c2} = 26.1$

3. CAD DESIGN

CAD design of supersonic RAMJET intake is designed for Mach number 2.4 at the altitude of 11,000 m.

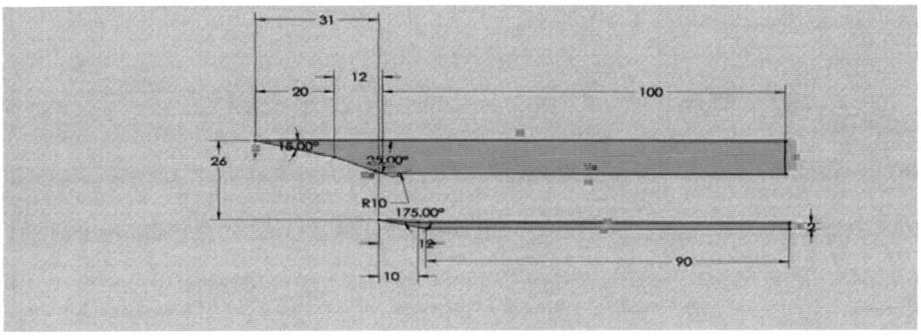

Fig. 1: 2-D design

3.1 GEOMETRIC PARAMETERS

15° – deflection angle of first wedge
10° - deflection angle of second wedge
132 mm – Total distance of the intake
20 mm – first ramp distance
12 mm – second ramp distance
26 mm – capture height

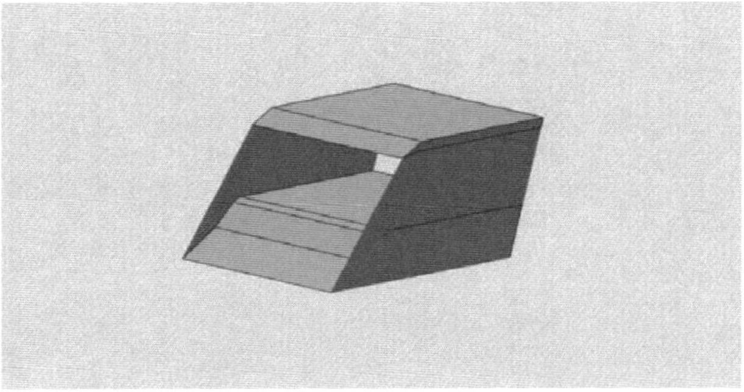

Fig. 2: 3D isometric view of the supersonic Ramjet intake

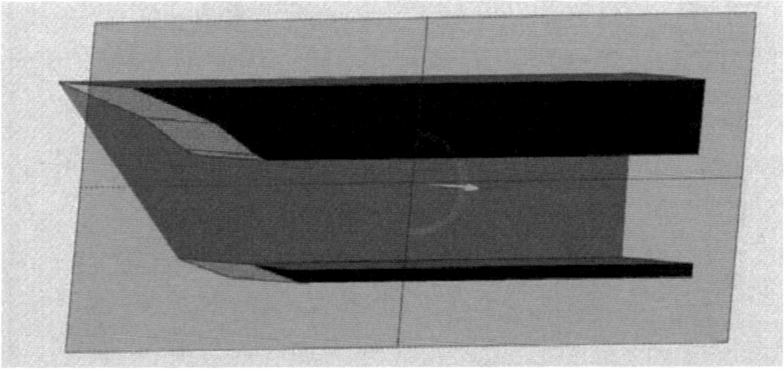

Fig. 3: Cross sectional view of the supersonic RAMJET intake

4. SOLUTION

ANSYS FLUENT is used for solving since 2D analysis is done. 2D analysis is selected. In this option, double precision is selected for a better and more stable solution. The processing option is selected as a parallel solver. Then the number of processers is given to start the boundary condition setup. In the general method solution setup, the solver is selected as density-based. This is because the flow is supersonic flow, so it will be a compressible flow and also turbulence flow will happen. Since it is a compressible flow, the density will vary. The flow happens in a 2D rectangle. So, 2D space is planar. In a density-based solver, the viscosity dissipation is calculated as in the energy equation. So, in general methods, the energy equation is turned on since it is a turbulent model because of supersonic flow.

In the General method, viscous model is selected as k-ω model, because it gives more stable solution. There are different types of turbulent models. Based on the requirements, the turbulent model was chosen.

5. BOUNDARY CONDITIONS

A condition which a quantity that varies throughout a given space or enclosure must fulfill at every point on the boundary of that space, especially when the velocity of a fluid at any point on the wall of a rigid conduit is necessarily parallel to the wall.

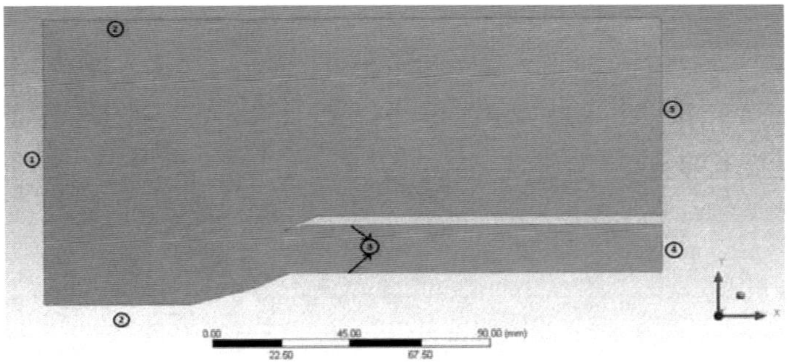

Fig. 4: BOUNDARY CONDITIONS

Boundary conditions set for our design are as follows:

1- INLET: Pressure far field
2- SIDE: Symmetry
3- WALL: Wall
4- OUTLET: Pressure outlet
5- OUTLET 2: Pressure outlet

6. RESULTS

6.1 FOR 5° COWL DEFLECTION

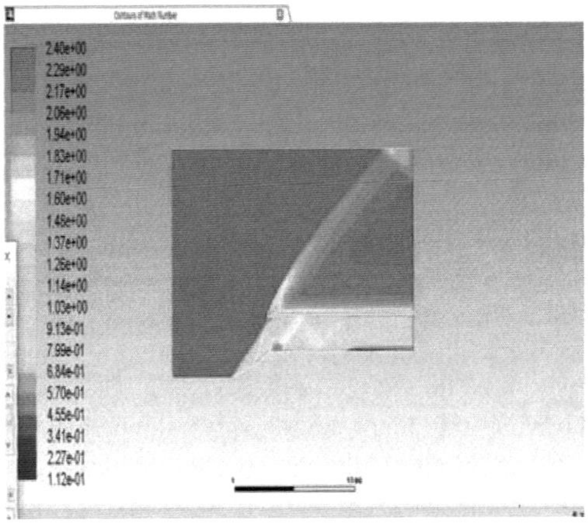

Fig. 5: Velocity contour of flow through intake with 5 degrees cowl deflection

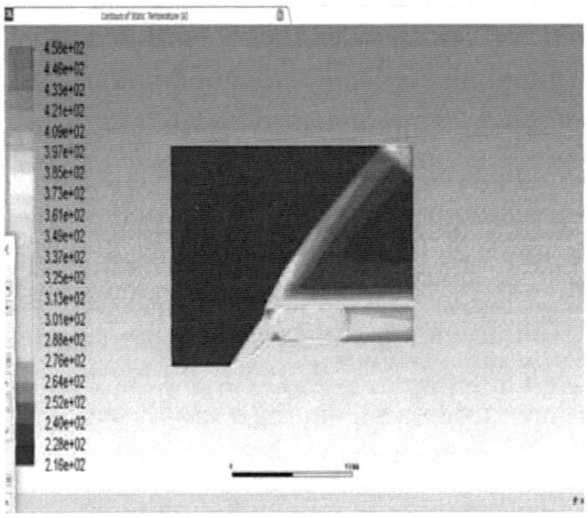

Fig. 6: Temperature contour of flow through intake with 5 degrees cowl deflection

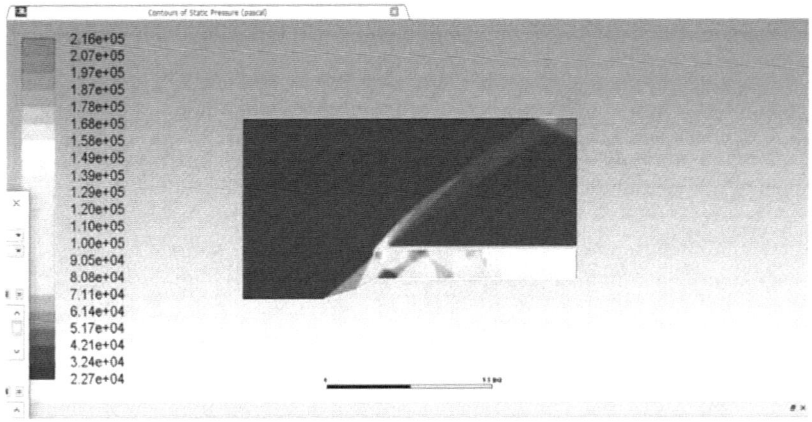

Fig. 7: Pressure contour of flow through intake with 5 degrees cowl deflection

6.2 FOR 10° COWL DEFLECTION

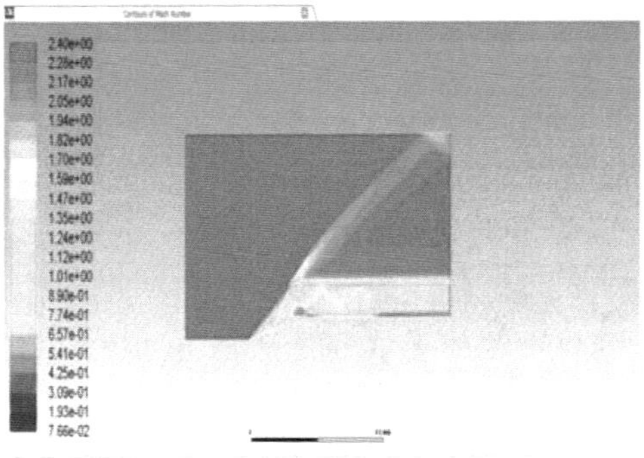

Fig. 8: Velocity contour of flow through intake with 10 degrees cowl deflection

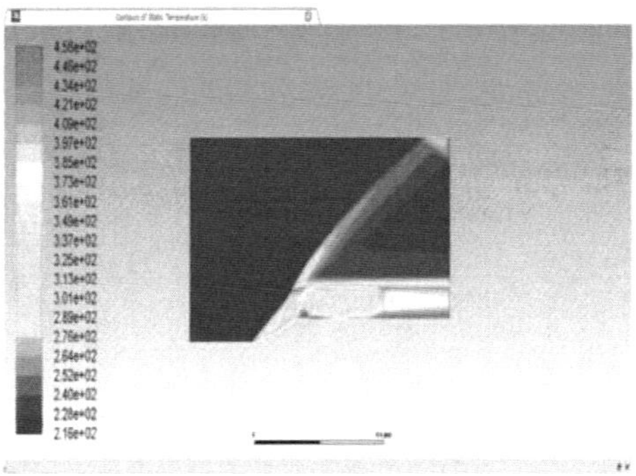

Fig. 9: Temperature contour of flow through intake with 10 degrees cowl deflection

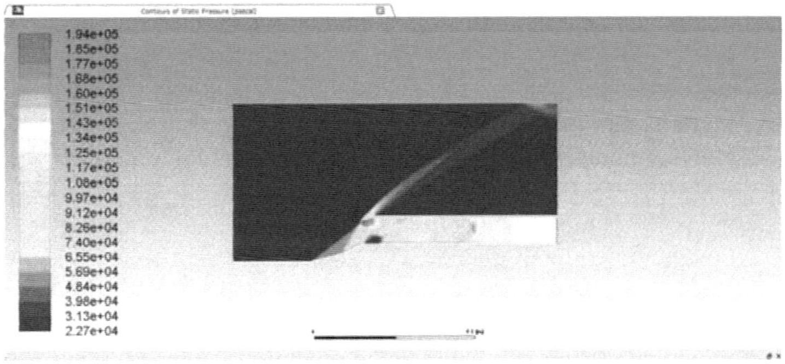

Fig. 10: Pressure contour of flow through intake with 10 degrees cowl deflection

6.3 FOR 15° COWL DEFLECTION

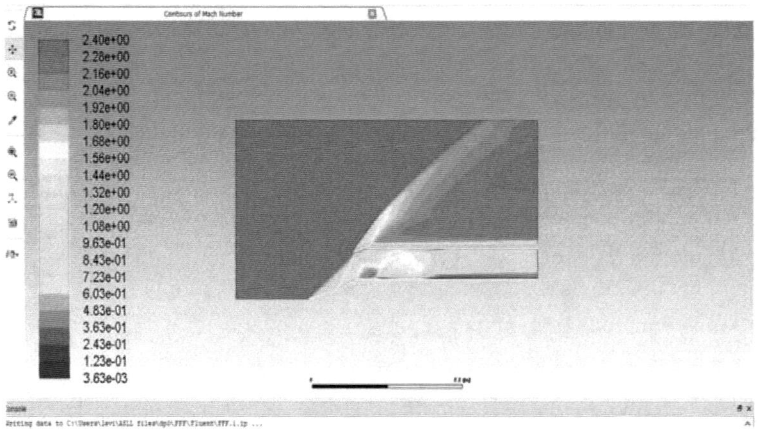

Fig. 11: Velocity contour of flow through intake with 15 degrees cowl deflection

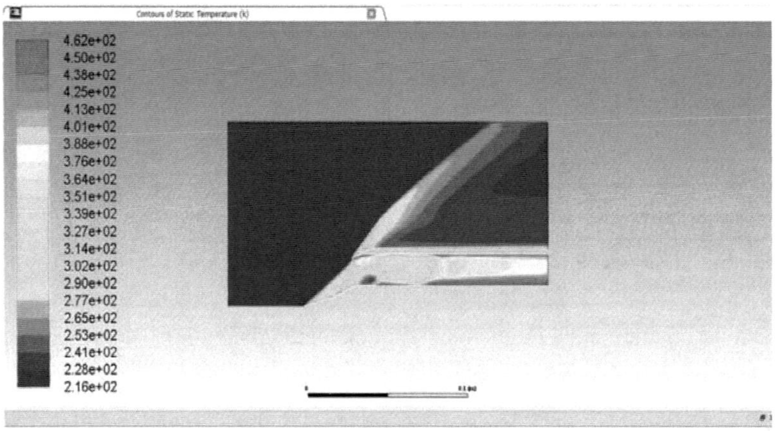

Fig. 12: Temperature contour of flow through intake with 15 degrees cowl deflection

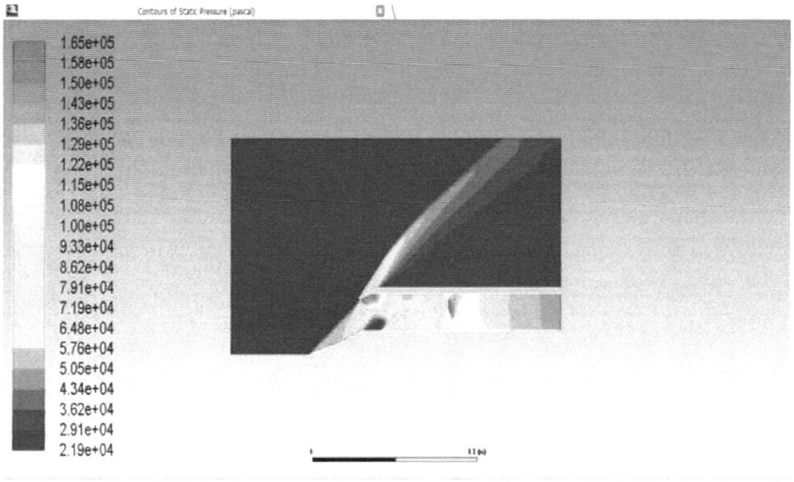

Fig. 13: Pressure contour of flow through intake with 15 degrees cowl deflection

Table 2. Comparison of total pressure recovery at various cowl lip angles.

Mach no.	Cowl deflection angle	Total Pressure recovery	Subsonic Mach no.
2.4	5°	0.476	0.6
2.4	10°	0.455	0.8
2.4	15°	0.43	1

7. CONCLUSION

A mixed-compression supersonic intake at Mach number 2.4 is designed at an altitude of 11,000 meters and is analyzed using a computational method. The terminal shock is formed as expected at a back pressure of 150 kPa. The effect of cowl lip angle deflection is studied, and the performance of the supersonic intake with different cowl lip angle deflections is analyzed. Analysis is done at cowl deflection angles of 5, 10 and 15 degrees, and the results for various cowl deflections at the same back pressure were obtained for comparison. Pressure recovery is observed to be maximum at 5 degrees of cowl lip deflection among the three angles analyzed. As the cowl deflection increases, supersonic flow is not compressed to subsonic flow at the subsonic diffuser region. The effective compression is found at 5° of cowl deflection.

8. REFERENCES

[1] AMES Research Staff, "*Equations, tables, and Charts for Compressible Flow*", NACA Report 1135, National Advisory Committee for Aeronautics, 1953.
[2] NEALE, M. C. & Lamb, P. S, "*Tests with a variable ramp intake having combined external / internal compression, and a design Mach number of 2.2*", Aeronautical Research Council – CP – 805,1962.
[3] Ione D. V. Faro, "*Supersonic Inlets*", North Atlantic Treaty Organization-Advisory Group for Aerospace Research and Development Fluid Dynamics Panel, May 1965.

[4] Philip G. Hill and Carl R. Peterson, "*Mechanics and thermodynamics of propulsion*", Pearson Education Publication, 1992.
[5] John D. Anderson, "*Modern Compressible flow*" -Third edition, Published by McGraw Hill Education, 2003.
[6] Hongjun Ran and Dimitri Mavris, "*Preliminary Design of a Supersonic Inlet to maximize Pressure Recovery*", Georgia Institute of Technology, AIAA Conference, September 2005.
[7] S. Das and J. K. Prasad, "*Cowl Deflection Angle in a Supersonic Air Intake, Birla Institute of Technology*", Mesra, Ranchi-835 215, March 2009.
[8] Ved Merchant, Jayakrishnan Radhakrishnan, "*Design and optimization of supersonic intake*", IOP Conference Series: Materials Science and Engineering, 2017
[9] Senthilkumar P, Sharan Chandran M, "*Effect of Cowl angle in the performance of scramjet air intakes*", International Journal of Mechanical Engineering and Technology, Volume 8, Issue 11, 2017
[10] B John, P Senthilkumar, "*Alternations of cowl lip for the improvement of the Supersonic-intake performance*", Journal of Applied fluid mechanics, Volume 11, 2018

Effect of Zn, Mg and heat treatment on Al base alloy on hardness using Taguchi Technique

Pruthvi H. M.[1], **H. Ghanashyam Shenoy**[2]

[1, 2] Department of Mechanical Engineering, Dr T Thimmaiah Institute of Technology, KGF, Karnataka, India.

Corresponding author email id: pruthvi@drttit.edu.in

Abstract. In the current investigation, effort has been made to develop the Al-Zinc-Mg alloy by changing weight percentage of Zn (5.1, 5.6 and 6.1) and weight percentage of Mg (2.1, 2.5 and 2.9) through liquid metallurgy process. Developed Al-Zn-Mg alloy is exposed to solutionzing treatment at 450°C for holding period of 3 hrs. followed by ageing at 180°C for 3 hrs and then Retrogression Reageing (RRA) at 150°C for a holding period of 3, 5 and 7 hrs. Then the modified Al-Zn-Mg cast alloys were cut and machined as per ASTM standard test size specimen. The microstructure shows the irregular grains with different shape and size resulting new grain boundaries. Formation of precipitation in the formed alloy was observed restricting dislocation of grains. The hardness of developed Al-Zn-Mg alloy increased with increase in weight percentage of Mg and Zn for all reageing duration. Further, design of experiments L_{27} orthogonal array was used to determine the signal to noise ratio, optimization and percentage contribution of each factor. Regression equation show that the experimental values are well within the acceptable range.

Keywords: *Al-Zn-Mg, Taguchi Technique, reageing duration, ANOVA*

1. Introduction

Al is one of the furthermost abundant materials in the earth's crust any working ore deposit must be eagerly amenable to benefice, so that a pure aluminum oxide can be obtained. However, physical benefice of the oxides has to been very successful [1]. Al and their alloys show the additional consideration which is due to their wide use in the applications of engineering sector [2]. These alloys are extensively used due to their machinability, durability, high strength and low density. In addition, these materials are effective in terms of cost and the intermittent mixture of properties cannot be attained by traditional materials [3]. The essential for these necessities has led to several enhancements in material design throughout the rare past periods over investigation on material science. When Al alloys are enhanced by thermomechanical or thermal process, a extensive variety of applications may be discovered [4]. Al–Zn based alloys are fundamentally used as structural apparatuses, particularly in the aerospace sector due to their outstanding hot workability, low density, high strength and fatigue resistance which depend upon type of thermal treatment [5]. Aluminium and its alloys (2xxx, 6xxx and 7xxx) can be age hardenable and when visible to atmosphere form a thin and supporter oxide defensive film. But then, the safety against corrosion by this film is inadequate to surroundings in which it is uneven [6].

Al-Zn [7] based alloy is representative heat treatable, and extensively used in transportation, aerospace, nuclear industry and other areas [8]. To fabricate bulk metal matrix composite, stir casting is being widely used technique because of cost effective, efficient, simple and flexible [9]. Al alloy [10] are important group of materials for wear characteristics due to their low density and high thermal conductivity [11]. In RRA heat treatment [12-13] the material is subjected to thermal treatment in fewer period and at low temperature and then aged under T6 [14] condition [15]. Ageing method takes place with the development of hard secondary segment in grain and its grain boundaries after the thermal treatment involving of adding to solution, quick cooling and ageing stages [16].
The above comprehensive literature review, it is seen that many researchers have worked on the fabrication of composites material and subjecting to various heat treatment state to find out the mechanical properties. Hence, for the present investigation modified Al-Zn-Mg alloy have been fabricated by using routing through stir casting method by varying wt.% of Zn and wt.% of Mg followed by RRA heat treatment for the machined test specimen. It is found that for Al-6.1 Zn-2.9 Mg alloy with 5hrs. of reageing duration exhibits higher hardness when compared to other composition. Further, L_{27} orthogonal array was used to investigate signal to noise ratio, optimization of process parameter, ANOVA and validation of experimental results using linear regression equation.

2. Experimentation Details

2.1 *Formation of modified Al-Zn-Mg alloy*

Calculated amount of aluminum, Zin, Magnesium and other elements were added into the pre-heated crucible (750°C) which is kept inside the casting furnace. The motel Al-Zn-Mg alloy is poured into the split die manually by using foundry tool and allowed to solidify for 15min as shown in Figure 1. The Figure 2 shows the modified Al-Zn alloy cast specimens after casting process.

Figure 1: Formation of Al-Zn alloy

Figure 2: Cast specimens of modified Al-Zn alloy

2.2 *Fabrication of cast specimen and RRA heat treatment*

The cast component was machined with suitable machining parameter by using CNC lathe machine tool (LT 20C) with approximate weight of 3500Kg as shown in Figure 3. The CNC machine has a maximum turning diameter of 320mm and maximum turning length of 310mm. The motor capacity of 7.5KW power having cutting speed of 3500rpm. Developed Al-Zn-Mg alloy is exposed to solutionzing treatment at 450°C for holding period of 3 hrs. followed by ageing at 180°C for 3 hrs. and then Retrogression Reageing (RRA) at 150°C for a holding period of 3, 5 and 7 hrs. by natural air cooling as quenching process. Figure 4 shows the machined hardness test specimens after RRA heat treatment process.

Figure 3: CNC machine (LT 20C)

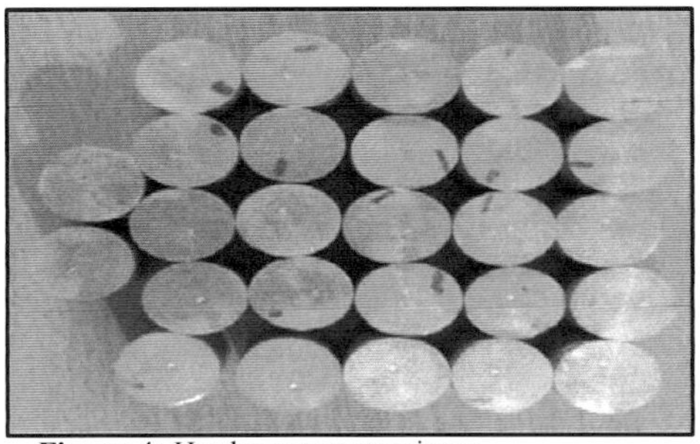

Figure 4: Hardness test specimen

2.3 Plan of Trials

In the present investigation, optimization of hardness of modified Al-Zn-Mg alloy have been carried out by using L_{27} orthogonal array. Signal to noise ratio was evaluated for the formed Al-Zn-Mg alloy by considering the 3 factors (wt.% of Zn, wt.% of Mg and reageing duration) at three level each.
Analysis of variance is a noteworthy investigating tool used to find the significance of each factor on the response output. ANOVA is carried out to know the percentage (Pr.) influence of individual factor on hardness of the modified Al-Zn-Mg alloy. To authenticate the experimental result, a regression model was generated by using Minitab software and assisted with the experimental results.
The control factors and their levels for the present investigation is shown in the Table 1.

Table 1. Control factors and their levels.

Sl. No.	Factors	Levels
A	Wt.% of Zinc	5.1, 5.6 and 6.1
B	Wt.% of Magnesium	2.1, 2.5 and 2.9
C	Ageing duration	3, 5 and 7 (hrs.)

3. Results and Discussion

3.1 Optimization and signal to noise ratio on hardness of modified Al-Zn-Mg alloy.

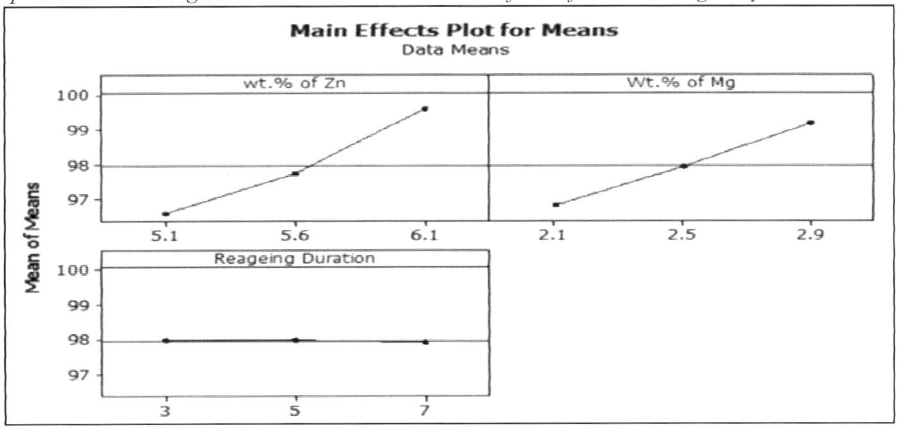

Figure 5. S/N ratio on hardness for modified Al-Zn-Mg alloy

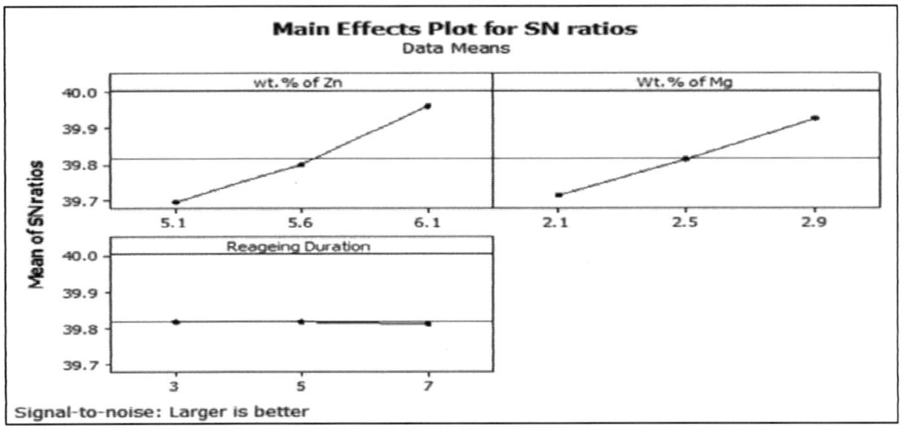

Figure 6. Means on hardness for modified Al-Zn-Mg alloy

Figure 5 and 6 represent the S/N ratio and optimization (Larger is better) on hardness for modified Al-Zn-Mg alloy. It is manifest that the optimum process parameter on hardness is 6.1 weight percentage of Zinc, 2.9 weight percentage of Magnesium and 5 hrs. of reageing duration within the factors selected for modified Al-Zn-Mg alloy.

The output response table for optimization and S/N ratio on hardness for modified Al-Zn-Mg alloy is shown in Table 2 and 3. It is manifest from the table that weight percentage of Zn is a leading factor for hardness property and hence it is important to control over the wt.% of Zn in achieving higher hardness.

Table 2. Output response table for optimization on hardness

Level	wt.% of Zn	Wt.% of Mg	Reageing Duration
1	96.56	96.79	97.97
2	97.72	97.91	97.99
3	99.59	99.17	97.91
Delta	3.03	2.38	0.08
Rank	1	2	3

Table 3: Output response table for S/N ratio on hardness.

Level	wt.% of Zn	Wt.% of Mg	Reageing Duration
1	39.69	39.71	39.82
2	39.8	39.81	39.82
3	39.96	39.93	39.82
Delta	0.27	0.21	0.01
Rank	1	2	3

3.2 ANOVA on hardness of modified Al-Zn-Mg alloy.

Table 4 represent the result of analysis of variance on hardness of modified Al-Zn-Mg alloy. Percentage contribution of each parameter on hardness is presented in last column of the table by conducting analysis of variance at assurance level of 95% and level of significance at 5%.

Table 4 revealed that weight percentage of Zn has a chief contribution on the hardness (47.05%). Hence the weight percentage of Zn plays a significant control factor which has to be controlled precisely and taken into consideration for achieving hardness value. Weight percentage of Zn is further followed by wt.% of Mg (28.44%), reageing duration (15.31%) on hardness of modified Al-Zn-Mg alloy.

Table 4: ANOVA for hardness (BHN)

Control Factors	Adj-SS	Adj-MS	Seq-SS	P-value	F-value	DoF	Percentage %
Weight percentage of Zn	42.14	21.07	42.14	0	51.25	2	47.05
Weight percentage of Mg	25.469	12.734	25.469	0	30.98	2	28.44
Reageing duration (hrs.)	13.716	6.858	13.716	0	16.68	2	15.31
Error	8.222	0.411	8.222			20	
Total			89.547			26	90.8

3.3 Regression equation on hardness of modified Al-Zn-Mg alloy.

Commercially existing Minitab V16 software is used for emerging the General Linear regression model. This equation gives the relationship between numerous controlled factors (weight percentage of Zn, weight percentage of Mg and reageing duration) and a response variable (hardness) to the experimental results.

General Linear (GL) regression model for hardness is:

Hardness (BHN) = 72.8439 + 3.03333 Zn + 2.97222 Mg + 0.138889 Re-Ageing hours (Eq.1)

R-Sq = 90.82%

Figure 7 represent the plot of experimental results verses predicted results for all the three levels of their factor. It is clear that the experimental results are very adjacent to the predicted values and hence hardness of modified Al-Zn-Mg alloy are validated.

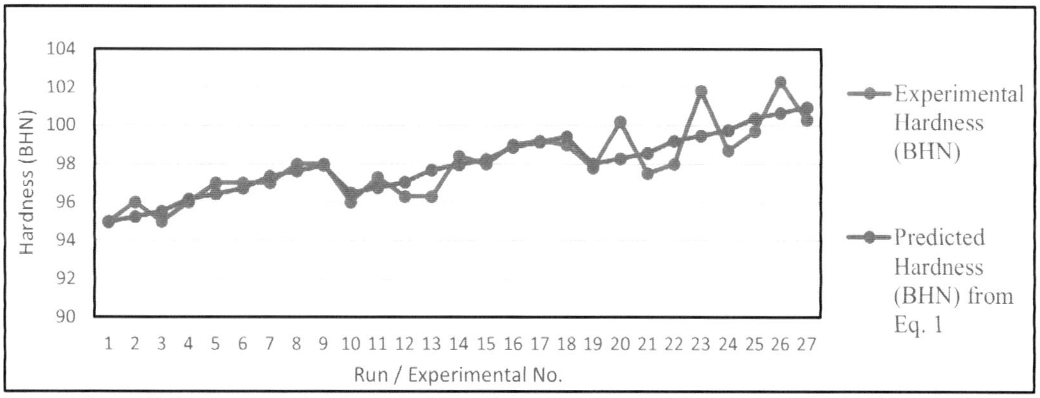

Figure 7. Assessment plot for experimental results verses predicted values for hardness

Table 5. Assessment results of experimental results and their regression model values on hardness.

Run / Exp. No.	wt.% of Zn	wt.% Mg	Re-Ageing hours	Experimental Hardness (BHN)	Hardness (BHN) from Reg. model (1)	Error	% Error
1	5.1	2.1	3	95	94.97	0.03	0.03
2	5.1	2.1	5	96	95.25	0.75	0.78
3	5.1	2.1	7	95	95.53	0.53	0.56

4	5.1	2.5	3	96	96.16	0.16	0.17
5	5.1	2.5	5	97	96.44	0.56	0.58
6	5.1	2.5	7	97	96.72	0.28	0.29
7	5.1	2.9	3	97	97.35	0.35	0.36
8	5.1	2.9	5	98	97.63	0.37	0.38
9	5.1	2.9	7	98	97.91	0.09	0.1
10	5.6	2.1	3	96	96.49	0.49	0.51
11	5.6	2.1	5	97.3	96.77	0.53	0.55
12	5.6	2.1	7	96.3	97.04	0.74	0.77
13	5.6	2.5	3	96.3	97.68	1.38	1.43
14	5.6	2.5	5	98.4	97.96	0.44	0.45
15	5.6	2.5	7	98	98.23	0.23	0.24
16	5.6	2.9	3	99	98.87	0.13	0.13
17	5.6	2.9	5	99.2	99.14	0.06	0.06
18	5.6	2.9	7	99	99.42	0.42	0.43
19	6.1	2.1	3	97.8	98.01	0.21	0.21
20	6.1	2.1	5	100.2	98.28	1.92	1.91
21	6.1	2.1	7	97.5	98.56	1.06	1.09
22	6.1	2.5	3	98	99.19	1.19	1.22
23	6.1	2.5	5	101.8	99.47	2.33	2.29
24	6.1	2.5	7	98.7	99.75	1.05	1.06
25	6.1	2.9	3	99.7	100.38	0.68	0.69
26	6.1	2.9	5	102.3	100.66	1.64	1.6
27	6.1	2.9	7	100.3	100.94	0.64	0.64

4. Conclusions

Based on the investigation, the modified Al-Zinc-Magnesium alloy are fabricated by routing through stir casting method. The hardness property has been evaluated experimentally by standard test process. L_{27} orthogonal array was used through design of experiments and found out that 6.1 wt.% of Zn, 2.9 wt.% of Mg and 5 hrs. of reageing duration are the optimum process parameter on hardness. The major percentage contribution on hardness is wt.% of Zn (47.05%) followed by wt.% of Mg (28.44%) and reageing duration (15.31%). Regression model developed shows that the experimental results are very adjacent to that of regression model values and hence the experimental results are validated. The average percentage error is found to be 0.6862%.

References

[1] Bharath L, Dr. M. Sreenivasa Reddy, Dr. Girisha H N and Dr. G. Balakumar, 2020 Influence of Process Parameters on the Mechanical Properties of Al-Cu alloy by Taguchi's Technique *Aegaeum J.* ***8*** 881-887.

[2] Bharath L, Dr. M. Sreenivasa Reddy, Dr. Girisha H N and Dr. G. Balakumar, 2021 Optimization of ductility and yield strength on Al2024/B_4C composite material using Taguchi Technique *IOP Conf Ser Mater Sci Eng.* ***1055***

[3] Bharath L, Dr. M. Sreenivasa Reddy, Dr. Girisha H N and Dr. G. Balakumar, 2021 Influence of process parameters on tensile strength and hardness of AW2024/B_4C composite using Taguchi's technique *Elsevier Mater Today Proc.*

[4] Arturo Abundez, Isa Pereyra, Bernardo Campillo, Sergio Serna, Enrique Alcudia, Arturo Molina,

Andrés Blanco, Jan Mayen 2016 Improvement of ultimate tensile strength by artificial ageing and retrogression treatment of aluminium alloy 6061 *Materi Sci Eng.* A **16** 201-207.
[5] D. Feng, X.M. Zhang, S.D. Liu, T. Wang, Z.Z. Wu and Y.W. Guo 2013 The effect of pre-ageing temperature and retrogression heating rate on the microstructure and properties of AA7055 *Materi Sci & Eng. A* **588** 34-42.
[6] Prasanta Kumar Rout, M.M. Ghosh and K.S. Ghosh 2015 Microstructural, mechanical and electrochemical behaviour of a 7017 Al–Zn–Mg alloy of different tempers *Materi Chara* **104** 49-60.
[7] Hagar Ashour Elkilany, Madiha A. Shoeib, Omar E. Abdel-Salam Infuence of hard anodizing on the mechanical and corrosion properties of different aluminum alloys *Metallography, Microstructure, and Analysis.*
[8] Rongde Li, Tong Liu, Ruiming Su, and Yingdong Qu 2018 Microstructure and Mechanical Properties of SprayFormed 7075 Alloy During Retrogression *JMEPEG* **27** 4437–4443.
[9] Thella Babu Rao 2017 An experimental investigation on mechanical and wear properties of Al7075/SiCp composites: effect of SiC content and particle size Journal of Tribology *ASME.*

[10] Pan Daia, Xian Luo, Yanqing Yang, Zongde Kou, Bin Huang , Jinxin Zang and Jigang Ru Thermal stability analysis of a lightweight Al-Zn-Mg-Cu alloy by TEM and tensile tests *Mater Chara, Elsevier* **153** 271-283.
[11] G. ozer, R. Gecu and A. Karaaslan 2019 The effect of retrogression and re-aging treatment on mechanical and tribological features of AA7075 aluminium alloy *Materialwiss. Werkstofftech,* **50** 761–768.
[12] Gokhan ozer and Ahmet Karaaslan 2018 Effect of RRA heat treatment on corrosion and mechanical properties of AA7075" *materials and corrosion* **70** 2064-2072.
[13] Yi Lin Wang, Qing Lin Pan, Li Li Wei, Bo Li and Ying Wang 2014 Effect of retrogression and reaging treatment on the microstructure and fatigue crack growth behavior of 7050 aluminum alloy thick plate *Materials and Design,***55** 857-863.
[14] Prasanta Kumar Rout, M.M. Ghosh and K.S. Ghosh 2015 "Microstructural, mechanical and electrochemical behaviour of a 7017 Al–Zn–Mg alloy of different tempers" *Mater Chara,* **104** 49-60.
[15] Gokhan ozer, Iski Kaya and Ahmet Karaaslan 2019 Effects of retrogression and reaging heat treatment on the microstructure, exfoliation corrosion, electrical conductivity, and mechanical properties of AA7050" *mater and corrosion,* 1-10.
[16] Musa Yildirim, Dursun Ozyurek and Metin Guru 2016 The Effects of Precipitate Size on the Hardness and Wear Behaviors of Aged 7075 Aluminum Alloys Produced by Powder Metallurgy Route *Arabian J. for sci. and Eng.* **41** 4273-4281.

Composite blades for lower extremity amputees

Viswanath Srinivasan[1], Praveen Kumar Govarthan[2], Om Prakash S[3], Dhanalakshmi Munirathinam[4*]

[1234]Department of Biomedical Engineering,

[1234]Sri Sivasubramaniya Nadar College of Engineering, Kalavakkam- 603110, Tamilnadu, India

(Email id: viswanathssrinivasan@gmail.com , praveenkumargovarthan@gmail.com, Sriprakash800@gmail.com, dhanalakshmim@ssn.edu.in)

ABSTRACT

The running blades used by the amputees are an advanced type of prosthesis or prosthetic limbthat are used as a replacement for a natural leg. The Principle behind the running blades is that itstores kinetic energy from the user's footsteps as potential energy, like a spring, allowing the user to run and leap. The major prevailing deficiencies of existing prosthesis are excess weight,lack of Indian manufacturers, high cost quoted by foreign manufacturers, lack of awareness of this technology among the local population. The major purpose of the project is to design and develop an affordable composite blade with unique design, which enables normal walking and running easier for the amputees. The methodology of the project includes selection of suitable composite from glass, carbon and hybrid fibers for blade fabrication through material testing, toevaluate the design using finite element analysis (FEA), to fabricate the actual blades and to analyze the walking pattern of the user's through gait analysis. Gait analysis results show that with blades the participant is able to follow normal gait pattern of initial contact, mid-stance andtoe-off as the blade supports normal knee and ankle biomechanics.

Keywords: Amputees, Running blades, Composite material, Carbon fiber, Gait analysis.

1. Introduction

The running blades are an advanced type of prosthesis, or prosthetic limb, which are used by lower limb amputees who are athletes. The two major categories in lower limb amputees are above and below knee amputees. In above-knee amputation (AKA), the leg is amputated at the thighbone above the knee-joint [1, 2, 3]. This type of amputation is also referred to as Transfemoral amputation. A below knee amputation (BKA), also known as a transtibial amputation, is an amputation done at the shinbone. In a BKA, as the knee-joint is spared, walking with a prosthesis is typically more successful. The Commercial blades are made from lightweight carbon fiber. Carbon fiber is a resilient material that is stiffer and around five times stronger than steel. The Principle behind the running blades is that it stores kinetic energy from the user's footsteps as potential energy, as a spring, allowing the users to run and jump [4, 5, 6]. For example, the Flex-Foot Cheetah is a prosthetic human foot replacement blade developed by biomedical engineer Van Phillips, who had lost a leg below the knee [7]. The other such available product is called the blades of glory. The major disadvantage with the above-mentioned prosthesis is that the same blades cannot be used for walking due to the lack of heel part that demands the user to be in a state of continuous motion throughout the blade usage, thus helpful in running activity alone. The other disadvantages with the commercially available blades such as Jaipur foot and Madras foot are heaviness, lack of local manufacturers and high cost quoted by the global manufacturers. The main aim of the project is to design and develop a unique low-cost lightweight blades that enables the amputees to walk normally [8, 9]. The project covers 3D modelling of blades, rapid prototyping and preparation of unique molds using different techniques such as 3D printing, iron casting and CNC cutting. It is followed by material testing, finite element analysis of the blades, fabrication of actual blades and gait analysis of the fabricated blades.

2. LITERATURE REVIEW

The Flex-Foot Cheetah is a prosthetic human foot replacement fabricated using carbon fiber. Figure 1, represents the product design of Flex-Foot Cheetah. Carbon fiber is actually a carbon-fiber-reinforced polymer, and is a strong, light-weight material used in a number of applications, including sports goods (such as baseball bats), car parts, bicycles and other equipment where rigidity and high strength-to-weight ratio is important. The polymer used for this equipment is normally epoxy, but other polymers are also used, depending on the application. Other reinforcing fibers may also be included. In the blade manufacturing process, sheets of impregnated material are cut into square sheets and pressed onto a mold to produce the final shape [10].

Figure 1. Flex-foot Cheetah

Base plate and blade design for a leg prosthetic: US 2016/0645337A1, Feb 18, 2016. A blade-type prosthesis assembly includes a blade body extending between a blade bottom and a blade top indicated in Fig 2. The blade top is securable to a socket on the limb and the blade bottom defines a convex surface formed by a single edge of the blade bottom curving towards the blade top. The prosthesis assembly further includes a base plate. The base plate includes a base plate top surface securable to the blade bottom, a base plate bottom surface opposing the base plate top surface, and a plurality of ground-engaging elements extending from the base plate bottom surface [11].

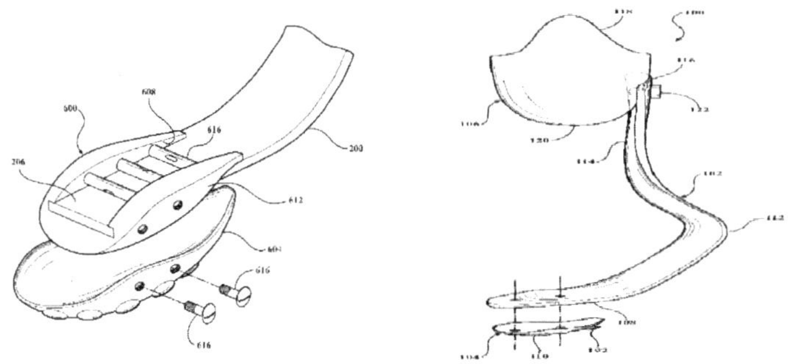

Figure 2. Base plate and blade design

Speed adaptive and patient prosthetic knee: US 6,610,101B2, Aug 26, 2003. The invention in Fig 3. relates to an automated speed-adaptive and patient-adaptive control scheme and system for a knee prosthesis. The control scheme and system utilize sensory information measured at a part of knee to the prosthesis, to automatically adjust stance and swing phase knee resistances to a user under a wide variety of locomotory activities. Substantially, no patient-specific information needs to be pre-programmed into the prosthetic knee by a prosthetist or the patient. The system can adapt to various types of disturbances once the patient leaves the prosthetist's facility because it is patient-adaptive and speed- adaptive [12].

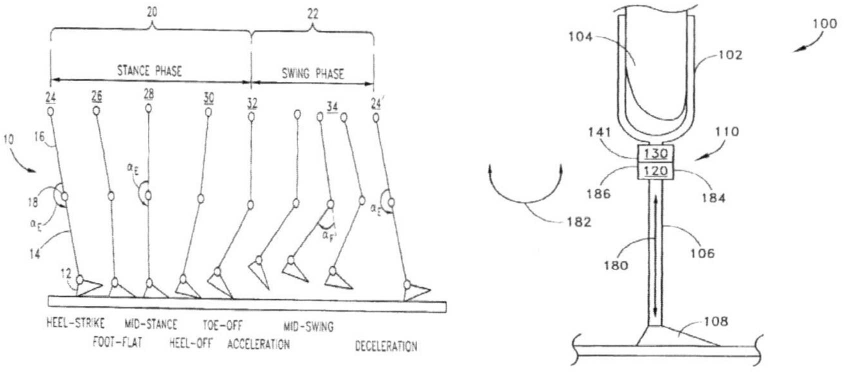

Figure 3. Speed adaptive and patient prosthetic knee

Foot prosthesis and method of making same: US 4,959,073 SEP 25, 1990. A foot prosthesis comprises of an elongate composite main member having a leg section and a toe section with one end of the leg section adapted to be connected to an amputation socket and the other end smoothly curving forward through an ankle section into the toe section and extending to a toe end, a heel member extending from the toe end generally along the toe section and then diverging from the toe section and extending to a heel tip. Fig 4 represents foot prosthesis with central composite structure covered with a resilient material, such as rubber, bonded between the toe section and heel member where the two are generally adjacent. The properties of the resilient material may be made adjustable by the user where the toe section and heel member diverge [13].

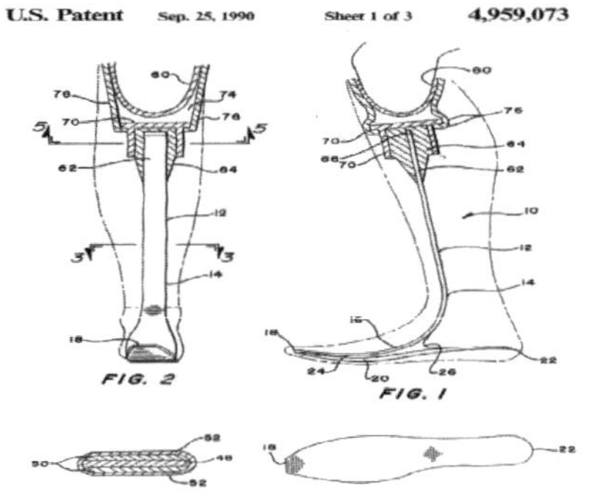

Figure 4. Foot prosthesis

IOP Conf. Series: Materials Science and Engineering 1258 (2022) 012044 doi:10.1088/1757-899X/1258/1/012044

3. METHODOLOGY

The methodology has been divided into eight following subdivisions namely, blade modelling, rapid prototyping, dye preparation, sample preparation, material testing, finite element analysis, blade fabrication and gait analysis as mentioned in Fig 5.

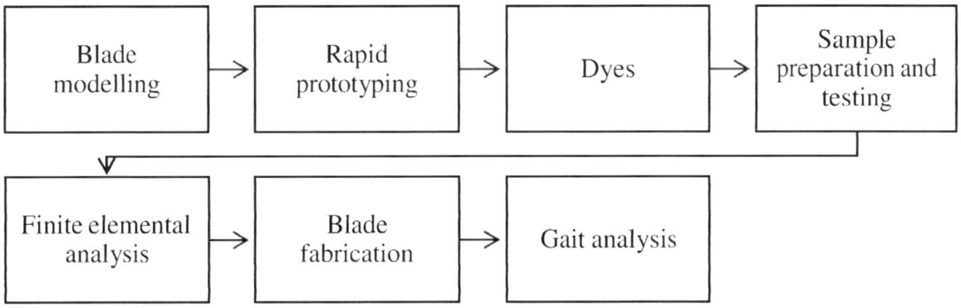

Figure 5. Block diagram

3.1 Blade modeling

The blade modelling section is subdivided into two categories, design principle and design description. Design principle discusses the principles used for designing of the blades and design description explains the reason for specific curves and arches in the blade design.

3.2 Design principle

The Principle behind the design is Bio-mimicry. The design is inspired from the foot mechanism of kangaroo. Stretchy tendons attach the muscle to the bone to provide power to the kangaroo's hop. Tendons compress with each hop and bound, releasing like a coiled spring and propelling the kangaroo into the air. Same Principle has been utilized in designing the prosthesis design that ails in generating maximum thrust with minimum effort [14, 15].

Hicks Windlass Effect: John Hicks (1954) was the first to describe this mechanism. During the gait cycle, the foot plays two major mechanical functions of acting as shock absorber (during initial contact) and helping in propulsion (during propulsion). According to Hicks, the bones and ligaments of the foot are arranged in a triangular arch form, during the initial contact, the arch of the foot flattens under the vertical load of the body, thus dispersing the weight in an organized fashion. Further flattening of the arch is restricted by the elastic property of the composite material. The windlass effect is also seen during the gait cycle just after the mid-stance as the foot prepares to lift off the ground [16].

3.3 Design description

The same principal is utilized in designing the composite prosthesis [17] that consists of two C-shape curves, one for the upper curvature and the other for lower curvature. The upper curve that is in contact with the socket, transfers the kinetic energy initiated from the user's footsteps to the lower curve. The lower curve generates upward thrust as a result of ground reaction [18]. During this push-off phase the heel is lifted upwards, which bends the toe part of the blade. The bending of the blade toes leads to winding of the virtual tendon. When the blade heel is moved further upwards it compresses all the joints in the blade. This compression helps the subsequent propulsion phase. Toe flexion is imperative for both shock absorption and propulsion of the foot during the gait cycle [19]. Figure 6 and 7 represents the Windlass mechanism assisting triangular arc and blade measurements respectively. Modelling involves graphical representation of the proposed blade design using SOLIDWORKS modelling environment with the actual shape and dimensions. Figure 8 (a) and (b) shows other views of the modeled blades along with their measurements. 3D modelling helps in actual visualization of the fabricated blades and reduces the chances of defective blade fabrication that in-turn causes a loss of manufacturing expenditure.

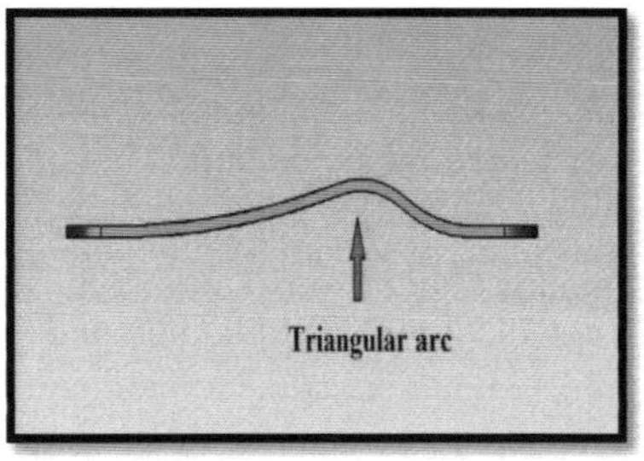

Figure 6. Lower curvature of the blade

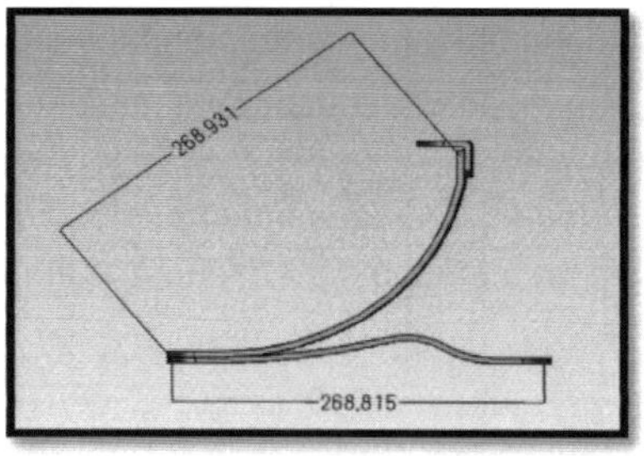

Figure 7. Blade measurement

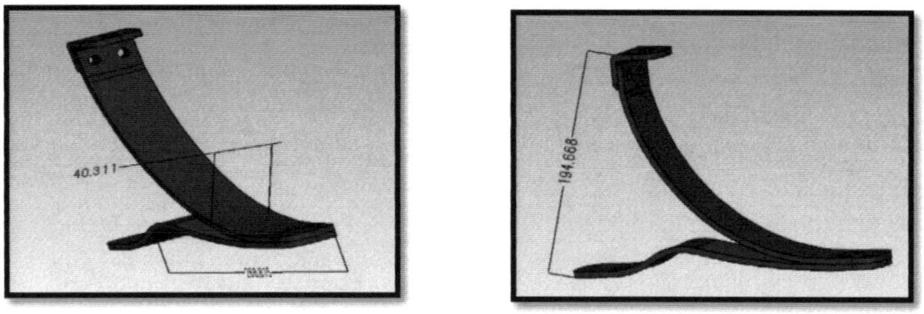

(a) (b)

Figure 8. Different views of 3D modelled blades

3.4 Rapid prototyping

The technique used for rapid prototyping is Fused Deposition Molding (FDM), which is a 3D printing process that uses a continuous filament of a thermoplastic material. Fig 5. is a picture taken during the actual FDM of the modeled blades. The material used for printing is Polylactic acid filaments (PLA), which is one of the more common materials for 3D printing [20, 21]. Filament is fed from a largecoil through a moving and heated printer extruder head. Molten thermoplastic is forced out of the printhead's nozzle and is deposited on the growing work. Fig 9. shows the 3D printed blades and Fig10. shows the actual 3D printed prosthetic blades.

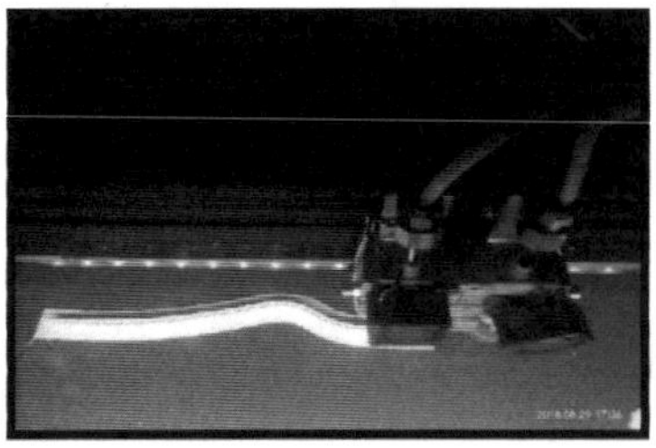

Figure 9. FDM process

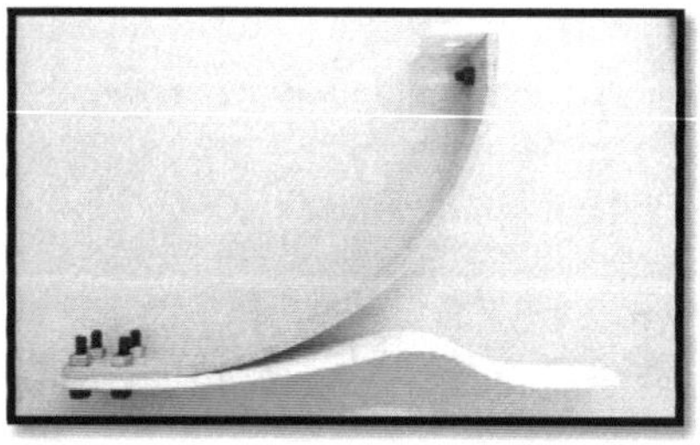

Figure 10. Blade prototype

3.5 Dyes

Dye preparation plays a crucial role in fabrication of the actual blades. In order to select a better fabrication technique, three different molds have been fabricated, this includes, PLA mold using 3D printing (Fig.11(a)), Iron mold using iron forging (Fig.11(b)) and Wooden mold (Fig.11(c)) using Computer numerical control (CNC) cutting [22]. The below depicted Table 1, provides the comparison of the above-mentioned molding techniques. The constraints chosen for comparison are edge precision, mold release agents, fiber alignment, resin leakage, reliability of molds and cost. On comparing the efficiency and reliability of the molds, the iron mold is selected for actual fabrication of the blades.

(a) (b) (c)

Figure 11. Different types of molds

Table 1. Comparison of molds

Constrains	**ABS mould**	**Iron mould**	**Wooden mould**
Edge cut-out	Requires machining	Requires machining	Precise edges
Mould release	Wax	OHP sheet	Teflon sheet/wax
Fibre alignment	Inconsistent	Inconsistent	Precise
Resin leakage	High	High	Low
Reliability	One time usage	Highly	Reliable
Cost	High	Low	High

3.6 Sample preparation and testing

A composite material is made up of two or more constituent materials with significantly different physical or chemical properties that when combined, produce a material with characteristics different from the individual components. Epoxy based thermoset and fiber-based reinforcement is used for composite fabrication [6]. Three different types of samples are prepared by Wet/Hand lay-up method to test mechanically their respective properties. They are glass fiber, Carbon fiber and hybrid fiber (mixed). Thermoset constituents such as Epoxy resin (LY556) and Hardener – HY 951 are used for the sample preparation. Sample preparation involves hand layup methods of arranging fibers and coating resin over it [23]. Arrangement is made on a wooden dye of 280x280mm square frame. OHP sheets are used for better surface finish. Wax is used as a releasing agent during the fabrication process. Multiple processing steps are involved in fabricating a sample for material testing as shown in Fig 12. and the resultant sample image is shown in Fig 13. Sample prepared are given for mechanical material testing to test its properties such that it can be given as input during the FEA step. Mechanical property assessed are tensile strength, Modulus of elasticity, Compression strength, density, hardness and flexural strength. The test values are incorporated in FEA for more realistic simulation results.

Step 1: Weigh the resin and hardener used for the fabrication based on calculation.
Step 2: Apply wax on the wooden frame.
Step 3: Fix OHB sheet over the sample base on the wooden frame.
Step 4: Spread the resin-hardener mixture for one fiber sheet on the sheet.
Step 5: Place the fiber sheet of same frame dimension on the matrix coat.
Step 6: Similarly repeat the above steps for sample thickness of 4mm.
Step 7: Close the wooden frame.
Step 8: Apply load for better compression and leave the sample for 24hrs.

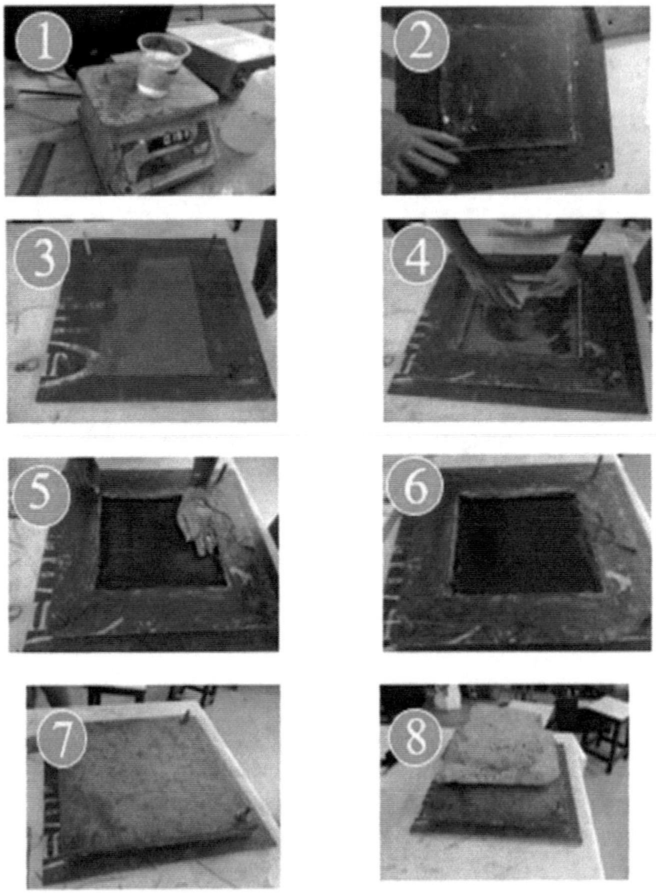

Figure 12. Test sample preparation

Figure 13. Resultant test samples

3.7 Finite elemental analysis

FEA involves a computer model of a design that is loaded and analyzed for specific results. The software used for static FEA is ANSYS. Analyses involve model importation, material assignment, meshing, analysis setting and result generation. During meshing, default element size and program controlled element order are used to generate 3495 nodes and 1546 elements. Properties of mechanical testing are declared in ANSYS materiallibrary. Boundary condition includes temperature of 22 °C, Force of 1000N and bottom surface of the lower curvature as fixed support for the loading operation. 1000N force is applied over the top face of upper blade curve with the time duration of 1s. Results are obtained as Total deformation, directional deformation, Stress intensity,Elastic strain intensity and Strain energy as shown in Figures 14, 15, 16 and 17 respectively. Red colorindicates the region of maximum parameter values and blue vice versa.

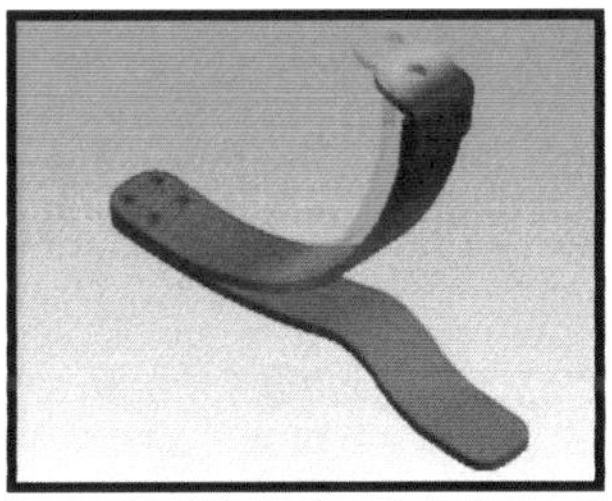

Figure 14. Directional deformation

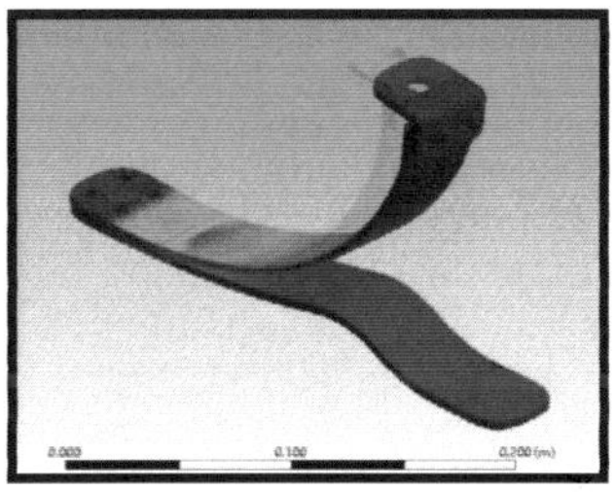

Figure 15. Strain intensity

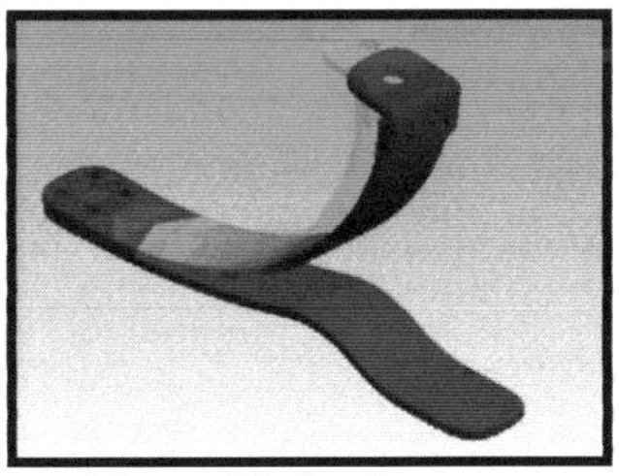

Figure 16. Strain energy

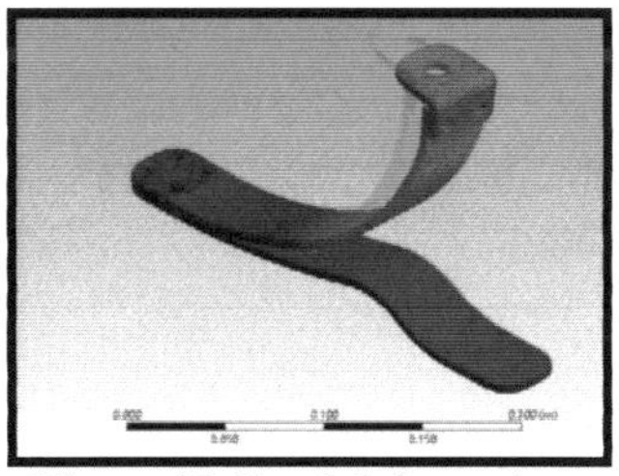

Figure 17. Total deformation

3.8 Blade fabrication

Three different Composite blades are fabricated using the discussed dye types. The fabrication involves arrangement of fiber and matrix in the prepared dyes. It requires necessary calculation of thickness to fiber count as well as weight of the resin required. Suitable releasing agent is also needed to be used for easy and safe removal of the blades from their respective dyes. Blades are made out of different materials such as glass fiber (Fig 18(a)), hybrid fiber (glass and carbon fiber) (Fig 18(b)) and carbon fiber (Fig 18(c)). Socket preparation involves taking the shape of the residual limb and filling the cavity with

Plaster of Paris along with a central stem. Later after solidification, the socket is covered using heated poly-propylene sheet to take the limb shape. Fabricated socket and blades are attached using below knee pylon kit, which offers a height adjustable stem. Fig 19., shows the complete lower extremity prosthesis with blades.

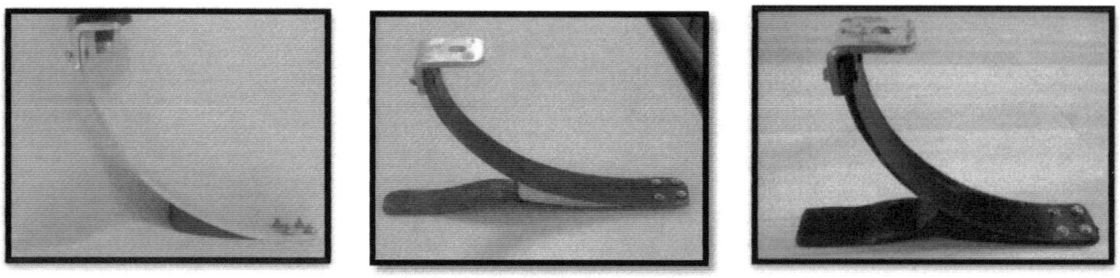

(a) (b) (c)

Figure 18. Fabricating composite blades

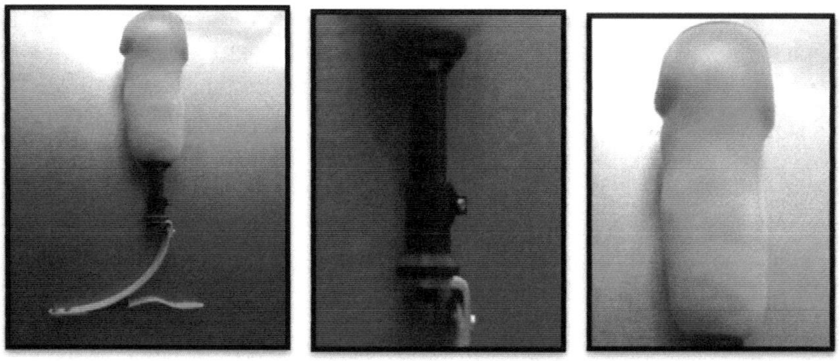

Figure 19. Assembly of entire prosthesis

3.9 Gait analysis

Human gait refers to locomotion achieved through the movement of human limbs. Four different analysis specific to walking has been performed with prior consent from the participant - Visual assessment; Plantar pressure analysis (P walk); Inertial Sensor analysis and Force analysis. Gait analysis and testing are conducted at Shoe and Product Design Centre, CLRI, Chennai.

3.9.1. Visual Assessment

Visual assessment is used to assess the gait pattern and prothesis visually. Visual assessment of the blades on user shows it is much sleeker and lighter in weight. Gait pattern of the user is also not distributed and better support is offered by the blades. Fig 20 (a) and (b) shows the pictures taken during the visual assessment phase in different views [24].

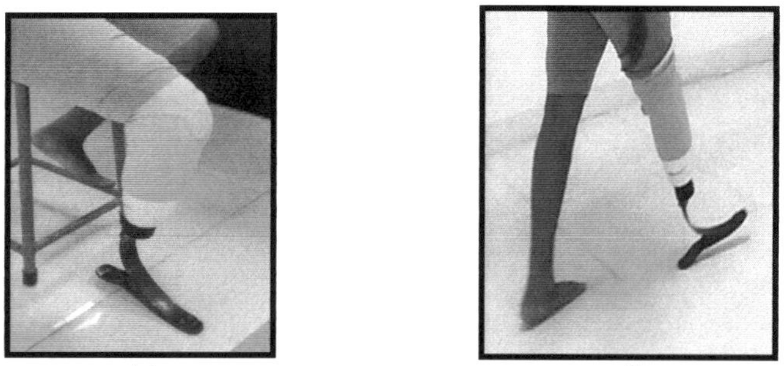

(a) (b)

Figure 20. Visual assessment in different views

3.9.2. Plantar pressure analysis (P walk)

Device used for plantar pressure analysis is BTS-P WALK of 2 m length shown in Figure 21. The Purpose of study is to measure temporal (time) and spatial (length) parameters and multi-chrome plantar pressure distribution map on motion. Figure 22. represents the p-walk system and patient walking over it [25, 26].

Figure 21. Pressure plates

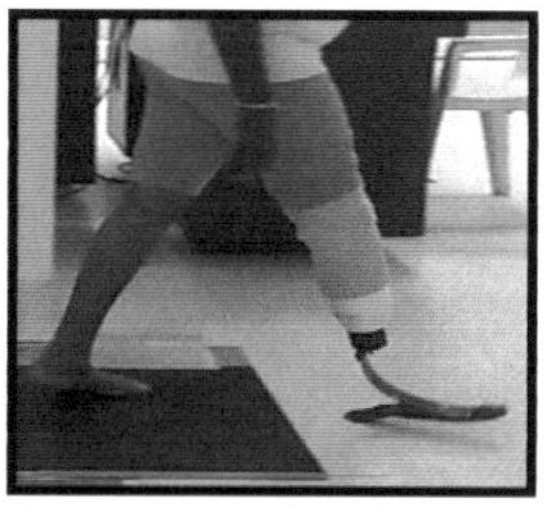

Figure 22. P-walk test

3.9.3. Inertial Sensor analysis (G-sensor)

G-sensor test is a solution for accurate and quick measurement of gait parameters. The wireless device consists of tri-axial accelerometer which gives the acceleration, tri-axial gyroscope which gives the motion and magnetometer which gives the speed of the subject. The G sensor shown in Fig 23. is positioned on L5 of lumbar region in human spine. The device Communicates with the computer using Bluetooth connection with a range up to 20 meters. It enables quick analysis and automatic report. The G sensor gives the spatial temporal parameters like, speed, cadence, stride length, Gait cycle duration, Stance phase duration, swing phase duration, double support duration, single support duration, right and left foot contacts, pelvic angles, and acceleration [27].

Figure 23. G-sensor test

3.9.4. Force analysis

Force analysis is performed using piezoelectric force plates and two AMTI strain gauge force plates. Force analysis is performed to measure the impact of the ground reaction force of the foot or the footwear to the ground. In sports industry, force analysis is used to measure the force applied by sportsmen during various activities with specific types of footwear [28]. Fig 24 and 25 shows static and dynamic force analysis performed.

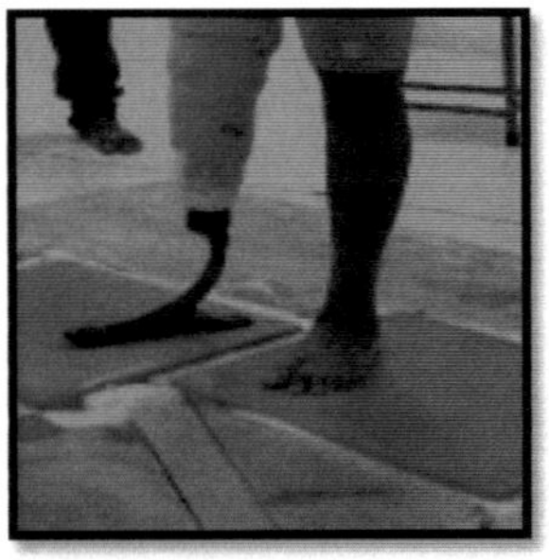

Figure 24. Force analysis- static

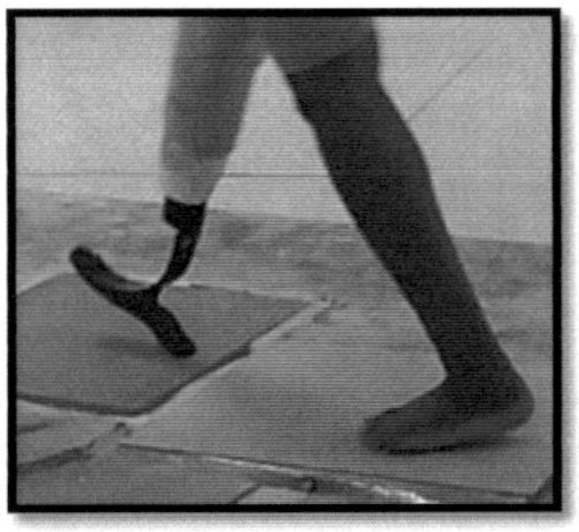

Figure 25. Force analysis- dynamic

4. RESULTS AND DISCUSSION

4.1. Material testing

As discussed in above sections, the following tests such as Tensile strength, Modulus of elasticity, Compression strength, Density, Hardness and Flexural strength are performed, and their respective values are depicted in Table 2. From the table one can infer that the out of the three composite samples prepared, carbon fiber reinforced composite sample shows the excellent tensile strength, compression strength and hardness. The density of carbon samples is less. Lower value of Young's modulus indicates the elastic property of the material and similarly decreased value in flexural strength indicates less stress developed during the time of fracture. These values are imported to ANSYS material library for finite element analysis. Non-destructive material testing using ANSYS is performed for all the three different material.

Table 2. Material testing results

S. No.	Parameters	Units	Glass fiber	Carbon fiber	Hybrid fiber
1	Tensile Strength	N/mm^2	320.34	403.72	341.49
2	Modulus of elasticity	Gpa	22.13	19	21.95
3	Compression strength	N/mm^2	489.9	519.47	475.14
4	Density	g/cm^3	1.569	1.239	1.5145
5	Hardness	HRR	66	62	64
6	Flexural Strength	N/mm^2	865.13	776.38	820.59

4.2. Finite element analysis
4.2.1 Static analysis

Finite element analysis tests of total deformation, directional deformation, stress intensity, elastic strain intensity and strain energy are performed and depicted in the Table 3. From the tabulation of finite elemental analysis carbon fiber composite shows superior property of elasticity by showing the maximum total and directional deformation due to the low young's modulus and density. With respect to the increased deformation carbon fiber shows higher value of stress, strain intensity and strain energy as well. The Obtained values indicate that carbon fiber composite offers higher elastic recoil force during walking. Literature evidences also show that carbon fiber provide better results in prosthesis and sports related applications [13, 14].

Table 3. FEA analysis results

S. No	Parameter	Material					
		Hybrid		Carbon		Glass	
		Min	Max	Min	Max	Min	Max
1	Total Deformation(m)	0	0.23053	0	0.26062	0	0.25254
2	Directional Deformation(m)	-0.12717	1.3831×10^{-5}	-0.14877	1.5638×10^{-5}	-014082	1.7054×10^{-5}
3	Stress Intensity(Pa)	11182	8.8276×10^{6}	11182	8.8276×10^{6}	9465	8.901×10^{8}
4	Elastic strain intensity (m/m)	6.1232×10^{-7}	4.834e-02	6.9222×10^{-7}	5.9647×10^{-2}	5.2138×10^{-7}	4.9031×10^{2}
5	Strain Energy (J)	1.6886×10^{-10}	1.9034	1.909×10^{-10}	2.1518	1.2033×10^{-10}	4.2379

4.3 GAIT ANALYSIS

4.3.1. Plantar pressure analysis

a. Static Plantar Pressure Analysis

Plantar Pressure distribution is found to be asymmetric between left foot (63.55%) and right (36.45%) during standing. There is increased average plantar pressure observed on the right foot (in the range of 85-105 kPa) than the left foot (in the range of 80-90 kPa) during quiet standing.

On the Right foot (Composite blade): Peak plantar (Normal/ reference foot) pressure is noted at both metatarsal (in the range of 104 – 116 kPa) and heel region (in the range of 104 – 114 kPa). Contact area with floor is observed in the range of 85 -95 cm2 which is comparatively lesser than the left foot (117 cm2) due to the design of the blade. Arch index in composite blade is 0% which indicates Heavy high arch foot design of prosthetic blade (0 – 7%).

On the Left foot: Peak plantar (Foot) pressure is observed at the medial heel region (111 kPa). Left static foot arch index is 34.09% which indicates Light flat Foot. Static plantar pressure analysis results are shown in Fig 26.

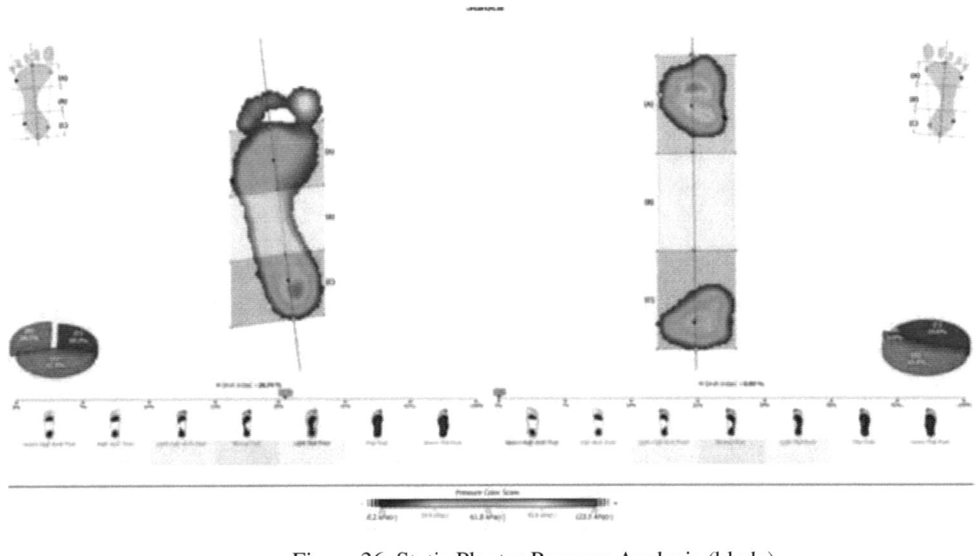

Figure 26. Static Plantar Pressure Analysis (blade)

b. Dynamic Plantar Pressure Analysis

Right foot (Composite blade):

Average plantar pressure is observed in the range of 83-90 kPa. Peak plantar (Foot) pressure is noted at the metatarsal region in the range of 240 – 310 kPa. Contact area of blade is observed in the range of 83-91 cm^2 which is much lesser than the left foot (130 cm^2). The center of path of pressure (COP) pattern of right foot travels from Heel to forefoot region in composite blade.

Left foot (Normal foot):

Average plantar pressure is observed in the range of 80-95 kPa. Peak plantar pressure (205-220 kPa) is noted at forefoot region while normal walking. Left foot contact area of 130 cm^2 is observed. The COP pattern of left foot starts from lateral heel to great toe in few steps through 2nd metatarsal with more time of contact at waist region. Dynamic plantar pressure analysis results are shown in Fig 27.

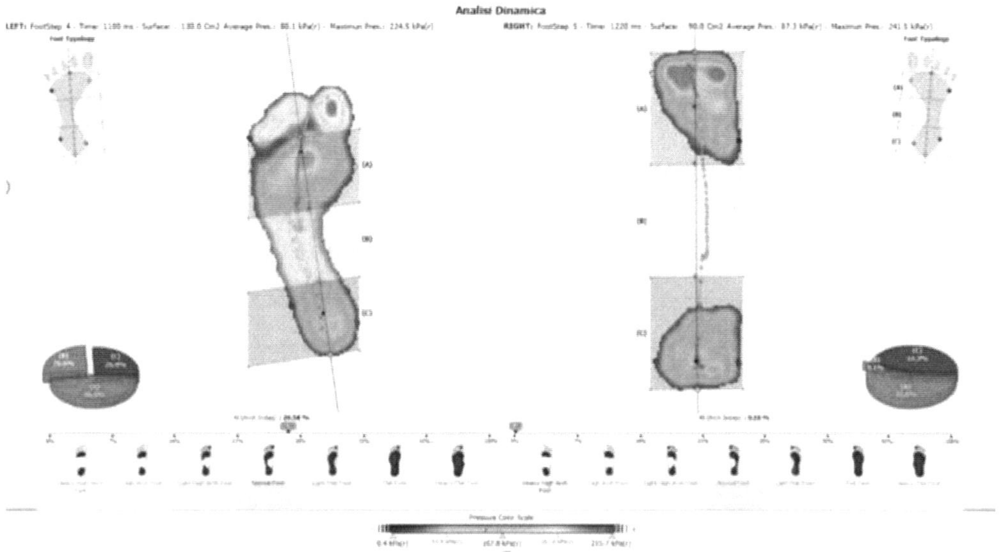

Figure 27. Dynamic Plantar Pressure Analysis (blade)

4.3.2. G sensor analysis

All spatiotemporal parameters are not meeting the normal values while using composite blade in right leg. Cadence (85.95±15.80 steps per minute), Stride length (0.59±0.03 meter) and Speed (0.41±0.0meters per second) are found to be less than normal as shown in Fig 28. Anterior pelvic tilt of 17° to 25° is noted and overall symmetry index is 74.5. Single support phase propulsion index is higher in right composite blade than left foot as shown in Fig 29.

IOP Conf. Series: Materials Science and Engineering 1258 (2022) 012044 doi:10.1088/1757-899X/1258/1/012044

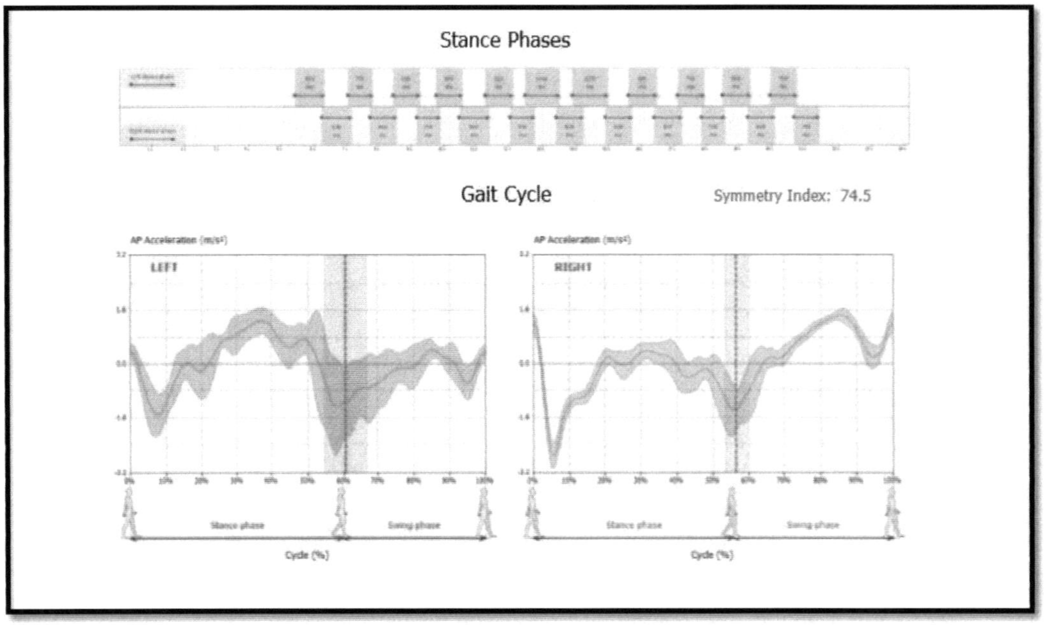

Figure 28. Walking analysis – Gait cycle

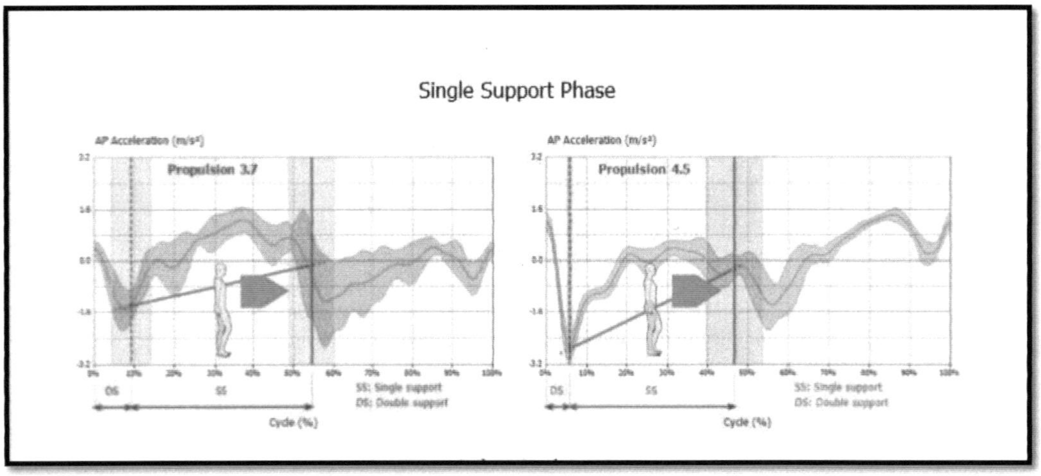

Figure 29. Walking analysis – Single support phase

4.3.3. Force analysis

Force analysis result shows all the values obtained from right leg with prosthetic blades are close to the average reference value of a healthy adult in comparison with the left leg. Further, equal load distribution of load on both legs were achieved as shown in Table 4.

Table 4. Force analysis

Test Condition	Right leg (Blades)	Left leg	Reference values
Force Profile Score	47	50.7	<20
Anterio-Posterior component (GRF)	104.4	192.4	<40
Medio-Lateral Component	67.4	60.3	<10
Vertical-Component	12	19.6	<15
Load balance	49.6	50.4	50:50

5. CONCLUSION

The uniquely designed composite blade is extremely light in weight i.e., around 250 grams and provides excellent propulsion during the gait cycle including normal walk pattern. Initially, to select thecomposite materials, the blades are fabricated using glass, carbon and hybrid fibers and mechanical testing is performed. Gait analysis is performed on a single limb amputee (right) using the fabricated carbon fiber blade. It was observed that the person had normal walking pattern. It is ensured that with the proposed blade the participant is able to follow normal gait pattern of initial contact, mid-stance andtoe-off as the blade supports normal knee and ankle biomechanics as observed from visual gait analysis.Increased peak plantar pressure with composite blade is due to less contact of foot with floor and it is observed at both the metatarsal and heel region which indicates poor posture and balance to the amputee. Also it may be attributed to the change from SACH foot (conventional foot worn by the participant prior to testing) to blade as the gait analysis is done on the first day of using blade. Increased pelvic tiltwith composite blade foot is also observed. Overall, as the design of artificial foot (blades) helps to follow normal gait pattern in right leg, the over strain and compensation gait pattern in the natural leg can be prevented. Further, walking analysis validates that leg with designed blades show better propulsion index over the other. Thus, it can be concluded that the proposed design of the blade offers superior support and enhanced normal gait. Composite blade with increased surface area and thicknesscould provide a greater load bearing with minimal stress. Further, the study can be extended by increasing the study population with simultaneously analyzing additional parameters such as joint anglestudy using 3D motion analysis, knee kinematics etc. The running characteristics of the blades can also be analyzed using treadmill-based sensors and EMG study.

6. REFERENCES

[1] Brüggemann G. P., Arampatzis, A., Emrich, F. and Potthast, Biomechanics of double transtibial amputee sprinting using dedicated sprinting prostheses, Sports Technology, 1, 220-227, (2008).

[2] Pailler, D., Sautreuil, P., Piera, J. B., Genty, M. and Goujon, H.: Evolution in prostheses for sprinters with lower-limb amputation, 47, 374-381, (2004).

[3] Upender et all; Design, Modeling and Optimization of a "Prosthetic runner blade", Indur Institute of Engineering & Technology; IJCESR 2018.

[4] Siobhan C. Strikea, Daniela Acronea, Michael Orendurffb, Running at submaximal speeds, the role of the intact and prosthetic limbs for trans-tibial amputees, Elsevier, Gait & Posture 62, 327–332 (2018).

[5] Scott, Lee, and Michele Stanten. The Walking Solution: Get People Walking for Results. Human Kinetics, 2019.

[6] Kapil, Shubhada. "Advanced Prosthetics: Changing the World of Sports." (2018).

[7] Van Philip; Flex-Foot cheetah, 2015.

[8] Radhakrishnan, Pradeep. "Developing a Scalable, Low-Cost Prosthetic Device for Below-Knee Amputations." (2020).

[9] Lecomte, Christophe. "Variable Stiffness Prosthetic Foot." (2022).

[10] Linda Patern`o, Michele Ibrahimi, Emanuele Gruppioni, Sockets for Limb Prostheses: A Review of Existing Technologies and Open Challenges, ieee transactions on biomedical engineering, vol. 65, no. 9, september 2018.

[11] Mackiewicz C, Prewitt MJ, Shinn KR, Shinn JR, Ikeda ES, Perrault JM, Hefner BD, Larry James Parker I, Hassberger BJ, Wettlaufer Jr EF, inventors; Altair Engineering Inc, assignee. Base plate and blade design for a leg prosthetic. United States patent application US 14/824,363. 2016 Feb 18.

[12] Herr HM, Wilkenfeld A, Bleck O, inventors; Massachusetts Institute of Technology, assignee. Speed-adaptive and patient-adaptive prosthetic knee. United States patent US 6,610,101. 2003 Aug 26.

[13] Merlette J, inventor. Foot prosthesis and method of making same. United States patent US 4,959,073. 1990 Sep 25.

[14] Weyand, P. G., Sandell, R. F., Prime, D. N. and Bundle, M.W.: The biological limits to running speed are imposed from the ground up, Journal of Applied Physiology, 108, 950-961, (2010).

[15] Hadj-Moussa, Firdous, Calvin C. Ngan, and Jan Andrysek. "Biomechanical factors affecting individuals with lower limb amputations running using running-specific prostheses: A systematic review." Gait & Posture 92 (2022): 83-95.

[16] Hiroaki Hobara; NIAIST; Running-specific prostheses: history, mechanics and controversy, 2014.

[17] Mark Bowkett and Kary Thanapalan, Comparative analysis of failure detection methods of composites materials systems, systems science & control engineering: An open access journal, vol. 5, 168–177 (2017).

[18] Fletcher, Jared R., Tessa Gallinger, and Francois Prince. "How Can Biomechanics Improve Physical Preparation and Performance in Paralympic Athletes? A Narrative Review." Sports 9, no. 7 (2021): 89.

[19] Baum, B.S., Tian, A., Schultz, M. P., Hobara, H., Linberg, A., Wolf, E. J., Shim, J. K.: Ground reaction force and temporal-spatial adaptations to running velocity when wearing running-specific prostheses, Proceedings of the 36th Annual Meeting of the American Society of Biomechanics, (2012).

[20] Lebedev SM, Gefle OS, Amitov ET, Zhuravlev DV, Berchuk DY, Mikutskiy EA. Mechanical properties of PLA-based composites for fused deposition modeling technology. The International Journal of Advanced Manufacturing Technology. 2018 Jul;97(1):511-8.

[21] Lay M, Thajudin NL, Hamid ZA, Rusli A, Abdullah MK, Shuib RK. Comparison of physical and mechanical properties of PLA, ABS and nylon 6 fabricated using fused deposition modeling and injection molding. Composites Part B: Engineering. 2019 Nov 1; 176:107341.

[22] Hayder saadi radaef, Firas Abdullah thweny al-saedi, Building a Three-Axis CNC Milling Machine Control System, International Journal of Artificial Intelligence and Mechatronics, ISSN 2320 – 5121, Volume 3, (2014).

[23] Yogeshwaran, S., L. Natrayan, S. Rajaraman, S. Parthasarathi, and S. Nestro. "Experimental investigation on mechanical properties of epoxy/graphene/fish scale and fermented spinach hybrid bio composite by hand lay-up technique." Materials Today: Proceedings 37 (2021): 1578-1583.

[24] Ridao-Fernández, Carmen, Elena Pinero-Pinto, and Gema Chamorro-Moriana. "Observational gait assessment scales in patients with walking disorders: systematic review." BioMed research international 2019 (2019).

[25] Buldt, Andrew K., Jamie J. Allan, Karl B. Landorf, and Hylton B. Menz. "The relationship between foot posture and plantar pressure during walking in adults: a systematic review." Gait & posture 62 (2018): 56-67.

[26] Zulkifli, Siti Syazni, and Wei Ping Loh. "A state-of-the-art review of foot pressure." Foot and Ankle Surgery 26, no. 1 (2020): 25-32.

[27] Rathore, R., A. K. Singh, and H. Choudhary. "Design, development, and calibration of bipedal force-plate for post prosthesis gait rehabilitation." Materials Today: Proceedings 44 (2021): 4873-4877.

[28] Caderby, Teddy, Jérémie Begue, Georges Dalleau, and Nicolas Peyrot. "Measuring foot progression angle during walking using force-plate data." Applied Mechanics 3, no. 1 (2022): 174-181.

A Comprehensive review on the current trends in Micro-Phasor Measurement Units

Adhitya Ravi[1], Madhusudan Saranathan[1*], Pa Hari Krishna Achuthan[1], M.C. Lavanya[2] and V. Rajini[3]
[1]Department of Electrical and Electronics Engineering, Sri Sivasubramaniya Nadar College of Engineering, Chennai, India.
[2]Deputy Director, R&D Division, National Institute of Wind Energy, Chennai, India
[3]Professor & Head, Department of Electrical and Electronics Engineering, Sri Sivasubramaniya Nadar College of Engineering, Chennai, India.

[*] madhusudan183001048@eee.ssn.edu.in (Corresponding Author)

Abstract. Micro-Phasor Measurement Units are invaluable data sources that improve visibility in a variety of domains of application, including power generation, transmission, and distribution. It started out with the objective of acquiring phasor values at specific points across the network and synchronizing them to a standard time frame. The technology has evolved into an all-encompassing unit with applications in the fields of protection, control, and automation. In light of the fact that these units are now being used as a standard benchmark, it is imperative that they be improved. This study presents a comprehensive review into the world of Synchro-Phasor technology, covering all aspects of the production of μPMUs and their role in the modern power system. In addition, the current deployment status of this technology has also been discussed to further emphasize its emergence.

1. Introduction

With the integration of renewable energy resources and electric vehicles to the distribution network, it has enormously changed the characteristics of the network. This has made supervision and control of the grid to be more complicated. Distribution networks are highly complex because of the presence of multiple nodes and being dynamic. There also exists a lot of uncertainties in the grid as there are a lot of parameters to focus on. Currently, there exists an architecture called the Supervisory control and data acquisition (SCADA) which receives data from remote terminal units located in various parts of the system and is being implemented in almost every substation. This acquisition system can measure magnitudes of voltage, current and reactive power flow, but it captures data at very low resolution (number of samples in unit time) and the measurements are unsynchronized. Thus SCADA only has the ability to observe the steady state of the distribution system.

In order to facilitate dynamic observability of the grid and to mitigate the complexities caused by SCADA, a distribution level phasor measurement unit or the micro - phasor measurement unit was designed and developed. Unlike SCADA, it can acquire a maximum of 120 samples per second and can observe the dynamic state of the distribution network. It also has the ability to measure the phase angle and time synchronize all the measurements by data stamping them. PMU facilitates wide area monitoring and control as opposed to local area monitoring in SCADA.

LA Lee et. al. compares the performance of PMUs and μPMU. A comparative analysis of both units was performed with respect to the IEEE C37.118.1 standard for PMUs. A PSL μPMU, PQube3 and a commercially available Artiber PMU, Model 1133 A was used to perform the comparative analysis. A Doble F2500 was used as a voltage supplier, to generate a 3-phase grid like low voltage.

IOP Conf. Series: Materials Science and Engineering 1258 (2022) 012045 doi:10.1088/1757-899X/1258/1/012045

Even though the units were used for different monitoring purposes within the grid, an equitable test bed was created to analyse their respective performances. This work showed that even though the μPMU had smoother magnitude and frequency tracking response, and cost less, the PMU performed better in the fields of phasor computation and response speed [4]. Furthermore, the significant differences between the PMU and the μPMU are, PMUs are employed in transmission systems while the μPMUs are employed in the Distribution systems. PMUs have an angle measurement accuracy of about ±1 % with ±0.1 % of angle resolution. On the other hand, μPMUs have an angle measurement accuracy of about ±0.01 % with ±0.002 % of angle resolution. In contrast to PMUs which have a data rate of 15 readings per second, μPMUs have a data rate of 100 – 120 readings per second [32].

The objective of this paper is to present a comprehensive survey on research based on micro-phasor measurement units. In this review paper, μPMU related literatures are classified into design and development, application and reports presenting the recent deployment of PMU in extensive electrical networks thus making this work novel.

This paper is organized as follows. Initially, a detailed description of the phasor measurement unit has been discussed followed by a discussion of numerous literatures focusing the design and development as well as papers that emphasize on the applications of μPMU. This paper concludes by discussing the deployment of PMU in major countries.

2. Phasor Measurement Units

The production and transmission of real-time data is done with the help of other peripheral devices by the PMU. The input for the PMUs is obtained in the form of stepped down voltage and current signals from the potential and current transformers which are connected to the respective measurement points. These parameters are then modified for potential analysis and data collection through the usage of electronic screening applications such as filters. The distinct processing algorithms are performed to obtain the final output, in the form of phasors which have both magnitude and phase angle components.

The process of time stamping is performed to make sure the data is in sync with the data which is obtained at the same moment at a different location. Making use of GPS (Global Positioning System) signals, the standard time reference is considered to timestamp the signal. The standard time reference usually considered is the Universal Time Reference (UTC). The peak of each sinusoid and a UTC unit reference is taken and the phase angle between the two is obtained. This phase difference is considered as the mark of the stamp. If the UTC reference is after the peak of the sinusoid, then the phase angle is positive and if the UTC reference is before the peak of the sinusoid, then the phase angle is negative [5].Depending on the unit UTC reference and the phase angle difference, the time at which the data was recorded can be accurately recovered, up to a microsecond of accuracy.

Generally, the μPMU usually comprises a processing unit, which receives the UTC and time stamps the analog signal, processes the phasor and then yields a digital value of that signal. It is also interfaced with a communications module to transmit the processed data. A GPS Receiver is used to obtain the UTC with the help of a GPS Satellite signal and a precision internal clock. The Input signal is usually obtained from a pair of current and potential transformers which step the signal down to an acceptable range. These analog signals are then passed through anti-aliasing filters to make sure the analog parameters are corrected to have the same phase and the same level of attenuation. This process makes sure that the incoming inputs at different instances have the same relative phase and magnitude. The processor can be configured to make sure that it contains a distinct phasor estimation technique to yield the phasor as determined by the user.

PMUs are designed according to standards set by the IEEE (IEEE C37.118.1-2011). This standard defines certain parameters that the synchrophasor is supposed to attain in terms of its performance. It defines a standard for frequency of operation, synchrophasors, and the Rate of Change of Frequency (ROCOF) measurements under all the operating conditions. These standards and requirements are defined for both steady-state and dynamic operating conditions. Other than this, the standard also defines the synchronization requirements and the time tag requirements to be fulfilled by both the PMU and incoming signals to the data concentrators.

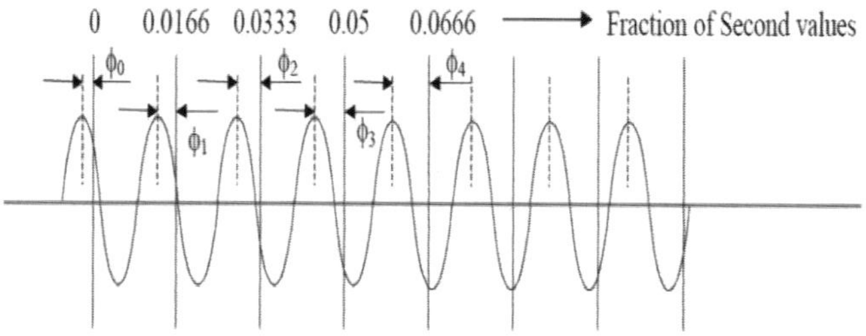

Figure 1. Sinusoid waveform with changing phase angles

The standard IEEE C37.118.1-2011 divides PMUs into two performance classes: P and M. The P class is designed for applications that require quick measurement response times, such as power system protections, but the M class should be considered where accuracy of the measurement is critical. At the very least, a standard compliant PMU should meet all of the standards for one class. The key distinctions between the two performance classes are the steady state performance test conditions and the dynamic performance criteria, particularly in terms of frequency and ROCOF estimation [6].

3. Block diagram of μPMU

The main components of μPMU include the current transformers (CT), potential transformers (PT), anti-aliasing filters, GPS module, phase locked oscillator, phasor processor and modem. Figure 2 depicts the block diagram of a phasor measurement unit.

- ***Instrumentation channel*:** The instrumentation channel's purpose is to provide isolation from the high voltage power system while reducing voltages and currents to standard instrumentation levels. It consists of current transformers, potential transformers, cables, burdens and analog to digital converter (ADC). The voltage and the current from the power system is stepped down by the PT and the CT respectively. The devices present in the instrumentation channel contribute to signal degradation [7]. An attenuator is used strengthen he signal for further processing.
- ***Anti – Aliasing Filter:*** Filters are utilized at the receiving end of the input sinusoid, where the stepped down analog parameter is altered to have the same phase and amount of attenuation. This ensures that the incoming inputs at each instance have the same relative phase and amplitude.
- ***GPS***: GPS delivers precise information about the GPS receiver's position and time (which might be local time or Universal Time Coordinate). Along with it, the satellite is capable of delivering 'one pulse per second' with an identity that is accurate to within one micro-second at any position on the planet.
- ***Phase Locked Oscillator*:** The satellite pulse signals are phase synchronized with the sampling clock. This function is performed by the phase locked oscillator. Phase Locked Oscillator splits the GPS signal's one pulse per second into the desired number of pulses per second for sampling.
- ***Analog to Digital Converter:*** It converts the analog signal from the filter to a discrete time and discrete amplitude signal. The sampling rate has to specify at which the analog signal is sampled.
- ***Phasor processor***: To get accurate estimates of all the current and voltage data, the microprocessor use the phasor estimation techniques. The output of the μPMU additionally includes estimations of frequency and rate of change of frequency recorded locally. Using two signals from the GPS receiver, the timestamp is generated. The time-stamp specifies the "universal time coordinated" (UTC) second and the moment that defines the limit of one of the power frequency intervals.

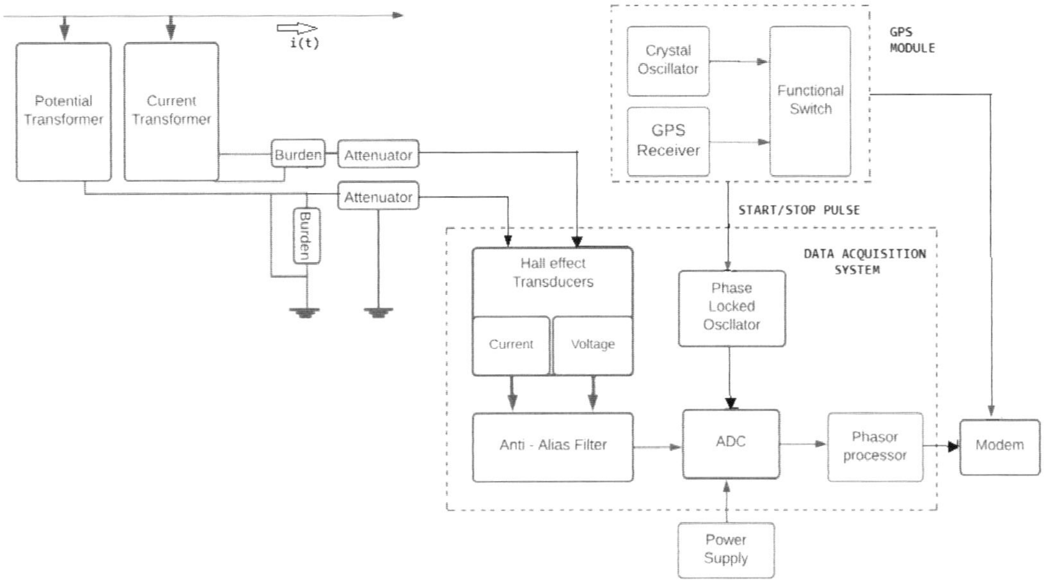

Figure 2. Block diagram of μPMU

4. Design and development of μPMUs

Jose R. Razo-Hernandez et. al., presented a new phasor estimation technique which was implemented using complex brick wall band pass filters which met all the requirements for both, P class and M class phasor measurement units. The weighted least squares method was used to obtain this filter with the help of a weighing function that aimed at fitting the passband and the rejection in the valleys. The performance of the PMU system was compared with a reference algorithm and validated with specific IEEE standards (IEEE Std. C37.118.1-2011 and its amendment) by estimating the frequency, rate of change of frequency and the phasor. This approach implements the FIR filter design throughout the online phasor estimation process thus making this algorithm less complex with simpler computations [8].

Hari Prasanna Das and Ashok Kumar Pradhan has presented a hardware implementation of a μPMU which is of significant importance for installation in distribution systems. Here, a TivaTM C series microcontroller was used for phasor estimation and signal acquisition and processing. An ESP 8266 WiFi development board has been used for receiving time information and two sets of probes for voltage and current are interfaced to the microcontroller. A predictive compensator block has also been installed to compensate for the system clock drift from the GPS clock. The main drawback of this implementation is the application of the microcontroller as the main processing unit, which has several drawbacks including the limited parallel processing capability (which is very pertinent to this application) and lower processing speeds [9].

Ahmed Fakhfakh et. al. presents a novel implementation of time stamping to overcome the drawbacks of utilizing GPS. Here IEEE 1588 PTP (Precision Time Protocol) is implemented instead of GPS to overcome the difficulty of having to have the units outdoors and to also increase time synchronization accuracy. The system is designed for both wired and wireless operation and can work indoors and outdoors which yields versatility to the unit on an unprecedented scale. The time of convergence and packet numbers for synchronization are reduced. The usage of embedded systems also reduces the encasement volume. This work is very valuable in the implementation of a smart grid and implementation of IOT in power systems. The increased timing accuracy can yield higher resolution for distribution level systems [10].

Andrea Angioni et. al. proposed a low-cost phasor measurement unit prototype which was meant for distribution grids. This prototype aimed at reducing the amount of computations thus reducing the cost, usage of robust, standard communication functionalities, and limited reporting rate by providing only the steady state information which is sufficient for the prescribed application. The steady state accuracy requirements and the total voltage error fulfilled the requirements. The author mentioned that the compensation of systematic errors will have to be looked at and could be experimented in future. This prototype does not make use of a microcontroller, DSP and it makes use of external pulse-per-second signal to enable real time operation thus making this work novel. The effect of noises and harmonics were also taken into consideration to provide an optimized result [11].

In [12], Paolo Romano et. al., presents an improved interpolated discrete fourier transform (e-IpDFT) method for synchrophasorestimation (SE) as well as a step-by-step procedure for creating a PMU prototype. The suggested e-IpDFT-based SE algorithm is a development of the standard IpDFT approach since it incorporates a scheme for compensating for the impacts of the spectrum's self-interaction between positive and negative image. This change enhanced the PMU's performance under static and dynamic conditions and also made it feasible for dumping the algorithm into the FPGA. The prototype exhibited unique characteristics aiding for its usage in transmission and distribution [12].

Talha Ahmed Bhatti et. al. provides a technique for developing PMUs that is not based on DFT and is compliant with IEEE standards C37.118.1 and C37.118.2. In a DFT based algorithm, as per Nyquist criterion the sampling rate for a 50 Hz system should be at least 100 Hz, but in order to avoid spectral leakage that can occur due to the non-integer frequencies (non-multiple of 50 Hz) present in the signal a high sampling rate will be required along with large window size which would increase the latency. Thus, a non - DFT based algorithm was chosen for the design of PMU. The suggested technique reduces processing time and sample rate, resulting in lower costs. In this literature, a PMU was implemented on the MTK3339 chipset with Raspberry Pi Model B+, ADS 1015, and GPS (Global Positioning System) model PA6H [13].

The open-source PMU project, OpenPMU, developed by QUB (Queen's University Belfast), was discussed by David M. Laverty et. al..OpenPMU began as a low-cost alternative to commercial devices for certain research applications that required quick synchrophasor estimates for real-time control. The primary impetus for the project is the need for a uniform, open platform for the development of PMU technology suitable for academic study. It is desirable to avoid duplication of effort caused by universities working on PMU technologies independently without sharing resources. The present LabView based OpenPMU was developed in this literature to produce a more modularized architecture. This unit's performance is measured against an IEEE standard and compared to a commercial device. Middleware designed for this purpose achieved compatibility with the IEEE C37.118.2 communications specification [14].

Shurong Li et. al., in this work proposes the extension of the topic of phasor estimation to not only to the fundamental phasor, but also to the harmonic components of the incoming AC signal. Citing low hardware utilization, the author has applied the same measurement algorithm used in fundamental phasor estimation for harmonic estimation and has compared the results from this method to the one computed using normal FFT estimation of phasors when there is a deviation from the rated frequency in the input AC signal. The results from the two methods were also processed to estimate the Harmonic Contribution Determination (HCD) and it was tested on a 18-bus distribution system to test HCD's accuracy [15].

E.M. Stewart et. al. discusses the use of data procured from two different projects (from a μPMU and from OpenPMU) in making grid management safer and more efficient. This research work also details various challenges in integrating the high fidelity data to the grid which include the visualization of the distribution grid, erroneous data and validation of the model. State estimation, one of the key applications of this data, predicts the voltage/current magnitude and phase angle using the existing values at every bus and also accounts for missing or distorted data. The effect of combining state estimation and power flow analysis was also discussed in detail by the author. [16]

5. Applications of μPMU

5.1. Fault Location

Lean Lorenzo F Blas presents a study where a μPMU is interfaced with an electric power system simulator to analyse and simulate a number of scenarios in a virtual substation setup. This setup also analyses transient events and operational behaviour. A compatibility study is also performed to test the viability of the setup to work in conjunction with a SCADA system. Fault studies are also performed to exhibit the potential for the system to be utilized as an event detection and state estimation platform for a distribution system. The work yielded promising results where high precision and frequent data combined with the granularity of the SCADA system provided a high resolution system which provides for a better perspective in transient observations and event detection [17].

D. Devaraj et.al. proposed a Non-Recursive DFT algorithm-based PMU on ARM based myRIO controller. This PMU was modelled and simulated in the LabView platform using a power system with a 150MW generating station, two transformers, one short transmission line and 110 MW load.
The results suggest that the created system can perform satisfactorily with the gathered real-time voltage signals. The simulated power system's faults were identified and detected using this method. In the sample power system, several sorts of faults were induced, and the fault current values acquired were recorded. Wavelet transformation was used to perform energy feature extraction from the stored values. The established Wide Area Monitoring System was then used to locate and identify the flaws using an Artificial Neural Network (ANN). The proposed setup was able to identify the fault within a few milliseconds thus signifying that this system will be able to support relaying systems for protection at high speeds [18].

Jonathan Lee describes how a PMU data stream can be used to detect faults accurately in a power system rather than depending on sensors that cost a lot. This work shows how synchronized voltage values measured by a PMU can help in precise identification of faults. The suggested technique is proved to be robust to variations in total load and uncertainty in the reaction of loads to voltage sags during a sample set of diverse fault scenarios using the IEEE 123 bus distribution feeder as a test case [19].

5.2. Event Detection

Alireza Shahsavari et. al. conducted a data driven analysis using the synchronized voltage measurements from the PMU to investigate lightning-induced contingencies at a real-life distribution system. A 7.5 MW PV farm was used for the study and few observations were made by the authors. One of them was the flow of power in the reverse direction (i.e. from the PV farm to the substation) because of the momentary short circuit caused due to the operation of the surge arrester [20].

Wenxuan Yao et. al. proposed a fast load control system that utilizes D-PMU for power mismatch detection where synchronization of data was done using GPS. Large power system disturbances were identified in real time using a frequency measurement threshold criterion. The event's starting point was computed using the rate of change of frequency values. Then, at an early point of the event, the power imbalance is evaluated using ROCOF. In order to improve the frequency stability, the authors were able to successfully create a communication between the PMU and a smart breaker for fast load control [21].

Huaiguang Jiang et.al proposed a steady state PMU-aided voltage security assessment for a wind power plant. This method aimed to identify the system locations having the greatest impact on the voltage at wind power plant's point of interconnection using thevenin equivalent matrix . The data from the PMU was used to evaluate the theveninmatrix. The fault impact of each bus was calculated using this matrix and a critical bus set was generated for the bus at wind power plant's interconnection. This approach produced an effective response compared to the dynamic simulation and required fewer computations [22].

Luigi Vanfrettiet.al developed a monitoring application for the detection of sub synchronous oscillations in power systems using the real-time measurements from phasor measurement units. Two algorithms were employed simultaneously to detect the level of energy in the oscillations as well as the frequency at which the oscillation occurs. This application was developed and tested in the framework of SmarTS Lab environment which is capable of Hardware in the Loop simulations.The monitoring tool was successful in detecting the strong oscillatory events at high frequencies as well as the injected oscillations at 10.83 Hz for modelling wind farms interactions at a faster speed with minimal delay [23].

Armin Aligholian et. al. aimed at utilizing the large volumes of μPMU data to capture and classify all distinct types of abnormalities in power distribution feeders. An unsupervised event detection algorithm, GAN, is used in this case which uses deep neural networks to train a model which can learn to characterize normal events and classify the abnormal events using a clustering method, based on origin and similarity. The results of the work exhibited a level of performance beyond that of conventional event detection and classification methods. [24]

A comprehensive event analysis was conducted from the μPMU data from a 4 MW solar farm distribution feeder. An unsupervised learning algorithm was utilized to determine event origin by impedance based analysis and classify locally caused and grid caused issues. These issues were also separately analyzed from the perspective of the solar farm, and its response to the dynamics was also analyzed. Multiple events' control dynamics which were classified were also separately analyzed. The findings revealed that more than 70% of the incidents happened when solar energy production was less than 30%. The study also gave insight into the equipment health on the farm.[25]

G. Sanitha et. al. presents an event classification and location identification, using wavelet transform and compensation theorem techniques respectively. Capacitor switching, Faults and load switching were among the events considered. μPMU data from two different units were used to compute the magnitudes and phase angles of voltage and current phasors.The results of simulations on the IEEE 34 bus test feeder were used to demonstrate the method's performance [26].

In the work proposed by Duan, Nan, and Emma M. Stewart, frequency abnormality events were recorded and categorized with the help of data recorded from μPMUs. These events were recorded from distribution feeders and involved analysis of the event's origin. It was also possible to identify if the reading was falsely produced by estimation algorithms such as IF (instantaneous frequency). Granger causality was utilized to identify if the event came from the transmission system. A sparse dictionary was used to identify physical origin of the event and an event-pair detector was used to identify if it was algorithm induced. [27]

An event locator was presented based on the results of an event-triggered distribution network state estimator by Dutta et. al.. Measurements from a small number of micro phasor measuring units (μPMUs) were used to estimate the state. The changes in system states were estimated using a state estimator which employs an L1 regularization. The results of the estimation were used to locate the event. This novel method was mainly suggested for radial and mesh topologies and was validated using an OPAL-RT real-time simulator in a 13 node testing distribution system. [28]

Anna L. Liao et. al. uses a network of three time-synchronized micro-Phasor measurement units at three different locations to identify the occurrence of any event at the transmission and the distribution side. Voltage sag was one event of concern in the paper. This event could be local or because of various external forces. Time synchronization of data would be highly useful to distinguish events similar to one discussed in this paper. This can be achieved by using a μPMU.[29]

Alireza Shahsavari et. al. initially proposed a technique for model-free event detection for extracting relevant data from μPMU data streams collected at a real-world test location in California. The identified events were then classified at two layers using a novel data-driven event labelling strategy paired with various classification methods. It was concluded that the classification accuracy increased when the classification features in the detection process were used. Furthermore, remote asset monitoring and distribution level analysis were performed on two real time datasets, whose results could be useful to utilities.[30]

Qiushi Cui and Yang Weng proposed a high impedance fault monitoring scheme. The proposed technique not only tackles the issue of significant labelling costs, but also shows how unlabeled data may be used to aid overall classification with high accuracy.[31]

Microgrid anomalies may be detected and allocated using just two PMUs that communicate with each other under an IoT-based umbrella using the effective data-driven stochastic method proposed by Abdollah Kavousi-Fard et al. For locating the precise position of an event, the compensation theorem and backward-forward voltage computations are used in conjunction with a minimum of two μPMUs. In order to enhance the precision of the findings, the phasor and magnitude measurements of the voltage of the μPMUs have been employed. Furthermore, the data transmission and computation complexity have been greatly reduced due to the use of IoT infrastructure by the μPMUs. [32]

Faults must be identified in real time and with high precision using the real-time high-resolution measurements made possible by the deployment of μPMUs. Debottam Mukherjee et al presented comparisons of several machine learning algorithms to classify faults such as L-G (Line to ground), LL-G (Double line to ground), and LLL-G (line to line to line to ground). The most and least common types of faults, L-G and LLL-G, produce an imbalanced dataset and thus a biased classifier. SMOTE-based synthetic data oversampling was used in this study to help alleviate this problem and create a machine learning classifier that was free of bias.[33]

Adjoints in PMUs and DGs integrated Distribution System was the basis for a fast and novel two-stage Loss-of-Mains (LoM) detection technique proposed by Yashasvi Bansal et al. Adjoint matrices and real-time PMU data are used to identify and locate the LoM using the difference form of Tellegen's Theorem. Fast and accurate LoM detection, economic feasibility, insensitivity to DGs or loads' transient actions, and the ability to distinguish LoM from other events are among the aspects of the proposed approach on the modified IEEE-13 bus distribution system.[34]

Using µPMUs, Mohammad Farajollahi et al developed an innovative method for locating the source of power distribution system events. The suggested method was based on the compensation theorem in circuit theory and utilizes voltage and current synchro phasors collected by µPMUs to create an equivalent circuit to represent the event. Additionally, since this approach makes significant use of both magnitude and synchronized phase angle data, it supports the use of µPMUs rather than conventional RMS-based voltage and current sensors. The suggested method was capable of processing data from as little as two µPMUs and its efficacy and robustness were demonstrated by computer simulations on the IEEE 123-bus test system [35].

5.3. Topology Status Verification

Guido Cavraro et. al. proposed a novel method in distribution networks for topology detection which was known as "Time-series signature verification method for topology detection" (TSV-top). Here measurement data from PMUs installed on distribution feeders were utilized to find common trends between time series data and network topology changes. The proposed algorithm was tested on a 33-bus test bed. This algorithm was capable of working efficiently even if the knowledge of the state of the grid was only partial. By making use of the measurement data, the algorithm understood not only the switching times but also the new topology [36].

G. Cavraro et. al. proposed a novel method to not only detect the switching actions but also the topology transition. This new method was data driven and it was based on an analysis in real-time of the time-series voltage measurement data obtained using PMUs. Using this data, specific patterns were observed which represent state transitions which include topology changes. This proposed algorithm was based on the cross validation of the measurement data and the signature library which was derived from distinct topology simulation. This algorithm on a standard test-bed of a IEEE 33-bus system [37].

In this work, Dutta, R., Chakrabarti, S., & Sharma, A. used bus voltage phase angle measurements available from Micro Phasor Measurement Units to investigate an M-estimator of scatter matrix for data-driven topology detection in Active Distribution Networks (ADNs). Testing on the IEEE-37 bus system revealed that the suggested technique is more reliable than the traditional inverse covariance matrix-based topology identification method. Even with the presence of bad quality measurements and noisy data, the method presented robust results.[38]

5.4. Adaptive Protection

Mohsen GhaleiMonfaredZanjani et. al. proposed an application of a µPMU to protect overcurrent relays in Microgrids. In this work, a thevenin equivalent was estimated using the µPMU data thus acquired. But uncertainties in the microgrid didn't have any effect on the thevenin parameters. Thus, a communication based method was proposed by using the capability of µPMUs to exchange data. The results of the suggested method's implementation show that it can detect uncertainties, reconfigure the main and backup relays, and change the relay settings [39].

Mohamed Salah Elbana et. al.proposed a smart protection scheme using a micro phasor measurement unit to obtain synchronized measurement data for purpose of protecting a microgrid. The author mentions that the tripping time iswithin 25 ms based on the usage of real-time P-type half cycle µPMUs which is accurate in measuring the phasors. The research work lists all the data measured by the µPMU for five different cases, i.e., normal mode, two cases for grid connected mode, three- and single-phase fault in islanded mode. The protection scheme was able to detect faults even when conventional schemes failed [40].

5.5. State Estimation

State estimation formulated as a quadratic programming problem was suggested by C.Lin et. al..To accommodate the autonomous operating mode among microgrids, as well as to safeguard information privacy between various operators, the problem was handled in a decentralized way. Here, µPMUs were used to record voltage and current measurements, including the injection current phasor, branch current phasor and the voltage phasor. The decentralized deployment of µPMUs exhibited the versatility of such a unity and also provided robust results in this study.[41]

Sagnik Basumallik et al. developed a non-linear state estimator that runs on the AoW (Analysis on the wire) framework with a massive volume, variety, and velocity of µPMU streaming data using real-time Deep Neural Networks. Monthly, around 4.45 billion data points were obtained from a variety of µPMUs in order to train the network. A large degree of variance in wind, sun, and demand data was shown to not affect the accuracy of the state estimator and the state estimator on the AoW framework was designed to be adaptable and scalable to accommodate future changes in the control strategy and the functionality of the device. [42]

5.6. *Voltage and Frequency regulation*

Alireza Shahsavari et. al. proposed a model which aimed at evaluating the adverse impact of load resource participation for frequency regulation on distribution system performance. Here, current and voltage data (both magnitude and phase angle) were collected at a sampling rate of 120 Hz µPMU, to study the transient load behavior to create a surge current profile. This profile was utilized to study the response of load resources during a regulation down event. The result of the analysis performed on the data obtained from the µPMUs exhibited the possible grid reliability issues with respect to having load resource frequency regulation [43].

Albuquerque, R. J. presents a novel and practical method for real-time per-phase voltage drop measurement in primary radial distribution feeders to improve voltage control. The author states that an accurate voltage drop and improved voltage regulation can be achieved by performing complex subtraction between voltage synchrophasors, even for scenarios involving heavy currents combined with leading power factors that may arise due to the high integration of renewable and Distributed Energy Resources. The IEEE 34-node radial test distribution feeder is used to generate numerical results, which are presented and debated.[44]

5.7. Protection and Islanding detection schemes

Kumar, G. P et. al. proposed a novel islanding scheme, whereby a linear correlation between phase difference and frequency difference is derived using Pearson's correlation coefficient formula in real time. A single threshold is created to remove the threshold redundancy and the power mismatch dependency conditions. The frequency and phase angle data is obtained from two different µPMUs at different locations to record the phase angle and frequency difference. [45]

In this work, Al-Maitah et. al. presents a modified novel WAMS (Wide Area Monitoring Scheme). It involved the ratio between the total of voltage phasors between two terminals of a transmission line and the amount of current phasors flowing in the respective line. The measurements were recorded by the µPMUs which were optimally deployed according to the scheme. The modified protection scheme also involved a revamped relay coordination which was tested with several different operational modes fault types, on a distribution system. The simulation results revealed that the scheme could robustly carry out isolation operations.[46]

Mahdi Jamei et. al. describes a security architecture that uses µPMUs to assess the actual condition of a distribution system at various points and in real time. Their proposed system analyses data from both the cyber and physical sectors to deliver a comprehensive, independent view of the condition of the distribution grid. The technique has three major benefits: it is resilient owing to its distributed structure; it can be used to both test existing cybersecurity systems and detect possible cyber threats; and it can be installed at existing utilities for a low cost.[47]

Soham Dutta et. al. proposed a method which uses a P-type µPMU to detect islanding for an active distribution network. The proposed technique does not require any additional hardware. The microprocessor in the µPMU is programmed to execute fortescue transform and send the required signal during an event of islanding using the unit's communication channel. Various non-islanding situations, fault scenarios, and other power system functions have no effect on the algorithm. Furthermore, the algorithm suggests smart islanding to avoid shutting down the critical load. [48]

5.8. Phase identification
Miles H.F. Wen et. al. proposed a novel method to identify the phase of a distribution network based on measurements from micro-synchrophasor units. Two µPMUs were used and their method was validated by simulating a IEEE 13-bus test system and using the field measurements. The author's approach is based on the fact that time-series voltage magnitudes on two ends of the same phase should have much greater correlations than those on two ends of different phases in an imbalanced three-phase system. Furthermore, after removing Delta-Wye transformer phase shifts, voltage angle changes between two ends of the same phase should be very minimal. The author aims to work on approaches based on voltage angle differences and magnitude correlation in future.[49]

5.9. Blockchain
New stochastic architecture based on unscented transform (UT) developed by Abdulaziz Almutairi et al. requires just a small number of PMUs to communicate in an encrypted blockchain-enabled structure that could allocate and identify irregularities and faults in the smart grid network. Blockchain-based techniques can be used to encrypt data transmitted among the PMUs in the smart grid, which includes a variety of distributed energy sources (DERs) and high impedance fault events, while taking into account the uncertainties associated with load. Three primary scenarios were simulated, and the results demonstrated that the suggested model was very accurate and effective in a variety of smart grid failure situations.[50]

6. Deployment of PMU across nations
The deployment of PMUs across various nations has been mainly concentrated on transmission line systems due to the increased cost and size of these units. In the 1980s, the first PMU prototype was introduced at Virginia Tech (Power System Research Laboratory), which was followed by the introduction of the first commercial unit in 1991 by Macrodyne. PMUs used in conjunction with IEDs instead of actual inputs from sensors for protection and monitoring systems have been more successful in automating the control process[51].
There are several applications for IED-PMU systems such as Photovoltaic monitoring systems, active and reactive power control, feeder management systems, online disturbance monitoring and complete automation of the control and protection systems.

Among major countries with extensive electrical networks, the need for Wide Area Management systems, greater visibility into electrical events and power distribution has increased by massive amounts. With the introduction of PMUs and µPMUs, this need is slowly being met but is hampered by the fact that the PMUs are either too bulky or expensive or that µPMUs are too simple and not efficient enough. So deployment across power networks has been slow to occur. Optimal deployment and placement of these PMUs has also been extensively studied.

China is a large country with an extensive electrical network and several power plants. WAMS are a major requirement for a network of that size and with its distributed generation units. To yield greater control, visibility and real time data exchange, 7 regional WAMS were set up with SPDCs in North China, East, Northeast, Northwest China, South and Central China power grids. Other than this provincial WAMs have also been set up in Guangdong, Shandong, Jiangsu, Guizhou, Shanxi and Yuannan power grids. With the increase in the number of commercial manufacturers for PMUs, standards were issued by the State Grid Company which supplements the transmission protocols, based on the IEEE std 1344-1995. The WAMS present in the provincial or regional dispatch centers had control, protection equipment and supplemental equipment for PMUs such as Phasor Data

Concentrators. Data transfers were performed using LAN connections to reduce time delays for the data[52].

In North America, Canada and the United States alone have deployed about 1700 PMUs in the transmission systems across their respective nations which has resulted in 100% bulk visibility. To facilitate the development of technology, several high-speed PMU data networks were also created. There are several localised PDCs present in both regional dispatch centres and Transmission line operators. Data sharing networks were set up between different reliability coordinators, and utility operators. PMU data is also being utilised for Special Protection Systems and Remedial Action Systems. Surge demand events were verified using ERCOT implemented using PMUs. They are also used for managing system islanding operations and black-out restoration[53].

The Power System Operation Corporation in India is building a network of 1700 PMUs to implement a nation-wide WAMS for real-time grid monitoring and to feed a suite of offline users. Consisting of separate regional grids which are interconnected using Asynchronous HVAC links, the national grid is widely distributed with several generating sources. To prevent cascading blackouts and to increase the visibility, the deployment of PMUs were planned. This implementation could also increase the level of automation in the operation and control segment which could reduce all dispatch and protection decisions taken by utility providers, to a time frame less than half a second. With the construction of this network in place, several pilot projects have been started to test the validity of the system, such as the deployment of PMUs in 8 selected sub-stations in the Northern grid and the deployment of 30 PMUs using optimal placement algorithms in the Western grid[54].

7. Conclusion

A comprehensive survey of current μPMU technologies was conducted which showed that existing technology can be extended on for a more efficient implementation of the technology. With the increasing penetration of renewable energy sources which are inconsistent in nature, we need applications which can monitor and provide real-time feedback regarding the operational parameters. These parameters can maintain power production stability and hence grid stability. A cheap and smaller unit is needed for high volume deployment across the fields of transmission, production and distribution. With increasing developments in the field of communications and electronics, we can achieve such a unit so as to not only increase the efficiency of current practices but also increase the level of automation in all aspects of power generation, transmission and distribution.

8. Acknowledgement

We thank the National Institute of Wind Energy for providing us with the opportunity to work on this project. Further, we would like to express our heartfelt gratitude to Dr. Rajesh Katyal, R&D Division, National Institute of Wind Energy, for providing us with this opportunity.

9. References

[1] E. Dusabimana and S.-G. Yoon, "A survey on the micro-phasor measurement unit in distribution networks," Electronics, vol. 9, no. 2, p. 305, 2020.

[2] A. Shahsavari et al., "A data-driven analysis of capacitor bank operation at a distribution feeder using micro-PMU data," in 2017 IEEE Power & Energy Society Innovative Smart Grid Technologies Conference (ISGT), 2017, pp. 1–5.

[3] A. Shahsavari, A. Sadeghi-Mobarakeh, E. Stewart, and H. Mohsenian-Rad, "Distribution grid reliability analysis considering regulation down load resources via micro-PMU data," in 2016 IEEE International Conference on Smart Grid Communications (SmartGridComm), 2016, pp. 472–477.

[4] L.-A. Lee and V. Centeno, "Comparison of μPMU and PMU," in 2018 Clemson University Power Systems Conference (PSC), 2018, pp. 1–6.

[5] P. Nanda, C. K. Panigrahi, and A. Dasgupta, "Phasor estimation and modelling techniques of PMU-a review," Energy Procedia, vol. 109, pp. 64–77, 2017.

[6] P. Castello, J. Liu, A. Monti, C. Muscas, P. A. Pegoraro, and F. Ponci, "Toward a class 'P+ M' Phasor Measurement Unit," in 2013 IEEE International Workshop on Applied Measurements

for Power Systems (AMPS), 2013, pp. 91–96.
[7] A. P. Meliopoulos et al., "Synchrophasor measurement accuracy characterization," North Am. SynchroPhasorInitiat. Perform. Stand. Task Team (Consortium Electr. Reliab. Technol. Solut., vol. 10, 2007.
[8] J. R. Razo-Hernandez, A. Mejia-Barron, D. Granados-Lieberman, M. Valtierra-Rodriguez, and J. F. Gomez-Aguilar, "A New Phasor Estimator for PMU Applications: P Class and M Class," J. Mod. Power Syst. Clean Energy, vol. 8, no. 1, pp. 55–66, 2019.
[9] H. P. Das and A. K. Pradhan, "Development of a micro-phasor measurement unit for distribution system applications," in 2016 National Power Systems Conference (NPSC), 2016, pp. 1–5.
[10] A. Fakhfakh and F. Derbel, "μPMU based on wireless IEEE 1588 PTP," in 2020 6th IEEE International Energy Conference (ENERGYCon), pp. 1030–1034.
[11] A. Angioni, G. Lipari, M. Pau, F. Ponci, and A. Monti, "A low cost PMU to monitor distribution grids," in 2017 IEEE International Workshop on Applied Measurements for Power Systems (AMPS), 2017, pp. 1–6.
[12] P. Romano and M. Paolone, "Enhanced interpolated-DFT for synchrophasor estimation in FPGAs: Theory, implementation, and validation of a PMU prototype," IEEE Trans. Instrum. Meas., vol. 63, no. 12, pp. 2824–2836, 2014.
[13] T. A. Bhatti, A. Raheem, T. Alam, M. O. Malik, and A. Munir, "Implementation of low cost non-DFT based Phasor Measurement Unit for 50 Hz power system," in 2016 International Conference on Computing, Electronic and Electrical Engineering (ICE Cube), 2016, pp. 120–125.
[14] D. M. Laverty, R. J. Best, P. Brogan, I. Al Khatib, L. Vanfretti, and D. J. Morrow, "The OpenPMU platform for open-source phasor measurements," IEEE Trans. Instrum. Meas., vol. 62, no. 4, pp. 701–709, 2013.
[15] S. Li et al., "PMU-based harmonic phasor calculation and harmonic source identification," in 2018 2nd IEEE Conference on Energy Internet and Energy System Integration (EI2), 2018, pp. 1–6.
[16] E. M. Stewart, S. Kiliccote, C. M. Shand, A. W. McMorran, R. Arghandeh, and A. von Meier, "Addressing the challenges for integrating micro-synchrophasor data with operational system applications," in 2014 IEEE PES General Meeting| Conference & Exposition, 2014, pp. 1–5.
[17] L. L. F. Blas, P. E. D. C. Bundoc, K. C. L. Melchor, and J. R. C. Orillaza, "Micro-Synchrophasors for Event Detection in an Electric Power System Simulator," in 2018 IEEE Innovative Smart Grid Technologies-Asia (ISGT Asia), 2018, pp. 178–183.
[18] D. Devaraj, M. K. Paramathma, and I. Rajalakshmi, "Simulation and implementation of phasor measurement unit for wide area monitoring system," in 2017 IEEE International Conference on Intelligent Techniques in Control, Optimization and Signal Processing (INCOS), 2017, pp. 1–6.
[19] J. Lee, "Automatic fault location on distribution networks using synchronized voltage phasor measurement units," in ASME Power Conference, 2014, vol. 46094, p. V002T14A008.
[20] A. Shahsavari, M. Farajollahi, E. Stewart, C. Roberts, and H. Mohsenian-Rad, "A data-driven analysis of lightning-initiated contingencies at a distribution grid with a PV farm using micro-PMU data," in 2017 North American Power Symposium (NAPS), 2017, pp. 1–6.
[21] W. Yao et al., "A fast load control system based on mobile distribution-level phasor measurement unit," IEEE Trans. Smart Grid, vol. 11, no. 1, pp. 895–904, 2019.
[22] H. Jiang, Y. Zhang, J. J. Zhang, and E. Muljadi, "PMU-aided voltage security assessment for a wind power plant," in 2015 IEEE Power & Energy Society General Meeting, 2015, pp. 1–5.
[23] L. Vanfretti, M. Baudette, I. Al-Khatib, M. S. Almas, and J. O. Gjerde, "Testing and validation of a fast real-time oscillation detection PMU-based application for wind-farm monitoring," in 2013 First International Black Sea Conference on Communications and Networking (Blackseacom), 2013, pp. 216–221.
[24] A. Aligholian, A. Shahsavari, E. Stewart, E. Cortez, and H. Mohsenian-Rad, "Unsupervised Event Detection, Clustering, and Use Case Exposition in Micro-PMU Measurements," IEEE

Trans. Smart Grid, 2021.
[25] P. Khaledian, A. Aligholian, and H. Mohsenian-Rad, "Event-Based Analysis of Solar Power Distribution Feeder Using Micro-PMU Measurements," in 2021 IEEE Power & Energy Society Innovative Smart Grid Technologies Conference (ISGT), 2021, pp. 1–5.
[26] G. Sanitha and R. M. Shereef, "Micro-pmu data based location identification and classification of events," in 2020 IEEE International Conference on Power Electronics, Smart Grid and Renewable Energy (PESGRE2020), 2020, pp. 1–6.
[27] N. Duan and E. M. Stewart, "Frequency event categorization in power distribution systems using micro pmu measurements," IEEE Trans. Smart Grid, vol. 11, no. 4, pp. 3043–3053, 2020.
[28] R. Dutta, S. Som, S. Chakrabarti, A. Sharma, and A. K. Srivastava, "Event Detection and Localization in Active Distribution Networks using μ PMUs," in 2020 IEEE Power & Energy Society General Meeting (PESGM), 2020, pp. 1–5.
[29] A. L. Liao, E. M. Stewart, and E. C. Kara, "Micro-synchrophasor data for diagnosis of transmission and distribution level events," in 2016 IEEE/PES Transmission and Distribution Conference and Exposition (T&D), 2016, pp. 1–5.
[30] A. Shahsavari, M. Farajollahi, E. M. Stewart, E. Cortez, and H. Mohsenian-Rad, "Situational awareness in distribution grid using micro-PMU data: A machine learning approach," IEEE Trans. Smart Grid, vol. 10, no. 6, pp. 6167–6177, 2019.
[31] Q. Cui and Y. Weng, "Enhance High Impedance Fault Detection and Location Accuracy via $\mu $-PMUs," IEEE Trans. Smart Grid, vol. 11, no. 1, pp. 797–809, 2019.
[32] A. Kavousi-Fard, S. Nikkhah, M. Pourbehzadi, M. Dabbaghjamanesh, and A. Farughian, "IoT-based data-driven fault allocation in microgrids using advanced μPMUs," Ad Hoc Networks, vol. 119, p. 102520, 2021.
[33] D. Mukherjee, S. Chakraborty, and R. Banerjee, "Power System Fault Classification with Imbalanced Learning for Distribution Systems," in 2020 IEEE 7th Uttar Pradesh Section International Conference on Electrical, Electronics and Computer Engineering (UPCON), 2020, pp. 1–6.
[34] Y. Bansal and R. Sodhi, "A novel μPMUs assisted loss-of-mains detection technique for active distribution systems," Electr. Power Syst. Res., vol. 202, p. 107578, 2022.
[35] M. Farajollahi, A. Shahsavari, E. M. Stewart, and H. Mohsenian-Rad, "Locating the source of events in power distribution systems using micro-PMU data," IEEE Trans. Power Syst., vol. 33, no. 6, pp. 6343–6354, 2018.
[36] G. Cavraro and R. Arghandeh, "Power distribution network topology detection with time-series signature verification method," IEEE Trans. Power Syst., vol. 33, no. 4, pp. 3500–3509, 2017.
[37] G. Cavraro, R. Arghandeh, K. Poolla, and A. Von Meier, "Data-driven approach for distribution network topology detection," in 2015 IEEE power & energy society general meeting, 2015, pp. 1–5.
[38] R. Dutta, S. Chakrabarti, and A. Sharma, "Robust Topology Detection of Distribution Network," IEEE Trans. Power Syst., vol. 36, no. 6, pp. 5967–5970, 2021.
[39] M. G. M. Zanjani, K. Mazlumi, and I. Kamwa, "Application of μPMUs for adaptive protection of overcurrent relays in microgrids," IET Gener. Transm. Distrib., vol. 12, no. 18, pp. 4061–4068, 2018.
[40] M. S. Elbana, N. Abbasy, A. Meghed, and N. Shaker, "μPMU-based smart adaptive protection scheme for microgrids," J. Mod. Power Syst. Clean Energy, vol. 7, no. 4, pp. 887–898, 2019.
[41] C. Lin, W. Wu, and Y. Guo, "Decentralized robust state estimation of active distribution grids incorporating microgrids based on PMU measurements," IEEE Trans. Smart Grid, vol. 11, no. 1, pp. 810–820, 2019.
[42] S. Basumallik, Y. Liu, D. Katramatos, and S. Yoo, "Real-Time Distribution State Estimation with Massive μPMU Streaming Data," in 2020 IEEE Power & Energy Society Innovative Smart Grid Technologies Conference (ISGT), 2020, pp. 1–5.
[43] A. Shahsavari et al., "Distribution grid reliability versus regulation market efficiency: An analysis based on micro-PMU data," IEEE Trans. Smart Grid, vol. 8, no. 6, pp. 2916–2925,

2017.
[44] R. J. Albuquerque, "Micro-PMU-based Voltage Drop Assessment for Improving Voltage Regulation in Primary Radial Distribution Feeders," in 2021 IEEE Power & Energy Society Innovative Smart Grid Technologies Conference (ISGT), 2021, pp. 1–5.
[45] G. P. Kumar and P. Jena, "Pearson's Correlation Coefficient for Islanding Detection Using Micro-PMU Measurements," IEEE Syst. J., 2020.
[46] K. Al-Maitah and A. Al-Odienat, "Wide Area Protection Scheme for Active Distribution Network Aided $\mu\text {PMU} $," in 2020 IEEE PES/IAS PowerAfrica, 2020, pp. 1–5.
[47] M. Jamei et al., "Micro synchrophasor-based intrusion detection in automated distribution systems: Toward critical infrastructure security," IEEE Internet Comput., vol. 20, no. 5, pp. 18–27, 2016.
[48] S. Dutta, P. K. Sadhu, M. J. B. Reddy, and D. K. Mohanta, "Smart inadvertent islanding detection employing p-type μPMU for an active distribution network," IET Gener. Transm. Distrib., vol. 12, no. 20, pp. 4615–4625, 2018.
[49] M. H. F. Wen, R. Arghandeh, A. von Meier, K. Poolla, and V. O. K. Li, "Phase identification in distribution networks with micro-synchrophasors," in 2015 IEEE Power & Energy Society General Meeting, 2015, pp. 1–5.
[50] A. Almutairi, O. Alrumayh, S. Alyami, N. Albagami, and M. Hossein, "A blockchain- enabled secured fault allocation in smart grids based on μPMUs and UT," IET Renew. Power Gener.
[51] M. Hojabri, U. Dersch, A. Papaemmanouil, and P. Bosshart, "A comprehensive survey on phasor measurement unit applications in distribution systems," Energies, vol. 12, no. 23, p. 4552, 2019.
[52] T. Hammons, K. P. Wong, and L. L. Lai, "Developments in power generation and transmission infrastructures in china," in 2007 IEEE Power Engineering Society General Meeting, 2007, pp. 1–5.
[53] J. Y. Cai, Z. Huang, J. Hauer, and K. Martin, "Current status and experience of WAMS implementation in North America," in 2005 IEEE/PES Transmission & Distribution Conference & Exposition: Asia and Pacific, 2005, pp. 1–7.
[54] M. Rihan, M. Ahmad, and M. S. Beg, "Phasor measurement units in the Indian smart grid," in ISGT2011-India, 2011, pp. 261–267.

Analog Performance Analysis of a Novel 5nm Stacked Oxide Top Bottom Gated Junctionless FinFET

L Gangwani[1] and S Hajela[1]

[1]Applied Physics Department, Delhi Technological University, New Delhi, India

E-mail: *lakshyagangwani1@gmail.com*

Abstract: This paper depicts the analog investigation of a Novel Stacked Oxide Top Bottom Gated Junctionless (TBG-JL) Fin-shaped Field Effect Transistor (FinFET) structure. The structure is designed in this way to enhance the switching performance and curtail the Short Channel Effects (SCEs). The study is done on Silvaco Atlas TCAD tools for the Novel Stacked Oxide FinFET device and analyzed against Top-Bottom Gated Junctionless FinFET and Traditional FinFET. The Novel device shows 280 times improvement in switching ratio when stacked against the Traditional JL FinFET at identical gate length at room temperature (300K). The results from device simulation affirm that the Novel device has better analog performance over Traditional Junctionless FinFET and diminishes the Short Channel Effects (SCEs) a cut above the Traditional Junctionless FinFET.

Keywords: *FinFET, Top-Bottom Gated, Junctionless, Stacked Oxide, Switching Ratio*

1. INTRODUCTION

New technologies are being developed every day. These progressions and types of developments are intended to serve the larger part and make daily tasks more efficient. Amidst such progressions it ends up being extensively more fundamental to work on the usefulness and resource use of our devices [1]. Various headways are being designed to achieve precisely that. This trend of downsizing the devices to achieve more portability and efficiency starting right from MOSFET devices till SOI and FinFET helps in developing far more capable devices which are handheld at the same time [2]. But reducing the device dimension comes at a cost. As we scale down, the dominance of certain undesirable attributes becomes more prominent. These undesirable attributes comprises of various short channel effects such as Velocity Saturation, Subthreshold Swing and DIBL. These undesirable attributes can cause delays and inefficient power use [3]. The new gate configurations for FinFET technology is a step in the right direction. These progressions are being embraced by different goliaths in the Chip and Semiconductor fabricating industry. Thusly in this research article, we have investigated the analog performance parameters of an Innovative Stacked Oxide Top-Bottom Gated (TBG) Junctionless (JL) FinFET [4]. The structure has been contrasted with the TBG-JL FinFET and a Traditional Junctionless FinFET structure to better glean the merits of the novel structure in terms of improved switching and reduced short channel effects.

2. DEVICE DESIGN

Figures 1, 2 and 3 outlines the Innovative FinFET structure. The oxide layer has been stacked i.e. one layer of Hafnium Oxide (HfO_2) placed over layer of Silicon Dioxide (SiO_2). Stacking is done to further develop the Ion/Ioff ratio and productivity of the gadget.

Gate is present only above the substrate in case for the Traditional Junctionless FinFET, wherein the Novel Stacked Oxide TBG JL FinFET has gates both above and below the substrate, subsequently working on the control of gate over the channel. This aides in improving the gadget efficiency [1].

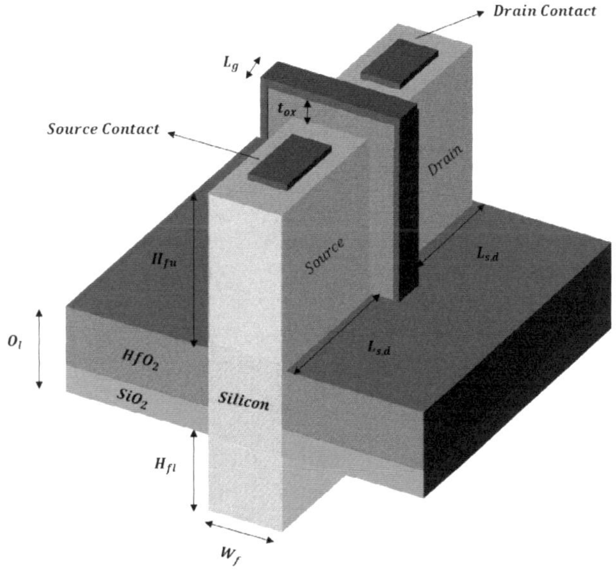

Figure 1. Systematic 3D structure of the Novel FinFET Structure

The Novel structure has two layers of oxide. 7nm HfO_2 oxide layer is placed on 3nm SiO_2 oxide layer. Hafnium Oxide (HfO_2) has dielectric constant in the range of 20 to 25 whereas for Silicon Dioxide (SiO_2) it is around 3.7 [5]. The dielectric constant of Hafnium Oxide (HfO_2) is taken as 25 for the simulation of this device structure using the TCAD tools [6]. This stacking of oxide layer alongside the fins reaching out on both the sides of the oxide helps in improving the analog performance and switching ratio of the device when stacked against the Conventional JL FinFET. Additionally it upgrades the gate controllability of the gadget when stacked against the TBG JL FinFET [2].

The Hafnium Oxide (HfO_2) has a band gap of 5.3- 5.7 eV and can behave as a dielectric insulator. HfO2 has high dielectric constant, great thermal stability and also lower values of leakage current [7].

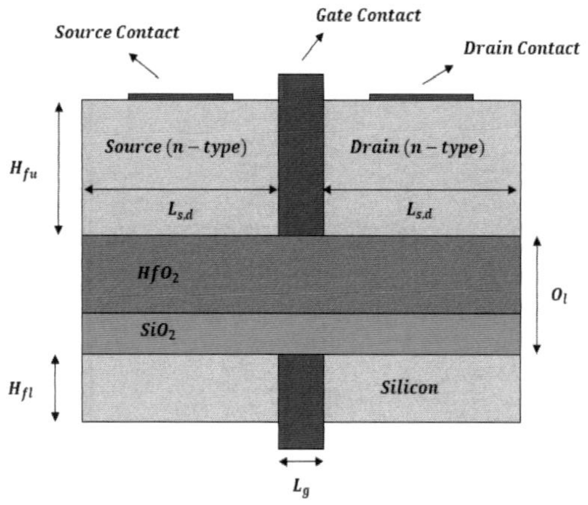

Figure 2. Horizontal View of the Novel FinFET Structure

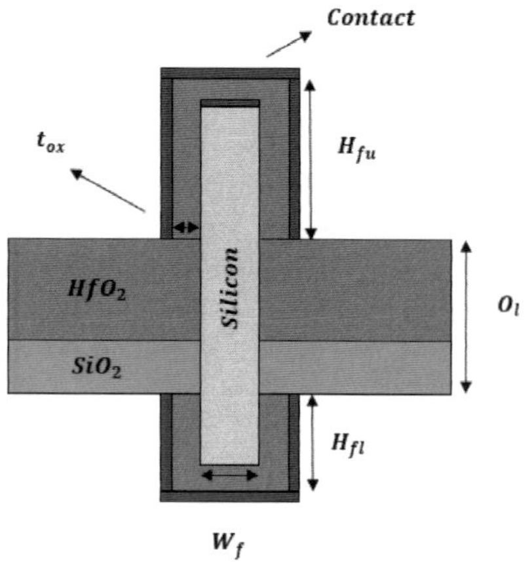

Figure 3. Vertical View of the Novel FinFET Structure

The Novel device structure simulated is of n-type with 5.0×10^{18} (cm^{-3}) doping concentration for all the regions of source, drain and channel.

So as we increase the doping from 5.0×10^{18} (cm^{-3}) to 5.0×10^{19} (cm^{-3}), the threshold voltage dips from 0.42V to 0.39V but the Switching ratio decreases by 16 times.

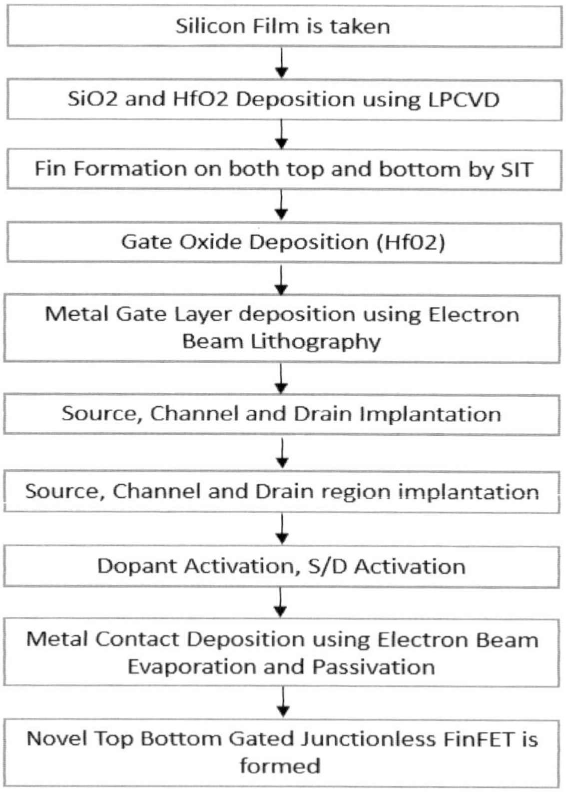

Figure 4. Fabrication Flowchart for Proposed TBG JL FinFET

Furthermore increasing the doping concentration shows improvement as far as Transfer Characteristics and Transconductance are concerned as depicted in the **figures 5 and 6**.

As we increase the doping concentration of the device, more charge carriers are available and hence the device exhibits better output current for same gate voltage as shown in **Figure 5** and similarly the Transconductance which varies directly with Output current increases as well as portrayed in **Figure 6**.

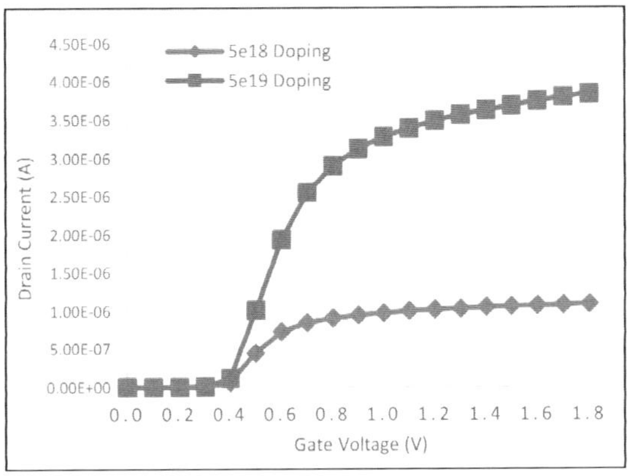

Figure 5. Variance of Transfer Characteristics with doping

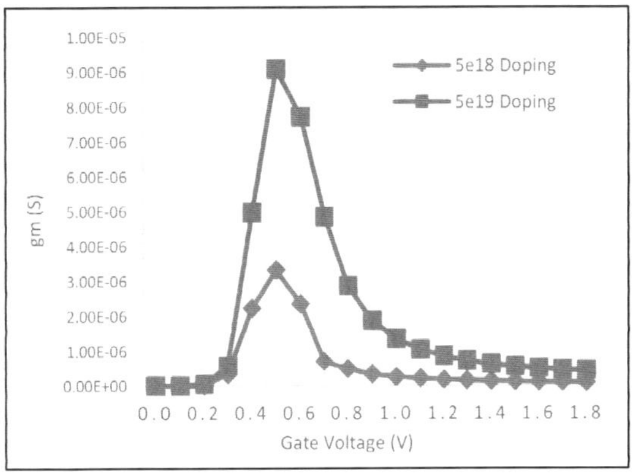

Figure 6. Variance of Transconductance with doping

The Top-Bottom Gated FinFET devices have fins on both sides of the oxide layer and thus have more device heights than the Traditional JL FinFET structure [2]. The Novel FinFET structure and TBG JL FinFET have device heights of 33nm due to presence of lower fin.

Introduction of High K HfO_2 gate dielectrics in FinFET devices prompts improvement in the analog efficiency and reduces power consumption [8]. The device parameters for the Novel structure are portrayed with the help of **Table 1**.

Table 1. Parameter Specification for Novel Stacked Oxide Top Bottom Gated FinFET

Parameter	Dimensions
Gate Length (L_g)	5nm
Upper Fin Height (H_{fu})	12nm
Lower Fin Height (H_{fl})	5nm
Fin Width (W_f)	5nm
Oxide Layer Thickness (O_l)	7nm HfO_2 + 3nm SiO_2
Substrate Width (W_s)	5nm
Source, Drain Length ($L_{s,d}$)	20nm

3. SIMULATION METHODOLOGY

The innovative gadgets have been simulated using the Silvaco Atlas TCAD tools [6]. A certain procedure is followed for developing FinFET devices using Silvaco tools. Above all the Construction Specification for the device is finished i.e. definition of the mesh grid structure for the FinFET lattice, definition of the doping concentration of regions and elucidating the regions itself. Secondly Definition of the Materials is done and physical models for the device are selected [6]. Finally the

Specification of Solution and Result Analysis is done for visualization of the results and analysis of various parameters [6].

4. RESULTS AND ANALYSIS

When a switch is on it has I_{ON} which is the ON state current and similarly it has I_{OFF} which is the off state current [9]. In an optimal circumstance, ON current is decided by the outside circuit while the OFF current should be zero [10].

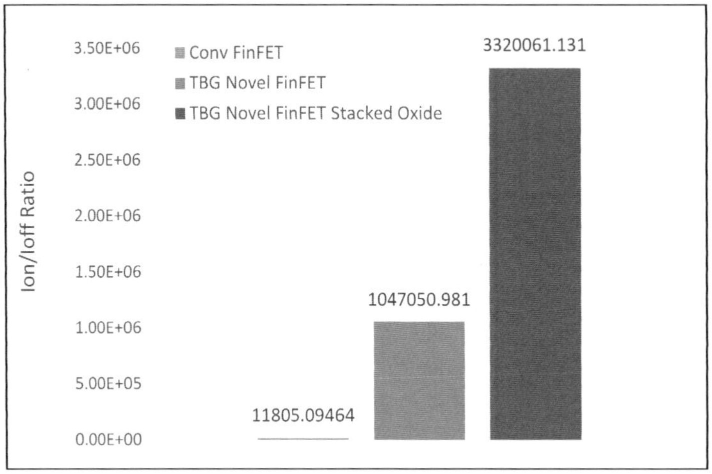

Figure 7. Switching Ratio for the three devices at 300K

For the common solid state technologies like MOSFET or FinFET that we use to emulate switches, the off current veers off from ideal zero and has a negligible value [11].

Figure 7 depicts the I_{ON}/I_{OFF} (Ion to Ioff) ratio of Traditional Junctionless FinFET, Novel TBG FinFET and the Innovative Stacked Oxide FinFET structure. It reveals that there is a great improvement in switching ratio for Innovative Stacked Oxide FinFET structure from that of Traditional JL FinFET and we observe that it improves by almost 280 times for Innovative Stacked Oxide FinFET structure when stacked against the Traditional JL FinFET and 3 times improvement as compared to TBGJL FinFET structure. Thus the proposed Stacked Oxide device has a higher on state current and better performance with reduced discharging ability due to the addition of high dielectric constant oxide. [10].

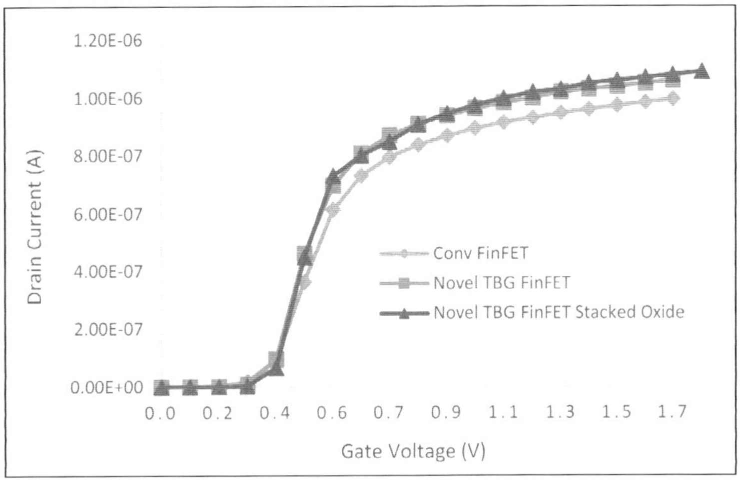

Figure 8. Transfer Characteristics Comparison for the devices at 300K

Figure 8 portrays the transfer characteristics of the Innovative Stacked Oxide FinFET structure and the same is shown for the other two devices for comparison [11]. It is clearly observed from the figure that the Innovative Stacked Oxide FinFET structure has improved yield current when stacked against the other two devices at the same given gate voltage at 300K. Thusly it can be inferred that the Innovative Stacked Oxide FinFET structure has incremented transfer characteristics [12].

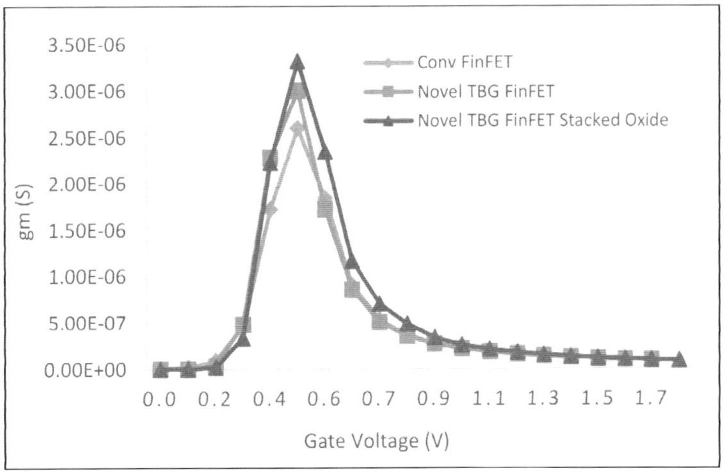

Figure 9. Transconductance Comparison for the devices at 300K

The transconductance as given by equation (1) (g_m- Input conductance) increments relatively with the linear gain in a gadget [3]. It connotes the variety in yield current with relating changes in input voltage V_{gs} (Gate Voltage) when V_{ds} (Drain Voltage) is kept unaltered [4].

The equation for Transconductance is:

$$g_m = \frac{\partial I_d}{\partial V_g} \tag{1}$$

IOP Conf. Series: Materials Science and Engineering 1258 (2022) 012046 doi:10.1088/1757-899X/1258/1/012046

Figure 9 gives the transconductance (g_m) of the three device. The Innovative Stacked Oxide FinFET structure gives incremented values of g_m for similar input voltage when stacked against the other two devices [13]. In this way the proposed gadget has better qualities with reference to transconductance. It moreover shows additionally created control of gate voltage over channel current [14].

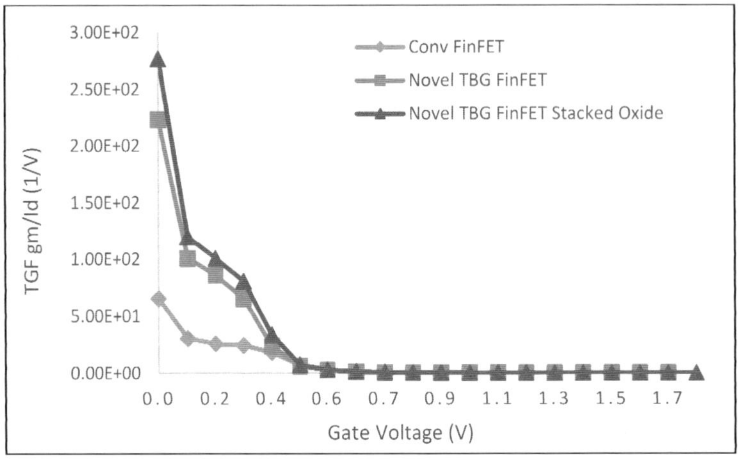

Figure 10. Transconductance Generation Factor Comparison for the devices at 300K

The ratio of g_m (Transconductance) to I_d (Drain current) is Transconductance Generation Factor (TGF) as depicted by equation (2) [15].

$$TGF = \frac{g_m}{I_d} \tag{2}$$

TGF demonstrates how successfully and deliberately the current is used to get a specific worth of transconductance [16]. The upside of high TGF is in the activity of circuits working at low supply voltage. And so we have superior performance [13]. **Figure 10** represents the variance in TGF as for changes in V_{gs} for the three devices. It tends to be seen that the Innovative Stacked Oxide FinFET structure has higher values of TGF when stacked against the other two devices and thus has more efficiency and performance values [16].

5. CONCLUSION

This work depicts the superiority of the Innovative Stacked Oxide TBG FinFET structure in terms of Analog Performance over the other two devices. 280 times improvement in switching ratio is observed for the Innovative Stacked Oxide FinFET structure when stacked against Traditional JL FinFET at 300K [1]. Further, enhanced analog performance is observed for the Innovative Stacked Oxide FinFET structure in terms of Transconductance and Transconductance Generation Factor when stacked against the other two devices [2]. Thus the device is an excellent choice for Analog Performance at Room Temperature (300K) and helps in diminishing Short Channel Effects a notch better.

References

[1] Gangwani L et al 2021 *J. Phys.: Conf. Ser.* **1921** 012100
[2] Gangwani L, Chakravarti R and Chaujar R 2021 *Devices for Integrated Circuit (DevIC) IEEE* 2021 pp. 1-5

[3] Kim S Y, Seo J, Yoon Y J, Yoo G, Kim Y J et al. 2014 *Journal of Semiconductor Technology and Science*. **14**. pp. 508-517
[4] Bughio A, Guerrieri S D, Bonani F and Ghione G 2017 *IEEE Electron Device Letters*, vol. **38**, no. 8, pp. 1004-07
[5] Kailasam M and Govindasamy M 2020 *International Journal Of Scientific & Technology*
[6] SILVACO I 2011 *ATLAS User's Manual* Santa Clara, CA, Ver., vol. **5**
[7] Boukortt N, Patanè S, and Crupi G 2019 *Silicon*, pp. 19–20, 2019
[8] Misra S, Mohan S B, Bara B, Kumar S S and Kumar S P *2018 IEEE Electron Devices Kolkata Conference (EDKCON)*, pp. 234-41
[9] Kaur N, Rattan M and Gill S S 2019 *Arab J Sci Eng* **44**, pp. 3101–16
[10] Singh M, Mishra S, Mohanty S S and Mishra G P 2016 *Advances in Natural Sciences: Nanoscience and Nanotechnology,* vol. **7**, no. 1, 2016
[11] Tripathi S, Gupta N, Tripathi P and Chaujar R 2019 *IEEE* 1-4
[12] Madan J and Chaujar R 2016 *Appl Phys A Material Science Process* **122**(11):973
[13] Kumar A, Gupta N and Chaujar R 2016 *Microsyst Technol* **22**, 2665–71
[14] Rai S, Sahu J, Dattatray W, Mishra R and Tiwari S 2012 *ISRN Electronics* 10.5402/2012/827452.
[15] Kranti A and Armstrong G 2007 *IEEE Trans. Electron Devices*, vol. **54**, no. 12, pp. 3308–16, 2007
[16] Kumar B, Kumar A and Chaujar R 2020 *IEEE VLSI Device, Circuit and System Conference* pp. 159–163

Design of Digital Meters for Industries using Data Acquisition System

J S Nancy Mary[1], A P Abishek Raj[2],A Inba Rexy[3], S Sathya Bharathy[4]

[1,3,4] Department of EEE, Loyola ICAM College of Engineering and Technology, Chennai, India

[2] Robert Bosch Engineering and Business Solutions. Pvt. Ltd., Coimbatore, India

nancymary.js@licet.ac.in

Abstract. Energy Monitoring is one of the most important routine that is carried out in every industry as a key process of Energy Management. The prime reason for energy metering is to ensure that the processes are done efficiently. Through energy monitoring, the energy consumption on a particular period is easily obtained thus helping to find the energy usage in a day and taking steps to improve energy efficiency. The energy consumption in a particular period of time is obtained using a digital meter. The meters in today's industries are digital meters and support serial communication. The energy meters present in the plant are connected to the server and are made to transfer their data directly to the server. This eliminates the man-hours wastage of every employee who performs this energy monitoring. The Man hours for a medium sized plant with 75 Digital meters spread across the plant was reduced to 2 hours/day.

Keywords—MAX 485 Converter, Arduino, Modbus RTU, RS- 485

1. Introduction

Electrical power has become vital to mortal existence and advancement. Beyond from efforts to fulfil thriving requirement, automation in the energy distribution is also inevitable to facilitate populace standardization. Digital meter reading by human operator is cumbersome to satisfy the emerging future industralization needs. So there is rising requirements for Automatic Meter Recording (AMR) systems which collects meter readings automatically. Collection of meter readings is also inefficient, because a meter reader has to physically be onsite to take the readings. This method of collecting meter readings becomes more troublesome and expensive when readings have to be collected from huge, and often coherent faraway areas. Manual meter readings are also time consuming that causes loss of energy and time. Hence, a system is proposed to eliminate this problem.[1]

Since 1983 for serial communication uses Recommended Standard #485 (RS-485) and it's been around since 1983.The device has terminal block connector and 9 pin D connector which uses serial ports. Nowadays, TIA-485-A standard maintained by Telecommunications Enterprises. The data transmission speed is around 10 Mbit/s up to 35m and 100 Kbit/s at 1200m. The cost is low as it uses two wire twisted pair cable and also enables high speed networks. The overheads of RS 485 cabling is little and also the cost is low. The speed of the network is 35 Mbit/s up to 10m and 100 Kbit/s at 1,200m cable but the cable length should not exceed 108 m such that it is not susceptible to electromagnetic interference wherein the factories will have electromagnetic inference sources. The cable uses differential twisted structure which makes it noise resilient. It has maximum 1200m as network length. [2,3]

IOP Conf. Series: Materials Science and Engineering 1258 (2022) 012047 doi:10.1088/1757-899X/1258/1/012047

2. Data Acquisition

Data acquisition is the procedure of sampling signals that would include actual global environment conditions and convert the resulting samples into digital numeric values that can be modified by a computer. DAS or DAQ - Data acquisition systems typically convert analog waveforms into digital values for processing. Data acquisition systems can be classified into PC (Personal Computer) Data Acquisition, USB (Universal Serial Bus) Data Acquisition, SCADA Systems. PC data acquisition systems and components download gathered database by hosting the computer. It employs Personal Computer Technique, DAQ Software, DAQ Hardware and DAQ device drivers were the parts of the data acquisition systems. USB data acquisition is a general technique conceptually employs a PC data acquisition. Supervisory Control and Data Acquisition (SCADA) is a control system architecture that uses computers, networked data communications and graphical user interfaces for high-level process supervisory management, but uses other peripheral devices such as programmable logic controller (PLC) and discrete PID (Proportional Integral Derivative) controllers to interface with the process plant or machinery.[4-6]

3. Types of Data Acquisition

The most common data acquisition processes are:
PC Data Acquisition
USB Data Acquisition
SCADA Systems

3.1. PC Data Acquisition

PC data acquisition is a phrase that encompasses all data acquisition systems and devices that need a interconnection to host computer for the appropriate download of gathered data and mainly function to control the plant. Most data acquisition systems employ some sort of PC method. It usually consists of
DAQ Hardware
DAQ device drivers
DAQ Software

3.1.1. DAQ Hardware. DAQ hardware is interconnects personal computer and signal through the modules which can be interfaced to the computer's ports (USB, serial, parallel etc.) or cards affined to slots [PCI-E (Peripheral Component Interconnect Express), S-100 (Standard 100) bus, PCI, Apple Bus, MCA (Micro Channel Architecture), ISA (Industry Standard Architecture) , etc.] in the motherboard. An external breakout is necessary as the space available on the back of the PCI card is little for complete interconnections. Shielding is required for the cables connecting the box and Personal Computer. DAQ Cards constitutes of variety of elements [ADC (Analog to Digital Converter), TTL-IO (Transistor Transistor Logic), DAC(Digital to Analog Converter), High Speed Timers, multiplexer, RAM(Random Access Memory)].Those are approachable through micro controller by small programs. A controller is less expensive compared to the CPU, more adaptable than a hard-wired logic, more admissible to block with simple polling loops. For example: Waiting for the ADC, to finish Waiting for a trigger, , move value to RAM, switch multiplexer, get TTL input, starting the ADC, looking up the time, let DAC proceed with voltage ramp.

3.1.2. DAQ device drivers. For the DAQ hardware to perform with personal computer, DAQ drivers are required. This driver executes low level register reads and writes on the hardware, while revealing API for establishing user applications in numerous programs,

3.1.3. DAQ Software. Specialized DAQ software might be dispatched with the DAQ hardware. Software tools employed for building large-scale data acquisition systems include EPICS (Experimental Physics and Industrial Control System). The programming environments used to build DAQ applications include MATLAB, Visual C++, LabVIEW, Visual Basic, and ladder logic.

3.2. USB Data Acquisition

USB data acquisition is a general technique conceptually employs a PC data acquisition because the USB device works by connecting the host computer straightaway. It employs a USB device is a serial bus for interconnecting and managing conservation amidst the host computers and data acquisition devices. Higher bandwidth (upto 12MB/s) and capability to supply power to incidental devices were the main exquisite characteristics of the USB data acquisition systems. It employs one cable to link the computer as the USB devices are used as power supply.

3.3. SCADA Systems

Supervisory Control and Data Acquisition (SCADA) is a control system architecture that uses computers, networked data communications and graphical user interfaces for high-level process supervisory management, but uses other peripheral devices such as programmable logic controller (PLC) and discrete PID controllers to interface with the process plant or machinery. The use of SCADA has been also considered for management and operations of project-driven-process in construction. SCADA Systems are complicated systems that engages a multiple of devices and programming to control, monitor and command systems. The operators can supervise massive numerous sites dispersed over long distances. For supervising and controlling they employs a variety of devices and software, consisting of computer ports (serial ports, USB ports, parallel ports, etc.), graphical user interfaces, and networked data communications. Programmable logic controllers or Discrete PID controllers are used to interconnect with machinery or process plants. Remote terminal units and network modules are linked with actuators and sensors to interpret real time calculations.

4. Existing System

In industries, different kinds of energy meters and multifunction meters are used depending on their applications. The existing system is a data acquisition system that focuses on data acquisition for a particular model of energy meters or multifunction meters only. In this system, the model that is more in number is identified and data is collected those meters. Usually, Energy meters show the power only whereas the Multifunction meters show a variety of values. Hence, the Multifunction Meters grab more attention in data acquisition.

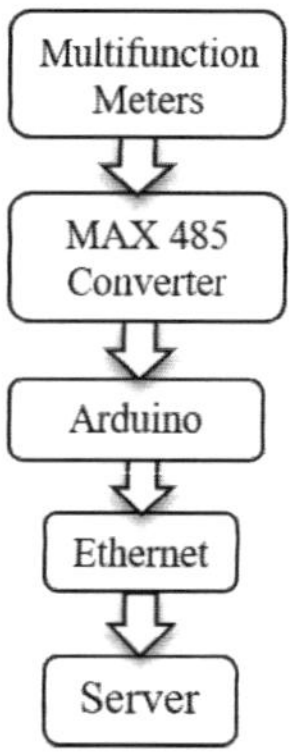

Figure 1. Block Diagram of Existing System

Multifunction meters sense the necessary data from the system. The existing system has a number of meters connected in a multidrop fashion(parallel). Since, these digital meters support RS-485 communication the associated Modbus RTU (Remote Terminal Unit) protocol is used. According to this protocol, a signal is sent from the master to the slave which makes the slave respond to the signal sent.

Here, the slave is usually the multifunction meter which sends the signal when signaled. The MAX 485 transceiver converts TTL to RS-485 and vice versa. The microcontroller processes TTL signals and not RS-485 signals and the master sends signals to the slave in the RS-485 format. This serves as a mediator between microcontroller and the slave. Arduino analyzes the signals and controls the overall process. The IC MAX 485 is specially designed for microcontroller-controlled conversion. The microcontroller initiates the data transfer by sending a control signal to the slave. The slave responds by sending the required data to the microcontroller and the data gets stored in the register map of the Arduino. Arduino sends the signals to the computer through the Ethernet Shield. The ethernet shield helps in data transfer as a LAN (Local Area Network) connector. The data is uploaded to the server using the ethernet shield. The server displays the current reading when initiated.[7-9]

5. Proposed System

In most of the industries, meters are scattered depending on their area of usage and the point of power output except those in the PCC (Power Control Centre). This project mainly focusses on collecting data from every meter available in the plant and uploading to it to the server for analysis and recording. The readings are periodically recorded. The main advantages are, the main hours required for recording the data is eliminated. Since, most of the data from the meters are utilized the exact values of power is obtained. The data can be accessed by any system within the network and has the IP (Internet Protocol) address.

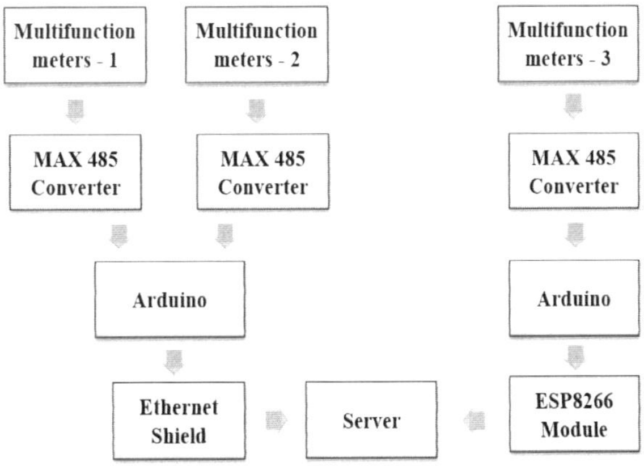

Figure 2. Block Diagram of Proposed System

The Data Acquisition System in this case works with RS-485 communication. The Remote Terminal Units with this type of communication is controlled with the Modbus-RTU protocol wherein the meters are connected in a multidrop fashion as in the existing system. The enhancement in this model is that, different meters are categorized based on their models initially. These meters are then contacted by the Arduino as per the Modbus-RTU protocol. The data is uploaded to the server either by the ethernet shield if the meters are in the PCC or by the ESP8266 Wi-Fi Module if the meters are in remote location within the plant. The difference is that the whole plant is given a particular data acquisition system to store its data by means of the Wi-Fi modules that connect to the server wirelessly and EEPROM (Electronic Erasable Programmable Read Only Memory)that enhances the data storage ability. The Arduino program is done to allow the microcontroller to send data in the form of packets which contain the slave id, address and the data to be stored into the slave. The Arduino sends data in the form of packets and receives back in the same way.[10,11]

5.1. Algorithm

Step 1 Start

Step 2 Import the necessary modules.
Step 3 Define the necessary constants and necessary elements.
Step 4 Define the functions used.
Step 5 Initialize server and packets.
Step 6 Configure connection between master and slave.
Step 7 Update values from the slave to a temporary memory.
Step 8 Store the data in the EEPROM.
Step 9 Start webpage transmission in HTTP(Hyper Text Transfer Protocol).
Step 10 Print the data in the webpage.
Step 11 Check the present number of packets sent.
Step 12 If the number of packets has reached maximum, skip to Step 14.
Step 13 If the number of packets is not maximum, skip to Step 7.
Step 14 If the digital meter read the values, skip to Step 16.
Step 15 If the digital meter does not read the values, skip to Step 5.
Step 16 Close connection with the server.
Step 17 Stop

5.2. Flow Chart

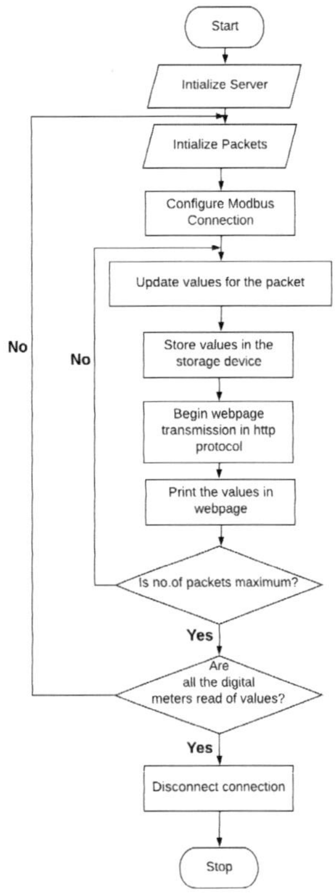

Figure 3. Flow Chart of Code Implementation

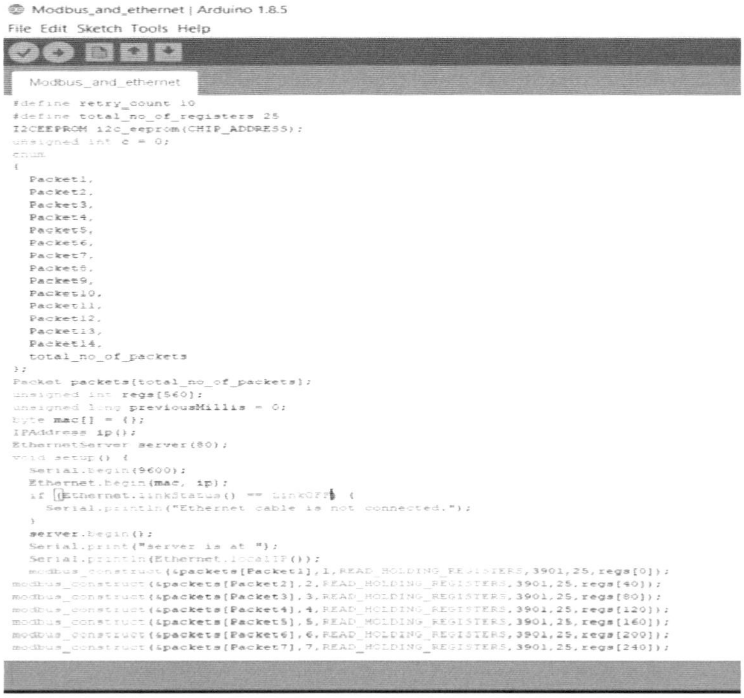

Figure 4. Screenshot of the program developed

5.3. Process Code Description

The flowchart of this project is clearly understood by understanding the modules that are added to it. The modules that are added to this Arduino Code are:

1. SimpleModbusMaster.h
2. Hardware Serial.h
3. Ethernet.h
4. ESP8266WiFi.h
5. I2CEEPROM.h

5.3.1. SimpleModbusMaster.h. This module is the core of the system which enables the use of Modbus RTU protocol in the RS-485 communication. To communicate with a slave, you need to create a packet that will contain all the information required to communicate to that slave. The goal of the project is to support all usable MODBUS commands on Arduino and expose all Arduino pins so you can use an Arduino as an advanced automation controller for both analog/digital in/out. The library also contains some counters There are numerous counters for easy diagnostic. These are variables already implemented in a packet. You can set and clear these variables as needed.

The library works on the basis of 3 functions:

- modbus_construct()
- modbus_configure()
- modbus_update()

modbus_construct():

modbus_construct is a function which has the details of the source and destination of the data, the slave id and the function.This function initializes the process.

modbus_configure():

modbus_configure() configures the packet with the details of the transmission and reception. This function is responsible for the proper functioning of the process.

modbus_update():

modbus_update() does the data transmission and reception.

5.3.2. Hardware Serial.h. This module supports Serial transmission from the microcontroller to the hardware.
The functions that are used in the program are:

- Serial.println()
- Serial.print()

And certain background functions.

5.3.3. Ethernet.h. The ethernet module operates the ethernet shield connected to the microcontroller. It contains functions like:

- client.available()
- client.connected()
- client.read()
- client.println()

5.3.4. ESP8266WiFi.h. The ESP8266WiFi.h module has the features and functions of a Ethernet.h module and controls a Wi-Fi shield.

5.3.5. I2CEEPROM.h This module is used to read and write in a EEPROM. Server is normally the database that holds the recorded data. In order to achieve the result for our system, server is usually networked. A working connection is established to the server to communicate the results. The Multifunction meters communicate in terms of packets of definite byte sizes. The packets contain the functional data: that is necessary to fetch the data from the meters and also the user data: that is supposed to carry the meter readings is setup.
The Modbus Connection shall be verified. The meters respond with acknowledgement upon right packets of signals. The packets are made of elements of Modbus RTU Frame. Slave Id corresponds to the given Device Id. Function Code is the operation that is to be done using the packet. Data is where the Multifunction Meter drops your message. CRC (Cyclic Redundancy Check) is the corresponding error check.[12,13]

		Protocol Data Unit (PDU)			
Start	Slave ID	Function code	Data	CRC error check	Stop
3.5 Bytes	1 Byte	1 Byte	n Bytes	2 Bytes	3.5 Bytes
	Application Data Unit (ADU)				

Figure 5. Modbus RTU Frame

Upon acknowledgement from Multifunction Meters, data packets are filled with necessary information and the packets are sent and meter readings are received.
The received information is stored in 2 different methods that are Persistent Memory (External Flash Memory) and Database (Server). In order to use the Persistent Memory, separate handling of External Flash memory through SPI with Arduino is done. Any operation with External flash memory is monitored and enabled with respect to the registers of the device. The primary operations with Flash memory in order to retain data are storing, reading and rearranging.
In Flash Memory, programming is done in terms of page and erasing is done in terms of sectors. To enable wear levelling and to ensure the device lifetime, the persistent memory is filled from start of flash to end of flash and the same is repeated. Once the persistent memory has no more space , in order to store further data, rearrangement of memory segments is done. Rearrangement of data is done in order to preserve the latest record of readings and not to miss any of useful information. The oldest sectors of the persistent memory are then erased and storing data in the erased sectors continue. This process is continued. In order to ensure industry standards, quality checks like CRC is done. Different algorithms of CRC like CRC32 and CRC16 are being used.

Data Transmission through Ethernet and WiFi are done using Arduino modules which enable simple transmission. The web page shall be available in a particular IP address that is private to the institution. Once data is stored in the webpage for one among many values, the register address in the multifunction meter's memory is changed for the next reading. The packet is then filled with necessary data for the next reading and the packet is again transmitted to the Multifunction meter (MFM).
The Arduino is the master and the MFMs are slaves. In the areas where multiple MFMs are available, the code shall perform multiple sequences of similar set of operations in a sequential manner. All the information are transmitted and stored in a first come first served basis. In case of time limit, all data can be cumulatively stored and transmitted through media.
At the end of operations, the server is disconnected. The whole setup works in timely manner. This setup also has a timer which allows time calculation and the necessary information can be obtained upon proper input of time. Hence, being totally automated and reducing man-hour requirements for recording power usage.

6. Results
Results show the successful implementation of the system in the industry with the following output results. The output is supposed to contain all the data from the meter.

S.No	Meter No.	Date	Time	Apparent Power	Active Power	Reactive Power	PF-Avg.	Line to Line Voltage	Line to Neutral Voltage
1	1	24/02/19	1000	36000	34445	35880.21	0.96	415	230.94
2	2	24/02/19	1000	3400	3100	3229.167	0.96	415	231
3	3	24/02/19	1000	6000	5100	5339.2	0.95	415	229
4	4	24/02/19	1000	2000	2100	2339.2	0.94	415	230

Figure 6. Output Format – 1stpart

Avg. Current	Frequency	Apparent Power (R-Phase)	Active Power (R-Phase)	Reactive Power (R-Phase)	R-Phase PF	R-Y Phase Voltage	R-Phase to Neutral Voltage	R-Phase current	Apparent Power (Y-Phase)
86.45833	50	35434	33000	34375	0.96	415	230.94	82.83133	35000
7.781124	50	3200	3000	3157.895	0.95	415	229	7.011	3300
12.93595	50	5600	5551	5657.9	0.95	415	230	11.98	5900
7.23	50	1600	1551	1657.9	0.95	415	230	6.735097	1900

Figure 7. Output Format –2ndpart

Active Power (Y-Phase)	Reactive Power (Y-Phase)	Y-Phase PF	Y-B Phase Voltage	Y-Phase to Neutral Voltage	Y-Phase current	Apparent Power (B-Phase)	Active Power (B-Phase)	Reactive Power (B-Phase)	B-Phase PF	B-R Phase Voltage	B-Phase to Neutral Voltage
34000	36170.21	0.94	415	230.94	83.831	35555	34521	36337.89	0.96	415	230.94
3021	3146.875	0.96	415	229	6.967	3003	2876	2905.051	0.99	415	231
5621	5746.9	0.96	415	230	11.5989	5700	5521	5546	0.95	415	229
1621	1746.9	0.95	415	230	6.4351	1700	1521	1546	0.95	415	230

Figure 8. Output Format – 3rdpart

7. Conclusion
The proposed system fulfils the need of the project which primarily is storing data in the server. The heart of the project lies in conversion, processing and transmission of the data from multifunction meters to the server. This project focusses on data processing and transmission which reduces manual calculations that might go wrong due to errors. This system has become flexible to any part of the plant that has a connection to the server through a MODEM (Modulator-Demodulator)and hence, from every multifunction meter data can be retrieved. The existing system was able to access only a particular set

of meters which only reduced half the work. The major requirement of this project was to obtain data from every meter available in the plant. The multifunction meters in the Power Control Centre were of 3 models and the data from them were collected by an Arduino by making use of the transmission and receiving pins of Arduino Mega to obtain signals from different converters at a time to make full use of the processing capacity of Arduino. The data converted is stored in the EEPROM to increase the storage capacity of the system in order to perform actions with the data obtained. The converted data is transmitted wirelessly through ESP8266 Wi-Fi Module thus increasing flexibility of the system. The proposed system is the implementation of data retrieval and storage whereas the existing had limitations in terms of wiring and transmission. The proposed system solved the problem using different transmission media.

8. References

[1] D. Peng, H. Zhang, Jiannian Weng, Hui Li and Fei Xia, 2009, "Design and Development ofModbus/RTU Master Monitoring System Based on Embedded Power PC Platform." *IEEE International Symposium on Indust. Electron.*, pp. 2148-2152.

[2] G. Gabor and G. Livint, "Solution for Monitoring and Remote Control Based on Modbus RTU and S7-1200 PLC," 2021 International Conference on Electromechanical and Energy Systems (SIELMEN), 2021, pp. 462-465, doi: 10.1109/SIELMEN53755.2021.9600411.

[3] K. Medrano, D. Altuve, K. Belloso and C. Bran, "Development of SCADA using a RTU based on IoT controller," 2018 IEEE International Conference on Automation/XXIII Congress of the Chilean Association of Automatic Control (ICA-ACCA), 2018, pp. 1-6, doi: 10.1109/ICA-ACCA.2018.8609700.

[4] L. Xinyun, L. Huidan, Y. Hang, C. Zilan, C. Bangdi and Y. Yi, "IoT Data Acquisition Node For Deep Learning Time Series Prediction," 2021 2nd International Conference on Big Data Analytics and Practices (IBDAP), 2021, pp. 107-111, doi: 10.1109/IBDAP52511.2021.9552096.

[5] H. Li, "Research on Big Data Analysis Data Acquisition and Data Analysis," 2021 International Conference on Artificial Intelligence, Big Data and Algorithms (CAIBDA), 2021, pp. 162-165, doi: 10.1109/CAIBDA53561.2021.00041.

[6] Y. Cai, "Adaptive Data Acquisition Technology for Android System Based on Big Data Analysis," 2020 International Conference on Advance in Ambient Computing and Intelligence (ICAACI), 2020, pp. 118-120, doi: 10.1109/ICAACI50733.2020.00030.

[7] M. S. S. Tanjim, A. M. Saleque and M. A. Rahman, "The Data Acquisition and Control Unit Design for Demand Side Low-Cost Customized SCADA," 2021 2nd International Conference on Robotics, Electrical and Signal Processing Techniques (ICREST), 2021, pp. 229-232, doi: 10.1109/ICREST51555.2021.9331095.

[8] L. D. Cuayo, J. Kerbee Culla, J. Gualvez, S. E. Padua and R. John Gallano, "Development of a Wireless Microcontroller-based SCADA RTU," TENCON 2018 - 2018 IEEE Region 10 Conference, 2018, pp. 2566-2570, doi: 10.1109/TENCON.2018.8650114.

[9] K. Y. Lim and H. M. Looe, "Design and Implementation of Automatic Transfer Function with RTU Programmable Logics," 2021 9th International Conference on Smart Grid and Clean Energy Technologies (ICSGCE), 2021, pp. 25-30, doi: 10.1109/ICSGCE52779.2021.9621381.

[10] A. V. Zinkevich, "ESP8266 Microcontroller Application in Wireless Synchronization Tasks," 2021 International Conference on Industrial Engineering, Applications and Manufacturing (ICIEAM), 2021, pp. 670-674, doi: 10.1109/ICIEAM51226.2021.9446411.

[11] Yoppy, R. H. Arjadi, H. Candra, H. D. Prananto and T. A. W. Wijanarko, "RSSI

Comparison of ESP8266 Modules," 2018 Electrical Power, Electronics, Communications, Controls and Informatics Seminar (EECCIS), 2018, pp. 150-153, doi: 10.1109/EECCIS.2018.8692892.

[12] G. Schatzberger, F. P. Leisenberger, P. Sarson and A. Wiesner, "High efficient low cost EEPROM screening method in combination with an area optimized byte replacement strategy which enables high reliability EEPROMs," 2018 IEEE 36th VLSI Test Symposium (VTS), 2018, pp. 1-6, doi: 10.1109/VTS.2018.8368631.

[13] P. Sarson, G. Schatzberger and R. Seitz, "Automotive EEPROM Qualification and Cost Optimization," 2013 22nd Asian Test Symposium, 2013, pp. 105-106, doi: 10.1109/ATS.2013.28.

CLCLP: Overlapping Community Detection in Networks Using a Combination of Link Community and Label Propagation Algorithm

Jayaprakash R[1]

[1]Department of Computer Science,
Nallamuthu Gounder Mahalingam College, Pollachi, Tamil Nadu, India

jpinfosoft@gmail.com

Abstract. Overlapping Community Detection is a tremendous method for accepting a crucial shape and complicated network software. As of overlapping communities descriptions of real-world networks must subsist quantified for community identification, this research suggested a method for discovering overlapping communities termed the Combination of Link Community with Label Propagation (CLCLP) algorithm. CLCLP converts a node separation issue into a link division issue because of its simplicity and competency of the label propagation technique, an improved label propagation algorithm that prioritizes connections over nodes to perceive communities has been developed. The proposed CLCLP then employs post-processing to develop discovered overlapping communities by removing over-overlapping and inaccurate fragile secures partitions. When tested on a large scale of synthetic and real-world networks, the proposed CLCLP approach attains good precision (i.e., accuracy) and NMI (Normalized Mutual Information) in recognizing overlapping communities in networks

Keywords: Propagation, Link, Detection, Cluster, Networks, Label.

1. Introduction

Nodes in many real-world networks could logically fit in to more than on communitiy, and hence the communities may overlap. In public networks, an individual be a part of a community of unit members, a society of associates, or a neighbourhood of coworkers. An internet page can wrap subjects that are related with atypical communities in an information network. Community detection methods that have been used in the past have not been successful in revealing community overlaps. Individuals who are unable to discern community be related in networks with logically overlapping communities are missing out on critical network structural information [1]. As a result, overlapping community detection techniques have a larger set of people paying attention to them. Overlapping communities can be identified using unique methods. Instead of partitioning the nodes, the followig ways is accroding to screening the transformation of a network's edges into communities [2, 3]. To recognise the communities in a network, a range of detection of a community techniques, commonly referred to clustering algorithms, have been suggested. Different communities are given up by particular community detection algorithms because they use various definitions of a community. On an actual network, Figure 1 shows that how two fundamentally different community recognition algorithms recognise the same communities.

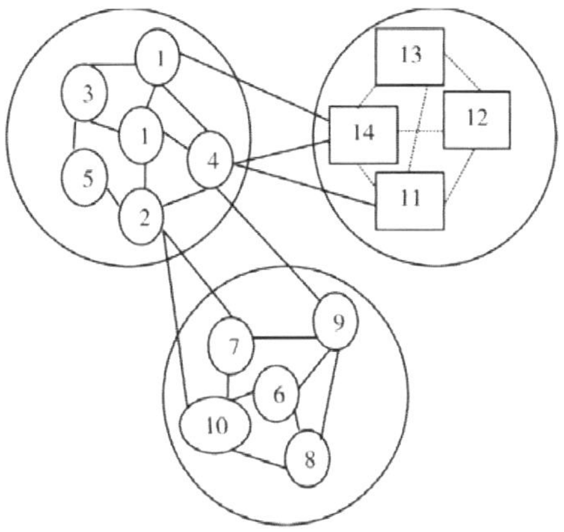

Figure 1. Zachary karate Club Network

Figure 1: The square and circular nodes in the Zachary karate club network represent two clusters of members. [4]

Previous community discovery algorithms assume that the entire network topology is known and accessible. To submit these algorithms as universal algorithms, because they requirement universal information from the whole network so as to categories of the communities therein networks. Because huge networks cannot have such information power, limiting algorithms are gaining popularity [5, 6]. By only looking at a small area of the network, local algorithms can extend a small number of known seed nodes into overlapping communities. Recognizing high-performing a large-scale real-world network communities are often computationally expensive and do not scale well. Parallelism is one way for processing the scalability of community detection. Parallelism can significantly boost up community discovery and furthermore required for dealing with massive amounts of data from the real world.

Local community identification methods are a quick and scalable tool for detecting communities. By just inspecting the neighborhood of 'seed nodes' in the network, local algorithms can finish calculations in parallel by starting with seed nodes and growing into communities. The enlargement of each node in the network into a community is a local method to limited community exposure. However, this method is computationally expensive and results in multiple replica communities. As a result, the goal of the test is to discover the ideal number of seeds to disperse into communities in order to cover the most popular nodes in a network.

The major information of community detection algorithms in difficult networks has been enlarged. Several of them can notice overlapping communities; a few can notice non-overlapping communities. Furthers can observe together overlapping and non-overlapping communities. To overcome the above exposed challenges, here an improved technique to recognize overlapping communities in a large scale network is proposed. The proposed method showed that by significant communities as partition of a network's links, and thus permitting entity nodes to appear in several communities, to calculate the extent to which every pair of communities in a network overlaps. This paper defines two actions of community overlap and applies them to the community formation of networks from special disciplines.

2. Related Works

(Wang, et.al., 2013) proposed the idea of local random walk and a novel similarity metric are introduced. Based on the novel similarity measurement, the dissimilarity index among every node of a network is considered firstly [9]. Then in order to maintain the novel distance node (similarity) to the greatest extent possible, the network formation is linked into low-dimensional space by the multidimensional scaling (MDS). Lastly, fuzzy c-means clustering is utilized to discover fuzzy communities in a network.

(C. Liu, J. Liu, and Z. Jiang, 2014) extended the unique similarity to the noticed similarity based on the social stability assumption [10]. Then, based on the signed similarity and the usual contradiction among negative and positive links, two goal functions are considered to form the problem of detecting communities in SNs as a multi-objective problem. Subsequently, the authors proposed a multi-objective evolutionary algorithm, called MEAs-SN. In MEAs-SN, to conquer faults of indirect and direct illustrations for communities, an indirect and direct pooled representation is considered.

(R. Aldecoa and I. Marin, 2014) reviewed that communities can be exactly distinguished by exploiting Surprise, a global network parameter [11]. Now, the authors presented "SurpriseMe", a tool that incorporating the results of seven of the finest algorithms obtainable to approximate the highest Surprise value. SurpriseMe also creates similarity matrices that permit which allow you to visualize the relationships among the algorithms' solutions.

(H.-J. Li and J. J. Daniels, 2015) discussed about combining the particular characteristics of actual society, the authors presented a framework to examine the importance of a social community [12]. The dynamics of social connections are formed by recognizing social leaders and matching hierarchical structures. Instead of a direct assessment with the standard outcome of a random model, they computed the similarity of a specified node with the leader by the amount of frequent neighbors.

(X. Yu, J. Yang, and Z.-Q. Xie, 2015) provided clustering algorithm for 'community discovery' based on the "Link-Field-Topic (LFT) model" recommended to handle the problem of a fixed number of communities [13]. Because the clustering technique is self-contained in terms of context sampling, the number of communities does not need to be predefined. They developed "Semantic Link Weight (SLW)" based on the LFT study to assess "semantic weight" of links for every sampling field in order to tackle the problem of "overlapping community detection."

(X. Wen, W. Chen, Y. Lin, and T. Gu, 2016) developed a "multi-objective evolutionary algorithm (MOEA) for overlapping community detection" based on maximal cliques (X. Wen, W. Chen, Y. Lin, and T. Gu, 2016) This programmed proposes a novel depiction scheme based on the well-known maximal-clique graph [14]. Because the maximal-group graph is constructed by employing the positions of maximum groups of a unique graph as nodes, and two maximal groups are allowed to assign the novel graph's identical nodes, overlap is a necessary attribute. Because of this attribute, the new representation method allows MOEAs to handle the overlapping community detection problem in a manner similar to divided community detection, hence simplifying optimization problems.

(Chakraborty, et.al., 2016) Presented a comprehensive structure could recognize together overlapping and non-overlapping communities, not including some previous contribution about the network or else its community division [15]. For the community division authors introduced a vertex-based metric, GenPerm, that computes how much a vertex belongs to every of its element communities.

3. Proposed Methodology

The proposed methodology is termed CLCLP and it works with a mix of "overlapping-community detection techniques" of Link Community and Label Propagation. The CLCLP demonstrates that link separation is theoretically common for difficulty of overlapping community recognition. The CLCLP converts the nodes separation problem into a link separation issue, before identifying communities using a label propagation technique that prefers links over nodes, despite the inefficiency of the label propagation algorithm. The CLCLP then does post-processing to advance found overlapping communities by avoiding overlapping and erroneous weak secure partitioning.

4. Network Model Creation

A "Graph G = (V, E), where V is the group of vertices and E is the group of edges" can be used to describe a given input of a network model. Overlapping community detection job is discovering densely connected overlapping sub-graphs in *SG*. The given network G= (*V*, *E*) problem is to expose vertex pair association within the local communities and their equivalent communities. While a vertex has the same multiple relationships to the local communities, the proposed CLCLP algorithm should connect it to only one of its corresponding local communities or communities.

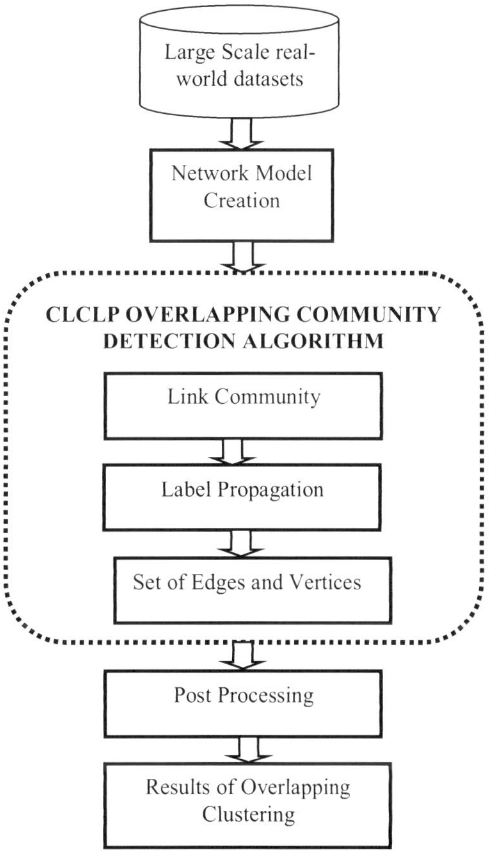

Figure 2: Improved CLCLP process Flow

5. Local community detection

The proposed CLCLP algorithm identifies local communities as follows: The counts of all vertices accessible in the network and arbitrarily chosen an initial vertex (v_i) and calculates its closeness. Then for both neighboring vertex pair in the (v_i), the y-close relative function is calculated. The y-close connected vertices are then clustered into their equivalent local communities. At every edge in the local community is labeled with its equivalent highest vertex close relative.

6. Improved CLCLP Overlapping Community Detection Algorithm

The suggested CLCLP (Combination of Link Community with Label Propagation) innovative overlapping community discovery technique consists of two stages to locate overlapping communities, as shown in Procedure 1. In the paper, an enhanced community discovery technique named CLCLP for large-scale networks to recognize overlapping communities in hard networks. Instead of considering all neighbours' in the present community, the suggested strategy only considers the neighbours of nodes introduced in the most recent expansion. CLCLP has linear time complexity with respect to network size, which is superior to most known overlapping community discovery techniques, according to theoretical study. CLCLP's efficacy and efficiency were demonstrated on large-scale synthetic and real-world networks, demonstrating the proposed algorithm's superiority over existing state-of-the-arts in terms of both effectiveness and efficiency. The proposed CLCLP method achieves good accuracy and NMI on finding communities that overlap in networks when tested on a vast scale of real-world and synthetic networks.

Procedure 1: CLCLP
Input: Given Network formation G = (V, E); *m* is the number of clusters to combine.
Output: Overlapping cluster SC of G as an output
Step 1: *LCLP* ← LinkCommunityLabelPropagation (G)
Step 2: SC ← Post Processing (*LCLP*)
Step 3: return S*C*

The CLCLP algorithm accepts a novel link Community with label propagation algorithm with favourite on links to separation into disjoint communities of nodes event on the connection fit in to the community which the connection members to.

Procedure 2: Link Community Label Propagation
Input: Given Network formation G = (V, E)
Output: ***SC***
Step 1: Set every communication with a single label
Step 2: Iterations = 100
Step 3: While Iterations > 0 **do**
Iterations --;
Node ← $\{ed|ed \in E\}$
For each ed ∈ Node **do**
if *ed* has an identical highest amount of labels of neighbouring edges in $Node_{nighbour}$ (*ed*) **then**
linkChoice(ed) = {*f* ∈ $Node_{nighbour}$ (*ed*) | max
linkSimilarity *LS*(*ed*, *f*)}
if isExist(linkChoice(*ed*) && nodeLabels.includes
(linkChoice(ed).label **then**
ed.label = *linkChoice*(*ed*).label;
else
ed.label = randomly choose form Node labels;
end if
else
label(ed) = label with maximum occurrence with
labels of neighbouring edges;
end if
end for
if each edge has a label that is shared by a high number of nearby edges **then**

quit;
end if
end while
Step 4: Keep edges with similar label into the similar cluster C_l;
Step 5: Into *C*, keep vertices that appear near the margins of the similar community in C_l;
Step 6: return C

The *LS* among two links ed_{ak} and ed_{bk} is follwed as,

$$LS(ed_{ak}, ed_{bk}) = \frac{|N(a) \cap N(b)|}{|N(a) \cup N(b)|} \quad (1)$$

Where *N(u)* is the formation neighbourhood of vertex u.

Procedure 1 and Procedure 2 are interred linked. After a link portions, the node be apt noticed as overlapping if its occurrence links fit in to unusual neighbourhoods. To conquer this issue noticed the overlapping community compositions improved, CLCP adopted a post-processing process encloses two stages: managing over-overlapping and combining related clusters. The process of post-processing was demonstrated in procedure 3.

Procedure 3: Post Processing

Input: *SC;* combine factor *m*
Output: ***OC*** (Overlapping Cluster)
Step 1: for every overlapping node *n* **do**
Neighbour (n) $= N(v) \setminus \{n\}$
for every Cluster $\in$ SC $\wedge$ n $\in$ Cluster **do**
IntraCluster (n) = $\{n_x | n_x \in$ Neighbour(n) $\wedge\, n_x \in$ Cluster$\}$
end for
end for
Step 2: Combine Clusters
for every Cluster $\in$ *C* **do**
if $C_l \in C \wedge C_1$.dimension() $\leq$ C.dimension() **then**
u = OC $\cup$ C_1
OC-Cluster;
OC $-C_l$;
OC $\cup$ u;
end if
end for
Step 3: return OC

7. Experimental Results

The experimental performance results of the proposed CLCLP algorithm perform on large-scale hard networks by evaluating with "overlapping-community detection algorithms". In the following experimental setting carried out through MATLAB, this comprises network parameters and evaluation metrics, and the test datasets.

Table 1. Simulation Parameters

Parameters	Symbol
Number of Links	L
Quantity of Nodes	N
Quantity of Communities	J
Quantity of links in community	Lc
Node Energy	E_{node}
Number of Links between the nodes	N_a *and* N_b

Figure 3 shows the Normalized Mutual Information (NMI) [16] rates of four "overlapping community detection algorithms" on the Lancichinetti–Fortunato–Radicchi (LFR)-N sets. From Figure 3, it can be seen that CLCLP performs significantly superior than OCLN (Overlapping-Community Detection Algorithm using Local-Neighbourhood Information), NISE (Neighbourhood-Inflated Seed Expansion), and LFM (Local Fitness Method) on the LFR-N set. The proposed CLCLP is superior than OCLN expressions of NMI.

Table 2. NMI values of the four methods compared

Nodes	0.9M	1M	1.1M	1.2M
NISE	0.5	0.62	0.68	0.72
LFM	0.75	0.77	0.78	0.79
OCLN	0.79	0.8	0.82	0.83
CLCLP	**0.79**	**0.83**	**0.86**	**0.91**

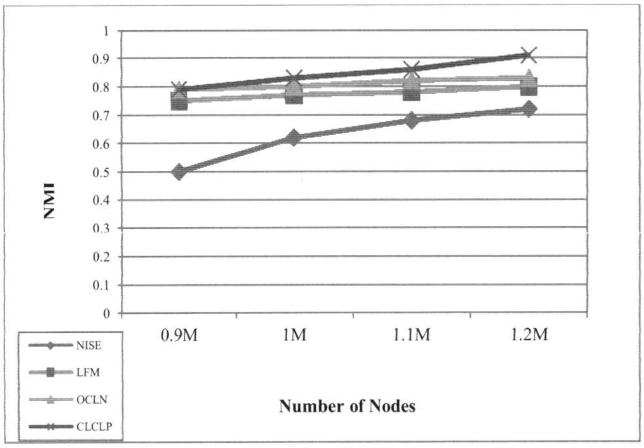

Figure 3. Chart of NMI Ratio

Figure 4 demonstrates the accuracy [17] measures of special "overlapping-community detection algorithms" on the LFR-N group. The LFR-N set consists of 12 LFR networks. Their node size n varies from 100000 (0.1 M) to 1200000 (1.2 M) with an interval of 100000.It can be practical of proposed CLCLP takes O(m) time forever attains a finest performance an expressions of accuracy using MATLAB.

Table 3. Comparison of Accuracy measures of the four algorithms.

Nodes	0.9M	1M	1.1M	1.2M
NISE	0.61	0.65	0.68	0.71
LFM	0.78	0.79	0.79	0.79
OCLN	0.87	0.88	0.89	0.89
CLCLP	**0.89**	**0.90**	**0.912**	**0.93**

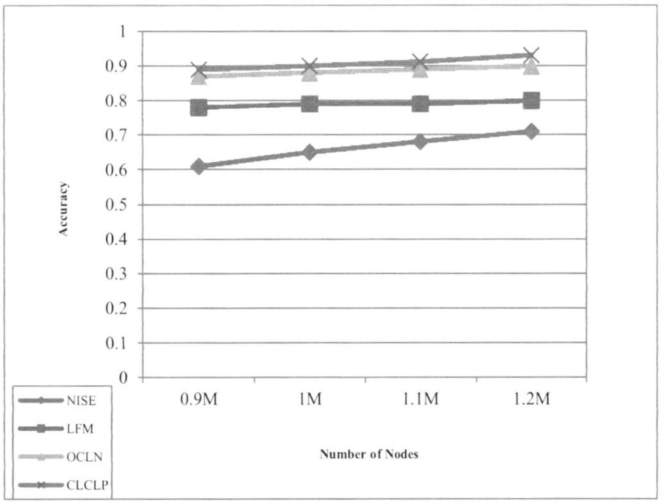

Figure 4. Chart of Accuracy Ratio

8. Conclusion

In this paper, an enhanced community discovery technique named CCLP was given to recognise overlapping communities in hard networks. Overlapping node partitioning is a problem was changed into overlapping link partition by using Link community based label propagation. The CLCLCP has innovatively utilised a new combination of CLCLP algorithm with favourite on connections and lastly locate the overlapping communities during post-processing technique to circumvent over-overlapping. The proposed CLCLP method achieves good accuracy and NMI on finding communities that overlap in networks when tested on a vast scale of real-world and synthetic networks.

References

[1] Ullas Gargi and Wenjun Lu. 2011 Large-Scale Community Detection on YouTube for Topic Discovery and Exploration *Proceedings of the Fifth International Conference on Weblogs and Social Media The AAAI Press*

[2] T. Evans and R. Lambiotte 2009 Line graphs, link partitions, and overlapping communities.

Physical Review E 80 1–8
[3] Yong-Yeol Ahn, James P Bagrow, and Sune Lehmann. 2010 Link communities reveal multiscale complexity in networks. *Nature* 466 7307 761–764
[4] W.W. Zachary 1977 An information flow model for conflict and fission in small groups *Journal of Anthropological Research* 33 452–473
[5] Michele Coscia, Giulio Rossetti, Fosca Giannotti, and Dino Pedreschi 2012 DEMON: a local-first discovery method for overlapping communities. *Proceedings of the 18th ACM SIGKDD international conference on Knowledge discovery and data mining - KDD '12* ACM Press, 615
[6] Aaron Clauset 2005 Finding local community structure in networks. *Physical Review E*, 72 2 026132
[7] Sucheta Soundarajan and John E Hopcroft 2014 Use of Local Group Information to Identify Communities in Networks *ACM Transactions on Knowledge Discovery from Data.*
[8] Joyce Jiyoung Whang, David F Gleich, and Inderjit S Dhillon 2013 overlapping community detection using seed set expansion. *Proceedings of the 22nd ACM international Conference on information & knowledge management - CIKM '13* ACM Press 2099–2108
[9] W. Wang, D. Liu, X. Liu, and L. Pan 2013 Fuzzy overlapping community detection based on local random walk and multidimensional scaling. *Phys. A, Stat. Mech. Appl.*, 392 24 6578–6586
[10] C. Liu, J. Liu, and Z. Jiang 2014 A multi objective evolutionary algorithm based on similarity for community detection from signed social networks. *IEEE Trans. Cybern.*, 44 12 2274
[11] R. Aldecoa and I. Marin 2014 SurpriseMe: An integrated tool for network community structure characterization using surprise maximization. *Bioinformatics*, 30 7 1041–1042
[12] H.-J. Li and J. J. Daniels 2015 Social significance of community structure: Statistical view. *Phys. Rev. E, Stat. Phys. Plasmas Fluids Relat. Interdiscip Top.*, 91 1 012801
[13] X. Yu, J. Yang, and Z.-Q. Xie 2015 A semantic overlapping community detection algorithm based on field sampling. *Expert Syst. Appl.,* 42 1 366–375
[14] X. Wen, W. Chen, Y. Lin, and T. Gu 2016 A maximal clique based multiobjective evolutionary algorithm for overlapping community detection. *IEEE Trans. Evol. Comput.*, 1–14
[15] T. Chakraborty, S. Kumar, N. Ganguly, A. Mukherjee, and S. Bhowmick 2016 GenPerm: A unified method for detecting non-overlapping and overlapping communities. *IEEE Trans. Knowl. Data Eng.*, 28 8 2101–2114
[16] A Lancichinetti, S. Fortunato, and J. Kertész 2009 Detecting the overlapping and hierarchical community structure in complex networks. *New J. Phys.*, 11 3 19–44

Confiner Based Video Captioning Model

J Vaishnavi[1*] and V Narmatha[2]
[1, 2] Annamalai University, Annamalai Nagar, India

[1] vaishnavijayabal22@gmail.com
[2] balaji.narmatha8@gmail.com

Abstract. Video captioning is the interesting task of encoding features and decoding the encoded features into natural language. Video captioning task is the perfect blend of computer vision and Natural language processing techniques. Video captioning process is a tentative mission since it must consider temporal features along with spatial features in order to generate appropriate captions. This task has the default framework as encoder-decoder. Ensuring the optimality of the generated caption for the particular video is the most important and essential thing which is not considered by most of the existing works. In order to ensure the optimality of captions for videos, our proposed work introduces confiner to the default framework. Confiner is planned to diminish the semantic gap among the generated caption and videos. Confiner is designed in our model with both LSTM and GRU separately as two different models. The actual work of the confiner is to inspect the visuals of the generated caption with the actual video. The proposed model has experimented with three different benchmark datasets as MSVD, MSR-VTT, and M-VAD. The performances of the models are measured by BLEU, METEOR, CIDEr evaluation metrics.

Keywords. Video captioning, Confiner, Attention mechanism, GRU, LSTM, Semantic gap.

1. Introduction

The process of video captioning is the attention-grabbing task of elucidating the visuals of a video in natural language sentences. It is the ideal mix of Computer vision along Natural language processing that instigates and magnetizes the attention of present researchers to work. In the early stages, the captioning task is mainly introduced for tagging purposes [1]. Image captioning [2, 3] is the vanguard for the video narration task. Image captioning is the task of explicating the association between objects, events, and activities in sentences. Spatial relationships between objects are considered to caption the image [4]. Calculating the semantic gap is a highly challenging task for captioning process. The major difference between image captioning and video captioning task is the temporal relationship. In video captioning tasks, temporal details [5] about the events, actions, and objects are essential to produce the optimal captions and are also considered as a challenging task compared to image captioning. Spatio-temporal [6] relationships are to be extracted to capture the temporal details from the input video. Video captioning task is considered as the stretching task for some major reasons [7] as complex nature, temporal dependencies, interconnectivity, diversified objects, events, scenes, actions, etc. video captioning task is designed to generate captions in SVO pattern [8, 9, 10, 11] about the visuals of the input video. Video captioning task has the advancement of profound learning [12, 13, 14, 15].

The video captioning task has the basic encoder-decoder framework [16] in which many existing models are designed. Currently, Attention mechanism techniques [17] overpower the task of describing videos. The most challenging task for video captioning which is not considered in many existing models is a semantic arrangement between videos and generated sentences. Our projected model is specifically intended to diminish the semantic gap between videos and sentences. The visual illustration of a certain object and activity is almost parallel to produce captions. The visual depiction of the produced caption is evaluated with the original visuals of the given video to trim down the semantic gap.

This article is assembled by the subsequences; section 2 explores the associated works of video captioning. Section 3 explicates the construction of the proposed model with encoder, decoder, and confiner. Section 4 shows the benchmark datasets used for the proposed model. Section 5 explains the preprocessing steps for the proposed approach. Section 6 discusses the evaluation metrics which is used to evaluate our model. Section 7 illustrates the proportional study of the proposed model through the state-of-art models. Section 8 terminates the paper.

2. Related Works

Video captioning is the inspiring task that processes videos to generate natural language sentences with the extracted features. The evolution of video captioning tasks starts from traditional methods toward machine learning techniques and at the present deep learning techniques dominate the field with exploiting various techniques such as LSTM, CNN, RNN, Attention mechanism, Transfer Learning, etc. Video captioning tasks in the early stages utilized template-based methods with SVO patterns to generate captions. Content hierarchies [18] of human activities are proposed by Atsuhiro et al. for extracting action-based features to generate captions. The semantic relationship [19] between videos and generated sentences plays an essential role for the superior performance of the model. Semantic relationships between various parts of sentences are proposed as a semantic hierarchy. Visuals of the video consist of a variety of objects, events, action, etc. the relationship between the contents of the visuals are modeled with Conditional Random Field (CRF) [20] in the existing model. Several events are presented in the video. In that particular, selected events [21] from the video are forwarded to generate captions. The event selection module in the model is proposed to select the particular event to generate consistent captions. The end-to-end masked transformer [22] technique is proposed to solve the problem of generating fewer standard captions for the video. This technique is designed based on the transformer model in machine translation [23].

A unified model is proposed for video-related tasks mainly for the mixture of computer vision and natural language tasks. For video captioning, a unified model [24] is considered to produce captions for the extracted features of the input video. Predefined templates lead to generating low-standard captions for video captioning. Neural machine translation techniques [25] frame encoder-decoder framework for video captioning. A recurrent neural network employs for the feature extraction process to convert the visual representations into sentences. The temporal relationship [26] of various frames is considered to generate high-quality captions with exact order. Yao et al. [17] design the model with temporal attention mechanism which picks the temporally allied segments from the video beside the temporal dynamics for superior extraction.

Hierarchical neural networks [27] are employed for the task of mining various features from video. Hierarchical Long Short Term memory (HLSTM) [28] is propounded for the production of corresponding captions. The temporal attention mechanism [29] considers both feature extraction and caption production for the video captioning model. Long Short Term Memory (LSTM) with the shifted semantic parameters are proposed for video captioning [30]. Dense video captioning with annotating the captions during the training part is proposed as supervised dense video captioning [31]. Informative frames are chosen based on the activities which are termed by the researcher as play-PicNet that leads to improving the performance with reducing computational cost [32]. Various features are fed by fusing together as one instead of feeding separately which is used to generate captions with multi-language generators as LSTM [33]. Bilinear pooling with an attention mechanism to introduce multi-modal reasoning is proposed for the video captioning tasks [34]. Convolutional Neural Network (CNN) [35] substitutes Recurrent Neural Network (RNN) in both encoders as well as a decoder to improve the performances by overcoming the problem of vanishing and exploding gradient. The above existing models influence the proposed model in some part with the attempt to employ attention-based generators for decoding the visual into appropriate and optimal captions.

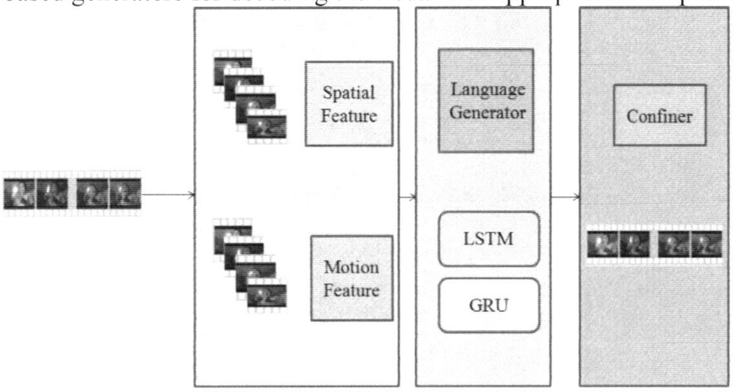

Figure 1. Structural design of the proposed work.

3. Proposed Work

In the proposed work, the task of video captioning is enriched by appending the confiner part in the existed Encoder-Decoder framework. The fore phase of the proposed model is an encoder that mines the diverse features from the input video. The second phase decoder engenders the captions from the extracted features. The proposed Confiner phase restrains the visuals from the videos by shrinking the semantic gap. The Confiner module is the chief part that meliorates the functioning of the proposed model. Fig. 1 explicates the structural design of the proposed work with encoder-decoder and confiner.

3.1. Encoder

In the encoder, different complementary features as spatial and motion are extracted for the finer performance of the model. Feature extraction is the foremost module in the framework of video captioning. N frames are extracted from every input video. Spatial features are about the location details of the visuals of the input video. Spatial features are extracted by feeding the frames into Inception v4 pretrained on the image dataset. The spatial features are denoted by I = {i1, i2, i3,..., in. Motion features are the moving details along with the temporal features which are extracted from the input video. Motion features are extracted by feeding the frames into ResNext101 pretrained on kinetic dataset which is represented as K = {k1, k2, k3, ..., kn}.

3.2. Decoder

In the decoding phase, the extracted features are modeled through a Temporal attention mechanism to generate captions. The temporal attention mechanism picks the content richer frame to generate captions by word through the word. Both Long Short Term Memory (LSTM) plus Gated Recurrent Unit (GRU) are exploited as the decoding techniques with Temporal attention mechanism in the proposed framework.

The temporal attention mechanism is modeled by the following Eqn. (1),

$$ag_{s,r} = x_g^R tanh(x_{gp}c_{r-1} + b_{gp}i_s + f_{gp}) \quad (1)$$

From Eqn. (1), x_g, x_{gp}, b_{gp} and f_{gp} are parameters for learning. (.) denotes the operation of concatenation. In each video frame, $ag_{s,r}$ denotes relative weight. The hidden fact is represented as cr-1 in Gated Recurrent Unit (GRU). Attention scores are calculated by normalizing the gained weights. Attention scores are formulated by Eqn. (2),

$$Pg_{s,r} = \frac{exp(ag_{s,r})}{\sum_g^N exp(ag_{g,r})} \quad (2)$$

As stated in attention scores, the spatial features of N inceptionV4 are merged together by temporal attention mechanism as follows in Eqn. (3),

$$e_r = \sum_{s=1}^{N} ag_{s,r} pg_{s,r} \quad (3)$$

Like spatial features, motion features of ResNext101 are merged together by the temporal mechanism by the following Eqn. (6),

$$ak_{s,r} = x_k^R \tanh\left(x_{kp}c_{r-1} + b_{kp}b_s + f_{kp}\right) \quad (4)$$

$$pK_{s,r} = \frac{exp(ak_{s,r})}{\sum_g^N exp(ak_{g,r})} \quad (5)$$

$$o_r = \sum_{s=1}^{N} ak_{s,r}\, pk_{s,r} \quad (6)$$

From Eqn. (4), (5) & (6), x_k, x_{kp}, b_{kp} and f_{kp} are parameters for learning. Here, the spatial features (e_r) and motion features (o_r) are merged together and form y_r as Eqn. (7),

$$y_r = [e_r, o_r] \quad (7)$$

GRU employs a decoding process and generates captions. The process of GRU is formulated here as Eqn. (8),

$$c_r = GRU(z_t, c_{r-1}) \quad (8)$$

LSTM is also employed for the decoding phase to generate captions. The formulation of LSTM is explained in Eqn. (9)

$$c_r = LSTM(z_t, c_{r-1}) \quad (9)$$

Gated Recurrent Unit employed with the default framework as it consists of two gates as update gate and reset gate to deal with the information. The operation of GRU is explained in Eqn. (10) & (11),

$$k_r = \sigma\left(x_k.\left[c_{r-1,}z_r\right]\right) \quad (10)$$

$$x_r = \sigma\left(x_y.\left[c_{r-1,}z_r\right]\right) \quad (11)$$

$$\tilde{c} = \phi(x_{\tilde{c}}.[k_r \odot c_{r-1}, z_r]) \quad (12)$$

$$c_r = (1 - y_r) \odot c_{r-1} + y_r \odot \tilde{c}_r \quad (13)$$

From Eqn. (9) to (10), σ represents a sigmoid function, ʘ represents element-wise product function, ϕ denotes hyperbolic tangent function. Multi-layer perceptron employs various layers for the further process. It consists of three different layers for different purposes; the foremost layer is the input layer for carrying the feature inputs. The subsequent layer is the hidden layer of the network. The activation function is offered in the final layer of the network. The output layer is activated with the soft activation function as the last layer for the process of the probability distribution of vocabulary words. The objective function is minimized by the following Eqn. (14),

$$d_h = \sum_{r=1}^{R} log\, p(x_r | x_{<r}, I; \theta) \quad (14)$$

Where r denotes the sentence length, θ denotes the parameters, x_r represents the predicted term in rth time step.

3.3. Confiner

Confiner is the final module of the proposed framework. It modeled to take the hidden states from the decoder as input to replicate the visual representation from the frames to diminish the semantic gap. Confiner is designed to resume the control over more visual data to increase the performance by arranging the semantic gap. Confiner is modeled with Gated Recurrent Unit (GRU) along with Temporal Attention Mechanism. The difference between the original visual representation and reproduces visual are calculated by the similarity loss to produce optimal results by reducing the gap. Confiner is the proposed module that helps to provide the appropriate captions by decreasing the gap. It plays a vital role in the proposed work.

$$l_r = \sum_{i=1}^{n} \beta_g^r c_g \quad (15)$$

$$k_r = GRU(l_r, k_{r-1}) \quad (16)$$

$$k_r = LSTM(l_r, k_{r-1}) \quad (17)$$

$$d_k = 1 \sum_{m=1}^{o} \psi(k_m, i_m) \quad (18)$$

In Eqn. (18), the Euclidean distance is denoted by the symbol ψ (..).
The loss of the proposed model is calculated by the following Eqn. (19),

$$d = d_h + \lambda d_k \quad (19)$$

IOP Conf. Series: Materials Science and Engineering 1258 (2022) 012049 doi:10.1088/1757-899X/1258/1/012049

In Eqn. (19), the trade-off parameter is predefined in the range of 0.1 and 0.5 and it is denoted by λ in the proposed structure.
Figure. 2 shows some examples of our approach.

Ground Truth Sentence: A Man packs his Laptop in a bag and walks out of the room.
Existing Model: A Man with laptop roaming in the room.
Proposed Approach: A Man sits on a bed and packs his laptop in a bag and walks out of the room

Ground Truth Sentence: A Girl eats and shares food with a dog.
Existing Model: A Girl is eating and shows something to a dog.
Proposed Approach: A Girl eats food and shares some food with a dog.

Figure 2. Shows some examples of our approach.

4. Dataset

The proposed work is demonstrated in three benchmark datasets as MSVD, MSR-VTT, M-VAD.
MSVD (Microsoft Research Video Description) [36, 38] is the benchmark dataset consisting of YouTube videos in total 1970 with an average duration of about 10 to 25 seconds and also consists of multi-lingual annotations by AMT workers for every video. This dataset is annotated with 16,000 new vocabularies. The data split to evaluate our model in the dataset is 1200 for training, 100 for validation, and 670 for testing.
MSR-VTT (MSR-Video to Text) [37] is also the popularly used dataset for video captioning task. It comprised 7180 videos with an overall duration of 4.2 hours. The captions consist of total 1.8 million words and 29316 individual words. The captions are annotated by AMT workers with 20 reference sentences. The data split for the evaluation of the model is 6513 for training, 497 for validation, and 2990 for testing.
M-VAD (Montreal Video Annotation) is a public video dataset with the collection of 92 DVD movies. It consists of 49,000 videos with annotations. The average period of a single video is 6.2 seconds and the overall duration is 84.6 hours. The sentence count for the entire video is 55,904. The data split for training is 38,949, validation is 4888, and testing is 5149.
These datasets are utilized separately to demonstrate and estimate the performance of the proposed model.

Table 1. Shows the comparative study of individual feature extractors with combined features.

Features	BLEU	CIDER	METEOR	ROUGH
Inception V4	56.3	79.1	38.5	76.1
ResNext 101	56.1	83.3	36.9	77.5
Inc + Res	60.3	94.5	40.7	80.2

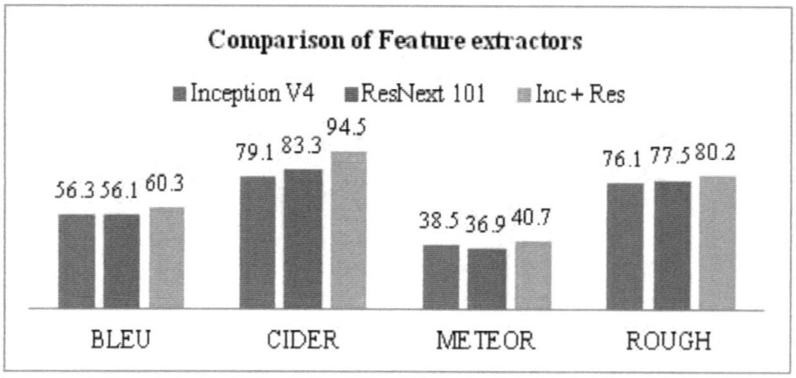

Figure 3. shows the comparison of feature extractors.

5. Preprocessing

The fixed amount of frames is taken as an input for every video in our proposed model. Here, the keyframes are mined using the Dynamic mode decomposition technique. 50 informative frames are mined by using the Dynamic mode decomposition technique from the video. Features are extracted by two different pretrained models as Inceptionv4 and ResNext101 from the selected frames. Reference sentences with more than 35 words are taken out of the process. Reference sentences are entirely changed into lower case and the special characters are removed before the process and then the encoding process is done with the one-hot vector. The pretrained model is stopped by the early stopping strategy. Semantic loss and final loss are intended to trim down the semantic gap among the ground truth sentences and produced captions.

Table 2. Shows the evaluation of two different generators.

Generator	BLEU	CIDER	METEOR	ROUGH
LSTM	55.2	83.9	37.9	77.3
GRU	60.3	94.5	40.7	80.2

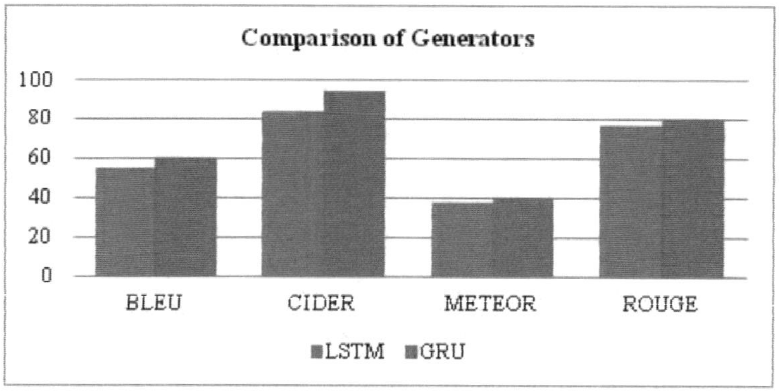

Figure 4. shows the comparison of two language generators

6. Evaluation Metrics

The proposed work is evaluated by the benchmark performance metrics such as BLEU, METEOR, and CIDEr. BLEU is the metric used to measure the similarity between reference sentences and generated captions. It considers the order and exact match of words in both ground truth sentences and generated sentences. It takes the unigram of sentences for the evaluation. The evaluation of BLEU is based on a precision score and not considered the Recall score. Brevity penalty is evaluated for the under lengthened sentences than the reference sentences. METEOR is evaluated based on both precision and recall scores and is specially meant for semantic matching. It calculates the synonyms and unigram of words between ground truth sentences and output sentences. Its basis is the harmonic mean of precision and recall scores. CIDEr is used for the assessment of consensus between reference sentences and generates captions. The ground truth sentences and predicted sentences are compared by using cosine similarity and also estimate the incidences of n-grams in both references and generated captions.

Table 3. Compares the proposed approach to existing models.

Model	BLEU	CIDER	METEOR	ROUGH
S2VT	39.6	66.6	31.2	67.4
GRU-RCN	43.3	68.2	31.6	-
MARN	48.7	92.2	35.2	71.8
Boundary aware model	42.3	63.5	32.4	-
RMN(Locate)	52.5	89.8	35.8	73.1
RMN (Relate)	52.6	89.8	36.1	73.1
RMN(S+L)	53.5	92.4	37.1	72.7
RMN (H+L)	56.6	94.4	34.5	73.4
PickNet	52.3	76.5	33.3	69.6
RecNet(SA-LSTM)	52.3	80.2	33.8	69.8
BP-LSTM	43	62.3	32.1	68.4
SA-LSTM	45.3	76.2	31.8	64.2
DMRM	51.2	74.8	33.6	-
MCNN+MCF	45.6	75.5	33.7	-
RES –ATT	53.5	73	34.4	-
ITS	34.2	65.5	25	59.9
MSGA	53.5	86.8	35.1	-
Proposed approach	**60.3**	**94.5**	**40.7**	**80.2**

7. Result Analysis

Table 1, explains the differentiation of various feature extractors in the proposed approach. In that, the combination of features shows the highest accuracy compared to individual features. Chart 1, explains clearly the highest performance of the combo. Table 2, shows the comparative study of two language generators. It shows that GRU scores the highest accuracy than LSTM. Chart 2, explains clearly the highest accuracy of GRU with LSTM. Table 3. Shows the comparative study of the various existing models with the proposed model. It deliberately shows the enhancement of the video captioning task as proposed approach with high accuracy. The proposed approach scores high accuracy on BLEU as 60.3 %, CIDER as 94.5 %, METEOR as 40.7 %, and 80.2% in ROUGH.

8. Conclusion

The proposed model is started with the default encoder-decoder framework with the additional confiner module. Spatial and motion features are extracted by Inceptionv4 and ResNext101. The encoded features are forwarded to the decoding phase. Captions are generated by employing the Temporal attention mechanism. Then confiner phase is employed with the Attention mechanism to progress the performance by shrinking the semantic gap between sentences and videos. Three benchmark datasets, MSVD, MSR-VTT, and M-VAD, are used to estimate and illustrate the performance of the proposed model. The model is evaluated using benchmark metrics like BLEU, METEOR, and CIDEr.

9. References

[1] Aradhye H, Toderici G, and Yagnik J 2009 Video2Text: Learning to annotate video content, in *Proc. Int. Conf. Data Mining* 144–151

[2] Wang Z, Feng B, Narasimhan K and Russakovsky O 2020 Towards unique and informative captioning of images, in Proc. 16th *Eur. Conf. Comput. Vis. Springer vol.* **12352** 629–644

[3] Xu G, Niu S, Tan M, Luo Y, Du Q and Wu Q 2021 Towards accurate text-based image captioning with content diversity exploration *arXiv:* 2105. 03236

[4] Li J and Qiu H 2020 Comparing Attention-based Neural Architectures for Video Captioning [Online]

[5] Su J 2018 Study of video captioning problem

[6] Agyeman R, Rafiq M, Shin H K, Rinner B and Choi G S 2021 Optimizing spatiotemporal feature learning in 3D convolutional neural networks with pooling blocks *IEEE Access vol.* **9** 70797–70805

[7] Rafiq M, Rafiq G and Choi S 2021 Video Description: Datasets & Evaluation Metrics *IEEE Xplore*, 10.11.1109/Access2021.3106565

[8] Barbu A, Bridge A, Burchill Z, Coroian D, Dickinson S, Fidler S, Michaux A, Mussman S, Narayanaswamy S, Salvi D, Schmidt L, Shangguan J, Siskind J M, Waggoner J, Wang S, Wei J, Yin Y and Zhang Z 2012 Video in sentences out, in Proc. 28th *Conf. Uncertainty Artif. Intell.* 102–112

[9] Guadarrama S, Krishnamoorthy N, Malkarnenkar G, Venugopalan S, Mooney R, Darrell T and Saenko K 2013 YouTube2Text: Recognizing and describing arbitrary activities using semantic hierarchies and zeroshot recognition in Proc. *IEEE Int. Conf. Comput. Vis.* 2712–2719

[10] Khan M and Gotoh Y 2012 Describing video contents in natural language, in Proc. *Workshop Innov. Hybrid* 27–35

[11] Krishnamoorthy N, Malkarnenkar G, Mooney R, Saenko K and Guadarrama S 2013 Generating natural-language video descriptions using text-mined knowledge, in Proc. 27th *AAAI Conf. Artif. Intell.* 541–547

[12] Venugopalan S, Rohrbach M, Donahue J, Mooney R, Darrell T, Saenko K 2015 Sequence to sequence-video to text in Proc. *IEEE Int. Conf. Comput. Vis.* ICCV 4534–4542

[13] Yang Z, Han Y and Wang Z 2017 Catching the temporal regions-of-interest for video captioning *In: ACM MM* pp. 146–153

[14] Baraldi L, Grana C and Cucchiara R 2017 Hierarchical boundary-aware neural encoder for video captioning *In: CVPR* 3185–3194

[15] Wang J, Wang W, Huang Y, Wang L and Tan T 2018 M3: multimodal memory modelling for video captioning *In: CVPR* 7512–7520

[16] Zhu L, Xu Z and Yang Y 2016 Bidirectional multirate reconstruction for temporal modeling in videos *In: CVPR* 1339–1348

[17] Yao L, Atousa T, Kyunghyun C, Nicolas B, Christopher P, Hugo L and Aaron C 2015 Describing videos by exploiting temporal structure in Proc. *IEEE Int. Conf. Comput. Vis.* ICCV 4507–4515

[18] Atsuhiro K, Takeshi T and Fukunaga K 2002 Natural language description of human activities from video images based on concept hierarchy of actions *International Journal of Computer Vision* 50(2) 171–184

[19] Guadarrama S, Krishnamoorthy N, Malkarnenkar G, Venugopalan S, Mooney R, Darrell T and Saenko K 2013 Youtube2text: recognizing and describing arbitrary activities using semantic hierarchies and zero-shot recognition Proc. *IEEE Int. Conf. Comput. Vis.* ICCV 2712–2719

[20] Rohrbach M, Qiu W, Titov I, Thater S, Pinkal M and Schiele B 2013 Translating video content to natural language descriptions Proc. *IEEE Int. Conf. Comput. Vis.* ICCV 433–440

[21] Xiong Y, Dai B and Lin D 2018 Move forward and tell: a progressive generator of video descriptions, in: Proc. *European Conf. Comput. Vis.* ECCV 468–483

[22] Yu J, Li J, Yu Z and Huang Q 2020 Multimodal transformer with multi-view visual representation for image captioning *IEEE Trans. Circuits Syst.* Video Technol. 30 4467–4680.
[23] Yu Z, Song Y, Yu J, Wang M and Huang Q 2020 Intra-and inter-modal multilinear pooling with multitask learning for video grounding *Neural Processing Letters* 1–17
[24] Ran X, Caiming X, Wei C and Corso J 2015 Jointly modeling deep video and compositional text to bridge vision and language in a unified framework *Twenty-Ninth AAAI Conf. Artificial intelligence* 1 29 2346–2352
[25] Venugopalan S, Huijuan X, Jeff D, Marcus R, Raymond M and Saenko K 2015 Translating videos to natural language using deep recurrent neural networks Proc. *Conf. North American Asso. Comput. Linguis.* 1494–1504
[26] Venugopalan S, Rohrbach M, Jeffrey D, Raymond M, Trevor D and Saenko K 2015 Sequence to sequence-video to text Proc. *IEEE Int. Conf. Comput. Vis. and Pattern Reg.* 4534–4542
[27] Baraldi L, Grana C and Cucchiara R 2017 Hierarchical boundary-aware neural encoder for video captioning, Proc. *IEEE Int. Conf. Comput. Vis. and Pattern Reg.* 1657–1666
[28] Song J, Gao L, Guo Z, Liu W, Zhang D and Shen H T 2017 Hierarchical LSTM with adjusted temporal attention for video captioning Proc. *Internat. joint Conf. Artificial intell.* 2737–2743
[29] Sepp H and Schmidhuber J 1997 Long short-term memory Neural Computation 9(8) 1735–1780
[30] Pan Y, Yao T, Li H and Mei T 2017 Video captioning with transferred semantic attributes. Proc. *IEEE Int. Conf. Comput. Vis. and Pattern Reg.* 6504–6512
[31] Duan X, Huang W, Gan C, Wang J, Zhu W and Huang J 2018 *Advances in Neural Information Processing Systems* 3059–3069
[32] Chen Y, Wang S, Zhang W and Huang Q 2018 Less is more: picking informative frames for video captioning, Proc. *European conf. Comput. Vis.* 358–373
[33] Jiang Y 2019 Multi-feature fusion for video captioning *International Journal of Computer Applications* 181(**48**) 975–8887
[34] Pan Y, Yao T, Li Y and Mei T 2020 X-Linear attention networks for image captioning Proc. *IEEE Int. Conf. Comput. Vis. and Pattern Reg.* 10971–10980
[35] Chen J, Pan Y, Li Y, Yao T, Chao H and Mei T 2019 Temporal deformable convolutional encoder - decoder networks for video captioning Proc. *AAAI Conf. Artificial intell.* **33**, 8167–8174.
[36] Chen S, Chen J, Jin Q and Hauptmann A 2017 Video captioning with guidance of multimodal latent topics Proc. 25th *ACM Intern. Conf. Multimedia* 1838–1846
[37] Jun X, Tao M, Ting Y and Rui Y 2016 Msr-vtt: A large video description dataset for bridging video and language Proc. *IEEE Int. Conf. Comput. Vis. and Pattern Reg.* 5288–5296.
[38] Yang Y, Zhou J, Jiangbo A, Yi B, Alan H, Shen H T and Yanli J 2018 Video captioning by adversarial LSTM *IEEE Transactions on Image Processing* 27(**11**) 5600–5611

IOP Conf. Series: Materials Science and Engineering 1258 (2022) 012050 doi:10.1088/1757-899X/1258/1/012050

PV Powered Neural Network Based SEPIC Converter Fed BLDC for Consumer Application

C.Infant Vinoth[1, a)]and **Dr.C.S.Ravichandran**[2, b)]

[1]Research Scholar, Department of Electrical and Electronics Engineering, Sri Ramakrishna Engineering College,Coimbatore- 641022, Tamil Nadu, India
[2]Professor, Department of Electrical and Electronics Engineering, Sri Ramakrishna Engineering College, Coimbatore-641022, Tamil Nadu, India

[a)] *Corresponding author: rockivinod@gmail.com*
[b)] *eniyanravi@gmail.com*

Abstract. The goal of this study is to investigate the utilization of a SEPIC in PV array-fed consumer applications. In order to run a centrifugal pump, a BLDC motor must be employed. It is possible to soften the start of a BLDC motor by using the INC-MPPT approach, which uses the SEPIC for management. . As an added advantage, the SEPIC can adjust the input voltage level at its output when used as a DC-DC buck-boost converter. An SPV array may be optimised at any voltage level, independent of the voltage level being utilised, thanks to this feature. Through the use of simulated results created in the MATLAB programme, the dynamic and steady state performance of a BLDC motor linked to a centrifugal water pump powered by a PV array-SEPIC is analysed and confirmed in this study.

I. INTRODUCTION

With its output gain flexibility, a SEPIC may be used to produce non-inverting output voltage, in which case it acts as a DC-DC regulator, adjusting its output voltage in line with its duty cycle [1]. With the other converters, the MPPT area is constrained [2]. This is not the case with the SEPIC, unlike the DC-DC stepup and stepdown converters. Because the switch control terminal on SEPIC is linked to ground, the gate-drive circuitry is simpler [3]. The negative voltage feedback sensing no longer need a splitting power source or optocoupler and associated circuit in place of a buck-boost or Cuk regulator with reversed outputs. This simplifies the system and speeds up reaction time [4]. A SEPIC with an input inductor has better MPPT accuracy because it reduces the pulsating of the input current.

The above-mentioned advantages of the SEPIC would be very beneficial to a consumer application system powered by an SPV array. To monitor the maximum power point (MPPT), the SEPIC has been employed in many PV-based applications, such as solar chargers and stand-alone PV systems [5-8]. SEPIC is not used in high-power applications to save money while still delivering decent performance, therefore making the SEPIC obsolete. The usage of an PV array-SEPIC-powered BLDC motor-driven water pump is recommended for this investigation. To keep the power supply and other components from being overloaded, the SEPIC runs in continuous conduction mode all the time. This technique may be used to create a switching pulse for the SEPIC's IGBT switch, which can then be utilised to run the SPV array at its optimal operating point.

For the VSI (Voltage Source Inverter), an electronic commutation is employed for the BLDC motor to produce switching sequences. There are several benefits to using a BLDC motor over a brushed DC motor, including great efficiency, reliability, low inertia and friction, and no need for maintenance. Maintaining a low starting current for the BLDC motor provides a gentle start. This may be achieved using the INC-MPPT approach by changing the increment size. Water may be pumped under any solar insulation variation thanks to the SPV array's design and BLDC motor's power rating.

System dynamics and steady-state behaviour in response to changing climatic conditions are shown in simulations using the Sim-powersystem toolboxes in the MATLB/SIMULINK environment.

II. PROPOSED CONFIGURATION OF PV FED SEPIC CONVERTER

Figure 1 depicts the system under consideration. After the SPV array, the SEPIC supplies the inverter with the drive feeding the water pump, which is a centrifugal type. For the SEPIC switch, the MPPT algorithm creates a switching pulse; while, through commutation generates switching signals. The position sensor on the BLDC motor provides the Hall signals that are logically transformed into the six pulses for the VSI's six switches. In the next sections, the control and design of each step of the setup depicted in Fig. 1 are explained.

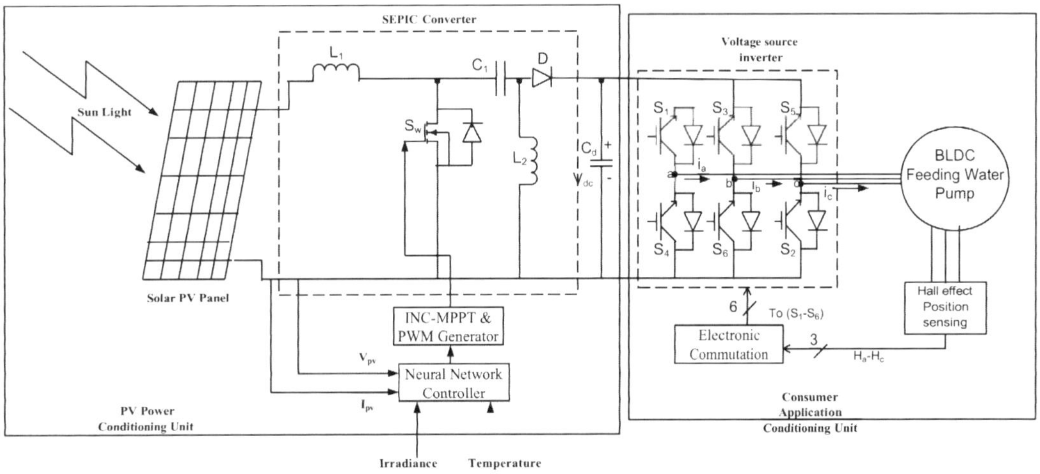

FIGURE 1. Proposed SEPIC Converter Fed BLDC For Consumer Applications with MPPT

In addition to the PV array, a SEPIC, a variable speed inverter, a brushless DC motor feeding a pump are included in the proposed system. The solar array, converter, and the pumping system have all been developed in accordance with the specifications of the planned solar energy system.

III. DC – DC SEPIC CONVERTER

As shown in Table I, SEPIC's architecture includes an input inductor (L_1); an output inductor (L_2); an intermediate capacitor (C_1); and a DC link capacitor (C_d).

TABLE 1. Design Parameters of DC-DC SEPIC Converter

S.No	Parameter	Design Equation	Value Chosen
1.	Input Inductor L_i	$L_1 = DV_{pv} / \{f_{sw}(\Delta I_{L1})\}$	$L_1 = 1.3mH$
2.	Output Inductor L_2	$L_2 = (1 - D)V_{dc} / \{f_{sw}(\Delta I_{L2})\}$	$L_2 = 1.5mH$
3.	Intermediate Capacitor C_1	$C_1 = D.I_{dc} / (f_{sw}(\Delta V_{c1})$	$C_1 = 43\mu F$
4.	DC Link Capacitor C_2	$C_d = I_{dc} / 6 * \omega * \Delta V_{dc}$	$C_d = 4780\mu F$

IOP Conf. Series: Materials Science and Engineering 1258 (2022) 012050 doi:10.1088/1757-899X/1258/1/012050

IV. CONTROL OF PROPOSED SYSTEM

Following is a division of the proposed system's control into two components. a) INC - MPP Tracking and b) Brushless Drive Commutation Control

Commutation Control of Brushless Drive: Powering brushless direct current motors using a voltage source pulse-width modulated inverter, which has a variable frequency output and generates variable voltage, is a typical practise. A switching logic pattern for VSI is devised on the basis of the rotor position information collected from the hall sensor and is implemented. A total of six switches are used in this inverter; depending on the power rating ofthe brushless DC motor, one of the switches may be an IGBT switch and the other a MOSFET switch, respectively [10]. Using a brushless direct current motor drive for a consumer application is shown in Figure 2.

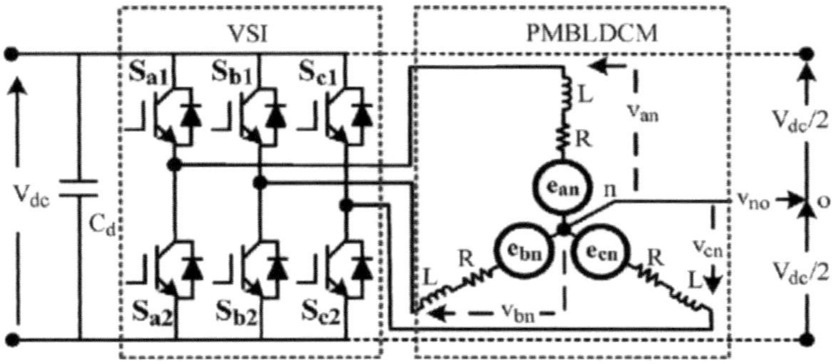

FIGURE 2. VSI Fed BLDC for Consumer Applications

Table 2: Electronic Commutator Output Based on Hall Signals

Hall Signals			***Switching Signals***					
H_a	H_b	H_c	S_{a1}	S_{a2}	S_{b1}	S_{b2}	S_{c1}	S_{c2}
0	0	0	0	0	0	0	0	0
0	0	1	0	0	0	1	1	0
0	1	0	0	1	1	0	0	0
0	1	1	0	1	0	0	1	0
1	0	0	1	0	0	0	0	1
1	0	1	1	0	0	1	0	0
1	1	0	0	0	1	0	0	1
1	1	1	0	0	0	0	0	0

INC - MPP Tracking: Controlling the solar PV array at its optimum operating point requires the use of an INC-MPPT algorithm with direct duty ratio control. The INC-MPPT approach is recommended because of its exact performance of tracking and the fact that it causes less confusion in dynamic circumstances [9]. The magnitude of the increment determines how often the MPP is inspected. There is a cost to consider when employing bigger instances for fast tracking, since the system may not operate exactly at the MPP and instead vary around it. A bigger increment size does not allow for soft beginning of the motor, thus it is vital to choose a smaller increment size in order to ensure that the system functions properly.

V. NEURAL NETWORK FOR MPP TRACKING

An artificial neural network (ANN) is used to extract the highest amount of electricity possible from a solar PV-powered consumer application system amid dynamic changes in weather conditions, including changes in sun temperature and irradiance. The suggested controller is divided into two stages. An artificial neural network is used to determine the reference maximum power point of a solar array for a given solar temperature and radiation. This is done in the initial step of the process. When comparing the reference maximum power point with real-time solar PV array power, this is achieved in a second phase, which is called the comparison phase. The controller uses the maximum power point to set the duty cycle for the DC–DC converter after analysing the data.

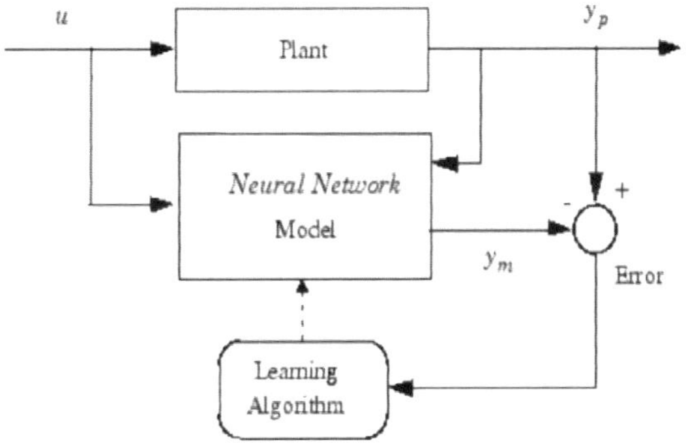

FIGURE 3. Generalized Model of Neural Network Based MPP Tracking Controller

VI. SIMULATION RESULTS AND DISCUSSION

In order to verify the suggested techniques, the MATLAB/Simulation tool is used, and the associated code is written. There are four components in this simulation diagram: the PV array model, the DC-DC sepic regulator, inverter, brushless drive, and the MPP controller for the consumer application.

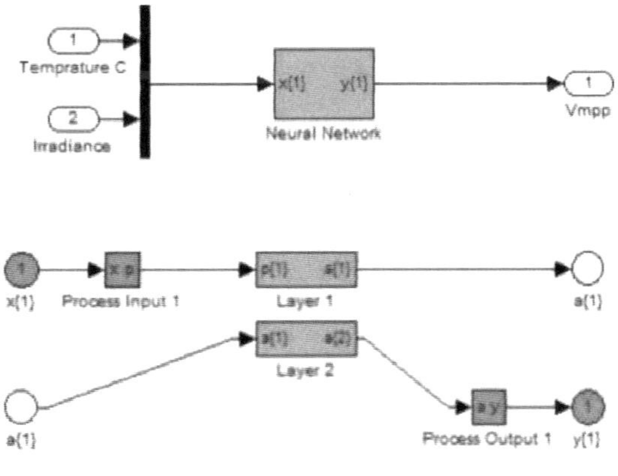

FIGURE 4. MPP Tracking Algorithm Using Neural Network

IOP Conf. Series: Materials Science and Engineering 1258 (2022) 012050 doi:10.1088/1757-899X/1258/1/012050

Figure 5 shows the output current, voltage, and power of the PV panels under a variety of irradiation levels.

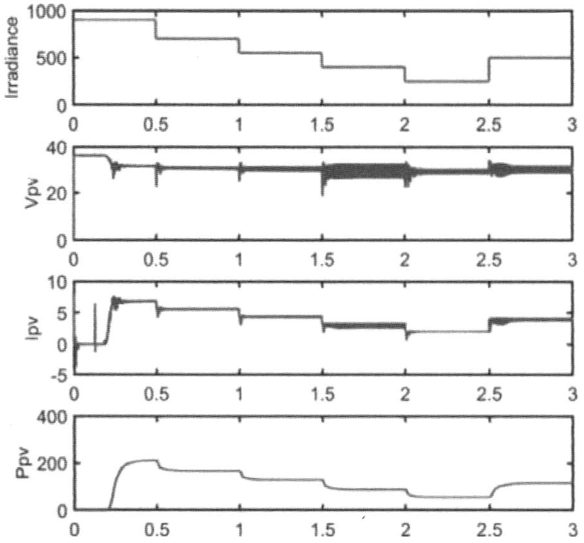

FIGURE 5. PV Results for steady state irradiation level with NN Based MPPT

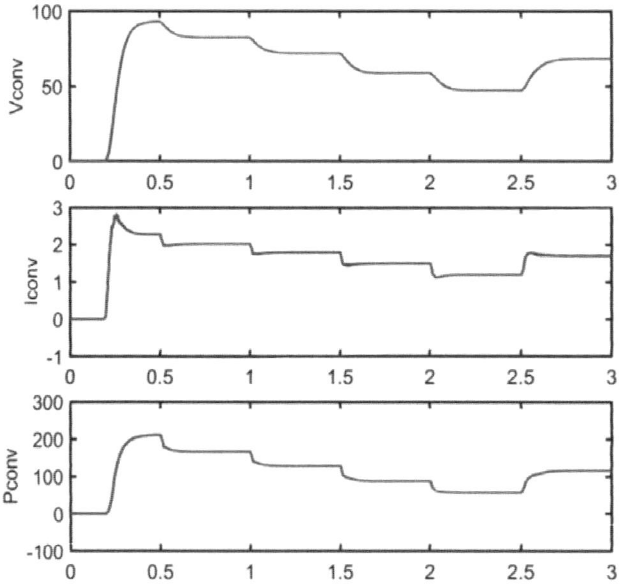

FIGURE 6. SEPIC Converter Results for steady state irradiation level with NN Based MPPT

High irradiance levels result in higher output power, whereas low irradiance levels result in decreased output power. Figure 6 depicts the zeta converter's output voltage, current, and power as a result of this.

Due to its ability to create multiple current and voltage outputs based on inputs such as light transmission or temperature, the MPPT controller is useful in a wide range of lighting applications. When the output voltage is compared to the measured voltage and current, an error signal is generated and amplified. For the next level of processing, a Proportional Integral controller handles the erroneous voltage and current data. It then performs a comparison between the reference and the real and, as a consequence, produces an error signal. A proportional integral controller (PIC) changes the duty cycle of the PWM generator to match the fault voltage or power. ThePWM generator generates pulses that the DC-DC SEPIC converter uses to extract maximum energy.

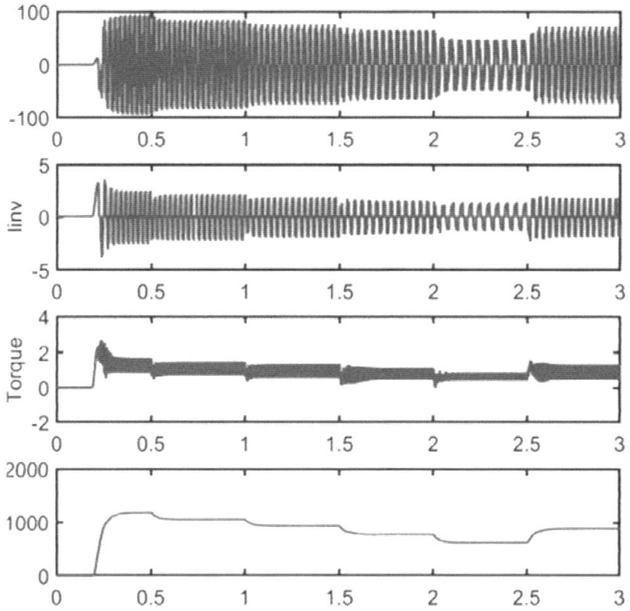

FIGURE 7. Results of BLDC for steady state irradiation level with NN Based MPPT

VII. CONCLUSION

Solar PV array with SEPIC-fed consumer application system BLDC motor driving was shown to be efficient. The suggested system is well-suited for daily usage by the typical person due to its dynamic and steady state properties. SEPIC's non-inverting polarity, a simple switch to control, and minimal input current pulses have helped to build a solar PV array water pumping system. Real-time PV panel data has been used to train the neural network. When compared to P&O, incremental conductance and fuzzy based P&O, the neural network MPP is shown to be the best. The findings show that the neural network MPP approach extracts the most power from the solar PV than other techniques studied. The solar PV array has been used to its greatest capacity to ensure that the water pumping system is running at maximum efficiency. The gentle starting of the BLDC motor may be achieved by correctly regulating the SEPIC with an MPPT algorithm. High frequency switching losses can be eliminated since the VSI is operating in 120° conduction mode. Spray irrigation and drinking water systems seem to benefit greatly from the suggested technology, which is based on a solar PV array.

VIII. REFERENCES

[1]. El Khateb, Ahmad H., Nasrudin Abd Rahim, and Jeyraj Selvaraj. "Fuzzy logic control approach of a maximum power point employing SEPIC converter for standalone photovoltaic system." Procedia Environmental Sciences 17 (2013): 529-536.

[2]. Coelho, Roberto F., Walbermark M. Dos Santos, and Denizar C. Martins. "Influence of power converters on PV maximum power point tracking efficiency." 2012 10th IEEE/IAS International Conference on Industry Applications. IEEE, 2012.

[3]. Taghvaee, M. H., et al. "A current and future study on non-isolated DC–DC converters for photovoltaic applications." Renewable and sustainable energy reviews 17 (2013): 216-227.

[4]. Lu, Dylan DC, and Quang Ngoc Nguyen. "A photovoltaic panel emulator using a buck-boost DC/DC converter and a low cost micro-controller." Solar Energy 86.5 (2012): 1477-1484.
[5]. Chiang, S. J., Hsin-Jang Shieh, and Ming-Chieh Chen. "Modeling and control of PV charger system with SEPIC converter." IEEE Transactions on industrial electronics 56.11 (2008): 4344-4353.
[6]. Mamarelis, Emilio, Giovanni Petrone, and Giovanni Spagnuolo. "Design of a sliding-mode-controlled SEPIC for PV MPPT applications." IEEE Transactions on Industrial Electronics 61.7 (2013): 3387-3398.
[7]. Moreno, Camilo, et al. "Design and implementation of a 140W charge regulator." 2013 IEEE 39th Photovoltaic Specialists Conference (PVSC). IEEE, 2013.
[8]. Muoka, Polycarp I., et al. "Modeling, simulation and hardware implementation of a PV power plant in a distributed energy generation system." 2013 IEEE PES Innovative Smart Grid Technologies Conference (ISGT). IEEE, 2013.
[9]. Esram, Trishan, and Patrick L. Chapman. "Comparison of photovoltaic array maximum power point tracking techniques." IEEE Transactions on energy conversion 22.2 (2007): 439-449.
[10]. Bist, Vashist, and Bhim Singh. "A reduced sensor PFC BL-Zeta converter based VSI fed BLDC motordrive." Electric Power Systems Research 98 (2013): 11-18.
[11]. Mokeddem, Abdelmalek, et al. "Performance of a directly-coupled PV water pumping system." Energy conversion and management 52.10 (2011): 3089-3095.

Smart helmet to start the motorbike and to prevent accidents

J Joy Mathavan[1]*, V K D Wijesekara[2], N Satheeskanth[3], W M U J Wanasinghe[4], M Maathushan[5], V V Wijenayake[6]

[1,2,3,4,5,6.] Department of Engineering Technology, Faculty of Technology, Kilinochchi Premises, Ariviyal Nagar, University of Jaffna, Sri Lanka.

joymathavan1991@gmail.com, Kaushi.781@gmail.com, sathees117@gmail.com, jayashanthaujith@gmail.com, mathushan.mahes@gmail.com, vihangavilthera@gmail.com,

Abstract. Most of the motorbike riders meet a fatal end due to severe head injuries during accidents. Usually in Asian countries, riders are reluctant to wear helmets while riding motorbikes due to carelessness and negligence. This research aims at reducing the accidents that may occur due to such carelessness and at the same time, immediately report the accident to the emergency contact numbers. This research work is aimed at making the necessary conditions essential to start the motorbike. The rider needs to wear the helmet, should be free from consumption of alcohol and should take off the side stand before igniting the vehicle; failure of any one of this condition would not ignite the engine. The proposed system detects the accident occurrence as well. SMS will be sent to the emergency numbers through the GSM module. The prototype works on start mode, running mode and accident mode with different functionalities in each mode. The rider needs to clear the start on mode to ignite the engine and a smart system will continuously monitor the bike and rider in the running mode. In case of unexpected interruptions during running mode, the proposed system enters into accident mode and enables GSM to report the accident to the pre-registered phone number.

1. Introduction

Motorbikes serve as the means of easy transportation compared to other vehicles. At the same time, safety while riding a motorbike is also an important issue to take care of. Out of around two million reported head injuries occurring every year, it is estimated that around 28% [1] of head injuries occur during motorbike accidents. Motorcyclists are not as protected on bikes as the driver or passenger in a car. These injuries may end up in life changing injuries or even death. World Health Organisation (WHO) indicates road accident injury as the 10th ranked global cause of death in 2016, which is predicted to be the 8th in 2030 [2]. The World Health Organisation has also shown that, the use of a helmet could reduce the risk of injury by 72% and the probability of death by 39% in case of motor bike accident [2]. Motorbike accidents have become a primary public hazard, which needs to be handled in a multi-disciplinary manner. Not wearing helmets, reckless driving, ignorance of traffic rules, drunkenness while driving and absence of a protective shield are some of the most significant reasons for these deaths. Mostly, the younger generation do not worry about safety when driving motorbikes. They may not have enough practice and attitude for being a good driver. With the introduction of the open economic policy to Sri Lanka in 1977 [3], a large number of three-wheelers and motorcycles were imported. The number of registered motorcycles in Sri Lanka in 2016 was 3,699,630 [3]. It was 55% of the total registered vehicle population at that time. Sri Lanka has seen a significant increase in its motorcycle population, from 2012 to 2019. The number of registered motorcycles increased by 45% with the growth rate of

11% per year [3] in Sri Lanka. This scenario shows the significant growth of the motorcycle population in Sri Lanka. This research work is aimed at making the necessary conditions essential to start the motorbike and to report the accident to a pre-registered phone number. Proper wearing of the helmet, lifting up the side stand and non-consumption of alcohol are considered as necessary factors in this project to start the motorbike. The objective of this study is to find a smart solution to minimise the accidents and to reduce the risk of injuries and number of deaths of bikers.

2. Literature Review

This research work is inspired from the recommendations of medical journals, which speak about the importance of wearing a helmet while riding a motorbike. The study of Bolbol SA and Zalat MM shows that 87.1% of the motorcyclists who succumbed to death in road accidents, do not have a proper driving licence and 98.1% of them do not wear a helmet [4]. This is an alarming situation considering the developing countries including India and Sri Lanka which are congested with huge numbers of vehicles on narrow roads. K.C. Mani Kulanthayan et al in their research has stated that, in Malaysia, helmets are mandatory for motorbike riders and it has reduced the fatality rate during motorcycle crashes by 30% [5]. J. Joy Mathavan et al have described the automatic side stand lift-up system while starting the motorbike. Once the rider is on the seat and the key is turned on, the side stand would be lifted up and when the key is turned off and rider gets off the seat, the stand retrieves back to the position [6]. This system is developed in order to prevent accidents by the position of the side stand. They clearly described how the position of the side stand can cause accidents on junctions. Nataraja N et al described about starting the motorbike, if the rider is not drunk. According to their work, an alcohol sensor is fixed on the helmet, and the bike would start if the rider is not drunk. Signals received from sensors attached to the ignition system and alcohol sensor are transferred to the microprocessor, which in turn enables the ignition of the motor bike [7]. Mohd Khairul Afiq Mohd Rasli et al in their attempt to enhance the functionality of the smart helmet, have incorporated Force Sensing Resistor (FSR) and Brushless Direct Current (BLDC) fan to estimate the speed limit and alarm the rider if the pre-determined speed limit is exceeded [8]. Rashmi Vashisth et al have described E-HELMET, which is used to inhibit engine ignition if not worn and incorporated sensors which serve as alcohol detectors, speed alarming system and tollgate auto payment E Wallet [9]. But, the system of integrating with tollgates for auto payment is practically tedious to implement in developing countries. Though a number of features are available in this research work, real time implementation integrating the system with the motor bike is not illustrated in their paper. L. Kanimozhi et al developed a smart helmet system that ensures the rider remains sober throughout the journey and also to monitor his heart beat rate to detect any health abnormalities while riding motorcycle [10]. Yet, the suggested prototype needs further improvements to make sure the helmet remains in contact with riders' head throughout the journey to protect him in case of any accidents. Abhinav Anand et al described an IR (Infra-red) sensor to detect the presence of a helmet using the infra-red rays emitted from the human body, and absence of which will not ignite the motorbike engine. He has also mentioned about pre-defined speed limits to prevent over speed of vehicles and inform them, if the speed exceeds the limit. His system was designed in a way to inform nearby hospitals in case of any accident as well [11]. In an attempt to improve the security and to prevent theft of automobiles, two factor authentication was described by O. Akinsanmi et al. It works on enabling keyless entry to ignite the engine using two-factor authentication. In addition to automobile keys, voice-based authentication can also be enabled to secure the vehicle from theft. If the automobile key is lost and found by an unauthorised person, the attempt of starting the engine can be prevented by implementing this voice-based authentication [12]. The working principle here is appreciable but brings practical difficulties when it comes to voice authentication. R.M. Vithlani et al described the possibility of a biometric automobile ignition locking system. It is mentioned that the automobile ignition system can be started only if the fingerprint authentication is provided [13]. Implementing such a biometric locking system is good for the use of single person, but may result in chaos in case of automobiles which are frequently used among inmates. S. Aliyu et al in their research work described a multi factor authentication technique using combination of password and fingerprints. In case if the vehicle is stolen, the current location of the vehicle could be communicated with the user using GSM / GPS (Global System for Mobile Communications / Global Positioning System) communication module [14]. However, with the advancement of technology, if the GSM module is removed after theft or if the automobile is taken to places where there are no proper mobile networks, the target of this research cannot be achieved. Researches [12-14] concentrated much on the security of automobiles rather than motorbikes. The research works [12–14] spoke about passing the robbery information or preventing the robbery of the vehicles. Mohammad Ali et al in their research work, analysed the accident-avoidance framework, in which, the proposed algorithm automatically decelerates the automobile in case of any dangerous terrains or road curves using automatic brake and steering control of individual wheels of the automobile [15]. J J Mathavan et al also carried out research regarding the accidents in automobiles. They described an automatic door unlocking system of the vehicle in case of any vehicle fire accident [16]. Chloe et al found that, during an accident, the riders of motor bikes and drivers of automobiles tend to blame the opponents for the cause of the accident [17]. It doesn't actually matter whether the accident involves two motorbikes or two automobiles or a motor bike and an automobile; the damage and loss is for both. Blaming the opponent after the accident will not restore the loss. Also, it hardly matters whether the accident involves two or more vehicle collisions or slip of a single vehicle. So, the best way to avoid accidents is by taking necessary precautions. Ching-Yao Chan has worked on sensor arrays during vehicular crashes. He observed the response time of those sensors for vehicular crashes happening at various speeds. His research concentrated on detecting suitable sensors for automobiles with the lowest response time with varying speeds [18]. Researches [15 - 18] are concentrated more on preventing accidents of automobiles.

Most of the required factors like starting conditions, safety while riding, informing accidents and preventing robbery are individually discussed by several authors. Individual solutions cannot solve this issue, since this subject is an interconnected matter of several factors. Eliminating one factor of accident while the others remain the same will not help to solve this problem. Therefore, the current research work is interested in solving all the available factors of accident in a single shot. The novelty of this research work resides in combining all the necessary factors for smooth and safe riding of a motorbike in a single small device, without making the existing system complex and implementing it in real time. Because, most of the research works failed to connect the solutions with the real time implementation for a developing country.

This research paper is written in way that, section one introduces the current scenario on the population of vehicles and accidents, section two describes the detailed literature review of the steps taken by different authors to solve the issue, section three describes the methodology of the solutions designed and developed in this research work to solve the above-mentioned issues, section four discuss the obtained results and the last section mention about the conclusion of this research work.

3. Methodology

The proposed smart helmet comprises overall functionalities as shown in Figure 1. The block diagram shown in Figure 2 shows the parts of the developed system such as alcohol sensor, IR sensor, GSM module, ignition relay, accelerometer and display. The flow diagram shown in figure 3 describes the necessary conditions to start the engine. Satisfaction of all the three conditions such as helmet worn properly, free of alcohol consumption and side stand lifted up will allow the ignition system to be activated. The flow diagram for the detection of accidents is shown in Figure 4. Once the accident is detected, a phone call and a message will be sent to the emergency number saved in the phone. The specifications of the hardware components used in the research work is shown in table 1.

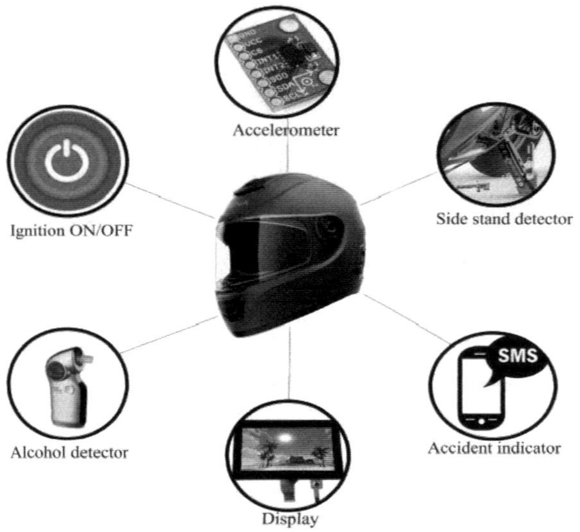

Figure 1. Integrated system of Smart Helmet

The Smart safety helmet is developed using IR sensors and it has two separate parts: receiver and transmitter. Receiver is fixed at the motorbike and the transmitter is fixed inside the helmet. The receiver fixed at the motorbike will control the ignition switch according to the signals received from the transmitter which is fixed inside the helmet. When the rider wears the helmet and ties the belt of it up, the IR sensor detects that the rider wears the helmet and then only it'll allow the bike to start. Simply wearing a helmet without tying the belt of the helmet will not work. When the rider wears the helmet, the mini switch placed inside the helmet detects the pressure and the switch would be turned ON. The transmitter looks to find the receiver which is fixed on the motorbike dashboard. Then the transmitter emits signals from Infra-Red Light Emitting Diode (IR LED) and which would be detected by the IR receiver. Then the receiver transfers this command to the controller. The helmet worn message will be displayed in the dashboard of the motorbike as shown in figure 5. Then as the next step, the controller will check for the position of the side stand.

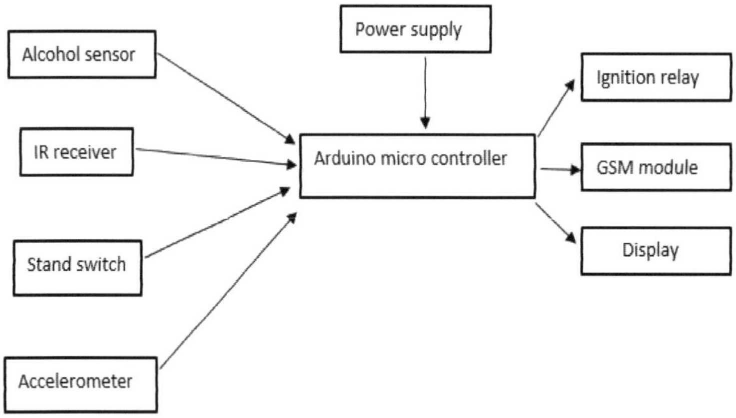

Figure 2. Block diagram indicating the overall function of smart helmet

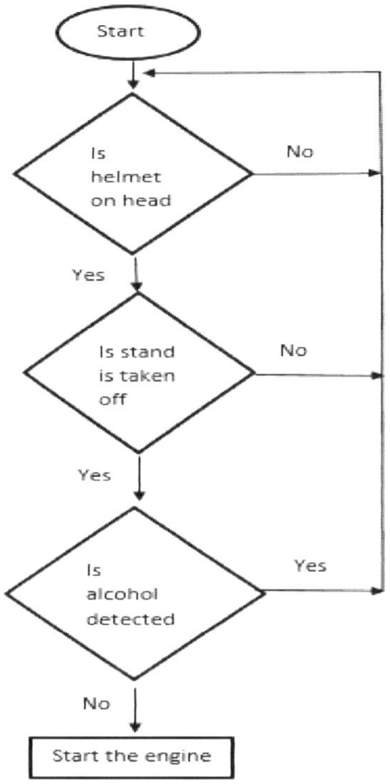

Figure 3. Flow diagram indicating the ignition of the engine

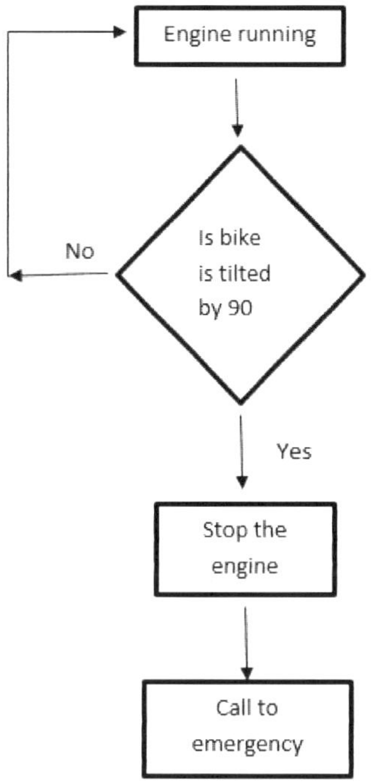

Figure 4. Flow diagram indicating the detection of accident

Table 1. Specification of the hardware components

Hardware co components	Specifications
Arduino UNO	Microcontroller: ATmega328P Operating voltage: 5V Flash memory 32KB Clock speed 16MHz Analog I/O pins: 6 Digital I/O pins:14
GSM module	Model type: SIM800L Power supply: 3.8V-4.2V SIM card socket: micro-SIM Supported frequencies: Quad-band (850/950/1800/1900 MHz)
Relay switch	Contact rating: 10A,240VAC/24VDC Max. carrying current: 14A Max. switching power: 2400VA, 240W Initial resistance: Max. 100mΩ at 6VDC, 1A
MQ 3 Sensor	Sensing resistance: 1MΩ – 8MΩ (0.4 mg/L alcohol) Operating voltage: 5V ± 0.1

	Load resistance: 200KΩ Operating temperature: 20^0C - 50^0C
Accelerometer	Model type: ADXL 345 Power supply: 1.8V – 3.6V Resolution: 13 bits Sensitivity: min 232 LSB/g, max 286 LSB/g Temperature range: -40 to +85℃
Display	Type: HD44780 Operating voltage: 4.7V – 5.3V Display: 2 × 16 characters Pixels: 5 × 8-pixel box Operating bits: 8-bit and 4-bit mode

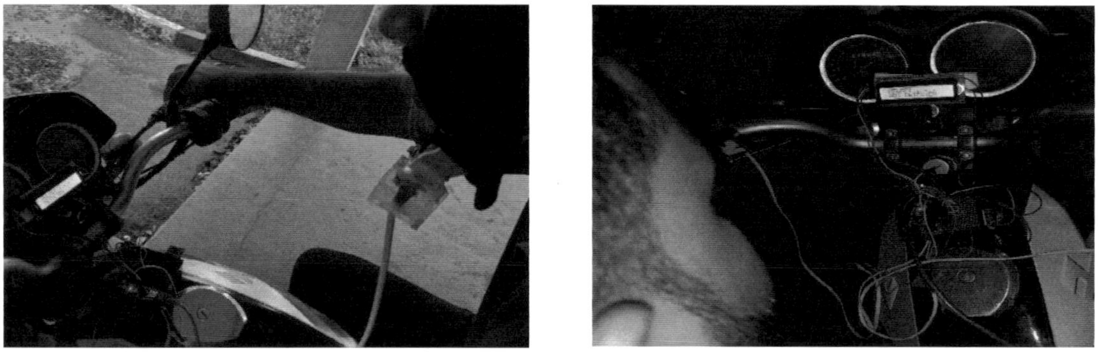

Figure 5. Detection of Helmet

Then the system detects the position of the side stand. If the stand is lifted, the toggle switch shown in figure 6 near the stand will be activated. Otherwise, the switch would not be pressed and the system would not allow the ignition of the vehicle. This scenario is shown in figure 7. In the similar way, when the side stand attached to the motorbike is released, the signals from the relay are transmitted to the controller to turn off the engine. Forgetting to lift the side stand is dangerous while taking left turns as it may hit the ground and the rider may lose his balance [6]. Once the controller detects that the side stand is lifted up, it looks for the signals from the alcohol sensor.

Figure 6. Arrangement to detect the position of side stand

IOP Conf. Series: Materials Science and Engineering 1258 (2022) 012051 doi:10.1088/1757-899X/1258/1/012051

Figure 7. Detection of side stand

Alcohol sensor (MQ3 sensor) is also fitted inside the helmet. It will be activated if the user has consumed alcohol. Under such conditions, the controller denies the ignition of the motorbike. If the rider is free of alcohol consumption, the alcohol sensor will not be activated and the motorbike ignition system can be started and it is shown in figure 8. The combination of all the signals from helmet, side stand and alcohol detector will allow the ignition of the motor bike engine to switch to start mode.

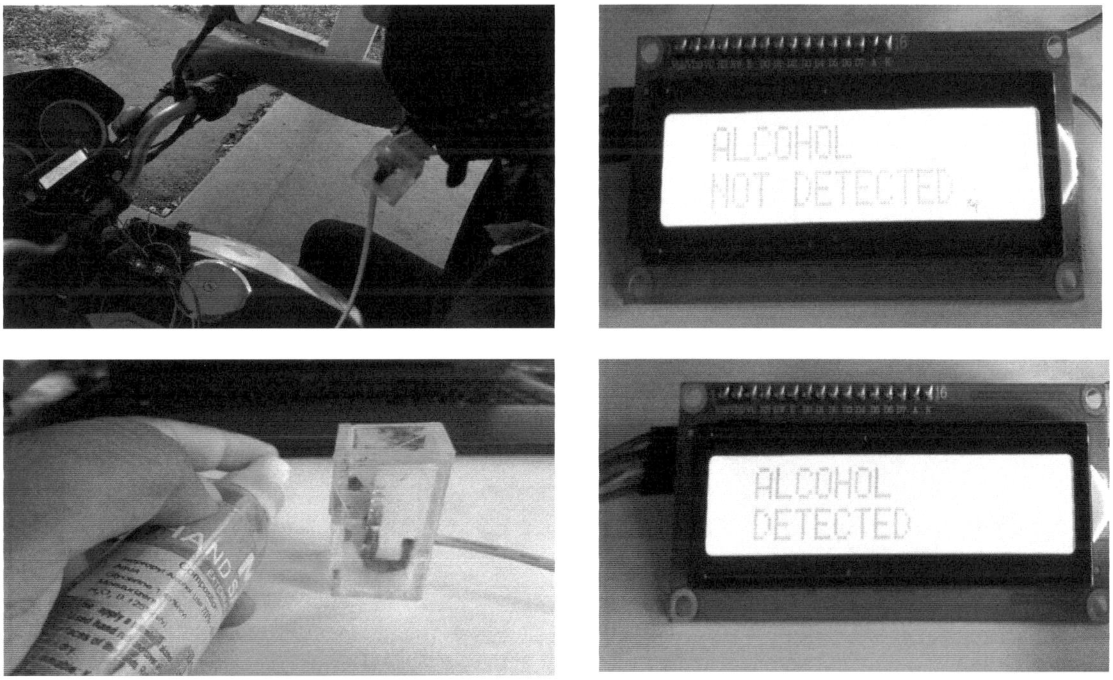

Figure 8. Detection of Alcohol

In running mode, a gyroscopic sensor will be monitoring the angle of inclination of the motor bike. If the gyroscope tilts above 90 degrees angle, then the accident mode would be enabled and the GSM

IOP Conf. Series: Materials Science and Engineering 1258 (2022) 012051 doi:10.1088/1757-899X/1258/1/012051

module gets activated as shown in figure 9. The GSM module sends the message "accident detected" to the pre-registered emergency contact numbers in case of any accidents, as shown in the fig 10.

Figure 9. Detection of accident

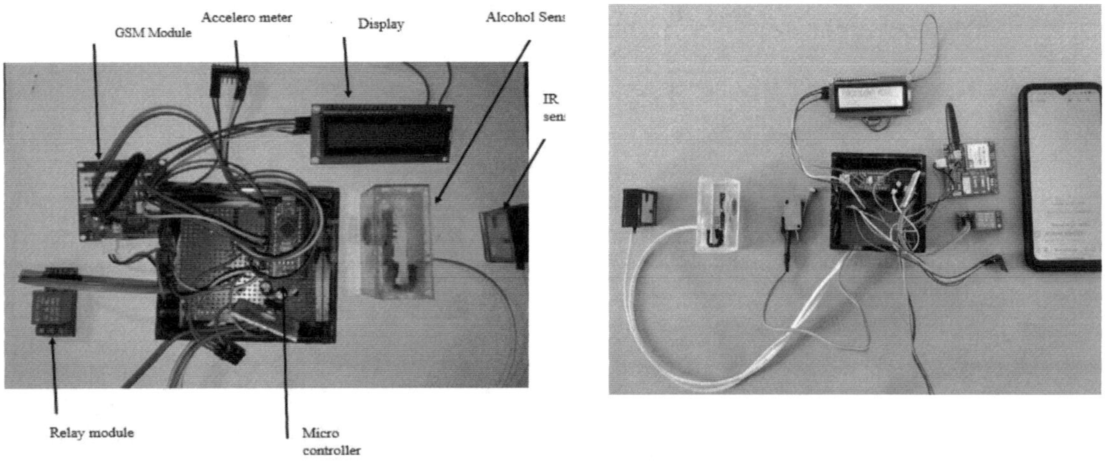

Figure 10. Proposed prototype for Smart Helmet

4. Results and discussion

The proposed smart helmet includes three modes namely Start ON mode, Running Mode and Accident mode. These three different modes of operations are designed using suitable sensors and tested in the real time environment.

In start ON mode, the controller will check whether the user has placed the helmet properly on his head and tie the belt of it, whether the side stand is taken off and whether any alcohol consumption is detected. If all these conditions are satisfied, then the controller sends signals to relay to ignite the engine. Even if any one of the above-mentioned conditions are not satisfied, relay will not be activated. Hence the engine would not be ignited.

In running mode, a gyroscopic sensor will be monitoring the angle of inclination of the motor bike. The ignition of the engine will be terminated when the bike is tilted above 90 degrees and which would be considered as an accident. 90 degrees here is a default set value. It can be customised to 60 or 70 degrees depending on user requirement. It can also be turned off if the user wants to travel to elevated terrain without getting unwanted accident alerts.

If the bike is tilted above 90 degrees, accident mode will be enabled where a message saying "accident detected" would be sent to the pre-registered phone number. This phone number can be changed time to time depending on the location where the rider is travelling. In other words, according to the place of movement of the rider, he can set the mobile number of a friend who is living nearby. So, in case of any accident, if the message is sent to that friend who is living nearby, he may arrive at the accident spot in a minimal time to assist the victim. It should not be forgotten that unnecessary delays during the arrival of emergency vehicles to and from a hospital due to traffic congestion in roads [19], may also increase the chance of life loss.

The proposed model is built and tested in a motorbike and found to work effectively. The modes of operations and emergency messages will be displayed in the Liquid Crystal Display (LCD) screen which can be fitted in the dashboard of the motorbike. Finalised prototype of the proposed smart helmet system is shown in Figure 10.

5. Conclusion and future scope

The smart helmet is designed and tested for real time application. All three modes of operations designed in the prototype namely Start ON mode, Running Mode and Accident modes are tested and found to work satisfactorily. Wearing a helmet, taking off side stand and negative in alcohol detection is mandatory during Start ON mode; without which the bike will not start. While riding the bike, gyro sensors will be continuously active and monitor the rider, and bike as well. In case if the gyroscope encounters an angle greater than 90 degrees, the engine will be off and will not be ignited and a message will be sent to the emergency number. This emergency number should already be set in the mobile phone.

Future scope of this project includes designing the proposed model using high-end controllers like raspberry pi and involvement in vehicle-to-vehicle communication (V2V). The automatic error detection in the vehicle and displaying those errors in the display to inform the user, integrating with the prompt audio messages are some advanced features which are currently under developing stage. Additionally, road signs are expected to be detected at a distance using image-processing software and a gentle alert can also be provided to the user while riding the bike.

6. References

[1] Bolbol S.A and Zalat M.M 2018 "Motorcycle riders' risky behaviours and safety measures: a hospital-based study", *Egyptian Journal of Occupational Medicine* **42** (3): 453-468
[2] World health statistics 2008 (www.who.int/whosis/whostat/2008/en/index.html)
[3] https://www.treasury.gov.lk/web/economic-phases/section
[4] https://www.transport.gov.lk/
[5] K.C. Mani Kulanthayan, Hui Fang Teow, Hari Krishnan Tamil Selvan, Kaviyarasu Yellappan and Vaidehi Ulaganathan 2020 "Determinants of standard motorcycle safety helmet usage among child pillion riders", *Transportation Research*, **74,** pp 408 – 417.
[6] J. J. Mathavan, K. G. L. Dinakara, P. G. S. G. Premathilaka and H. M. T. D. Herath, "Arduino based Hardware Implementation of Automatic Side Stand for MotorBike," *2020 Fourth International Conference on Computing Methodologies and Communication (ICCMC)*, 2020, pp. 627-630, doi: 10.1109/ICCMC48092.2020.ICCMC-000116.
[7] N. Nataraja, K. S. Mamatha, Keshavamurthy and Shivashankar, "SMART HELMET," *2018 3rd IEEE International Conference on Recent Trends in Electronics, Information & Communication Technology (RTEICT)*, 2018, pp. 2338-2341, doi: 10.1109/RTEICT42901.2018.9012338.

[8] M K A Mohd Rasli, N K Madzhi and J Johari, "Smart helmet with sensors for accident prevention," *2013 International Conference on Electrical, Electronics and System Engineering (ICEESE)*, 2013, pp. 21-26, doi: 10.1109/ICEESE.2013.6895036.

[9] R Vashisth, S Gupta, A Jain, S Gupta, Sahil and P Rana, "Implementation and analysis of smart helmet," *2017 4th International Conference on Signal Processing, Computing and Control (ISPCC)*, 2017, pp. 111-117, doi: 10.1109/ISPCC.2017.8269660.

[10] L Kanimozhi, R Sambasivam, M Pragathi and M Ranjith 2019 "Smart Helmet with Accident-Avoidance System ", *cikitusi journal for multidisciplinary research*, **6** Issue 10.

[11] Abhinav Anand, Kumar Harsh, Kushal Kumar and Sourav Gouthi 2015, "Microcontroller based smart wear for driver safety", *International Journal of Research in Engineering and Technology*, **04** Issue: 05.

[12] O Akinsanmi, A D Usman, A Abdulraheem, G D Obikoya and B G Bajoga 2015 "Two-Factor Authentication Based Automobile Keyless Entry System", *International Journal of Engineering and Applied Sciences (IJEAS)* ISSN: 2394-3661, **2**, Issue-8.

[13] R M Vithlani, Sagar Shingala, Dr. H N Pandya 2016 "Biometric automobile ignition locking system", *International Journal of Electronics and Communication Engineering and Technology (IJECET)* **7**, Issue 5, pp. 28–37.

[14] S Aliyu, Umar Abdullahi, Majeedat Pomam, Mustapha Hafiz, Adeiza Sanusi, and Sodiq Akanmu 2019 "Smart Protection of Vehicle using Multi Factor Authentication (MFA) Technique", *3rd International Engineering Conference IEC* pp 701 - 710.

[15] Mohammad Ali, Paolo falcone, Claes Olsson and Jonas Sjoberg. 2013. "Predictive prevention of Loss of Vehicle Control for Roadway Departure Avoidance". *Trans. Intell. Transport. Sys.* **14**, 1, pp 56-68. Doi.org/10.1109/TITS.2012.2206584

[16] J. J. Mathavan, A. W. Faslan, N. U. A. Basith and W. V. S. D. Wanigasinghe, 2020, "Hardware Implementation of Fire Detection, Control and Automatic Door Unlocking System for Automobiles, "*4th International Conference on Trends in Electronics and Informatics (ICOEI) (48184)*, pp. 125-129, doi: 10.1109/ICOEI48184.2020.9142990.

[17] Chloe J. Robbins, Harriet A. Allen, Peter Chapman, 2018. "Comparing car drivers' and motorcyclists' opinions about junction crashes", Accident Analysis & Prevention, 117, pp 304-317

[18] Ching-Yao Chan, "On the detection of vehicular crashes-system characteristics and architecture," in *IEEE Transactions on Vehicular Technology*, vol. **51**, no. 1, pp. 180-193, Jan. 2002, doi: 10.1109/25.992078

[19] Satheeskanth N, Mathushan M, Mathavan J J, Kunaraj A, Daisan G. (2022). "Smart Traffic Controller Based on Traffic Density and Prioritised Emergency Vehicle Clearance". In: Bindhu, V., Tavares, J.M.R.S., Du, KL. (eds) *Proceedings of Third International Conference on Communication, Computing and Electronics Systems. Lecture Notes in Electrical Engineering*, vol **844**. Springer, Singapore. https://doi.org/10.1007/978-981-16-8862-1_33

A review on electrical transport properties of thin film Schottky diode

S D Dhruv and D K Dhruv*
Natubhai V. Patel College of Pure and Applied Sciences, Charutar Vidya Mandal University, Vallabh Vidyanagar-388120, Anand, Gujrat, INDIA

Corresponding author e-mail ID: dhananjaydhruv@rediffmail.com
Contact number: +91-9427422470

Abstract. The article outlines an inclusive list of thin film Schottky diodes (TFSD) references. The review audits the fabrication and characterization of the TF metal-semiconductor (MS) diode, a TFSD. The work functions of metal (ϕ_m) and semiconducting material (ϕ_s) determines whether the established MS contact is ohmic or rectifying. Current-voltage ($\mathrm{I}-\mathrm{V}$) and capacitance-voltage ($\mathrm{C}-\mathrm{V}$) characterizations are essential electrical transport measures of TFSDs. The $\mathrm{I}-\mathrm{V}$ and $\mathrm{C}-\mathrm{V}$ outcomes are conferred, and archetypal results are parroted. The TFSD device is emblematic of a heterojunction diode (HJD). The rectification ratio (RR), saturation current (I_0), ideality factor (n), Schottky barrier height ($\emptyset_\mathrm{B}$), and carrier concentration may all be calculated using $\mathrm{I}-\mathrm{V}$ and $\mathrm{C}-\mathrm{V}$ data. Thermionic emission (TE) and/or space charge limited conduction mechanisms (SCLC) may conduct electricity in TFSDs. Anderson's model can theoretically be used to construct a TFSD energy band diagram.

Keywords: Thin film Schottky diode; $\mathrm{I}-\mathrm{V}$ characterization; $\mathrm{C}-\mathrm{V}$ characterization; Work function (ϕ); Schottky barrier height (ϕ_B).

1. Introduction

Contact is required for all electronic devices to communicate with one another. When metal comes into contact with semiconductors, several electrical behaviours can be observed based on the applied voltage (*V*). The connection behaves like a heterojunction diode (HJD) in the first type, rectifying the current (*I*). On the other hand, the ohmic contact transmits *I* equally regardless of the *V* polarity of the source. Microwave diodes, photodetectors (PDs), field-effect transistors (FETs) and solar cells (SCs) are all examples of thin film (TF) semiconducting devices based on the Schottky diode (SD) technology [1]. The performance and reliability of an SD are impacted by the eminence separating metals from semiconductors. [2]. The superiority of the metal-semiconductor (MS) interface has a remarkable impact on the electrical properties of Schottky contacts. Factual SDs necessitate using the capacitance (*C*) technique for characterization [3]. Diverse deposition technologies are employed to fashion TFSDs, such as flash evaporation [4], sputtering [5,6], pulsed laser deposition [7,8], etc. SDs have been shown much reverence in the electronic industry, and they are actively involved in developing gas sensors, solar cells and switching devices. [9–11].

Current research inspects the development of TFSD and its allied parameters such as rectification ratio (RR), saturation current (I_0), ideality factor (n), Schottky barrier height (ϕ_B) and carrier

IOP Conf. Series: Materials Science and Engineering 1258 (2022) 012052 doi:10.1088/1757-899X/1258/1/012052

concentration derived from I − V and C − V measurements. The TFSDs' I conduction mechanisms are explained by the author. Using Anderson's approach, the authors stress the MS energy band diagram.

2. Fabrication and characterization of TFSD

2.1. Fabrication of TFSD

The thermal evaporation/sputtering system will deposit a thin metal film of ≃100 nm thickness on a meticulously cleaned substrate. Many substrates, such as glass, aluminium foil, alumina (Al_2O_3), silicon (*n-/p-* type), sapphire (*SiC*), and others, can be used to deposit the TF. Semiconducting TF of ≃300 nm thickness is then grown on a pre-grown metal TF at a substrate temperature (Ts) of ≃573 K. In the final step, a metal contact of ≃100 nm thickness will be made on top of the semiconducting TF using the thermal evaporation/sputtering technique to fashion electrical contact. Indium (*In*) or silver (*Ag*) paste is commonly used to make electrical connections. A junction area of ≃0.25 cm^2 can be realized through an apposite masking arrangement. A standardized thermocouple (*Cr-Al*) measures T_S. All temperature-related measurements can be carried out under ≃1.333 Pa (low vacuum) to dodge TF deterioration owing to environments. The specified TF T_S and thickness values are symptomatic; they can be altered/optimized to lift the device's sensitivity. The schematic depicted in figure 1 exemplifies the TFSD [12].

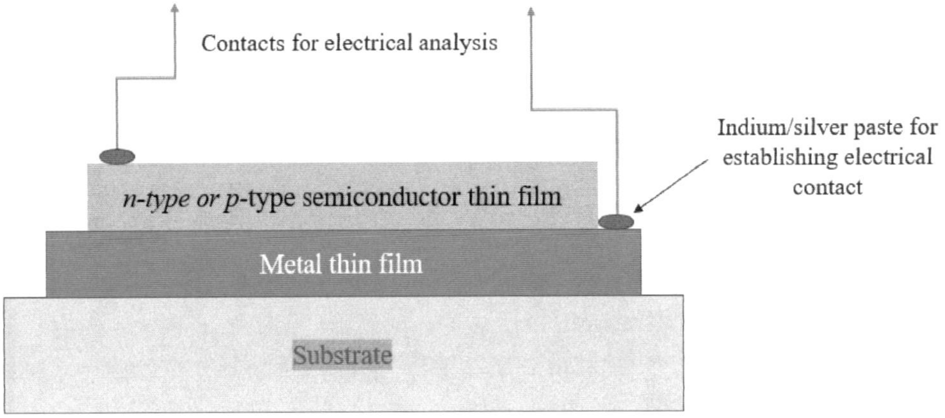

Figure 1. The schematic of TFSD.

2.2 Characterization of TFSD

An abundance of characterization methods makes it simpler to investigate TFs [13–15]. X-ray diffractometer (XRD) is used to identify crystalline phases, measure crystallite sizes, lattice parameters, orientation, quantitative phase analysis, and atomic coordinates of TFs. The elemental analysis and surface morphological study of TFs can be carried out with the help of a scanning electron microscope (FEG-SEM). The morphology, crystal structure, particle size, quantitative elemental analysis, and elemental mapping study of TF can be done using a transmission electron microscope (HR-TEM). The optical properties of TF can be studied with a UV-Vis-NIR help spectrophotometer. The type of semiconductor (*n-/p*-type) and its mobility and carrier concentration of TFs can be determined using the Hall effect characterization system. The electrical transport portrayals of the TFSD devices can be carried out using an electrometer and a computer-controlled semiconductor characterization system. There are advantages and disadvantages to each characterization method. As a result, a sole tactic is insufficient to depict a TF device. Thus, it is possible to use more than one practice to scrutinize TF electronic devices.

A barrier is moulded at the MS edge when metal is connected to a semiconductor, pedalling I conduction and C. Schottky-Mott [16] explicates the rudimentary science of ϕ_B; the conception of the ϕ_B depends uniquely on the kind of semiconductor (*n-/p*-type), its work function (WF) (ϕ_s) and metal

IOP Conf. Series: Materials Science and Engineering 1258 (2022) 012052 doi:10.1088/1757-899X/1258/1/012052

WF (ϕ_m). The WFs rheostats ohmic or rectifying behaviour of the established MS contact, ϕ_s and ϕ_m (table 1).

Table 1. Correlation amid the diode's WF and its conduction nature.

WF	*n*-type semiconductor	*p*-type semiconductor
$\phi_m > \phi_s$	Rectifying	Ohmic
$\phi_m < \phi_s$	Ohmic	Rectifying

Electrons from the metal can reach the semiconductor when ϕ_m is less than ϕ_s, resulting in a line of Fermi levels being aligned between the two materials, resulting in a positive charge (hole) behind the electron that transfers to the metal, confirming a thicker interface[17]. ϕ_B is the potential difference between the Fermi energy of metal and band edge, where the majority charge careers are sited. A more extensive ϕ_B grades better-rectifying appearances.

3. Electrical analysis of TFSD

3.1 $I-V$ characterization

In TF electronic devices, many issues plague I transference mechanisms. The interface between ***1)*** a metal electrode (used for electrical connection) and TF, and ***2)*** metal and semiconductor (MS contact) plays an optimistic role. The assortment of metal (electrode) for founding electrical contact and metal for SD is thorny, as the WF of metal has an indispensable role. A metal used for electrical connection should bid an ohmic plot with little contact resistance when it comes into contact with the metal/semiconductor in the MS junction, which is the most perilous consideration. The metal used in MS junctions, on the other hand, should exhibit rectifying behaviour [18]. Thermal transport over the ϕ_B, tunnelling through the barrier, carrier recombination in the depletion area, and carrier recombination in the neutral region governs the I induced by carriers crossing semiconductor to metal and vice-versa [19]. Semiconductors with high mobility are commonly used to create a TFSD [20].

Diodes are known for their non-linear behaviour. The RR should be high enough for the superlative diode; RR is the forward-to-reverse I ratio at a given bias[21]. When operated at different bias levels, the TFSD reveals two conduction mechanisms: thermionic emission (TE) and space charge limited conduction mechanisms (SCLC) [12]. Following TE theory [22], I can only flow if the carrier's energy exceeds the conduction band energy at the MS interface. It is agreed upon by equation (1) that I is dependent on V [23,24],

$$I = I_0 \left[\exp\left(\frac{qV}{nkT}\right)\right] \quad (1)$$

The terms used in this equation have their typical connotations.

Figure 2's $\ln(I)$ versus V (for low bias) rectilinear plot advocates that TE is the foremost conduction mechanism and can be used to infer I_o (from the intercept, $\ln(I_0)$) and n (from the slope, $\frac{q}{nkT}$) [12]. Though SD and HJD trail identical $\mathrm{I-V}$ relations (equation 1), I_0 perceived in SD is mostly $10^3 - 10^8$ times larger than HJD [25]. The n value advanced than one intitles the aberration of the diode device from its archetypal nature, which can be explicated due to the recombination of charge carriers in the depletion section or swelling diffusion I [26,27]. At higher bias, the SCLC is evinced by witnessing the linear plot of I as a function of the square of V (the plot is not incorporated) [12,28]. Materials selected for MS contacts dictate the overriding conduction mechanism. Table 2 pageants a few TFSDs airing TE and SCLC leading conduction mechanisms.

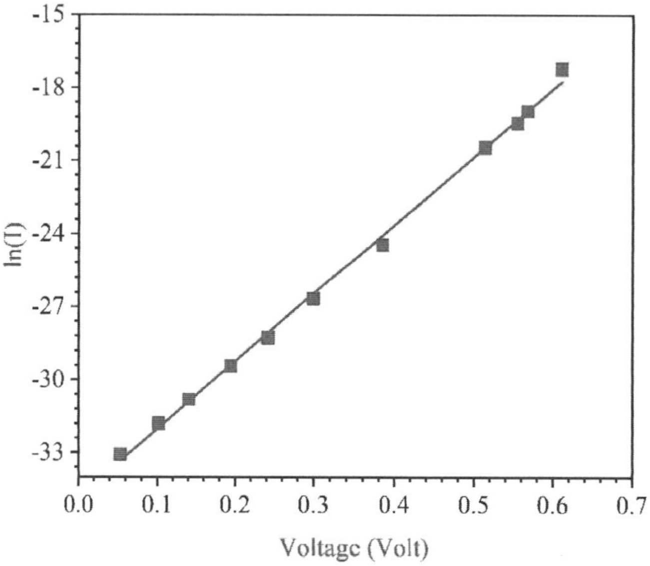

Figure 2. Archetypal $\ln(I)$ versus V plot for TFSD.

The equation (2) is plotted in figure 3 as a function of $\ln\left(\frac{I_s}{T^2}\right)$ verses $\frac{1000}{\mathrm{T}}$ to derive the Richardson's constant (A^*) and ϕ_B values [29],

$$I_0 = A\,A^*T^2\exp\left(\frac{-\phi_B}{kT}\right) \quad (2)$$

Here $A^* = \frac{4\pi m^* k^2}{h^3}$, m^* is the effective mass of semiconducting material.

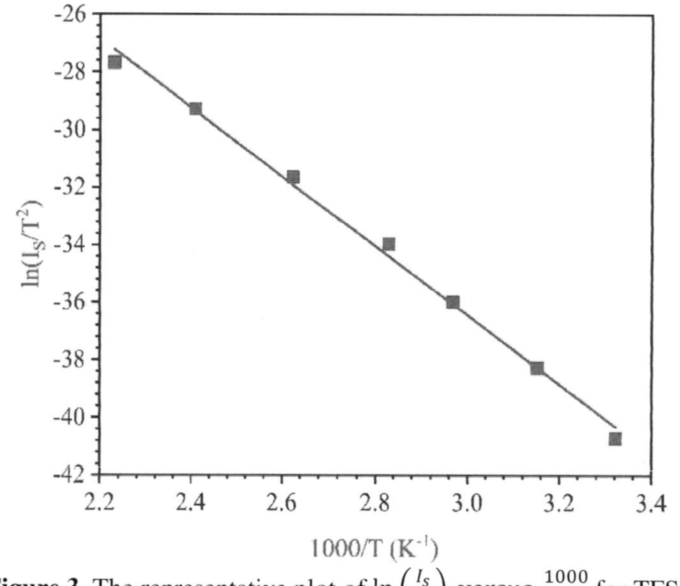

Figure 3. The representative plot of $\ln\left(\frac{I_s}{T^2}\right)$ versus $\frac{1000}{\mathrm{T}}$ for TFSD.

Table 2. The dominant conduction mechanisms in TFSDs.

Dominant conduction mechanism			
TE		SCLC	
TFSDs	Ref	TFSDs	Ref
Au/V_2O_5/n-Si	[30]	Au/Si_3N_4/p-GaAs	[31]
Bi doped p-ZnO	[32]	$Pt/BiFeO_3/SrRuO_3$	[33]

Ag//ZnO/Al	[34]	Al/p-$ZnIn_2Se_4$	[12]
PEDOT:PSS/ZnO	[35]	Ag/CuPc/Au	[36]
Pt–ZTO	[37]	Pt/n-Ge/Bi_2Te_3	[38]

Table 3 conscripts several diode parameters for a few TFSDs.

Table 3. Electrical parameters of various TFSDs.

Sr no	TFSDs	RR	I_0 (μA)	n	Ref
1	Cu/p-SnSe	097.63	0.245	1.38	[39]
2	Al/p-Si	736.39	-	2.12	[40]
3	Ag/p-Cu_2NiSnS_4	-	8.240	2.23	[41]
4	Pd/Zno	113.00	0.0876	5.79	[42]
5	Au/ZnO	093.00	0.0014	6.00	[43]

3.2 $C-V$ characterization

The reverse bias potential applied to the TFSD device causes a modification in depletion area width, W to W + dW, due to movement of charge carriers into the space charge sheet [18,19]. When a bias is applied to a TFSD, instigating a change in ϕ_b due to changes in movement of electrons.

The alteration in junction C by reverse alternating bias (V) is given by equation (3) [44],

$$C = A\left(\frac{q\varepsilon_0\varepsilon_r N_d}{2(V+V_0)}\right)^{\frac{1}{2}} \tag{3}$$

Here V_0 is zero bias built-in potential, A is the active cross-sectional area of the diode, ε_r is a relative dielectric constant of semiconducting material and N_d is the carrier density of the donor.

The diffusion potential (V_D) can be inferred by using the equation (4) [12],

$$V_0 = V_D - \frac{kT}{q} \tag{4}$$

Figure 4's straight-line plot amid $\frac{1}{C^2}$ and V_R proposes that donor concentration N_d (for n-type material) or acceptor concentration N_a (for p-type material) is constant through the depletion area. The negative slope is a telltale sign of an n-type semiconductor (figure 4). When the electric field in the semiconductor is zero, the attained ϕ_B is identified as flat-band ϕ_B, $\phi_{B(C-V)}$, which can be inferred using equation (5) [45],

$$\phi_{B(C-V)} = V_D + E_F - \Delta\phi_B \tag{5}$$

Equation (6) is employed to deduce Fermi energy (E_F) [46],

$$E_F = \left(\frac{KT}{q}\right)\ln\left(\frac{N_V}{N_D}\right) \tag{6}$$

Where $N_V = 2\left(\frac{2\pi m_p kT}{h^2}\right)^{\frac{3}{2}}$ is the effective density of state in the valence band of semiconducting material, m_p is the effective mass of semiconducting material.

The ϕ_B dropping can be inferred using equation (7) [12],

$$\Delta\phi_B = \left[\frac{qE_m}{4\pi\varepsilon_0\varepsilon_s}\right]^{\frac{1}{2}} \tag{7}$$

Here E_m is extreme conceivable electric strength at the MS junction surface. equation (8) derives E_m [12],

$$E_m = \left[\frac{2qN_A V_0}{\varepsilon_0\varepsilon_s}\right]^{\frac{1}{2}} \tag{8}$$

Values of V_D, E_F and $\Delta\phi_B$ are placed in equation (5) to find $\phi_{B(C-V)}$.

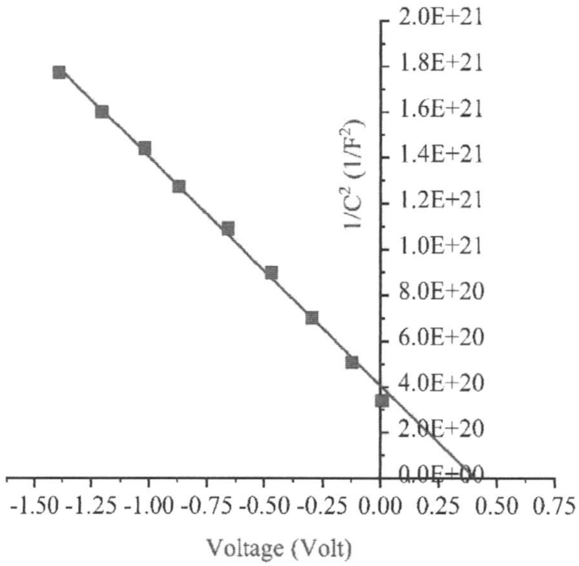

Figure 4. A representative plot of change in C with V_R for TFSD.

The ϕ_B attained from $\mathrm{I-V}$ and $\mathrm{C-V}$ is typically contrary. The ϕ_B resulting from $\mathrm{C-V}$ characterization is more advanced than derived from the $\mathrm{I-V}$ due to diverse natures of measurement approaches. Table 4 portrays ϕ_B values ensuing from $\mathrm{I-V}$ and $\mathrm{C-V}$ techniques.

Table 4. Comparison of ϕ_B values derived from $\mathrm{I-V}$ and $\mathrm{C-V}$ techniques.

Sr. No.	SD	$\boldsymbol{\phi_{B(I-V)}}$ (eV)	$\boldsymbol{\phi_{B(C-V)}}$ (eV)	Ref
1	Pd/Ti/n-InP	0.35	0.73	[47]
2	Pt/Ru-n-GaN	0.73	1.04	[48]
3	Ti/p-Si	0.75	1.04	[49]
4	Ni/Cu/n-InP	0.52	0.78	[50]
5	Au/β-Ga_2O_3	1.38	1.98	[51]

4. Energy band diagram construction for TFSD

The thermal equilibrium energy band diagram for TFSD can be shaped notionally founded on Anderson's model, which is shown in figure 5 [52]. The barrier height is the variance amid metal and semiconducting materials' WFs.

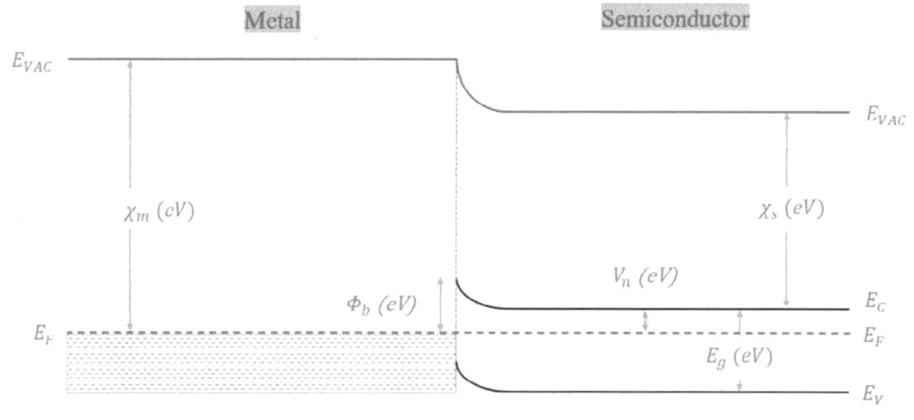

Figure 5. The band diagram of TFSD.

Conclusion

TF MS diode' fabrication and characterization are thoroughly reviewed. The WFs of metal (ϕ_m) and semiconducting material (ϕ_s) decides MS contact's behaviour. TFSDs are typically tested using $I-V$ and $C-V$ techniques for their electrical transport measurements. The TFSD device's nature is emblematic of a HJD. The plentiful physical parameters like RR, I_s, n, ϕ_B and carrier concentration of TFSD can be derivative from electrical transport measurements. The electrical conduction in TFSD may occur by TE and/or SCLC. The Anderson model can be used to construct a theoretical energy band diagram for the TFSD.

Acknowledgement

One of the authors, S.D. Dhruv is thankful to SHODH-ScHeme of Developing High-Quality Research-MYSY (202010820029) for a scholarship.

References

[1] Ali G M 2020 *Journal of Alloys and Compounds* **831** 154859
[2] Courtin J, le Gall S, Chrétien P, Moréac A, Delhaye G, Lépine B, Tricot S, Turban P, Schieffer P and le Breton J C 2019 *Nanoscale Advances* **1** 3372
[3] Ukraintsev V A, Potts F R, Wallace R M, Magel L K, Edwards H and Chang M-C 2009 *AIP Conference Proceeding* **449** 736
[4] Dhruv D K and Patel B H 2015 *International Journal of Innovative Research in Electrical, Electronics, Instrumentation and Control Engineering* **3** 9
[5] Yu S, Zhang C, Yang P, Wu M, Sun Y and Li L 2019 *Journal of Materials Science: Materials in Electronics* **30** 18287
[6] Varma T, Periasamy C and Boolchandani D 2017 *Superlattices and Microstructures* **112** 151
[7] Ortiz W, Ramirez N J, Barrionuevo D, Bhattarai M K and Feng P 2021 *Nano Express* **2** 010020
[8] Durmuş H and Karataş Ş 2019 *International Journal of Electronics* **106** 507
[9] Mohd Azol M E, Shafiei M, Leow P L, Foo K L, Hashim U and Arsat R 2013 *Sensors and Transducers* **149** 49
[10] Mukhamale S v., Chavan A R, Lokhande R M and Khirade P P 2020 *Solar Energy* **211** 866
[11] Fan Y S and Liu P T 2014 *IEEE Transactions on Electron Devices* **61** 1076
[12] Dhruv D K and Patel B H 2016 *Materials Science in Semiconductor Processing* **54** 29
[13] Caglar Y, Caglar M and Ilican S 2018 *Optik* **164** 424
[14] Ryu S R, Ram S D G, Lee S J, Cho H D, Lee S, Kang T W, Kwon S, Yang W, Shin S and Woo Y 2015 *Applied Surface Science* **347** 793
[15] Çakici T, Güzeldir B and Saǀlam M 2015 *Journal of Alloys and Compounds* **646** 958
[16] Kandemir B B, Gören S, Erdem M, Cengiz A, Şale Y, Fedotov A K, Mammadov T G and Seyidov M Y 2020 *Semiconductor Science and Technology* **35** 125010
[17] Dhruv S D and Dhruv D K 2022 *Materials Today: Proceedings* **55** A1
[18] Rhoderick E H 1982 *IEE Proceedings I: Solid State and Electron Devices* **129**
[19] Crowell C R and Sze S M 1966 *Solid State Electronics* **9** 1035
[20] Pecunia V, Banger K, Sou A and Sirringhaus H 2015 *Organic Electronics* **21** 177
[21] Chetia S K, Das A K, Ajimsha R S, Sahu V K and Misra P 2020 *Physica Status Solidi (A) Applications and Materials Science* **217** 2000440
[22] Türüt A 2021 *Turkish Journal of Physics* **44** 302
[23] Patel H, Patel K, Patel A, Jagani H, Patel K D, Solanki G K and Pathak V M 2021 *Journal of Electronic Materials* **50** 5217
[24] Jang J, Song J, Lee S S, Jeong S, Lee B J and Kim S 2021 *Materials Science in Semiconductor Processing* **131** 105882
[25] Khediri A H, Talbi A, Benamara M A and Benamara Z 2021 *Journal of Nano- and Electronic Physics* **13** 04002-1
[26] Gayen R N and Bhattacharyya S R 2016 *Journal of Physics D: Applied Physics* **49** 115102

[27] Touati R, Trabelsi I, Rabeh M ben and Kanzari M 2017 *Journal of Materials Science: Materials in Electronics* **28** 5315
[28] Darwish S 2004 *Physica B: Condensed Matter* **349** 269
[29] Agarwal L, Singh B K, Tripathi S and Chakrabarti P 2016 *Thin Solid Films* **612** 259
[30] Mahato S, Biswas D, Gerling L G, Voz C and Puigdollers J 2017 *AIP Advances* **7** 085313
[31] Buyukbas-Ulusan A and Tataroglu A 2020 *Journal of Materials Science: Materials in Electronics* **31** 9888
[32] Singh B K and Tripathi S 2018 *Superlattices and Microstructures* **120** 288
[33] Lee D, Baek S H, Kim T H, Yoon J G, Folkman C M, Eom C B and Noh T W 2011 *Physical Review B - Condensed Matter and Materials Physics* **84** 125305
[34] Ali G M, Khalid A K and Swadi S M 2020 *Journal of Semiconductors* **41** 102103
[35] Hernandez-Como N, Rodriguez-Lopez A, Hernandez-Cuevas F J, Munguia J, Garcia R, Baca-Arroyo R and Aleman M 2016 *Semiconductor Science and Technology* **31** 115007
[36] Shah M, Sayyad M H and Karimov K S 2011 *Journal of Semiconductors* **32** 044001
[37] Schlupp P, von Wenckstern H and Grundmann M 2017 *Physica Status Solidi (A) Applications and Materials Science* **214** 1700210
[38] Ahmad F, Kandpal K and Kumar P 2021 *Journal of Materials Science: Materials in Electronics* **32** 8106
[39] Jagani H S, Gupta S U, Bhoraniya K, Navapariya M, Pathak V M, Solanki G K and Patel H 2022 *Materials Advances* **3** 2425
[40] Aldemir D A, Esen M, Kökce A, Karataş S and Özdemir A F 2011 *Thin Solid Films* **519** 6004
[41] Ghediya P R, Palan Y M, Bhangadiya D P, Nakani I I, Chaudhuri T K, Joshi K, Sumesh C K and Ray J 2021 *Materials Today Communications* **28** 102697
[42] Somvanshi D and Jit S 2014 *Journal of Nanoelectronics and Optoelectronics* **9** 21
[43] Rajan L, Periasamy C and Sahula V 2016 *Perspectives in Science* **8** 66
[44] Riazimehr S, Belete M, Kataria S, Engström O and Lemme M C 2020 *Advanced Optical Materials* **8** 66
[45] Hudait M K, Venkateswarlu P and Krupanidhi S B 2001 *Solid-State Electronics* **45** 133
[46] Moloi S J and McPherson M 2013 *Radiation Physics and Chemistry* **85** 73
[47] Reddy D S, Reddy M B, Reddy N N K and Reddy V R 2011 *Journal of Modern Physics* **2** 113
[48] Reddy N N K and Reddy V R 2012 *Bulletin of Materials Science* **35** 53
[49] Asıl Uğurlu H, Çınar Demir K and Coşkun C 2021 *Journal of Materials Science: Materials in Electronics* **32** 15343
[50] Munikrishana Reddy Y, Nagaraj M K, Siva Pratap Reddy M, Lee J H and Rajagopal Reddy V 2013 *Brazilian Journal of Physics* **43** 13
[51] Jiang K, Lyle L A M, Favela E, Moody D, Lin T, Das K K, Popp A, Galazka Z, Wagner G and Porter L M 2019 *ECS Transactions* **92** 71
[52] Liberto G di and Pacchioni G 2021 *Journal of Physics Condensed Matter* **33** 415002

Bending Assessment and On-body Investigation of a Textile Antenna for Body-Centric Wireless Communication Applications

D Ram Sandeep*, Habibulla Khan, B T P Madhav, M Gayathri , A Kalyana Priyatham, M Nikhilvarma, S Chandra Sekhar Kumar and S Salma

ALRC, R&D, Department of ECE, Koneru Lakshmaiah Education Foundation,

AP, India.

***E-mail:** askram91@hotmail.com

Abstract: This article presents the design and investigation of a textile-based antenna for body wireless communication applications. The Christmas tree structure inspired the antenna design, and a staircase-like defective ground element supported it. The antenna is tiny with a footprint of 30×20×1 mm^3 and is developed on polyester fabric. This material has good grades of flexibility and has the most negligible affinity to moisture, so it was chosen as a suitable substrate for the developed wearable antenna. The radiating elements and grounds conductivity is actualised using the adhesive conductive copper tape. The propounded antenna operates in a treble band of Telemetry, Industrial Scientific and Medical (ISM), and X-band at 4.5, 5.8, and 8 GHz frequencies. The developed textile antenna is intended for body applications, so it was methodically examined by bending it at various vertical and horizontal bending angles. This analysis shows that the antenna's resonating frequencies are not substantially varied in the vertical and horizontal bending aspects. The antennas on body performance were also examined by placing the prototype directly on the subject's skin. These results confirm the developed antennas' applicability for body wireless communication applications.

Keywords: Textile antenna, conductive copper tape, polyester fabric, Telemetry, ISM and X-bands

1. Introduction

In the era of modern, flexible electronics, there has been a massive demand for wearable and durable conformal antennas [1]. The conventional printed circuital board microstrip antennas are not suitable for modern wearable gadgets, owing to their inflexibility limitations. At the same time, not skin and eco friendly; wearing these rigid antennas is practically impossible due to their sharp edges that severely infuriate the human skin. When it comes to contemporary wearable antennas, they are of a unique kind and made with conformal materials that facilitate being worn. The wearable antennas have massive demand in emergency and rescue operations, medicine, military, on-body wearable communication systems and bio-medical radiofrequency system applications. Personal on-body wearable gadgets are connected through a shared network called Wireless Body Area Networks (WBAN). The wearable antennas are the crucial components that will establish a stable connection between these devices. They are also responsible for the possibility of variable communications like off-body, on-body, and in-body communications [2].

Wearable antennas are generally made with different materials and methodologies. However, the textile patch antennas are the best class of wearable antennas because they are made from conformal and skin-friendly textile materials. They won't cause any irritation or discomfort while worn; simultaneously, they are visually concealed. Usually, while fabricating the textile antennas, the insulation elements of the antennas are made with dielectric fibres. The conductive radiating and ground elements are actualised using several techniques like inkjet printing [3] and screen printing [4], conductive tapes or fibres [5-6], embroidering [7], and wet-etching [8]. Modern advanced manufacturing methods help fabricate these wearable antennas in huge volumes and help attain a compact size, dual-polarization (linear and circular) and low profile. The choice of selecting the above fabricating methods also depends on several factors like antennas size, textile material used, the thickness of the substrates, flexibility, and conformability while being worn as a piece of outfit, manufacturing scales and cost. Usually, these microstrip textile antennas (textennas) are available in several forms like circular, elliptical, triangular, square and rectangular patterns. However, many natural inspired shapes are being evolved to design an efficient, flexible and perfect safe wearable antenna that can facilitate the operations even in different bending circumstances. The other fabrication factors like fabrication precision, maintaining the accurate dimensions and structural deformation while bending also need to be considered while developing these antennas.

The antenna developers of [9] designed a textenna for on-body communication applications, and they incorporated the embroidering method for fabricating the textenna. The Shieldex Plated Nylon fibres are utilised for embroidering, and the cotton textile was used as a proper substrate. This antenna is operating at 2.4 GHz for ISM band applications. Another type of antenna was designed on Pellon material, and conductive fabrics were utilised to actualise the radiating and ground elements [10]. This fabrication method achieved compact dimensions of 57×32.1 mm^2 by operating the antenna at 2.45 GHz. The antenna engineers of [11] developed a dual-band texteena on Denim material and operated at 2.45, 5 frequencies. The conductivity is drawn using the conductive stitching threads. At the same time, the wearable antenna researchers proposed textile antennas on various materials like silk, jute, leather and cotton [12-17].

This study presents the development of a polyester-based wearable textile antenna through a simple and easy fabrication method that can facilitate the production of wearable antennas on large scales. The developed antennas' functional behaviour is also analysed on-human body and under different bending circumstances. This paper is structured as follows; Antenna design section deals with the antenna design and fabrication methodology. The bending assessment section analyses the antenna's performance in on-body and various bending circumstances. Finally, a conclusion is drawn with the obtained results in different on-body and bending aspects.

2. The Antenna Design

The developed textile antenna design is inspired by a Christmas tree shape, and a staircase-like defective ground element supported it. The textenna is actualised on a polyester-based textile substrate with a footprint of 30×20×1 mm^3, as shown in figure 1. Ansys High-Frequency Structure Simulator (HFSS) version 19 software was used to model the antenna's physical parameters. The overall measurements of the proposed antenna are H4=3, h3=3, h2=3, h1=3, FL1=3.5, Gh=15, L6=5, L5=2.5, L4=2.5, L3=3.5, L2=3.5, L1=4.5, Lf=15, Ls=30, Ws=20. The simulated antenna is practically fabricated on a polyester-based material, and the radiation parts (patch and ground) are actualised using adhesive conductive copper tape. The propounded antenna operates in treble bands of Telemetry, ISM and X-band at 4.5, 5.8

and 8 GHz frequencies. The prototype model is illustrated in figure 2 and practically validated resonating frequencies are shown in figure 3.

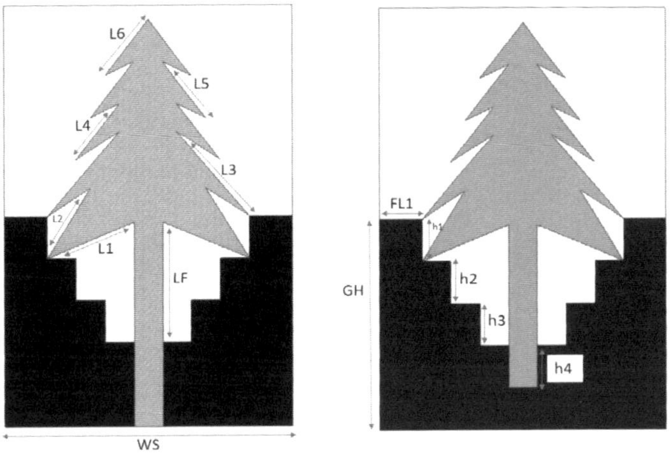

Figure 1. The propounded antennas complete view and entire topology.

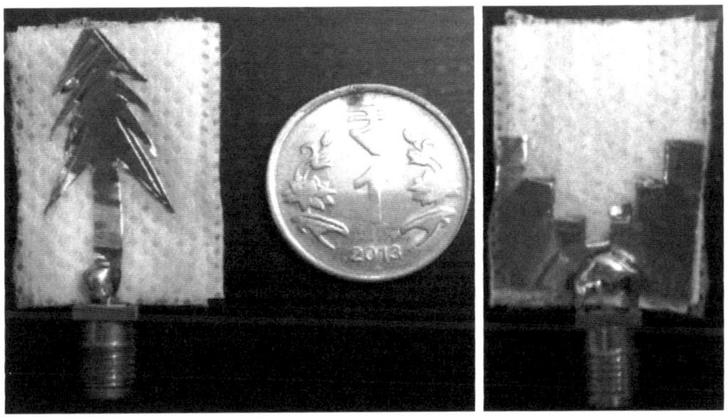

Figure 2. The fabricated prototype on polyester material.

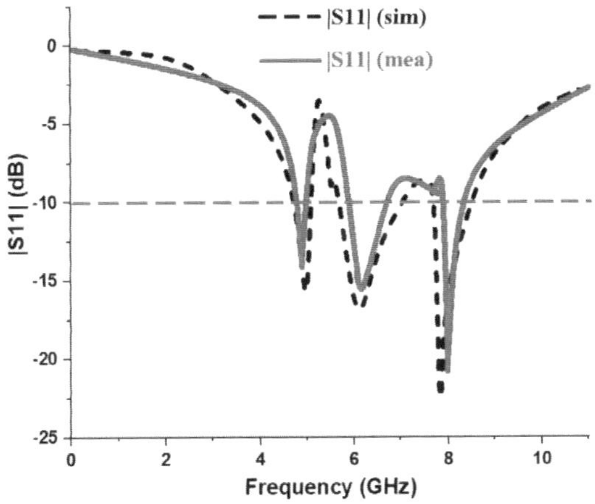

Figure 3. The simulated and validated resonating frequencies.

The designed textenna is meant for on-body wireless communication, so the textenna performance while positioning it on the human body and functionality performance concerning bending it in various dimensional aspects are detailed in the below section.

3. Bending Analysis

3.1. Modelling of Textenna Bending

Several methodologies exist to assess the bending consequences on resonating characteristics of textennas: (a) Keeping the textile antenna dimensions constant as in a deflated form and assessing the impacts of flexing without affecting the textennas size. (b) By adjusting the size of the patch, the frequencies shift can be reverted to the flat circumstance resonant frequency state. (c) The patch of the textenna must be tuned up at regular intervals using adaptable methods such as reconfigurable approaches for every flexing position. In the present bending assessment analysis, approach-1 must be considered because it can provide the significance of bending on the textile antenna's performance without altering the overall dimensions of the antenna. The bending angle is fully responsible for the textenna's curving during bending. It's the curving angle created by stretching the antenna's textile substrate over a cylinder with a radius of Rc. Each bending angle creates a unique bending radius for a texteena with distinct measurements. The cylinder structure bending radius is calculated using equation 1.

$$Rc = \frac{(T\times 360°)}{(\theta\times 2\pi)} \tag{1}$$

Where T is the antennas' substrates length.

This study analyses the propounded textile antennas' functional behaviour by bending in both horizontal and vertical dimensional aspects. While wearing this antenna as a piece of clothing, both the vertical and horizontal bending circumstances happen due to the human body's heterogeneous nature. Practical measurements are taken by moulding the proposed antenna on a cylindrical foam surface with radius Rc, as illustrated in figure 4.

Figure 4. Proposed antenna mounted on cylindrical foam realising horizontal bending conditions.

3.2 Textile antenna: Different types of Bending Analysis

Two types of bending characteristics are investigated in this work. They are divided into two categories: (1) vertically perspective turning (bending with respect to length) and (2) horizontal perspective turning (i.e., bending with respect to width). In both situations, the bending position ranged from 0 to 45 degrees. The measurement setup of the textenna while bending in an anechoic chamber is illustrated in figure 5.

3.3 Vertical bending analysis

Vertical bending situations arise when the textenna is twisted in its length plane. The propounded polyester antennas vertical flexibility response is evaluated in the following section by twisting vertically at various bending angles ranging from 0-45 degrees with a step size of 15 degrees.

3.4 Resonating frequencies behaviour concerning bending angles

The propounded textile antenna operates at 4.3-4.55, 5.6-7, and 7.65-8.2 GHz in a normal flat state. When it was bent 15 degrees vertically, the resonating frequencies shifted slightly towards the higher frequencies and functioned at 4.35-4.57, 4.63-7.19, and 7.72-8.34. When the antenna is furtherly bent 30 degrees in the vertical aspect, the operating frequencies are shifted further towards higher bands and resonated at 4.41-4.61, 4.72-7.31, 7.8-8.45. When the antenna is bent over 45 degrees, unlike in other bending circumstances, they are shifted towards the lower bands and function at 4.19-4.5, 5.49-6.84, 7.56-8.11. A comparative analysis of the antenna functioning behaviour in different bending aspects are tabulated in table 1.

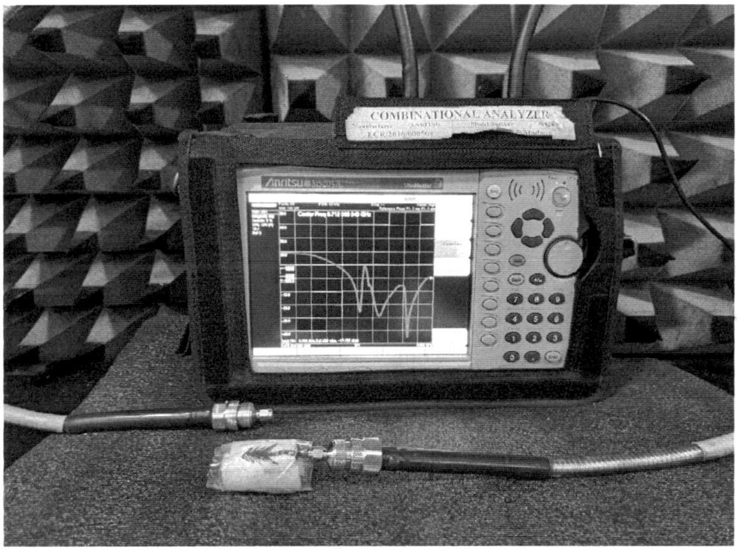

Figure 5. Measurement setup of the polyester antenna when mounted on cylindrical foam.

Table 1. A comparative analysis of flat state and different vertical bending circumstances.

S. No.	State /Bending angle	Operating Frequencies
1	Flat	4.3-4.55,5.6-7,7.65-8.2
2	15-Degrees vertical	4.35-4.57,4.63-7.19,7.72-8.34
3	30-Degrees vertical	4.41-4.61,4.72-7.31,7.8-8.45
4	45-Degrees vertical	4.19-4.5,5.49-6.84,7.56-8.11

3.5 Horizontal bending analysis

Horizontal bending situations arise when the textenna is twisted in its width plane. The propounded polyester antenna horizontal flexibility response is evaluated in the following section by twisting horizontally at various bending angles ranging from 0-45 degrees.

3.6 Resonating frequencies behaviour concerning bending angles

The propounded textile antenna operates at 4.3-4.55, 5.6-7, and 7.65-8.2 GHz in a normal flat state. When it was bent 15 degrees horizontally, the resonating frequencies shifted slightly towards the lower frequencies and functioned at 4.3-4.55, 5.6-7, and 7.65-8.2. When the antenna is furtherly bent 30 degrees in the horizontal aspect, the operating frequencies are shifted further towards lower bands and resonated at 4.19-4.49, 5.48-6.81, 7.42-8.01. When the antenna is bent over 45 degrees, unlike in other bending circumstances, they are shifted towards the higher bands and function at 4.36-4.62, 5.71-7.3, and 7.73-8.34. A comparative analysis of antenna functioning behaviour in different bending aspects is tabulated in table 2.

Table 2. A comparative analysis of flat state and different horizontal bending circumstances.

S. No	State/Bending angle	Operating Frequencies
1	Flat	4.3-4.55,5.6-7,7.65-8.2
2	15-Degrees horizontal	4.21-4.52,5.51-6.92,7.59-8.13
3	30-Degrees horizontal	4.19-4.49,5.48-6.81,7.42-8.01
4	45-Degrees horizontal	4.36-4.62,5.71-7.3,7.73-8.34

4. On-body Functioning Behaviour

The antenna's resonating behaviour was assessed by directly positioning it on the subject's body. As seen in figure 6, the textenna was placed directly on the human body and examined in an anechoic chamber. While placing the prototype polyester antenna on the human body, a slight variation in the resonating frequencies is reported, operating at 3.45, 4.93, and 5.78 GHz. These reported variations are due to human tissue lossy and conductive behaviour.

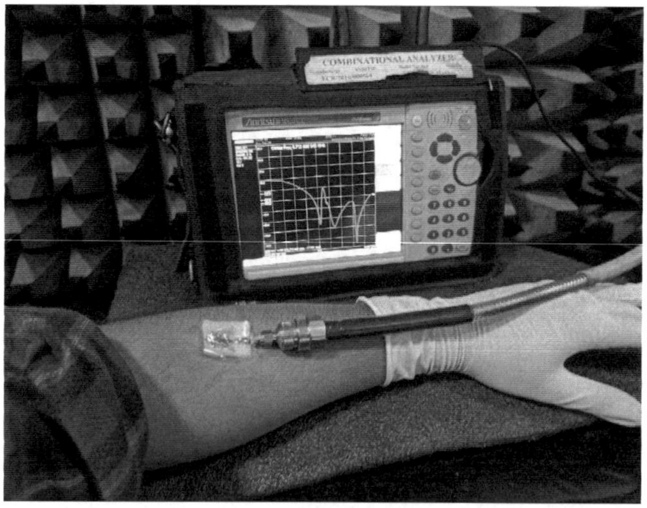

Figure 6. On-body measurement setup of the propounded polyester textenna.

Table 3. A comparative analysis of the proposed antenna with some existing literature

S.NO.	Antenna material	Dimensions (mm^2)	Operating Frequencies (GHz)	Bending analysis	Direct on-body evaluation
2	Condura	120×80	0.9, 1.8	No	No direct contact; was tested while embedded in a coat
6	Cotton	56×50	2.46	Yes	No direct testing on the human body
9	Felt	60×20	2.45	No	No
11	Denim	70×65	2.45, 5	No	No
12	Jean	64.1×50	2.45	No	No
17	Felt	72×66.46	2.45	No	No
This work	Polyester	30×20	4.5, 5.8, and 8	Yes	Yes

Table. 3 provides a comparative analysis between the proposed textenna with some existing literature. The analysis concluded that the propounded textenna is compact, operates with treble bands, and can function in various vertical and horizontal bending aspects. These results furtherly confess the excellent performance of the textenna for body wireless communication applications.

5. Conclusion

In this communication, a compact Christmas tree-shaped textile antenna is proposed on polyester material. The textennas radiating and ground layers conductivity is drawn with the help of adhesive copper tape. This simple fabrication method is found helpful in fabricating textile antennas on large scales. The developed antenna operates in a triple band of Telemetry, ISM and X-band at 4.5, 5.8 and 8 GHz frequencies. The prototype was methodically examined as a body-worn antenna by bending it in various vertical and horizontal bending angles and tested directly on the subject's skin. A marginal variation of resonating frequencies is reported in both the horizontal and vertical bending examination; however, they even fall under the limits of Telemetry, ISM and X-band applications. So the antenna has shown good grades of flexibility and is also compatible for on-body communication at the application bands of Telemetry, ISM and X-band.

References

[1] Kumar, Jayendra, Banani Basu, Fazal Ahmed Talukdar and Arnab Nandi 2018 Multimode-inspired low cross-polarisation multiband antenna fabricated using graphene-based conductive ink *IEEE Antennas Wirel. Propag. Lett.* **17** 1861-1865
[2] Loss, Caroline, Ricardo Gonçalves, Catarina Lopes, Pedro Pinho and Rita Salvado 2016 Smart coat with a fully-embedded textile antenna for IoT applications *Sensors* **16** 938
[3] Mohamadzade, Bahare, Raheel M Hashmi, Roy BVB Simorangkir, Reza Gharaei, Sabih Ur Rehman and Qammer H Abbasi 2019 Recent advances in fabrication methods for flexible antennas in wearable devices: State of the art *Sensors* **19** 2312
[4] Stempien Z, E Rybicki, A Patykowska, T Rybicki and M I Szynkowska 2018 Shape-programmed inkjet-printed silver electro-conductive layers on textile surfaces *J. Ind. Text.* **47** 1321-1341
[5] Agu, Daniel, Rachel J Eike, Allyson Cliett, Dawn Michaelson, Rinn Cloud and Yang Li 2021 Investigation of e-textile dipole antenna performance based on embroidery parameters *Text. Res. J.*

[6] Anbalagan, Abirami, Esther Florence Sundarsingh, Vimal Samsingh Ramalingam, Aadesh Samdaria, David Ben Gurion and Karthik Balamurugan 2020 Realisation and analysis of a novel low-profile embroidered textile antenna for real-time pulse monitoring *IETE J. Res.* 1-8
[7] Simorangkir, Roy BVB, Duc Le, Toni Björninen, Abu Sadat Md Sayem, Maxim Zhadobov and Ronan Sauleau 2018 Washing durability of PDMS-conductive fabric composite: Realising washable UHF RFID tags *IEEE Antennas Wirel. Propag. Lett.* **18** 2572-2576
[8] Ivsic, Branimir, Davor Bonefacic and Juraj Bartolic 2013 Considerations on embroidered textile antennas for wearable applications *IEEE Antennas Wirel. Propag. Lett.* **12** 1708-1711
[9] Gil, Ignacio, Raul Fernández-García and José Antonio Tornero 2019 Embroidery manufacturing techniques for textile dipole antenna applied to wireless body area network *Text. Res. J.* **89** 1573-1581
[10] Alemaryeen, Ala and Sima Noghanian 2019 On-body low-profile textile antenna with artificial magnetic conductor *IEEE Trans. Antennas Propag.* **67** 3649-3656
[11] Wiri Abraham T 2021 Stitched Ground Planes for Textile Antenna Application: An Experimental Study *Open J. Antennas Propag.* **9** 11-26
[12] Kavitha A and J N Swaminathan 2019 Design of flexible textile antenna using FR4, jeans cotton and teflon substrates *Microsyst. Technol.* **25** 1311-1320
[13] Sandeep, D Ram, N Prabakaran, B T P Madhav, K L Narayana and Y Pratapa Reddy 2020 Semicircular shape hybrid reconfigurable antenna on jute textile for ISM, Wi-Fi, Wi-MAX, and W-LAN applications *Int. J. RF Microw. Comput.-Aided Eng.* **30** e22401
[14] Lajevardi, Mir Emad and Manoochehr Kamyab 2017 Ultraminiaturized metamaterial-inspired SIW textile antenna for off-body applications *IEEE Antennas Wirel. Propag. Lett.* **16** 3155-3158
[15] Prabakaran N and Madhav Boddapati TP 2020 Circularly Polarized Jute Textile Antenna for Wi-MAX, WLAN and ISM Band Sensing Applications *Appl. Comput. Electromagn. Soc. J.* **35** 1493-1499
[16] Sandeep D Ram, N Prabakaran, B T P Madhav, K L Narayana and P Rakesh Kumar 2020 Systematic investigation from material characterisation to modeling of jute-substrate-based conformal circularly polarised wearable antenna *J. Electron. Mater.* **49** 7292-7307
[17] Ullah, Irfan, Mahmoud Wagih, and Steve P Beeby 2022 Design of Textile Antenna for Moisture Sensing *Eng. Proc.* **15** 1-4

Hybrid Multi-Level Source Inverter Design and Validation in Solar Photo-voltaic Renewable Energy System

Thirumalini P[1], M Murali[1] and Arulmozhiyal R[1]
[1]Department of Electrical and Electronics, Sona College of Technology, Salem - 5, India
E-mail: muralimunraj@gmail.com (M Murali)
E-mail: arulmozhiyal@sonatech.ac.in (Arulmozhiyal R)

Abstract. This paper proposes the hybrid multi-level source inverter in light of staggered coordinated framework for boosting photo voltaic power. Hybrid multi-level inverter (MLI) proposed to design less number of switches compare with other MLIs of diode clamped multi-level inverter, the smallest number of switches are required to accomplish the highest steps in staircase AC output voltage. HMLIs additionally include variable info voltage and carries out load guideline to keep up with consistent result voltage. The HMLIS intends to operate in multiple stages during shoot through stage exchanging beat width (PWM) adjustment keeps up with to work in buck and lift mode. The activity of proposed work implemented in MATLAB and discuss effectiveness with diminished sounds took care of inverter operation. Model of converter is planned and accomplished the most extreme effectiveness of 95.4% as information voltage goes from 10-60 volts and result of half breed staggered source inverter keeps up with steady of 230v dc. Trial results approve the exhibition of proposed half and half staggered source inverter for exchanged mode power supplies utilized in PC.

Keywords: Multilevel, Inverter, Hybrid, PWM, Solar, Controller

1. Introduction

Global energy consumption is expected to grow at a rate of 1.6-1.7 percent each year. By 2050, worldwide consumption will have increased to 28 TW. As the world's energy demand grows, the majority of it is met by burning fossil fuels, which contributes to global warming and pollution. Coal power stations emit 59 percent of total sulphur dioxide and 18 percent of total nitrous oxide emissions each year. Only about 15% of coal is used to generate electricity. The total carbon emitted to produce one unit of power is 0.8 kg. Renewable energy is a newly developed energy production method that generates pollution-free, environmentally friendly energy from natural sources such as sunlight, wind, geothermal, and tidal energy. Renewable energy, by supplying clean energy, will help to the worldwide effort to face the difficulties of climate change.

PV energy is increasing popularity as a result of its widespread availability. In India, the sky is clear for 280 days out of the year. The earth's surface has about 90PW of PV power that can be converted to electric energy. One hour of PV energy received by the planet can provide all of the world's energy needs for a year. The proper use of this energy can make a significant contribution to global energy output. The major goal of this project is to model and develop a system that efficiently supplies continuous PV to a single phase grid. Low voltage PV cells are linked in a series to get a high DC output voltage to interface solar energy to the grid [2]. This method necessitates high-rated voltage inversion equipment and a step-up transformer, which increases the system's losses, cost, weight, and dimensions [3]. The transformer can be removed to solve these problems.

As a result, a multilayer inverter (MLI) integrates the staircase waveform near a sine waveform using the power of semiconductor device sources. In an SPV system, the multilayer inverter is the best solution for the majority of power output. According to [4,5], typical 2-level inverters suffer from high switching voltage stress, low efficiency, and poor power quality. Among the accessible saddling frameworks of, the proposed framework has been ended up being more effective. The traditional tackling framework comprises of DC-DC converter, charge regulator and an inverter. There are a few phases of change. Along these lines it builds the intricacy, size and cost of the framework [6-7]. The proposed framework comprises of staggered Z source (PV) inverter and control circuits with single-stage to deal with an inconsistent number of PV strings with various electrical boundaries, areas, and directions is presented and hence lessening the expense and decrease absolute symphonies distortion[8]. Crossover staggered source inverter collects the most elevated conceivable power from each string freely, despite the fact that they may be at divergent irradiance levels and working temperatures [9-10]. In this venture, PV boards at wanted electrical circumstances is re-enacted utilizing MATLAB/SIMULINK and an itemized examination is done utilizing hybrid multi-level source inverter and fuzzy rationale regulator progressively exploratory setup.

2. PV Power Generation

PV source

A PV module is made up of solar cells that are joined together as shown in Figure 1. PV modules are typically grouped in series or parallel, depending on the requirements. PV modules of various ratings are commercially available in the market. The shunt resistance (RSH) in this variant is extremely high. As a result, it has no effect on the circuit. As a result, it is overlooked. Consider only a current source (I), a diode, and series resistance (Rs). As a result,

I = Isc – Id Id = Io (eqVd/kT - 1) (1)

I is the output current obtained from the solar array. Where Io is the diode's reverse saturation current, q denotes the electron charge, Vd denotes the voltage across the diode, and k denotes the Boltzmann constant (1.38 * 10-19 J/K). T - Kelvin Junction Temperature (K).

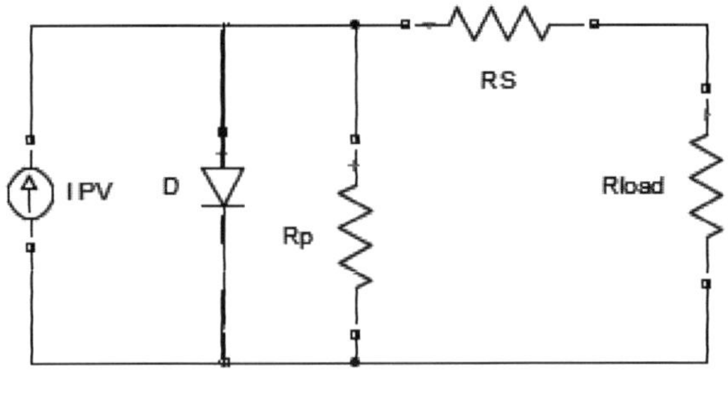

Figure 1. Solar Photo Voltaic cell

I =Isc – Io (eq((V+IRs)/nkT)- 1) (2)

Where I- denotes photovoltaic cell current, V denotes photovoltaic cell voltage, T denotes temperature (in Kelvin), and n is the Diode Ideality Factor. T is Variations in the environment's weather, on the other hand, have a direct impact on the PV system's production. To get the most power out of the PV array as source in need to performs active tracking system.

Control feedback system

Feedback mechanism developed for control and operates hybrid multi-level inverter that attributes that allows the output to be compared to the reference input to provide an error signal that may be used to determine the appropriate controlling action as shown in Figure 2.

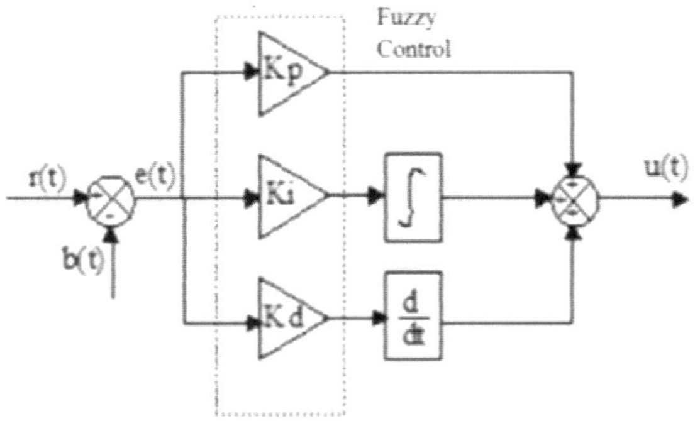

Figure 2. Feedback Control for HLI

The controller is given a generic closed loop feedback control system as part of the output, which is determined by the sensor and feedback element.

3. Simulation Result and Discussion

The fuzzy controller is simulated in MATLAB and THD data are provided for comparison between linear and nonlinear loads as single phase grid loads. Fuzzy logic is used as a controller drive PV power generating system to increase the output performance of the. Figure 3 shows a simulation schematic of a hybrid multi-level source inverter with fuzzy logic control for PV power generation. The proper gate pulses for the switching devices of the inverter are created using this approach, which uses the errors of the load current and input current of the hysteresis controller.

The simulated waveforms of the fuzzy sliding mode controller outperform the fuzzy logic controller, as seen in the accompanying figures, where Fig.4 and Fig.5 respectively indicate load voltage and load current observed when the fuzzy sliding mode controller is used.

The load voltage was measured at 229V and the load current was measured at 2.40 amps, with harmonics levels of 3.95 percent and 3.95 percent of the linear load, respectively, as shown in Fig.5 and Fig.6.

IOP Conf. Series: Materials Science and Engineering 1258 (2022) 012054 doi:10.1088/1757-899X/1258/1/012054

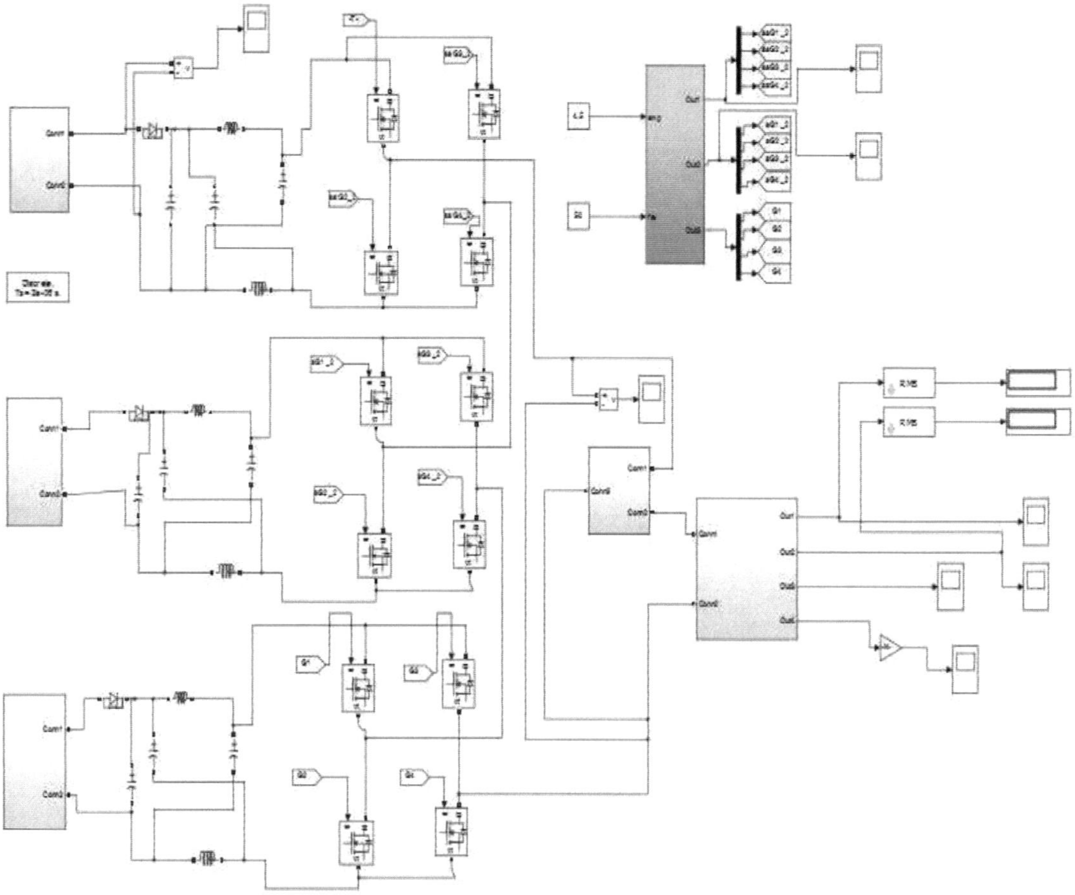

Figure 3. Simulation diagram of hybrid multi-level source inverter using fuzzy controller for PV power generation

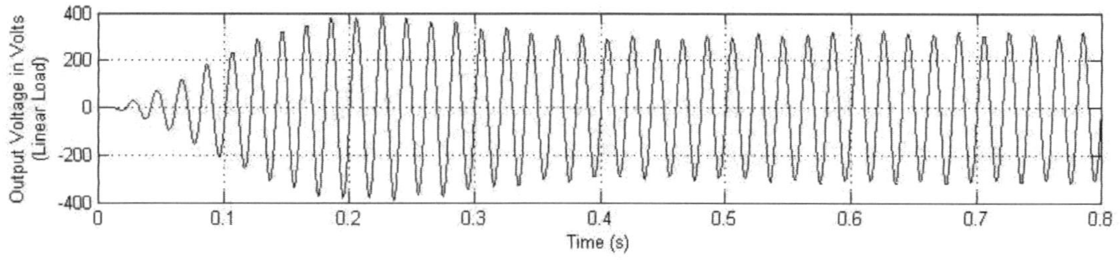

Figure 4. Output voltage of the linear load

IOP Conf. Series: Materials Science and Engineering 1258 (2022) 012054 doi:10.1088/1757-899X/1258/1/012054

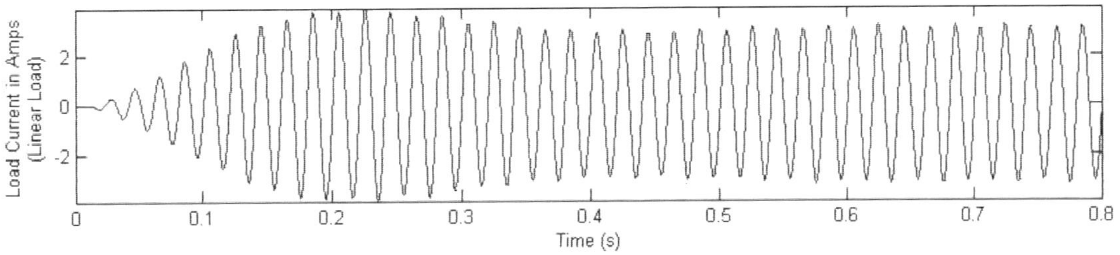

Figure 5. Load current of the linear load

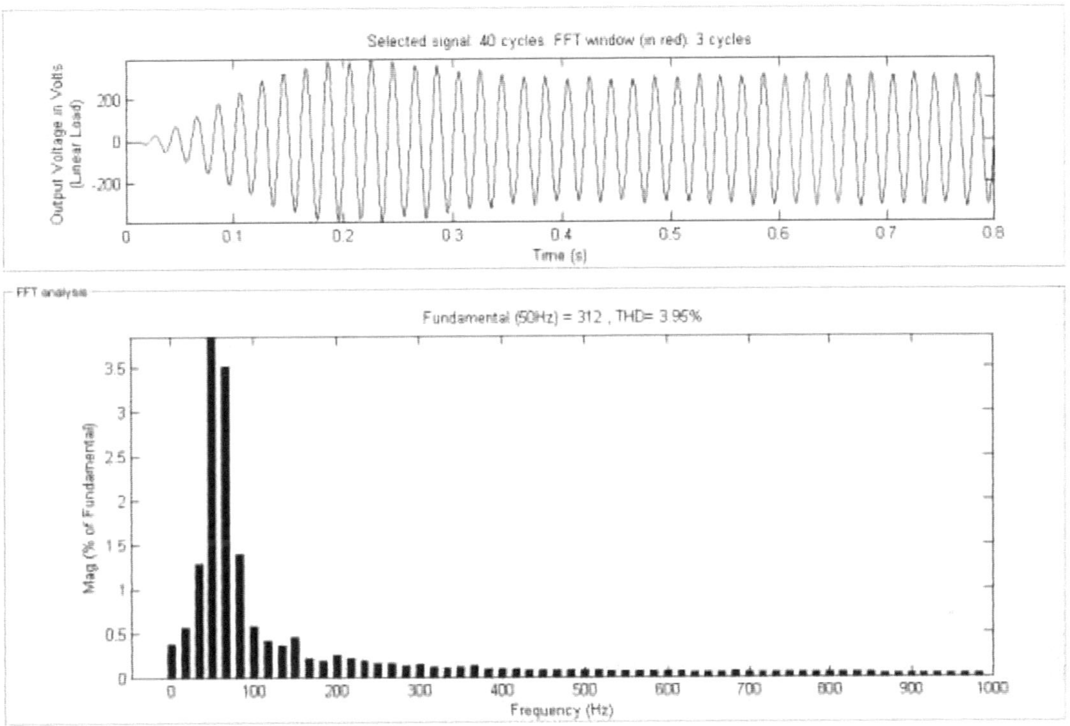

Figure 6. Total Harmonic Distortion for load voltage

IOP Conf. Series: Materials Science and Engineering 1258 (2022) 012054 doi:10.1088/1757-899X/1258/1/012054

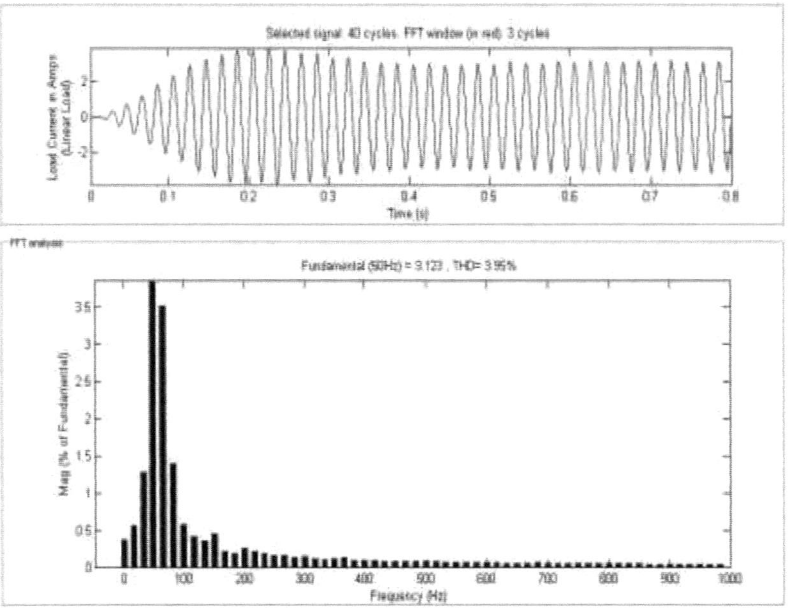

Figure 7. Total Harmonic Distortion of the load current

With observation of the characteristics of the nonlinear load by looking at the simulated waveforms of the output voltage and load current of the PV power generating system with fuzzy. The simulated waveforms of the output voltage and load current show only minor disturbances, but they perform better when compared to simulated waveforms of the nonlinear load with the fuzzy logic controller, as shown in the following figures, where Fig.8 and Fig.9 show the load voltage and load current observed when the fuzzy sliding mode controller is used with a nonlinear load, respectively.

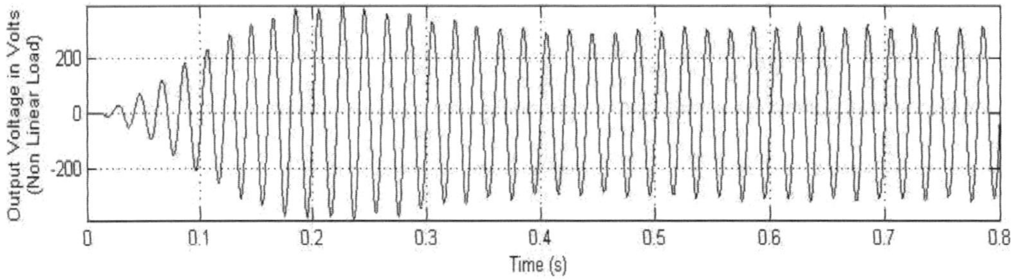

Figure 8. Output voltage of the nonlinear load

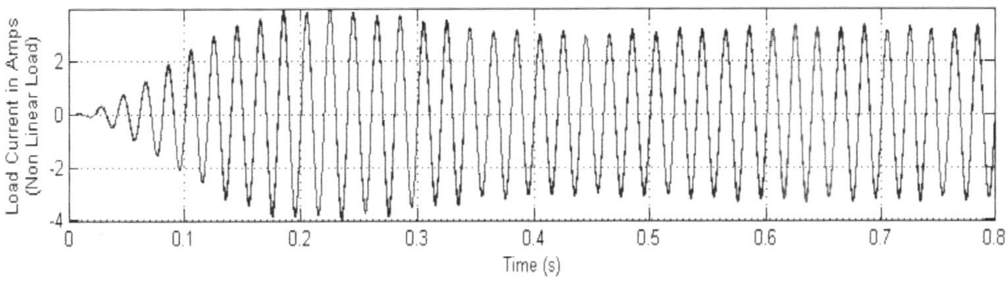

Figure 9. Nonlinear load current

The load voltage was estimated at 226V and load current was estimated at 2.33 amps with sounds level of burden voltage and burden current is 4.00 % and 5.55% separately of the straight burden as displayed in the Fig.10 and Fig.11.

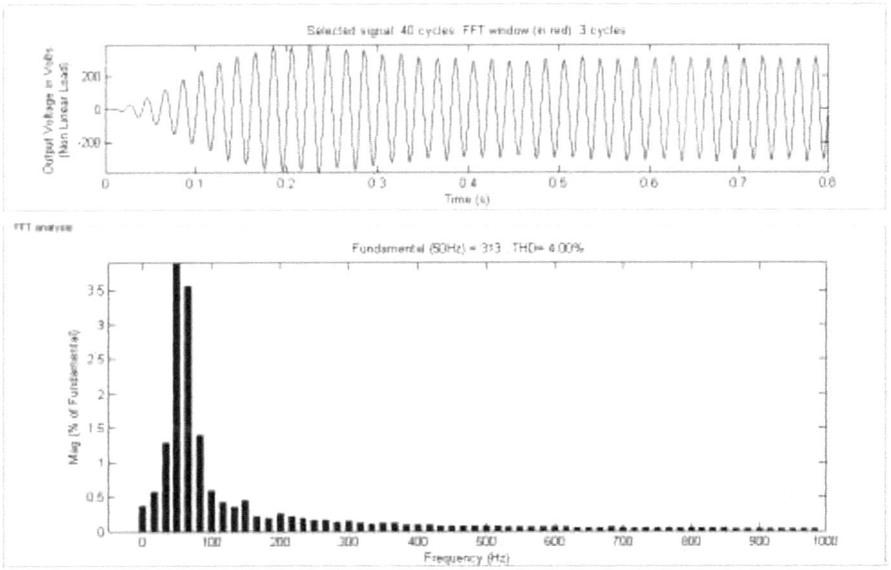

Figure 10. Total Harmonic Distortion for voltage

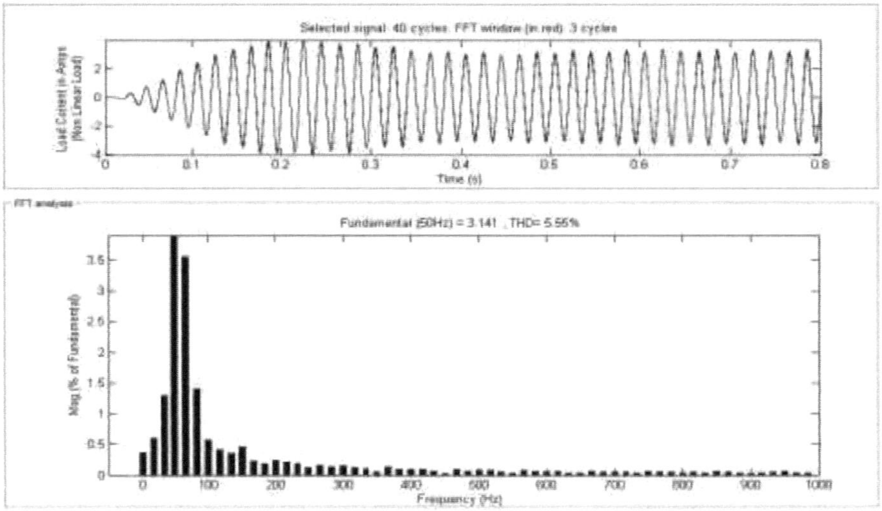

Figure 11. Total Harmonic Distortion of the load current

Table.1 THD value of output voltage and load current for linear and nonlinear load.

Type of Load (W)	$V_{O(rms)}$ Volts	Load Current(I_{load}) amps`	% THD for V_O	%THD for I_{load}
Linear	229	3.4	3.95	3.95
Nonlinear	226	3.3	4.00	5.55

Table no. 1 shows the overall harmonic distortion of output voltage and load current for linear and nonlinear loads. Because of its nature, the THD value of current waveforms is higher in nonlinear loads, and this value is not acceptable per IEEE standards.

4. CONCLUSION AND FUTURE SCOPE

The proposed work constructed with reduced switches, diodes, voltage sources, fundamental voltage, and THD %, the proposed work compared with similar designs of multi-level inverter. The efficient hybrid multi-level source inverter with PV power generation fed single phase power grid is proposed in this research. In addition, to coordinate hybrid multi-level source inverter switching, a complete control technique based on fuzzy logic control is devised. The simulation model was constructed in the MATLAB environment, and the simulation results and THD analysis show that current ripples have been reduced. A linear load and a nonlinear load were simulated and the results were given to illustrate the efficacy of the suggested system. The harmonics are significantly decreased, lowering the distortion factor and overall harmonic distortion (THD). The output waveform of the H-MLI clearly exhibits less THD. As a result of the lower harmonics, the output power quality improves. Further the proposed work will be implemented in dSPACE controller to evaluate performance of inverter in real time with solar PV as input source.

References

[1] Khajesalehi J, Sheshyekani K, Hamzeh M and Afjei E, 2015 High-performance hybrid photovoltaic -battery system based on quasi-Z-source inverter: application in micro grids IET Generation, Transmission & Distribution 9 895-902.

[2] Carlos Meza, Domingo Biel, Juan Martinez and Francesc Guninjoan, 2005 Boost-Buck inverter variable structure control for grid-connected photovoltaic systems IEEE International Symposium on Circuits and Systems 2 1318-1321.

[3] Mario Fortunato, Alessandro Giustiniani,Giovanni Spagnuolo and Massimo Vitelli 2008 Maximum Power Point Tracking in a One- Cycle-Controlled Single-Stage Photovoltaic Inverter' IEEE Transactions on Industrial Electronics, 55, 2684-2693.

[4] Domingo, Cortes, Nimrod Vázquez & Jaime Alvarez-Gallegos 2009 Dynamical Sliding-Mode Control of the Boost Inverter IEEE Trans. Industrial Electronics 56 3467-3476.

[5] Guzman R, Garcia de Vicuna, Camacho L, Matas A, Castilla J & Miret M 2013 Active damping control for a three phase grid connected inverter using sliding mode control Industrial Electronics Society, IECON 2013 - 39th Annual Conference of the IEEE, Vienna 382 – 387.

[6] Murali M, Arulmozhiyal A & Sundaramoorthy P 2014 Design and Analysis of Voltage Source Inverters for Renewable Energy Applications Telkomnika Indonesian Journal of Electrical Engineering 12 8114-8119.

[7] Thirumalini P, Murali M & Arulmozhiyal R 2016 Implementation of Multilevel Z-Source Inverter using Solar Photovoltaic Energy system for Hybrid Electric Vehicles Computation and Communication Technologies,2016.

[8] Liu H, Li Y, Zhou Z, Wang W and Xu D, 2019 A Family of Low-Spike High-Efficiency Y-Source Inverters IEEE Transactions on Industrial Electronics 66 9288-9300.

[9] Ayala-Chauvin, M, Kavrakov, Buele B.S, Varela-Aldás J 2021 Static Reactive Power Compensator Design, Based on Three-Phase Voltage Converter Energies 14.

[10] Thirumalini P, Arulmozhiyal R and Murali M 2015 Cascade Multilevel Z-Source Inverter for Photovoltaic System Telkomnika Indonesian Journal of Electrical Engineering 16 473-479.

Accurate and High Speed Garbage Detection and Collection Technique using Neural Network and Machine Learning

Usha S M[1], **Mahesh H B**[2]
[1]JSS Academy of Technical Education, Bangalore, Karnataka India
[2]PES University, Bangalore, Karnataka, India

Abstract. Due to the increasing development of cities' populations, which has resulted in massive garbage output, waste management systems in urban areas are confronting issues. The ravage of possessions can be employed powerfully with the incorporation of the internet of things (IoT), TensorFlow based deep learning model, as conventional ravage managing system are extremely uneconomical. The major goal of this study is to create a smart waste management system based on a deep learning model that optimizes trash isolation and allows for bin status monitoring in an IoT context. Yolo real time object detection algorithm is employed and educated with a dataset that includes paper, cardboard, glass, metal, and plastic for garbage sorting and grouping. Yolo algorithm enhances the detection speed and yields precise findings with low background noise. Yolo uses convolutional neural network to detect the object. The camera module detects garbage and the servomotor linked to a plastic board, categorizes the waste into the appropriate waste cubicle using the educated model on TensorFlow Lite and Raspberry Pi 4. The garbage fill is monitored by an ultrasonic sensor, and the latitude and longitude are obtained in real time by a GPS module. The smart bin's LoRa module transmits the bin's status to the LoRa receiver at 915 MHz. The smart bin's electronic mechanisms are safeguarded by an RFID-based locker that can only be opened with a registered RFID badge for maintenance or upgrades. This work is framed out of the technologies such as Robotics, neural network, Internet of Things and deep learning algorithm. The garbage detection system is more precise and faster than the other existing methods. The YOLO algorithm can predict objects in real time, which speeds up detection. It's a prediction method that produces exact results with little background noise. The algorithm has outstanding learning capabilities, allowing it to learn and apply object representations to object detection.

1. Introduction

According to the World Bank, roughly 2.01 billion tonnes of trash were generated in 2016 as a result of population and economic expansion in metropolitan areas, and this number is expected to rise to 3.40 billion tonnes by 2050. According to EUROSTAT, the European Union recycled 423 million tonnes of waste in 2016, accounting for 56% of locally produced excess. Waste management, often known as waste collection, entails a series of activities and movements to effectively achieve excess discarding, including waste collection, transportation, monitoring, and regulation. Waste management procedures differ between urban and rural settings. In most cases, the best way to deal with gathered waste is to reprocess and recondition it. The expense of good surplus authorization, on the other hand, is substantial, necessitating collaboration from both authorities and users. Although the government or authorities have made significant attempts to enhance the waste management system, it remains a major issue in every country, particularly in urban areas. If the waste is collected on time, one of two things happens: either the bins aren't entirely filled, or they overflow. The waste overflow would pollute the environment, resulting in air pollution and the spread of dangerous diseases. Recycling garbage is another way to cut down on waste generation and help the environment. However, because of the consumers' incompetence in appropriately categorizing the garbage, this strategy does not yield favorable outcomes. The rapid advancement of the digital world has had a significant impact on technological advancements, particularly by allowing intelligence to be integrated into current technology [1]–[3], commonly known as the Internet of Things (IoT). IoT-enabled technologies have taken numerous fields, such as engineering; to a whole new level of development. IoT market is predicted to grow at a rapid pace. The figure will be 1.1 trillion by 2023.

The present unwanted waste material organization system can be improved with using Internet of things. Many areas of the organization system can be enhanced, including the price of resources user- friendly, manageable and cost-effective. Garbage recycling is a good way to get rid of it. A trash management system that is connected to the internet (IoT) is designed. Aside from IoT, "machine learning" is additional discipline which has grown and took a significant effect. Machine learning allows computers or machines to learn unsupervised from labelled or un-labelled data. Machine learning entails the use of facsimiles and algorithms which predict the results of methodical research. Machine learning has reached its pinnacle since it provides the most astounding computational features. According to recent figures [4], the market for machine learning and AI-based technologies was 1.4 billion dollars in 2016, and it is expected to grow to 59.8 billion dollars by 2025. Such figures demonstrate C utility of machine learning-based enactment. In several applicable target contests in machine learning and pattern recognition, profound neural network results revealed enhanced accuracy.

Deep learning is a machine learning technique that allows computers to learn from their failures [6]. Using several layers of processing, deep learning allows computational models to grasp a data presentation with multiple layers of abstraction. These technologies help us produce cutting-edge technology in areas including speech recognition, visual objects, medicine research, and genomics. Convolutional neural networks (CNNs) are a type of deep learning that may be used to interpret images, recognize objects, and perform other tasks [8]. CNN can greatly improve trash classification and categorization accuracy in a smart waste management system, saving resources and cutting global waste output. A smart waste management system is presented in this study that detects and categorizes various sorts of garbage before depositing them in a specified compartment.

In this work, a waste management system is designed making use of technologies such as Robotics, convolution neural network-based Yolo algorithm, deep learning and deep internet of things. This paper describes the related work.

2. Related work

Several writers have proposed several ways for monitoring the quantity of garbage in the dustbin, including an infrared sensor that measures detachment by reproducing light impressions and an ultrasonic sensor that measures distance using the standard of reproduced sound waves. In this paper presented a strategy to lower the cost of waste applications while increasing their efficiency [3]. A trashcan is connected to a microcontroller-based scheme that includes infra-red wireless schemes that shows the current waste status. A GSM electronic monitoring system that sends a message to the expert when the trashcan is completely occupied, indicating that the trash car should be dispatched [4].The amount of trash in the trashcan was detected using ultrasonic sensors, and the GPRS module was utilized to deliver message on the status of the dustbin. A vitality-effectual telecommunication etiquette that portable long spaces can be implemented into the shrewd waste management scheme to screen the prominence of the container. There are many distinct types of telecommunicationprotocols, each with its own set of strengths and weaknesses in particular contexts. Smart garbage bins [5] use ultrasonic sensors and a GSM-enabled microprocessor to relay data to a control unit to provide information about the amount of filling. Another study [6] uses various sensors to maintain the surplusand state of the container, including an ultrasonic sensor, a moisture sensor, and a gas sensor. The garbage level is monitored using an ultrasonic sensor. Moisture sensors can detect wet trash, while poisonous waste can be detected with toxic waste sensors. Gas sensors can be used to detect gases.

The microcontroller collects data from sensors and sends it over a great distance via the ZigBee transmitter. In addition, the microcontroller transmits SMS messages to the cellular phone through global system mobile. Apart from that, in 2016, a transducer node equipped with RFID technology for monitoring the level of garbage bin filling was presented; this might be a viable explanation due to its resilience and cheap [7]. Because 2G communication is costlier, it cannot be perfect solution for the future applications. [8], leading 2G communications to cease operations. This will make it impossible to use IoT devices that communicate using GSM standards. LoRa offers a free spectrum under 1 GHz and can broadcast up to 15 kilometers. In intelligent waste segregation scheme, a system to look after the levels of the container is critical, since it allows authorities to more easily manage the entire conditions of all dustbins. A cloud-based trash management system [10-11] was presented in this paper. Due to the longer range supplied by ultrasonic sensors compared to IR sensors, ultrasonic sensors are utilized to intelligence the status of garbage in the dustbin in their experiment. Aside from that, sunshine, object color, and object hardness have been discovered to impact IR sensors. Their method can detect the amount of waste created as well as the quantity of biogas produced in the publicregion [4]. The data collected by instruments is transferred to a central system over the internet, where it is warehoused and managed. This information is then utilized to track the garbage container, and the appropriate waste container is selected for collection. Daily waste levels are monitored and accordingly containers are arranged. Aside from that, the scheme will forecast upcoming events based on experience, such as the accessibility of automobiles nearby the location and other variables.

In this paper an IoT-based technique for intelligent waste monitoring in urban areas [9], this system monitors the status of the container through graphical user interface. The level of garbage is detected and monitored through a GUI. A GUI is designed to track the needed details pertinent to the waste for several selected places in order to have greater control over garbage disposal. This mechanism ensures that trash in the bins is collected as soon as the amount of garbage reaches a certain level. Furthermore, by recognizing bogus reports, this technology eradicates exploitation in the whole garbage monitoring scheme. The number of rubbish collection visits by vehicle has been cut, lowering the overall waste collection costs. When the waste bins are nearly or completely full, a smart waste management system requires automation to notify the authorities about the condition of the bins.A smart garbage alarm system is proposed in this study, which is an integrated system consisting of anArduino Uno, GSM Module, Ultrasonic sensor, and LED light [6]. The ultrasonic sensors are used to detect two threshold levels: 70% and 90%, respectively, of the bin height. When the first threshold level is reached, the green LEDs will come on to inform the inhabitants on that floor, and a first warning message will be sent to the municipality.

3. The proposed waste management system hardware module

Traditional garbage bins are normally divided into categories based on the nature of surplus, such as biodegradable and non-decomposable surplus. The garbage is again classified as paper, metal and plastic. The old traditional garbage collection method is quite successful; here four distinct bins are adequately prepared for public use, and they are frequently the only ones available. In traditionalbins, since there is no segregation in the dustbin. The public will through the mixed garbage to the bin.As a result, the traditional bins have demonstrated their unsuccessfulness in the view of the public. A GPS segment keeps track of the bin's site. The location and status of the bin's filling level are relayedto the server for the purpose of observing via a LoRa module. For maintenance needs the dustbin hasto be authorized, for this purpose proposed garbage management includes an RFID module that allows authorization. Figure 1 and Figure 2 depicts the block diagram of the proposed system with Arduino Uno and Raspberry Pi boards respectively. The selected container is adequately arranged for public use, this work proposes a explanation to this problem by separating waste sections to explanation for various forms of trash, food waste paper plastic, metal, and other garbage are all acceptable. Different sorts of garbage must be identified and separated

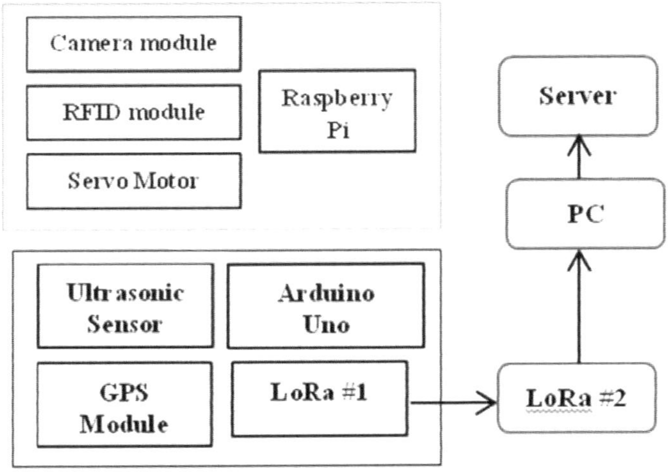

Figure 1. The proposed system with Arduino Uno.

The LoRa module is connected to Arduino using a multiprotocol radio shield, which acts as an interconnection shield for Arduino and allows two communication modules to be connected simultaneously. Because bins are often a few meters to a few kilometers apart, LoRa is an excellent fit for this system because it can broadcast data over long distances while using less power. The LoRa module is connected to Arduino with the help of a multiprotocol radio shield, which acts as an interconnection shield for Arduino and may connect two communication modules at the same time.

TensorFlow framework is utilized to train the detection model. The detection of waste and status of filling is monitored by an ultrasonic sensor. The garbage segregation assortment and dustbin level observing are all important aspects of developing intelligent garbage managing scheme. The garbage assortment and cataloguing system uses a CNN-based YOLE algorithm to detect the real time objects, Raspberry Pi, camera unit, ultrasonic sensor, and servo motors as well as other hardware. Besides, the bin's monitoring system is based on Arduino and includes ultrasonic sensors, a GPS module, and a LoRa communication module. Arduino IDE is used to for obtaining real-time bin information from a different location. In addition, the RFID-based locker system is linked to an Arduino to protect thebin's electronic components.

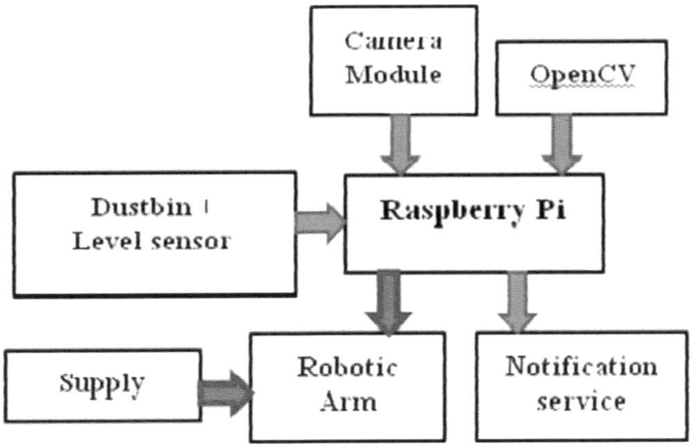

Figure 2. The proposed system with Raspberry Pi.

The database controller will have a powerful CPU, but the robot computers can be very basic. The robots will essentially take photos and upload them wirelessly to the server. The image processing willbe done on the computer (resulting in extremely quick results), and the result will be sent back to the robot. After that, this same machine should decide whether or not to pick up these things. In our company, the latter has been implemented. Installing a high-end processor on the server will be more expensive. Nevertheless, the reality that standalone robot processors will indeed be substantially less expensive, as well as the fact that prediction accuracy will be much faster, will make this a viable option. The robot will dispose of the article if it is categorized as trash. Bring yourself closer to the objective. The robopet that is positioned on the chassis will pick up the garbage of the material whenit is approached. If the images aren't rubbish, the robot will sidestep the obstacle as continue under its own course.

4. Design methodology and Implementation

The system is broken down into several modules. Each one of these components is unique and will be examined in detail:

4.1. Trajectory and motion module

The robot goes in a predetermined path, wrapper the area The Raspberry delivers data to the DC engine drivers, which control the engines in various capacities, including ahead, backward, and left, right turn, and rest. In a domain-agnostic manner, the robot car continues to go forward, for example. The path can also be progressed by considering numerous aspects in order to traverse the least amount of distance while still accomplishing its goal. It will act after recognizing the impediment, depending on whether it is garbage or not. If there is no trash, the robot will travel in its intended path after keeping a strategic position. The algorithm for trajectory and motion module is shown below.

- Robot starts
- Check for obstacle, if obstacle not detected, repeat the process and check for the obstacle.
- if obstacle detected, activate camera, capture the image and send the image to the server.
- Classify the garbage images using machine learning algorithm.
- Send the classified images to the robot.
- If garbage is detected, lift the objects using robot arm and place the object in collector.
- Continue in trajectory.
- If garbage is not detected, continue in trajectory.
- Repeat the process.

4.2. Object detection module

YOLO is a machine-learning real-time object identification technology. Because of its speed and precision, this algorithm is very popular. In a variety of applications, and has been used to recognize traffic lights, pedestrians, parking meters, and animals. This method improves recognition rate by predicting objects in the scene. YOLO is a forecasting process that generates exact performance with minimum noisy environment. The acronym YOLO stands for "You Only Look Once." This is a method to detect and recognizing various objects in an image (in real-time). Object identification in YOLO is carried out in the following way: The YOLO approach uses convolution neural network (CNN) to identify objects in the image. The method utilizes a specific upwards transmission through neural network to detect objects, as the name implies. This signifies that the full image isforecasted using a single algorithm run. At the same time, the CNN is utilized to forecast variousregion proposals and class probabilities. The YOLO algorithm comes in a variety of forms. Two prominent examples are Tiny YOLO and YOLOv3.

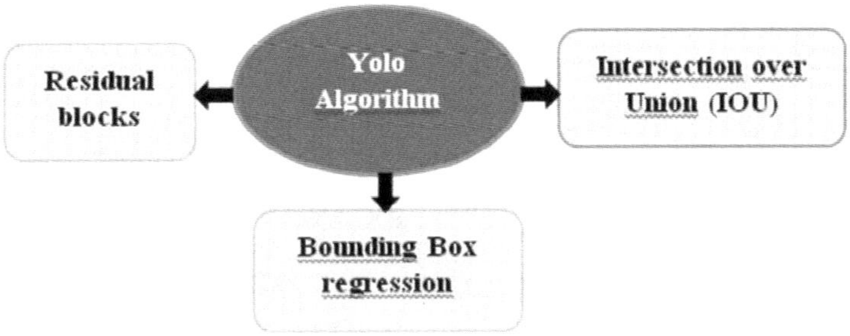

Figure.3. YOLO algorithm approach.

The image is divided into many panels initially. Each grid has the dimensions SxS. Each grid cell will accurately detect that occur within its boundaries. If, for example, an item center appears within a given grid cell, that battery will be useful for sensing it. A bounding box is an envelope in a photograph that draws attention up to a specific object. ISM bands at 915 MHz have multiple channels in each frequency band. There are three bandwidth options: 125 kHz, 250 kHz, or 500 kHz.It is decided to use a higher bandwidth. For speedier transmission, more bandwidth is preferred, for quite a wide distance, a smaller throughput is preferred. The data stream is received and decoded usingReal Term. The message was sent from the LoRa node. Star topology of LoRa module.

Every frame in the image contains the following attributes:

- Dimensions (bw)
- a specific height (bh)
- Class is represented by the letter c. (for example, person, car, traffic light, etc.).
- Located in the centre of the structuring element (bx, by)

YOLO uses a single bounding box regression to predict the length, height, center and category of object. In object detection, the interconnection over union (IOU) notion depicts how containersoverlap. Table 1. Depicts the characterization of the training set.

Table 1. Characterization of the training set.

class competition dataset	1000
input size	224x224
convolutional layers	20
input resolution	448x448 pixels
batch size	64
momentum	0.9
decay	0.0005
epochs	135
Learning rate schedule:	from 0.001 to 0.01

4.3. Picture capturing module

If an obstacle is detected, the camera takes a single photo, raises the flag, and communicates the image to the web via the correspondence module. It also holds everything together till the server responds.

The camera, which is installed on the robot's chassis, remains dormant until the IR sensor detects an item by processing a single image rather than continuously processing every frame of a live video feed, this method saves a lot of time, memory, and power. As a result, we may achieve this goal by using low-end processors, lowering the hardware cost significantly.

4.4. Classification module

When it receives a photo, the web conducts characterization or location. The computation is being used more frequently now. As a result, the computation stream should be carried out with precision in mind. Fast has been used as a verification tool. It is concerned with the highlight standard. Coordination between the preparation and information photographs demonstrates the alternatives, such as high-complexity pixels from previously produced images and those that match the information havea picture. Nearest Neighbor is a snappy approximation. The calculation consumes very little processing power and produces very precise results. The picture classifier for garbage trash is one of the optional calculations that can be used. YOLO (only live once) object discovery using the TensorFlow API, and so forth.

5. Results discussion and comparison

This segment presents the arrangement of equipment and framework design of the given RC robot framework as shown in Figure 4. The robot is built on a square under carriage that is level and has four-wheel drives attached to the body. A special waste or picking arm is attached on the front side, and it is used to remove the trash when it is detected. The case is built of a sturdy aluminum sheet with precise dimensions, which is attached to a tight, immediate to the RC vehicle robot body. It is connected to four wheels by four different DC engines. The repository is installed on either side of the rover's broadness with the use of two servo engines. The robot moves in a pre-programmed path, casing the area or region for which it is accountable.

Figure 4. Robotic arm movement in different direction.

Images can be classified in two ways. Use the robot processor to do image processing: This will necessitate the presence of a low-cost processor on the robot's body. Because the processor isn't as powerful, there may be some latency in the output. However, the cost will be lower for the same reason. Secondly, utilize a high-end processor on the central server to send photos from numerous robots. The experiment is carried out in a safe environment. The challenge revolves around the concepts of identification and assortment. The garbage identification is completed with the help of OpenCV. The Raspberry Pi camera will take a photo in a defined region and save it as the default image. The Raspberry Pi camera will continue to take photographs and compare them to the default picture.

Figure 5. When an object is discovered, the camera takes a single image and sends it to the server for categorization as garbage or non-junk using image processing

The object will be identified as trash, and the garbage will be collected using a Robotic Arm. The rover moves in a pre-programmed pattern that covers the area that has to be cleaned. The camera, which is installed on the robot's chassis, remains dormant until the IR sensor detects an item.

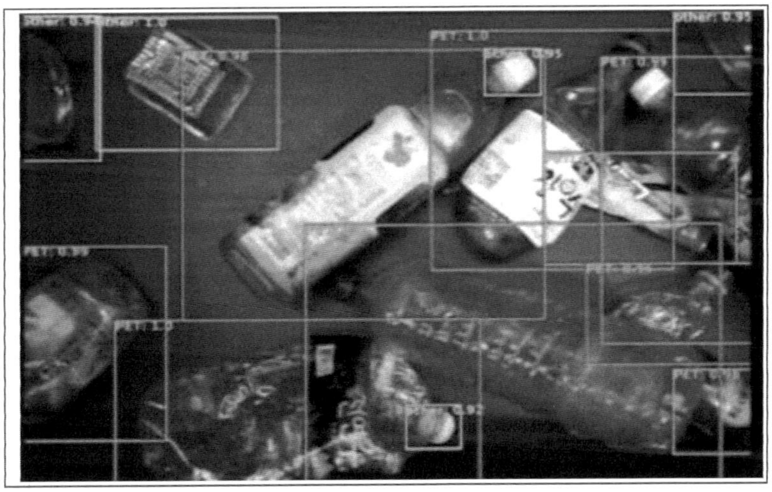

Figure 6. Different types of waste materials.

The camera, which is positioned on the robot's chassis, is dormant until the IR sensor detects an item. After an entity is discovered, the camera takes a solitary image and directs it to the server for categorization as garbage or non-junk using image processing. If it's garbage, the garbage will be collected by a robotic arm. The trash is collected and deposited in the bin, and the robot continuesto the next collection point.

Figure 7. The rc vehicle robot travels on to grab more items.

To some extent, the project's goal was accomplished. Garbage was collected in a timely and efficient manner. The robot moves at a steady pace. When the rubbish is 20cm away from the trash, it is detected. The project is currently being worked on to get the best results possible with a few more tweaks.

5.1. Comparison

The proposed work is based on Yolo algorithm and this work is compared with SVM algorithm and results obtained are given in table [2] below. As per the comparison, Yolo is better and faster than the SVM algorithm. 20 convolutional layers. The objects detected are classified in to three categories, such as positive, negative and false positive. The positive indicates accurate detection and negative indicates no object detection. The other category of false positive indicates detecting other than the listed objects such as plastic, glass, Iron material, stone and soil.

Table 2. Comparison of Yolo algorithm vs. SVM algorithm.

	Yolo	SVM
Positive	257	154
Negative	42	112
False positive	4	84
Accuracy	85.7%	57.8%
Frames per second	48 fps	4fps

6. Conclusion

This work is based on the usage of neural networks, machine learning and internet of things. Yolo algorithm is used to detect the waste in real time. The characterization of the training set consists of

1000 competition data set with 135 epochs. Computer vision and image processing are used to spot and process many images simultaneously. This algorithm identifies and detects the objects such as plastic, glass, Iron material, stone and soil etc. The other algorithms are used to identify objects are fast R-CNN, Retina-Net and SVM. Although these methods have overcome the limitations of data restriction and parameterization in object tracking, they are not capable of finding objects in a single run. The YOLO algorithm has gained popularity as a result of its higher performance over the above-mentioned object detection techniques. The proposed work with Yolo algorithm is having accuracy of 85.7% whereas SVM has 57.8%.

7. Limitation and Future Work

Even though Yolo algorithm is faster, it is having a false positive rate and negative rate. The resolution of this algorithm is lower compared to SVM algorithm. SVM detects the objects with higher resolution rate. However higher resolution data would require more computation and therefore the performance would be worse. In future, the performance of image classification networksis improved a lot with the use of refined training procedures such as Image mix-up with geometry preserved alignment, using cosine learning rate scheduler, synchronized batch normalization, data augmentation and label smoothing.

8. References

[1] H. Robinson 2007. Commun. Waste Resource Manage. **8** 1 1932.
[2] M. A. Al Mamun, M. A. Hannan, and A. Hussain 2014. Jurnal Kejuruteraan. **26** 1519
[3] Waste Management Indicators. Accessed: May 2021.[Online]. Available: https://ec.europa.eu/eurostat/statistics-explained/index.php?title=Waste_management_indicators#Overview
[4] T. J. Sheng, M. S. Islam, N. Misran, M. H. Baharuddin, H. Arshad, M. R. Islam, M. E. H. Chowdhury, H. Rmili, and M. T. Islam 2020. IEEE Access. **8** 148793–148811
[5] M. Shahidul Islam, M. T. Islam, M. A. Ullah, G. KokBeng, N. Amin, and N. Misran 2019. IEEE Access. **7** 127850–127861
[6] L. Atzori, A. Iera, and G. Morabito 2010. Comput. Netw., **54** 15 2787–2805
[7] S. Zavare, R. Parashare, S. Patil, P. Rathod, and V. Babanne 2017. Int. J. Comput. Sci. Trends Technol., **5** 3, 7478
[8] J. Bobulski and M. Kubanek 2019. in Proc. 25th ACM SIGKDD Conf. Knowl. Discovery Data Mining
[9] S. Han, F. Ren, C. Wu, Y. Chen, Q. Du, and X. Ye, 2018. ISPRS Int. J. Geo-Inf., **7** 4 120,
[10] M. S. Medvedev and Y. V. Okuntsev 2019. J. Phys. Conf., **1399**
[11] Talahua, J.S.; Buele, J.; Calvopina, P.; Varela-Aldás, J. 2021. Sustainability **13** 6900 https://doi.org/10.3390/su13126900

IOP Conf. Series: Materials Science and Engineering 1258 (2022) 012056 doi:10.1088/1757-899X/1258/1/012056

An overview of load frequency control for grid using various control techniques

Vinitha JC[1], Usha Rani P[2]

[1]*R.M.D. Engineering College, Kavaraipettai, Tiruvallur, Tamil Nadu, India*
[2]*R.M.D. Engineering College, Kavaraipettai, Tiruvallur, Tamil Nadu, India*

E-mail: vinithajc.eee@rmd.ac.in, pur.eee@rmd.ac.in

Abstract. Nowadays Electrical power generation using renewable energy systems brings more novel ideas to satisfy the power demand worldwide. Microgrid is one of the ideas that ensures stability and reliability of the power system. Due to seasonal or climatic effect, the power generated by the renewable energy sources has issue in providing a constant power output and hence produces the power imbalance between the source to load. Furthermore, the system frequency cannot be kept constant in all cases. In this case, an electric vehicle (EV) can serve as a power generation source to support the power grid and meet the Load Frequency Control requirement (LFC). Due to their fast-regulating characteristics, electric vehicles provide frequency regulation services. Electric vehicles are used in both primary and secondary frequency management in power systems to quickly suppress frequency oscillations caused by changes in load. The control centre gathers and updates real-time Electric Vehicle data, such as information regarding state of charge and the state of each Electric Vehicle, in order to decide on load frequency regulation dispatch to Electric Vehicles. This paper compares and [illegible] the performance of Load frequency control through Conventional plants and Renewable energy systems. The graphical abstract of power system grid is shown in Figure.1

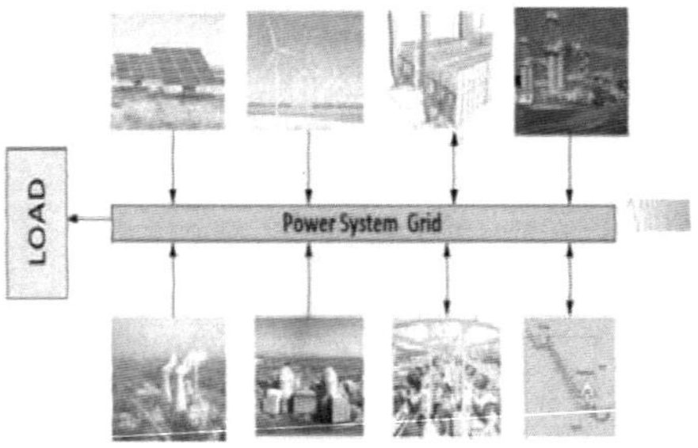

Figure 1. Graphical abstract of Power system Grid

1. INTRODUCTION

Electric vehicles play a major role in the upcoming generations to overcome the fossil fuel issues in society. The ecology and ecosystem are two aspects that are impacted by fossil fuel consumption. As a result, engineers take great care of these challenges as part of their profession. To address this problem, extensive research has been performed in several areas of EVs, particularly in the integration of EVs into power grids for bidirectional operation in order to better dampen grid oscillations. The government also offered numerous programs to help people adopt EV. International Energy Agency has anticipated that the sales of passenger light-duty plug-in hybrid EV's will surpass 100 million/year worldwide by 2050.

Renewable energy technologies are well renowned for being pollution-free. Renewable Energy systems have been increasingly popular in recent years for power generation, with Solar and Wind being the most common. Electric vehicles and renewable energy sources have attracted a lot of attention as they are integrated into the power grid. The system frequency deviates from its ideal value due to different disruptions in the power system, such as unexpected load shifts and the intermittent nature of renewable sources like solar and wind power. By optimizing solid state devices in power systems and electrical power generation, these risks are reduced. Load Frequency Control (LFC) is used in the power system to keep the system frequency within a certain range [1, 2, 3, 4]. Grid system frequency can be continuously monitored in real time utilizing IoT and remote devices or local control rooms.

The main parameter in a power system is frequency, which must be managed and kept within legal limits by regulating the power sources. EVs are used in power systems to achieve frequency regulation due to their fast-regulating qualities. Adaptive loop control is used to create EVs with primary frequency regulation [5,6]. The frequency response in an islanded system is studied using EVs, and this article looks at frequency regulation in PFC and SFC with and without EVs [7,8]. The load frequency is regulated in this case by the participation factor(α) [9]. This research [10] explains the economics and dispatching tactics of EVs participating in frequency regulation.

When EVs contribute to LFC, the V2G control approach regulates the frequency of the interconnected system by taking into account the SOC levels of the EV owners [11]. With electric vehicles, power to the network is delivered effectively and reliably, improving system stability. As a result, the system will run at a low cost and provide optimum convenience to EV owners [12]. Load Aggregators with appropriate charging control are offered for minimal cost in order to take use of the benefits of EV [13,14]. The V2G approach in EVs uses a BESS to control the charging and discharging of the batteries to maintain the generating and load side frequency [15,16,17].

LFC for electric vehicles When considering SOC of EV, the expectations of EV owners are largely dependent on the participation factor of various loads linked in the system to minimize grid frequency oscillation. By lowering the likelihood of load frequency oscillations, grid frequency measures can be maintained efficiently [18,19,20,21].

2. LOAD FREQUENCY CONTROL USING VARIOUS CONTROLLERS

2. 1. Load frequency control by incorporating Electric Vehicles(EVs) and Heat Pump Water Heater)(HPWH)

The combination of various sources along with its generator model and control system for the regulation of LFC is shown in the below Figure.2.

The LFC of a single-area system is investigated with the involvement of electric vehicles and other conventional energy sources. The LFC signal is sent to several conventional sources, such as EVs, BESS, and HPWH, depending on its speed and control capacity. The EV battery's adjustable SOC range is 85-5 percent, and the battery is routinely charged up to that point. The lower and upper limits of battery SOC are defined by EV owner's expectations and the battery's lifetime. It is not possible to charge the battery to 100 percent SOC. As a result, the EV user must select whether or not they need to engage in LFC.

The frequency deviation for the previous 24 hours is shown in table 1. Three scenarios are examined.
Case 1: EVs and HPWHs are not taken into account for LFC.
Case 2: EVs are taken into account for LFC but HPWHs are not.
Case 3: EVs and HPWHs are taken into account for LFC.

Equation 1 gives the formula for calculating the frequency deviation (RMS Value).

$$frms = \frac{1}{N}\sum_{i=1}^{N} \Delta f^2 (1)$$

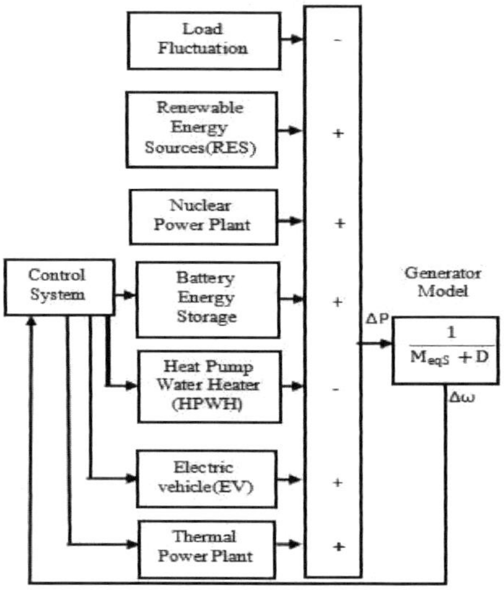

Figure 2. Block Diagram of LFC with EV and HPWH

Table 1. Frequency Deviation with and without EV and HPWH

	Controller	Max (Hz)	RMS (Hz)
Case1	Without EV or HP	0.439	0.0353
Case2	With EV	0.251	0.0272
Case3	With EV and HPWH	0.225	0.0263

It is obvious from the aforementioned findings that the presence of EVs and HPWH effectively reduced frequency fluctuations [10].

2.2 Load frequency control by incorporating EV's with time delay

Figure 3 shows a model of an EV with a time delay for controlling the primary and secondary frequencies of a power system. EVs are used in this model to increase the reliability and flexibility of power grids [21,22]. The use of time delays affects system performance, which leads to power system instability and, as a result, the frequency cannot be kept constant. H∞ control approach for LFC has been established in this model to alleviate this drawback [23,24,25,26].

Different cases has been taken for analysis in this system:

1) Thermal turbine LFC
2) SFC plus EV
3) EV with PFC

4) SOC Calculation for EV

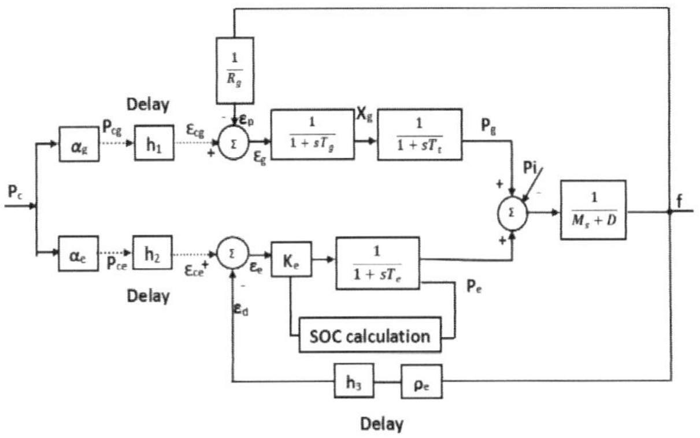

Figure 3. Block Diagram of Load Frequency Control of EV with Time delay

The output power from the thermal turbine and EVs is controlled by generating LFC signals using PFC and LFC loops in order to keep the system frequency at a certain level. EVs can serve comparable functions to turbine generators by adopting a droop control, however in the case of SFC, EVs are also employed to generate power to satisfy the demand of power plants to maintain LFC [27,28,29,30].

The control centre considers the participation factors when sending control signals (i.e. incremental changes in power) to thermal plants and EVs Pcg(t) and Pce(t). The participation factors for thermal turbines and electric vehicles are αg = 0.8 and αe =0.2, respectively. Controlling the PFC of electric vehicles aids the thermal turbine in achieving zero frequency deviation [9].

2.3 Load frequency control by incorporating Heat Pump Water Heaters (HPWH)

This research examines the SLFC with a variable load of a Heat Pump Water Heater (HPWH). For the SFC frequency in a two-area power system, a model predictive controller is used [31,32,33,34,35,36]. Various factors, such as off-load shifts, intermittent renewable power generation, and parameter adjustments, are used to calculate the LFC for HPWH. The system's power consumption is determined by sending an appropriate input signal between the main and local centres. The load frequency is adjusted within the range of 90-100% of the rated power in this model with HPWH. HPWH uses an MPC with AGC for a two-area power system for SFC. The outcomes of AGC with MPC and AGC with PID controller are compared.

Table 2. Computation of peak Overshoot with PID and MPC Controller

Controller Type	Signal	Peak Overshoot(pu)
AGC-MPC with HPWH-MPC	Δf_1	-0.0103
	Δf_2	-0.0123
	ΔP_{12}	-0.0020
AGC-MPC	Δf_1	-0.0198
	Δf_2	-0.0238
	ΔP_{12}	-0.0038
AGC-PID	Δf_1	-0.0390
	Δf_2	-0.0449
	ΔP_{12}	-0.0074

Case 1: Change in Load

A 0.01pu load change is applied to area1. In comparison to AGC with MPC and AGC with PID Controller, HPWH with MPC controller has better oscillation dampening, overshoots, and settling time.

The peak overshoot of area1 with HPWH-MPC has been reduced nearly half of the time with AGC-MPC, as shown in Table.2. With the merger of HPWH and MPC, the similar frequency decrease will occur for area2, as well as a reduction in tie line power.

Case 2: Change in input wind power and random load change

The two-area network receives changeable wind power and unpredictable changes in demand as inputs. The output power departure from a given point is determined by the change in input. The change in electricity owing to the wind farm is 0.03pu, while the random load change is 0 to 0.05pu. The findings are compared to three different types of controllers, as indicated in Table 3.

Table 3: Computation of maximum Frequency Deviation with PID and MPC Controller

Controller Type	Signal	Maximum Deviation (pu)
AGC-MPC with HPWH-MPC	Δf_1	-0.0507
	Δf_2	0.0791
	ΔP_{12}	0.0139
AGC-MPC	Δf_1	-0.0825
	Δf_2	0.121
	ΔP_{12}	0.0203
AGC-PID	Δf_1	0.2257
	Δf_2	0.1726
	ΔP_{12}	-0.0377

In comparison to AGC-MPC and AGC-PID controllers, the maximum frequency variation observed in AGC-MPC with HPWH-MPC is relatively minimal (see table.3).

2.4 Load frequency control by incorporating EV's with time delay in an Islanded system

There is a power imbalance between the source and the load side when using renewable energy sources in the power system. This power imbalance easily influences the islanded system, resulting in power system frequency instability [37,38,39,40]. Controllable loads such as EVs and V2G are employed to restore the system frequency to a steady state. Because of the fast-regulating features of electric vehicles, it was employed primarily in PFC, although EVs can also participate in SFC in addition to PFC.A single EV with V2G capability is explored in this frequency control model to inject power back into the grid as needed. EVs can absorb or inject power depending on the frequency change. EVs have the ability to charge and discharge quickly at the request of the client. This attribute of EV is utilized to balance the power in an islanded network when the output or frequency changes [41].

Three scenarios are examined when using EVs to control frequency.
(i) No EVs are involved in frequency regulation
(ii) EVs are exclusively involved in PFC
(iii) EVs taking part in the PFC and SFC

Simulation results for all the three cases are given in Table.4.

Table 4. Simulation Results with and without EV Participation

Type	Steady-state Frequency (Hz)
No EV	49.89
EV with PFC	49.85
EV with PFC and SFC	49.92

The simulation results suggest that having an electric vehicle with PFC and SFC reduces the load frequency to 49.92.

2.5 Load Frequency Control using Coefficient Diagram Method

This method investigates load frequency regulation in the presence of solar and thermal energy. To mitigate frequency variations induced by renewable energy sources, this system combines heat pumps and electric vehicles. The disturbance in the power system and load caused by changeable solar and thermal is also decreased with this technology. The Coefficient Diagram Method [42,43,44,45] is an algebraic approach for analyzing polynomial loops in the power system. The CDM approach gives information on the systems stability condition. The block diagram for the CDM Controller is shown in Figure 4.

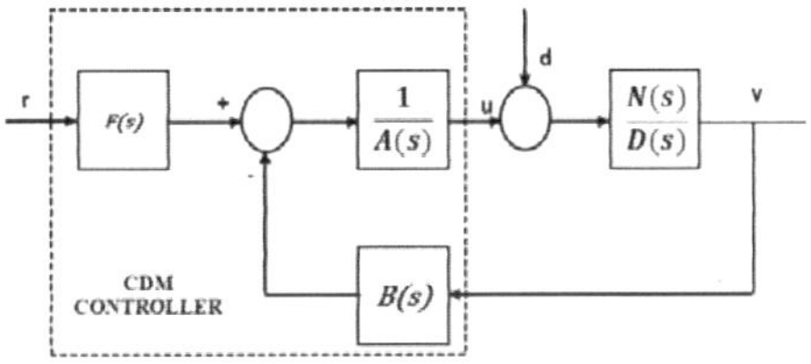

Figure 4. Block Diagram for CDM Controller

MATLAB/SIMULINK software is used to calculate the system's LFC. The HPWH and EV controllers are compared to the CDM controller used in solar and thermal power generation with variable load demand. The simulation results indicate how the inclusion of HPWH and EV with integrated controller and CDM controller changes the frequency.

Table 5. Max values of frequency deviation [Hz]

Type of Controllers	Maximum value of Frequency Deviation (Hz)
HPWH and EV with Integral Controller	0.013
HPWH and EV with CDM Controller	0.007

The transfer function of the system is given by

$$Y = \frac{N(s)F(s)}{P(s)} r + \frac{A(s)N(s)}{P(s)} d$$

Where,
Y-Output of the of the system
N(s)-Numerator Polynomial of the system
D(s)-Denominator Polynomial of the system
A(s)-Forward denominator Polynomial of the system
F(s)-Reference Numerator
B(s)-Feedback Numerator
r-Reference signal
u-Controller signal
d-External signal

In comparison to HPWH and EV with Integral Controller, Table.5 shows that the variation in frequency is suppressed to a very low value in HPWH and EV with CDM controller.

2.6 Load frequency Control in Isolated Micro-grids with the coordination of EVs

Frequency instability in micro-grids is caused by the rapid expansion of renewable energy sources and their intermittent nature. Micro-grids can function in either a linked or islanded mode. The frequency cannot be kept constant throughout this autonomous operation due to the existence of RES's. As a result, LFC is essential to maintain system stability between the source and load sides [46,47,48,49].The charging and discharging of electric vehicles can be calculated using an EVPC power controller [26]. Depending on the availability of energy, electric vehicles can be used as a source or as a load. Whether energy is given or absorbed to or from the network, the two modes are V2G and G2V. EVs may charge and absorb electricity in G2V mode, and discharge and inject power into the network in V2G mode. EVs can thus operate as a load or a generator, depending on the energy level. When the load is low, the EVs help with frequency management to charge, and when the generation is high, the EVs help with discharge. Thus, the load frequency control and stability of the system is obtained by optimally scheduling the charging and discharging process of EV. In an isolated mode, frequency stability is analyzed in two modes i.e., without considering EVPC and with EVPC and the results are tabulated in Table.6

Table 6. Frequency Calculation by EVPC Controller

Controller	**Min frequency (Hz)/24 hours**	**Max frequency (Hz)/24 hours**	**Frequency Deviation(Hz)/24 hours**
With EVPC	59	61.4	2.4
Without EVPC	< 57	61.4	4.4

2.7. Participation of EV's in Load frequency control using H2/H∞ Controller.

The H2/H∞-SOF controller is used to evaluate the performance of electric vehicles in conjunction with traditional energy sources for LFC. The key benefit of employing this controller is that it regulates the load frequency, which improves system stability [50]. Using a combined H2/ H∞-SOF controller, the time delay caused by power system components is also investigated [51,52]. Various controller models have been presented, but they have not been applied in practice due to higher order [53]. To address this, the H2/H∞-SOF controller is presented, which produces a simple lower order controller with performance that is nearly identical to that of a dynamic controller [54]. The system's load frequency is investigated under various conditions such as load change, time delay, and so on. The outcomes of this technique are compared to H∞ controller for a single and three area system.

A single area system is subjected to a load change with and without time delay. The responses for with and without time delay are described, and the frequency drops to -0.23Hz due to a sudden rise in load, which is
prevented by the Primary Frequency Controller. The main cause of this variation in frequency was system
inertia. In this single-area system, the rate of change in frequency (Δf) is 0.178Hz/s. The system frequency is
reduced from -0.23Hz to zero using the H2/H∞-SOF controller.

When there is no time delay, the system with H∞ controller is stable, but when there is a 5s time delay, the system becomes unstable. The frequency error for both single and three area systems is almost negligible when the system operates without time delay. The H∞controller performs better than the H2/H∞ controller in this no-delay state, which is mostly due to the turbines' high inertia. The H2/H∞-SOF controller is used in the system to offset this disadvantage.

Table 7. Comparison of H∞ controller and H2/H∞ controller for a singlearea system

Performance Criteria	**H∞ controller with and without time delay**		**H2/H∞ controller with and without time delay**	
	0s	**5s**	**0s**	**5s**
Sum of Squared Errors	0.306	1.6×10^5	0.57	1.2
Mean Absolute Error	0.0037	8.06	0.016	0.0206
Mean Squared Error	0.00039	203	0.0014	0.0027

The performance of the H2/H∞-SOF controller is clearly superior to that of the H∞ controller with time delay, as seen in the table above.

Table 8. Comparison of H∞ controller and H2/H∞ controller for a three-area system

Performance Criteria	**H∞ controller with and without time delay**		**H2/H∞ controller with and without time delay**	
	0s	**5s**	**0s**	**5s**
Sum of Squared Errors	0.16	1.08× 1013	0.57	1.55
Mean Absolute Error	0.011	1.73 × 1004	0.015	0.025
Mean Squared Error	0.0004	1.6 × 1009	0.0014	0.00311

3. Conclusion

The research of load frequency regulation of a power system network utilizing various control approaches, including EV, is presented in this paper. The load frequency control is investigated for a variety of sources, including traditional power systems, electric vehicles, heat pumps, battery energy storage systems, and renewable energy systems. RES-induced frequency changes in the system can be suppressed utilizing a variety of control techniques. The frequency fluctuation of the system is much reduced when EV and HPWH are present, as well as the battery's SOC. To mitigate the influence of load disturbances in the power system, H∞ control techniques for LFC are employed in EVs with time delay systems. With the use of a Model Predictive Controller, secondary load frequency management can be done while only considering HPWH(MPC). The frequency fluctuation is reduced in an islanded time delay system by using V2G as a configurable parameter. When EVs and HPWH are present, the system is disrupted by intermittent sources, and traditional units are managed using the Coefficient Diagram approach. In isolated micro-grids, load frequency management is performed with an EV Power Controller and an H2/H∞-SOF controller. For various modes of operation, LFC for big parking areas of EV is regulated by centralized and local control centres. Under various conditions, the LFC control of a four-area power system utilizing PID and MPC controllers is compared and studied. According to the comparison, V-G bidirectional control for frequency variations is an effective way for suppressing oscillations.

REFERENCES

[1] Ota Y, Taniguchi H, Nakajima T, Liyanage K M, Baba J, and Yokoyama A, 2012 Autonomous distributed V2G (vehicle-to-grid) satisfying scheduled charging IEEE Trans. Smart Grid, **3**, no.1, pp. 559–564.

[2] Shimizu K, Masuta T, Ota Y, and Yokoyama A, 2011 A New Load Frequency Control method in power system using vehicle-to-grid system considering users convenience 17th Power Syst. *Computation Conf. Stockholm Sweden.*

[3] Yilmaz M and Krein P T, 2013 Review of the impact of vehicle-to-grid technologies on distribution systems and utility interfaces *IEEE Trans. Power Electron.,* ***28,*** *no. 12, pp. 5673–5689.*

[4] Meng J ,Yunfei Mu , Hongjie Jia , Jianzhong Wu , Xiaodan Yu , Bo Qu 2016 Dynamic frequency esponse from electric vehicles considering travelling behaviour in the Great Britain power system *Appl. Energy,* ***162****, pp. 966–979.*

[5] Izadkhast S, Gonzalez P G, and Frias P, 2015 An aggregate model of plug-in electric vehicles for primary frequency control *IEEE Trans. Power Syst,* ***30****, no. 3, pp. 1475–1482.*

[6] Neofytou N, Blazakis K, Katsigiannis Y, and Stavrakakis G 2019 Modeling vehicles to grid as a source of distributed frequency regulation in isolated grid with significant RES penetration.

[7] Hui Liu, Hu Z, Song Y & Lin J 2013 Decentralized vehicle-to-grid control for primary frequency regulation considering charging demands, IEEE Trans. on Power syst, 28(3), 3480 – 3489.

[8] Athanasios Z, Mattia M, Chresten T, Katarina K, Pete Bach A 2014 A dynamic behaviour analysis on the frequency control capability of electric vehicles , *9th Int. Universities Power Eng. conf.*

[9] Pham T N, Nahavandi S, Hien Le V, Trinh H, Kit Po Wong, 2017 Static output feedback frequency stabilization of time-delay power systems with coordinated electric vehicles state of charge control, *IEEE Trans. on Power Syst., 32, No.5.*

[10] Masuta T, Yokoyama A, 2012 Supplementary load frequency control by use of a number of both electric vehicles and heat pump water heaters, *IEEE Trans. on Smart Grid,* ***3(3)****, 1253–1262.*

[11] Peng C, Zou J,Lian 2017 Dispatching strategies of electric vehicles participating in frequency regulation on power grid: A review,*Renewable and Sustainable Energy Reviews,* ***68****,147–152.*

[12] Hui Liu, Junjian Qi, J Wang, Peijie Li, Canbing Li, 2018 EV dispatch control for supplementary

frequency regulation considering the expectation of EV owners, *IEEE Trans. on Smart Grid, 9(4), 3763–3772.*

[13] Dogan A, Bahceci S, Daldaban F, Alci M, *2018* Optimization of charge/discharge coordination to satisfy network requirements using Heuristic algorithms in vehicle-to grid concept, *Advances in Electr. and Computer Eng,* ***18****, no1.*

[14] Han S & Sezaki K. 2010, Development of an optimal vehicle-to-grid aggregator for frequency regulation, *IEEE Trans. Smart Grid,* ***1(1),*** *65–72.*

[15] Shimizu K,Masuta T, Ota Y, *2010* and A Yokoyama, Load frequency control in power system using vehicle-to-grid system considering the customer convenience of electric vehicles, *Int. Conf. on Power Syst. Technol.*

[16] Pillai J R & Bak-Jensen, B 2010, Integration of vehicle-to-grid in the western danish power system, *IEEE Trans. on Sustainable Energy.*

[17] Iqbal S, Xin A, Jan M U, H UR Rehman, Masood A, Rizvi S A A and Salman S 2018, Aggregated electric vehicle-to-grid for primary frequency control in a microgrid- A Review, *IEEE 2nd Int. Electr. and Energy Conf.*

[18] Irie H, Yokoyama A, and Tada Y 2008, Modeling for frequency control analysis of power system with a large penetration of wind power generation by a lot of controllable heat pump systems and battery systems, in Proc. *Int. Conf. Power Syst. Technol., New Delhi, India,*

[19] Masuta T, Yokoyama A, and Tada Z *2010* System frequency control by heat pump water heaters on customer side based on statistical HPWH model in power system with a large penetration of renewable energy sources, in Proc. *Int. Conf. Power Syst. Technol., Hangzhou, China.*

[20] Nishizaki Y, Irie H, Yokoyama A, and Tada Y, *2010* coordinated control of blade pitch angle of wind turbine generators and battery for frequency regulation and the battery capacity reduction, in Proc. *Int. Conf. Electr. Eng., Okinawa, Japan.*

[21] Yilmaz M and Krein P T, 2013 Review of the impact of vehicle-to-grid technologies on distribution systems and utility interfaces, *IEEE Trans. Power Electron.,* ***28****, no. 12, pp. 5673–5689.*

[22] Kempton Wand Tomic J, 2005 Vehicle-to-grid power implementation: From stabilizing the grid to supporting large-scale renewable energy, *J. Power Sources,* ***144****, no. 1, pp. 280–294.*

[23] Yu X and Tomsovic K, 2004 Application of linear matrix inequalities for load frequency control with communication delays, *IEEE Trans. Power Syst.* ***19****, no. 3, pp. 1508–1515.*

[24] Dey R, Ghosh S, Ray G, and Rakshit A, 2012 H∞ load frequency control of interconnected power system with communication delays, *Int. J. Elect. Power Energy Syst.* ***42****, no. 1, pp. 672–684.*

[25] Pham T N, Trinh H, Hien L V, and Wong K P, 2016 Integration of electric vehicles for load frequency output feedback H∞ control of smart grids, *IET Gener., Transm., Distrib.* ***10****, no. 13, pp. 3341– 3352.*

[26] Zang C K, Jiang L, Wu Q H, He Y, and M Wu, 2013 Delay-dependent robust load frequency control for time delay power system, *IEEE Trans. Power Syst., vol. 28, no. 3, pp. 2192–2201.*

[27] Ota Y, Taniguchi H, Nakajima T, Liyanage K M, Baba J, and Yokoyama A 2012 Autonomous distributed V2G (vehicle-to-grid) satisfying scheduled charging, *IEEE Trans. Smart Grid,* ***3****, no. 1, pp. 559–564.*

[28] Pahasa J and Ngamroo I 2015 PHEVs bidirectional charging/discharging and soc control for microgrid frequency stabilization using multiple MPC, *IEEE Trans. Smart Grid,* ***6****, no. 2, pp.526–533.*

[29] Yang H, Chung C Y, and Zhao J 2013 Application of plug-in electric vehicles to frequency regulation ased on distributed signal acquisition via limited communication, *IEEE Trans. Power Syst.,* ***28****, no. 2, pp. 1017–1026.*

[30] Vachirasricirikul S and INgamroo I 2014 Robust LFC in a smart grid with wind power penetration by coordinated V2G control and frequency controller, *IEEE Trans. Smart Grid,* ***5****, no. 1, pp. 371–380.*

[31] Oshnoei A, Khezri R and Muyeen S M 2019 Model predictive-based secondary frequency control considering heat pump water heaters.

[32] Shiroei M, Ranjbar A M 2014 Supervisory predictive control of power system load frequency control *Int. J. Electr. Power Energy Syst.* ***61****, 70*–80.

[33] Ersdal A M, Fabozzi D Imsland, L Thornhill N F *2014* Model predictive control for power system frequency control taking into account imbalance uncertainty, *IFAC Proc.* ***47****, 981–986.*

[34] Ersdal, A M , Cecılio I.M, Fabozzi D, Imsland L, Thornhill N F 2010 Applying model predictive control to power system frequency control, *In proceedings of the 2013 IEEE Innovative smart grid technol. Europe, Copenhagen, Denmark, 6–9.*

[35] Mir A S, Senroy N 2018 Adaptive model predictive control scheme for application of SMES for load frequency control, *IEEE Trans. Power Syst,* ***6****, 526–533.*

[36] Mohamed T H, Bevrani H, Hassan A A, Hiyama T 2011 Decentralized model predictive based load frequency control in an interconnected power system, *Energy Convers, Manag.,* ***52****,1208–1214.*

[37] Baccino F, Marinelli M, Massucco S, and Silvestro F 2012 Low voltage microgrid under islanded operation: Control strategies and experimental tests, Power *Gener, Transm., Distrib* and Energy conversion, *8th Mediterranean Conf. pp.1-7, Cagliari, 1-3.*

[38] Zoka Y, Tonoda R, Mashima Y, Sasaki Y, and Yorino N 2012 An on-demand control system for demand and supply control of small independent power grids, *Universities Power Engineering Conf. 47th Int., pp.1-6, London, 4-7.*

[39] Martinenas S, Marinelli M, Andersen P B &Traholt C 2014 Implementation and demonstration of grid frequency support by V2G enabled electric vehicle, 49th *Int. Universities Power Eng. Conf.*

[40] Sumithra J, Vinitha J C and Jesintha T 2018 Controller design for single area load frequency control with time delays using genetic algorithm, *Int. J of Appl. Eng. Research ISSN 0973-4562* ***13****, No 15, pp. 11909-11916.*

[41] Lopes A P, Soares F J and Rocha Almeida P M 2009 Using vehicle-to- grid to maximize the integration of intermittent renewable energy resources in islanded electric grids *Int. conf. on clean Electr. power renewable energy resources impact, Capri, Italy, 9-11.*

[42] Manabe S, Coefficient diagram method 1998 *14th IFAC Symp on Automatic control in Aerospace, Seoul.*

[43] Bernard M Z, Mohamed T H, Mitani Y& Qudaih Y S 2013, CDM application on power system as a load frequency controller, *IEEE Electr. Power & Energy Conf., doi: 10.1109/epec.2013.6802920.*

[44] Mohamed T H, Shabib G, Ali H 2016 Distributed load frequency control in an interconnected power system using ecological technique and coeffificient diagram method, *Electr. power and energy syst.* ***82****, 496–507.*

[45] Kumar B, Adhikari S, Datta S, & Sinha N 2020 Real time simulation for load frequency control of multisource microgrid system using grey wolf optimization based modified bias coefficient diagram method controller, *J of Electr. Eng & technol. Doi:10.1007/ s42835-020-00596-2.*

[46] Dahraie M V, Kermani H R, Najafifi A, Moghaddam A and J M. Guerrero 2017 Coordination of EVs participation for load frequency control in isolated microgrids,*Appl.Sci.7,539; doi:10.3390/app7060539.*

[47] Veronica A J, Kumar N S, & Longatt F 2020 Design of load frequency control for a microgrid using D-partition method, *Int. J of emerging electr. power syst.*

[48] Aghamohammadi M R, Abdolahinia H 2014 A new approach for optimal sizing of battery energy storage system for primary frequency control of islanded microgrid, *Int. J Electr Power Energy Syst. 54:325–33.*

[49] Jafari H, Mahmadi M, Rastegar H 2012 Frequency control of micro-grid in autonomous mode using model predictive control, *In: IEEE Iranian Conf. on Smart Grids, 1–6*

[50] Khana M, Suna H, Xiangb Y, Di Shib 2021, Electric vehicles participation in load frequency control based on mixed H2/H∞ *Electr. Power and Energy Syst., 125,106420.*

[51] Bevrani 2014 H. Robust power system frequency control. 2nd Edn. Cham: Springer, Cham.

[52] Scherer CW 1995 Multiobjective H2/H∞ control. *IEEE Trans. Automat Contr.40(6):1054–62.*

[53] El-Sherbiny MK, El-Saady G, Yousef AM,2002 Efficient fuzzy logic load–frequency controller, *Energy Convers Manage,**43(14):**1853–63.*
[54] Asghar, R., et al.2020, Load frequency control for EVs based smart grid system using PID and MPC. *3rd Int. Conf. on computing, mathematics and eng. technol. IEEE.*

Dual Threshold Voltage Pseudo Domino Logic - Based Buffer with Reduced Power

K A Verma[1], M Kumar[2], and M Deo[3]

[1,3]Dept. of Electronics & Communication Engineering, MMMUT, Gorakhpur, U.P
[2]Dept. of EEE, SCE, Sasaram, Bihar (Earlier with ECE Dept, MMMUT, Gorakhpur, U.P)
Email: [*1]anjaliv1311@gmail.com, [2]er.manish.k@gmail.com, [3]manishiec2006@gmail.com

Abstract. In this paper, a new technique of power reduction in pseudo domino logic-based buffer is proposed with dual threshold voltage. The proposed method employs the dual threshold voltage technique to reduce the power. During the precharge stage, we have used the high threshold voltage transistor in the non-critical path of the circuit, to reduce the leakage current due to which the average power dissipation is minimized. In critical path, low threshold voltage transistors are used. The low threshold voltage transistors are used in evaluation stage, for the performance enhancement of the circuit. Keeper circuit is used with two PMOS transistor which helps in maintaining voltage of dynamic node and improves the noise immunity and speed of the circuit. Also, a NMOS transistor, which is operated by the dynamic node is used which minimizes the leakage by creating a stacking impact. In addition, the clock inverter is enabled via low voltage NMOS which significantly reduces the current in the inverter due to direct connection between the supply voltage (V_{dd}) to ground.

Key words: Domino, Dual threshold Voltage, Keeper circuit, Evaluation, and Precharge.

1. Introduction

Power consumption has become a primary concern in VLSI circuits with scaling down in technology into the nanometre regime. Domino circuits are widely used in many applications such as arithmetic circuits, microprocessors and this logic provides advantages in terms of power consumption and chip area. Domino logic operates in two stages: the first stage is precharge and the second stage is evaluation, both stages are controlled by clock. When clock is low, the standard domino logic works in precharge stage, and the dynamic node is charged to supply voltage (V_{dd}) using PMOS transistor [1]. And when clock is high, the standard domino logic works in evaluation stage in which PMOS is OFF, NMOS evaluation transistor turns ON, and the output is evaluated by the applied inputs at pull-down network, in this phase dynamic node discharges to ground. Otherwise, it will remain in the precharge stage.

The average power consumption of domino logic circuit is calculated and can be written by the equation given as:

$$Pavg = Pswitching + Pshortckt + Pleakage \tag{1}$$

Here,
$P_{switching}$ = Switching power because of the charging and discharging the circuit capacitance at the output.
$P_{shortckt}$ = Short circuit power due to the short path between the V_{dd} to ground.
$P_{leakage}$ = Leakage power present in MOS transistor.

Low power as well as design reliability must be always considered at all levels of development, according to the pioneering computing system experts. Static CMOS and domino circuitry are inadequate for forthcoming computing needs. Static logic and dynamic logic are two

IOP Conf. Series: Materials Science and Engineering 1258 (2022) 012057 doi:10.1088/1757-899X/1258/1/012057

fundamental CMOS circuit architectures. Static CMOS seems to be more energy-efficient and resilient, but it is extremely sluggish, when employed in important or enormous circuits. Domino logic is speedy, but still, it uses a lot of power and isn't robust. Furthermore, domino logic grows inadequately, restricting its speed advantage.

2. Literature Review

Different domino circuits [5-10] like Conventional Logic, Pseudo Logic, DOIND logic, C-3D logic, M-TSPC logic, CDDK logic, Stacked Pseudo domino, DVT logic and Charge sharing work have been studied.

Dual Threshold MOS Technique: Dual threshold MOS technique utilizes low threshold voltage transistor in critical path of CMOS based design and these transistors can be activated only during the evaluation stage, whereas high threshold voltage transistors are employed in the non-critical path of the circuit and can be activated only in precharge stage [2]. In this, a standby switch is used to switch OFF all the high threshold voltage transistors [3], boosting the efficiency of high threshold voltage CMOS based technology in terms of reducing sub-threshold leakage current.

Modified Keeper Circuit: Figure 1 shows the modified keeper circuit, which contains two keeper transistors MK1 and MK2. The gate terminal of MK1 transistor is operated by the output of inverter, whereas the gate terminal of MK2 transistor is driven by dynamic node of the circuit [4]. The keeper transistor MK1, which is controlled by the inverted output, helps in maintaining the voltage of dynamic node and the keeper transistor MK2 helps to improve the noise immunity and the speed of the circuit.

In recent years, different logic styles have been proposed to implement domino logic which can be further used to implement several circuits including the memory designs [5-10].

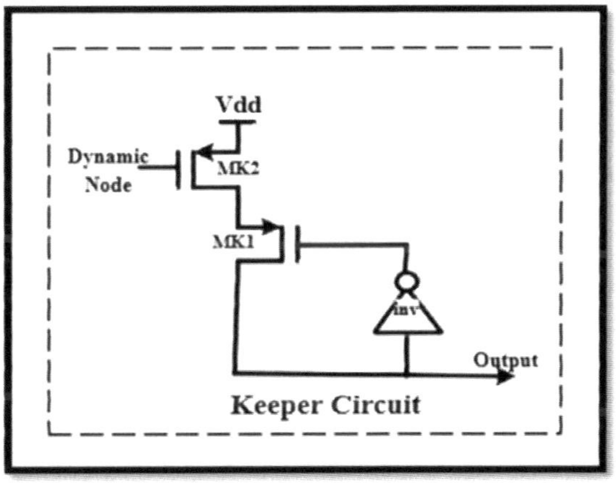

Figure 1. Keeper Circuit

Researchers offer an optimized pseudo domino methodology with lower leakage, lesser power, as well as an elevated noise tolerance.

3. Proposed Circuit

In the proposed circuit, Figure 2 shows the dual threshold voltage pseudo domino logic-based buffer. The Pull-down network (PDN) along with two PMOS M2 and clocked high V_{th} transistor M4 is used. The dynamic node is connected to the source of PMOS M2 and M4 transistor. The PMOS transistor M2 and NMOS transistor M3 constructs an inverter, which is operated by the voltage of dynamic node. In precharge phase, stacking effect will found with the evaluation due to NMOS transistor M5.

With the inverted clock footer transistor, the NMOS transistor M5 restrains the N_inv of the output buffer in generating precharge stage by preserving the N_foot at higher capacity.

The stacking effect increases the V_{th} at N_foot due to switching off the NMOS transistor, and this will lead to rise in noise immunity of the circuit. At footer transistor, an inverted clock is used which helps to discharge the previous charged N_foot in precharge phase and it helps in avoiding the wrong output generation.

In the proposed circuit we have used low threshold voltage NMOS transistor M6 and M7. The NMOS transistor M6 whose gate terminal is operated by the dynamic node, minimise the leakage by creating the stacking impact. Depending on the applied input to the evaluation network transistor M_eval switches ON, holding or discharging the dynamic node to V_{dd}. In addition, the clock inverter is enabled via transistor M7, which significantly reduces the current in the inverter due to direct connection from Vdd to ground.

This proposed dual threshold voltage pseudo domino-based buffer circuit works in two modes: precharge mode and evaluation mode.

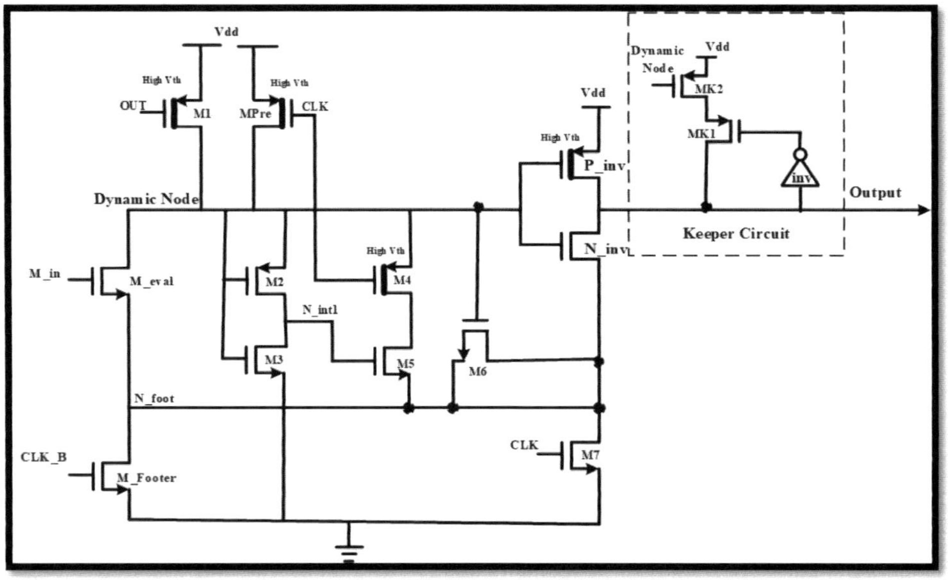

Figure 2. Proposed Pseudo Domino Logic-Based Buffer with Dual Threshold Voltage

In precharge stage, the input M_in = 1, clock CLK = 0 and clock-bar CLK_B = 1 then both transistor M1 and more are on and make the dynamic node logic 1. CLK_B = 1 is applied to the M_Footer NMOS transistor so the footer transistor also turns on. During this stage, the NMOS transistor M3 and PMOS transistor M4 are in high stage due to the connection of gate terminal with the dynamic node = 1 and CLK = 0 respectively. Low Vth NMOS is used to make the N_inv1 node low because it acts as an inverter with the PMOS transistor M2. Since N_inv1 become low the transistor M5 become off. It forms a stacking effect which lowers the voltage at N_foot and increases the voltage between source to gate of the NMOS, which is the static inverter PDN network, present at the output. It shows the dominating nature at the prior evaluation stage output as high-rise.

In the evaluation stage, when the clock CLK = 1 and clock-bar CLK_B = 0, the high Vth PMOS M1 & MPre are available between the evaluation network, which makes the M5 transistor on due to which the discharging of voltage of dynamic node starts from one to zero. At the same time there is a reduction in voltage at N_foot which guides the switch on the NMOS of the inverter which makes output logic 1.

The proposed technique provides lesser power consumption at the cost of speed, delay & energy is slightly higher than the previous circuit.

4. Simulated Results and Discussion

The simulation for pseudo-domino logic circuit and proposed dual threshold pseudo domino logic design is performed using the software cadence virtuoso gpdk 45nm tool at 1V power supply at 27^0C temperature. The previous circuit has high speed, but it has maximum power about 11.72uW. By comparing with the proposed circuit power of the circuit is minimized or reduced, which can be applicable for low power application.

Table 1. Performance analysis of proposed domino circuit using dual threshold voltage

PARAMETER	PSEUDO-DOMINO BUFFER WITH INVERTED OUTPUT [6]	PROPOSED PSEUDO DOMINO LOGIC
Average Power (uW)	11.72	5.663
Delay (ps)	10.81	185.9
Energy ((E-15) *J)	12.19	12.29
Power Delay Product (f)	0.126	1.052

Table 1 shows the comparison between the proposed design with the reference circuit in terms of power, delay, energy, and power delay product.

The average power dissipation for the proposed circuit is also analyzed by varying the supply voltage (Vdd) from 0.8V to 1.2V at 27^0C temperature which is shown in Figure 3.

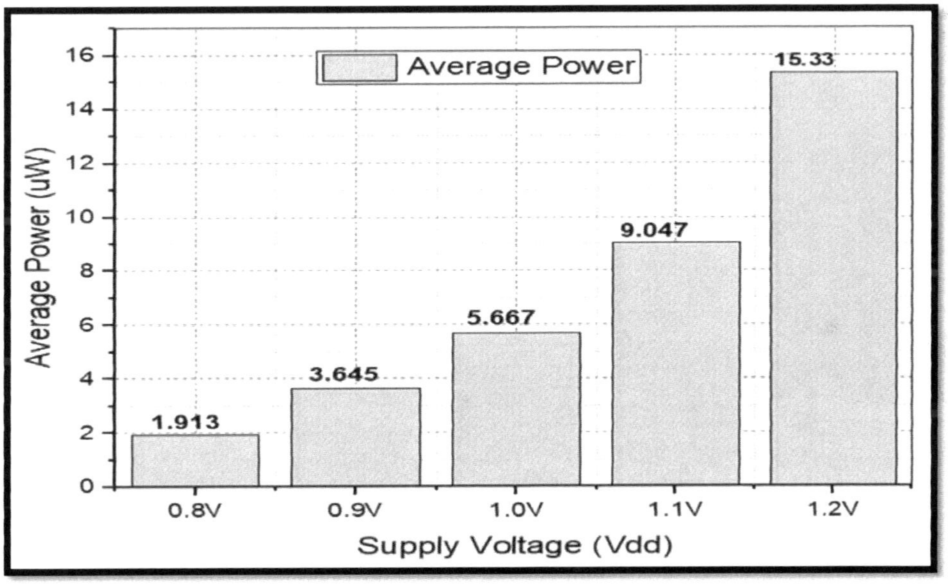

Figure 3. Average power in buffer circuit using proposed domino circuit (change in Vdd)

The output waveform of the proposed circuit is shown in Figure 4. The output waveform contains clock signal CLK, input signal A. Corresponding to the clock signal and input signal, the output signal and power signal of proposed circuit is obtained.

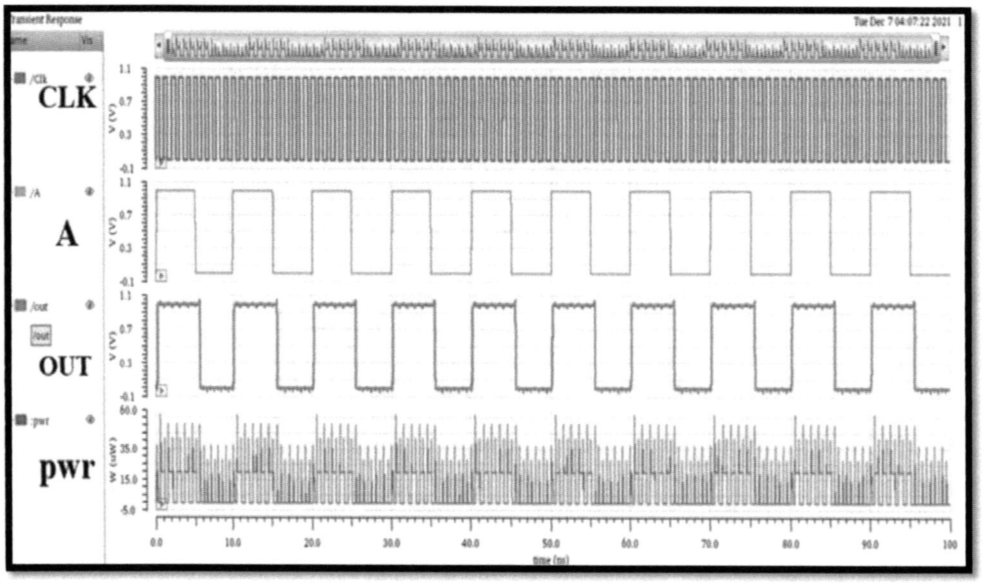

Figure 4. Transient waveform of proposed domino logic with dual threshold voltage

The current-temperature analysis at varying temperature 0^0C to 110^0C of the proposed circuit is shown in Figure 5.

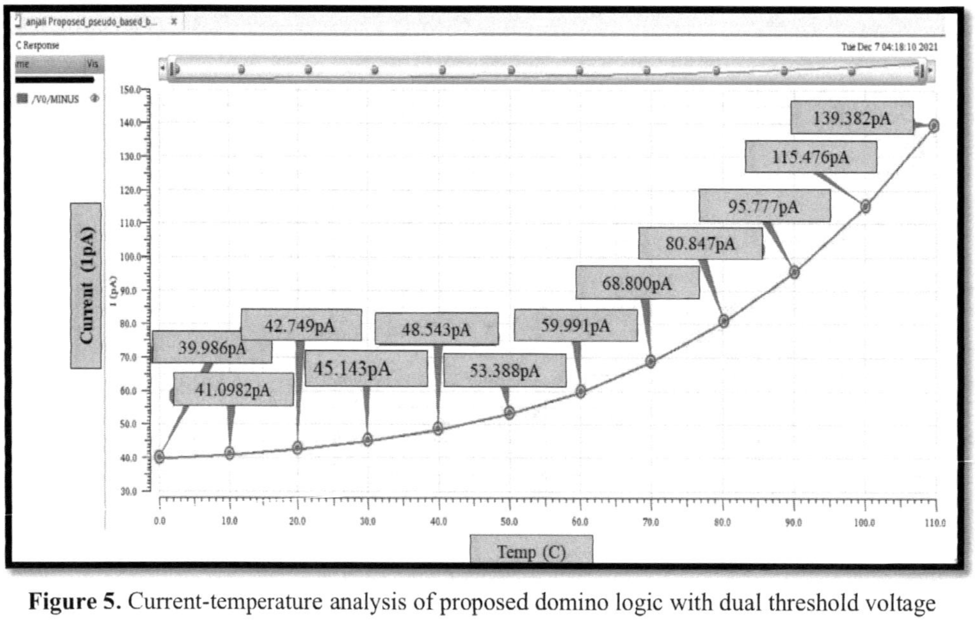

Figure 5. Current-temperature analysis of proposed domino logic with dual threshold voltage

5. Conclusion

The proposed pseudo domino logic-based buffer is designed using dual threshold voltage. The proposed domino circuit shows less power dissipation about 51.68% as compared to the existing circuit. We have calculated different parameter like power, delay, energy, and power delay product to analyze the circuit. A significant average power reduction is observed in the proposed circuit by using dual threshold voltage. The delay of the proposed circuit is higher than the previous existing circuit, so it has limitation in terms of speed of the circuit. This circuit is applicable for low power application.

6. References:

[1] Kannan R, Rangarajan R. Low power noise immune node voltage comparison keeper design for high speed architectures. Microprocessors and Microsystems, 2020, 103192.
[2] Shah A P, Neema V, Daulatabad S. A novel leakage reduction DOIND approach for nanoscale domino logic circuits. 2015 Eighth International Conference on Contemporary Computing (IC3), 2015.
[3] Liu Z, Kursun V. Sleep switch dual threshold voltage domino logic with reduced subthreshold and gate oxide leakage current. Microelectronics Journal, 2006, 37(8), 812–820.
[4] Panday A K, Gupta T K, Gupta A, Pandey D. Keeper Effect on Nano Scale Silicon Domino Logic Transistors. Silicon, 2021.
[5] Sah N, Mittal E. An improved domino logic. 2017 International Conference on Energy, Communication, Data Analytics and Soft Computing (ICECDS), 2017.
[6] Ghimiray S R, Meher P, Dutta P K. An improved charge-sharing elimination pseudo-domino logic. International Journal of Circuit Theory and Applications, 2020.
[7] Ghimiray S R, Meher P, Dutta P K. Ultralow power, noise immune stacked-double stage clocked-inverter domino technique for ultradeep submicron technology. Int J Circ Theor Appl, 2018, 46(11): pp. 1953–1967.
[8] Bansal D, Nagar B C, Singh B P, Kumar A. A New Keeper Domino Logic Based Full Adder for HighSpeed Arithmetic Circuits. Micro and Nanosystems, 2021
[9] Angeline A A, Bhaaskaran V S K. Design impacts of delay invariant high-speed clock delayed dual keeper domino circuit. IET Circ Devices & Syst. 2019;13(8):1134-1141.
[10] Shah A P, Neema V, Daulatabad S, Singh P. Dual threshold voltage and sleep switch dual threshold voltage DOIND approach for leakage reduction in domino logic circuits. Springer Microsystem Tech. 2017;25(5):1-14.
[11] Ayala-Chauvin M, Kavrakov B S, Buele J, Varela-Aldas J. Static Reactive Power Compensator Design, Based on Three-Phase Voltage Converter. Energies, 2021, 14, 2198.

Design and Implementation of Bridgeless AC/DC PFC SEPIC Converter with Valley-Fill Circuit

B Lakshmi Praba[1,3] and R Seyezhai[2]

[1] Research scholar, Renewable Energy Conversion Laboratory, Department of EEE, Sri Sivasubramaniya Nadar College of Engineering, Kalavakkam, India

[2] Professor, Renewable Energy Conversion Laboratory, Department of EEE, Sri Sivasubramaniya Nadar College of Engineering, Kalavakkam, India

Abstract: The continuous requirement to improve power quality has motivated the proposal of several PFC circuit topologies. The conventional AC/DC SEPIC has increased conduction loss due to the presence of bridge rectifier. In order to reduce the conduction loss and to improve the efficiency this research work deals with a bridgeless SEPIC (single ended primary inductance converter) AC/DC PFC integrated valley-fill circuit (VFC). The number of components is reduced compared to the existing topologies. The bridgeless topology results in reduced conduction loss, reduced current stress and improved system performance compared with the traditional SEPIC and the existing topologies. The circuit configuration of the suggested topology has been simulated using MATLAB/SIMULINK. The functional parameters of the projected AC/DC integrated VFC is compared with the conventional SEPIC (Con-SEPIC), con-SEPIC integrated VFC and bridgeless SEPIC in terms of passive component count, output voltage ripple, input current ripple, supply power factor, Total Harmonic Distortion. From the simulation outcomes, it is concluded that the bridgeless topology provides high power factor, reduced ripple and reduced supply current harmonics. The results are validated.

Keywords: Bridgeless SEPIC, Ripple, PFC (Power Factor Correction), THD, Valley-Fill (VFC).

1. Introduction

Nowadays, LED lightings have occupied over other lightings in all probable applications owing to developments such as excessive lifespan, enhanced efficiency and adaptability in size, shape and reachable in various colour. For this LED applications the proposed work fully deals with the AC/DC PFC SEPIC converter. A 24W LED is chosen for this proposed work [1-3].

The continuous requirement to attain improvement in power quality factors has inspired the suggestion of several PFC circuit topologies for LED applications [4]. They are prevalent to reduce total harmonic distortion (THD) at AC input, retain a constant DC output voltage, decrease losses, and enhanced power factor (PF) for supply current. These PFC circuits regulate the waveforms of the current in the supply side to enhance the power factor, so that extreme power can be strained from the supply side [5,

[3] *Corresponding Author*
email: lakshmiprabab@ssn.edu.in

6]. Numerous types of converter are employed in power factor correction, such as buck, boost, buck-boost and flyback converters for DC side and STATCOM for AC side [7]. All these above mentioned circuits are analyzed in two basic modes of processes named as, Continuous Conduction Mode (CCM), Discontinuous Conduction Mode (DCM).

This research work fully deals with the SEPIC topologies that are operated in DCM [8-10], with better power factor modification capability. Different SEPIC PFC topologies have been studied in this research work, namely AC/DC Conventional PFC SEPIC, Bridgeless AC/DC SEPIC, bridgeless PFC and VFC derived SEPIC PFC discussed in [11, 12]. From the literature study, it is observed that these topologies suffer with higher output voltage ripple and THD. To overcome the limitations, a new Bridgeless AC/DC PFC SEPIC integrated VFC is proposed. The projected AC/DC PFC SEPIC is used to drive a load of 24W LED. This work offers a different topology of single power MOSFET switch bridgeless AC/DC PFC SEPIC that gives reduced THD and also improved PF with fewer numbers of components. The proposed bridgeless AC/DC PFC SEPIC topology contacts the bridge rectifier and DC/DC converter in single stage with one active switch that results in reduced cost. [13-16]. AC/DC Conventional PFC SEPIC integrated VFC

A comparison is made among the proposed topology with AC/DC Conventional PFC SEPIC, Bridgeless AC/DC SEPIC, bridgeless PFC, VFC derived SEPIC PFC and modified AC/DC Conventional PFC SEPIC integrated VFC [17]. From comparison it is found that, the nonexistence of an input rectifier bridge helps in the reduction of conduction loss. Also, the projected converter has reduced supply current THD, improved power factor and contains two additional diodes to eliminate a short circuit condition. In addition to that, the Valley-fill PFC circuit is used to separate the input PFC current and output current, which reduces the switch current stress and improves system thermal management compared with the traditional SEPIC converter. Therefore, this novel VFC SEPIC topology is suitable for LED lighting applications.

2. AC/DC PFC SEPIC Converter Topologies

The various AC/DC PFC SEPIC topologies analyzed in this research work are as follows:

- Con-AC/DC PFC SEPIC
- Con-AC/DC PFC SEPIC integrated VFC
- Bridgeless AC/DC PFC SEPIC
- Bridgeless AC/DC PFC SEPIC integrated VFC

2.1. Con-AC/DC PFC SEPIC circuit

From Fig.1, when the power MOSFET switch S_1 is turned on, current raises at a fewer rate and drives in negative polarity. When the power MOSFET device is in off state, the current through capacitor is identical to the current via the inductor. A SEPIC is functioned in discontinuous-conduction mode if current through the inductors L1 or L2 is permissible to fall to zero throughout an operating cycle.

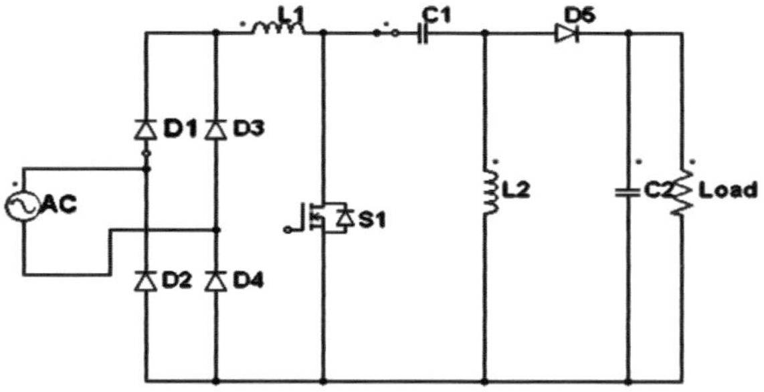

Figure 1. Circuit for Con-AC–DC PFC SEPIC

2.2. Con-AC/DC PFC SEPIC integrated VFC

The con-SEPIC integrated VFC is depicted in Fig.2. The VFC involves 2 capacitors and 3 diodes. The capacitors that are present in VFC are gets charging in series and shunt to the load. The SEPIC integrated VFC functions in DCM mode of operation. The power MOSFET switch and also the passive devices are supposed to be in ideal. The salient advantage of VFC is that, which offers enhanced power factor and high efficiency by the addition of diodes and capacitors.

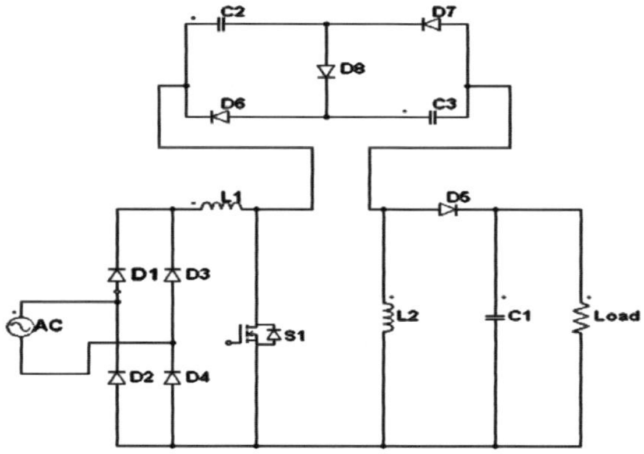

Figure 2. Circuit for AC/DC PFC SEPIC integrated VFC

2.3. Bridgeless AC/DC PFC SEPIC

A bridgeless SEPIC PFC is depicted in Figure 3. which has single power MOSFET switch, it makes this circuit simple and is easy to develop the control scheme and the switching losses are also lower compared to conventional AC/DC SEPIC PFC. The switch S_1 is used and it transfers by an electrical cycle. Once S_1 turns on, blocking diodes D_2 and D_3 are mandatory to avoid confliction of on (positive), off (negative) half cycles.

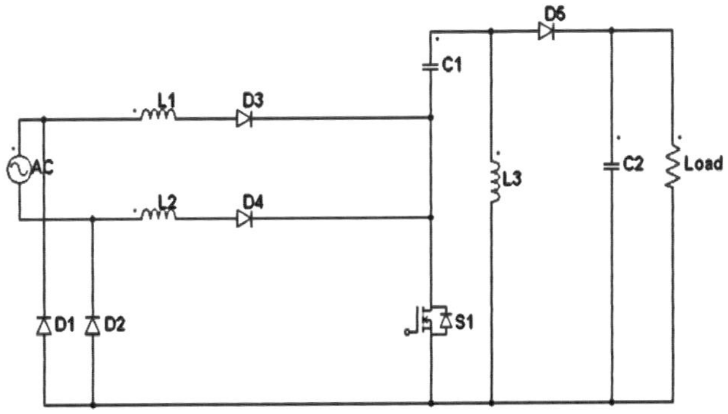

Figure 3. Circuit for Bridgeless AC/DC PFC SEPIC

2.4. Proposed Bridgeless AC/DC PFC SEPIC integrated VFC

This proposed topology deals with minimized current stress through diodes and supply capacitor. Moreover, this projected one offers a single switch and also the continuous assembly among the input and the output processed voltages over the diodes D_1 D_2, hence reducing the common-mode electro-magnetic interference (EMI). However, it contains similar quantity of diodes like conventional-SEPIC. The VFC is not only used for increasing supply PF; the primary role of VFC is used to shrink the ripple in load voltage. The basic principle of this VFC is to increase the time that rectifier consumes input line current.

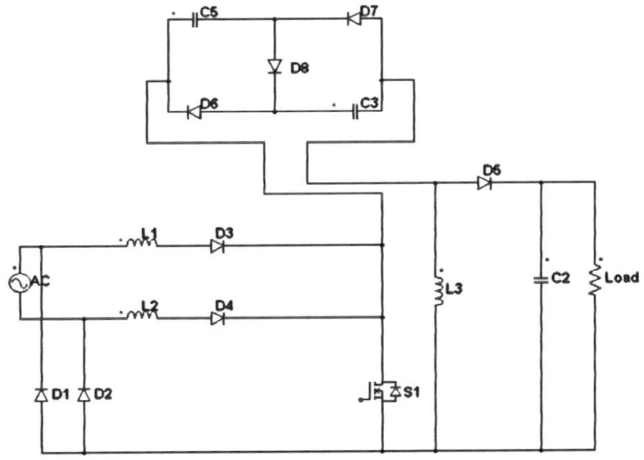

Figure 4. Circuit for Bridgeless AC/DC PFC SEPIC integrated VFC

3. Design Procedure

Following are the typical design equations of the Con-SEPIC converter is specified in Eq. 1 – 3,

$$\text{Conversion gain of SEPIC is } \frac{V_o}{V_{in}} = \frac{D}{1-D} \tag{1}$$

$$\text{The inductor value of SEPIC is } L1 = L2 = L3 = L_c = \frac{DV_{in}}{\Delta I_0 2f_s} \tag{2}$$

$$\text{The output capacitor value is } C_C = \frac{DI_0}{\Delta V_0 2f_s} \tag{3}$$

Where,

V_0 is load voltage,
V_{in} is supply voltage
D is duty ratio.
f_s is switching frequency,
$\Delta I_0 = 30\% - 50\%$ in I_0.
$\Delta V_0 = 1\% - 5\%$ of V_0

Eq. 1 - 3 holds good for all the topologies.

4. Simulation Results

The simulation parameters are calculated using the design equations that are discussed in sec.3 and depicted in Table.1

Table 1. Simulink modelling parameters

Sl.No	Parameters	Values
1.	Input voltage	40V
2.	Switching frequency	50KHz
3.	Duty ratio	0.45
4.	Capacitor	14.18μF
5.	Filter capacitor	700μF
6.	Resistor	40Ω

MATLAB Simulink results for all the four AC/DC PFC SEPIC topologies are discussed below,

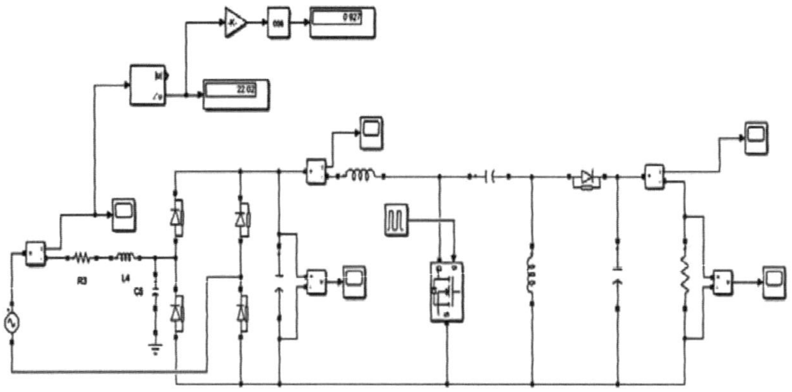

Figure 5. MATLAB Simulink diagram of Con-AC/DC PFC SEPIC

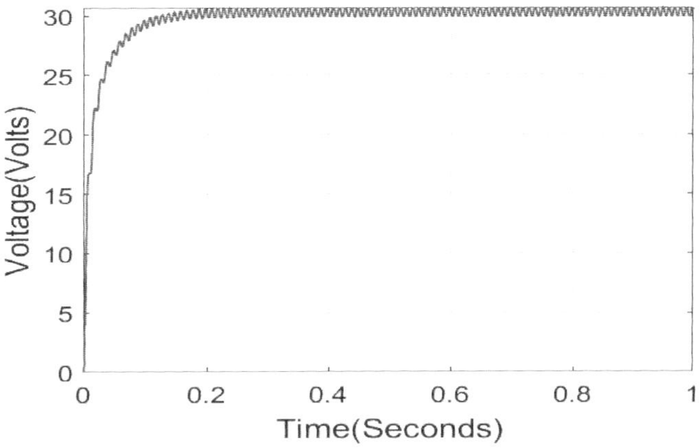

Figure 6. Simulation output waveform for Con-AC/DC PFC SEPIC

The output voltage for con-AC/DC PFC SEPIC is described in Figure 6, from that the output voltage attained is around 30.7V.

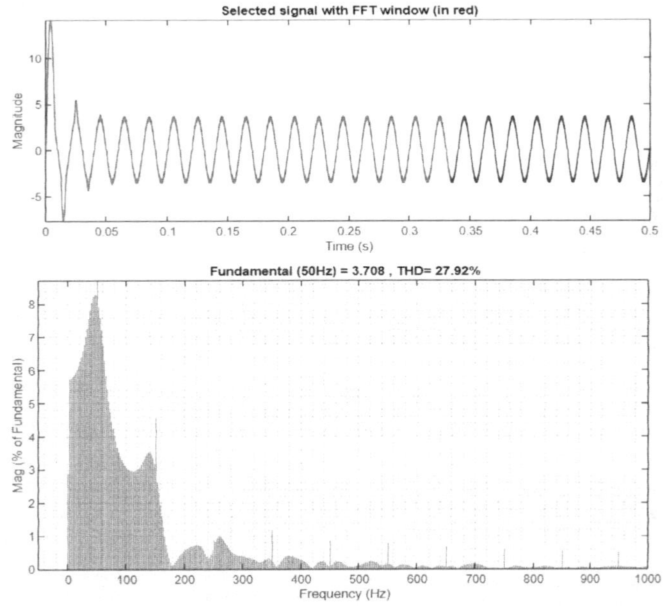

Figure 7. THD value of Con-AC/DC PFC SEPIC

From Figure 7, THD spectrum shows that the THD value for con-AC/DC PFC SEPIC is about 28%.

IOP Conf. Series: Materials Science and Engineering 1258 (2022) 012058 doi:10.1088/1757-899X/1258/1/012058

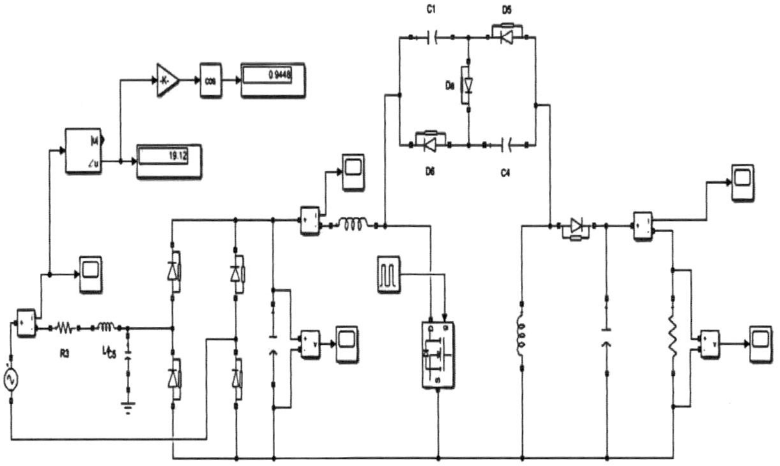

Figure 8. MATLAB Simulink diagram of con-AC/DC PFC SEPIC integrated VFC

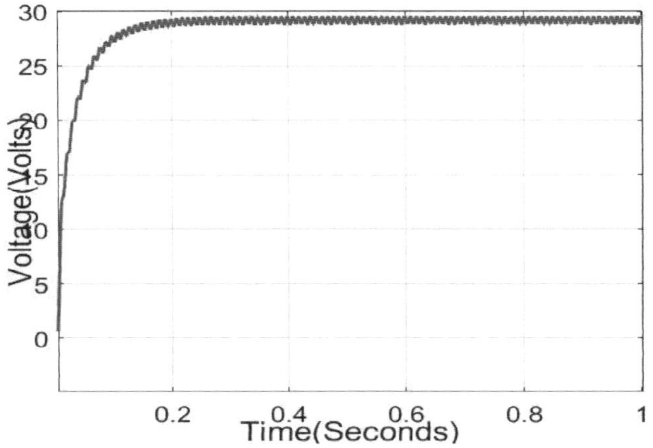

Figure 9. Simulink output for con-AC/DC PFC SEPIC integrated VFC

The output voltage for con-AC/DC PFC SEPIC integrated VFC is depicted in Figure 9, where the output voltage attained is around 29.5V.

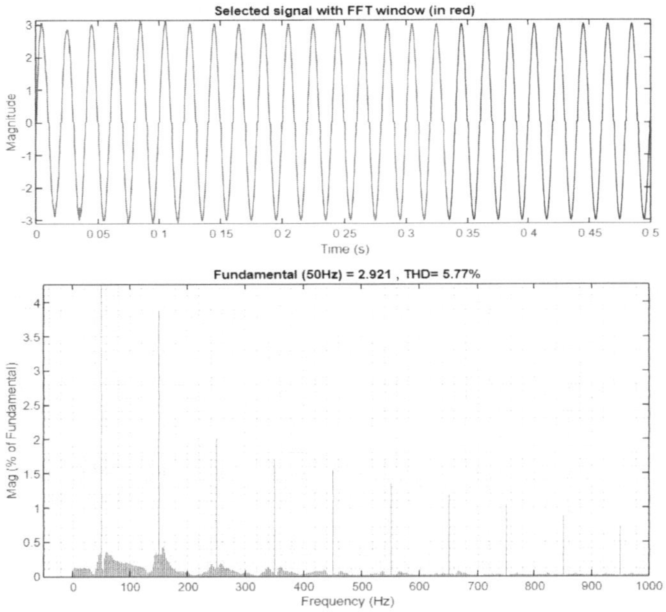

Figure 10. THD for con-AC/DC PFC SEPIC integrated VFC

From Figure 10, THD spectrum shows that the THD value for con-AC/DC PFC SEPIC integrated VFC is about 5.77%.

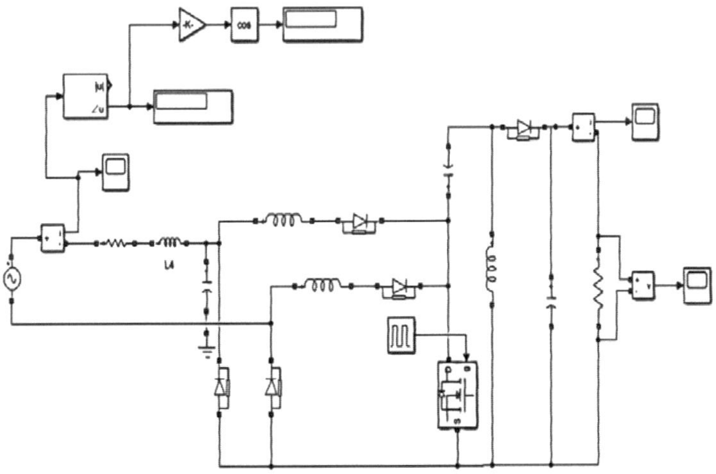

Figure 11. MATLAB Simulink diagram of Bridgeless AC/DC PFC SEPIC

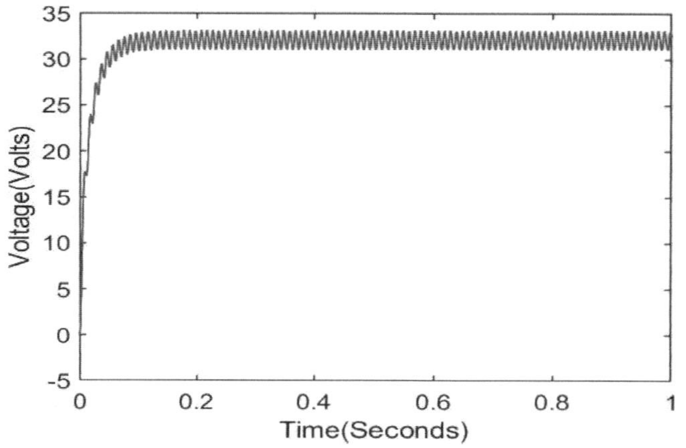

Figure 12. Simulink output for Bridgeless AC/DC PFC SEPIC

The output voltage for Bridgeless AC/DC PFC SEPIC is depicted in Figure 12, where the output voltage attained is around 33.1V.

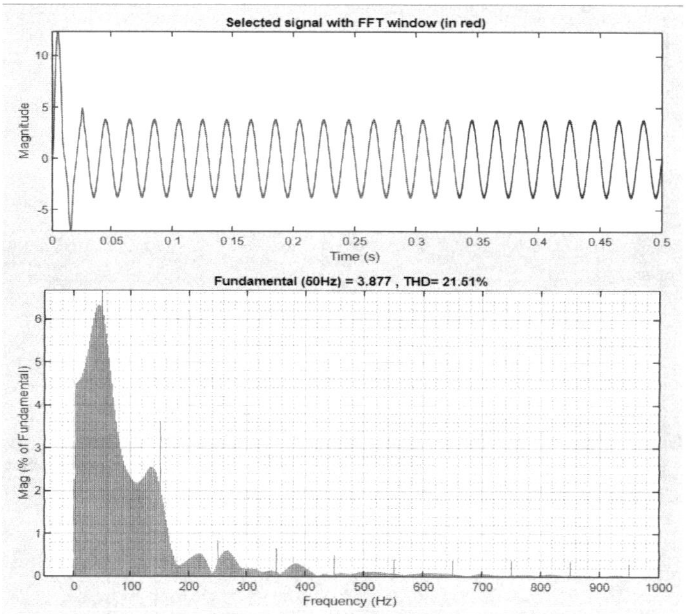

Figure 13. THD waveform for Bridgeless AC/DC PFC SEPIC

From Figure 13, THD spectrum shows that the THD value for Bridgeless AC/DC PFC SEPIC integrated VFC is about 21.51%.

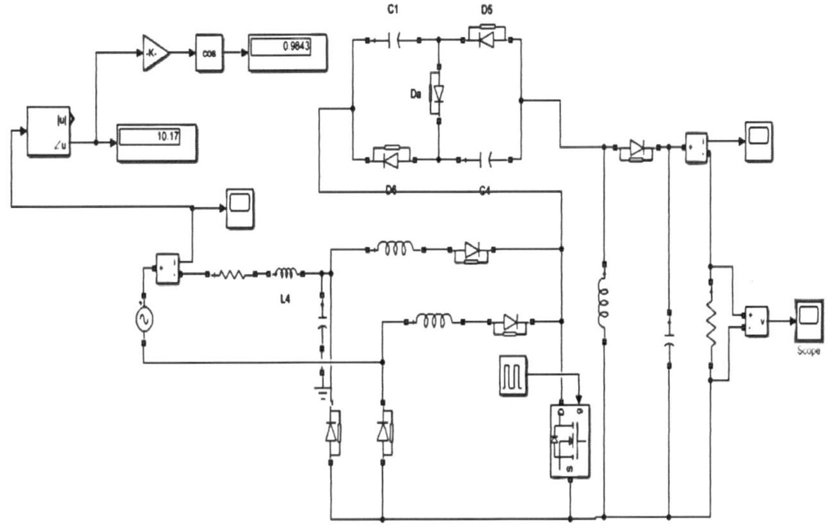

Figure 14. MATLAB Simulink diagram of Bridgeless AC/DC PFC SEPIC integrated VFC

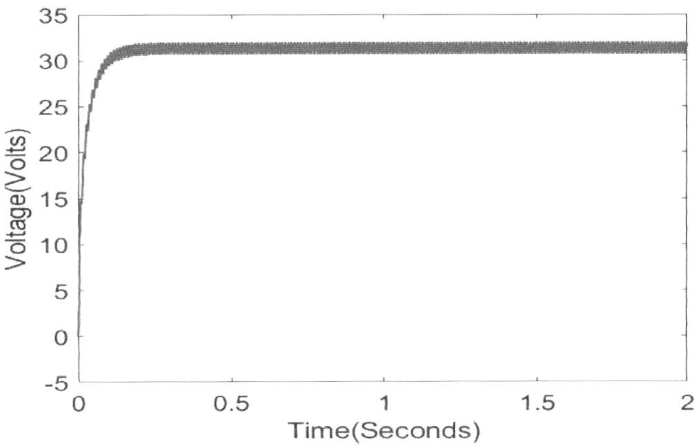

Figure 15. Simulink output for Bridgeless AC/DC PFC SEPIC integrated VFC

The output voltage for Bridgeless AC/DC PFC SEPIC integrated VFC is depicted in Figure 15, where the output voltage attained is around 32.1V.

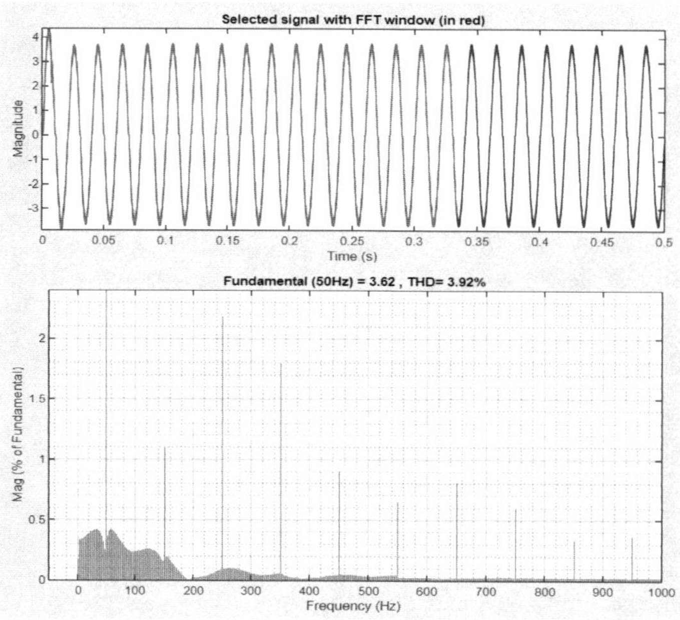

Figure 16. THD waveform for Bridgeless AC/DC PFC SEPIC integrated VFC

From Figure 16, THD spectrum shows that the THD value for Bridgeless AC/DC PFC SEPIC integrated VFC is about 3.92%

Table 2. Comparison of Topologies

Topologies	Output Voltage (V)	Output Voltage Ripple (%)	THD (%)	PF
1. Con-AC/DC PFC SEPIC	30.7	8.53	27.92	0.921
2. Con-AC/DC PFC SEPIC integrated VFC	29.5	4.14	5.77	0.985
3. Bridgeless AC/DC PFC SEPIC	33.1	7.15	21.51	0.952
4. Proposed Bridgeless AC/DC PFC SEPIC integrated VFC	32.1	3.62	3.92	0.995
5. A New PFC converter using bridgeless SEPIC discussed in [11]	24.99	8	17.36	0.999
6. A Valley-Fill SEPIC-Derived Power Factor Correction Topology discussed in [12]	11.908	-	28.794	0.969

From Table 2, Concluding that the proposed Bridgeless AC/DC PFC SEPIC integrated VFC has enhanced performances like THD, Voltage Ripple and Power Factor.

5. Conclusion

In this research work a bridgeless AC/DC PFC SEPIC circuit integrating with the VFC converter is examined. The comparison of the conventional bridged SEPIC, bridgeless SEPIC, existing AC-DC SEPIC PFC topologies and the proposed bridgeless with VFC SEPIC converters are offered to illustrate the merit and demerit of the proposed converter. From Table.2, it is concluded that the Bridgeless AC/DC PFC SEPIC integrated VFC converter has reduced output voltage ripple, enhanced power factor and better harmonic spectrum for the sinusoidal supply current associated with the Con-AC/DC PFC SEPIC converter. Moreover, the projected bridgeless SEPIC integrated VFC will be suitable for LED applications. This proposed topology is limited to low current applications. The research work can be extended with the reliability analysis for the projected topology in future.

6. References

[1] A. N. Padmasali and S. G. Kini, 2020, Accelerated degradation test investigation for life-time performance analysis of LED luminaires, *IEEE Trans. Compon., Packag., Manuf.Technol.*,vol.**10**, no.4,pp.551–558,doi:10.1109/TCPMT.2958852.

[2] A. N. Padmasali and S. G. Kini, 2020, a lifetime performance analysis of LED luminaires under real-operation profiles, *IEEE Trans. Electron Devices*, vol. **67**, no. 1, pp. 146–153, doi: 10.1109/TED. 2950467.

[3] A. N. Padmasali and S. G. Kini, 2020, a generalized methodology for predicting the lifetime performance of LED luminaire, *IEEE Trans. Electron Devices*, vol. **67**, no. 7, pp. 2831–2836, doi: 10.1109/TED.2020.2996190.

[4] Bayan Hussein, Nima abdi and Ahmed massoud, 2021, Single-Phase SEPIC-based Power Factor Correction for Electric Vehicles Charging, 11th IEEE Symposium on Computer Applications & Industrial Electronics (ISCAIE)|978-1-6654-03IEEE,|DOI: 10. 1109/ ISCAIE517 53.9431 82438-2/21.

[5] Khadiza Akter, Gholam sarovar, 2018, Modeling and Simulation of Input Switched AC-DC SEPIC Converter with PFC Control for Optimized Operation, 4th international conference electrical engineering, information and communication technology, 978-1-5386-5541-2/18/$31.00, IEEE.

[6] Kazi Khurshidi Haque Dia1, Mohammad Ali Choudhury, 2016, A Single Phase Two Switch SEPIC Power Factor Corrected Rectifier, 9th International Conference on Electrical and Computer Engineering, IEEE, 978-1-5090-2963-1/16.

[7] Ayala-Chauvin, M.; Kavrakov, B.S.; Buele, J.; Varela-Aldás, J. Static Reactive Power Compensator Design, Based on Three-Phase Voltage Converter. Energies 2021, 14, 2198. https://doi.org/10.3390/en14082198

[8] Zeynep Bala Duranay, 2021, Output Voltage Regulated Sepic Converter for Photovoltaic Systems, 13th International Conference on Electronics, Computers and Artificial Intelligence (ECAI)|978-1-6654-2534-6/21/$31.00, IEEE |DOI: 10.1109 /ECAI52376.9515157.

[9] G. M. Soares, J. M. Alonso, and H. A. C. Braga, 2018, Investigation of the Active Ripple Compensation Technique to Reduce Bulk Capacitance in Offline Flyback-Based LED Drivers, *IEEE Trans. on Power Electron.*, vol. **33**, no. 6, pp. 5206–5214.

[10] Thiago G. Neves, Luiz C. G. de Freitas, Antônio O. C. Neto, Vítor F. Barbosa, Gustavo B. de Lima, João Batista Vieira Júnior, 2021, Analysis and Comparison of Bridgeless SEPIC Topologies operating in Discontinuous Conduction Mode for Power Factor Correction in Aircraft Power Systems, 14th IEEE International Conference on Industry Applications (INDUSCON)|978-1-6654-4118-6/21/$31.00,IEEE|DOI:10.1109 /INDUSCON51756.9529622.

[11] M. R. Devi and L. Premalatha, 2015, "A new PFC converter using bridgeless single-ended primary induction converter (SEPIC)," International Conference on Control, Instrumentation, Communication and Computational Technologies (ICCICCT), 2015, pp. 321-324, doi: 10.1109/ICCICCT.2015.7475297.

[12] Hariprasath.S and Dr.Balamurugan.R, 2013, A Valley-Fill SEPIC-Derived Power Factor Correction Topology for LED Lighting Applications Using One Cycle Control Technique,"2013 International Conference on Computer Communication and Informatics (ICCCI -2013), IEEE, 978-1-4673-2907-1/13/$31.00.

[13] R. Lv and D. Dai, 2020, Analysis and Design of Bridgeless Continuous Input Current Charge Pump PFC Converter, *IEEE Access*, vol. **8**, pp. 133734–133745.

[14] Radha Kushwaha and Bhim Singh, 2020, An Isolated Bridgeless Cuk-SEPIC Converter Fed Electric Vehicle Charger, IEEE Industry Applications Society Annual Meeting,978-1-7281-7192-0/20/$31.00, IEEE|DOI:10.1109/IAS44978.9334823.

[15] Jitendra Gupta and Bhim Singh, 2020, Single Stage Isolated Bridgeless Charger for Light Electric Vehicle with Improved Power Quality, IEEE International Conference on Power Electronics, Drives and Energy Systems (PEDES) | 978-1-7281-5672-9/20/$31.00, IEEE|DOI: 10. 1109/PEDES49360. 9379852.

[16] Umang Singh Patel, S.Shiva kumar, Vinaya Sagar Kommukuri, 2019, Power Factor Correction using Continuous Conduction Mode based Bridgeless SEPIC Converter for Improved THD and Voltage Regulation, Conference on Vision towards Emerging Trends in Communication and Networking (ViTECoN), IEEE 978-1-5386-9353-7/19/$31.00.

[17] Dylan Dah-Chuan Lu, 2007, Analysis of an AC-DC Valley-Fill Power Factor Corrector, *ECTI Transactions on Electrical Eng., Electronics, and Communications* Vol.**5**, No.2.

Tversky Similarity based Deep Neural Learning Classification for Engineering Alloys

P M Siva Raja[1], S Vidhya[2], R.P. Sumithra [3] and K Ramanan[4]

[1,2,3]Department of Computer Science & Engineering, Amrita College of Engineering and Technology, Nagercoil, Tamil Nadu, India, 629901
[4]NPR College of Engineering and Technology, Dindigul, Tamil Nadu"

Abstract. Integrated Computational Materials Engineering (ICME) is an environment friendly technique used for performing cloth discovery and design. Computational methods introduced a new deep studying classification approach to display screen the candidate cloth designs. During the product designing stage, the ingredients are customised to meet particular needs. In ICME processes, there is always a degree of uncertainty in the process, structure, and property components. Uncertainties may be quantified, reduced, and propagated via structure–property links using the Tversky Similarity based Deep Neural Learning Classification (TS-DNLC) Method. In TS-DNLC Method, number of compound data are considered as input and given to the input layer. An input compound data is given to hidden layer 1. In that layer, regression is employed for performing the compound data analysis with structure–property linkages. After that, the regression coefficient value is sent to the hidden layer 2. In that layer, Tversky similarity function is used to identify the similarity between the regression coefficient value of training compound data and threshold value. Tversky similarity value varies from 0 to 1 and the results are transmitted to the output layer. By this way, TS-DNLC Method improves the performance of structure–property linkages. The computational cost of proposed TS-DNLC Method is higher than conventional uncertainty quantification.

1. Introduction

A data-driven approach was designed in [1] for estimation of coarse scale basis function. Using solution samples and a neural network predictor, we generated future basis functions with low computing overhead. A non-intrusive method was used in [2] to establish multi-scale uncertainty quantification (UQ) and uncertainty propagation. Using the top-down sampling strategy, the non-stationary and continuous spatial variations of nested random fields were used There is a relationship between polycrystalline material microstructure unpredictability and fatigue life scatter, as shown by a fatigue crack initiation and life prediction model developed in [3]. For the development and modelling of material systems and structures, a brand-new data-driven computational framework was proposed in [4].

Tversky similarity function identifies the similarity between the regression coefficient value of training compound data and threshold value. Ttversky similarity value varies from 0 to 1 and the results are transmitted to output layer. TS-DNLC Method improves the performance of structure–property linkages.

2. Related Works

Engineering design (ED) is the process of solving technical problems within requirements and constraints to create new artifacts. Data science (DS) is the inter-disciplinary field that uses computational systems to extract knowledge from structured and unstructured data [1]. The synergies between these two fields have a long story and throughout the past decades, ED has increasingly benefited from an integration with DS

A generalized framework for the digital generation of composite microstructures using filter-based approaches that can devise and utilize a wide variety of cost functions reflecting the desired targets on geometrical and statistical measures. The use of filter-based approaches leads to remarkable computational advantages compared to the conventional approaches used currently for microstructure generation [2]. The framework provides a highly modular and flexible approach to generate stochastic ensembles of microstructures meeting user-defined microstructural characteristics.

Microstructure evaluation is one of the key element to determine the properties and performance of a material which has been great eternal towards humanity and industrial applications. The present research work aims to study the grain growth kinetics influenced by various second phase particles and also homogeneity of the grains using Potts model under the Monte Carlo simulation method [3]. A 2D square lattice with optimum matrix size (N), Q-States (Q), and MCS was employed. The mean grain size (Rmean) and maximum grain size R(max) was determined for various surface fractions, implying average grain growth and microstructure consistency

As an important factor of dual-phase polycrystalline materials, dual-phase morphology distribution significantly affects its deformation mechanism due to the complex biphasic interactions. However, it was rarely investigated, especially for titanium alloy. Although crystal plasticity modeling has been regarded as an important method to study the deformation behaviors, the current models do not fully consider the actual grain scale information of dual-phase materials. In this paper, a pseudorandom dual-phase crystal plasticity finite element (CPFE) model is established, which fully considers the actual grain size and orientation information, volume fraction, and constitutive behaviors of the dual-phase grains [4] . The model is calibrated by uniaxial tensile experiments, and validated from micro and macro levels respectively.

When the grain cores are in their original state, a 2D dual-phase polycrystalline microstructure has been developed in [5]. Different phases were created to have identical grain growth velocities so that the grain growth velocity variations may be linked to each other. The Uncertainty Quantification (UQ) technique was initially introduced in the AM process in [6] for calculating the uncertainty in material attributes owing to uncertainty sources [7]. The ultimate tensile strength of the structure was achieved by laser sintering of nanoparticles.

Structural dynamic model and crowd behaviour uncertainty was quantified in [7] using a comprehensive method. Uncertainty was communicated from input parameters to the construction of the footbridge. For laminated composite plates, [8] suggested a stochastic natural frequency approach using an artificial neural network (ANN) model. The ANN model was introduced as a surrogate model using Latin hypercube sampling. Complexity was, however, not lessened. In [9], a methodology for comparing fatigue resistance among different microstructures with microstructurally tiny fracture development regimes was developed.

3. Research Methodology

Integrated Computational Materials Engineering is used to design the products and materials through linking the material model at different length scales. The key objective is on materials that how process creates the material structure, how structures respond to material properties and how to select the materials for particular application. All four of these components play an important role in the whole process. ICME's goal is to speed up the materials design and development process by integrating modelling and simulation, resilient design, and information technologies. By establishing structural–property links, microstructure statistical uncertainty is able to quantify output property uncertainties robustly, mitigate them, and disseminate them. SVE (Statistical Volume Elements) are used to calculate Fatigue Indicator Parameters (FIPs) for each element in the discretized microstructure, with the volume averaged across a user-specified finite damage process zone. There was no longer any sensitivity to mesh size, and the numerical solution now accurately reflected both the physical damage and the extent of the damage. To aid in the comparison of fatigue resistance, the average volume is fixed throughout many simulations. As part of the elasto-plastic shakedown, the FIPs are calculated using the third of three complete cycles of completely reversed, strain-controlled uniaxial loading.

3.1. Uncertainty

This uncertainty is expressed in terms of a probability density function (PDF) or interval limits. The use of intervals is a good way for describing uncertainty when the unknown values fall within a certain range without extra details on fluctuation, frequency or preference. The numerical mesh refinement or convergence criteria are connected to the epistemic uncertainty. Materials' microstructure and mechanical characteristics depend on beginning circumstances and process factors that are subject to uncertainty. MSU was transmitted statistically to the uncertainty of macroscopic mechanical characteristics of material. In the setting of uncertainty propagation, it is difficult to establish accurate cause-and-effect relationships between processing, structure and characteristics. Using QFD technologies, designers may focus on the demands of their clients. In order to translate customer demands into technical characteristics, QFD tools give an orderly means of doing so. It is a comparable system design diagram that aids material engineers in identifying the most critical connections between various components.

By propagating microstructure attributes uncertainty from the structure column of PSPP maps, design engineers may construct procedures and evaluate the uncertainty in important material qualities. When assessing the fatigue performance of polycrystalline materials based on crystal plasticity, brute force PDF propagation with high-fidelity is the most computationally intensive method. By use of computer models and the Monte Carlo (MC) technique, uncertainty is disseminated. To produce output distribution, PDF files are sampled and then reassembled. In order to get the output statistics, MC uses a huge number of samples. Covariance functions and kernel objects are used as prior distributions in the inner function of the surrogate model, which is optimized by training data. The SP connections are estimated with the help of the Matern kernel function. The infinitely differentiable radial basis function kernel is a generalization known as the matern kernel. Using the generalization, the user was able to figure out whether a kernel function was differentiable or not

ICME is a daring and transformational programme that calls for a fundamental change in the way materials design and development are treated. Traditional material science approaches stress a sequential approach to material creation that spans discovery. It is not necessary to create custom materials for specific purposes in order to use existing design processes and engineers. The ICME expedited the materials design and development process by integrating modelling and simulation within the uncertainty environment, strong design, and information systems.

3.2. Modeling Fatigue

Deformation gradient is multiplicatively decomposed in the microstructure-sensitive FEM model, and the slip system is governed by a power law flow rule with a rate sensitivity. UMAT, or User MATerial, is the method through which models in ABAQUS are shown to the user. To handle the primary alpha-phase and the alpha–beta-colony phase, the original Ti-64 crystal plasticity model is developed to address discrete three-dimensional slip geometries for each phase. Crystallographic texture and dislocation structure are both improved with its help. The dislocation interaction between distinct microstructure properties is captured by the constitutive model for monotonic and cyclic loads. The alternating laths of secondary alpha- and beta-phase were composed of grains in the alpha–beta-colony phase phase.

Hall–Petch connection is employed with threshold stress in primary and secondary alpha-phases. Grain boundaries are seen as an impermeable surface that facilitates grain compatibility. The CPFEM model has a varied model form and model parameter ambiguity that contribute to the uncertainty. The model's primary goal is to measure and disseminate the uncertainty associated with the statistical distribution of

IOP Conf. Series: Materials Science and Engineering 1258 (2022) 012059 doi:10.1088/1757-899X/1258/1/012059

important microstructure properties. In future efforts, the developed model will be supplemented with a crystal plasticity constitutive crystal plasticity model to reduce uncertainty. Fatigue fracture development and tiny trans granular expansion were both surrogated by FIP. For fractures at the slip band/matrix interface, a substantial link between FIP and Fatemi–Socie was found. Analogy is used to detect the driving factor for minor fatigue crack formation. To counter fatigue fracture development and subsequent growth, FIP was used as a stand-in strategy.

3.3. Tversky Similarity based Neural Classification (TS-DNLC) for Structure–Property Linkage Modelling.

Deep Learning is a kind of machine learning approach used to system the statistics and imitate reasoning process. Deep getting to know makes use of the layer to technique the records and apprehend the human discourse.

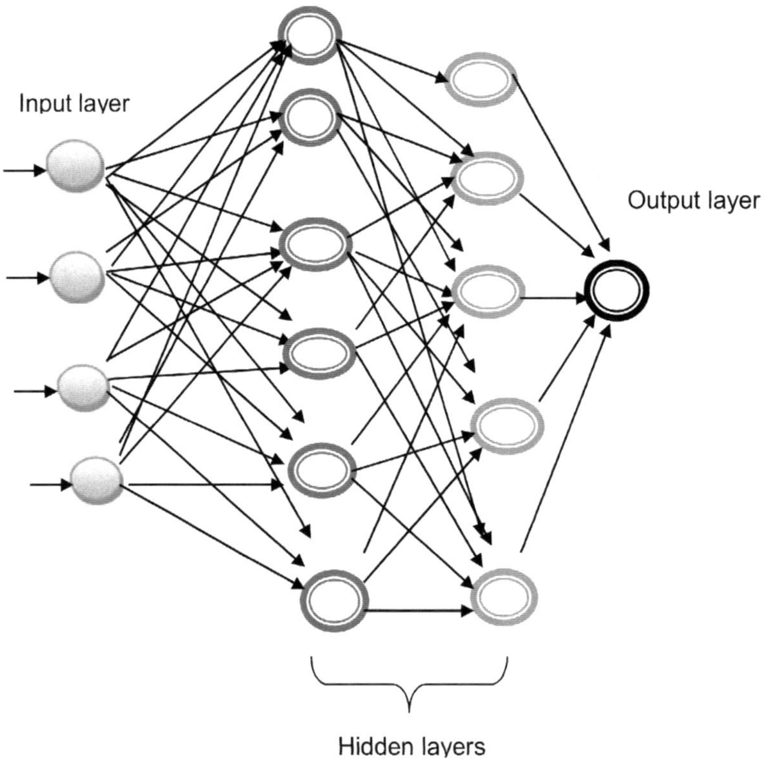

Figure 1. Construction of Deep Neural Learning Classifier.

The deep neural studying classification approach comprised 4 layers, specifically enter layer, two hidden layers, output layer for modelling the structure–property linkages with Fatigue Critical Engineering Alloys. The shape of deep neural mastering is illustrated in fig. 1.

The output of the hidden layer is symbolized as,

$$Hidden(t) = [\sum_{i=1}^{n} data_i * weight_i] + [weight_{ih} * hidden_{t-1}] \quad (1)$$

From (5), '$Hidden(t)$' denotes the output of hidden layer at time period 't'. '$hidden_{t-1}$ represent the previous hidden layer output, '$data_i$' denotes the input compound data, '$weight_{hh}$' symbolizes the weight between the input and hidden layer. Tversky similarity coefficient presents the output results from '0' to '1'. The error rate is computed at output of hidden layer for improving the classification accuracy performance and reducing the error rate. The difference between the actual and expected results is multiplied by the square root of the mistake. To put it another way,

$$E = (Actual\ value - Obtained\ value)^2 \tag{2}$$

From (6), the error 'E' value is computed. When network not exactly classifies the particular compound data, proposed algorithm changed the weights and identified minimum error. It is attained as,

$$\arg\min E \tag{3}$$

From (7), 'arg min' denotes the argument of minimum function. As a result, the values of similarity are transformed into deep neural network output layer inputs. The output layer compares the similarity data to a threshold value. According to the formula used to create the output layer

$$Output(t) = \{Hidden(t) * weight_{ho}\} \tag{4}$$

From (8), '$Output(t)$' denotes the output layer result. '$weight_{ho}$' shows how much weight is given to the output layer against the concealed layer. By this way, an efficient structure–property linkage modelling is carried out with fatigue critical engineering alloys. The algorithmic process of the proposed TS-DNLC Method is illustrated as,

\\ **Tversky Similarity based Deep Neural Larning Classification**
Input: Number of compound data
Output: Improves structure–property linkage modelling performance
Begin
1. Collect number of compound data at input layer
2. The input layer transmits the alloy data to the hidden layer 1
3. **For** each data
4. Hidden layer 1 uses the piecewise regression to examine and identify the data
5. Hidden layer 2 uses the Tversky similarity function
6. Determine the similarity between training compound data and threshold at the hidden layer 2
7. Estimate error value
8. Update initial weights
9. Find arg min $error$
10. Output layer displays the result
11. **End for**
End

Algorithm 1 describes the algorithmic process of TS-DNLC Method for improving the structure–property linkage modelling performance. TS-DNLC Method considers the compound data as input for the Deep Neural Leaning Classifier. The hidden layer 1 examines the compound data by using piecewise regression classifier. After that, the result is sent to the hidden layer 2. In that layer, Tversky similarity function is employed to determine the similarity between input compound data and threshold data. When the error is minimal, the result is transmitted to the output layer. Finally, the output layer displays the final result for enhancing the structure–property linkage modelling performance.

4. Results and Discussion

FIP distributions with extreme values are reduced to a level that allows for a suitable trade-off between computational effort and forecast accuracy. Training and Validation of Gaussian Process Surrogate Models" teaches GP models as a replacement for the CPFEM. GP models are used to assess the decrease in epistemic uncertainty that may be achieved by training models with reduced uncertainty. From structure to responses and characteristics, aleatory microstructure statistical uncertainty is propagated by use of GP surrogates in "Propagation of Aleatory Microstructural Statistic Uncertainty." FIP is calculated using all three complete cycles of three fully inverted, strain-controlled, uniaxial computational strain cycles with periodic boundary conditions set at strain amplitude at CPFEM simulation. The settings are set at ambient temperature and quasistatic strain rate. Local quantities are calculated at each integration point/element in the digital microstructures to represent the distribution of fatigue fracture development locations. To eliminate the possibility of a localised characterisation, only one FIP sample is obtained from each grain. It is necessary to use SBA FIP from every grain in order to prevent counting SBA FIPs twice in the same grain. An SBA FIP is a metric used to describe a wide range of values. In order to compare the trends in the extreme value distribution of FIP across various SVE ensembles, the Fréchet extreme value distribution is utilised.

Table 1. Computational Simulation Parameters.

Parameter	Value
Strain rate	0.001 s^{-1}
# of applied cycles	3
Load ratio	-1
Temperature	300K
Element per SVE	30^3
Boundary conditions	Periodic

4.1. Gp Model Prediction Of Yield Strength For Random And Transverse Textured Ti-64"

In order to diminish the noise in the extreme value FIP response of Ti-64, the synthetic microstructures are implemented. 7.5 m is the predefined standard deviation for grain size, including maximum and minimum cutoffs.

Grain size and volume percent of main alpha-phase are calculated using the statistical data. These 60 SVEs were generated using microstructure statistics. A total of 12 distinct SVE sizes, ranging from 100 to 300 grains per SVE, were included in the ensembles. Each of the five ensemble SVEs, namely SVEs 10, 15, 20, 25, and 30, had 200 grains, resulting in a total of 20 SVEs per ensemble. For each ensemble size, the 99th percentile EV FIP response prediction is simulated five times.

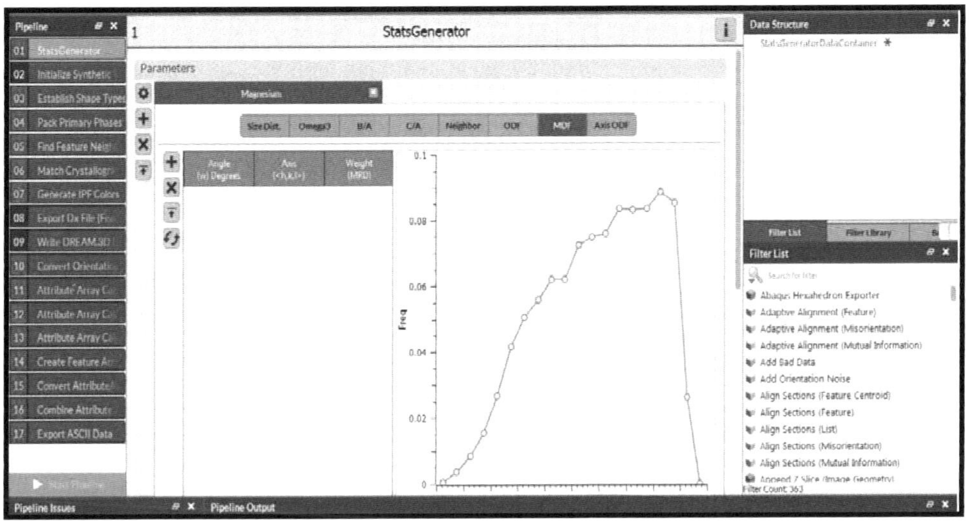

Figure 2. 2D Surface plots of GP model prediction of yield strength for random and transverse textured Ti-64.

4.2. Training and Validation of Gaussian Process Surrogate Models."

High-fidelity Using CPFEM simulations, fatigue performance of materials may be anticipated by the quantification of the projected EV FIP response. With Ti-64, it is critical to reduce the computational cost of prediction in order to convey the uncertainty with microstructure statistics to output parameters such as FIP99 and yield strength. The CPFEM simulation data for FIP99 were used to train a Gaussian Process (GP) regression model. A wide range of input parameters, including grain size and main alpha-phase volume percent, were used to create the yield strength values, ranging from 10 to 60 m and 0.3–0.6, respectively. The average grain size and volume percentage of the most relevant alpha-phase are sampled using an LHS method. A total of ten samples from the input space are used for the first training set for each texture. FIP output unpredictability is seen in Figure 3.

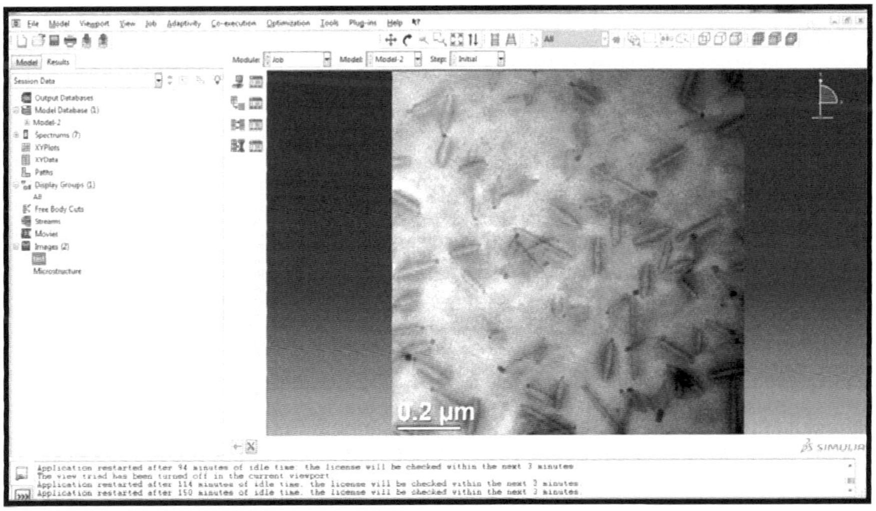

Figure 3. EV FIP and projected yield strength output uncertainty distributions.

When training the GP surrogate, a Matérn kernel and white noise kernel are mixed together in a linear fashion. With two input and two output dimensions, GP models may be used for any texture. The previous and posterior FIP99 distributions from Ti-64 were fed into a GP surrogate model. The volume percentage of primary alpha-phase varies along the x-axis, as should be noted. Readers may understand how the GP surrogate model handled UQ for increased uncertainty associated to its ft via the use of this figure. The grey colouring that surrounds the black curve reveals the degree of uncertainty. Since training, the level of confidence in GP forecasts has risen. It is easy to see that there are potentially an endless number of possible variations of the curve represented by the oscillating multi-colored curves. One standard deviation from the mean was represented by a grey shaded region in all directions on the black prediction curve. With the remaining training samples, the model is trained to predict the desired output value.

4.3. *Propagation of Aleatory Microstructure Statistic Uncertainty.*

Microstructure statistics are conveyed across structure–property relationships by using a surrogate model. As a statistical measure, the normal distribution and uncertainty in microstructure properties of primary alpha-phase grains are used to characterise the statistical distributions. Using data from several batches of the same manufactured material, an OEM constructed histograms based on the real microstructure statistics, which were derived from the histogram. The uncertainty distribution of microstructure statistics is communicated via SP links associated with the output distributions. To train the yield strength surrogate model, elastic moduli are directly measured. Input category crystallographic texture alters the elasticity modulus and phase information, but the elasticity modulus remains constant over grain size. Uncertainty is propagated via CPFEM without a GP surrogate model by using computational time.

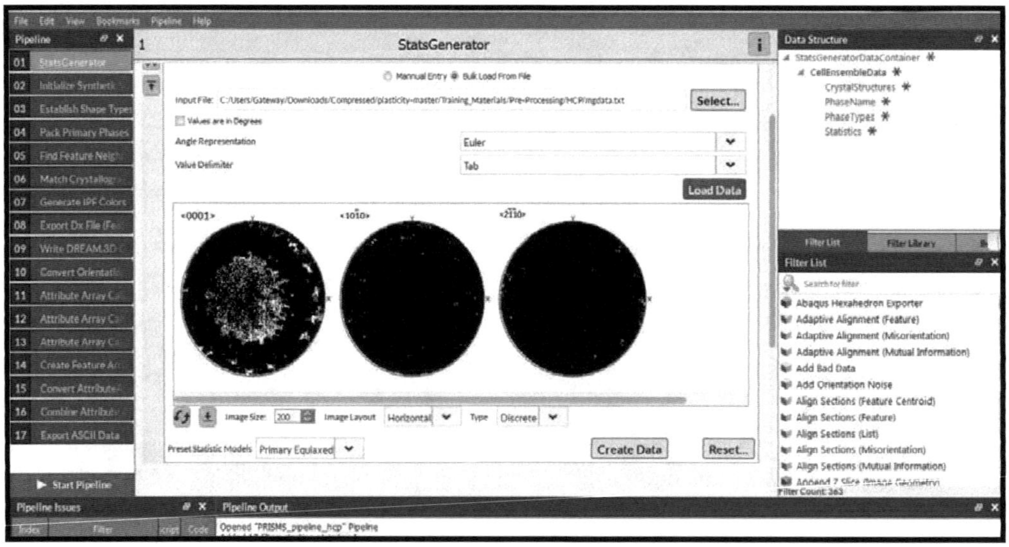

Figure 4. Normal distributions for average grain size.

The typical distribution of grain size is seen in Fig. 4. Using SP connections and the whole process, we may calculate and transmit the uncertainty in microstructure characteristics in order to estimate the uncertainty in relevant properties. EV FIPs are used to quantify the yield strength and fatigue life of surrogate models and to re-establish the SP links.

5. Conclusion

A new technique termed TS-DNLC Method quantifies reduces and propagates the uncertainties thru structure–property linkages for strong uncertainty quantification. An enter compound information is regarded as input. The piecewise regression performs the compound facts evaluation with the structure–property linkages. Tversky similarity feature identifies the similarity between the regression coefficient fee of coaching compound information and threshold value. Ttversky similarity fee levels from zero to 1 and effects are despatched to the output layer. TS-DNLC Method enhances the overall performance of structure–property linkages. The computational price of proposed TS-DNLC Method is greater than traditional uncertainty quantification.

References

[1] Filippo Chiarello Paola Belingheri Gualtiero Fantoni "Data science for engineering design:State of the art and future directions" , Computers in Industry, Volume 129, August 2021, 103447

[2] Ahmet Cecen, Berkay Yucel and Surya R. Kalidindi "A Generalized and Modular Framework for Digital Generation of Composite Microstructures", Journal of Composites Science, 2021

[3] Rajendra P, Phaneesh K R, Prakrathi S, Madeva Nagaral "Simulation of Two-Dimensional Polycrystalline Material by using Monte Carlo Approach 'Journal of Mechanical Engineering Research and Developments", Vol. 44, No. 9, pp. 205-220

[4] WentongCai.QinghuaSong, HansongJi ZhanqiangLiu " Dual-phase morphology distribution effects on mechanical behaviors of Ti6Al4V via pseudorandom crystal plasticity modelling", Journal of Materials Research and Technology , Volume 17, March–April 2022, Pages 2897-2912

[5] Shing Chan and Ahmed H. Elsheikh, "A machine learning approach for efficient uncertainty quantification using multiscale methods", *Journal of Computational Physics, Elsevier*, Volume **354**, 1 February 2018, p 493-511.

[6] Ramin Bostanabad, Biao Liang, Jiaying Gao, Wing Kam Liu, Jian Cao, Danielle Zeng, Xuming Su, Hongyi Xu, Yang Li and Wei Chen, "Uncertainty quantification in multiscale simulation of woven fiber composites", *Computer Methods in Applied Mechanics and Engineering, Elsevier*, Volume **338**, 15 August 2018, p 506-532.

[7] Saikumar R. Yeratapally, Michael G. Glavicic, Christos Argyrakis and Michael D. Sangid, "Bayesian uncertainty quantification and propagation for validation of a microstructure sensitive model for prediction of fatigue crack initiation", *Reliability Engineering & System Safety, Elsevier*, Volume **164**, August 2017, p 110-123.

[8] M. A. Bessa, R. Bostanabad, Z. Liu, A. Hu, Daniel W. Apley, C. Brinson, W. Chen, Wing Kam Liu, "A framework for data-driven analysis of materials under uncertainty: Countering the curse of dimensionality", *Computer Methods in Applied Mechanics and Engineering, Elsevier,* Volume **320**, June 2017, p 633-667.

[9] Guowei Cai and Sankaran Mahadevan, "Uncertainty Quantification of Manufacturing Process Effects on Macro-scale Material Properties", *International Journal for Multiscale Computational Engineering*, Volume **14**, Issue 3, January 2016,p 1-15.

[10] Zhen Hu and Sankaran Mahadevan, "Uncertainty quantification in prediction of material properties during additive manufacturing", *Scripta Materialia, Elsevier*, Volume **135**, 1 July 2017, p 135-140.

[11] Xinxin Wei, Jin-Cheng Liu and Sifeng Bi, "Uncertainty quantification and propagation of crowd behaviour effects on pedestrian-induced vibrations of footbridges", *Mechanical Systems and Signal Processing*, March 2022, p 1-12.

[12] Sudip Dey, Tanmoy Mukhopadhyay, Axel Spickenheuer, Uwe Gohs and Sondipon Adhikari, "Uncertainty Quantification in Natural Frequency of Composite Plates - *An Artificial Neural Network Based Approach", Advanced Composites Letters, Elsevier*, Volume **25**, Issue 2, February 2016, p 43–48.

[13] Gustavo M Castelluccio and David L. Mc Dowell, "Mesoscale modelling of microstructurally small fatigue cracks in metallic polycrystals", *Materials Science and Engineering: A, Elsevier*, Volume **598**, 26 March 2014, p 34-55.

Electronic transport properties of atomic wires on monolayers

K S Yadav[1*], A K Srivastava[2], N Behera[3], B S Bhadoria[1] and S Saxena[4*]

[1]Department of Physics, Bundelkhand University, Jhansi,Uttar Pradesh – 284128 India
[2]Department of Physics, School of Science, Indrashil University, Rajpur, Mehsana Gujarat – 382740, India
[3]Department of Applied Physics, Adama Science and Technology University, Adama PO Box1888 Ethiopia
[4]Department of Physics, Sardar Vallabhbhai National Institute of Technology, Surat, Gujarat – 395007, India

ksaurabhyadav@yahoo.com, swastisaxenaa@gmail.com

Abstract. In recent years, several scientific organizations have focused on the electrical characteristics of atomic wires on monolayers. The usage of MoS_2 in many technologies requires contacts between metallic atomic wires and MoS_2. Using first-principles density functional theory, we investigate the interaction of a variety of metallic atomic wires on MoS_2, including Al and Pt. The electronic properties of atomic wire sandwiches containing MoS_2 monolayers such as aluminium (Al) and platinum (Pt) have been thoroughly investigated using density functional theory calculations. Despite significant lattice deformation, structural investigations show that all sandwich systems almost maintain the original MoS_2 structure type. By inserting flat impurity bands within the band gap, the electronic transport property is improved. The electrical structure of the system reveals the binding processes and trends in binding energy. On MoS_2, we show that monitoring monomer adhesion energy and mobility on the substrate may be used to predict metal nanoparticle growth morphologies. The indirect bandgap of layered atomic wires sandwiched with MoS_2 becomes a direct band gap in the visible spectrum, making them appropriate for optoelectronics devices.

1. Introduction

Since its development using an exfoliation technique [9] identical to that seemed to create graphene, two-dimensional molybdenum disulfide (MoS_2) has engrossed a great deal of interest [1–8]. The bulk MoS_2 has a hexagonal structure with the space group P_{63}/mmc [10,11], with a layer of Mo atoms sandwiched between two layers of S atoms under ambient circumstances. A monolayer is the name given to this atomic trilayer structure [11]. The MoS_2 monolayer's individual atomic sheets are kept together by a weak van der Waals force, but the monolayer's atoms are covalently bound. The monolayer has a direct gap, despite the fact that the majority of MoS_2 has an indirect gap [12–14]. MoS_2 monolayers exhibit a number of controllable features that are useful in nanoelectronics [15–17]. They also complement graphene in applications that need thin transparent semiconductors, and because to their high surface-to-volume ratio, they are projected to offer excellent electrochemical sensors properties [18-21].
Creating substitutional solutions, in which atoms with a greater or lesser number of electrons are placed as dopants for the creation of semiconductors of type n or p, or forming covalent bonds between organic molecules and atoms at the materials' edges were all common methods for achieving electronic properties of atomic wires in 2D materials [22–24]. On the other hand, these interactions are essentially unstable, impacting the crystalline structure of 2D materials and drastically altering their intrinsic qualities and stability. Physical adsorption of organic molecules or metallic atoms has recently been proposed as a cost-effective and damage-free method of modifying 2D material properties [25–27].

Lewis' acid-base concepts would be used to utilize the occurrence of nonbonding electron pairs on the surface of InSe [28]. As a consequence, basic acid-base procedures might be utilized to make complexes like Lewis acids and two-dimensional materials. The Fermi level and band gap of a material can be controlled depending on the acid strength [28].

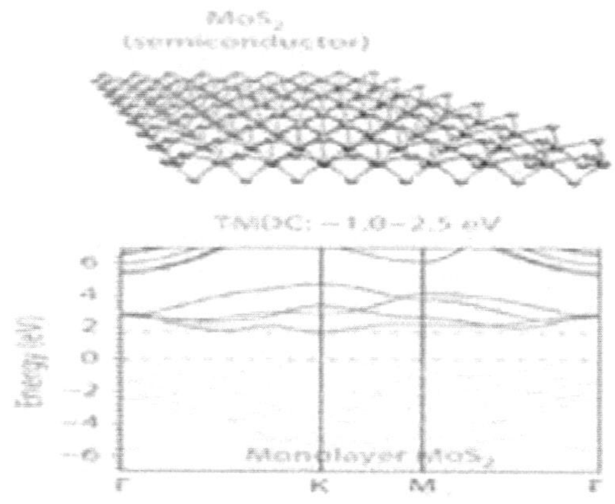

Figure 1: Band structure of the monolayers

This opens up a wide range of uses for these materials in the construction of solar devices. Electronic devices for next-generation applications are frequently made of a mix of conducting and insulation material, which leads to improved performance and adaptability. For example, Al nanowires on graphene increased the functionalized graphene's conductivity significantly. As a result, the Al–graphene combination has greater promise for high-performance, flexible energy conversion and storage systems [29]. In addition, chemical and microwave integration of Al nanoparticles on MoS_2 drastically altered the physical and chemical properties.

We're interested in the role of metal nanostructures in changing the electrical characteristics of a monolayer substrate based on existing research observations. Is the impact just due to nanostructures, or is the interfacial chemistry between nanostructures and monolayers also important? We'll look at the durability and electronic properties of Al ($3s^2$ $3d^1$) and Pt ($6s^1$ $5d^9$) monoatomic wires anchored on the MoS_2 monolayer, as well as density function theory computations (DFT). In the break-junction tests and on the substrate, metallic monoatomic wires including Al and Pt were formed.

DFT was recently utilized to study the interactions of atomic wires with MoS_2 by one group [30]. They observed that the adsorption energies of hollow (hexagonal centre) and Mo sites were almost identical, and that both were higher than those of S sites. Isolated Al atoms on the MoS_2 surface have recently been shown to allow electrons to flow out of the 2D material, hence boosting the surface-hole concentration [31]. According to DFT theoretical study, changes in the coordination sphere might impact the MoS_2-Al binding energy. The MoS_2-Al bond is two times stronger when one coordination site on the Al atom is occupied by one S atom on the surface of MoS_2 and the other three by Cl anions, the MoS_2-Al bond is two times stronger. One group recently used DFT to investigate the interactions of atomic wires with MoS_2 [30]. They discovered that the adsorption energies of hollow (hexagonal centre) and Mo sites were almost comparable, and that both were greater than the S sites.

Isolated Al atoms on the MoS_2 surface have recently been shown to enable electrical current to flow out of the 2D material, hence boosting the surface-hole concentration [31]. According to DFT theoretical study, changes in the coordination sphere might impact the MoS_2-Al binding energy. The MoS_2-Al bond is two times more powerful when one coordination site on the Al atom is occupied by one S atom on the surface of MoS_2 and the other three by Cl anions. Ultimately, the obtained patterns are explained using crystal-field stabilization energy, absolute hardness of metal wires, and standard potentials for MoS_2 conduction and valence band reductions.

2. Computational Method

Before and after functionalization, we tracked the same area of the same flake. For each MoS_2-Al combination and Pt concentration, at least two distinct flakes were measured. The density functional theory is used to investigate the equilibrium structure, stability, and electrical structure of atomic wires on

MoS_2 (DFT). For all DFT computations, the system's ab-initio calculations [31] are done using the VASP [32]. We employ extended gradient approximation and spin-polarized density functional theory (GGA). The generalized gradient approximation (GGA) as implemented by Perdew-Burke-Ernzerh is used to address the exchange-correlation aspect of electron–electron interactions (PBE). A plane wave foundation with a 540 eV energy cutoff is utilized. The computations were carried out in a 3 x 3 unit cell supercell framework.

3. Results and Discussion

The upper or hollow areas can also be used to accept responsibility for the orientation of atomic wires on the surface of the MoS_2 monolayer; the former refers to the case where the metal atoms are positioned between the two S atoms, thus directly above the monolayer's underlying Mo atoms, whereas the latter refers to the case where the metal atoms are positioned between both S atoms, thus effectively above the monolayer's underlying Mo atoms (Fig. 1).

The preferred binding domains were determined by analysing the total amount of energy of the produced monolayers while adjusting the spacing between the wire and the monolayer for both the top and hollow locations. In Al and Pt, the difference between the hollow and top sites is minimal, yet the hollow placement is favoured over the top site. Pt, on the other hand, clearly prefers the MoS_2 monolayer's top location. The generated monolayers' anticipated order of stability at the top location is Pt.

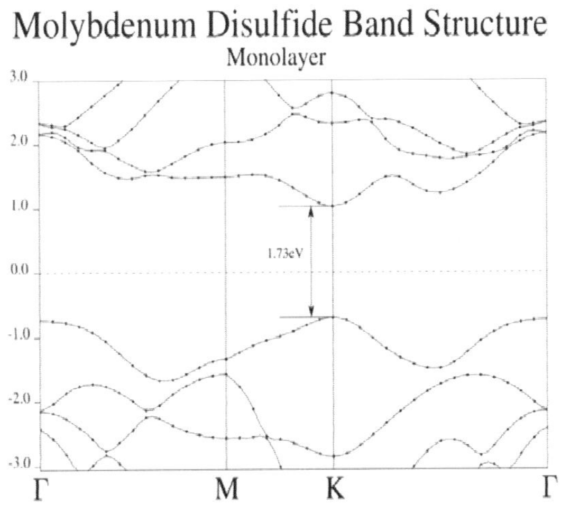

Figure 2: Band structure for Molybdenum Disulfide

The electronic band structures of the produced monolayers and virgin wires are shown in Figure 3. Contact with metallic atomic wires significantly alters the electrical properties of the semiconducting MoS_2 monolayer.

The created system's metallic feature can be seen in projected density of states, where the metallic atoms of the atomic wires are connected to Fermi level electronic states. It's worth noting that the metal wires in the monolayers generated appear to have retained their bands (Fig. 3). At Fermi energy, the pristine monolayer is semiconducting, with a direct gap of 1.55 eV at point K (Fig. 3); mixed Mo-d and S-p states dominate the top of the valance band, while Mo-d states control the bottom of the conduction band. On the other hand, the generated monolayers feature partly populated electrical bands at the Fermi level.

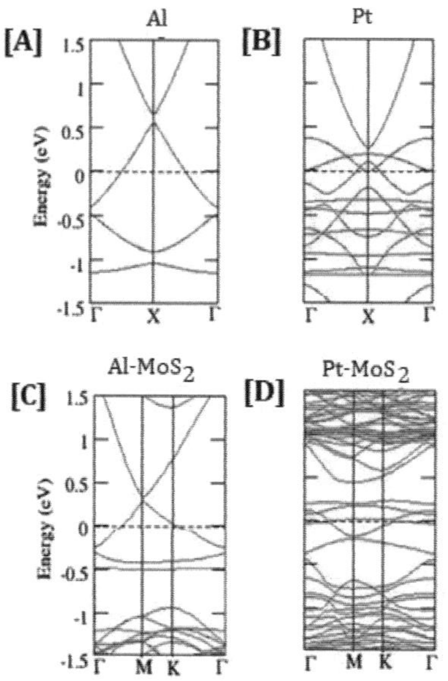

Figure3 : Electronic band structures of the pristine monolayer (top panel) and wires of the constructed MoS_2 monolayer (bottom panel).

The patterns of coordination on the MoS_2 surface will be investigated using coordination chemistry.The creation of coordination complexes on the material's surface produces coordinates on the 2D Monolayer surface [34]. As a consequence, comparing the molecular orbital theory for coordination complexes with 2D Monolayer solid-state physics can describe the patterns and properties observed after functionalization. Coordination chemistry and the hard-soft acid-base concepts can help us better understand and control the chemical, electric, and optical properties of 2D materials.

4. Electron transport properties

The metallic nature is predicted to turn up in the combined monolayer's higher conductivity rather than the semiconducting pristine monolayer's conductance. To quantify such improvements, we now use a model setup employed in scanning tunnelling microscopy (STM) studies to measure the current–voltage (I–V) characteristics of the generated monolayers [35]. A vacuum barrier width of 5 mm is considered to isolate the point from the sample for STM investigations, simulating a non-bonded tip arrangement. A cage-like atom Al cluster represents the probe tip's cap structure. Throughout this model configuration, the tunneling current was calculated using the Bardin, Tersoff, and Hamman (BTH) formalism of electron tunneling. When the sample is coupled to a positive potential and electrons tunnel from the tip to the sample, the bias is positive. The tunneling current is determined by the distance between the tip and the sample.

As a result, the tip-sample separation determines the amplitude of the tunneling current, but it should have no bearing on the expected trend for the monolayers under consideration. The predicted tunnelling properties of the pristine and assembled monolayers are shown in a graph. When the bias voltage, V bias, is raised, the production of finite Density of State at the Fermi level of the wire–MoS_2 system is predicted to be the source of an increase in tunneling current since tunneling current is proportional to the convolution of DOS between the tip and the sample. The connected monolayers show an ohmic response at low forward bias, despite the fact that the virgin MoS_2 monolayer is a semiconductor.

With the exception of Pt–MoS_2, the monolayers generated are asymmetric, with approximately constant current for reverse bias. Because the states associated with atomic wires lined up around the Fermi energy are the conducting routes under an applied bias, the variation in current–voltage characteristics can be explained in terms of the density of states of the constructed monolayers. As a conclusion, the performance of Pt–MoS_2 and other metallic wire–MoS_2 monolayers may differ due to the unique nature of Pt–MoS_2 bonding, as previously mentioned. A computational investigation of electron transport across the MoS_2 monolayer linked with Al and Pt contacts was done with the goal of determining the selection of metallic connections for MoS_2-based systems.

5. Conclusion

In summary, metallic monoatomic wires have structural stability, electrical structure, and electron transport qualities. A side view of the generated monolayers' valance band charge density, which is 0.005 e below the 0.5 eV Fermi threshold. The current–voltage (I–V) properties of MoS_2 monolayers, both pure and manufactured, are examined. The Al and Pt wire monolayers are stable, according to the calculations. States linked with metallic atomic wires arise in the Fermi level region, generating electron conduction channels, according to the electronic band structure and density of states of the generated monolayers. Pt–MoS_2 has quantum conductance due to partly filled Pt-d states, whereas Al–MoS_2 has quantum ballistic conductance. The constructed monolayers' transverse current was computed. With the exception of Pt–MoS_2, which has a fundamentally different kind of bonding at the created monolayer interface, asymmetric current–voltage characteristics are expected for the produced monolayers.

According to our findings, the MoS_2 monolayer interacts strongly with the metallic mono atomic wires, changing its electrical characteristics. In the MoS_2 monolayer, Al wires provide an n-type doping effect, whereas Pt wires produce a p-type doping effect that may be used to reflect changes in the monolayer's electrical structure caused by metal wires. The results of this work are expected to contribute in the development of new applications for MoS_2 monolayer as a suitable material for catalysis and optoelectronic devices.

References

[1] Xu M, Liang T, Shi M, Chen H2013 *Chem. Rev.***113**(5) 3766.
[2] WangQ H, Kalantar-Zadeh K, Kis A, Coleman J N, Strano M S 2012*Nat. Nano-technol.***7**(11) 699.
[3] ButlerS Z, Hollen S M, Cao L, Cui Y, Gupta J, Gutiérrez H R, Heinz T F, Hong S S, Huang J, Ismach A F, Halperin E J, Kuno M, Plashnitsa V V, RobinsonR D, Ruoff R S, Salahuddin S, Shan J, Shi L, Spencer M G, Terrones M, Windl W, Goldberger J E 2013 *ACS Nano***7**(4)2898.
[4] Huang X, Zeng Z Y, ZhangH2013 *Chem. Soc. Rev.***42**(5) 1934.
[5] SongX F, HuJ L, ZengH B2013 *J. Mater. Chem. C***1**(17) 2952.
[6] EnyashinA, Gemming S, SeifertG2007 *Eur. Phys. J. Spec. Top.***149**103.
[7] KumarA, AhluwaliaP K2014''*Chapter 3: Tunable Electronic and Dielectric Properties of Molybdenum Disulfide*'' under book title '*MoS₂: Materials, Physics, and Devices*', *Springer International Publishing, Switzerland***21** 53.
[8] SagynbaevaP P M, Yunguo L, AhujaM R A R2014 *Nanotechnology***25** 165703.
[9] NovoselovK S, JiangD, SchedinF, Booth T J, KhotkevichV V, MorozovS V,GeimA K 2005 *Proc. Natl. Acad. Sci.***102**(30)10451.
[10] WilsonJ, YoffeA D 1969 *Adv. Phys.***18**(73)193.
[11] KumarA, AhluwaliaP K 2012 *Mater. Chem. Phys.***135**(2–3)755.
[12] MakK F, Lee C, Hone J, Shan J, Heinzet TF2010 *Phys. Rev. Lett.***105**(13) 136805.
[13] Splendiani A, Sun L, Zhang Y, Li T, Kim J, Chim C Y, Galli G, Wang F 2010 *Nano Lett.***10**(4) 1271.
[14] Kumar A, AhluwaliaPK 2012 *Eur. Phys. J. B***85**(6)186.
[15] Kumar A, Ahluwalia P K 2012*Phys. B***407**(24)4627.
[16] Kumar A, AhluwaliaP K 2013 *Mater. Sci. Eng.***21**(6) 065015.
[17] Kumar A, AhluwaliaP K 2013 *Phys. B***419** 66.
[18] Saxena S, Srivastava A K, Srivastava R, Kheraj V2019*European Journal of engineering Science and Technology***2**(3) 71.
[19] Saxena S, Verma A L 2014 *Advanced Material Letters***5**(8) 472.
[20] Saxena S, VermaA L 2013 *AIP Conf Proc.***1536**298.
[21] Verma A L, SaxenaS, SainiGSS, Gaur V, JainVK 2011 *Thin Solid Films***519** 8144.
[22] Bhimanapati G R, Lin Z, Meunier V, Jung Y, Cha J, Das S, Xiao D, SonY, Strano M S, Cooper

V R, Liang L, LouieS G, Ringe E, Zhou W, Kim S S, Naik R R, Sumpter B G, Terrones H, Xia, F, Wang Y, Zhu J, Akinwande D, Alem N, Schuller J. A, Schaak R. E, Terrones M, Robinson J A2015*ACS Nano***9** (12)11509.
[23] Zhang T, Fujisawa K, Zhang F, Liu M, Lucking M C, Gontijo R N, Lei Y, Liu H, Crust K, Nakajima T G, Terrones H, Elías A L, Terrones M2020 *ACS Nano***14** 4326.
[24] Zheng J, LebedevK, Wu S, Huang C, Ayvalı T, Wu T S, Li Y, Ho P L, Soo Y L, Kirkland A, Tsang S C E2021 *J. Am. Chem. Soc.* **143** (21)7979.
[25] Sahin H, Peeters F M 2013 *Phys. Rev. B - Condens. Matter Mater. Phys.***87** 085423.
[26] Zhou Y, Reed E J 2015 *J. Phys. Chem. C***119**(37) 21674.
[27] Lin X, Ni J 2012 *Phys. Rev. B - Condens. Matter Mater. Phys.***86**075440.
[28] Lei S, Wang X, Li B, Kang J, He Y, George A, Ge L, Gong Y, Dong P, Jin Z, Brunetto G, Chen W, Lin Z T, BainesR, Galvão D S, Lou J, Barrera E, Banerjee K, Vajtai R, Ajayan P2016 *Nat. Nanotechnol.***11** (5) 465.
[29] Chen J 2013 *ACS Appl. Mater. Interfaces***5**(4) 1408.
[30] Gangwar R, Pandey D, Kancharlapalli S, Raychaudhuri D, Chakrabarti A, Banerjee A, Ghanty T K2021 *J. Phys. Chem. C***125** (2)1493.
[31] Liu H, Grasseschi D, Dodda A, Fujisawa K, Olson D, Kahn E, Zhang F, Zhang T, Lei Y, Nogueira R B, Elías A L, Silva R C, YehY. T, Maroneze C M, Seixas L, Hopkins P, Das S, de Matos C J S, Terrones M 2020 *Sci. Adv.***6** 49.
[32] Srivastava A K, Saxena S 2021 *AIP Conf Proc***2369** 020007.
[33] Giannozzi P, Baroni S, Bonini N et al. 2009 *J. Phys. Condens. Matter***21**(39) 395502.
[34] Hartel P, Rose H, Dinges C 1996 *Ultramicroscopy***63** (2)93.
[35] Tersoff J, HamannD R 1983 *Phys. Rev. Lett.***50**(25) 1998.

Electricity Consumption Forecasting System Using Deep Learning

[1] R.Sujatha, [2]Gowtham Sukumar, [3]Rajesh.V, [4]Sathish Kumar.R, [5]Judeson David

Department of Information Technology,
M. Kumarasamy College of Engineering, Karur, Tamil Nadu, India

sujathar.it@mkce.ac.in, gowthamsukumarravi@gmail.com, rajeshrishi267@gmail.com sathishkumarramsk2000@gmail.com judesondavid1998@gmail.com

Abstract. Energy modeling in Smart Buildings (SB) and planning and operating the generation of power based on the information extracted are key components of the Smart Grid's (SG's) energy management system. In the world, buildings use a significant amount of energy that contributes to energy efficiency programs. Additionally, excessive utilization power generation appliance also including air conditioners and heater, breathing, and climate control (HVAC) units, improper microclimate control, and inappropriate start-up and ordering of power equipment waste a lot of heat pumps. The utility can mitigate energy generation costs when it anticipates electrical loads and schedules generation resources in accordance with the demand. To estimate electricity usage at varying tiers of utility grid systems, a range of techniques have already been used. The goal of this study will be to create a hybrid deep learning model can predict resource utilization in infrastructures. Model building and data cleaning are the two stages of the proposed framework. Data cleaning involves pre-processing techniques and adding additional lag values to raw data. This hybrid deep learning (DL) approach is made up of a series of completely connected layered and linear Long Short-Term Memory (LSTM) sections layered over bi-directional Long Short-Term Memory (LSTM) components and is based also on collected information. You could incorporate the dependence structure of electricity usage on regressors and increase computation efficiency, training time, as well as computational complexity by using the results obtained.

Keywords: Smart Grid, Deep learning, LongShort-Term Memory, Smart Building, SVM.

1. INTRODUCTION

Electricity is a critical commodity for human survival. Heating, cooking, ironing, and other key components of daily life are all done with it in homes. It is used in industry to power factories and manufacturing. It is also used in offices for daily operations including as powering computers, running air conditioning, and so on. It comes from a variety of sources, including thermal, coal, nuclear, and hydro. Manage the spined words as you want. The presence of an adequate supply of strength aids in a rustic's monetary increase and balance. The cost of manufacturing is greatly influenced by the type of energy used, such as hydroelectric power plants, petroleum products, nuclearpower, and wind power. Energy consumed has been one of the primary challenges due to its effects on the financial realms. In Saudi Arabia, a number of reasons result in increased energy use in non-residential properties. The CoC has a slew of difficulties that contribute to growing demand. Heavy rainfall is also one of the major concerns, as it raises average temperatures and enhances power usage due to the constant usage of air conditioning. In the winter, repeated the use of such a heater increases energy consumption, depending on the kind, size, operating time, and watts consumed per hour. In broad sense, the number of gadgets puts a majorstrain on energy; as the number grows, so does energy usage.

2. LITERATURE REVIEW

The complication of the prediction problemexpanded as more modern products became popular, making voltage increases planning a real problem with in power industry. In addition, as time passes, several approaches for optimization have been developed, ranging from traditional procedures to soft techniques. In the electronic industry, several of them are employedin a variety of ways. Advanced approaches are required for reliable and realistic power monitoring system in the smart grid. Smart-grid investments and innovations have also introduced new difficulties to the focuses on delivering sector, such as load forecasting and forecasting of renewable power. The smart-grid age hasgiven new life to the century-old discipline of energy forecasting. Numerous computer algorithms areemployed, but each has advantages and disadvantages that are discussed in this paper. A good anticipatedresult must include a variety of parameters and have a lower MAPE. Industrial Electronics on Distributed Generators, Vol.16, No.3, "Mild situations prediction employing large dataset simulation with peak energy estimating capability" (August 2001). Time series analysis: A succession of data is used in time series analysis, which is normally at uniform intervals. This strategy attempts to comprehend the actual data and forecasts upcoming events given previous ones. TSA is widely employed in the future for a small space of time.D.C. Parkaeia, M.A. El-Sharkawi, R.J. Marks II,L.E. Atlasae, and M.J. Damborgo, "Electric load forecasting with an artificial neural network," IEEE Transactions on Electrical Engineering, volume (1991). "Electric load forecasting with an artificial neural network," by D. C. Parkaeia, M.A. El-Sharkawi, R.J. Marks II, L.E. Atlasae, and M.J. Damborgo, During That on Electrical And electronic engineering, volume (1991). ANN (Artificial Neural Network): An artificial neural network (ANN) is a soft approach utilised in a number of optimization procedures. This approach is capable of dealing with complicated modelling and adapting. It is not necessary to assume that load and weather factors have any significant link. We can change the ANN by feeding it fresh information. Other electricity system failures being studied with the ANN include security testing, harmonic load recognition, alert processing, troubleshooting, and geometrical testability.

H. Mori and S. Tsuzuki, "Breaker Panel Computational Geometry Availability Evaluation But use a Neural Network Model," Proc. of 2nd Sym. on Required Successive to Electromagnetic Systems, July 1989, pp.385-391. The System of Experts: A computer softwarethat can act as an expert is known as an expert system. This means that as situations change, this computer software can reasoning, interpret, and extend its knowledge base. The load forecast model is built with thehelp of an expert in the field's knowledge of the load forecast domain. The "Knowledge Engineer" extracts this knowledge from the acquisition module component of the expert system, which is called the load forecast (domain) expert. Christiani Crowleye and Frederick L. Joutz are the eighth and ninth authors. Modeling and Forecasting of Weather Effects on Electricity Loads 12th of December, 2005 Report Final Weather Effects on Electricity Loads, according to the EPA. Disturbances at Random: The unceremonious stimulation of any mayor's public radio show, any curriculum to be staged in almost any region, and so on are examples of random load demanddisturbances. Beside this, several sorts of customers, suchlike home, agribusiness, and manufacturers, make up the power sector. The resource generating power source primarily consists of inductive loads such as electrical machines, lightning, and so on, the turning on and off of which causes a sudden surge. Sanjeev Kumar Aggarwal, 1, Lalit Mohan Saini A review and evaluation of electricity price forecasting in deregulated markets. Economic Aspects: The way to generate income has a lot more influence on protracted demand forecasting, although it also affects the SLTF. Environmental economics have a serious influence on peak demand. Several factors influence system load increase, namely regional airport demographic, manufacturing athletic performance, change in the rural economy, the form and extent of equipment coverage, governmental backdrop advances, and, more loosely, economic cycles. Besides, load is affected by consumer software's also including frequency design adjustments and requirement prevention strategies.

3. EXISTING SYSTEM

The conventional machine learning method

Support vector machine (SVM), k nearest neighbour (KNN), random forest (RF), extreme random forest (ERF), Support Vector Regression (SVR), Artificial Neural Network (ANN), Autoregressive Integrated Moving Average (AIMA), Multiple Linear Regression (MLR) are eight traditional ml methods used only to construct power tool energy utilization projections (MLR).

SVM (Support Vector Machine)

In supervised learning, SVM is a sort of generalisationclassification algorithm used for data stream sorting.
To create home appliances overall energy statistical models,eight classic machine learning approaches are applied.

- Support Vector Machine (SVM)
- K nearest neighbor (KNN)
- Random Forest (RF)
- Extreme Random Forest (ERF)
- Support Vector Regression (SVR)
- Artificial Neural Network (ANN)
- Autoregressive Integrated Moving Average(AIMA)
- Multiple Linear Regression
- Support Vector Regression (SVR)
- Electrical Expenditure Analytical Technology for Integrated Dwellings based on the Internet of Things

Deployment of a smart utility meter using IoT and prepaid transactions. Remote ON/OFF Control of Utility Grid and Real - timemonitoring and Storage System for Electricity Consumption.

ARIMA (Autoregressive Integrated Moving Average) : ARIMA, the very well principal components for regressionanalysis, is capable of modeling and forecasting complicated patterns in univariate data. The characters p, d,and q, which stand for auto - regressive, integrative, andexponential smoothing factors, respectively, are used to designate three important factors in the ARIMA model operation. The ARIMA (p, d, q) model is represented as follows in general:

$$\Phi p\ (BO)(1\text{-}BO)^{\wedge}d\ Yt = \theta(BO)et+\sigma$$

4. PROPOSED METHODOLOGY

When comparison to something like the sequence alignment models, a hybrid model (HSBUFC) is generated that is primarily focused also on arrangement ofnon - linear and non and single - mode LSTMs, coupled by fully linked dense layers, to implement forecasts overall accuracy and low error probabilities. The representations of unidirectional and bidirectional LSTMsare described. Bidirectional LSTMs recognize intrinsic patterns in data sets combining past and forthcoming inputs by using forward and reverse passes.

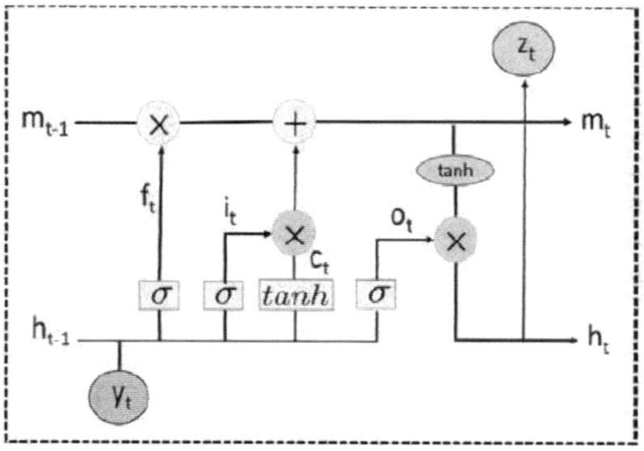

Figure 1. HSBUFC Model 1

UNI-DIRECTIONAL LSTMS is an Abbreviation forLong-Term (OR LSTMS)

The LSTM is a type of training algorithm that has been usedto process, analyze, and generate genomic information in the past. Depending on the latest time step on the ladder intake and the specific time step's output, the RNN model makes predictions. LSTMs have memory cells that aggregate steps throughout projection sequences, ability to perform daily better with long-term dependent tasks such as power generation forecasting, in addition to being able to use the information from back propagation to past time step outcomes. LSTMs also have gates and m more complicated recurrent units, controlling the content that is fed through and resist vanishing gradients.

LSTMS WITH TWO DIRECTIONS

A more precise assessment of total energy on the site will have a substantial impact on estimating wind velocity and producing capacity at a construction site. By merging LSTMs, RNN groups, and linear Word embeddings, the current study presents a unique domain adaption model. LSTMs with resonant frequencies that comprehend how electric energy is consumed both uphill and downhill can beused to accurately estimate energy usage. The accuracy of the measurement is unaffected by the load specificity or probabilistic reasoning of a building. By encouraging students to participate in the planned hybrid, early pause was used to lessen computational load. In a variety of applications, particularly natural language processing, bi- directional LSTMs have been shown to outperform unit-directional LSTMs.

5. PROPOSED MODEL ARCHITECTURE

The suggested hybrid stacked macro - and micro postgrad LSTM with completely integrated convolution (HSBUFC) photographer's concept is shown. There are three levels in the HSBUFC model: bidirectional LSTM layers, stacked unidirectional LSTM layers, and fully integrated strands layers. As previously mentioned, bi-directional LSTMs use forward as well as retrograde dependence. Mostly during characteristic educational process, the initial layer ofbidirectional LSTM retrieves the spatial long-term dependency of something like the energy demand numbers in two directions. Then, in the top part, LSTM pieces are utilised, which absorb the deliverables from the bottom layerupon learning from either the gathered detailed andcomplicated features and are economical in forward interactions.

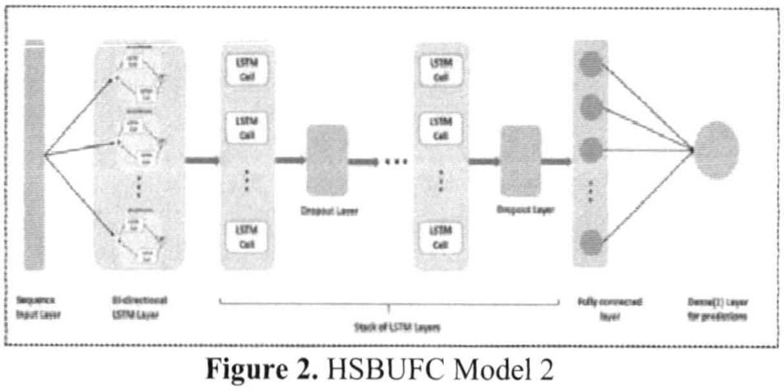

Figure 2. HSBUFC Model 2

IOP Conf. Series: Materials Science and Engineering 1258 (2022) 012061 doi:10.1088/1757-899X/1258/1/012061

6. ADVANTAGES OF PROPOSED SYSTEM

a. The prediction methodology is intended to help to effective electric-system budgeting by estimating monthly power demand, who represents for the bulk of overall consumption of electricity.
b. The capacity to accurately estimate annual energy demand can aid energy staff and interested parties in making prolonged power planning decisions, ensuring effective development and resource utilization.
c. This exact need plenty prediction model might is also used to solve various problems
including regression analysis, energy process optimization, resource efficiency, and long-term energy resources.

7. RESULT AND DISCUSSION

Several tests were carried out in this section to get extremely good estimations of infrastructure electricity usage using the provided HS UFC model, as well as to analyze their performance to that of several fundamental algorithms and efficient and promising computer programs. The foundations handles include linear regression, ELM, neural networks, fundamentally two LSTM, stacking LSTM models, and bidirectional LSTM frameworks. The observed and projected equipment consumption using the conceptual approach on a spontaneous deign the test sample are shown in Graph 1. The independent axis of the graph depicts the time during the day, although the dependent axis represents applianceelectricity usage. As you can see from the graph, the anticipated values closely align with the actual energy usage figures. The real and predicted values using the proposed forecasting model for global active power on a subset of records from case study 2 are demonstrated in Graph 2. The proposed model delivers substantially perfect accuracy performance at both spikes and troughs in the illustration.

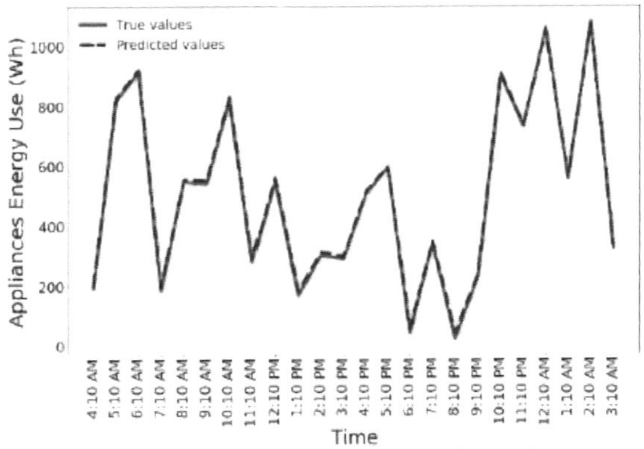

Graph 1: study 1 actual v/s predicted values for appliancesenergy use.

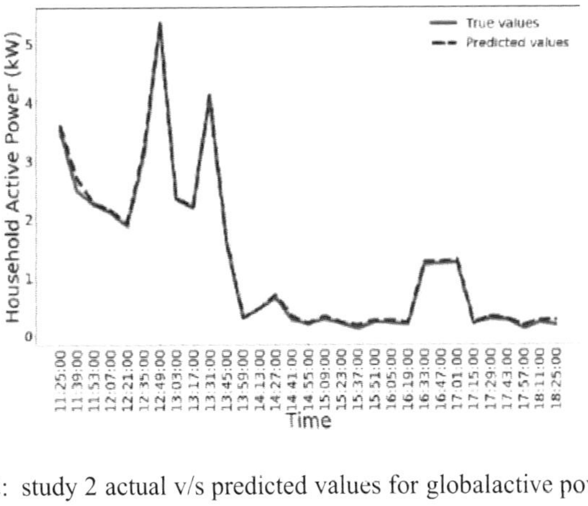

Graph 2: study 2 actual v/s predicted values for globalactive power.

8. CONCLUSION

A more precise estimate of total energy somewhere on a construction site will now have a substantial impact on wind velocity efficiency and system generation capacity. By combining the properties of linear Word embeddings, bilateral LSTMs, and RNN grouping, this work presents a new domain adaption model. In order to accurately estimate energy consumption numbers, LSTMs with resonant frequency grasp how electric is used both uphill and downwards. Probabilistic reasoning or uncertainty regarding specific building loads have no effect on the accuracy achieved. Early pause was employed to reduce computational load by inviting students to engage in the planned hybrid.

REFERENCES

[1] W. Kong, Z. Y. Dong, Y. Jia, D. J. Hill, Y. Xu, and Y. Zhang, ``Short-term residential load forecasting based on LSTM recurrent neural network,'' IEEE Trans. Smart Grid,vol. 10, no. 1, pp. 841851, Jan. 2019.M. Massaoudi, S. S. Refaat, I. Chihi, M. Trabelsi, F. S. Oueslati, and H. Abu-Rub, ``A novel stacked generalizationensemble-based hybrid LGBM-XGB-MLP model for short-term load forecasting," Energy, vol. 214, Jan. 2021, Art. no.118874.

[2] Y. Dong, X. Ma, and T. Fu, ``Electrical load forecasting: A deep learning approach based on K-nearest neighbors," Appl. Soft Comput., vol. 99, Feb. 2021, Art. no.106900.

[3] S. M. J. Jalali, S. Ahmadian, A. Khosravi, M. Shae- khah, S. Nahavandi, and J. P. S. Catalao, ``A novel evolutionary-based deep convolutional neural network model for intelligent load forecasting," IEEE Trans. Ind.Informat., vol. 17, no. 12, pp. 82438253, Dec. 2021.

[4] M. Ibrahim, A. Alsheikh, Q. Al-Hindawi, S. Al-Dahidi, and H. ElMoaqet, ``Short-time wind speed forecast using articial learning-based algorithms," Comput. Intell. Neurosci., vol. 2020, pp. 1-5, Apr. 2020.

[5] A new method for short-term electricity load forecasting, Jing-Min Wang and Li-Ping Wang, Transactions of the Institute of Measurement and Control 30, 3/4 (2008), pp. 331–344.

[6] "Weather Sensitive Method for Short-Term Load Forecasting in Serbian Electric Power Utility," Ruzic, A.Vuckovic, and N. Nikolic, IEEE Transaction on Power Systems, 18:1581–1586, 2003.

[7] "Regression Based Peak Load Forecasting Using Transformation Technique," T. Haida and S. Muto. IEEE Transactions on Power Systems, vol. 9, no. 9, pp. 1788– 1794, 1994.

[8] P. Van Olinda, M.S. Chen, and W. Charytoniuk. IEEE Transactions on Power Systems, "Nonparametric RegressionBased Short-Term Load Forecasting."

[9] Short term load forecasting using time series modelling with peak load estimation capability", IEEE Transactions onPower Systems, Vol.16, No.3 August 2001.Mohamed Mohandes, Support vector machines for short- term electrical load forecasting International Journal of Energy Research Int. J. Energy Res. 2002; 26:335}345 (DOI: 10.1002/er.787).

[10] Wei Chu, S. Sathiya Keerthi, Chong Jin Ong, a general formulation for support vector machines, Proceedings of the 9th International Conference on Neural Information Processing(ICONIP'OZ) , Vol. 5.

[11] Study report on electricity demand curve and system peak reduction, Public Utilities Commission of SRI LANKA, December 2012.

[12] Christian Crowley and Frederick L. Joutz Weather Effects on Electricity Loads: Modeling and Forecasting 12 December 2005 Final Report EPA Weather Effects onElectricity Loads.

[13] M. S. Kandil, S. M. El-Debeiky, Senior Member, IEEE,and N. E. Hasanien , Long-Term Load Forecasting for Fast Developing Utility Using a Knowledge- Based Expert System.

IoT based guided fire fighting vehicle

A R M Raafeek[1, a], **N Satheeskanth**[2, b], **J Joy Mathavan**[3, c *], **A Kunaraj**[4, d]

[1,2,3,4] *Department of Engineering Technology, Faculty of Technology, University of Jaffna, Kilinochchi premises, Ariviyal Nagar, Sri Lanka.*

[a]*arm.raafi69@gmail.com*, [b]*sathees117@gmail.com*, [c]*joymathavan1991@gmail.com*, [d]*kunaraj1984@gmail.com*

Abstract. Fire accidents are occurring frequently nowadays in domestic and industrial environments with unpredictable climatic changes and human negligence. The proposed prototype is designed to detect such fire accidents and to extinguish the fire by remote operation of the user, and thereby reducing the risk of fire fighters involved. The movement of the vehicle is dictated by the combination of sensor-based inputs and user guidance. When the fire is detected by the fire detection sensor, the signal is immediately passed to the controller. GSM modules will be activated alerting the user about the fire accident. Once the alert is received, the user can guide the vehicle to the required location with the help of Blynk software & in-built camera module to extinguish the fire. The servo motor connected with the firefighting hose can control the direction and volume of water to be sprayed. The controls of the servo motors are with the user. The proposed firefighting vehicle is designed to function using Internet of Things (IoT), thereby enabling the user to control the fire from anywhere.

Keywords: *ESP32 cam, Arduino, Flame sensor, IoT, Blynk software, Guided vehicle, Fire hazard*

1. Introduction

Nowadays, domestic and industrial fire accidents are getting frequent. It increases the amount of damage caused by fire and also the risk to human lives [1]. A smart home automation system [2, 3] could be an easy solution to prevent small scale fire accidents from occurring in a home or to extinguish such fire hazard. But once it is spread out to the entire complex, the supporting factors for the fire cannot be controlled by the smart home automation system since the system is designed to control the fire only to that particular home. So, it won't control the fire on the entire building. Under such a situation, a fire extinguishing vehicle is necessary to control the fire throughout the entire complex. These needs paved the path to this research work. The proposed system aims at building a guided vehicle that can be installed in a domestic or industrial environment, to alert the user in case of fire accidents. This vehicle can be guided by the user through mobile phone control to the intended location where the fire accident is occurring and water can be sprayed to extinguish the fire via mobile phone control. The velocity, volume and the direction of water while spraying can also be controlled by the user. The developed vehicle can be observed in the screen of the mobile phone from far away to guide the motion of the vehicle and thereby the user can communicate with it to effectively extinguish the fire. IoT (Internet of

Things) integration enables the guided vehicle to organise, analyse and process data, allowing the user to make optimum decisions on a real-time basis. By this way, the damage that can be caused by fire before the arrival of the fire extinguishing squad can be drastically reduced and the risk of involvement of fire extinguishing squad can also be minimised. The chance of life loss and / or injury in a normal automobile fire accident can be reduced by automatic door unlocking system [4]. But fire extinguishing squads are the people who are dedicated to stopping the fire even at the expense of their lives. So, this research work targets to minimise human involvement to stop any fire accidents.

2. Literature review

Automatic Guided Vehicles (AGV) are increasingly utilised in the logistics in smart warehouse [5] operations due to their increased efficiency [6]. Hiroshi Yoshitake et al in their research, proposed a system using AGV that works on real time holonic scheduling algorithms to improve the efficiency in warehouses. This AGV transports both inventory shelves and sorting shelves to the picker point simultaneously, preventing the movement of the picker across the warehouse. At the same time, the role of the AGVs in smart logistics [7] are also increasing steeply. The use of computer vision in vehicle automation [8] for guiding AGV in complicated environments are also discussed in this research work. The automatic guided vehicle follows the magnetic tape [9] attached to the floor or desired predefined trajectory [10] and identifies the destination, and load or unload based on the RFID (Radio Frequency Identification Definition) tag. This system works well for loading and unloading where guided systems have to travel only through well predefined paths. Kajan. K et al [11] have described an automatic guided vehicle which is controlled by PLC (Programmable Logic Controller) and integrated with Arduino using Transmission Control Protocol (TCP) communications to connect any third-party devices with the PLC. However, these prototypes are developed to work in the industrial environment with predefined paths for AGV movement. Kelber et al [12] in their research paper described automated guided vehicles that can be guided by human machine interface such as joystick and mobile phone in contrast to the conventional pedal and steering wheel arrangements. The automation was made using the technology available in 2003 and still inspired us for the proposed research work. Butdee and Suebsomran [13] in their research work, described an automatic guided vehicle that can be guided based on the colour tape sensed by the Charge Coupled Device (CCD) camera sensor. Edge detection to identify the red tape against its background is performed by the Laplacian operator and the control is provided by the PLC controllers. However, for firefighting, this method of guiding the robot will not be effective as the sources of fire incidents are unpredictable and therefore it will be difficult to move based on colour tape. Ratul Ahmed Rahat et al [14] developed a firefighting mechanism which can be operated by both manual and automatic mode. Water is sprayed in case of manual operation and CO_2 is sprayed in automatic operation. The robot was built using obstacle avoidance [15, 16] algorithm and detection of fire was performed by fire and smoke sensors. Once the fire is extinguished, it is important to patrol [17] the region to make sure that the fire is properly extinguished. Rakib and Sarkar [18] introduced a firefighting autonomous robot controlled by a PID (proportional–integral–derivative) controller that can throw water using a centrifugal pump once fire is detected. However, guiding mechanisms are not well developed in these prototypes. A clear guiding mechanism is designed by Kunaraj et al to be a guided robot which can be operated by an operator from a distance; but their objective is landmines detection in war-torn regions [19]. The idea of using hybrid fuzzy PID [20] controllers is also utilised in mobile robots to improve the performance of the robot and increase the dynamic response of the robot. The robot can avoid obstacles [21], detect flames and track through to extinguish the fire. Robots are not only utilised to extinguish the fire but also using a block of robots [22] the fire or nuclear radiation can be prevented. Water or lead solution pumped through those chains of robots can act as shields to protect humans from any harmful flares and radiations. Yet reliable models that can work in the robust and real time environment are still in the developing stage.

3. System description

When a fire accident occurs from different sources, the fire detection sensor senses the fire and sends the signal to the Arduino controller. According to that signal, the Arduino controller sends commands to the GSM (Global System for Mobile communication) module as shown in figure 1. The GSM module in turn will send a message as "Emergency fire alert" to the user or owner of the particular domestic / industrial complex. Based on the signal, the user may guide the fire fighting vehicle using Blynk software. While guiding, he can consider the observations obtained from the inputs of the camera module. Once the target is reached, then the user can activate the servo motor to spray the water to extinguish the fire. The outline of the proposed prototype is shown in figure 2.

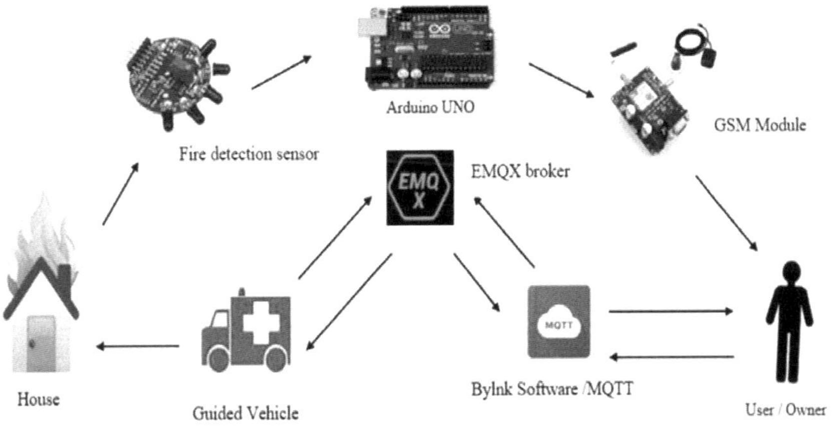

Figure 1. Block diagram of the proposed work

Figure 2. Outline of the proposed firefighter

Meanwhile, using ESP camera module, damages caused by the fire can also be monitored by IoT (Internet of Things) from far away. GSM modules are attached as a backup option to reach out the user through SMS or emergency calls in case of any network issues. The flow diagram of the operation of the fire sensing module is shown in figure 3 and the function of the fire fighting vehicle is shown in figure 4 respectively.

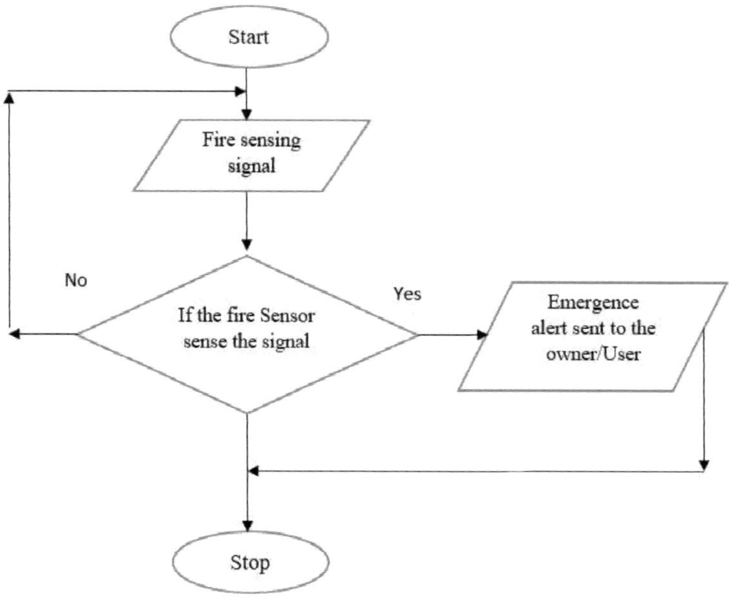

Figure 3. Flow diagram of fire sensing module

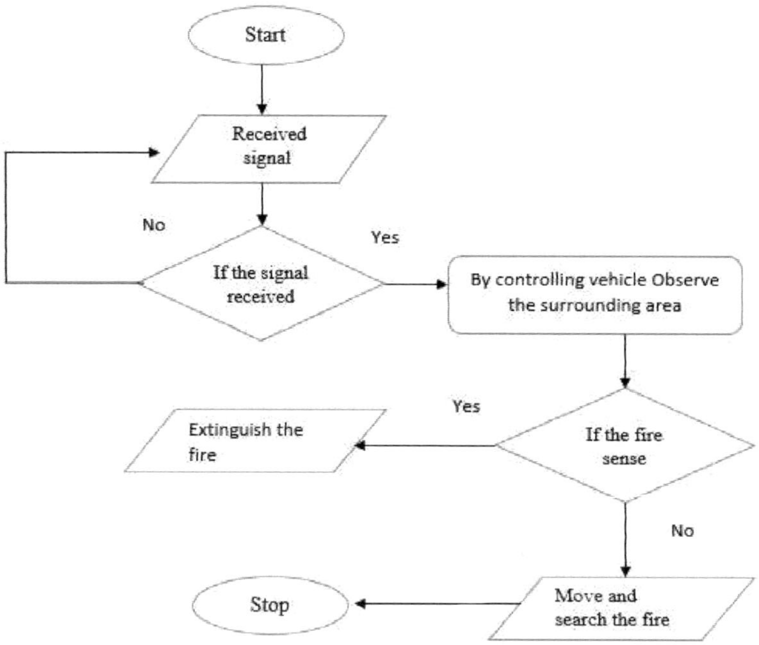

Figure 4. Flow diagram of firefighting guided vehicle

Table 1. Specification of the hardware components.

Hardware components used	Specification
Arduino UNO	Microcontroller: ATmega328P Operating voltage: 5V Flash memory 32KB Clock speed 16MHz Analog I/O pins: 6 Digital I/O pins:14
ESP Camera	802.11b/g/n/Wi-Fi Operating voltage: 3.3V Input voltage: 5V Flash memory: 4 MB Clock speed: 160MHz GPIO ports: 9
Ultrasonic sensor	Power supply: 3.3V-5V Operating current: 8mA Range distance:3cm-350cm Measuring angle:15 degree
Blynk Software	All supported hardware and devices Connection mode: Wi-Fi, Bluetooth, Ethernet, USB (serial), and GSM
Motor	Rated voltage: 3-6V Operating speed(6V): 200+/-10% RPM Gear ratio: 1:48 Shaft: dual-shaft
Motor drive controller	Diver model: L298N Power supply: 5V-35V Peak current: 2A Operating current range: 0-36mA Maximum power: 25W
GSM module	Model type: SIM800L Power supply: 3.8V-4.2V SIM card socket: microSIM

	Supported frequencies: Quad-band (850/950/1800/1900 MHz)
DC water pump	Voltage range: 2.5V-6V Operating current: 130-220 mA Hydraulic head 40-110cm Power: 0.4-1.5W
Servo motor	Speed: 0.12sec/60 degrees(6V) Torque:1.5kf-cm (4.8V), 2.0kgf-cm(6V) Voltage:4.8V-6V

3.1 *Fire sensing system*

In the proposed system, once the fire incident is sensed by the fire detection sensor, the controller sends the user an emergency fire alert message through the GSM module. In this system, SIM800L is used as transmitter considering its affordability and its GPRS (General Packet Radio Service) transmission capability to send and receive SMS (Short Message Service) and Voice call. The schematic diagram shown in figure 5 depicts the connection between Arduino uno fire sensor and GSM sim 900D module.

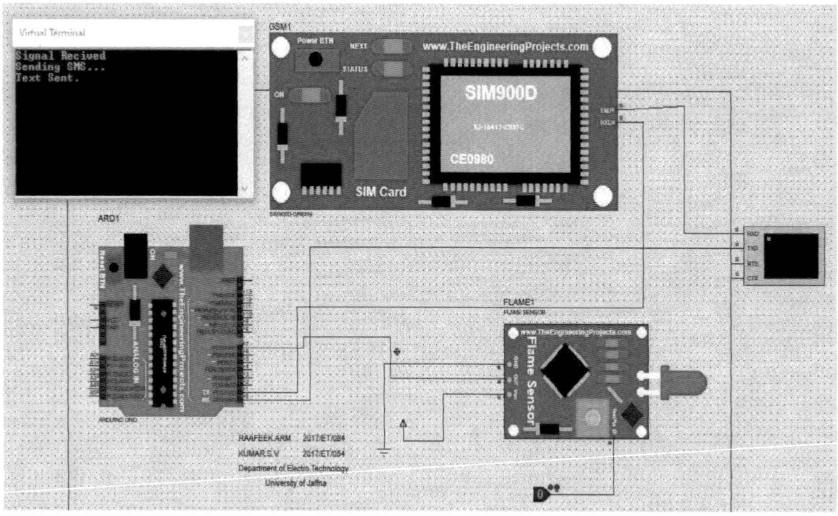

Figure 5. Emergency fire alert message sending mechanism

3.2 *Vehicle controlling system*

Vehicle controlling system comprises several parts such as video streaming, obstacle detection system, lighting facility, water level observation system, and water spraying system. Forward, backward and sideways movement of the vehicle is controlled using the L298N motor driver. Users can control the movement of the vehicle using the ESP32 camera module attached to the vehicle. Once the ESP32 camera module is connected to the internet, through the Blynk application installed in the mobile phone of the user, the vehicle can be operated. The schematic diagram showing the connection between Arduino uno motor driver and stepper motors are shown in the figure 6.

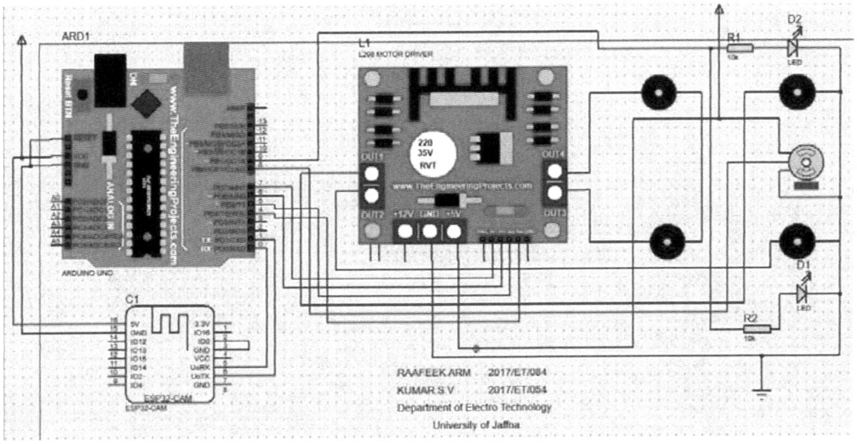

Figure 6. Vehicle controlling system

3.3 Water level observation

The proposed system uses water to control the fire. Therefore, it is mandatory to monitor the level of water in the tank. The level of water is monitored by the water level sensor which acts like a potentiometer to vary the electric resistance according to the level of water. When the level of water in the tank falls below predefined critical value, it sends an alarm signal to the user to refill the water tank. The schematic diagram showing the connection between Arduino uno and water sensor is shown in figure 7.

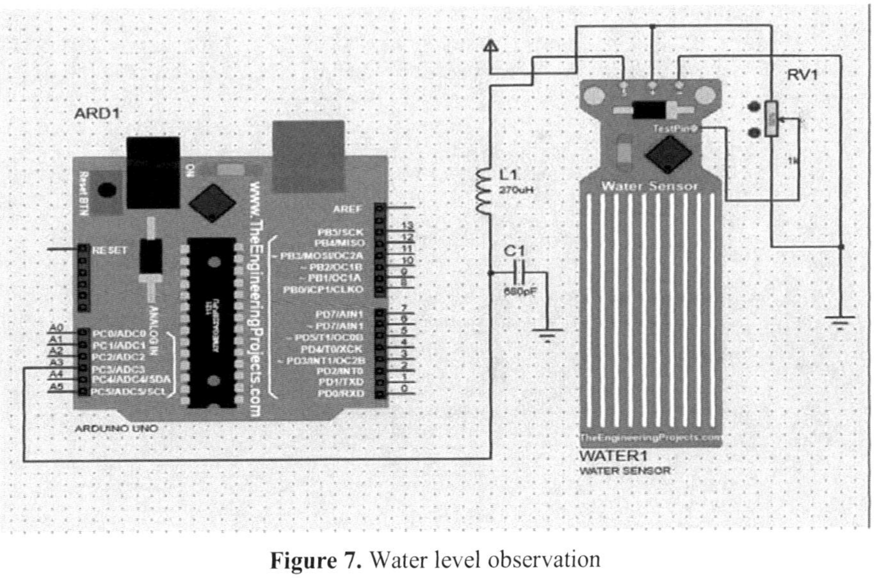

Figure 7. Water level observation

3.4 Obstacle detection system

The presence of an obstacle detection system in the guided fire fighting vehicle is to identify the obstacles on its way and to alert the user. Once the obstacle in the path of the robot is identified using an ultrasonic sensor, a sound buzzer is activated alerting the user about the presence of obstacles. This obstacle detection mechanism will be helpful while fumes created by fire block the view of the camera. The schematic diagram of the obstacle detection system is shown in figure 8.

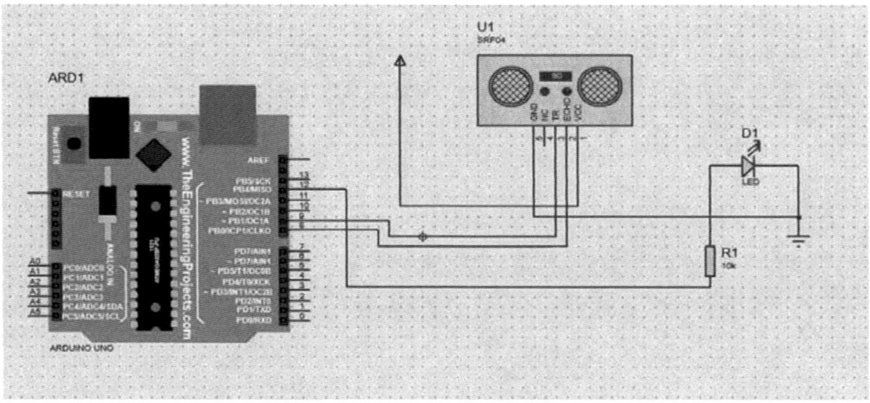

Figure 8. Obstacle detection system

3.5 ESP32 cam with Arduino configuration system

FTDI or Arduino can be used to upload code for ESP32 Cam boards. Arduino is used in this research as the external programmer for uploading the code to the ESP32 Cam board, since the connection port is unavailable in ESP32. The method of uploading the code using an external programmer is described below. Initially, configure the Arduino 5V to ESP32 Cam board power pin and Arduino ground pin to the ground pin of ESP32 module as shown in figure 9. The Arduino ground pin is again connected to the reset pin of Arduino. Arduino TX pin is connected to UOT pin of ESP32 cam and Arduino RX pin is connected to UOR pin of ESP32 cam. Once the configurations are completed, Arduino UNO can be used as a programmer to ESP32 module.

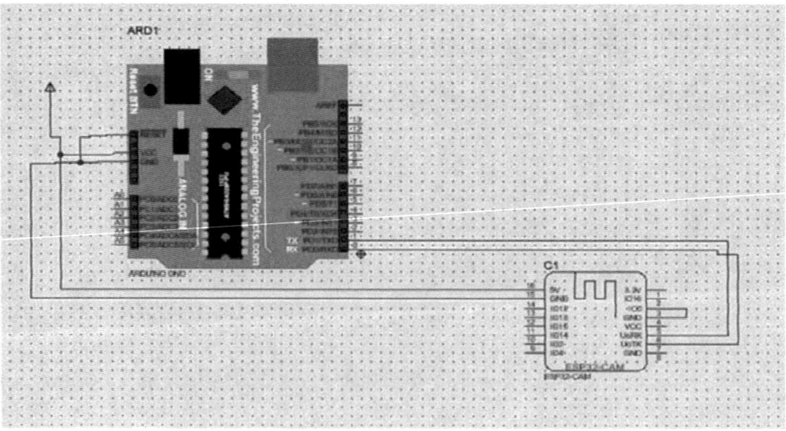

Figure 9. ESP32 cam with Arduino configuration

3.6 Software controlling system

To control and monitor the surroundings, the Blynk application shown in figure 10 is used. Based on the alarms received from the vehicle, it should be guided using the Blynk application which is integrated with the ESP32 cam board, telecasting video streaming using internet connection. This application contains several features to connect the device. It has Wi-Fi, Ethernet, serial port connection and several widgets such as button, slider, joystick and video streaming. In this system, control switches are provided for controlling the headlight, water spray and water nozzle adjustment. Joystick is provided to control the forward, backward, left and right motions of the vehicle and video streaming is enabled for sharing the live video streaming.

Figure 10. Vehicle controlling system

3.7 LAN to WAN IP address Forwarding system

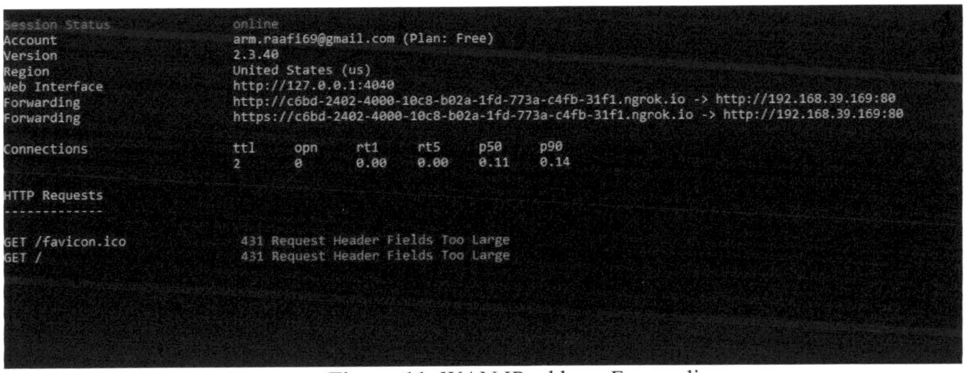

Figure 11. WAN IP address Forwarding

In order to access the ESP32 remotely, IP addresses can be forwarded using the NGROK application as shown in Figure 11. Using the IP address, the ESP32 cam board can be accessed from any part of the world.

4. Results and discussion

An IoT based guided fire fighting vehicle was constructed and tested. The final prototype designed based on the aerodynamic requirements are shown in figures 12 and 13. The proposed IoT-based firefighting-guided vehicle includes fire sensing elements, water level sensing elements, obstacles sensing elements, lighting elements, application control and live video streaming. For each of these applications, suitable sensors are employed and integrated with the controller. For the purpose of fire detection, heat detection sensor is used; for obstacle detection, ultrasonic sensor is used; and for the live video streaming, ESP32 camera module is used.
The fire detection sensor senses the fire and sends the signal to the Arduino controller. According to that signal, the Arduino controller sends commands to the GSM (Global System for Mobile communication) module. The GSM module in turn will send an alert message to the user. Based on the signal, the user may guide the fire fighting vehicle using Blynk software, considering the observations obtained from the inputs of the camera module. The fire fighting vehicle is designed to move forward, backward and sideways as directed by the commands from the operator. Once an obstacle is detected, a buzzer is triggered to alert the user. Once the fire is detected water can be pumped directly towards the fire at varying speed as determined by the command centre to put away the fire. Meanwhile, using ESP camera module, damages caused by the fire can also be monitored by IoT from far away. GSM modules are attached as a backup option to reach out the user through SMS or emergency calls in case of any network issues. The designed prototype functions effectively to detect the fire.

Figure 12. Side view of the prototype fire fighting vehicle

Figure 13. Front view of the prototype fire fighting vehicle

5. Conclusion and Future scope

In the proposed IoT-based firefighting guided vehicle, all the inbuilt devices were controlled using the Blynk application specifically designed for this purpose. The user can connect either PC or mobile phone to any network as the module. Therefore, the exchange of signals takes place frequently which can be monitored from any part of the world. Proposed firefighting guided vehicle would be useful for extinguishing fire when no one is around a domestic or industrial complex. The IoT based firefighting guided vehicle proves to be effective especially for the people who are away while having the loved ones in their home with no one to take care of them. It is also useful in corporate offices when the workers are away from the office building. The status of the place can be monitored and controlled from anywhere easily.
In the future, several features can be added to this system like providing first aid, notification to emergency units and identifying the human inside smoke. This system is built using an Arduino UNO controller which includes limited features and functionalities. For effective control and increased

efficiency, a wireless control of ESP32 can be used. Proposed plan of the IoT-based firefighting guided vehicle is entirely adaptable and can be effortlessly extended and applied to bigger structures by expanding to high torque motors, high-capacity water tanks and enhanced control devices.

6. References

[1] National Fire Protection Association 2019, "Firefighter Fatalities in the United States – 2018," June.

[2] Satheeskanth N, Marasinghe S D, Rathnayaka R M L M P, Kunaraj A and Joy Mathavan J 2022 IoT-Based Integrated Smart Home Automation System. In: Karuppusamy P, Perikos I., García Márquez F.P. (eds) *Ubiquitous Intelligent Systems. Smart Innovation, Systems and Technologies*, vol **243**, pp 341-355. *Springer, Singapore*. https://doi.org/10.1007/978-981-16-3675-2_26.

[3] Kunaraj, A., Mathavan, J.J., Mathushan, M., Kamalesan, G.M. (2021). "Hardware Implementation of Smart Home Automation System Using Aurdino UNO ". In: Sengodan, T., Murugappan, M., Misra, S. (eds) *Advances in Electrical and Computer Technologies. Lecture Notes in Electrical Engineering*, vol **711**. Springer, Singapore. https://doi.org/10.1007/978-981-15-9019-1_48

[4] Mathavan J J, Faslan A W, Basith N U A and Wanigasinghe W V S D 2020 "Hardware Implementation of Fire Detection, Control and Automatic Door Unlocking System for Automobiles" *4th International Conference on Trends in Electronics and Informatics (ICOEI)* (48184), 2020, pp. 125-129, doi: 10.1109/ICOEI48184.2020.9142990.

[5] Koster R de, Le-Duc T, and Rood bergen K J 2007 "Design and control of warehouse order picking: A literature review", *Eur. J. Oper. Res.*, **182**, no. 2, pp. 481–501.

[6] Wang biao, Xiwei Z 2017, "Design of Intelligent Logistics System for Power Plant Based on AGV[J]", *Automation Applications*, **11**: 142-144, 152

[7] Schmidtke N, Behrendt F, Wagner M, Alina B. Rettmann and Theresa Ansorge, 2020, *5th International Conference on Logistics Operations Management (GOL)*, pp. 1-9, doi: 10.1109/GOL49479.2020.9314741.

[8] Kelly A, Nagy B, David S and Unnikrishanan R 2007 "An Infrastructure-Free Automated Guided Vehicle Based on Computer Vision", *Robotics and automation magazine*, **14** (3).

[9] Xing Wu, Peihuang L, Qixiang C, Chidong Z, Shen ke and Jin che 2012 'Design and control of material transport system for automated guided vehicle', *UKACC International Conference on Control*, Control pp 765 -770.

[10] Gupta N, Singhal A and Rai J K, 2014 "Path Tracking of Automated Guided Vehicle", 2014, *Seventh International conference on contemporary computing (IC3)* pp. 260-264, doi: 10.1109/IC3.2014.6897183.

[11] M. Kajan, J. Šovčik, F. Duchoň, L. Mrafko, M. Florek and P. Beňo, 2014, "Sensoric subsystem of automated guided vehicle: TCP communication between SIMATIC S7 PLC and Arduino," *23rd International Conference on Robotics in Alpe-Adria-Danube Region (RAAD)*, pp. 1-6, doi: 10.1109/RAAD.2014.7002258

[12] Christian R Kelber, Rubem Sprenger Dreger, Guilherme Klein Comes, Daniel Webber, Juliano Schirmbeck, Roberto Hoffmann Netto and Diogo Albert Borges; 2003 "Cell-phone guided vehicle, an application based on a drive-by-wire automated system", *IEEE Intelligent Vehicles Symposium, Proceedings*, pp. 645–649.

[13] Butdee S and Suebsomran A 2009 'Automatic Guided Vehicle Control by Vision System', 2009 *IEEE International Conference on Industrial Engineering and Engineering Management* pp. 694–697.

[14] M. R. A. Rahat, M. M. Rashid, M. C. Dey and T. Banik, 2020, "Design and Implementation of a Safety Fire Fighter (SAFF) Robot with Dual Controlling Mechanism," *2020 IEEE Region 10 Symposium (TENSYMP)*, pp. 937-940, doi: 10.1109/TENSYMP50017.2020.9230709.

[15] Tong feng, Xu Lufeng and Tong Daoling, 2002, "An ultrasonic obstacle avoidance system for firefighting robot," *Proceedings of the 4th World Congress on Intelligent Control and Automation (Cat. No.02EX527)*, 2002, pp. 1219-1222 vol.2, doi: 10.1109/WCICA.2002.1020775.

[16] Yamamoto, M., Ushimi, N and Mohri, A 1999 Development of sensor-Based Navigation for Mobile Robots Using Target Direction Sensor, *Journal of Robotics and Mechatronics* Vol.**11** No.1 pp. 39-44.
[17] Mittal S, Rana M K, Bhardwaj M, Mataray M and Mittal S 2018 "Cease Fire: The Fire Fighting Robot", *International Conference on Advances in Computing, Communication Control and Networking (ICACCCN)*, Greater Noida (UP), India pp. 1143-1146.
[18] Rakib T and Sarkar M A R 2016 "Design and fabrication of an autonomous firefighting robot with multisensory fire detection using PID controller", *5th International Conference on Informatics, Electronics and Vision (ICIEV)*, Dhaka pp. 909-914.
[19] A. Kunaraj, M. Mathushan, J. J. Mathavan and G. M. Kamalesan, "Sensor Controlled Defence Purpose Robot for Landmine Detection," *2020 International Conference on Smart Electronics and Communication (ICOSEC)*, 2020, pp. 42-47, doi: 10.1109/ICOSEC49089.2020.9215338.
[20] Teng-Yi Wang and Chia-Der Chang 2018 "Hybrid Fuzzy PID Controller Design for a Mobile Robot", *IEEE International Conference on Applied System Invention (ICASI)*, pp. 650-653, doi: 10.1109/ICASI.2018.8394340.
[21] Gao, Shang et al. "Vision and Infra-Red Sensor Based Fire Fighting Robot." *2018 IEEE 61st International Midwest Symposium on Circuits and Systems (MWSCAS)* pp: 873-876.
[22] Amano H, Osuka K and Iwano Y 2005 "Development of robotic wall system to protect victims", *ICAR '05. Proceedings, 12th International Conference on Advanced Robotics*, Seattle, WA, pp. 579-586.

Survey and analysis long term deflection of cantilever bridge in Vietnam

Duy Tien Nguyen[1], Xuan Tung Nguyen[1], Van Minh Ngo[1].
[1] Faculty of Civil Engineering, University of Transport and Communications, 3 Cau Giay, Lang Thuong, Dong Da, Hanoi, Vietnam
E-mail address: ngxuantung@utc.edu.vn;

Abstract. The deflection and camber of bridge, especially of the box girder bridge as well as their effected factors have been important research topics since the end of the last century. In Vietnam, the box girder bridge constructed by cantilever method had been introduced since the 1990s and become popular in the last twenty years. After more than two decades of exploitation, these bridges face various problems such as over deflection, cracks... To clarify the actual deflection of the cantilevered box girder bridge and the effected factors such as creep and shinkage, the survey of long-term deflection of such box girder bridge (Hung Ha bridge) has been carried out and compare with the numerical analysis. The results show the significant effect of creep and shinkage to the long-term deflection of the cantilevered box girder bridge.

Keywords: Deflection, camber, box girder bridge, cantilevered bridge.

1. Introduction

Control of camber/deflection of large span of prestressed concrete bridge constructed by cantilever method is one of the main focused contents mentioned in the design and operation standards of bridge [1]. This issue has been considered in the design stage and has been observed carefully since the cantilever method was applied to prestressed concrete bridge structures in Vietnam (since 1995). However, camber/ deflection is the result of many long-term effects such as shrinkage, creep of concrete; relaxation and loss of prestressing, temperature; damage of bearings, changing in strength and properties of concrete over time... In order to evaluate the effects of these impacts on bridge construction, long-term monitoring and inspection are required.

In some of cantilever bridges in Vietnam such as Phu Luong (1996), Song Gianh (1996), Tan De (2002), Tram Bac (2003), Kien and Khe (2004), after a long time of exploitation, the signs of degradation have been observed through the appearance of cracks on girders and the change in the geometry of the span such as girder camber/deflection over time [2]. Finding out the cause of this change in bridge's deflection is a complicated problem. The deflection of the cantilever bridge is the result of many influenced factors including deterioration of the material (concrete, prestressed steel), changes in the operating mechanism (e.g. additional asphalt concrete layers on the bridge deck), changes in boundary conditions (loss of serviceability of bearing) or due to the effects of long-term loads such as shrinkage, creep of concrete and loss of prestressing as mentioned above.

In some bridges, the survey showed that the actual deflection of the bridge during operation was significantly larger than the calculation in design stage and it effect to the exploitation capacity of the bridge. In this research, the effected factors for deflection/camber will be reviewed and the deflection of the large span of prestressed concrete bridge in Vietnam will be surveyed and analyzed.

2. Effected factors of long-term deflection/camber

According to Machelsky, the mechanism of camber of prestressed concrete box girders constructed by the cantilever method was divided into two stages: construction and exploitation stage [3]. The proposed formula for calculating the camber/ deflection for beams in these two phases is based on beam theory.

However, during the construction stage, since the bridge diagram is static, short-term loads do not have a long-term influence on the deflection. The deflection is thus calculated according to the formula related mainly to the bending moment and prestressing force as follows:

$$w_v = \int_0^{\frac{L}{2}} \frac{M_v x}{EI_x} dx \quad (1)$$

$$M_v = M_{DL} - S(v_g - a) \quad (2)$$

Where:

M_v: the bending moment at the considered section due to the dead load moment minus the prestressing moment

M_{DL}: moment due to dead load

S: prestressing force

v_g: the distance from the top of beam to the center of cross-section

a: the distance from prestressed force of cover to the top of beam

x: the distance from the considered section to the restrained end.

The deflection equations (1) and (2) show that the main long-term deflection effects are static loads and prestressing. The effects of these long-term loads by time (shrinkage) continue along with the service life of the bridge and further develop the camber/deflection from the initial state.

In the cross section of the box girder bridge, the areas of the top and the bottom plate are different and so their shrinkage and creep. This can lead to an increasing in bending moment and deflection by time. To investigate this difference, Kristek et al, [4], created creep and shrinkage camber/delflection for different concrete slab thicknesses according to Bazant's B3 model, [5]. The results showed that the different deflection between the top anh the bottom plate due to the effect of the shrinkage could be up to 28mm at about 5 to 7 years, hence this effect on the total deflection of the box girder is significant. This mechanism has also been experimentally verified for the Zvikov Bridge crossing the Vltava River built in 1962, the result showed that this different deflection due to shrinkage of the 85m-span-cantilever bridge could be up to 25mm after 1300 days (about 3.5 years).

In the box girder, large shear stresses are transferred from the vertical girder ribs to the flanges, causing shear strain in the plane of the flange. Due to the influence of this shear strain, the longitudinal tensile or compressive stresses in the top and bottom plates of the girder decrease with the distance to the girder. This stress reduction is known as shear lag. This stress causes the longitudinal displacement of the flange center area to be different from that of the flange near the web girder. A simple example of the influence of shear lag on two-span continuous box girder deflections under evenly distributed loads is given by Kristek, [6]. Calculation results on 2-span box girder bridge showed that under the influence of shear lag, the negative bending moment on the top of the center decreased by about 5% while the positive moment increased by the same amplitude. This caused the static deflection increase significantly.

Prestressing helps to create camber in the box girder. However, prestressing loss by time also reduces girder deflection. Jayaseelan, [7], proposed a time step method to show that the top strand with the mild steel can work to reduce short term cambers, which marginally improves the constructability of prestressed concrete bridges. A. Abasi et al, [8], prove that deformation appears in the first three years and mostly caused by time-dependent properties of material.

3. Survey the deflection of cantilever bridge in Vietnam

In recent years, the bridge construction industry in Vietnam has made significant changes in investment in construction technology, one of which is "Construction technology of reinforced concrete continuous box girders by the balanced cantilever method" (referred to as cantilever technology) applied to construction of long-span prestressed concrete structures.

Span structure of large span precast concrete bridge is usually constructed from 1 pier or multiple piers by means of a formwork support device (casting vehicle). Normally, the construction of girder structure is started from the box girder on the top of the pier, then the girder segments from 3m to 5m are constructed symmetrically from both sides. Construction of each span is finished with a segment in the middle of that span.

In order to research the camber development of girder by time, the 5-span continuous (97.5m+3*150m+97.5m) cantilever bridge, Hung Ha bridge, was selected to carry out the survey. Hung Ha Bridge cross the Red River anh connect Hung Yen city and Phu Ly city in the North of Vietnam as shown in figure 1. The construction of the bridge started on December 27, 2015 and officially opened on January 26, 2019.

Figure 1. Location of Hung Ha Bridge

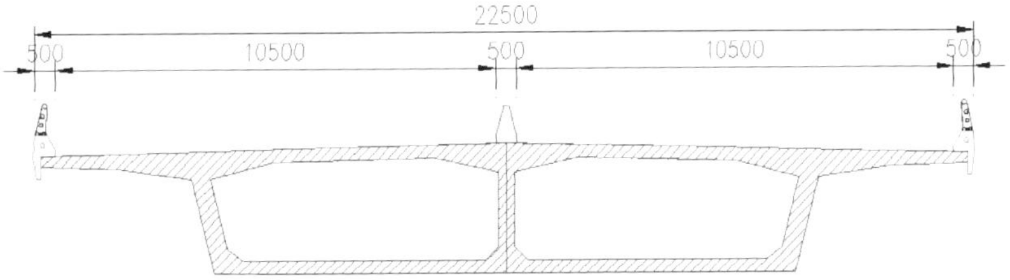

Figure 2. Cross section of Hung Ha Bridge

For investigating the camber of Hung Ha bridge, the surveys were deployed in 3 times on June, December of 2020 and June of 2021, corresponding to 18, 24 and 30 months after the bridge was put into operation.

The measured arrangement is conducted as follows: there are 5 measured points on 97.5m span with 25m space and 7 measured points on 150 m span with 25m space. There were 27 measuring points in the entire length of the main span. The height is measured by hydrometer and a 5m-rod. The elevation at the T22 pillar is measured to calculate the elevation of the detailed points.

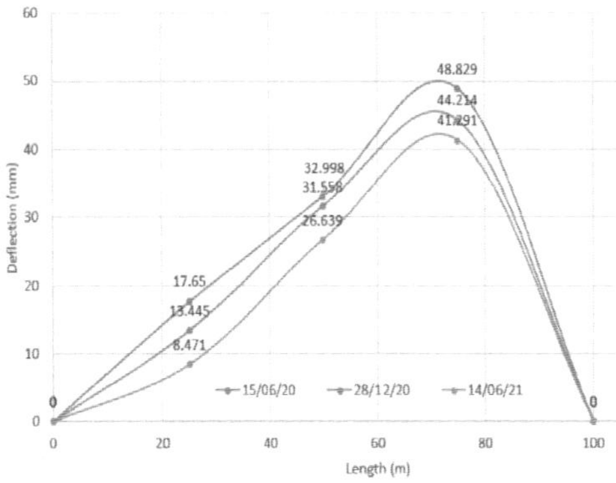

Figure 3. Deflection diagram of N23 span

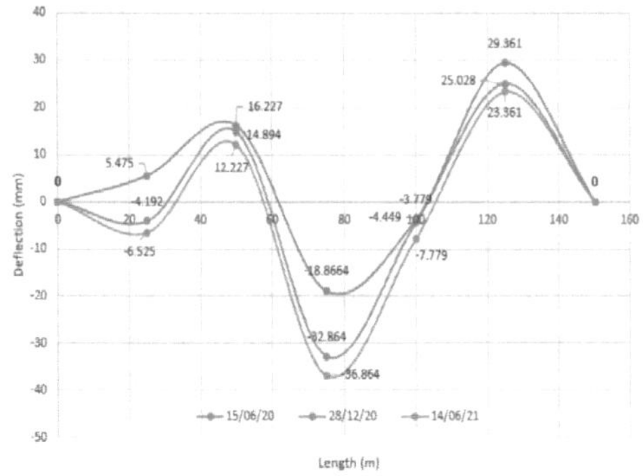

Figure 4. Deflection diagram of N24 span

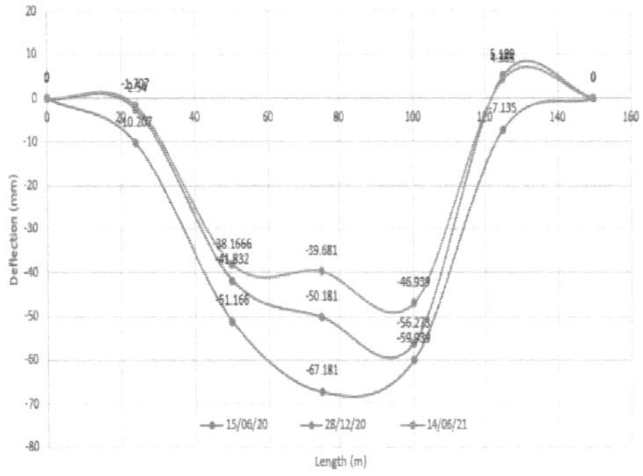

Figure 5. Deflection diagram of N25 span

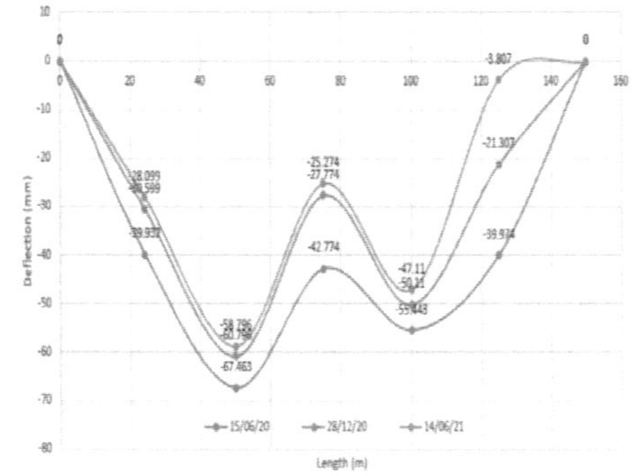

Figure 6. Deflection diagram of N26 span

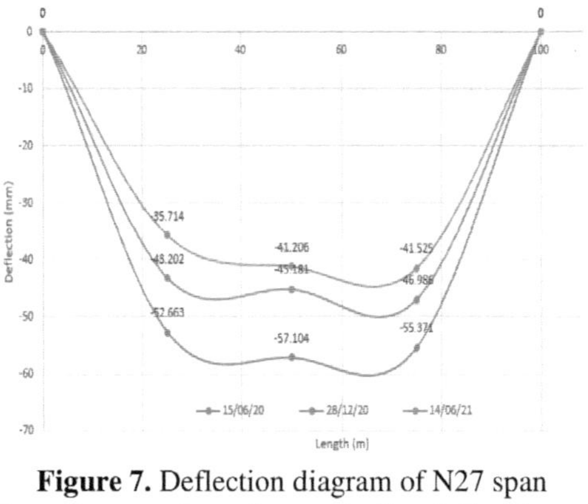

Figure 7. Deflection diagram of N27 span

The survey results on the bridges show a clear change in the deflection of the span structure as shown from Figure 3 to Figure 7.

In the side spans of 97.5m, N23 and N27, although the length and structure are completely similar, the formation and development of the deflection are significant different. On span N27, after 18 months of exploitation, the deflection was -57mm. This deflection gradually decreased to -45mm and -41mm at 24 and 30 months of exploitation. Meanwhile, on span N23 of the same length, the deflection of girder are 41, 44 and 49 mm, respectively, at 18, 24 and 30 months of exploitation.

Similarly in the 150m-side middle spans, spans N24 and N26, although both have the same length and structure, their deflection values at the middle of the span do not coincide. In N24 span, there is a tendency to gradually increase the deflection in the middle section over time with deflection values of -18, -32 and -34mm at the time of 18, 24 and 30 months of exploitation, respectively. While at span N26 and deflection decreased from -42mm after 18 months to -27mm after 24 months and finally after 30 months it reached the value of -25mm. At these two spans, the effect of positive prestressing on the formation of deflections over time can be observed clearly with geometrical "breaks" in the girder at locations with and without positive prestressing.

At mid-span N25, the girder's deflection decreases over the time with values of -67mm, -50mm and -40mm at the 18, 24 and 30 months of exploitation, respectively. At this span, the effect of positive prestressing is also clearly shown through the geometrical fractures recorded on the survey data.

4. Analysis of the deflection of cantilever bridge in Vietnam

In order to analyze the influence of factors on the development of camber, the cantilever bridge is simulated with the effects including shrinkage, creep of frequent loads (Dead load, prestressing) and prestressed losses. The bridge selected for simulation is Hung Ha bridge with the parameters introduced in previous paragraph. The Hung Ha bridge was modeled using Midas/Civil software with the construction steps as designing as shown in Figure 8.

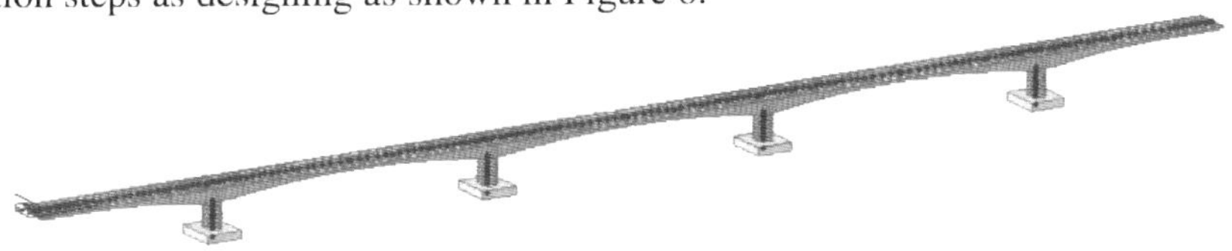

Figure 8. Hung Ha bridge modelling by Midas/Civil software

To evaluate the influence of shrinkage, creep and prestressing loss on the formation of deflection, the loads considered during construction and operation include creep, shrinkage and prestressed loss.

To clarify the difference between the models, the development of camber and deflection of girder according to the model results were taken at 18 months, 24 months and 30 months after the bridge was put into operation (similar to survey time).

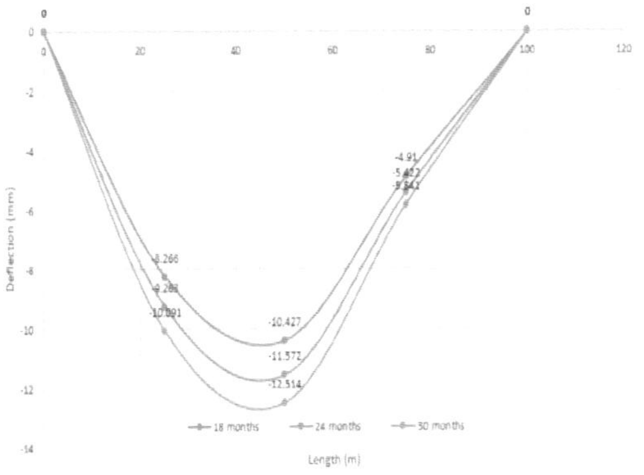

Figure 9. Deflection diagram of N23 span by Midas/Civil

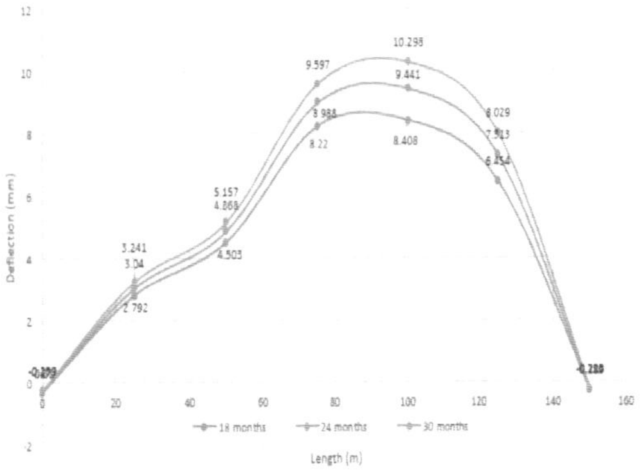

Figure 10. Deflection diagram of N24 span by Midas/Civil

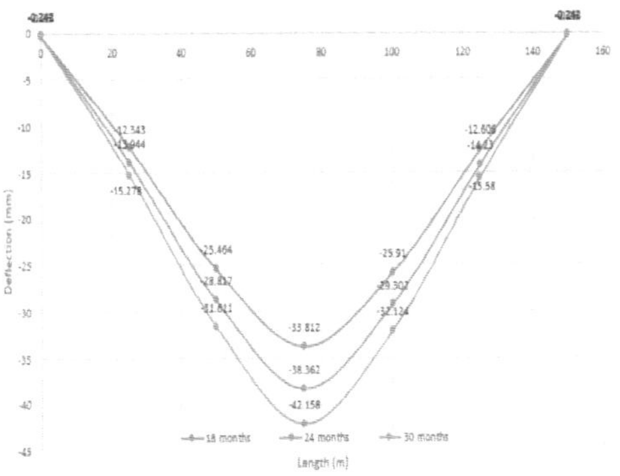

Figure 11 . Deflection diagram of N25 span by Midas/Civil

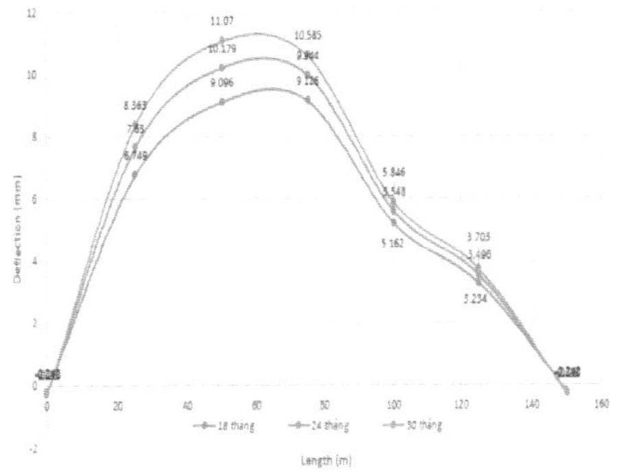

Figure 12. Deflection diagram of N26 span by Midas/Civil

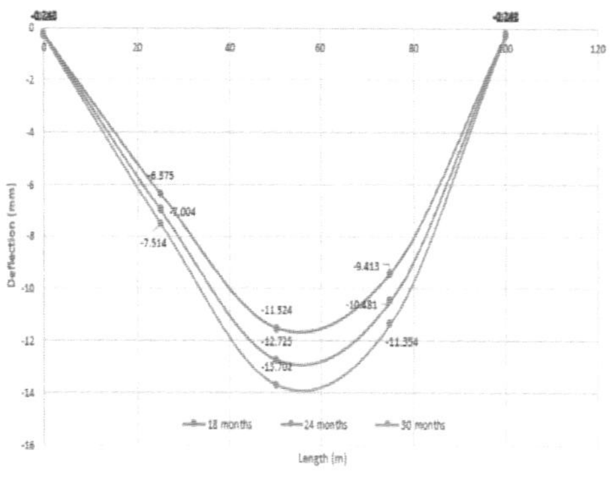

Figure 13. Deflection diagram of N27 span by Midas/Civil

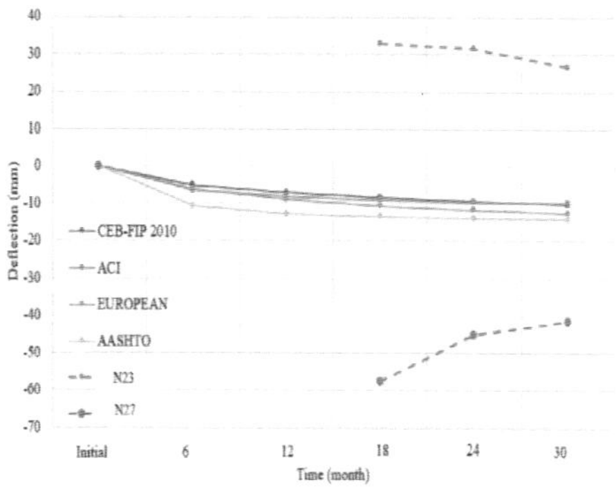

Figure 14. Deflection diagram at mid span of N23 and N27 span by modeling and survey

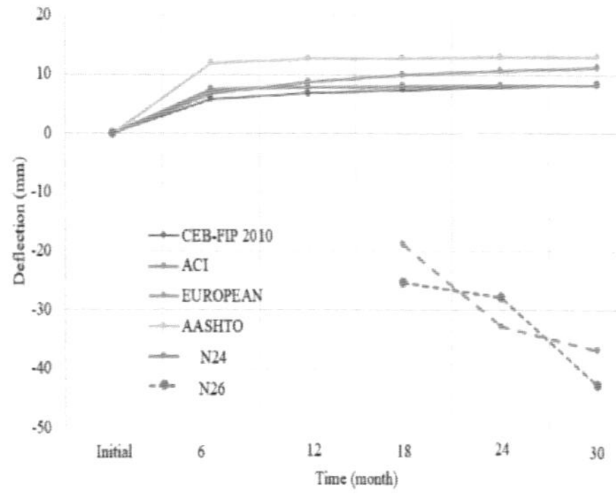

Figure 15. Deflection diagram at mid span of N24 and N26 span by modeling and survey

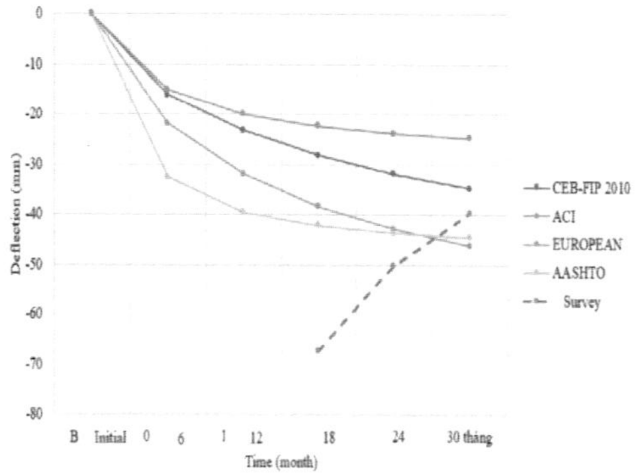

Figure 16. Deflection diagram at mid span of N25 span by modeling and survey

The modeling results are shown from Figure 9 to Figure 13. The shrinkage and creep model is also analyzed using different standards (CEB-FIP 2010, ACI, European and ASSHTO) for comparison of the surveyed results as shown from the Figure 14 to Figure 16.

As seen in Figure 9 to Figure 13, the shrinkage - creep models in the standard do not accurately predict both the trend and the deflection of bridge.

At the middle span (span N25, 150m long): the deflection after 18 months of exploitation is -67.18mm, this deflection decreases over time with values equal to -50.18mm and -39.68mm at the 24th and 30th months, respectively. The increasing trend shows no sign of stopping. Calculation results for the middle span according to different standards (Figure 14 to Figure 16) show that the girder will be deflected, however the deflection develops gradually over time and the rate of increase of the deflection decreases gradually. Shrinkage - creep model according to AASHTO standards (equivalent to TCVN 11823-2017) and EUROPEAN have the largest results (about - 48mm at 30 months), while the deflection according to ACI's model is smaller (about -25mm after 30 months).

At the side spans N23 and N27 (L = 97.5m), the survey results showed opposite data on deflection which must be theoretically similar due to the use of the same type of concrete and same operating conditions. In fact, span 23 and span 27 have completely opposite development of deflection. At span N23, the camber tend to decrease gradually (33mm, 31.59mm and 26.64mm in month of 18, 24 and 30) while at span 27, the deflection tend to decrease gradually (-57.66mm to -45.18mm and -41.525mm at months 18, 24 and 30). Meanwhile, the model results using shrinkage - creep models according to the standards give the results that the deflection is stable over time. The deflection according to AASHTO's model is largest, -14.04mm, while the ACI's model is the smallest, -9.83mm, at 30 months of exploitation.

At the span N24 and N26, the model also completely predicts both the magnitude and the development trend of the camber/deflection over time. In fact, the survey showed that the deflection of both spans of 24 and 26 tend to increase quickly (from about -20mm at 18 months to -30mm at 24 months and -40mm on average at 30 months) and the model results show that the camber reaches a stable value at about 10mm after 30 months.

5. Conclusion

From the above results, some conclusions can be drawn as follow:

- The effect of shrinkage and creep is significant to the camber of the continuous cantilever bridge;
- Even when the construction conditions and materials are similar, accurate prediction of camber over the time is not possible due to the great variability of material properties. It is clarified by the measurement results for span N23 and span N26, using the same type of concrete and same construction conditions, with the same structure but opposite deflection/camber.
- All theoretical models show that the curve/deflection tends to be stable over time, however, they are not consistent with the actual survey at spans 24, 25 and 26 of Hung Ha bridge, the camber/deflection of the span structure is still progressing and developed strongly. The trend has not yet stopped after 30 months from the exploitation time.

Acknowledgments
This research is funded by Ministry of Education and Training (MoET) under grant number B2020 – GHA – 06.

6. References

[1] Vietnamese standard 11823:2017 - Highway Bridge Design Standard.
[2] N. V. Minh, B. T. Thanh. N. H. Thuan. H. X. Son, "Damages on the pretresed large span concrete: cause and repairing method", Construction Publisher, 2017.
[3] C. Machelski, "The Effects of Dead Loads in Cantilever Concreting Bridges," Archit. Civ. Eng. Environ., vol. 12, no. 1, pp. 109–120, 2019, doi: 10.21307/acee-2019-010.
[4] V. Křístek, L. Vráblík, and V. Hrdoušek, "Causes of long-term deflections of large-span pre-stressed concrete box girders and recommendations on how to avoid these," Transactions on Transport Sciences, vol. 1, no. 3. pp. 97–102, 2008, doi: 10.5507/tots.2008.014.
[5] Z. P. Bazant and S. Baweja, "Creep and shrinkage prediction model for analysis and design of concrete structures: Model 83," American Concrete Institute, ACI Special Publication, vol. SP-194. pp. 1–83, 2000.
[6] V. Křístek and Z. P. Bažant, "Shear Lag Effect and Uncertainty in Concrete Box Girder Creep," J. Struct. Eng., vol. 113, no. 3, pp. 557–574, Mar. 1987, doi: 10.1061/(ASCE)0733-9445(1987)113:3(557).
[7] H. Jayaseelan and Bruce W.Russel, "Prestress losses and the estimation of long-term deflections and camber for prestressed concrete bridges", Feb.2007; PCI Convention and National Bridge Conference.
[8] A. Abasi, A. Rahai, A. Shokoohfar, "The Expectation of Long-Term Deformation of Concrete Cantilever Bridges: Case Study"; Dec. 2019; 5th International Conference on Bridges (5IBC2019)

A Critical Review on the application of coal Bottom Ash inconcrete: An Indian Prospective

Kiruthiga P.[1], Dr. Niragi dave[2], Dr. Ramesh Guduru[3]

[1] Civil Department, Pandit Deendayal Energy University, Gandhinagar, Gujarat, India.
[2] Civil Department, Pandit Deendayal Energy University, Gandhinagar, Gujarat, India.
[3] Mechanical Department, Pandit Deendayal Energy University, Gandhinagar, Gujarat, India.

Abstract. Coal Bottom Ash (CBA) is one of the by-products of Coal sourced Thermal Power Plants. Unlike Fly ash most of the CBA are dumped as landfills rather than utilized. Replacement of CBA in concrete reduces the burden on waste management and reduces the demand for the river sand and Portland cement. Themain and major composition of CBA is Silica hence it has pozzolanic properties. The utilization of CBA in concrete is done by adopting several treatment and process in order to its effective replacement as sand and cement. The process involved in the replacement of cement is Grinding or Pulverization. Pulverization increases the Surface area and fineness of CBA and is comparable with the cementhence pulverized CBA can be used as a cement replacement. Whereas sieving is done to replace as a sand. The focus of this study is to prove that CBA can be used in concrete without any significant loss in the strength, durability properties and the optimum dosage of CBA to improve the performance of the concrete. This review compares the results of several researches on mechanical and durability properties and determine the areas where the research has to be carried out further.

Keywords: Coal Bottom ash, Pulverization, EDX, Consistency and Setting time, Compressivestrength, Permeability and ASR Expansion.

1. INTRODUCTION

India is the largest producer and consumer of electricity. The main source of power generation is from thermal power plants. The total installed generation capacity as on 31.10.2021 is 3, 90,791 MW. Nearly 50% of the thermal power plants are coal fueled with the generation capacity of 2, 01,995 MW [43]. The by-products of thermal power plants are Fly Ash (FA) and Bottom Ash (BA). BA contributes 25% of the total Ash generation and FA contributes 75% [1]. FA is widely used now-a-days in the concrete and also in bricks whereas the BA are dumped as landfills which is a threat to Environment. India has taken initiative in the ash utilization by notifying the brick units to utilize 25% of ash by weight and ash brick should be used in the construction [2]. The coal ashes both fly ash and bottom ash can beused in the field of construction as fills or back fills [3]. Because CBA's chemical composition and characteristics are comparable to those of FA, researchers are drawn to work on BA. Several studies have been conducted on the application of these CBA in the construction field. Due to its silica concentration, which has pozzolanic qualities, CBA can be utilized as a sand and cement replacement in concrete, according to the findings. The use of CBA minimizes the demand for river sand and cement,as well as reducing landfill space. The use of CBA instead of cement reduces the amount of cement used, lowering CO2 emissions during cement production. CO2 emissions are also harmful to the environment since they contribute to global warming. Bottom ash from coal-fired thermal power plantscan be integrated into concrete to lessen the environmental impact of CO2 emissions from cement manufacturing plants and bottom ash from coal-fired thermal power plants. As a result, using CBA as asand and cement replacement in

concrete is a good idea. This review examines and contrasts the physical, chemical, mechanical, and durability properties of CBA in concrete from prior investigations.
According to prior research, the replacement of CBA in concrete is suitable and does not have any significant disadvantages; rather, it improves the durability properties of the CBA concrete and its strength after 28 days of curing. The CBA concrete outperformed the control concrete in terms of resistance to harsh environments.

2. COAL FLY ASH

The Coal Fly ash (FA) is fine and spherical by product produced from the coal powered thermal power plants. These fine dust particles are collected using Electrostatic precipitators from the chimneys.These particles are finer than cement and found to have pozzolanic properties hence can be used as a binder in concrete. The main composition of FA are Silica and Alumina. FA is classified into class C and class F categories. FA contains hazardous materials like metals, organic impurities and radioactive materials hence disposal of it affects the environment [4]. For safe disposal and cementing properties ofFA, it has been widely used in bricks and concrete now-a-days. Fine particles of FA in concrete improved the strength of the concrete [5]. The major chemical composition of FA are Si, Al and Fe [6].The effective utilization of FA in concrete as cement depends on the optimum dosage. The strength increases with increase in addition of FA in concrete up to optimum dosage beyond that limit the strengthtends to decrease [7].

3. COAL BOTTOM ASH

Bottom ash is the coarse, granulated, and incombustible coal residue found at the bottom of the furnace or boiler, whereas fly ash is the fine dust that flew up and was collected by electrostatic precipitators. Because of its high silica, alumina, and iron oxide concentration, fly ash is utilised in concrete, but the majority of CBA is deposited as landfill, combined with water to produce slurry for ash ponds, and only a little amount is used as an aggregate to replace river sand. Cement clinker now contains fly ash as a component. Thermal power plant of 2000MW capacity can produce 2200 tons of bottom ash [8]. In India the thermal power plants produce 25 million tons of coal bottom ash which contributes 20% of total coal ash [9]. The disposal of such huge amount of BA is challenging and has adverse impact on the Environment. In order to avoid this, the utilization of the BA as a cement and aggregate replacement plays a vital role. The physical properties and chemical composition of CBA areinfluenced by the sources of raw coal used in the thermal power plant.

3.1 CHEMICAL COMPOSITION

The chemical composition of the CBA varies with the type of coal burnt and the furnace conditions which influences the performance of the CBA in the concrete [10]. The chemical composition is determined either by X-ray Energy Dispersive Spectrometry (EDS) or X-ray fluorescence (XRF) and itis found that the main compounds are Silica, Alumina and Iron [2]. From several reports the sum of percentages of Silica, Alumina and Iron varies from 70% - 90% [11]. There are two types of classes, ifthe sum of (SiO_2 + Al_2O_3 + Fe_3O_2) is less than 70 % but greater than 50 % is class C whereas if the sumof (SiO_2 + Al_2O_3 + Fe_3O_2) is greater than 70% then is class F. The high Silica content is responsible forthe formation of C-S-H gel during hydration [12]. The chemical composition of CBA is similar to FA hence it is evident that CBA can be used as a Pozzolanic material similar to FA [13]. The percentage ofchemical composition of some of the references are given in the Figure (1). The chemical composition of FA is given in Figure (2). From the figures it is understood that the composition of CBA and FA varies with every thermal power Plants and the sources of coal. XRD pattern of FA and CBA are shownin Fig (3) [14].

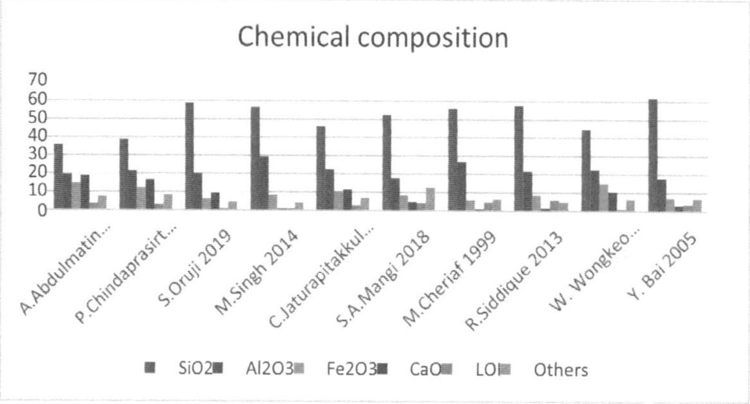

Figure 1. Chemical composition of Coal Bottom Ash in percentage from several studies [13][14][15][9][16][17][18][19][20][21]

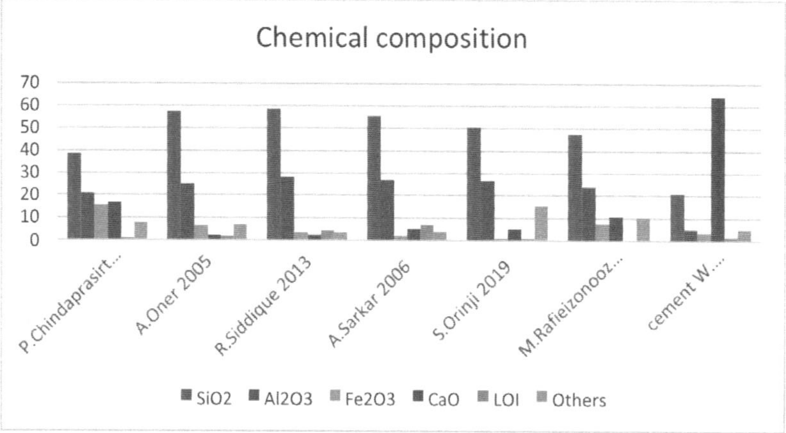

Figure 2: Chemical composition of cement [20] and FA [14][7][19][22][15][23]

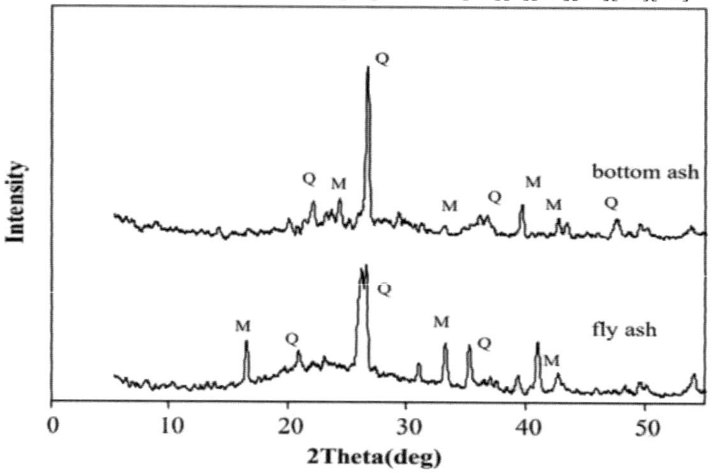

Figure 3. XRD patterns of Fly ash and Bottom Ash Q= Quartz, M= Mullite [14]

3.2 PHYSICAL PROPERTIES

An extensive interpretation on Physical properties is necessary for the better and effective utilization of CBA in concrete. The appearance of CBA is greyish, rough, spherical or irregular and porous [2]. The performance of the CBA concrete depends on their physical attributes like specific gravity (SG), specific surface area (SSA), fineness modulus (FM) and moisture absorption (ω). FM governs the utilization of CBA as a fine aggregate. FM value of CBA should be in the range of 1.6-2.2if it has to be used as a sand. The fineness can be improved by adopting Pulverization technique or grinding CBA [12]. During grinding the FM increases which in turn increases the specific gravity and reduces the porosity [13]. SSA can also be increased significantly by adopting sophisticated grinding techniques or longer duration of grinding which results in the replacement of cement in concrete with ground CBA [12]. The effect of grinding process on SSA is also shown in Figure 4. The sieve analysis of CBA shows that it resembles the natural sand as 80% -90% particles pass through 4.75mm sieve [10],[16]. CBA has higher amount of fine particles and lower amount of coarse to medium coarse particles [17]. The specific gravity of CBA is directly proportional to iron oxide content [1]. SG increases with increase in grinding time [10]. The physical attributes of several researches are listed in the Figure 5,6.The water requirement for the original CBA binder is more than the cement binder due to its high porosity but the water requirement for the ground CBA is less than the cement mortar [16]. The particlesize distribution and SEM image of CBA are shown in Figure 7,8 [13].

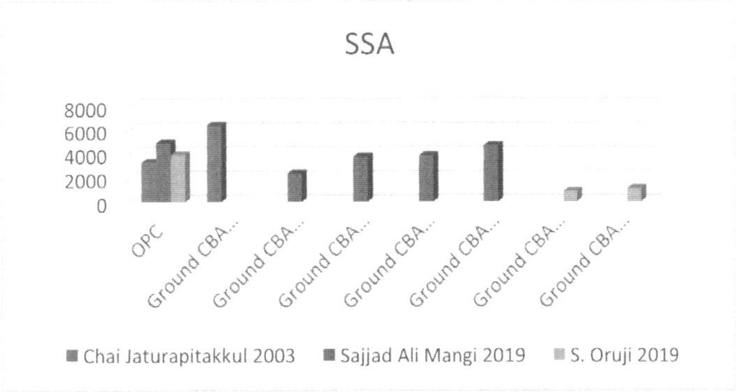

Figure 4: Specific surface area of coal bottom ash in various grinding periods in cm^2/g [16][26][15]

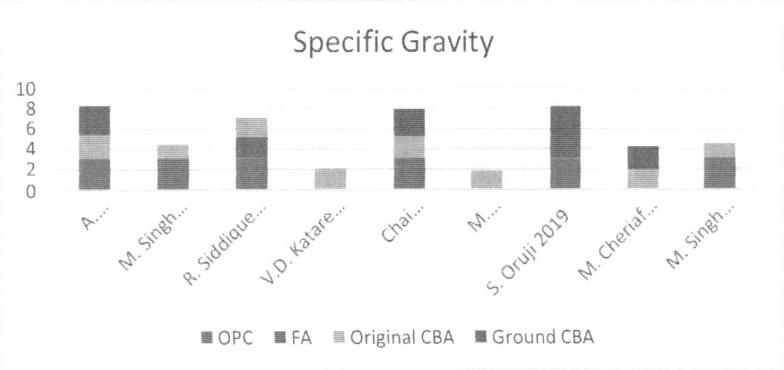

Figure 5. Specific gravity of Coal Bottom Ash from various studies[13][27][19][16][23][15][18][9]

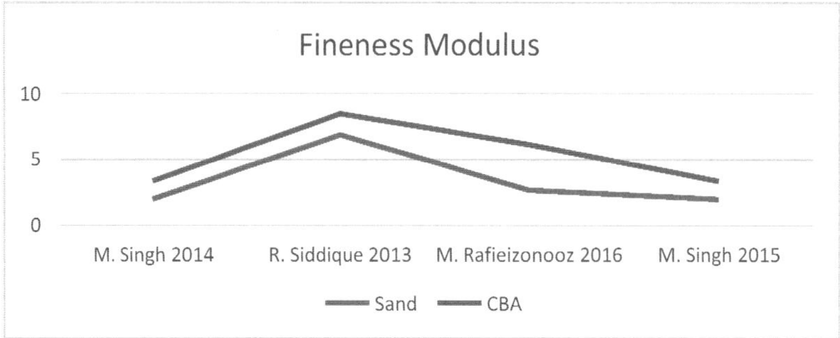

Figure 6. Fineness Modulus of CBA from various studies [27][19][23][9]

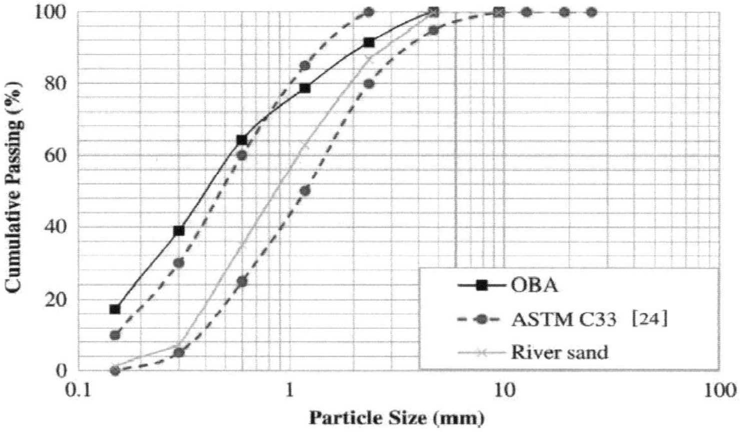

Figure 7. Particle size distribution of CBA compared with river sand [13]

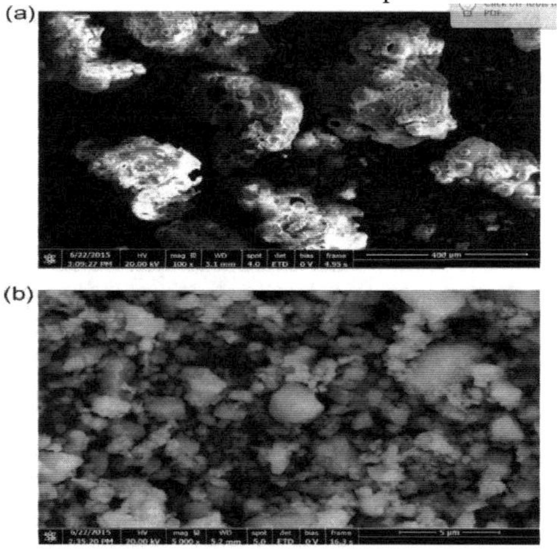

Figure 8. SEM image of Original CBA (a) and ground CBA(b) [13]

3.3 MICROSTRUCTURE

The CBA from thermal power plants are generally rough, irregular and porous from the SEM results but grinding will alter its porosity as well as smoothens the rough texture [13]. The microstructure of CBA is influenced by the combustion temperature and cooling rate during incineration of coal [10]. According to V. Asokbunyarat et al. the CBA under SEM (Scanning Electron Microscope) resembles aggregates with irregular structure, polycrystalline, amorphous and glassy appearance. X-ray diffraction (XRD) analysis of the CBA samples indicated the crystalline phases like mullite, feldspar, magnetite and quartz [28]. The predominant crystalline phases are quartz and mullite [14]. L.B. Andrade et al. confirms the porosity of CBA through SEM and put forward that the determination of exact water cement ratio is difficult due to its high porosity [29].

3.4 PROCESSING OF CBA

The processing of CBA is necessary to improve the desired mechanical and durability properties as the usage of raw CBA is unsuitable. There are two types of processing physical and chemical. The physical treatment includes soaking, burning, grinding and sieving whereas the chemical treatment includes the addition of alkaline activators or chemical additives during mixing in order to improve the pozzolanic reactivity of CBA [12]. The wetted CBA was oven-dried at 110°C for 24 hours [9] then sieved to retain on sieve no 4 to eliminate larger particles and then ground and sieved to retain on sieve no 325 then usedas a cement replacement [13], [16],[26] - [30]. CBA was dried before used in concrete as a sand replacement [27], [31]. M. Rafieizonooz et al. sieved CBA in 4.75 mm sieve before used as a replacement of sand in concrete [23] S. Oruji et al. ground CBA in vibratory mill to get fine pulverizedCBA which then is used as a replacement of cement [15], [32].In another research the CBA was subjected to grinding and then thermal treatment as a result the color of the CBA changes from dark grey to light grey due to the melting of carbon content [33]. H. Kurama, M. Kaya used low grade lignite coal where the amount of unburnt coal content is reduced from 8.9% to 4.65% by sink and float test, particle size classification, and electrostatic separation methods as the concrete with high carbon content is inferior to low carbon content [34].

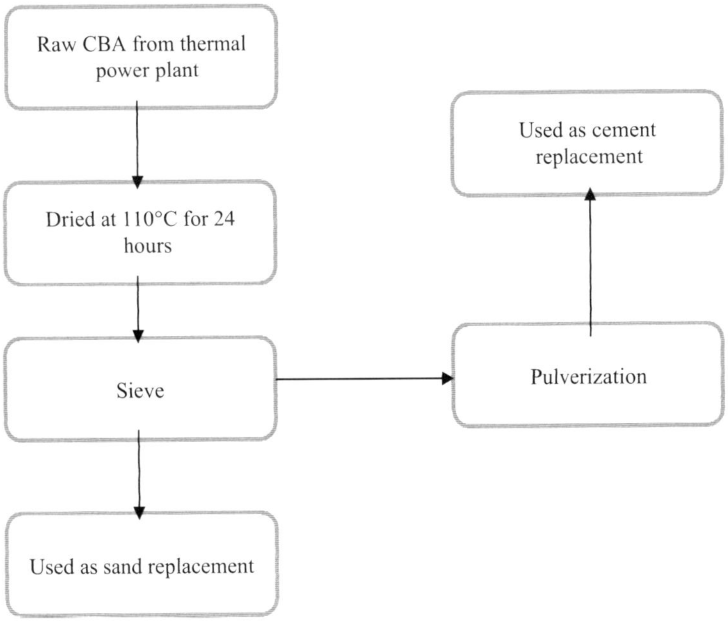

4. COAL BOTTOM ASH IN CONCRETE

CBA replaces sand and cement in concrete because of its high silica content and also its pozzolanic reactivity which improves the mechanical and durability properties. The pozzolanic reactivity is the chemical reaction between the calcium hydroxide and silica to produce C-S-H gel. The pozzolanic reactivity can be improved my its fineness [2]. The improvement of Fineness result in high energy consumption during grinding hence the amount of fineness is confined with the strength activity index and water requirement. For good pozzolanic material the grinding should be carried out till the particles retained on sieve no 325 should be at least 25% by weight and their strength activity index is more than 80% at 7 or 28 days [13]. Pozzolanic Activity Index (PAI) qualifies the pozzolanic behavior of CBA and it was found that PAI increases with increase in grinding durations. The minimum satisfactory PAIrequirement is 75% [33]. Cheriaf et al. studied the calcium hydroxide consumption it was 5%, 15%, 37% and 60% at 7, 14, 28 and 90 days respectively. The pozzolanic activity of CBA occurs after 14 days and its calcium hydroxide consumption is remarkable at 90 days [18].

4.1 CBA AS A SAND REPLACEMENT

There is a huge demand for the river sand and the sources are depleting hence an alternative to river sand needed the attention. The particle distribution on CBA is similar to river sand hence it can be a replacement of sand [21]. The CBA and river sand were oven dried and sand was replaced with CBA by mass in concrete [9],[27], [21]. CBA with FM 1.6 was used as a sand replacement in Self-CompactingConcrete (SCC) [19]. M. Singh et al. used oven dried CBA as a sand replacement and studied the workability of CBA concrete [35]. FA and CBA both used as a replacement of cement and sand respectively where CBA is sieved through 4.75mm sieve before replacement [23]. The strength and durability properties of concrete containing FA as a cement replacement, raw CBA and GBFS (Granulated Blast Furnace Slag) as a sand replacement without grinding [36].

4.2 CBA AS A CEMENT REPLACEMENT

The ground CBA is a pozzolan like fly ash with high silica content which react with calcium hydroxideto form C-S-H gel which is responsible for the strength and durability properties. The chemical composition of CBA and FA are similar but the raw CBA is undesirable to use as a cement replacementas it is porous and has larger particles. To make CBA as a cement replacement the particle size and porosity need to be improved. A. Abdulmatin et al. ground CBA till the particles pass through sieve no 325 (45 mm opening) and then used in concrete as a cement replacement [13]. W. Wongkeo et al. groundCBA for 6 hours in the ball mill to gain pozzolanic property and then used in concrete as a cement replacement in autoclaved light weight concrete [20]. Grinding the CBA improves the inactive pozzolanic reaction and can be used as a cement replacement in mortar and concrete [16]. The CBA with SSA 3835.75 cm^2/g ground for 20 and 30 hrs was the energy efficient and effective pozzolan as binder [26], [37]. The ultrafine pulverized CBA is three times finer than cement was used as a cement replacement and studied their ASR (Alkali-Silica Reaction) expansion and strength index [15], [32]. The CBA concrete was tested for short term effects of chloride and sulphate where the CBA was groundfor 2 hrs then sieved through 300 micron and again ground for 20 hrs to match the size of the cement [17]. Haldun Kurama studied the effects of pulverized low grade lignite CBA in mechanical propertiesup to 25% by mass as a cement replacement where the CBA had high amount of unburnt carbon contentwhich were then reduced by particle size classification, heavy medium separation and electrostatic separation [34].

5. FRESH PROPERTIES OF CBA MORTAR AND CONCRETE

5.1 WORKABILITY

Workability is defined as the ease with which concrete can be compacted hundred per cent having regard to mode of compaction and place of deposition. Some of the factors affecting workability are particle size, surface texture and use of admixtures [38]. Workability is determined by slump flow and the slump flow of Self Consolidating Concrete (SCC) was studied with CBA as a sand replacement. ForSCC the slump flow value should be in the range of 500 – 700 mm beyond that the concrete tends to segregate [39]. R. Siddique studied the slump value of SCC incorporating CBA as a sand replacementat 10% - 30% and found the slump value for all the mixes were within the range hence workability is not significantly affected by CBA [19].When CBA and FA as a sand and cement replacement respectively, the workability of the concrete decrease with increase in CBA but at 25% and 50% replacement of CBA the workability exceeds control mix due to the 20% of FA content [23]. The workability not only decreases with increase in CBA content as a cement replacement but also decreases with increase in grinding period as the water requirement increases with increase in SSA due to grinding[37][30]. According to Y. Bai et al. the workability increased with increase in CBA content as a sand replacement, because of the spherical and rounded particles which passed through 5mm sieve [21]. O¨. O¨ zkan et al. used CBA as sand replacement along with GBFS (Granulated Blast Furnace Slag) and found that concrete containing CBA and GBFS showed reduction in the workability with increased in sand replacement [36]. The workability of CBA concrete with 10% CBA as a cement replacement decreased by 10% than control mix as more water is absorbed by CBA particles [17]. S. Oruji et al. compared the results of workability for cement, FA and pulverized CBA mortar and found that the workability increased with the addition of fine pozzolanic material but ultrafine CBA mortar lagged than FA because of the fineness [32].

Table 1. Workability of concrete with different percentage of CBA

Percentage of CBA replacement	Effect on Workability	Reference
10% - 30% as sand	Workability is not affected	R. Siddique [19]
25% - 50% as sand	Workability exceeded control mix due to 20% FA as a cement replacement	M. Rafieizonooz [23]
10% as cement	Workability decreased by 10%	S. A. Mangi [17]
30% - 100% as sand	Workability increased with increase in CBA passing through 5mm sieve due to its spherical shape	Y. Bai [21]
5% - 30% as sand	Workability reduces with increase in sand replacement	Ö. Özkan [36]
10% - 30%	Workability decreases with CBA content and also decreases with increase in grinding periods	S. A. Mangi [37]
10% - 30% as cement	Workability decreases with increase in replacement	S. A. Mangi [30]
10% - 70% as cement	Workability of mortar increased with increase in fine pozzolanic material but lagged with ultrafine pozzolanic material	S. Oruji [32]

5.2 NORMAL CONSISTENCY AND SETTING TIME

Consistency is the degree of fluidity or the degree of mobility of the mortar or concrete. The setting time is divided into initial and final setting time. The initial setting time is regarded as the time elapsed between the moment that the water is added to the cement, to the time that the paste starts losingits plasticity. The final setting time is the time elapsed between the moment the water is added to the cement, and the time when the paste has completely lost its plasticity and has attained sufficient firmness to resist certain definite pressure [38]. A. Abdulmatin et al. studied that the normal consistency of the cement paste is 25% whereas the CBA paste (cement + CBA) has the lesser consistency of 26%-27%. The setting times of the CBA paste retards as there is decrease in C_3A and C_3S due to 20% replacementof cement with CBA. The setting time is affected by CBA fineness, original CBA has the setting time of 113 and 185 min whereas the ground CBA has the range of 120-125 min and 190-200 min [13]. Chai

Jaturapitakkul compared the normal consistency and setting times of control mix and CBA concrete (replacement of cement at 10%-30%) both original and ground CBA and found that there was no considerable difference in the consistencies which ranged from 24.7% - 25%. The initial setting time ofCBA concrete was longer than cement by 9 – 23 minutes as C_3S decreased because of cement replacement whereas the final setting times of original CBA and ground CBA were 195 and 210 min [16]. The normal consistency of concrete containing CBA (as a cement replacement where the CBA was ground for 20 hrs till 75% pass through 63micron sieve) at 10% - 30% replacement was in the rangeof 32%- 36% , the initial setting time was 10 – 45 minutes and the final setting time was 280- 315 minutes [30]. Both the initial and final setting times retards or increased with increase in CBA content,3 hrs pulverized CBA mortar was relative with concrete as the fineness decrease the setting times [32].

5.3 BLEEDING AND SEGREGATION

The effect of bleeding on CBA concrete was explained by M. Singh, R. Siddique, according to that,the dry pores available in the CBA particles as a sand replacement absorbed the excess water available hence reduction in the bleeding. the bleeding decreased with increase in CBA content [31].

6. HARDENED PROPERTIES OF CBA MORTAR AND CONCRETE

6.1 COMPRESSIVE STRENGTH

The compressive strength of the CBA mortar was compared with control mortar and it was found the early strength (3- 14 days) of the CBA mortar is decreased due to 20% replacement which reduces the hydration process but later (after 28 days) the pozzolanic reactivity initiated and the strength tends to increase. The compressive strength of the CBA mortar containing CBA particles retained on sieve no 325 of 5% by weight at 90 days was 49.8 MPa but the cement mortar had 47.2 MPa which is due to thefiner particles of CBA than cement [13]. CBA has also used as a cement replacement in Light Weight Concrete (LWC) at a rate of 20% and 30% by weight and was found that the compressive strength of bottom ash LWC at 7 days exceeds 10MPa which is higher than control concrete which is due to Tobermorite formation [20]. The strength of CBA mortar with CBA as a cement replacement showed reduction in strength with increase in CBA content [34]. M. Singh, R. Siddique used oven dried unground CBA as a sand replacement at various amount of replacement (30%-100%) and found that the compressive strength decreases with increase in CBA content and also the early strength is lower that control mix but it gradually increases after 28 days due to pozzolanic activity [9]. At 90 and 365 days the compressive strength exceeds the control mix due to the spread of C-S-H gel in the pores [27]. Thecompressive strength of SCC with 10% - 30% sand replacement with CBA showed decrease in strengthwith increase in CBA content but strength increased with aging [19]. But at fixed slump value (as the workability increased with CBA content [21]) where the water demand is reduced then the strength result was similar, with no significant loss than the control mix [21]. When the CBA concrete used alongwith FA as a cement and CBA as a sand replacement there was no significant changes in the compressive strength hence CBA can be used as a sand replacement [23][36]. The compressive strength of the mortarand concrete for the specimens containing CBA as a cement replacement looks identical, the strength increases with aging but decreases with increase in CBA content. Ground CBA showed better results than original CBA. The early strength is lower than control but it exceeded after 56 days due to pozzolanic reactivity which was observed in ground CBA [17]. Hence ground CBA has to be used as a cement replacement [16]. The strength increases with increase in grinding period [37].The optimum dosage of ground CBA as a cement replacement is 10% [30]. The compressive strength of the CBA concretes at 28 days with two grinding durations 30 mins and 3 hrs of CBA was studied and found thatwith 9% replacement of cement with 3 hrs grounded CBA exceeded the control mix with strength activity 111% [15]. When the cement mortar compared with the CBA mortar the strength increased withcuring period but decreased with dosage. At 28 days strength exceeded the control mortar, CBA mortarwith 3 hrs grinding showed more strength that 30 mins grinding as pozzolanic reactivity increased withfineness [32].

Table 2. Compressive Strength of concrete with different percentage of CBA

Percentage of CBA replacement	Effect on compressive strength	Reference
30% - 100% as sand	Strength decreases with increase in CBA content. Early strength is lower than control specimen but gradually increases after 28 days.	M. Singh [9]
20% as cement	Early strength may be lower but exceeds control after 28 days.	Abdulmatin [13]
9% - 41%	Concrete with 9% CBA ground for 3 hrs exceeds control at 28 days.	S. Oruji [15]
20% - 100% as sand	Strength exceeds the control mix due to the spread of C-S-H gel in the pores.	M. Singh [27]
10% - 30% as sand	Strength decreases with increase in content.	R. Siddique [19]
20% - 100% as sand	After 90 days strength exceeds control specimen.	M. Rafieizonooz [23]
10% as cement	Lower early strength but exceeds control after 56 days due to its finer particles.	S. A. Mangi [17]
5% - 30% as sand	Strength decreased with increase in content but when 20% FA replaced cement then strength exceeded control after 90 days.	Ö. Özkan [36]
10% - 30% as cement	Strength exceeds cement mortar after 28 days for ground CBA.	C. Jaturapitakkul [16]
10% - 30% as cement	Strength of concrete with 10% CBA ground for 30hrs showed satisfactory result at 28 days.	S. A. Mangi [37]
10% - 30% as cement	Strength of concrete with 10% CBA ground for 20hrs showed satisfactory result at 28 days.	S. A. Mangi [30]
9% - 41%	Strength of mortar increased with curing period but decrease with increase in CBA.	S. Oruji [32]
5% - 25% as cement	Reduction in strength with increase in CBA content.	Haldun Kurama [34]
20% and 30% as cement	Strength exceeds control concrete.	W. Wongkeo [20]

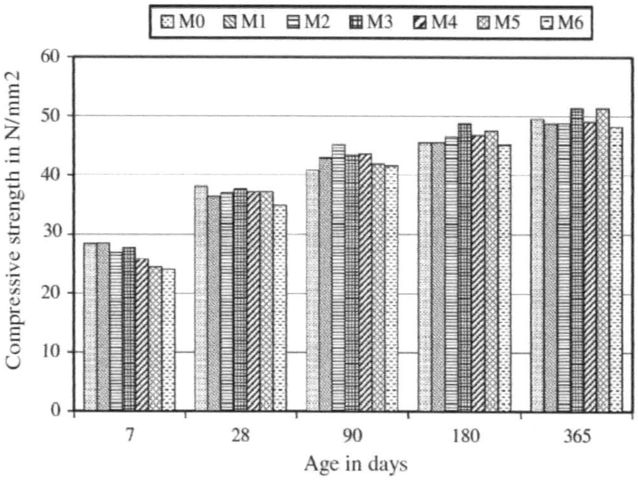

M0- control mix M1- 20% CBA as sand M2- 30% CBA as sand M3- 40% CBAas sand M4- 50% CBA as sand M5- 75% CBA as sand M6- 100% CBA as sand

Figure 9. Compressive strength of CBA concrete at various sand replacement levels [27]

6.2 SPLIT-TENSILE STRENGTH

The split-tensile strength of the SCC with CBA as a sand replacement was similar to compressivestrength decreased with increase in CBA content but strength increased with aging [19]. The split-tensilestrength of concrete with CBA as cement replacement shows similar result as that of compressive strength, the strength decreases with increase in CBA content but increases with grinding. The concretecontaining 10% of 20 hrs ground CBA showed less strength than control but 10% of 30 hrs ground CBAexceeded the control [37][30]. According to M. Rafieizonooz et al. the split tensile strength of concretewith FA and CBA as a cement and sand replacement respectively at 7 and 28 days were lower than thecontrol except the concrete with 75% CBA replacement. After 90 days all the FA and CBA concrete exceeded control but by 24.07% and 12.68% higher than control concrete with 75% and 100% replacement of CBA at 180 days [23].

6.3 FLEXURAL STRENGTH

The flexural strength of the LWC with Al is similar to compressive strength, strength increased with increased in CBA content due to increased bulk density of concrete [20]. The flexural strength of the concrete containing CBA as a cement replacement showed that the strength decreases with increasein the CBA content and the optimum dosage of CBA in concrete for better flexural strength was found as 10% replacement [37]. M. Rafieizonooz et al. studied the flexural strength and found that the strengthincreased with aging but decreased with CBA content which is similar to compressive strength with 5.51% and 3.05% higher strength than the control concrete for 75% and 100% replacement respectivelyat 180 days curing [23]. The flexural strength of the CBA mortar containing treated lignite coal bottom ash as a cement replacement was similar with compressive strength increased with aging but decreased with CBA content addition and the optimum dosage was found as 10% CBA of cement replacement [34].

6.4 MODULUS OF ELASTICITY

The mechanical behavior of structures to stress and strain are determined by its modulus of elasticity. The concrete containing CBA as a sand replacement showed the modulus of elasticity is nearly 98% of the control concrete up to 50% sand replacement [35].

Table 3. Tensile Strength of concrete with different percentage of CBA

Percentage of CBA replacement	Effect on tensile strength	Reference
10% - 30% as sand	Strength decreases with increase in content.	R. Siddique [19]
20% - 100% as sand	After 90 days strength exceeds control specimen.	M. Rafieizonooz [23]
10% - 30% as cement	Strength of concrete with 10% of 30 hrs ground CBA exceeded the control.	S. A. Mangi [37][30]

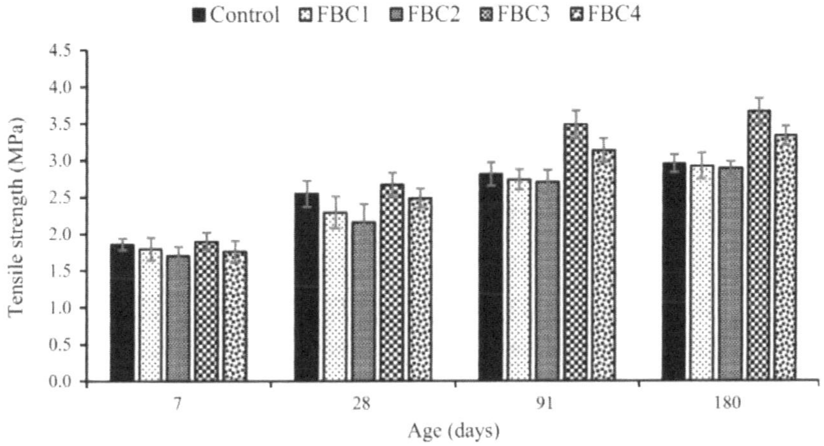

FBC1- 20% FA 25% CBA FBC2- 20% FA 50% CBA FBC3- 20%FA 75% CBA FBC4- 20% FA 100%CBA

Figure 10. Tensile strength of CBA concrete at various levels of sand replacement [23]

Table 4. Flexural Strength of concrete with different percentage of CBA

Percentage of CBA replacement	Effect on flexural strength	Reference
20% - 100% as sand	Strength increased with aging but decreased with CBA content	M. Rafieizonooz [23]
10% - 30% as cement	Strength decreases with increase in the CBA content	[37]
5% - 25% as cement	Strength increased with aging but decreased with CBA content	Haldun Kurama [34]
10% - 30% as cement	Strength increased with increased in CBA content due to increased bulk density of concrete	W. Wongkeo [20]

7. DURABILILITY PROPERTIES OF CBA CONCRETE

7.1 CHLORIDE PENETRATION

The chloride attack occurs in the marine structure and the structures exposed to salty environment. The chloride ion penetration is due to the presence of large size of voids and microcracksavailable in the concrete and also due to the alkalinity of pore solution in the concrete. M. Singh, R. Siddique conducted the experiment on the concrete containing CBA as a sand replacement at various replacement levels (20%, 30%, 40%, 50%, 75% and 100% of sand) where the total charge is passed through the CBA concrete and found that the charge passed through the CBA concrete was lower than the control concrete and after 90 days the chloride resistance increased with increase in CBA content [27]. The above results were also similar for SCC concrete having CBA as fine aggregates at 10%, 20%,30% of sand increased CBA content increased the chloride resistance which is the evident the microstructure becomes denser and gains strength [19]. The short term effect of 5% sodium chloride onCBA concrete (10% CBA as a cement replacement) was studied and stated that the compressive strength of CBA concrete was lower than control concrete, strength declined with age in control concrete whereas the CBA concrete gained strength with age [17]. The chloride migration coefficient and chloride diffusion coefficient were determined by Norwegian standard and found that the 10% to 25% cement replacement by CBA showed decrease in both the coefficients with increase in ground CBA replacement [41].

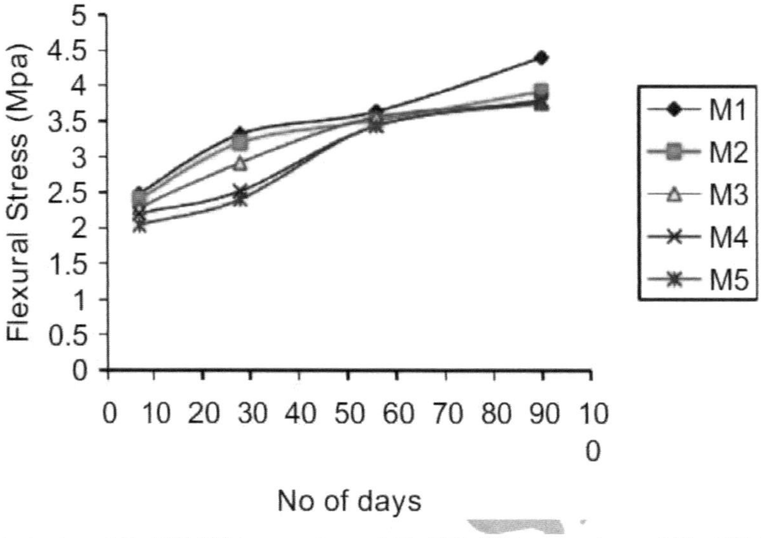

M1- control mix M2- 20% CBA as sand M3- 30% CBA as sand M4- 40% CBA as andM5- 50% CBA as sand

Figure 11. Flexural Strength at various levels of sand replacement [40]

Table 5. Chloride attack of concrete with different percentage of CBA

Percentage of CBA replacement	Effect of chloride attack	Reference
20% - 100% as sand	Resistance to chloride penetration increase with increase in CBA content.	M. Singh [27]
10% - 30% as sand	Increased chloride resistance due to dense microstructure.	R. Siddique [19]
10% as cement	CBA concrete is resistant to chloride as the strength is not declined due to chloride attack.	S. A. Mangi [17]
10% - 25% as cement	chloride migration coefficient and chloride diffusion coefficient decrease with increase in CBA replacement.	C. Argiz [41]

7.2 ALKALI SILICA EXPANSION

Alkali Silica expansion occurs when reactive aggregates react with alkali cement to form sodium silicate hydrate gel which swells up by absorbing water and results in the expansion and deterioration of the concrete structures. S. Oruji et al. studied the ASR (Alkali Silica Reaction) Expansion on the concrete with FA and CBA as a cement replacement and compared with control concrete. According to that, the ASR expansion of the FA and CBA concrete was significantly lower than the control and the expansion decreased with increased in the replacement fig. Comparing FA and CBA concrete, FA concrete showed better resistant to ASR again comparing the CBA concrete with 30 minus and 3 hour pulverization, the concrete contained 3 hours pulverized CBA showed much less expansion than the former CBA specimen due to its fineness [15].

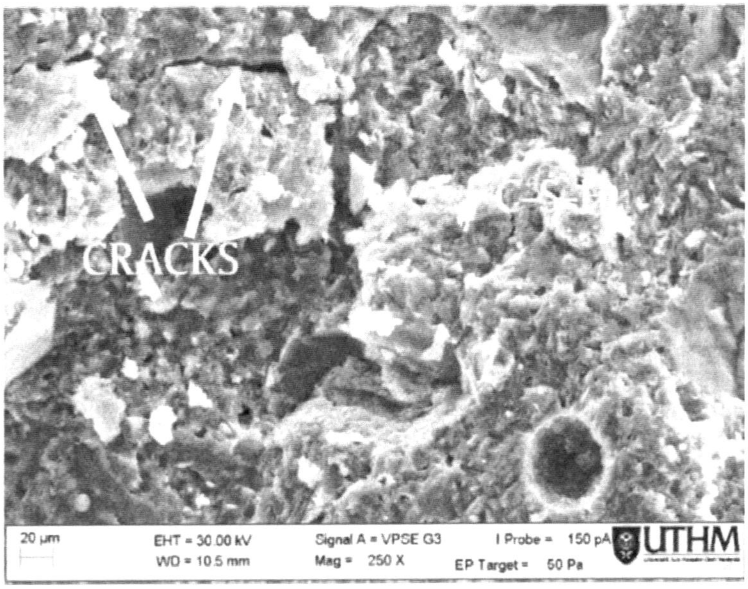

Figure 12: Short term effect of chloride on CBA concrete at 56 days [17]

7.3 SULPHATE ATTACK

The concrete should be resistant to sulphate attack to prevent the loss of compressive strength, volume expansion and mass of concrete by the external sulphate attack. The concrete becomes resistantto sulphate attack if it becomes impermeable. Magnesium sulphate has a severe impact on concrete hence an experiment was conducted on the concrete containing (CBA as a sand replacement ranged from 20% - 100% of CBA) immersed in 10% magnesium sulphate and found that there was no loss in the mass for all mix. The expansion of the CBA concrete is more than the control mix but the loss of compressive strength was lower than control mix even when the specimen was immersed for 180 days [27]. Low permeability is responsible for the better performance of the CBA concrete under sulphate environment, no sign of external attack was also observed through SEM and XRD of the specimen immersed for 180 days [27]. V. Sata et al. studied the sulphate resistance of lignite BA geopolymer mortar on 5% sodium sulphate solution and found that the expansion due to sulphate attack was less in BA geopolymer than the Portland cement [42]. The effect of sodium sulphate on compressive strength of concrete containing 10% CBA as a cement replacement was studied and found that there was no significant variation on the strength at early stage for short term exposure but after 28 days the CBA concrete performed better with higher strength of about 4.7% and 11.32% at 56 and 90 days respectively than the control concrete [17].

Table 6: Sulphate attack of concrete with different percentage of CBA

Percentage of CBA replacement	Effect of sulphate attack	Reference
20% - 100% as sand	CBA concrete performs better than control as there is no loss of compressive strength even after 180 days of immersion in sulphate environment.	M. Singh [27]
10% as cement	CBA concrete performed better in sulphate environment than control concrete.	S. A. Mangi [17]
40% and 100% as binder	Less attack on CBA geopolymer mortar than control mortar.	[42]

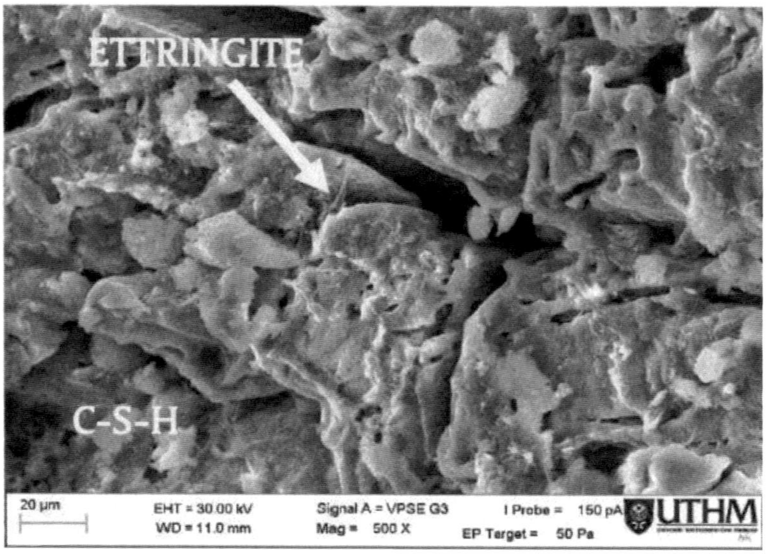

Figure 13. Short term effect of sulphate on CBA concrete at 56 days [17]

7.4 DRYING SHRINKAGE

The hardened concrete undergoes drying shrinkage. The concrete containing CBA as a sand replacement experienced lower shrinkage strain than control concrete and also the rate of shrinkage decreased with increase in CBA content because the porous CBA particles had absorbed more water atthe time of mixing the ingredients [27]. The drying shrinkage of the FA and CBA concrete was lower than the control concrete as the pores available in the CBA particles absorbed more water and preventedthe shrinkage during drying process. At 180days drying period, the shrinkage of FA and CBA concrete for various amount of replacement varied from 4.26%, 15.77%, and 37.89% lower than the control concrete which implies the rate of shrinkage decreases with increase in CBA content [23]. The drying shrinkage of both series A and B (fixed W/C and fixed slump respectively) was experimented by Y. Bai et al. and found, for A series the CBA concrete showed lower shrinkage than control for all ages and shrinkage decreased with increase in CBA sand, for B series the shrinkage increased with increase in CBA sand beyond 30% replacement [21].

Table 7. Shrinkage measurement of concrete with different percentage of CBA

Percentage of CBA replacement	Effect of drying shrinkage	Reference
20% - 100% as sand	Lower shrinkage strain of CBA concrete than control concrete as the excess water is absorbed by the porous CBA.	M. Singh [27]
20% - 100% as sand	CBA has lower shrinkage than control concrete.	M. Rafieizonooz [23]
30% - 100% as sand	For fixed W/C, shrinkage decreased with increased in CBA sand.	Y. Bai [21]

8. CONCLUSION

CBA can be used as a sand replacement in concrete without compromising the strength and durability properties. The raw CBA cannot be used as a cement replacement due to its lower pozzolanicreactivity. The pozzolanic reactivity can be improved by pulverization or grinding in a ball mill. Thus, ground CBA has good pozzolanic property and can be used as a cement replacement with improved strength and durability properties. The performance of the CBA concrete or mortar escalate with lengthening the grinding time. The longer grinding time is not economical as the cost of operating the grinding mill makes the CBA concrete expensive. The ideal grinding time of the CBA to attain economical and durable CBA concrete from the above researches range from 3hrs to 30hrs. The optimum replacement level of CBA as a cement without significant loss in strength and durability is 30% of cement. The early strength might be comparable or lower than control concrete but it increases with ageing. Similar results were observed for CBA as a sand. The durability properties like drying shrinkage, chloride and sulphate attack were discussed but no results were stated regarding carbonization and corrosion. Likewise, strength, durability properties also improve with increase in CBA content up to 30% replacement levels. Several fields related to the research like cost analysis, corrosion, bond strength, carbonization of CBA concrete remains unexplored hence researches need to focus on the above durability properties to fill the gaps in the CBA concrete research.

ACKNOWLEDGEMENT

I would like to acknowledge GUJCOST funding agency for providing the funding forresearch on nanomaterials

REFERENCES

[1] M. Singh and R. Siddique, "Effect of coal bottom ash as partial replacement of sand on properties of concrete," *Resour. Conserv. Recycl.*, **72** 20–32, 2013.

[2] A. K. Mandal and O. P. Sinha, "Review on Current Research Status on Bottom Ash: An Indian Prospective," *J. Inst. Eng. Ser. A*, **95** 277–297

[3] B. Kim and M. Prezzi, "Compaction characteristics and corrosivity of Indiana class-F fly and bottom ash mixtures," *Constr. Build. Mater.*, **22** 694–702, 2008

[4] V. C. Pandey, "Fly ash application in reclamation of degraded land: opportunities and challenges," *Phytomanagement Fly Ash*, 167–193.

[5] R. Siddique, *Waste materials and by-products in concrete*. 2008.

[6] V. C. Pandey and N. Singh, "Impact of fly ash incorporation in soil systems," *Agric. Ecosyst. Environ.*, **136** 16–27 2010.

[7] A. Oner, S. Akyuz, and R. Yildiz, "An experimental study on strength development of concretecontaining fly ash and optimum usage of fly ash in concrete," 35. 1165–1171.

[8] B. Ash, "Management of Ash Disposal," 2010.

[9] M. Singh and R. Siddique, "Properties of concrete containing high volumes of coal bottom ash as fine aggregate," *J. Clean. Prod.*, **91** 269–278

[10] N. Singh, Shehnazdeep, and A. Bhardwaj, "Reviewing the role of coal bottom ash as an alternative of cement," *Constr. Build. Mater.*, **233** 117276.

[11] K. Muthusamy, M. H. Rasid, G. A. Jokhio, A. Mokhtar Albshir Budiea, M. W. Hussin, and J. Mirza,"Coal bottom ash as sand replacement in concrete: A review," *Constr. Build. Mater.*, **236** 117507.

[12] S. Gooi, A. A. Mousa, and D. Kong, "A critical review and gap analysis on the use of coal bottom ash asa substitute constituent in concrete," *J. Clean. Prod.*, **268** 121752.

[13] A. Abdulmatin, W. Tangchirapat, and C. Jaturapitakkul, "An investigation of bottom ash as a pozzolanic material," *Constr. Build. Mater.*, **186**, 155–162.

[14] P. Chindaprasirt, C. Jaturapitakkul, W. Chalee, and U. Rattanasak, "Comparative study on thecharacteristics of fly ash and bottom ash geopolymers," *Waste Manag.*, **29** 539–543.

[15] S. Oruji *et al.*, "Mitigation of ASR expansion in concrete using ultra-fine coal bottom ash," *Constr. Build. Mater.*, **202** 814–824.

[16] C. Jaturapitakkul and R. Cheerarot, "Development of Bottom Ash as Pozzolanic Material," *J. Mater. Civ. Eng.*, **15**. 48–53.

[17] S. A. Mangi, M. H. Wan Ibrahim, N. Jamaluddin, M. F. Arshad, and R. Putra Jaya, "Short-term effectsof sulphate and chloride on the concrete containing coal bottom ash as supplementary cementitious material," *Eng. Sci. Technol. an Int. J.*, **22** 515–522.

[18] M. Cheriaf, J. C. Rocha, and J. Péra, "Pozzolanic properties of pulverized coal combustion bottom ash," *Cem. Concr. Res.*, **29** 1387–1391.

[19] R. Siddique, "Compressive strength, water absorption, sorptivity, abrasion resistance and permeability of self-compacting concrete containing coal bottom ash," *Constr. Build. Mater.*, **47** 1444–1450,

[20] W. Wongkeo, P. Thongsanitgarn, K. Pimraksa, and A. Chaipanich, "Compressive strength, flexuralstrength and thermal conductivity of autoclaved concrete block made using bottom ash as cement replacement materials," *Mater. Des.*, **35** 434–439.

[21] Y. Bai, F. Darcy, and P. A. M. Basheer, "Strength and drying shrinkage properties of concretecontaining furnace bottom ash as fine aggregate," *Constr. Build. Mater.*, **19**,. 691–697.

[22] A. Sarkar, R. Rano, G. Udaybhanu, and A. K. Basu, "A comprehensive characterisation of fly ash from a

thermal power plant in Eastern India," *Fuel Process. Technol.*, **87** 259–277.
[23] M. Rafieizonooz, J. Mirza, M. R. Salim, M. W. Hussin, and E. Khankhaje, "Investigation of coal bottomash and fly ash in concrete as replacement for sand and cement," *Constr. Build. Mater.*, **116** 15–24.
[24] M. Singh, *Coal bottom ash*. Elsevier Ltd, 2018.
[25] N. Singh, M. Mithulraj, and S. Arya, "Influence of coal bottom ash as fine aggregates replacement on various properties of concretes: A review," *Resour. Conserv. Recycl.*, **138** 257–271.
[26] S. A. Mangi, M. H. Wan Ibrahim, N. Jamaluddin, M. F. Arshad, S. A. Memon, and S. Shahidan, "Effectsof grinding process on the properties of the coal bottom ash and cement paste," *J. Eng. Technol. Sci.*, **51** 1–13.
[27] M. Singh and R. Siddique, "Compressive strength, drying shrinkage and chemical resistance of concrete incorporating coal bottom ash as partial or total replacement of sand," *Constr. Build. Mater.*, **68** 39–48.
[28] V. Asokbunyarat, E. D. Van Hullebusch, P. N. L. Lens, and A. P. Annachhatre, "Coal bottom ash as sorbing material for Fe(II), Cu(II), Mn(II), and Zn(II) removal from aqueous solutions," *Water. Air. Soil Pollut.*, **226.**
[29] L. B. Andrade, J. C. Rocha, and M. Cheriaf, "Influence of coal bottom ash as fine aggregate on fresh properties of concrete," *Constr. Build. Mater.* **23** 609–614.
[30] S. A. Mangi *et al.*, "Influence of ground coal bottom ash on the properties of concrete," *Int. J. Sustain. Constr. Eng. Technol.*, **9** 26–34.
[31] M. Singh and R. Siddique, "Effect of coal bottom ash as partial replacement of sand on workability and strength properties of concrete," *J. Clean. Prod.*, **112** 620–630.
[32] S. Oruji, N. A. Brake, L. Nalluri, and R. K. Guduru, "Strength activity and microstructure of blendedultra-fine coal bottom ash-cement mortar," *Constr. Build. Mater.* **153** 317–326.
[33] V. D. Katare and M. V. Madurwar, "Design and investigation of sustainable pozzolanic material," *J.Clean. Prod.* **242** 118431.
[34] H. Kurama and M. Kaya, "Usage of coal combustion bottom ash in concrete mixture," *Constr. Build. Mater.***22** 1922–1928.
[35] M. Singh and R. Siddique, "Effect of coal bottom ash as partial replacement of sand on workability and strength properties of concrete," *J. Clean. Prod.* **112** 620–630.
[36] Ö. Özkan, I. Yüksel, and Ö. Muratoğlu, "Strength properties of concrete incorporating coal bottom ashand granulated blast furnace slag," *Waste Manag.* **27** 161–167.
[37] S. A. Mangi, M. H. Wan Ibrahim, N. Jamaluddin, M. F. Arshad, and P. J. Ramadhansyah, "Effects ofground coal bottom ash on the properties of concrete," *J. Eng. Sci. Technol.* **14** 338–350.
[38] M. S. Shetty, *Concrete Technology: Theory and Practice*, **55**.
[39] T. R. Naik, R. Kumar, B. W. Ramme, and F. Canpolat, "Development of high-strength, economical self-consolidating concrete," *Constr. Build. Mater.* **30** 463–469.
[40] P. Aggarwal, Y. Aggarwal, and S. M. Gupta, "Asian Journal of Civil," **32** 1–12.
[41] C. Argiz, A. Moragues, and E. Menéndez, "Use of ground coal bottom ash as cement constituent in concretes exposed to chloride environments," *J. Clean. Prod.*, **170** 25–33.
[42] V. Sata, A. Sathonsaowaphak, and P. Chindaprasirt, "Resistance of lignite bottom ash geopolymermortar to sulfate and sulfuric acid attack," *Cem. Concr. Compos.* **34** 700–708.
[43] Power Sector at a Glance ALL INDIA | Government of India | Ministry of Power (powermin.gov.in)

Effect of outrigger-belt truss system on the storey drift and maximum displacement of high-rise building

G Venkatraman, V Vanathi, S Veeraraghavan and K Sornamugi*

Department of Civil Engineering, PSG College of Technology, Coimbatore, India

***E-mail:** 20mc06@psgtech.ac.in (K Sornamugi)

Abstract: High-rise structures are becoming more common around the world, posing new challenges that necessitate ongoing technological foresight. When a structure's height rises, it's anticipated to have a lateral load-resisting system other than shear walls in place to negate the effect of lateral loads. However, advances in structural design and engineering have enabled building systems to withstand lateral forces such as wind and seismic stresses. There are numerous structural solutions tailored to resist lateral forces in the structure without adding to the plan area of the building to stabilize it against lateral load. The Outrigger-belt truss system is a common structural solution for successfully managing excessive drift caused by the lateral load. Hence, in this project, an attempt is made to investigate the characteristics of an Outrigger-Belt truss system. The seismic analysis is limited to the linear static method. A comparative analysis is done using STAAD Pro V8i SS6. Finally, the storey drift and maximum displacements obtained are calculated and the results are compared with the moment-resisting frame and the values are tabulated.

Keywords: Linear Static analysis, Lateral Loads.

1. Introduction

In the modern era, high-rise structures are conspicuous due to the factors such as population density, lack of available land, economic growth, etc. especially the major cities have provoked the structural designers to move towards a vertical structural system [4]. A high-rise structure, also known as an MDU (Multi-Dwelling Unit), is a tall continuous dwelling structure with multiple floors. This type of structure is considered to possess a serious impact on evacuation [1]. High-rise buildings were introduced in practice after the invention of elevators, and they became less expensive due to the invention of more abundant building materials. It is generally used as a residential building, official building, or for miscellaneous purposes such as retail, hotel, or multi-purposes combined [14].

The structural system of tall buildings for all the types namely, steel buildings, reinforced concrete buildings, or composite structures may be classified as,

1. Braced frame and shear walled system
2. Outrigger structural system
3. In-filled frame structural system
4. Tube structural system
5. Rigid frame systems

An Outrigger-Belt truss system shown in figure 1 is one of the lateral load resisting systems in the tall buildings, where the concept of stiffening is accustomed. The major objective of this type of structure is to reduce the hampered space compared to the conventional building. The outrigger system is one of the major types of structural system where the structure acts like a cantilever-shaped horizontal member connected to the structure's inner core and outer columns. It is an efficient system to provide resistance against lateral loads with or without columns on the internal phase of the structure [15].

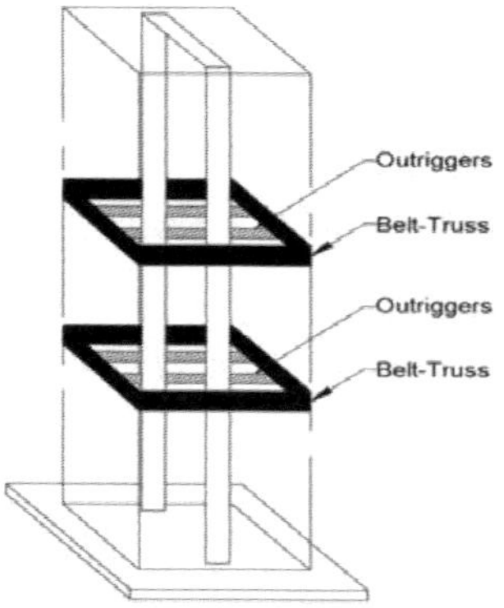

Figure 1. Outrigger-Belt truss system.

The outrigger and belt truss system comprises an internal core linked to the exterior periphery of columns using relatively rigid horizontal members known as outriggers. Outriggers can either be of concrete beams or trusses [12]. The core can either be in the centre, or the centre may be located on the sides of the building with the outriggers extending to the building columns from the other side. The core in the structure acts as a cantilever and the outriggers are provided to reduce the overturning moment in the core and to transfer the moment from the core of the building to the outer periphery of the columns by connecting both core and the columns [2]. The exterior periphery of the column consists of dense and closely placed columns which are connected by deep spandrel beams through moment connections. The columns on the exterior periphery are tied together using trusses or bracings named as Belt truss [11].

Initially, when the structure is subjected to lateral loads, the column on the exterior framing tends to pose like a strong structural wall which in turn provides increased resistance against the lateral forces [8]. The external moment is expected to be resisted by the combined action of both bending of the core and the axial tension and compression of the exterior periphery of the columns connected to the core using stiff outriggers [13]. As a result of this effect the ability of the structure to resist bending is increased when the core acts as a vertical cantilever development of tension in the windward columns and compression in the leeward columns is observed [6]. The belt truss and the outrigger combined to increase the stiffness of the structure and alleviate the rotation of the core, storey drift and storey displacement. Since a considerable number of forces are resisted by the exterior framing, the interior framing takes care of gravity loads thus acting like a simply framed structure. Generally, it is not mandatory to provide columns in the interior part of the structure because gravity loading is considerably low and can be taken care of by the core or shear wall of the structure. Besides its structural integrity, an outrigger system provides the interior floor plan without core bracing and heavy columns, enhancing the usage of the net floor area [3].

The basic concept of this system is very fundamental and simple. The ships of the past used outriggers to resist the lateral forces during their sails. Like the ship shown in figure 2, the core in high rise building can be related to the mast of the ship and the outrigger acting like the

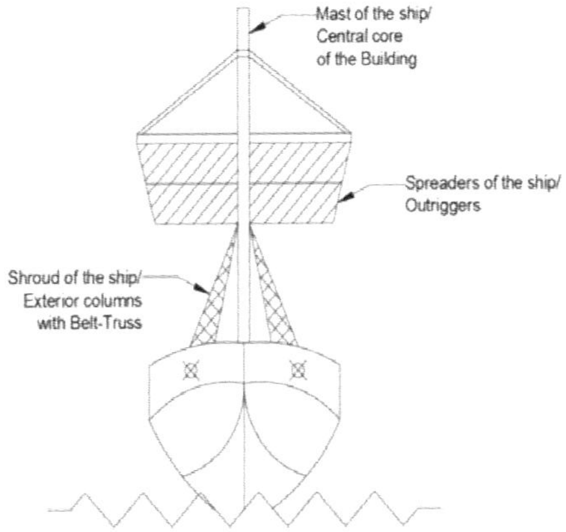

Figure 2. Concept of the outrigger-Belt truss system.

spreaders where the exterior columns perform the role of the shroud in the ship [9]. From its conception the Outrigger systems may be classified into two types:

1. Conventional Outrigger system
2. Virtual Outrigger system

In the "conventional" outrigger concept, the outriggers are connected directly to shear walls using deep beams or trusses at the core and to columns. In most cases, the columns are at the outer edges of the building [7].

In the "virtual" outrigger, the same transfer of overturning moments from the core to structural elements is achieved, but in this case, there is no direct connection between the outrigger trusses and the core. The basic idea behind the virtual outrigger concept is to use floor diaphragms, which are very stiff and strong in their plane [10].

In the present study, a conventional outrigger system is used. The outriggers provided were half the storey height, and length and breadth were like that of the particular storey. The structure was divided into three portions concerning their height and the outriggers and belt trusses were provided at the end of each portion.

2. Building and Material Configuration

In the present study, the number of stories has been taken as 30 and the height of the storey has been kept at 3.96m, thus making the total height of the building 118.8m and as specified in table 1. The exterior dimensions are kept unchanged. The materials considered are concrete of grade M-30, rebar of yield strength Fe 500, and structural steel of yield strength Fe-250 as mentioned in table 3. The specification and the size of the beams, columns, slab, and core wall are mentioned in table 2.

Table 1. Specifications of the Building.

Built-up area	200ft x 120ft
No. of Stories	30
Storey to Storey Height	3.96m
Total Height	118.8 m
Location	Coimbatore
Type of Building	Commercial

The Plan view of the Conventional Moment resisting frame and the Outrigger-Belt truss system are shown in Figures 3 & 4. The conventional building plan is a conventional grid type structure with ten bays along X-Direction and six bays along Z-Direction each of distance of 6.096m as shown in figure 3, whereas in the outrigger concept the exterior column to column spacing is the same with a central core of dimension 24.384m x 12.192m as shown in figure 4.

Table 2. Specifications of Beam, Column, and Slab.

Dimension of Beam(0-10 Floors)	0.51 x 0.84m
Dimension of Beam (10-20floors)	0.41 x 0.84m
Dimension of Beam (20-30floors)	0.31 x 0.84m
Dimension of Column (0-10floors)	0.90 x 0.81m
Dimension of Column (10-20floors)	0.90 x 0.51m
Dimension of Column (20-30floors)	0.90 x 0.41m
Depth of Slab	0.15m
Thickness of Structural Wall	0.15m
Thickness of Brick Wall	0.23m

Table 3. Material Properties.

Density of Concrete	25 kN/m^3
Grade of Concrete	30
Grade of Steel	Fe-250
Grade of Rebar	Fe-415
Poisson's Ratio	0.15

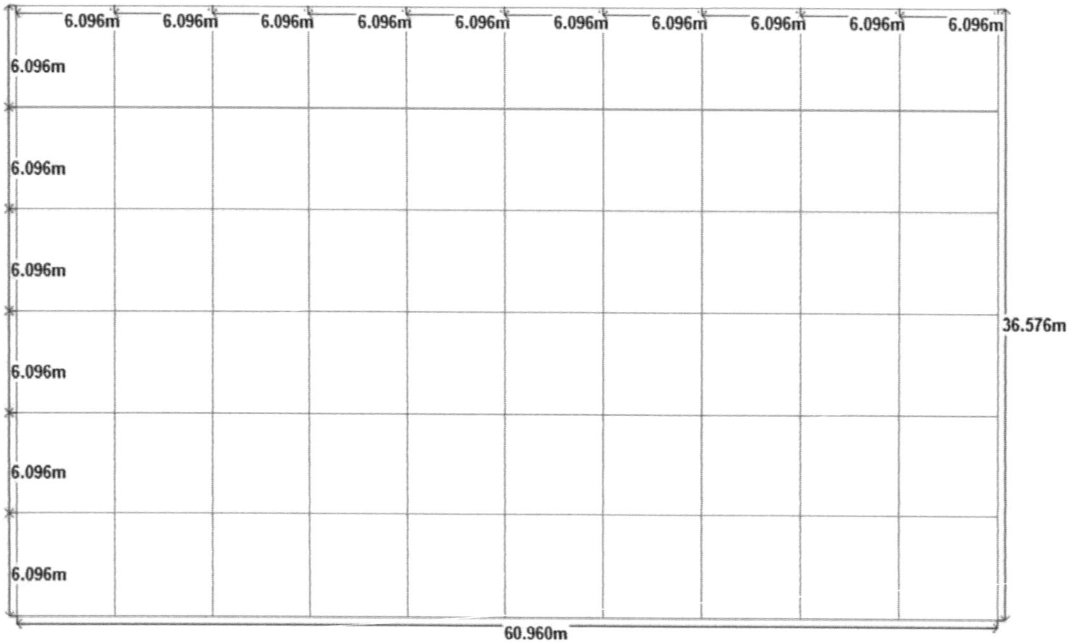

Figure 3. Plan of Moment Resisting Structure.

For better comparison of results, the plan dimension of each building is considered as 60.96m x 36.576m. The center-to-center column spacing is assumed to be 10 feet.

Inverted V bracing or Chevron bracing have been used as belt truss in the Outrigger belt truss system [5]. Chevron bracings have been considered because it resists the vertical unbalanced force in the beam.

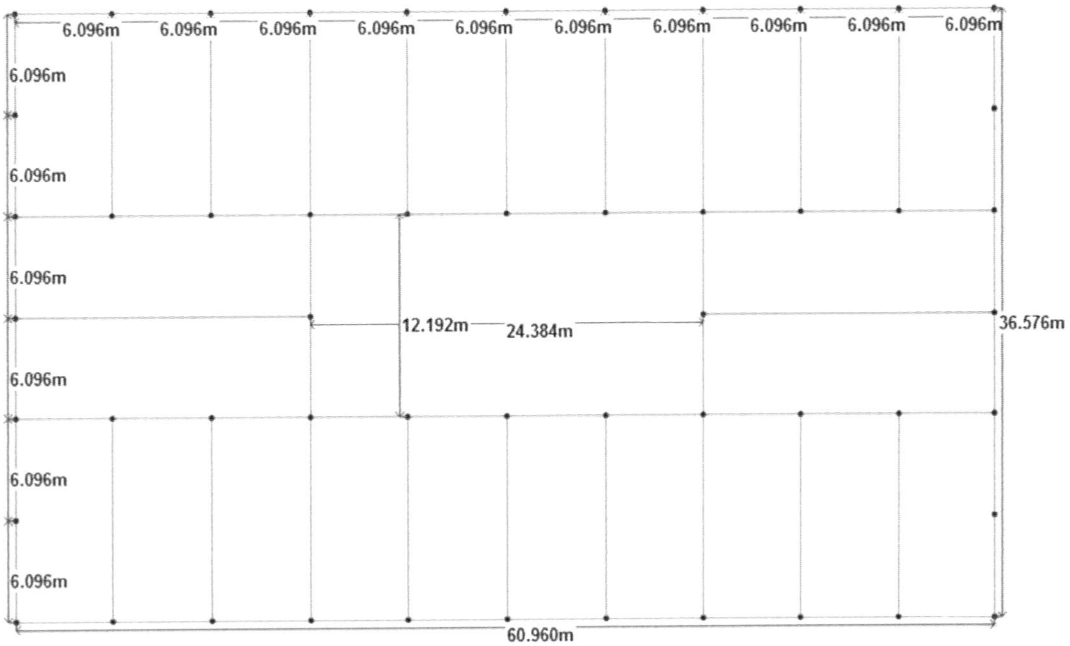

Figure 4. Plan of the Outrigger-Belt truss system.

Chevron bracings are provided at every $1/3^{rd}$ of the total building height. The structural steel used was of ISA: 200x150x15 and yield strength of 250 grades was used.

The structure is in Coimbatore; therefore, it comes under Zone III. The type of soil is assumed as medium soil. The approximate natural period of the structure in both axes is calculated as per IS 1893:2016, for these the period and response factors were calculated and applied. Loading conditions taken for analysis are given in table 4.

Table 4. Loading Conditions.

Total Wall Load	15 kN/m
Total Parapet Load	4.5 kN/m
Total Floor Load	5 kN/m^2
Terrace Load	9.25 kN/m^2
Live Load Floor Slab	4 kN/m^2
Basic Wind Speed	39 m/s

3. Analysis of the Structure

The structures were analysed using STAAD Pro V8i SS6, a powerful tool used for analyzing structures. The basic information regarding the structure was mentioned earlier. Concerning the data, the structure is analyzed and the average displacement & storey displacement for each storey is determined and calculated.

As per IS: 1893:2016, for structural analysis of RC structures the moment of inertia can be taken as 70% of I_{gross} for columns and 35% I_{gross} for beams.

The following assumptions are considered before the analysis of the structure,

1. The columns on the external periphery of the column are designed to resist lateral forces and storey shear acting due to the forces.
2. The column flexural strength is assumed to be less than the beam flexural strength.
3. The slabs are assumed to be simply supported and made of concrete as a structural material.
4. The floor slabs are rigid.
5. The effects of secondary structural elements and non-structural elements are considered very small in number and hence those values are negligible.
6. Deformations that are relatively small and which have a low impact on structure are neglected.

4. Results & Discussions

The analysis of the structure was carried out based on the data collected regarding the project and the assumptions mentioned above. The load applied was calculated and applied as explained earlier. The Load combinations were selected under the Indian standards. Seismic analysis of structures is done using Response-Spectrum analysis. The zone factors and soil type were identified using the data obtained from the location of the structure. From the analysed results post-processing is done and the values of displacements and storey drifts are obtained.

This study is limited only to these two factors because the rest of the considerations were neglected with respect to the assumptions made.

4.1 Storey Displacement

For a structure, it is important to ensure whether the structure maintains its integrity and it meets the functional requirements of the structure i.e., Serviceability limit state (SLS). The displacement check should be checked to determine the performance against the serviceability limit state [16].

If the structure fails due to displacement, then it needs to be checked. Displacements along X and Z directions are depicted in Figures 5 and 6 respectively for both structures.

4.2 Storey Drift

Storey drift is the lateral displacement of one level relative to the level above or below along X and Z directions are depicted in Figures 7 and 8. Storey drifts can be computed using the maximum inelastic response displacement and shall consider P-Delta effects. The calculated storey drift shall not exceed 0.025 times the storey height for structures having a natural period of fewer than 0.7 seconds. For a natural period, greater than 0.7 seconds the storey drift shall not exceed 0.020 times the storey height. The Structures have been modelled and analysed using Staad.Pro V8i. The average percentage variations in the structures have been discussed.

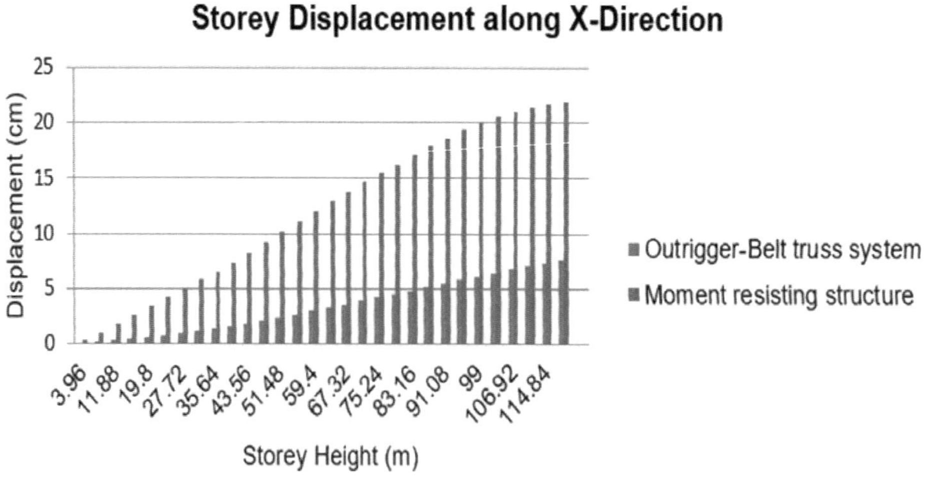

Figure 5. Storey Displacement along X-Direction

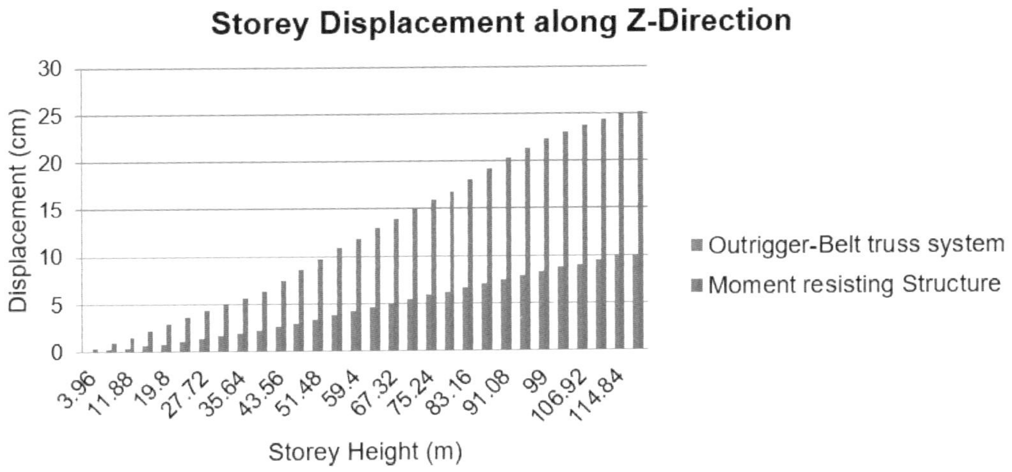

Figure 6. Storey displacement along Z-Direction

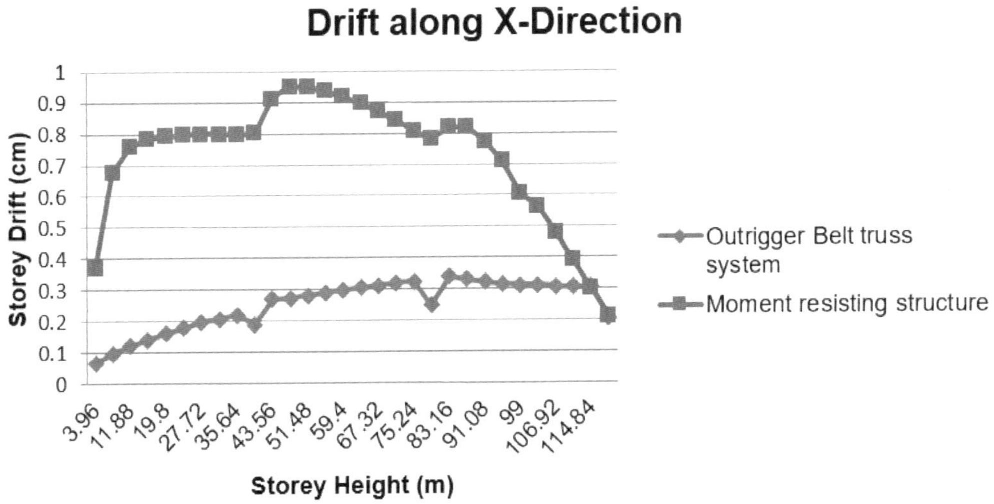

Figure 7. Storey Drift along X-direction.

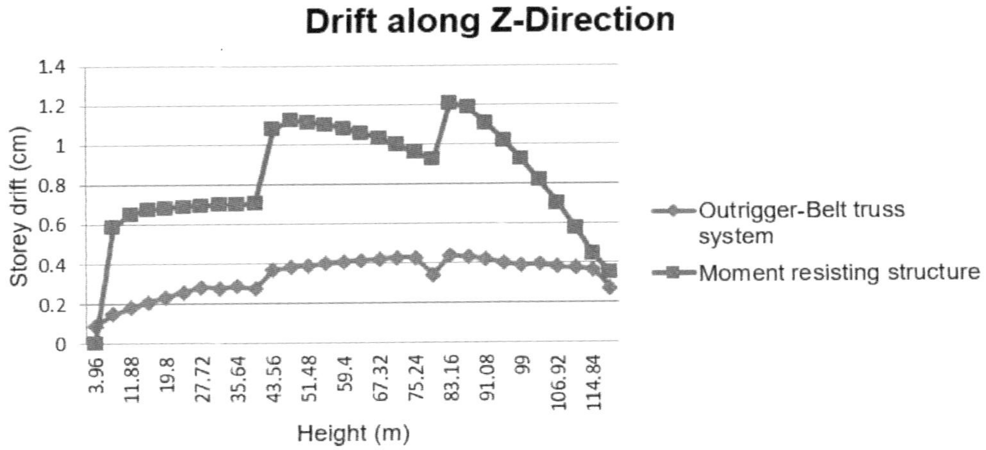

Figure 8. Storey Drift along Z-direction

5. Conclusion

The types of structural systems in tall buildings have been studied and discussed. Following the literature reviewed and references a suitable methodology and a draft has been prepared. For the project information and draft, the structure has been modelled. The structure has been analyzed with respect to the project information, collected data and codal provisions. The obtained results have been processed and the following points are observed,

- The dimensions of structural members are considered the same for both the structures, the effectiveness of Outriggers-Belt truss structures posed substantial resistance against the load conditions.
- The Outriggers belt truss system resisted the gravity loads without any interior columns, thus making the structure economical and cost-efficient.
- The Outrigger-Belt truss system strongly supports the earthquake design philosophy of Strong Column and Weak Beam which maximizes the efficiency of the structure.
- Without any increase in the plan dimension, the structural system resisted the lateral loads and gravity loads effectively when compared to Moment resisting frames.
- The Outrigger-Belt truss system shows reduced average displacement of about 24.81% and 34.74% along X and Z directions respectively as compared to the displacement of the Moment resisting frame for seismic zone III.
- The story drift for the Outrigger-Belt truss system is reduced by about 38.41% and 42.02% along X and Z directions respectively for zone III in comparison to the Moment resisting frame.

6. Acknowledgement

The authors wish to record their sincere and heartfelt thanks to the Principal in Charge, Head of the Department, Former Head of the Department, Department of Civil Engineering, PSG College of Technology, Coimbatore.

References

[1] Salman K, Kim D, Maher A and Latif A 2020 *Journal of Structural Integrity and Maintenance* 2470–5314.

[2] Hussini M and Nasier S 2020 *International Journal of Engineering Research & Technology (IJERT)* **9** 2278–0181.

[3] Prof N G and Gore 2018 *International Journal of Engineering Trends and Technology* **64.**

[4] Kamgar R and Rahgozar R *International Journal of Steel Structures* 2017–2017.

[5] Sundar R S and Gore N 2017 *International Journal of Engineering Research & Technology (IJERT)* **6** 2278–0181.

[6] Akshay A, Khanorkar M S V and Denge 2016 *IJSTE - International Journal of Science Technology & Engineering* **2** 2349–784.

[7] 2016 *International Journal of Engineering Research & Technology (IJERT)* **5** 2278–0181.

[8] Mulla A K and B S 2015 *International Journal of Engineering Research & Technology (IJERT)* **4** 2278–0181.

[9] R V K G M, Manohar B C and Study 2015 *International Journal of Engineering Research & Technology (IJERT)* **4** 2278–0181.

[10] Badami M R S and Suresh 2014 *International Journal of Engineering Research & Technology (IJER)* **3** 2278–0181.

[11] Nanduri P M B R K, Suresh B and Hussain M I 2013 *American Journal of Engineering Research* 76–89.

[12] Kamgar R and Saadatpour M M 2011 *Applied Mathematical Modelling* **36** 4918–4930.

[13] Liu N X, Zhao X, Sun H H, Zheng Y M and Ding J M 2011 *The Twelfth East Asia-Pacific Conference on Structural Engineering and Construction* **14** 2503–2510.

[14] Jie Z, Zhong-Xian Z, Wen-Guang Z, Hong-Ping Z and Chun-Sheng Z 2007.

[15] Kian F P S and Siahaan T 2001 *Dimensi Teknik Sipil* **3.**

[16] Sharma P and Singh G 2018 *International Journal of Civil Engineering and Technology (IJCIET)* **9** 976–6308.

Seashells to lime: swotting of a vernacular method and needs for revival

Haritha B Nair[1] **and Binumol Tom**[1]

[1]Department of Architecture, Rajiv Gandhi Institute of Technology, Kottayam, Kerala.
E-mail: harithabnair24@gmail.com

Abstract. Lime is the versatile mineral. Various forms of lime are used in environmental, metallurgical, construction, chemical/industrial applications, and more. It has been used for masonry mortars and plastering for centuries. Lime is usually obtained from lime stone. Places like Kerala where lime stone is scarce, lime is majorly obtained from seashells so they developed their own traditional method to extract lime. Shell lime was one of the major components in construction of most of the houses, temples, bridges, dams etc. in Kerala during the earlier days. People used shell lime not only as a replacement to cement but also as a plastering agent which helps to reduce temperature in the interiors. This dissertation is retrospection into the vernacular method of making lime mortar from seashells by visiting the workspace. The study also intends to analyse the challenges faced by this method and provide with necessary recommendations.

Key words – Challenges, Seashells, Recommendations, Labour

1. Introduction

Our set of experiences makes an extraordinary effect on moulding our future. The old innovations embraced by our own special progenitors are very notable. During the early times, there was a biological equilibrium kept up with among the human and common habitat. They believed in amalgamating the nature with the building to create a picturesque scenario so they did not harm the natural beauty of the environment. India has been perceived worldwide for its variation culture and its commitment to the equivalent. If each and every heritage structure are taken in account, a common factor that seems to be distinctive is its construction technique and structural stability which ensures itsexistence even till date in spite of witnessing calamities, manmade disasters and negligence.

While searching in our past we can get many such techniques which are developed specifically for our region and needs locally available materials. And the main advantage of using such technique is that, these methods are widely used and proven for thousands of years. And even today also evidences can be Seen and shown proudly.

No literature studies are available about the vernacular method which resulted in creating a big knowledge gap in this particular field. The main objective of this paper is

- To understand the process of making lime from seashell in the traditional way by visiting the workspace.

- To analyze the advantages of using lime mortar in buildings through live case studies.
- To establish the opportunity of its revival.
- To identify the knowledge gap and provide necessary information.

1.1. Aim
To study and document the vernacular method of making lime from seashells and analyze the opportunity of its revival.

1.2. Scope
Traditional knowledge and wisdom about this method is only known to the Craftsmen. It is of utmost importance to document and identify the potential of using this method. It is essential to make this technique popular among the common man.

1.3. Research questions
- What are the advantages of shell lime construction?
- Are seashells available in abundance at present?
- How did this method become obsolete?
- What are requirements for traditional extraction of lime from seashells?

2. Literature review

2.1. Paper 1: Role of Additives in Mortars: Historic Precedents Anjali Acharya, *Shubhjeet Archroy and Vasudha A Gokhale (2017)*
This research is an attempt to provide a review about history of lime as a building material. The study intends to explore the significance and sustainability of lime mortars and their use in architectural conservation and analyse the use of additives in historic buildings in India and abroad [7].

The Research Methodology employed in this study is Quantitative Research based on available literature resources.

The findings outline the role of additives in lime mortar, especially in Indian historic buildings. The study also reveals that the components of original lime mortar which remained in ancient sites is of fundamental significance particularly in terms of role of additives. Historical examples demonstrated in the journal depicts the effective use of additives which are taken from nature and are sustainable. These additives include jaggery, powdered shells, cow dung, fenugreek, lentils, yoghurt, and the pulp of the wood-apple or Bael, mustard oil, incense etc [7].

2.2. Paper 2: Traditional Lime Mortar and Plaster – Reconstruction with emphasis on durability Balksten Kristin (2007)
Through this study, the author wants to identify the main parameters affecting the durability of different types of lime mortar. And to explore the influence of craftsmanship on the material propertiesof lime mortar and also to study traditionally produced mortar and lime putty [9].

The Research Methodology employed in this study is qualitative in nature. They prepared lime mortar by themselves and explored the various stages.

The outcome of the study depicts that Lime as a binder receives different properties depending on the raw material chosen as well as on the burning, slaking and storing processes. The choice of raw material has an impact on the chemical composition (hydraulic/non-hydraulic) while burning, slaking and storing has an impact on the crystal structure and the workability of the lime [9].

2.3. Review on Herbs used as Admixture in Lime Mortar used in Ancient Structures Thirumalini P, And S. K. Sekar (2013)

The objective of this study is to identify the various herbs used in traditional construction and its role in modifying the fresh and hardened properties of lime mortar. The author also aims to retrieve the traditional concept of addition of admixture to concrete [8].

The Research Methodology employed in this study is Quantitative Research based on available literature resources.

Through the studies conducted on ancient structures it is evident that organic proteins and carbohydrates are used in concrete to enhance the fresh and hardened properties of concrete. The sticky rice paste, potato starch, black gram and other pulses, rubber milk (latex) viscous liquidobtained from elm shavings in water, pulses, molasses, boiled stems and leaves of banana plants, oils, egg whites, cashew nut shell, liquid resin, gluey fluid from cactus plants and powdered shells were some of the natural materials used in the traditional construction. The buildings in ancient times are constructed not only to withstand the external forces but also wellbeing of inhabitants. The ingredients used in plaster and mortars makes the environment healthy because the building itself breathe like an organism [8].

2.4. Summary

Studies prove that Lime is one of the most important building materials used since ancient times. Old mortar remaining on historical buildings is an important source of knowledge that can be used in the reconstruction of similar mortar using the materials available today. Today lime mortar is mainly used as a repair mortar for historic buildings. Most of the limestone quarries in existence produce stone only for aggregate use. The uses of lime and its durability have been proven by centuries of use and this ancient technology has shown it can withstand earthquakes, floods, and environmental changes. Limitation with the literature studies is None of the studies show how lime was prepared traditionally in the past and what is the ratio of adding the materials, this manifests the demand of doing a paper explaining the traditional method of making lime.

3. Primary study

3.1. Case studies

A total of 4 live case studies was done to understand the advantages of using shell lime in construction. Vadakumnathan temple, varikassery mana, tirurmunsif court, kadavathuparambil residence are visited. The munsiff court as shown in figure 1 which I visited was about 110 years old and was constructed using shell lime instead of cement and the walls were plastered using lime. None of the walls had felt any weakness till date. We can experience the temperature difference inside the building even in our first visit. Likewise, all the buildings I visited have never experienced any structural issues due to the usage of lime. The externals walls of Varikassery mana as shown in figure2 and Kadavathuparambil residence as shown in figure 3 clearly shows the perfect finish obtained from lime mortar.

Figure 1. Munsiff court.

- Age of the building – 110 years
- Constructed using Shell Lime mortar instead of cement
- Properly functioning, no walls are felt unstable.
- Walls are plastered using lime, still have a smooth finish

Figure 2. Varikassery mana.

- Age of the building – 100+ years
- Constructed using Shell Lime mortar instead of cement
- Still have no structural issues

Figure 3. Kadavathuparambil Residence.

- Age of the building – 75 years
- Constructed using Shell Lime mortar instead of cement.
- Properly functioning

3.2. Documentation

Step 1 - A portion of workspace floor is cleaned to mix wood coal and seashells. Wood coal is spread as shown in figure 4(a) at first and water is sprinkled above it (big pieces of wood coal are kept aside which is later used in creating fire in round furnace), then seashell is laid above it as shown in figure 4(b). And again, wood coal is laid on top as shown in figure 4(c).

(a) (b) (c)

Figure 4. (a) Spreading wood coal at floor, (b) Seashells laid above it, (c) Putting wood coal at top.

Wood coal and seashells are mixed in a ratio 1:2, here 7 basket of wood coal are used to mix 14 baskets of seashells.

Step 2 – Mixing of wood coal and seashells are done using a wooden spade as shown in figure 5. The materials are mixed properly and are kept for some time

Figure 5. Mixing of seashells and wood coal.

Step 3 – As shown in figure 6(a) and (b) generate fire using coconut husk and carry it to the furnace as shown in figure 6(c), which already have big pieces of wood coal that helps to uplift the fire. Oncethe fire is spread, we will add one basket of mix (seashells and wood coal) as shown in figure 6(d) in to it

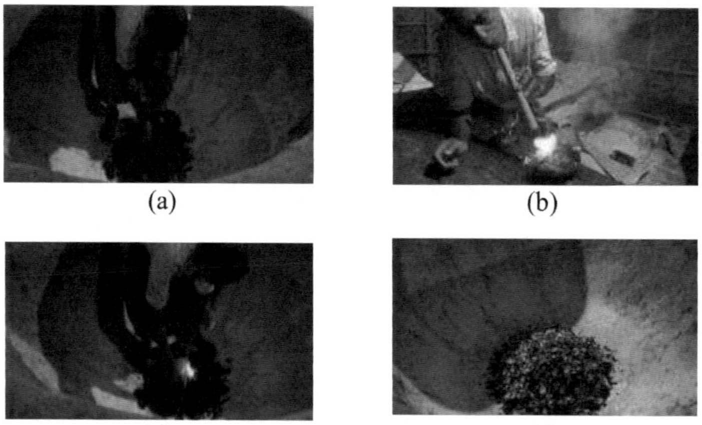

(a) (b)

(c) (d)

Figure 6. (a) Creating fire using coconut husk, (b) Wood coal laid in round furnace, (c) Creating fireinside the round furnace, (d) One basket of seashells in furnace.

Step 4 – Switch on the motor as shown in figure 7(a) to provide sufficient oxygen for the combustion. At that point itself white smoke will come out as shown in figure 7(b) and (c) from the furnace and seashells will start burning. Earlier days when there was no electricity this process was done by rotating the spinning wheel manually.

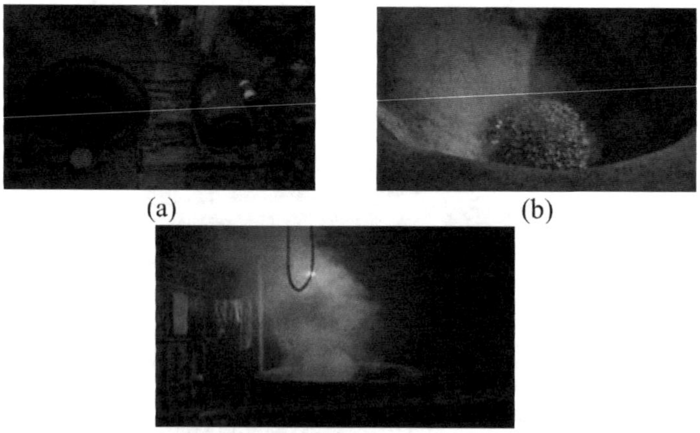

(a) (b)

(c)

Figure 7. (a) Motor, (b) Seashells getting burned, (c) White smoke coming out from furnace.

Step 5 - The mixture is added from four sides as shown in figure 8(a) and (b) at frequent intervals of 15-20 mins to the furnace. This process is repeated until the mixture is completed. White fumes will start coming as shown in figure 8(c). It will take almost 2.30 hours to complete the entire process.

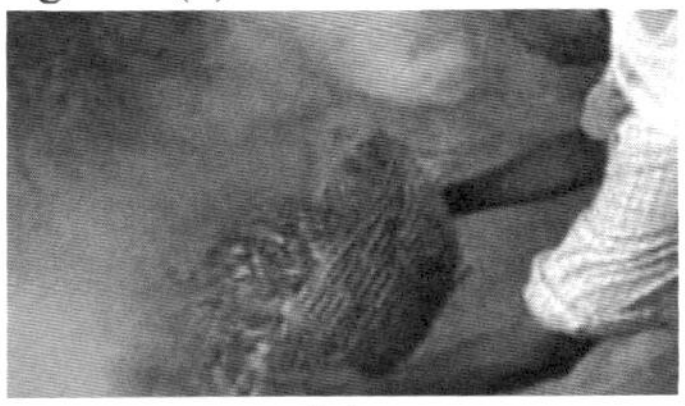

(a)

(b)

(c)

Figure 8. (a) Adding mixture to furnace, (b) Adding mixture to furnace, (c) Burning of mixture.

Step 6 – Once the entire mixture is added to the furnace, the next step is to lay jute over the mixture to accelerate the rate of burning as seen in figure 9(b). The smoke emission will increase during that time as shown in figure 9(a).

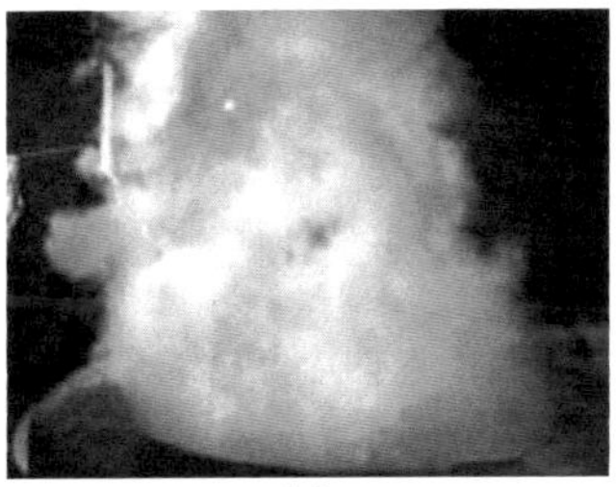

(a)

(b)

Figure 9. (a) Heavy smoke coming out from furnace, (b) Fire generated after jute is laid.

The process of burning this mixture is usually carried out at night. The mixture is kept burning for 4-5 hours, post which the mixture is left as is overnight. The Calcination process is now completed.

Steps 7 – In the morning, when the mixture (Quick lime) is cooled as seen in figure 10(a) and (b) it is packed and weighed as shown in figure 10(c) and sent for various purposes.

(a)

(b)

(c)

Figure 10. (a) Mixture kept in the furnace, (b) Mixture, (c) packed and weighed.

Step 8 – Water is added to quick lime as shown in figure 11(a) to produce lime powder as seen in figure 11(f). The process of changing from shells to powder can be seen in figures 11(b),(c),(d) and (e).

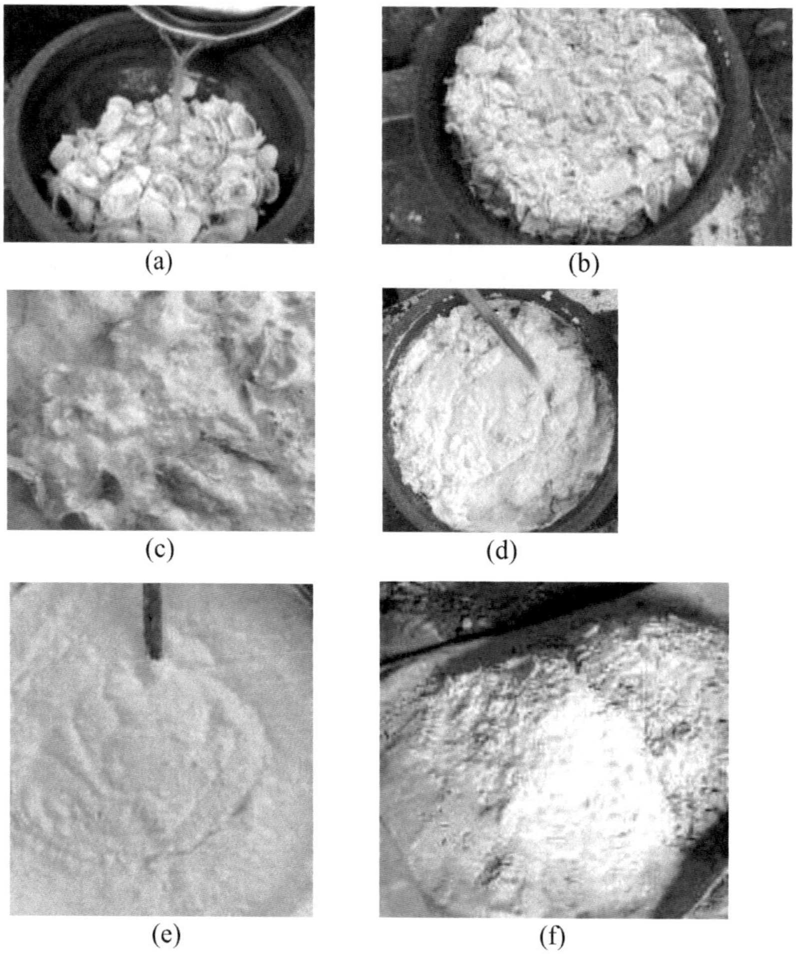

Figure 11. (a) Adding water to quick lime, (b) Quick lime changing to lime powder, (c) Quick lime changing to lime powder, (d) Quick lime changing to lime powder, (e) Quick lime changing to lime powder, (f) lime powder.

Step 9 – Adding jaggery, fat of snake head fish, gluey fluid from cactus plant to this mixture, we will get lime mortar suitable for construction [7].

3.3. Interview

Figure 12. Craftsman Devan near his workspace

Devan, a 63-year-old man as seen in figure 12 who is pursuing this profession for past 45 years clearly remembers the older times when people used to construct their own houses using locally available materials, without causing harm to the nature and even exploring their own methods in construction.

Due to the arrival of cement and other materials which are easily available and can manufacture quickly, people have forgotten about this traditional technique which require skilled labourers. "I am pretty sure that the new generation have no idea that a method like this exists and that the dams, bridges etc. which they still use were made by these materials" says Devan.

By interviewing the craftsman, I got to know how they practiced this method earlier and what all changes they came across. He also explained the benefits and drawbacks of this process.

He strongly recommended for the revival of this method which have several benefits including the inexistence of environmental impacts created by cement (which is used as an alternative for lime) which is one of the main producers of carbon dioxide - a potent greenhouse gas, lesser energy spent in the making process and inexistence of waste produced after production.

This interview assisted in contemplating the answers to the research questions proposed in this paper.

3.4. Finding the challenges faced by this vernacular method, analyzing and providing recommendation

3.4.1. Lack of skilled labourers

Recommendations:

- Training programs should be conducted to make people/workers understand every step of this method and encourage them to follow this.
- Implement improvisations in Labour oriented technology which in turn will benefit the betterment of the society who work in this field.
- If the Government can announce incentives for constructions using this method, as it is eco-friendly and promote environmental sustainability, which will attract more people to learn these skills.

3.4.2. Majority of the clients are unaware of this method and material.

Recommendations

- Educate and make the clients aware of these material and method.
- Create awareness about the benefits of this method.
- Marketing and publicity and awareness programs can increase the use of this product.
- Use of this method in public projects can increase the value of the product.
- If the curriculum in architecture schools in India would encourage its students to document vernacular ways of construction processes and usage of traditional building materials, it will help improve the visibility of this method among the new generations.

3.4.3. Lack of Marketing.

Recommendations:

- Tie-up with contractors and architects and encourage them to use this material in their projects, this will accelerate the reach of this method.
- Reduced tax rates for such products can increase the market value so that it can compete with cement in case of economic feasibility
- Advertise about the benefits of this method with the help of media influencers to create publicity.

3.4.4. Lack of modern inputs:

Recommendations:

- Make the workers aware of the changes happening in the field of architecture like the introduction of technology, equipment etc., so that they can experiment with modern methods, which are simple and ideal for modern world.

3.4.5. Unavailability of raw materials
Recommendations

- Tie up with suppliers who provide these raw materials currently, in order to widen the area of distribution.

3.4.6. Time consuming
Recommendations

- Analyze the areas in this process which can be improved with the help of modern technological advancements and implement them.

4. Conclusion

Lime is eco-friendly comparing to cement and when it is combined with sand and natural additives, we get a plaster and mortar every bit as satisfactory as cement mortar and often with many advantages. Yet, we have allowed this plentiful sustainable product and technique almost to disappear from the building world altogether. Very few people realize that the cost of establishing a cement factory is a hundred times that of starting a similar sized lime factory, which uses seashell as raw material. Furthermore, there is in fact no advantage in building very huge lime- producing factories, small inexpensive production units can be established at comparatively very little cost and spread over wide areas to simplify labour distribution and cut down transit and delivery costs. We have already forgotten that many of our big old irrigation dams, railway bridges etc. which still serve us efficiently, were built with this lime mortar and knew nothing about this new-fangled cement. By developing economic, simple, wide-spread lime production units we could solve many unemployment problems and produce fine, efficient, versatile building material which is a much greener option for society.

The opportunity to document a vernacular technique is extremely prestigious. This gives us a chance to preserve our ancestral skills and expertise before it is long-lost.

5. References

[1] Carran, Dorn, John Hughes, Alick Leslie, and Craig Kennedy 2012 A short history of the use of lime as a building material beyond Europe and North America *Int. J. Archit. Herit.* **6** 117-146
[2] Singh, M 1993 Analysis and characterization of Charminar lime plaster *Curr. Sci.* **64** 760-764
[3] Thirumalini Selvaraj, Ramadoss Ravi, S K Sekhar, and M Nambirajan 2015 Knowing from the past – Ingredients and technology of ancient mortar used in vadakumnathan temple, Trisssur, Kerala, India *J. Build. Eng.* **4** 102-112
[4] Boffey, Geoffery and Elizabeth Hirst 1999 The Use of Pozzolans in Lime Mortars *J. Archit. Conserv.* **5** 34-42
[5] Pavlik, V., and M. Uzakova. 2016 Effect of curing conditions on the properties of lime, lime-metakaolin and lime-zeolite mortars *Constr. Build. Mater* **102** 14-25
[6] Kennedy, Craig J., William A. Revie, Lore Troalen, Matthew Wade, and Tim J. Wess 2013 Studies of hair for use in lime plaster: Implications for conservation and new work *Polym. Degrad. Stab.* **98** 894– 898
[7] Anjali Acharya, Shubhjeet Archroy and Vasudha A Gokhale 2017 Role of Additives in Mortars: Historic Precedents *Int. j. eng. res. appl.* **7** 07-12
[8] Thirumalini, P., and S. K. Sekar 2013 Review on herbs used as admixture in lime mortar used in ancient structures *Indian J. Appl. Res 3* **8** 295-298.
[9] Balksten Kristin 2007 Traditional Lime Mortar and Plaster: Reconstruction with emphasis on durability *Doctoral dissertation, Chalmers Tekniska Hogskola* **7** 7-12.

Modelling Approach to Develop a Multimodal Transportation Network

K. Aravinthan[1]*, Deepak. D[2], Chethan[3], Borudregowda[4]
[1234] Department of Civil Engineering, Rajarajeswari College of Engineering, Bengaluru, Karnataka
aravinthank444@gmail.com

Abstract. Multimodal transportation (MMT) refers to a system that utilizes various means of transportation to transfer people & goods from one place to another. Transit operations are measured as the maximum number of passengers that can be carried past a fixed route, in a given period. Investing in and using multiple modes of transportation does not only bring efficiencies but also goes a long way to reduce time & delay, as well as the overall cost of transportation. This paper presents the efficient mode of transportation system at selected routes in the Bengaluru city, the route stretch selected is 12.2 km in length and it has 8 major intersections. The type of pavement surface consists of both bituminous and Reinforced cement concrete (R.C.C). The complete study stretch is a 2-way 4-lane road. This study provides walkability-index, traffic volume studies in terms of passenger car unit (PCU), time and delay study, and Cost & time analysis. Four modes of transportation are used in this study to find the effective multimodal-transportation system namely, 2-wheeler, 4-wheeler, para-transit [15], and buses/metro. The outcome of this paper is to arrive at the shortest distance for reducing delay & time, and cost-effective travel using multi-mode transport. While focusing on the present condition of the para-transit transport facilities. The time, delay & cost during MMT and to provide a safe, cost-effective & convenient way to travel or transport. The result shows the reduction in time, cost & delay of transportation by using the MMT model. This MMT system could be adopted as an efficient mode of transportation on this route. Similar rural-urban junction routes around the world could use this approach for effective transit by making a modal study.

Keywords: Multi-modal transportation system, walkability index, time & delay study, Cost & time analysis, para-transit.

1. Introduction

Rural and urban liveability [5] can be improved by transportation settings and commuters strongly agree that public transport improvement is necessary for wellbeing. The Multimodal transportation system (MMTS) comprises one trip that involves two or more two different modes of transportation [13] like bus, metro, car, tram, etc. Whether it might be government or privately operated, the passengers have to transfer into another mode. Even imports and export of any county can be improved by making unimodal a multimodal transportation network [19]. The potential benefits of transit, social benefits, and economic impact of MMTS can be measured [2] in the required region. This technique can be used for regional public, and similar regional public areas [1]. It also promotes the road transportation department for decision making and planning the investment [14,17]. The network congestion on roads can be minimized and helps in providing alternative fixed, frequent, and scheduled public transportation options [11]. The independence of the personnel vehicle as a major mode of road transportation reduces traffic [3] and indirectly promotes public transportation in urban areas. A good transportation system facilitates passenger interface, provides flexibility in the use of alternative mode choice, minimizes transportation costs, CO_2 emission, and time and is cost-effective [8]. Numerous countries (Denmark, Germany, Netherland, etc,) gauge their public transport system by various assessment methods [6,16]. In Paris and suburban routes advanced traveller information systems [7] for a public transit network and a multi-criteria routing algorithm are used for a transit program to end in a reasonable travel planning program. The congestion on roads can be minimized by making the V/C ratio nearly one. Bangalore is a highly-populated area and there is no space for expansion of roads, hence increasing the capacity of roads is the costliest solution. The volume of traffic should be decreased by encouraging fast, fixed, scheduled and economical public transport which includes buses (BMTC), metro rail, etc. Recent studies [4] show four transportation modes such as automobile, bus, rail, roads, and a mixed model for developing transportation networks. Further, the time-table should be designed in such a way that the time of arrival of the bus at a certain area should be at least 5- 10 minutes from the time of departure of a metro of the same area so that there is no wastage of time and thus the owners of private vehicles will be encouraged to use the public transport facilities.

There is still space for filling the gap in research made. It might be the models used [4] that cannot be directly transferred to other models due to the modal choice. Specific benefits in monetary terms for rural areas are considered [2] and the small urban areas were not covered. An overall transportation survey was only carried out [5] and the multimodal transportation system is not being considered. Tools that are used to gauge [6] the transport assessment do not describe positive and negative aspects. Some other modes of transportation [7] are not considered like bike and car-sharing. The area chosen for this study would fill up the research gaps and suit well for the targeted modal area.

2. Methodology

The study area begins from Ramohalli Cross (Bengaluru South) to Mysore Road (Bengaluru North). The total distance of the selected stretch is 12.2 km as shown in figure 1. This stretch is further divided into two zones. Zone 1 is the rural part of Bangalore and zone 2 is the urban part of Bangalore.
Zone 1 – Ramohalli Cross to Kengeri Bus Terminal
Zone 2 – Kengeri Bus Terminal to Mysore Road
The reason for making it two zones is this road is a major highway connecting the city and the outer road. This transition part of the route has a considerable high traffic volume. A substantial amount of people working in Bangalore are travelling from the rural parts and they travel to the city for major activities. The problem in the selected area is carefully identified such that a perfect solution for the problem is obtained. To quantify the traffic, suitable traffic survey data is obtained such as Origin – destination survey [9], walkability-index, public transport accessibility, traffic volume study, time & delay study, and Cost & time analysis. The data which is collected is been properly charted and analyzed. Analysis of data is of prime importance as it is very essential to generate a solution. A most feasible solution could be generated by analyzing the collected data. Before implementation of the solution, it has to be evaluated for maximum development and complete abolishment of the problem.

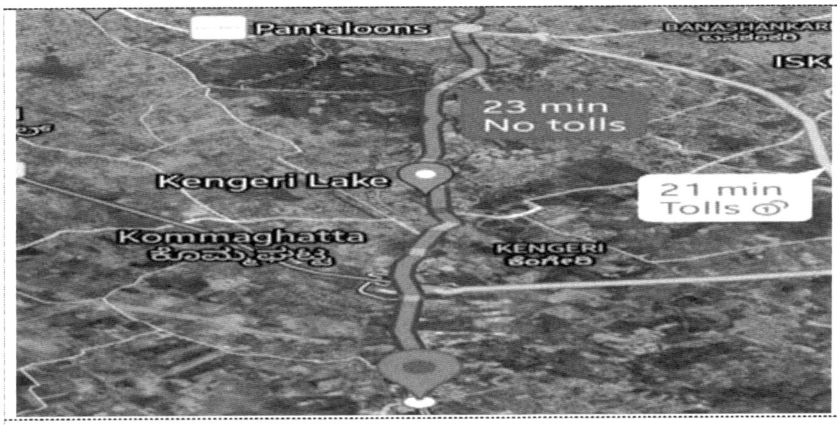

Figure 1. Origin & destination route map.

3. Walkability Index

3.1. Walkability Index

An aggregate rating of 0 to 1 has been formulated to rate the walkability index which aggregates the footpath availability and visual observation rating. Visual observations were conducted for every bus stop at every 250 m of stretch for each zone. Visual observations were conducted for every bus stop of the stretch for each zone. The odometer is used to identify the length of the footpath. The obtained above rating is averaged for both the zones.

Table 1. Walkability index parameters in zone - 1.

Bus stop junctions	Ramohalli Cross - Anchepalya	Anchepalya – R.R Medical College	R.R Medical –Check-post	Checkpost – Kengeri	Kengeri – KBT
Total Length in m	1000	700	800	2000	1100
Footpath Length in m	0	0	0	1900	1050
Footpath Availability	0	0	0	0.86	0.95
Visual Observation Rating	0	0	0	0.75	0.75
Walkability Index	0	0	0	0.805	0.85
Total Length in m	1000	700	800	2000	1100

Table 2. Walkability index parameters in zone- 2.

Bus stop junctions	KBT – R.V College	R.V College – University Gate	University Gate – R.R Gate	R.R Gate – Nayandahalli (Mysore Road)
Total Length in m	2000	1900	700	1800
Footpath Length in m	1850	1800	650	1650
Footpath Availability	0.925	0.95	0.93	0.92
Visual Observation Rating	0.5	0.75	0.5	0.75
Walkability Index	0.713	0.85	0.715	0.835
Total Length in m	2000	1900	700	1800

The rating is done as mentioned below:
0 = There is no Footpath.
0.25 = Footpath condition is poor.
0.5 = Footpath Condition is fair.
0.75 = Footpath condition is good.
1 = Footpath condition is very good.

Footpath availability = Footpath Length in m / Total length in m (1)

Walkability Index = (Footpath Availability + Visual observation rating) / 2 (2)

Table 3. Cumulative Walkability index.

Zone	Footpath length (m)	Total length (m)	Footpath Availability	Visual Observation Rating	Walkability Index
1	2950	5800	0.51	0.3	0.405
2	5950	6400	0.93	0.625	0.775

3.2. Public Transport Accessibility (PTA)

Accessibility can be utilized to discover the purposes of land use and interdependency of various transport frameworks. Accessibility is the simplicity of the transportation framework. When accessibility is good, the journey time will be reduced as the walking time is reduced. Equation 3 provides the formulation of the PTA index. In table 4, the different bus stop under the selected route is provided with the minimum walking distance to the nearest bus stop and PTA index. Footpath length is measured by using the odometer.

Public Transport Accessibility Index = 1 / (Average distance in meters to the nearest transport systems like bus stop/metro station) x 1000 (3)

Table 4. Cumulative Public transport accessibility Index.

Bus stop	Identification	Walking distance to the bus stop (m)	PTA index
Ramohalli Cross	Stop 1	45.40	22.02
Anchepalya	Stop 2	25.78	38.76
R.R Medical College	Stop 3	52.36	19.08
Checkpost	Stop 4	161.18	06.20
Kengeri	Stop 5	62.24	16.06
Kengeri Bus Terminal	Stop 6	90.12	11.09
R.V College	Stop 7	60.03	16.65
University Gate	Stop 8	95.15	10.50
R.R Gate	Stop 9	101.83	09.82
Nayandahalli	Stop 10	35.33	28.30
Mysore Road	Stop 11	61.32	16.31

3.3. Traffic Volume Study

Traffic volume data provides the number of vehicles in various modes passing through the study area region (i.e., from Ramohalli Cross to Mysore Road). In traffic-engineering studies, there are many volumes such as daily volume, hourly volume, and peak hour volume [10]. In addition, volumes of a day or an hour can vary greatly, depending on the different day of the week or different periods of a day. It also determines the traffic volume number, movements, and classifications of roadway vehicles at a given location. The table-5 provides the sample of traffic volume data. The traffic volume data consists of two-wheeler, three-wheeler (Autorickshaw), four-wheeler (Car), mini-bus, Bus, different types of trucks, tractors, bicycles, and other modes of transportation. The individual category of the vehicle is counted manually and the traffic volume is tabulated at every 30minutes interval. Both individual counts of the vehicle and the PCU (Passenger car unit) are being found out for further understanding. It is observed that more than 75% of traffic volume is due to two-wheeler and four-wheelers. The three-wheeler, Bus, and two-axel truck volume is around 4% each.

IOP Conf. Series: Materials Science and Engineering 1258 (2022) 012067 doi:10.1088/1757-899X/1258/1/012067

Table 5. Sample of Traffic Volume Data

Time	2W	PCU 0.75	3W	PCU 1.2	4W	PCU 1	Mini Bus	PCU 2.2	Bus	PCU 2.2	LCV	PCU 1.4	2-axel	PCU 2.2	Multi-axel Truck	PCU 2.2	Semi-axel Truck	PCU 2.2	Tractor	PCU 4	Cycle	PCU 0.4	Others	PCU 4	Total Vehicles	Total PCU
7.00-7.30	1073	804.8	153	183.6	847	847	72	158.4	148	325.6	71	99.4	152	334.4	148	325.6	28	61.6	0	0.0	0	0.0	0	0.0	2692.0	3140.4
7.30-8.00	1204	903.0	144	172.8	808	808	69	151.8	203	446.6	56	78.4	172	378.4	172	378.4	25	55.0	0	0.0	0	0.0	2	8.0	2855.0	3380.4
8.00-8.30	1612	1209.0	176	211.2	1976	1976	116	255.2	168	369.6	88	123.2	236	519.2	196	431.2	36	79.2	2	8.0	3	1.2	1	4.0	4610.0	5187.0
8.30-9.00	2024	1518.0	184	220.8	1608	1608	188	413.6	152	334.4	118	165.2	248	545.6	188	413.6	44	96.8	5	20.0	8	3.2	0	0.0	4767.0	5339.2
9.00-9.30	2348	1761.0	212	254.4	1228	1228	167	367.4	188	413.6	77	107.8	196	431.2	220	484.0	80	176.0	9	36.0	12	4.8	1	4.0	4738.0	5268.2
9.30-10.00	2044	1533.0	196	235.2	1648	1648	88	193.6	132	290.4	83	116.2	200	440.0	195	429.0	28	61.6	6	24.0	9	3.6	0	0.0	4629.0	4974.6
10.00-10.30	1284	963.0	184	220.8	948	948	77	169.4	144	316.8	89	124.6	165	363.0	200	440.0	4	8.8	1	4.0	3	1.2	0	0.0	3099.0	3559.6
10.30-11.00	1740	1305.0	128	153.6	1584	1584	208	457.6	109	239.8	94	131.6	184	404.8	128	281.6	9	19.8	0	0.0	0	0.0	0	0.0	4184.0	4577.8
11.00-11.30	2212	1659.0	116	139.2	1096	1096	80	176.0	137	301.4	75	105.0	120	264.0	189	415.8	22	48.4	2	8.0	0	0.0	2	8.0	4051.0	4220.8
11.30-12.00	1420	1065.0	232	278.4	1209	1209	86	189.2	180	396.0	63	88.2	155	341.0	160	352.0	28	61.6	0	0.0	4	1.6	0	0.0	3537.0	3982.0
12.00-12.30	1236	927.0	176	211.2	1224	1224	91	200.2	136	299.2	48	67.2	132	290.4	97	213.4	18	39.6	0	0.0	0	0.0	0	0.0	3158.0	3472.2
12.30-1.00	1376	1032.0	183	219.6	1184	1184	48	105.6	112	246.4	74	103.6	143	314.6	85	187.0	14	30.8	0	0.0	1	0.4	0	0.0	3220.0	3424.0
1.00-1.30	1576	1182.0	127	152.4	1247	1247	63	138.6	84	184.8	95	133.0	84	184.8	113	248.6	11	24.2	0	0.0	0	0.0	0	0.0	3400.0	3495.4
1.30-2.00	1548	1161.0	161	193.2	836	836	82	180.4	97	213.4	84	117.6	108	237.6	88	193.6	32	70.4	0	0.0	1	0.4	0	0.0	3037.0	3203.6
2.00-2.30	1120	840.0	148	177.6	1088	1088	89	195.8	209	459.8	71	99.4	95	209.0	136	299.2	21	46.2	0	0.0	0	0.0	0	0.0	2977.0	3415.0
2.30-3.00	860	645.0	125	150.0	920	920	108	237.6	116	255.2	52	72.8	99	217.8	76	167.2	28	61.6	0	0.0	1	0.4	0	0.0	2385.0	2727.6
3.00-3.30	844	633.0	164	196.8	860	860	94	206.8	141	310.2	68	95.2	116	255.2	89	195.8	22	48.4	0	0.0	0	0.0	0	0.0	2398.0	2801.4
3.30-4.00	838	628.5	180	216.0	884	884	75	165.0	122	268.4	103	144.2	128	281.6	144	316.8	40	88.0	2	8.0	0	0.0	0	0.0	2516.0	3000.5
4.00-4.30	884	663.0	284	340.8	1473	1473	115	253.0	105	231.0	108	151.2	124	272.8	87	191.4	45	99.0	3	12.0	0	0.0	0	0.0	3228.0	3687.2
4.30-5.00	928	696.0	167	200.4	1581	1581	97	213.4	119	261.8	87	121.8	155	341.0	73	160.6	58	127.6	0	0.0	0	0.0	1	4.0	3266.0	3707.6
5.00-5.30	1228	921.0	112	134.4	1912	1912	106	233.2	203	446.6	112	156.8	138	303.6	69	151.8	63	138.6	0	0.0	1	0.4	0	0.0	3944.0	4398.4
5.30-6.00	1320	990.0	145	174.0	1040	1040	79	173.8	154	338.8	44	61.6	148	325.6	87	191.4	35	77.0	4	16.0	0	0.0	0	0.0	3056.0	3388.2
6.00-6.30	1411	1058.3	173	207.6	1098	1098	65	143.0	192	422.4	94	131.6	122	268.4	77	169.4	18	39.6	0	0.0	0	0.0	1	4.0	3251.0	3542.3
6.30-7.00	1188	891.0	156	187.2	1196	1196	88	193.6	133	292.6	97	135.8	155	341.0	75	165.0	11	24.2	0	0.0	0	0.0	0	0.0	3099.0	3426.4
7.00-7.30	1196	897.0	102	122.4	1240	1240	66	145.2	161	354.2	66	92.4	177	389.4	108	237.6	33	72.6	0	0.0	1	0.4	0	0.0	3150.0	3551.2
7.30-8.00	1228	921.0	122	146.4	1160	1160	101	222.2	157	345.4	101	141.4	108	237.6	110	242.0	20	44.0	0	0.0	0	0.0	0	0.0	3107.0	3460.0
8.00-8.30	1124	843.0	72	86.4	987	987	104	228.8	116	255.2	73	102.2	112	246.4	95	209.0	14	30.8	0	0.0	0	0.0	0	0.0	2697.0	2988.8
8.80-9.00	1112	834.0	87	104.4	1012	1012	88	193.6	87	191.4	57	79.8	128	281.6	88	193.6	11	24.2	0	0.0	0	0.0	0	0.0	2670.0	2914.6
Total	37978		4409		33894		2710		4005		2248		4100		3493		798		34		44		8		93721	
Percentage	40.5		4.7		36.2		2.9		4.3		2.4		4.4		3.7		0.9		0.0		0.0		0.0		100.0	

3.4. Time and Delay Studies

The Floating Car Method is used for calculating the travel time & delay. This route has a 12.2 Km distance and the journey time is 27.46 minutes to 29.46 minutes. The number of parameters that decides the vehicle speed and travel delay are traffic volume, the number of vehicles overtaking, the number of vehicles overtaken, and vehicles from the opposite way. These parameters are considered while travelling in-between each bus stop and cumulative trip parameters are indicated in table 6. This table provides the to-and-fro operational data

Table 6. Cumulative Public transport accessibility Index.

Direction of Trip	From-To	Journey Time min: sec	Stopped Delay	No. of vehicles overtaking	No. of vehicles overtaken	Vehicles from the opposite way
E – W	Mysore Road-Ramohalli Cross	27:46	08:54	36	67	1596
W – E	Ramohalli Cross-Mysore Road	29:34	12:58	45	82	1862

3.5. Cost & Time Analysis

The cost& time analysis is done by driving the respective vehicle on the proposed route for finding the cost and time taken during travel. The distance travelled from each mode of transport is a 12.2 km stretch of the study area. This survey is on the peak traffic time which is arrived in the traffic volume study. The Yamaha RX-135 4-speed 2-stroke, petrol engine which has a mileage of 24 kmpl is used to find the 2-wheeler travel time and cost of travel. For 4-wheeler study Hyundai i-10: 4-seater car: petrol engine with the mileage capacity of 19 kmpl is used. In the case of para-transit, a car-pooling application (Uber) is used, time travel and cost are indicated in table 7. The next mode of transportation is the bus Bangalore metro transport corporation (BMTC), and an average of 5 trips is considered. For multimodal the bus fare till KBT is Rs. 15 and the metro charge from KBT to Mysore Road is Rs.17.

Table 7. Cumulative Public transport accessibility Index.

Type of mode	Time (minutes)	Cost (Rs)
Two-wheeler	26 min	50
Four-wheeler	32 min	62.5
Para-transit	34 min	125
Bus	48 min	22
Multi-Modal (Bus + Metro)	24 min	32

4. Results and Discussion

4.1. Walkability Index

Figure 2 shows the walkability index of zone-1 and zone-2 as 0.405 and 0.775 respectively. In Zone-1, the walkability index is a below-average condition, and it needs to be improvised for better usage for commuters. Under Zone-2 the walkability index is very good in condition and is very useful for commuters.

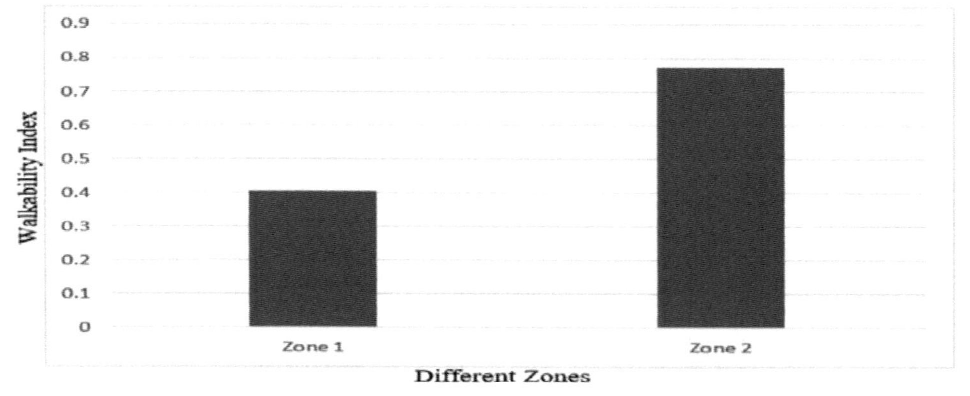

Figure 2. The Walkability Index.

4.2. Public Transport Accessibility

When looking into table 9 the range from 0 to 10 falls under poor levels of PTA index and two such stops are under this level, namely Check-post and RR Gate. This indicates that there is a need for para-transit, own vehicle, etc. to get good transportation accessibility in these areas. These are the weak links to be filled by the multi-modal transportation system. The rest of the stops come under moderate to excellent conditions of PTA levels. In figure 3 every stop stretch is highlighted with PTA. There is a non-linearity between the stretches which creates a hindrance to the pedestrians while traveling to the nearest bus stop. This affects the people in the respective bus-stop zones and also, it is difficult for multi-modal transportation system users.

Figure 3. Public transport accessibility Index.

4.3. Traffic Volume Study

The traffic volume study is done by finding the Peak one-hour volume. This value is compared with the capacity of the road according to the Indian Road Congress (IRC) guideline i.e., the volume of vehicles concerning the capacity of the road. The Peak Hour Factor (PHF) is determined by using equation 3. The number of vehicles and PCU data is provided in table 8 for the peak 30 minutes and one hour. This value is nearly 0.736 hence, it is under the prescribed limit. The capacity to the volume of vehicles on the road is safe and sufficient to avoid traffic delays.

Peak Factor = Volume in PCU during Peak hour / 4 x Highest volume in PCU during 30 minutes interval of peak hour (4)

Table 8. Cumulative Public transport accessibility Index.

Description	Vehicle	PCU
Peak 30 minutes	4767	5339.2
Peak One Hour	9505	10607.4
Peak Hour Factor		0.5

Peak One-hour PCU = 10607
Volume = 2651.75
For one day PCU = 10607 / 4 = 2651.75
Capacity from IRC = 3600 (for 4-line divided two way)
V / C = 0.736

4.4. Time and Delay Study

Due to the higher number of vehicles overtaking and from the opposite way, there is a delay in travel time and speed as shown in table 9. The E-W direction of the trip was done at 11:00 am and the W-E direction of the trip was done at 6:00 pm. By comparing the above-obtained values, it is concluded that the delay is more during the evening time than morning time. Nearly 30% of time and speed delay is observed during the analysis. This delay is due to an increase in traffic volume. Here, multi-modeling transportation helps reduce the time and delay by using partial metro services.

Table 9. Time and Speed analysis

Description	E-W direction	W-E direction
Average Journey Speed	26kmph	23.79kmph
Average Running Speed	36.95kmph	41kmph
Average Journey Time	28.71min	30.85min
Average Running Time	19.81min	17.9min

4.5. Cost and Time Analysis

The cost & time analysis is being carried out for the selected route from Ramohalli Cross to the Mysore Road bus stop. This route was travelled by different modes such as four-wheeler, two-wheeler, bus, para-transit, and combination of the above (multimodal transportation). It is observed that time and money are saved during the multi-model transportation system. The two-wheeler and four-wheeler are owned transportation systems that are nearly equal to the MMTS. When compared to public transportation, i.e., bus/metro the cost of travel is reduced comparing any mode of transportation. Due to the delay in each stop and the traffic volume of that route, the time is higher than any of the other modes. The MMTS provides a cost and time-effective solution. The para-transit is the costliest mode and the time is almost the same as the owned vehicle. Hence, multi-modal transport is convenient in both the travel-cost ratio and travel-time ratio.

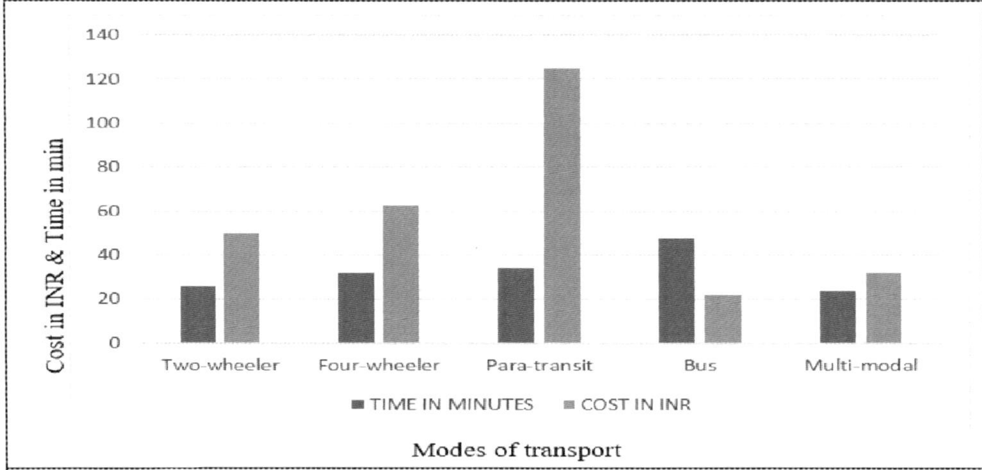

Figure 4. Cost and time analysis.

5. Conclusions

In this paper, the stretch of road is selected and analyzed for five different traffic analyzing parameters they are walkability index, traffic volume studies in terms of passenger car unit (PCU), time and delay study, and Cost & time analysis. Four modes of transportation are used in this study to find the effective transportation system namely 2-wheeler, 4-wheeler, para-transit, and buses/metro. A new schedule for representing a multimodal transportation network has effectively worked in this traffic, both in cost and duration of travel. The multimodal-transportation system works well here. The assessments also verify that the computational an online journey planning system may not work well in poor walkability index zones. This lags the connection between one mode of transportation to another mode also, this is the reason for vehicle side friction. The traffic flow that is disturbed in this zone may be at a particular time. Soon, the traffic volume is increased in that zone. Further studies could be made to assess the increase in traffic due to passenger side friction. The time delay and cost could be reduced when using the multimodal transportation system. Other vehicles such as commercial and e-commercial [18] transportation of goods through this MMTS should function efficiently for comprehensive transportation planning. An intellectual multimodal transportation system helps in the monitoring and management of cargo transportation [12]. Integrating such modes into the model is therefore one of the future goals.

References

[1] Joel Hansson Fredrik Petterson and Helen Svensson, Anders Westland 2019 Preferences in regional public transportation *European Transport Research Review*.

[2] Jeremy Mattson 2020 Measuring the Economic Benefits of Rural and Small Urban Transit Services in Greater Minnesota, Upper Great Plains *Transportation Institute North Dakota State University*.

[3] Litman, T. 2018 Evaluating Public Transit Benefits and Costs: Best Practices Guidebook. *Victoria Transport Policy Institute*.

[4] Mattson, J. 2016 Innovative Approach to Estimating Demand for Intercity Bus Services in a Rural Environment. University of South Florida, Tampa*: National Center for Transit Research.*

[5] Godavarthy, R., J. Mattson, J. Brooks, J. Jain, L. Quadrifoglio, I. Sener, and C. Simek 2018 Transit and Livability: Results from the National Community Livability Survey SURLC 16-004. *Montana State University, Bozeman: Small Urban and Rural Livability Center.*
[6] Eric Bruun 2017 Assessment Methods from Around the World Potentially Useful for Public Transport Projects *Journal of Public Transportation*, Vol. 20, No. 2, pp. 103-130.
[7] Omar Diba, Marie-Ange Manierb, Laurent Moalicb, Alexandre Caminada 2017 A multimodal transport network model and efficient algorithms for building advanced traveler information systems *Transportation Research Procedia* 22, 134–143
[8] Athakorn Kengpol Sopida Tuammee Marakku Tuomine 2014 The Development of framework for route selection in multimodal transportation *The International Journal of Logistics Management* vol 25 3 pp 581-610.
[9] Paola Moesti and Anna Sciommachen 2014 A Utility for finding multi-objective shortest paths in urban multimodal transport *European Journal of Operational Research* vol 111 issue 3 495-508 1998.
[10] Agarwal M K 1996 Urban Transportation in India, *Allied Publishers Ltd* New Delhi.
[11] Simpson J Berry 2017 Urban Transport Today *E & FN Spon* London.
[12] Anatolii Asaul Igor Malygin and Vladimir Komashinskiy 2017 The project of intellectual multimodal transport system *Transportation Research Procedia* 20 25-30.
[13] Nes V R 2002 Design of Multi-Modal Transport Networks *Ph.D. dissertation Department Civil Engineering, Delft University Press Delft.*
[14] Lizzette Perez-Lespier 2013 The Multimodal transportation system A systems dynamic approach *Thesis dissertation Master of Science in Engineering Management, Missouri University of Science and Technology.*
[15] Embarq and Taruu 2005 A Study on Para-transit system in Indore city. https://wrirosscities.org/sites/default/files/Study-Para-Transit%20System-Indore-City_0.pdf
[16] Institute of Urban Transport 2004 Sustainable transport: A sourcebook for policy makers in developing city *Urban Transport Development Training course material IUT.*
[17] Govt. of India 2006 National Urban Transport Policy Ministry of Urban Development, *Government of India New Delhi* September.
[18] Aman Dua and Deepankar Sinha 2019 Assessment of Quality of Multimodal Transportation for Containerized Exports *IIM Kozhikode Society & Management Review* 8(1) 10–22.
[19] Aman Dua and Deepankar Sinha 2015 The Multimodal Transportation: Research Trend and Literature Review *Udyog Pragati* Vol. 39, No. 4, October – December 1-11.

A Review on Relative Linear and Nonlinear Strains Performances of Laminated Composite Shell Roofs

Arpita Saxena[a]**, Kaustav Bakshi**[b]

[a,b] Civil Engineering Department, Indian Institute of Technology Indore, Madhya Pradesh 453552, India.

E-mail: *[a]phd2101104001@iiti.ac.in, [b]bakshi.kaustav@gmail.com*

Abstract. This paper comprehensively reviews research articles on linear and nonlinear performances of laminated shells. The study on composite surfaces gained popularity since the last decade. The composite shell problems for complex forms are difficult to solve analytically. Hence, the numerical solutions became popular. The invent of fast computers popularized the numerical approaches further among practicing engineers. Thin shell surfaces are widely popular to cover long unsupported spans in civil engineering; however, these are vulnerable to instability especially under transverse overloading. The stiffened surfaces addressed such problems. The detailed review on bare and stiffened surfaces highlights gaps in literature which can help the researchers to find out scopes of future research problems.

Keywords: Laminated composite shells, geometrically nonlinear strains, numerical solutions, finite element formulation, stiffened shells.

1. Introduction

In civil engineering, shell structures are widely used to cover long unsupported plan areas in building structures such as in retail malls, airport terminals, auditoriums, and parking facilities. The shells are also popular in aircraft, spaceships, rockets, submarines, boats, and storage tank manufacturing. Owing to the curved geometry, shells combine axial and bending rigidities to sustain the external load. Hence, shells offer greater load carrying capacity than flat plates for a given plan dimensions and material thickness. Due to the growing demand to develop light-weight structures there was appreciable advancements in material science and engineering. Hence, the laminated composites and functionally graded shell structures has increased in the last decade. Laminated composites have been used for shell fabrication since the last century. The composites were popular in civil engineering applications because of their high specific strength, lesser fabrication cost, and better durability than conventional reinforced concrete. As a result, the composite shells offer lesser seismic and foundation forces than their isotropic counterparts. The stiffness of composite shells can be modified by changing the lamina fabrication sequences, which is a desirable flexibility for practicing civil engineers.

The conoidal (Fig.1a) and cylindrical (Fig.1b) shells offer stiff, aesthetically pleasing, singly ruled geometry. Aesthetics and greater rigidity are the attributes for wide popularity of hypar (Fig.1c), saddle (Fig.1d) and spherical shells (Fig.1e). The cylindrical, conoidal and hypar shells are stiffer than flat plates and easier to fabricate at the same time. Such shells also allow air and sunlight. Moreover, these shells reduce operational costs of industrial units as the artificial ventilation is not needed due to availability of the natural air and light. Naturally, a large number of studies on laminated composite shells began to appear. The detailed review on such composite shell panels is organized systematically in the following manner:

1. Review of literature on unstiffened shell roofs studied using geometrically linear strains on furnished in Section 1.1.

2. Review of literature on stiffened shell roofs studied using geometrically linear strains on furnished in Section 1.3.
3. Review of literature on unstiffened shell roofs studied using geometrically nonlinear strains on furnished in Section 1.2.
4. Review of literature on stiffened shell roofs studied using geometrically nonlinear strains on furnished in Section 1.4.

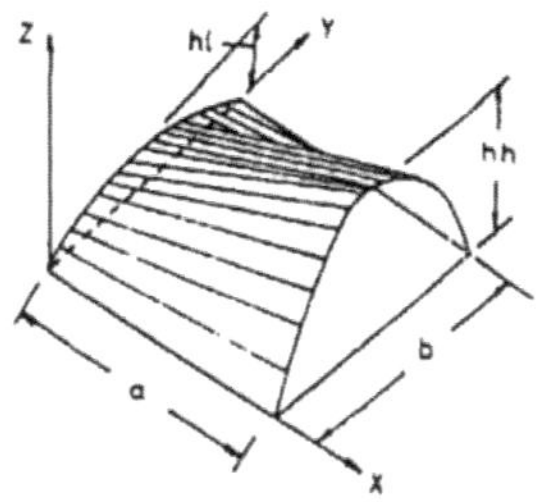

Figure 1 (a) Conoidal Shell

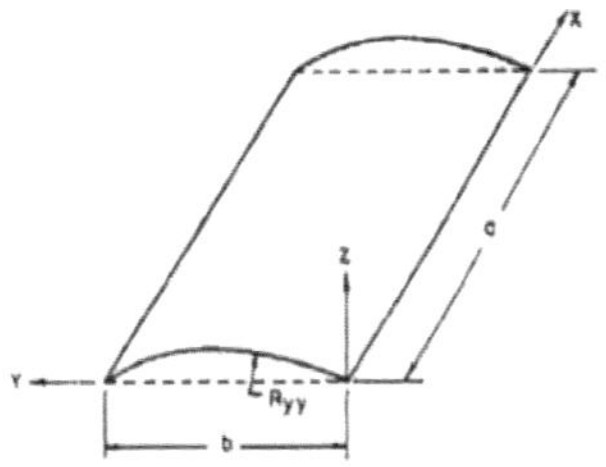

Figure 1 (b) Cylindrical Shell

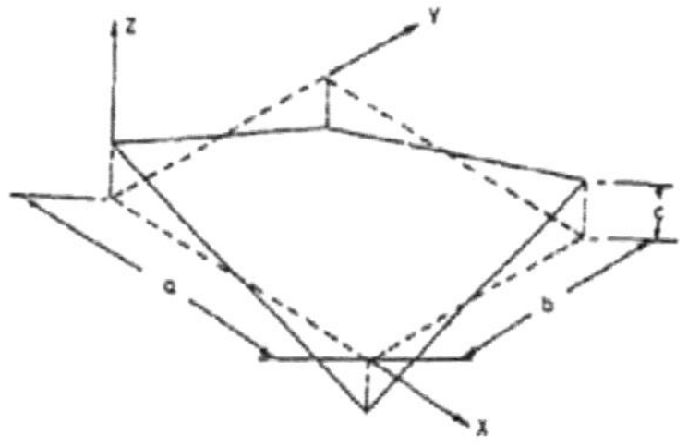

Figure 1 (c) Hypar Shell

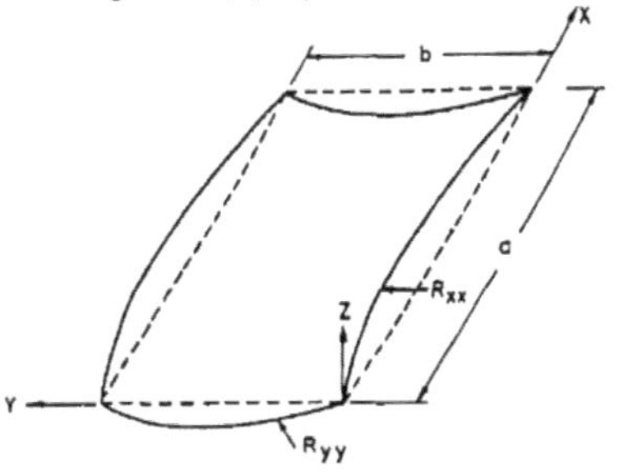

Figure 1 (d) Saddle Shell

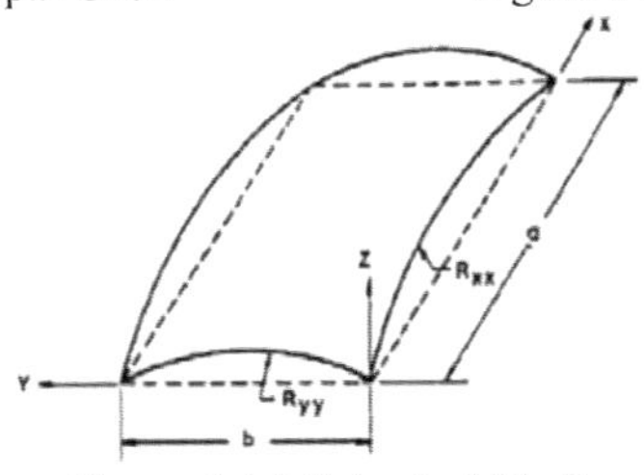

Figure 1 (e) Spherical Shell

Figure 1. Shell Configurations

1.1 Review of literature on unstiffened shell roofs studied using geometrically linear strains

Several studies explored the dynamic and static of composite laminated singly curved shell roofs using linear FEM owing to the widespread appeal of singly ruled curved surfaces, and a large number of papers are accessible. There are both isotropic and anisotropic shell structures in these works. Reddy [1] adopted the shear deformation theories of first and higher-order (FSDT and HSDT) to study vibration of isotropic and composite laminated cylindrical shell under complicated loading. The exact closed form deformations and fundamental frequencies reported by the author acted as benchmark for many numerical findings. Chakravorty et al. [9], Bardell et al. [12], Ganapathi et al. [23] and Dogan et al. [40] worked on dynamics of cross ply cylindrical shells. The authors adopted isoparametric finite element approaches in their studies. Isotropic conoidal shells were the focus of several studies. Ghosh and Bandyopadhyay investigated the static behaviour of isotropic truncated conoidal shells [3 and 7].The authors adopted different numerical methods like Galerkin approach and isoparametric finite element approach. Das and Chakravorty [47] worked on composite shells and demonstrated the bending response of the conoidal shell without stiffeners. The authors evaluated the relative performance of several shell alternatives under different boundary conditions along with different laminations. Furthermore, the authors presented appropriate ways for selecting the optimal shell option from all those considered. Chakravorty et al. [14] used FEM for vibration problems of composite cylindrical, spherical, saddle, hypar and conoidal shells with cut-out. The transient responses along with dynamic magnification factors of composite doubly curved shells were reported by the authors. The numerical approaches got popular to solve the complicated shells forms like conoid, hypar and elliptic paraboloid as seeking analytical solutions for such shells involves greater computational efforts. Among different numerical approaches, the FEM became popular due to its versatility with relatively easier to solve approach. The popularity of FEM can also be attributed to the invent of high-speed computers. Das and Chakravorty [31 and 35] adopted isoparametric finite element approach to report behaviors of conoids which is made of laminated composite for a variety of boundary conditions. The authors worked considering FSDT. Pradyumna et al. [36] adopted HSDT for getting static deflections and fundamental frequencies of composite laminated hypar, conoid, and hyperbolic paraboloid shells. Das and Chakravorty [41] also worked on bending of full conoidal shells which is made of laminated composite having lower height of the shell as zero. The instability of laminated conoids were investigated by Pradyumna and Bandyopadhyay [50]. Bakshi and Chakravorty [57] reported bending of laminated conoidal shells. The vibration responses of laminated conoids for three different time-dependent forces were reported by the authors [58] later. The isoparametric finite elements were combined with FSDT by the authors to study the conoidal shell forms. Bakshi and Chakravorty [65] worked on laminated composite conoids and reported the relative shell performances in terms of static deflections, fundamentals frequencies and forced vibration deflections. The authors also reported the dynamic magnification of deflections due to the dynamic loads.

The laminated composites are weak in transvers shear capacity. Hence, such shells under transverse overloading may get delaminated. The delamination can also initiate due to fabrication deflects. A number of authors realized that and worked on delaminated shell forms. The vibration frequencies and mode morphologies of cylindrical shells with separated laminae were reported by Acharyya et al. [44] using isoparametric FEM. The bending deformations of such conoids was described by Kumari and Chakravorty [45 and 49]. The authors also reported the deformed shapes of shell roofs, although the damage initiation was not studied. The laminated composites may start to fail at any inner lamina or interfaces of laminae. Such latent damage may progress gradually within the shell which ultimately culminates to collapse of the shell roof. Hence, the failure initiation in composite shells must be studied in detail to confidently apply such surfaces in industrial applications. The failure initiation in laminated conoidal shells was studied in 59 and 64. The authors considered isoparametric FEM along with FSDT to solve the shell static problems. The lamina stresses were solved using the laminate constitutive relationship. The authors [68] later reported the failure initiation of shells under concentrated loads. Sengupta et al. [69] employed an isoparametric shell elements to perform progressive crack analysis of laminated cylindrical roofs under transverse uniform pressure. The failure behavior of a laminated ellipsoidal shell withheld to internal pressure was studied by Gohari et al. [70]. To explore crack initiation in laminated composites made from woven cloth, the scientists used a linear interpolation technique. Kumar et al. [71] used higher-order zig-zag strains to study the failure of laminated and sandwich single curved panels. For the analysis of a doubly curved shell which is made of composites, Thakur et al. [76] proposed a HSDT based FEM model. That model was able to calculate the shear stress over the thickness direction with great accuracy. Bakshi and Chakravorty [82] reported the crack initiation in composite conoidal shells under central concentrated load, while Ghosh et al. [90] studied crack progression in laminated hypar shells using the FEM in linear and elastic ranges under uniform pressure.

The researchers investigated difficult vibration problems such as impact responses in addition to failure concerns under bending pressures. For various impactor velocities, Neogi et al. [51] examined the low energy impact of composite hypar shells under the central impact. Kumar et al. [56] used an efficient 2D FE model on zigzag theory to analyse the vibration of laminated and sandwich shells (HOZT). This was the first FE version of the HOZT to deal with the vibration of shells using Sanders' approximations, which addressed all three curvature radii as well as the influence of cross curvature.

Despite the fact that the researchers were working on complex shell problems, a few of them were still interested in bending deformations and fundamental frequencies. However, such research is carried out for criteria that the researchers had not previously examined. Sahoo [81] investigated the natural frequencies of hypar shells. The author worked on tangential / rotary inertia on those shells. Chaubey et al. [80] adopted thick and moderately thick conoids for a variety of boundaries. Arefi et al. [85] used nonlocal elasticity theory and HSDT to work on the bending of a piezoelectric nano shell. The motion equations were created using the virtual work principle. They investigated how the piezoelectric nano-mechanical shell's and electrical components were affected by the electric potential, a nonlocal parameter. By using the C^0 FEM based on HSDT to studied the behavior of laminated shells is examined by Thakur et al. [95].

1.2 Review of literature on stiffened shell roofs studied using geometrically linear strains

The thin structural elements may get unstable under transverse overloading. The stiffened surfaces can address that problem. A number of authors worked on stiffened surfaces and compared the shell performances with those for unstiffened ones. The stiffened surfaces show significantly improved performances like reduced deformations and increased fundamental frequencies. Among the numerical approaches adopted by different researchers the finite element method was mostly adopted to evaluate shell performances like fundamental frequencies for differing stiffener geometries, fiber alignments and boundaries. Mustafa and Ali [2] used structural symmetry approaches for vibration analysis of closed cylindrical and conical shells. Samanta and Mukhopadhyay [26] worked on vibration of triangular stiffened shells. Vibration of cylindrical, paraboloid, hypar and conoidal stiffened shells were worked by Nayak and Bandyopadhyay [21,22 and 27]. The stiffened composite hypars were studied by Sahoo and Chakravorty [29] and Shi et al. [73] worked on the vibration of curvilinear shallow shells which was doubly curved using the Ritz method.

Cheng and Dade [4], Srinivasan and Krishnan [5], Sinha and Mukhopadhyay [10 and 13], and Prusty and Satsangi [19] published on stiffened cylindrical and elliptic shells and concluded that the stiffener can be placed anywhere with suitable transformations. In many regions, conoidal shells with a singly ruled surface are chosen over other shell forms because of their attractive aesthetics, ease of building, and efficiency in allowing natural ventilation. As a result, of the several shell configurations accessible in the literature, these are the most chosen for industrial applications. However, compared to stiffened cylindrical and elliptic shells, there is a dearth of evidence on the dynamics of such conoidal shells. To fill up this void in literature, Nayak and Bandhopadhyay [28] worked on such conoidal shells using the FEM. They reported displacements of stiffened conoidal shells under transient load cases. A number of parameters were varied by the authors which includes varying stiffener depth, number and orientations.

Anisotropic shells have a combination of unique displacement modes due to their associated axial and bending rigidities, resulting in significant pre-buckling deformation. As a result, the researchers observed that determining the critical load requires a linearized stability study. Furthermore, the same was used in post-buckling analysis to identify the significant load and stability modes.

Prusty and Satsangi [19] presented a FEM for stiffened plates and cylindrical shells based on FSDT. For FE analysis of stiffened shells, Goswami and Mukhopadhyay [8] used nine noded Lagrange element and heterosis element. Sahoo and Chakravorty [39] worked on the static reactions of stiffened hypar shells, including deflections and stresses, under various geometries, boundary conditions, and stacking sequences.

Das and Chakravorty [53] demonstrated the bending response of the stiffened conoid by coupling a three-noded element with an eight-noded element, and deflections and force/moment resultants were comprehensively studied. Pradhan and Jena [55] investigated the static bending of a stiffened conoidal shell under focussed strain. Several shell actions were calculated by varying stiffener dimensions. The cut-out shells were examined using stiffeners. Hota and Chakravorty [33] created a similar FE code. The authors looked at natural vibration of a stiffened conoids having cut-outs. The frequencies and mode forms of laminated stiffened conoids having cut-out were studied by Sahoo [61].

Sahoo [75] worked on the stiffened hyperbolic paraboloids having cut-out using FEM based on FSDT. Chaudhuri et al. [86] and Sahoo [88] worked on stiffened hypar shells using the finite element approach and reported natural frequencies and mode shapes. Chaudhuri et al. [89] investigated first five mode shapes laminated hypars with cut-out and found that free vibration is primarily determined by boundary conditions.

The failure of composite laminates is progressive in nature. It starts with failure of the weakest lamina which could be an inner one. Hence, unlike metals, the cracks in laminated composites may not be visible at the surface. Realizing the importance, Prusty et al. [18] studied failure of stiffened shells. Rout et al. [79 and 84] worked on vibration responses of twisted composite stiffened shell with separating layers.

1.3 Review of literature on unstiffened shell roofs studied using geometrically nonlinear strains

When exposed to transverse overloading, the thin shell roofs are prone to significant deformations. Deflections of thin shells under time-varying stresses can be equal to or greater than the material thickness. It is also possible to experience dynamic snapping or dynamic buckling. As a result, it is critical to consider substantial displacement/rotation in the vibration of thin shells. The geometrically linear FE formulation may be unable to accurately predict such shell responses, requiring the use of nonlinear analysis. Because the geometrically nonlinear approach is more difficult to execute than the linear approach, there are fewer research articles on nonlinear shell performances than there are on linear shell performances.

Tsai and Palazatto [6], Chang and Chang [16] and Toorani [24] adopted geometrically nonlinear finite element formulation to work on vibration of cylindrical shells. The authors [6] adopted HSDT to report the natural frequencies, while Przekop et al. [25] proposed FEM for large amplitude vibration analysis of triangular shallow panels. The nonlinear frequencies of shallow shells were reported by the authors. Nanda and Praduymna [52] combined FE approaches to report the nonlinear frequencies of composite singly and doubly shells having initial imperfections and being treated for hygrothermal condition variations. Bich and Nguyen [54] explored the nonlinear vibrations of shells exposed to mechanical loads using the displacement method. The differential equations were formulated taking into account kinematic nonlinearity and ignores shallowness of cylindrical shells. For dynamic analysis of shells Volmir's assumption, Galerkin and Runge–Kutta method were adopted to obtain the frequencies. Hashemian and Jam [46] employed a mix of FE and Lindstedt–Poincaré perturbation methods to study the vibrations of laminated cylindrical panels. Ribeiro [38] used a p-version FE study geometrically non-linear transient nature of cylindrical shell panels and Singh and Panda [66] demonstrated the large amplitude nonlinear behavior of hyperboloid and ellipsoid shells.

Owing to the greater specific strength, the vibration of shells with cut-outs was of tremendous interest to designers. Cut-outs in the shell panels offer natural ventilation, passages for electrical and communication cables and altering the resonating frequency. The shell surface near the cut-outs is expected to show relatively greater deformations than solid parts of the same shell. Hence, the nonlinear formulation was important in precise vibration behaviour prediction of such structural elements.

Nanda and Bandhopadhyay [32 and 37] used nonlinear finite elements to analyse vibration of composite singly curved panels with cut-outs. The authors looked at the influence of plan dimension ratios, laminations, and elastic parameters on nonlinear fundamental frequencies of cut-out shells. Forced vibration responses of perforated laminated shells and early dimensional flaws were also studied by Nanda and Bandhopadhyay [42]. The authors investigated square-cut-out shells and addressed the shell problem using a combination of finite elements and geometrically nonlinear stresses. Jin et al. [60] proposed an analytical model for transient predictions of thick laminated cylindrical shells based on FSDT, and Kurpa [43] worked on multi-layered spherical shells with complex bases. To formulate the shell problem, the author used HSDT and Green strains. Chen et al. [101] investigated the post-critical behavior of sandwich shells surrounded by elastic media. The authors created the governing equations by combining Reddy's HSDT with simplified nonlinearity.

To correctly anticipate the failure loads of shells subjected to static overloading, geometrically nonlinear strains are required. Bakshi and Chakravorty [67] and Bakshi [87] demonstrated failure of laminated, fairly thin shell panels using nonlinear technique. Failure loads, failure modes, and tendencies, as well as the areas where failures began, were all reported. Ghosh and Chakravorty [77 and 83] explored the failure of composite skewed hypars. Nonlinear failure loads of conoids were observed in 72 and 78. Failure loads were provided for a variety of laminations, stacking sequences, and shell boundary conditions. Bakshi and Chakravorty [91, 92, and 93] provided delamination and first ply failure predictions for conoids using FSDT and simplied nonlinearity. To propose load carrying capacities, Dona et al. [96] offered design guidelines for skewed laminated composite hypar shells with partially open borders, taking into account both the material failure and serviceability requirement.

The critical load of curved panels was studied by Sahu and Datta [17]. Different loads like partial and intense edge loadings. Soares and Goncalves [74] employed ABAQUS to study the conoids. Pinho et al. [100] used nonlinear analysis to investigate thin shells.

1.4 Review of literature on stiffened shell roofs studied using geometrically nonlinear strains

Section 1.2 reports that the research papers on stiffened surfaces are not available in good numbers compared to the ones published on unstiffened ones. Such papers are even scanty while considering the geometrically nonlinear strains. Goswami and Mukhopadhyay [11] and Samanta and Mukhopadhyay [15] adopted geometrically nonlinear strains to study cylindrical shells to trace the nonlinear equilibrium path. Samanta and Mukhopadhyay [15] worked on isotropic shell while Goswami and Mukhopadhyay [11] adopted $90^0/0^0/0^0/90^0$ shell only. For different stiffener properties like orientations, eccentricities, numbers, depth and varying boundary condition the mode shapes and fundamental frequencies of laminated curved stiffened shells were presented by Bakshi [97].

Patel et al. [30, 34, and 48] reported parametric critical natuer of isotropic stiffened doubly curved spherical shells. The authors studied the influence of angle of stiffener with boundaries and plan dimension ratio. Bakshi [94] looked at the maximum transverse displacements of laminated composite stiffened curved surfaces for industrially significant parametric changes. Using Sanders' first approximation theory and simplified nonlinearity, the author constructed an iso-parametric C^0 FEM for the composite stiffened one. The formulation was expanded by Saxena and Bakshi [98] to accommodate composite stiffened conoidal shells.

Duc [62] examined the dynamic buckling and non-linear vibrations of eccentrically stiffened functionally graded panels ignoring temperature effects. Li et al. [63] devised layer-wise/solid-element (LW/SE) approach for stiffened composite shells, which was based on the layer-wise theory and standard FEM package. The authors reported static bending deformations and frequencies for of vibration. Along with such results, the authors also reported the mode shapes and deformation shapes of the cylindrical shells having uniform boundary conditions. The critical load and failure of laminated stiffened cylindrical panels were proposed by Keshav et al. [99]. The analysis was carried out in ABAQUS.

2. Conclusions

This study aimed to (i) enable researchers working in the field of shell structures with the necessary references for their work, and most importantly (ii) to locate the gaps in the field of shell research which can be filled in future. The authors put their best efforts to include as many research articles as possible and that with sufficient clarity and discussions on individual papers. However, it is undeniable that there is always a probability to have a few remain unattended. The authors sincerely apologies for such unintentional left outs. The following points may help to identity the research gaps in the literature.

- There are a number of research papers available on different behavioural aspects of laminated composite shell structures. The researchers mostly adopted simpler to implement geometrically linear strains to study cylindrical, conoidal, hypar and elliptical shell surfaces.
- Although the isotropic shells are still adopted by a few researchers specifically for its simplicity to solve the shell problem, most researchers now-a-days work on laminated composite shell surfaces due to the advantages offered by the material. The stiffened surfaces are preferred in industrial units to avoid the instability of thin shell surfaces under transverse overloading. However, such studies generally adopt the geometrically linear strains.
- The relatively greater involving geometrically nonlinear formulations were also adopted by the researchers however, such studies are mostly limited to simpler geometries like cylindrical and spherical surfaces. The advent in fast computers enabled the researchers to work on complex shell forms having cross-curvatures like conoidal, hypar and saddle shells.
- The solutions of stiffened shell problems using geometrically nonlinear strains are scanty in the literature. The stiffened shell problems demand a lot more attention from the researchers where the transient responses, instability, random vibration, impact responses can be studied alongside bending and free vibration. Recently, a few researchers worked on this area, but the simpler cylindrical surfaces were solved. The complex shell surfaces are not yet reported by any researchers which can become an active research area in near future.

References

[1] Reddy JN 1984 Exact solutions of moderately thick laminated shells *Journal of Engineering Mechanics* **110(5)** pp.794-809.

[2] Mustafa BAJ, Ali R 1987 Prediction of natural frequency of vibration of stiffened cylindrical shells and orthogonally stiffened curved panels *Journal of Sound Vibration*

113(2) pp.317-27.
[3] Ghosh B, Bandyopadhyay JN 1989 Bending analysis of conoidal shells using curved quadratic isoparametric element. *Computers & structures* **33(3)** pp.717-28.
[4] Cheng SP, Chuang D 1990 Dynamic analysis of stiffened plates and shells using spline gauss collocation method *Computers & structures* **36(4)** pp.623-29.
[5] Srinivasan RS, Krishnan PA 1990 Response of orthogonally stiffened cylindrical shell panels *AIAA journal* **28(6)** pp.1144-45.
[6] Tsai CT, Palazotto AN 1991 On the finite element analysis of non-linear vibration for cylindrical shells with high-order shear deformation theory *International journal of Non-linear Mechanics* **26(3-4)** pp.379-88.
[7] Ghosh B, Bandyopadhyay JN 1994 Bending analysis of conoidal shells with cut-outs *Computers & structures* **53(1)** pp.9-18.
[8] Goswami S, Mukhopadhyay M 1994 Finite element analysis of laminated composite stiffened shell *Journal of Reinforced Plastics and Composites* **13(7)** pp.574-616.
[9] Chakravorty D, Bandyopadhyay JN and Sinha P 1995 Finite element free vibration analysis of point supported laminated composite cylindrical shells *Journal of Sound and Vibration* **81(1)** pp.43-52.
[10] Sinha G, Mukhopadhyay M 1995 Transient dynamic response of arbitrary stiffened shells by the finite element method *Journal of Vibration and Acoustics* **117** pp.11-16.
[11] Goswami S, Mukhopadhyay M 1995 Geometrically nonlinear analysis of laminated composite stiffened shells *Journal of reinforced plastics and composites* **14(12)** pp.1317-36.
[12] Bardell NS, Dunsdon JM and Langley RS 1997 Free and forced vibration analysis of thin, laminated, cylindrically curved panels *Composite structures* **38(1-4)** pp.453-62.
[13] Sinha G, Mukhopadhyay M 1997 Static, free and forced vibration analysis of arbitrary non-uniform shells with tapered stiffeners *Computers & structures* **62(5)** pp.919-33.
[14] Chakravorty D, Sinha PK and Bandyopadhyay JN 1998 Applications of FEM on free and forced vibration of laminated shells *Journal of Engineering Mechanics* **124(1)** pp.1-8.
[15] Samanta A, Mukhopadhyay M 1999 Finite element large deflection static analysis of shallow and deep stiffened shells *Finite elements in analysis and design* **33(3)** pp.187-208.
[16] Chang TP, Chang HC 2000 Nonlinear vibration analysis of geometrically nonlinear shell structures *Mechanics research communications* **27(2)** pp.173-180.
[17] Sahu SK, Datta PK 2001 Parametric instability of doubly curved panels subjected to non-uniform harmonic loading *Journal of Sound and Vibration* **240(1)** pp.117-129.
[18] Prusty BG, Ray C and Satsangi SK 2001 First ply failure analysis of stiffened panels–a finite element approach *Composite structures* **51(1)** pp.73-81.
[19] Prusty BG, Satsangi SK 2001 Finite element transient dynamic analysis of laminated stiffened shells *Journal of Sound and Vibration* **248(2)** pp.215-33.
[20] Prusty BG, Satsangi SK 2001 Finite element buckling analysis of laminated composite stiffened shells *International Journal of Crashworthiness* **6(4)** pp.471-84.
[21] Nayak AN, Bandyopadhyay JN 2002 Free vibration analysis and design aids of stiffened conoidal shells *Journal of Engineering Mechanics* **128(4)** pp.419-27.
[22] Nayak AN, Bandyopadhyay JN 2002 On the free vibration of stiffened shallow shells *Journal of Sound and Vibration* **255(2)** pp.357-82.
[23] Ganapathi M, Patel BP, Patel HG and Pawargi D 2003 Vibration analysis of laminated cross-ply oval cylindrical shells *Journal of Sound and Vibration* **262(1)** pp.65-86.
[24] Toorani MH 2003 Dynamics of the geometrically non-linear analysis of anisotropic laminated cylindrical shells *International journal of non-linear mechanics* **38(9)** pp.1315-35.

[25] Przekop A, Azzouz MS, Guo X, Mei C and Azrar L 2004 Finite element multiple-mode approach to nonlinear free vibrations of shallow shells *AIAA journal* **42(11)** pp.2373-81.

[26] Samanta A, Mukhopadhyay M 2004 Free vibration analysis of stiffened shells by the finite element technique *European journal of mechanics-A/solids* **23(1)** pp.159-179.

[27] Nayak AN, Bandyopadhyay JN 2005 Free vibration analysis of laminated stiffened shells *Journal of engineering mechanics* **131(1)** pp.100-105.

[28] Nayak AN, Bandyopadhyay JN 2006 Dynamic response analysis of stiffened conoidal shells *Journal of Sound Vibration* **291(3-5)** pp.1288-97.

[29] Sahoo S, Chakravorty D 2006 Stiffened composite hypar shell roofs under free vibration: Behaviour and optimization aids *Journal of sound and vibration* **295(1-2)** pp.362-77.

[30] Patel SN, Datta PK and Sheikh AH 2006 Buckling and dynamic instability analysis of stiffened shell panels *Thin-walled structures* **44(3)** pp.321-33.

[31] Das HS, Chakravorty D 2007 Design aids and selection guidelines for composite conoidal shell roofs—A finite element application *Journal of Reinforced plastics and Composites* **26(17)** pp.1793-1819.

[32] Nanda N, Bandyopadhyay JN 2007 Nonlinear free vibration analysis of laminated composite cylindrical shells with cutouts *Journal of reinforced Plastics and Composites* **26(14)** pp.1413-27.

[33] Hota SS, Chakravorty D 2007 Free vibration of stiffened conoidal shell roofs with cutouts *Journal of Vibration and Control* **13(3)** pp.221-40.

[34] Patel SN, Datta PK and Sheikh AH 2007 Dynamic instability analysis of stiffened shell panels subjected to partial edge loading along the edges *International Journal of Mechanical Sciences* **49(12)** pp.1309-24.

[35] Das HS, Chakravorty D 2008 Natural frequencies and mode shapes of composite conoids with complicated boundary conditions *Journal of Reinforced plastics and Composites* **27(13)** pp.1397-1415.

[36] Pradyumna S, Bandyopadhyay JN 2008 Static and free vibration analyses of laminated shells using a higher-order theory *Journal of Reinforced Plastics and Composites* **27(2)** pp.167-186.

[37] Nanda N, Bandyopadhyay JN 2008 Nonlinear transient response of laminated composite shells *Journal of engineering mechanics* **134(11)** pp.983-90.

[38] Ribeiro P 2008 Forced large amplitude periodic vibrations of cylindrical shallow shells *Finite Elements in Analysis and Design* **44(11)** pp.657-74.

[39] Sahoo S, Chakravorty D 2008 Bending of composite stiffened hypar shell roofs under point load *Journal of engineering mechanics* **134(6)** pp.441-54.

[40] Dogan A, Arslan HM 2009 Effects of curvature on free vibration characteristics of laminated composite cylindrical shallow shells *Scientific Research and Essays* **4(4)** pp.226-38.

[41] Das HS, Chakravorty D 2009 Composite full conoidal shell roofs under free vibration *Advances in Vibration Engineering* **8(4)** pp.321-28.

[42] Nanda N, Bandyopadhyay JN 2009 Geometrically nonlinear transient analysis of laminated composite shells using the finite element method *Journal of Sound and Vibration* **325(1-2)** pp.174-185.

[43] Kurpa LV 2009 Nonlinear free vibrations of multilayer shallow shells with a symmetric structure and with a complicated form of the plan *Journal of Mathematical Sciences* **162(1)** pp.85-98.

[44] Acharyya AK, Chakravorty D and Karmakar A 2010 Natural frequencies and mode shapes of composite cylindrical delaminated shells by finite element *Journal of reinforced plastics and composites* **29(2)** pp.226-37.

[45] Kumari S, Chakravorty D 2010 On the bending characteristics of damaged composite

conoidal shells-a finite element approach. *Journal of Reinforced Plastics and Composites* **29(21)** pp.3287-96.
[46] Hashemian AH, Jam JE 2010 Nonlinear free dynamic response of laminated compressible cylindrical shell panels *Mechanics of Composite Materials* **46(1)** pp.15-28.
[47] Das HS, Chakravorty D 2010 A finite element application in the analysis and design of point-supported composite conoidal shell roofs: suggesting selection guidelines *The Journal of Strain Analysis for Engineering Design* **45(3)** pp.165-177.
[48] Patel SN, Datta PK and Sheikh AH 2010 Effect of harmonic in-plane edge loading on dynamic stability of stiffened shell panels with cutouts *International Journal of Applied Mechanics* **2(04)** pp.759-85.
[49] Kumari S, Chakravorty D 2011 Bending of delaminated composite conoidal shells under uniformly distributed load *Journal of engineering mechanics* **137(10)** pp.660-68.
[50] Pradyumna S, Bandyopadhyay JN 2011 Dynamic instability behavior of laminated hypar and conoid shells using a higher-order shear deformation theory *Thin-walled structures* **49(1)** pp.77-84.
[51] Neogi SD, Karmakar A and Chakravorty D 2011 Impact response of simply supported skewed hypar shell roofs by finite element *Journal of reinforced plastics and composites* **30(21)** pp.1795-1805.
[52] Nanda N, Pradyumna S 2011 Nonlinear dynamic response of laminated shells with imperfections in hygrothermal environments *Journal of Composite materials* **45(20)** pp.2103-12.
[53] Das HS, Chakravorty D 2011 Bending analysis of stiffened composite conoidal shell roofs through finite element application *Journal of composite materials* **45(5)** pp.525-42.
[54] Bich DH, Nguyen NX 2012 Nonlinear vibration of functionally graded circular cylindrical shells based on improved Donnell equations *Journal of Sound and Vibration* **331(25)** pp.5488-5501.
[55] Pradhan N, Jena J 2012 Static characteristics of stiffened conoidal shell roofs under concentrated load. *International Journal of Advances in Engineering & Technology* **4(2)** p.195.
[56] Kumar A, Chakrabarti A and Bhargava P 2013 Vibration of laminated composites and sandwich shells based on higher order zigzag theory. *Engineering Structures* **56** pp.880-88.
[57] Bakshi K, Chakravorty D 2013 Static and dynamic characteristics of composite conoidal shell roofs *Advances in Acoustics and Vibration* **539817**.
[58] Bakshi K, Chakravorty D 2013 Relative static and dynamic performances of composite conoidal shell roofs *Steel and Composite Structures* **15(4)** pp.379-97.
[59] Bakshi K, Chakravorty, D 2013 First ply failure study of composite conoidal shells used as roofing units in civil engineering *Journal of failure analysis and prevention* **13(5)** pp.624-33.
[60] Jin G, Ye T, Ma X, Chen Y, Su Z and Xie X 2013 A unified approach for the vibration analysis of moderately thick composite laminated cylindrical shells with arbitrary boundary conditions *International Journal of Mechanical Sciences* **75** pp.357-76.
[61] Sahoo S 2013 Dynamic characters of stiffened composite conoidal shell roofs with cutouts: design aids and selection guidelines *Journal of Engineering*.
[62] Duc ND 2013 Nonlinear dynamic response of imperfect eccentrically stiffened FGM double curved shallow shells on elastic foundation *Composite Structures* **99** pp.88-96.
[63] Li D, Qing G and Liu Y 2013 A layerwise/solid-element method for the composite stiffened laminated cylindrical shell structures *Composite Structures* **98** pp.215-27.
[64] Bakshi K, Chakravorty D 2014 First ply failure study of thin composite conoidal shells subjected to uniformly distributed load *Thin-Walled Structures* **76** pp.1-7.

[65] Bakshi K, Chakravorty D 2014 Relative performances of composite conoidal shell roofs with parametric variations in terms of static, free and forced vibration behavior *International Journal of Civil Engineering* **12** 299-311.

[66] Singh VK, Panda SK 2014 Nonlinear free vibration analysis of single/doubly curved composite shallow shell panels *Thin-Walled Structures* **85** pp.341-49.

[67] Bakshi K, Chakravorty D 2014 Geometrically linear and nonlinear first-ply failure loads of composite cylindrical shells *Journal of Engineering Mechanics* **140(12)** p.04014094.

[68] Bakshi K, Chakravorty D 2015 First ply failure loads of composite conoidal shell roofs with varying lamination *Mechanics of Advanced Materials and Structures* **22(12)** pp.978-87.

[69] Sengupta J, Ghosh A and Chakravorty D 2015 Progressive failure analysis of laminated composite cylindrical shell roofs *Journal of Failure Analysis and Prevention* **15(3)** pp.390-400.

[70] Gohari S, Sharifi S, Vrcelj Z and Yahya MY 2015 First-ply failure prediction of an unsymmetrical laminated ellipsoidal woven GFRP composite shell with incorporated surface-bounded sensors and internally pressurized *Composites Part B: Engineering* **77** pp.502-18.

[71] Kumar A, Chakrabarti A, Bhargava P and Prakash V 2015 Efficient failure analysis of laminated composites and sandwich cylindrical shells based on higher-order zigzag theory *Journal of Aerospace Engineering* **28(4)** p.04014100.

[72] Bakshi K, Chakravorty D 2015 First Ply Failure of Laminated Composite Conoidal Shell Roofs Using Geometric Linear and Nonlinear Formulations In *Advances in Structural Engineering* Springer pp. 113-24.

[73] Shi P, Kapania RK and Dong CY 2015 Free vibration of curvilinearly stiffened shallow shells *Journal of Vibration and Acoustics* **137(3)**.

[74] Soares RM, Gonçalves PB 2016 Buckling and Nonlinear Analysis of Conoidal Shells *International Design Engineering Technical Conferences and Computers and Information in Engineering Conference* American Society of Mechanical Engineers **50206** p. V008T10A017.

[75] Sahoo S 2016 Performance evaluation of free vibration of laminated composite stiffened hyperbolic paraboloid shell panel with cut-out *International Journal of Engineering and Technologies* **7** pp.1-24.

[76] Thakur SN, Ray C and Chakraborty S 2017 A new efficient higher-order shear deformation theory for a doubly curved laminated composite shell *Acta Mechanica* **228(1)** pp.69-87.

[77] Ghosh A, Chakravorty D 2017 Failure analysis of civil engineering composite shell roofs *Procedia engineering* **173** pp.1642-49.

[78] Bakshi K, Chakravorty D 2017 Geometrically nonlinear first ply failure loads of laminated composite conoidal shells *Procedia engineering* **173** pp.1619-26.

[79] Rout M, Bandyopadhyay T and Karmakar A 2017 Free vibration analysis of pretwisted delaminated composite stiffened shallow shells: a finite element approach *Journal of Reinforced Plastics and Composites* **36(8)** pp.619-36.

[80] Chaubey AK, Kumar A and Chakrabarti A 2018 Novel shear deformation model for moderately thick and deep laminated composite conoidal shell *Mechanics Based Design of Structures and Machines* **46(5)** pp.650-68.

[81] Sahoo S 2018 Simply supported composite hypar shells under free vibration –some observations *Material Science & Engineering International Journal* **2(4)** 139–142.

[82] Bakshi K, Chakravorty D 2018 Relative Performance Study of Composite Conoidal Shell Roofs in Terms of First Ply Failure Loads *INCOM18: Proceedings of the 1st International Conference on Mechanical Engineering* **INCOM18-054**.

[83] Ghosh A, Chakravorty D 2018 First ply failure analysis of laminated composite thin hypar shells using nonlinear finite element approach *Thin-Walled Structures* **131** pp.736-45.

[84] Rout M, Hota SS and Karmakar A 2018 Free vibration characteristics of delaminated composite pretwisted stiffened cylindrical shell *Proceedings of the Institution of Mechanical Engineers, Part C: Journal of Mechanical Engineering Science* **232(4)** pp.595-611.

[85] Arefi M, Rabczuk T 2019 A nonlocal higher order shear deformation theory for electro-elastic analysis of a piezoelectric doubly curved nano shell *Composites Part B: Engineering* **168** pp.496-510.

[86] Chaudhuri PB, Mitra A and Sahoo S 2019 Higher Mode Vibration of Composite Stiffened Hypar Shell with Cut-Out for Varying Boundary Conditions and Ply Orientation *i-Manager's Journal on Structural Engineering* **8(2)** p.37.

[87] Bakshi K 2019 A Nonlinear Finite Element Failure Study of Laminated Composite Cylindrical Shell Panels *64th Congress of Indian Society of Theoretical and Applied Mechanics (ISTAM 2019)*.

[88] Sahoo S 2019 Characteristics of vibrating composite stiffened hypars with cut-out at different modes *Engineering Transactions* **67(1)**, pp.75-99.

[89] Chaudhuri PB, Mitra A and Sahoo S 2019 Mode frequency analysis of antisymmetric angle-ply laminated composite stiffened hypar shell with cutout *Mechanics and Mechanical Engineering* **23(1)** pp.162-71.

[90] Ghosh A, Chakravorty D 2020 Fem Analysis of Progressive Failure for Composite Hypar Shells *Strength of Materials* **52(4)** pp.507-20.

[91] Bakshi K, Chakravorty D 2020 First Ply Failure Study of Laminated Composite Conoidal Shells Using Geometrically Nonlinear Formulation *Recent Advances in Theoretical, Applied, Computational and Experimental Mechanics* Springer, Singapore pp. 119-131.

[92] Bakshi K, Chakravorty D 2020 A finite element prediction of first ply failure and delamination in composite conoidal shells using geometric nonlinearity *Sādhanā* **45(1)** pp.1-9.

[93] Bakshi K, Chakravorty D 2020 Numerical Study on Failure of Thin Composite Conoidal Shell Roofs Considering Geometric Nonlinearity *KSCE Journal of Civil Engineering* **24(3)** pp.913-21.

[94] Bakshi K 2021 A numerical study on nonlinear bending performance of transversely loaded composite singly curved stiffened surfaces *The Journal of Strain Analysis for Engineering Design* **56(7)** pp.430-42.

[95] Thakur SN, Ray C 2021 Static and free vibration analyses of moderately thick hyperbolic paraboloidal cross ply laminated composite shell structure *Structures* Elsevier **32** pp. 876-88.

[96] Chatterjee D, Ghosh A and Chakravorty D 2021 Finite element prediction of first-ply failure loads of composite thin skewed hypar shells using nonlinear strains *Thin-Walled Structures* **167** p.108159.

[97] Bakshi K 2021 A numerical study on nonlinear vibrations of laminated composite singly curved stiffened shells *Composite Structures* **278** p.114718.

[98] Saxena A, Bakshi K 2021 Nonlinear Bending Performances of Laminated Composite Stiffened Conoidal Shell Panels *66th Congress of ISTAM.*

[99] Keshav V, Patel SN and Kumar R 2021 Non-linear stability and failure of laminated composite stiffened cylindrical panels subjected to in-plane impulse loading. *Structures* Elsevier **29** pp. 360-72.

[100] Pinho FAXC, Del Prado, ZJGN and da Silva FMA 2022 Nonlinear static analysis of thin shallow and non-shallow shells using tensor formulation *Engineering Structures* **253** p.113674.

[101] Chen X, Shen HS and Xiang Y 2022 Thermo-mechanical postbuckling analysis of sandwich cylindrical shells with functionally graded auxetic GRMMC core surrounded by an elastic medium *Thin-Walled Structures* **171** p.108755.

AUTHOR INDEX